W9-CLS-018

Annual Review of Neuroscience

Production Editor: Jennifer E. Mann
Managing Editor: Linley E. Hall
Bibliographic Quality Control: Mary A. Glass
Electronic Content Coordinators: Suzanne K. Moses, Erin H. Lee
Illustration Coordinator: Douglas Beckner

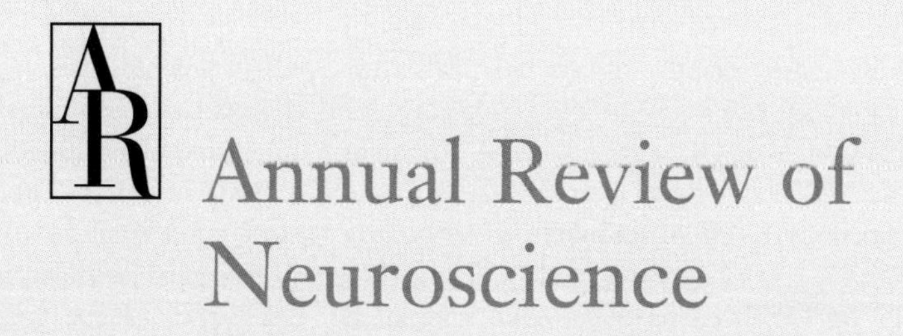

Annual Review of Neuroscience

Volume 36, 2013

Steven E. Hyman, *Editor*
Broad Institute of MIT and Harvard

Thomas M. Jessell, *Associate Editor*
Columbia University

Carla J. Shatz, *Associate Editor*
Stanford University

Charles F. Stevens, *Associate Editor*
Salk Institute for Biological Studies

Huda Y. Zoghbi, *Associate Editor*
Baylor College of Medicine

www.annualreviews.org • science@annualreviews.org • 650-493-4400

Annual Reviews
4139 El Camino Way • P.O. Box 10139 • Palo Alto, California 94303-0139

Annual Reviews
Palo Alto, California, USA

International Standard Serial Number: 0147-006X
International Standard Book Number: 978-0-8243-2436-0
Library of Congress Control Number: 78643473

∞ The paper used in this publication meets the minimum requirements of American National Standards for Information Sciences—Permanence of Paper for Printed Library Materials, ANSI Z39.48-1992.

TYPESET BY APTARA
PRINTED AND BOUND BY SHERIDAN BOOKS, INC., CHELSEA, MICHIGAN

Annual Review of Neuroscience

Volume 36, 2013

Contents

Indexes

Errata

An online log of corrections to *Annual Review of Neuroscience* articles may be found at http://neuro.annualreviews.org/

Related Articles

From the ***Annual Review of Biochemistry***, Volume 82 (2013)

Molecular Architecture and Assembly of the Eukaryotic Proteasome
Robert J. Tomko Jr. and Mark Hochstrasser

The Voltage-Gated Calcium Channel Functions as the Molecular Switch of Synaptic Transmission
Daphne Atlas

From the ***Annual Review of Biomedical Engineering***, Volume 14 (2012)

Quantitative Imaging Methods for the Development and Validation of Brain Biomechanics Models
Philip V. Bayly, Erik H. Clayton, and Guy M. Genin

From the ***Annual Review of Cell and Developmental Biology***, Volume 28 (2012)

The Membrane Fusion Enigma: SNAREs, Sec1/Munc18 Proteins, and Their Accomplices—Guilty as Charged?
Josep Rizo and Thomas C. Südhof

LINE-1 Retrotransposition in the Nervous System
Charles A. Thomas, Apuã C.M. Paquola, and Alysson R. Muotri

Axon Degeneration and Regeneration: Insights from *Drosophila* Models of Nerve Injury
Yanshan Fang and Nancy M. Bonini

From the ***Annual Review of Clinical Psychology***, Volume 8 (2012)

Default Mode Network Activity and Connectivity in Psychopathology
Susan Whitfield-Gabrieli and Judith M. Ford

From the ***Annual Review of Genomics and Human Genetics***, Volume 13 (2012)

The Genetics of Substance Dependence
Jen-Chyong Wang, Manav Kapoor, and Alison M. Goate

From the ***Annual Review of Medicine***, Volume 64 (2013)

Glioblastoma: Molecular Analysis and Clinical Implications
Jason T. Huse, Eric Holland, and Lisa M. DeAngelis

Treating the Developing Brain: Implications from Human Imaging and Mouse Genetics
B.J. Casey, Siobhan S. Pattwell, Charles E. Glatt, and Francis S. Lee

Genetic Basis of Intellectual Disability
Jay W. Ellison, Jill A. Rosenfeld, and Lisa G. Shaffer

From the ***Annual Review of Pathology: Mechanisms of Disease***, Volume 8 (2013)

Progressive Multifocal Leukoencephalopathy: Why Gray and White Matter
Sarah Gheuens, Christian Wüthrich, and Igor J. Koralnik

Pathological and Molecular Advances in Pediatric Low-Grade Astrocytoma
Fausto J. Rodriguez, Kah Suan Lim, Daniel Bowers, and Charles G. Eberhart

From the ***Annual Review of Pharmacology and Toxicology***, Volume 52 (2012)

Chronic Pain States: Pharmacological Strategies to Restore Diminished Inhibitory Spinal Pain Control
Hanns Ulrich Zeilhofer, Dietmar Benke, and Gonzalo E. Yevenes

Molecular Mechanism of β-Arrestin-Biased Agonism at Seven-Transmembrane Receptors
Eric Reiter, Seungkirl Ahn, Arun K. Shukla, and Robert J. Lefkowitz

Addiction Circuitry in the Human Brain
Nora D. Volkow, Gene-Jack Wang, Joanna S. Fowler, and Dardo Tomasi

From the ***Annual Review of Physiology***, Volume 75 (2013)

Functional Insights from Glutamate Receptor Ion Channel Structures
Janesh Kumar and Mark L. Mayer

The Impact of Adult Neurogenesis on Olfactory Bulb Circuits and Computations
Gabriel Lepousez, Matthew T. Valley, and Pierre-Marie Lledo

Pathophysiology of Migraine
Daniela Pietrobon and Michael A. Moskowitz

Perspectives on Kiss-and-Run: Role in Exocytosis, Endocytosis, and Neurotransmission
AbdulRasheed A. Alabi and Richard W. Tsien

Understanding the Rhythm of Breathing: So Near, Yet So Far
Jack L. Feldman, Christopher A. Del Negro, and Paul A. Gray

Metabolic and Neuropsychiatric Effects of Calorie Restriction and Sirtuins
Sergiy Libert and Leonard Guarente

From the ***Annual Review of Psychology***, Volume 63 (2012)

Annual Reviews is a nonprofit scientific publisher established to promote the advancement of the sciences. Beginning in 1932 with the *Annual Review of Biochemistry*, the Company has pursued as its principal function the publication of high-quality, reasonably priced *Annual Review* volumes. The volumes are organized by Editors and Editorial Committees who invite qualified authors to contribute critical articles reviewing significant developments within each major discipline. The Editor-in-Chief invites those interested in serving as future Editorial Committee members to communicate directly with him. Annual Reviews is administered by a Board of Directors, whose members serve without compensation.

Annual Reviews of

Analytical Chemistry
Animal Biosciences
Anthropology
Astronomy and Astrophysics
Biochemistry
Biomedical Engineering
Biophysics
Cell and Developmental Biology
Chemical and Biomolecular Engineering
Clinical Psychology
Condensed Matter Physics
Earth and Planetary Sciences
Ecology, Evolution, and Systematics
Economics
Entomology
Environment and Resources
Financial Economics
Fluid Mechanics
Food Science and Technology
Genetics
Genomics and Human Genetics
Immunology
Law and Social Science
Marine Science
Materials Research
Medicine
Microbiology
Neuroscience
Nuclear and Particle Science
Nutrition
Pathology: Mechanisms of Disease
Pharmacology and Toxicology
Physical Chemistry
Physiology
Phytopathology
Plant Biology
Political Science
Psychology
Public Health
Resource Economics
Sociology

SPECIAL PUBLICATIONS
Excitement and Fascination of Science, Vols. 1, 2, 3, and 4

Active Properties of Neocortical Pyramidal Neuron Dendrites

Guy Major,[1] Matthew E. Larkum,[2,*] and Jackie Schiller[3,*]

[1]School of Biosciences, Cardiff University, Cardiff, CF10 3AX, United Kingdom; email: majorg@cardiff.ac.uk

[2]Charité University, Neuroscience Research Center (NWFZ), D-10117 Berlin, Germany; email: matthew.larkum@gmail.com

[3]Department of Physiology, Technion Medical School, Haifa 31096, Israel; email: jackie@techunix.technion.ac.il

Annu. Rev. Neurosci. 2013. 36:1–24

The *Annual Review of Neuroscience* is online at neuro.annualreviews.org

This article's doi:
10.1146/annurev-neuro-062111-150343

*These authors contributed equally to this work.

Keywords

spike, action potential, NMDA receptor, NMDAR, synaptic integration, computational subunit

Abstract

Dendrites are the main recipients of synaptic inputs and are important sites that determine neurons' input-output functions. This review focuses on thin neocortical dendrites, which receive the vast majority of synaptic inputs in cortex but also have specialized electrogenic properties. We present a simplified working-model biophysical scheme of pyramidal neurons that attempts to capture the essence of their dendritic function, including the ability to behave under plausible conditions as dynamic computational subunits. We emphasize the electrogenic capabilities of NMDA receptors (NMDARs) because these transmitter-gated channels seem to provide the major nonlinear depolarizing drive in thin dendrites, even allowing full-blown NMDA spikes. We show how apparent discrepancies in experimental findings can be reconciled and discuss the current status of dendritic spikes in vivo; a dominant NMDAR contribution would indicate that the input-output relations of thin dendrites are dynamically set by network activity and cannot be fully predicted by purely reductionist approaches.

Contents

CURRENT OVERALL UNDERSTANDING OF THE NEOCORTICAL PYRAMIDAL NEURON

The layer 5 pyramidal neuron is the largest neuron of the cerebral cortex, extending its dendrites for more than 1 mm across the cortical layers. This review assesses our knowledge of this important cell type and of other neocortical pyramidal neurons, in terms of the electrical properties and information-processing capabilities of their dendrites (first and second sections). We also provide a framework for understanding the role of dendrites in the cortical network (third and fourth sections). The complex and strikingly beautiful shape of a pyramidal neuron is a fundamental determinant of signal flow within it (Larkman et al. 1992, Rapp et al. 1996, Rhodes & Llinas 2001, Spruston et al. 1994, Stafstrom et al. 1984), as well as the pattern and types of synaptic input it can receive (Cauller & Connors 1994, Dantzker & Callaway 2000, Larkman 1991, Petreanu et al. 2009). However, the active properties of dendrites also have a profound influence (Johnston et al. 1996, Mel 1993, Migliore & Shepherd 2002, Rhodes 1999, Spruston 2008). Understanding the transformation of synaptic inputs to output action potentials (APs) is therefore an incredibly complex task involving the combination of input patterning, dendritic architecture, signal transformation, and regenerative properties.

Dendrites Under the Linear Regime

EPSP size: location-dependent somatic impact of single synaptic inputs. The first major question concerns the impact of the passive dendritic architecture on the size and spread of electrical signals impinging on different portions of the basal, oblique, and tuft trees (Agmon-Snir & Segev 1993, Gulledge et al. 2005, Larkman et al. 1992, Magee 2000, Major et al. 1994, Mel 1994, Stuart & Spruston 1998, Williams & Stuart 2003). Direct patch pipette recordings of miniature excitatory postsynaptic potentials (EPSPs) in basal, apical trunk, and tuft dendrites reveal that distal basal, apical, and tuft EPSP peak voltages attenuate similarly, by ~30-fold or more, as they spread from the synapse to the closest downstream (more proximal) intrinsic spike initiation zone (**Figure 1**), either at the soma/axon initial segment or around the main apical bifurcation point (Larkum et al. 2009, Nevian et al. 2007, Williams & Stuart 2002). In contrast to the strong attenuation of peak voltage, the loss of current along the input dendrite itself is comparatively minor (Agmon-Snir & Segev 1993, Williams 2004), reflecting the fact that its membrane resistance is around an order of magnitude higher than its axial resistance over much of the likely biological parameter range.[1] In the case of basal and oblique branches, whatever the synapse site, most of the charge injected flows rapidly via the soma to the rest of the cell's membrane capacitance before more slowly leaking out through the membrane resistance; over the distal ~80% of these trees, input location per se makes only a relatively minor contribution to variations in postsynaptic potential (PSP) amplitude at the soma (Hardingham et al. 2010, Stratford et al. 1989).

Dendritic integration. Dendritic integration (how dendritic inputs combine to influence the membrane potential) has been extensively reviewed (Rall 1977, Stuart et al. 1999). One school of thought is that passive/linear or sublinear summation is the dominant mode of neocortical dendritic integration (Cash & Yuste 1999, Priebe & Ferster 2010). This would be more likely if inputs are activated in a distributed and sparse manner onto the dendritic tree (Jia et al. 2010, Varga et al. 2011) (see third section, below). The result is an "every input for itself," non-cooperative type of integration; the final input-output transformation is dominated by the passive properties of the dendritic tree (although linear summation is not synonymous with passive behavior).

Passive summation is far from boring or trivial. For example, PSP rise times at the soma differ as they spread from different locations in the dendritic tree (Agmon-Snir & Segev 1993, Major et al. 1994, Rinzel & Rall 1974). This could be exploited for computing the direction of input sequences lasting within the proximal-to-distal range of EPSP rise times (Rall 1964). Because their peaks coincide, distal-to-proximal sequences of synaptic inputs will result in a larger voltage surge at the soma than would the same inputs in reverse order. I_h (hyperpolarization-activated conductance) in the distal apical trunk and tuft may selectively reduce the duration of PSPs originating from these branches, while having much less effect on basal dendritic PSPs (Berger et al. 2001, Berger & Lüscher 2003, Kole et al. 2006, Williams & Stuart 2000).

Regenerative Mechanisms and Spikes in Dendrites

Both in vitro recordings and modeling studies have shown that dendrites can switch to a highly supralinear regime, which in principle can lead to so-called dendritic spikes (Ariav et al. 2003, Häusser et al. 2000, Judkewitz et al. 2006, Larkum & Nevian 2008, Losonczy & Magee 2006, Mel 1993, Polsky et al. 2004, Rhodes 2006, Schiller et al. 1997, Schiller & Schiller 2001, Spruston 2008). Do pyramidal cell dendrites actually fire spikes? Are voltage-sensitive channels in dendrites activated in an all-or-none manner during normal brain function, or do they simply boost synaptic inputs in a graded fashion? What does all-or-none actually mean (see sidebar, Terms)? This topic remains contentious despite decades of evidence suggesting the existence of all-or-none dendritic events under certain conditions (Amitai et al. 1993, Llinás et al. 1968, Schiller et al. 1997, Schwindt & Crill 1997, Stuart & Sakmann 1994, Wong et al. 1979, Yuste et al. 1994, Xu et al. 2012). The question is perhaps better restated as whether we should expect

[1] During the awake state or network upstates, however, membrane conductance and hence current loss may increase several-fold.

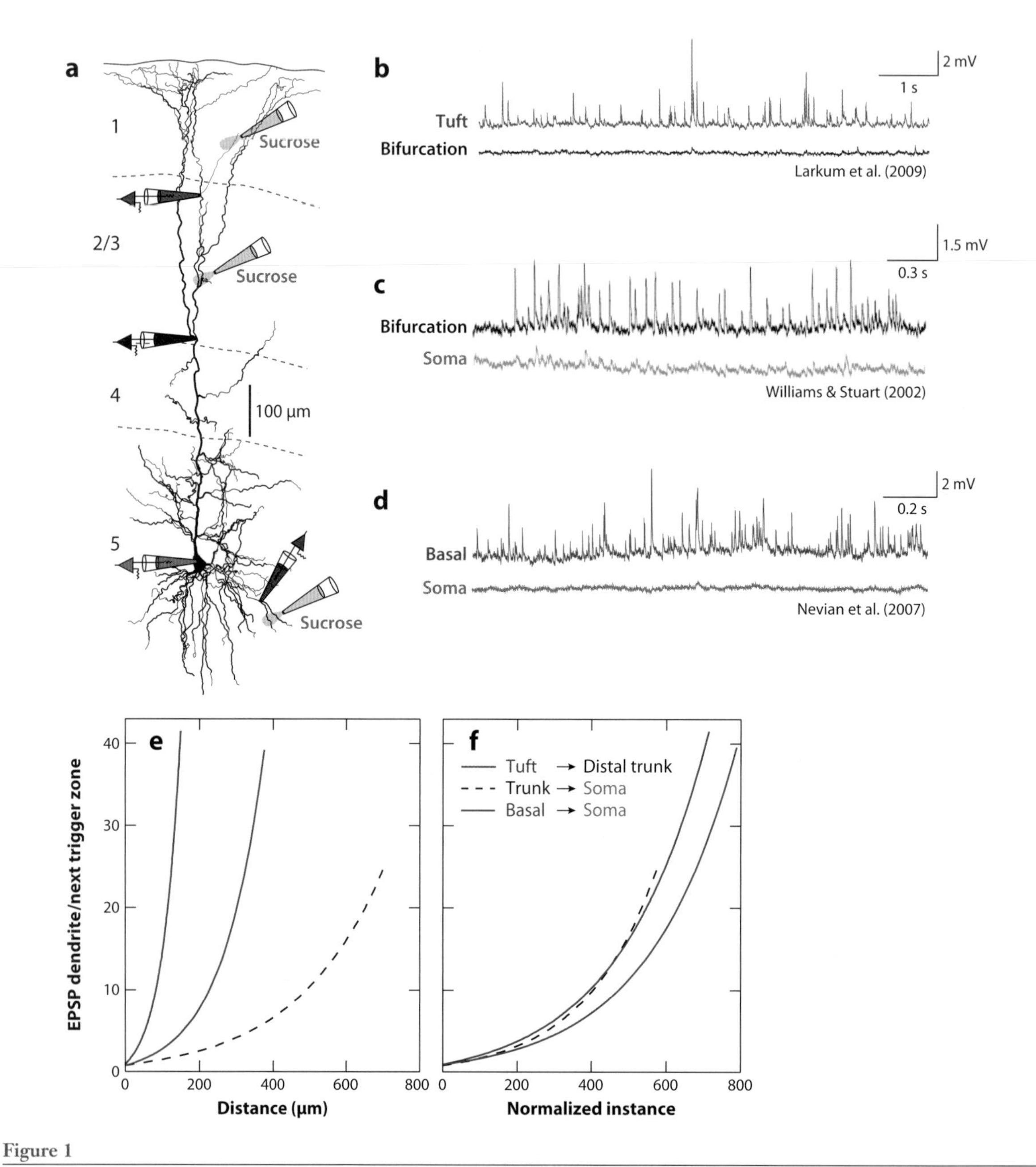

Figure 1

Attenuation of miniature excitatory postsynaptic potentials (mEPSPs) across the dendritic tree. (*a*) Sucrose-evoked mEPSPs at different locations in the L5 pyramidal neuron. Whether evoked in the distal tuft and recorded (*b*), or evoked near the apical bifurcation and recorded at the soma (*c*), or evoked in the basal dendrites and recorded at the soma (*d*), most individual propagated mEPSPs were barely detectable: in terms of attenuation, the two spike trigger zones (both the calcium initiation zone near the apical bifurcation and the sodium initiation zone near the soma) appear electrically remote from the sites distal to them providing the bulk of their synaptic input. (*e*) Attenuation of EPSPs across different regions of the dendritic tree (peak V near synapse/peak V near next downstream spike trigger zone). (*f*) Distances normalized to average total length of the respective dendrites.

dendrites to spike under relevant physiological conditions. We argue here that this question can be approached in principle in terms of the net current-voltage (I-V) relationships of dendritic compartments. In this context, many of the currently confusing and disparate results can easily be reconciled into a unified framework. We do not attempt to archive exhaustively all the details of active dendritic properties in neocortical pyramidal neurons, which have been addressed in several other reviews (London & Häusser 2005, Migliore & Shepherd 2002, Spruston 2008).

One dominant ion? Not necessarily. The word spike has been a convenient term for most of the history of neuroscience, typically being used to denote the firing of a sodium AP, usually recorded at the soma but actually arising in the axon (Kole & Stuart 2008, Stuart et al. 1997). The discovery that dendrites are capable of electrogenesis (Amitai et al. 1993, Llinás et al. 1968, Wong et al. 1979) undermined this once simply understood terminology. Although all dendritic spikes are somewhat mixed in nature, one can differentiate between three major types of spikes—Na^+, Ca^{2+}, and *N*-methyl-D-aspartate (NMDA)—according to the main underlying class of conductance. The wide range of electrical properties, channel types, and densities gives rise to regenerative events in dendrites whose rise times and durations differ by orders of magnitude. Nonetheless, dendritic spikes do conform to the classical description of APs in as much as they have a threshold, and some types have a refractory period and propagate actively for some distance (Larkum & Zhu 2002, Schiller et al. 1997).

All-or-none? Yes. Stereotyped? No. The term all-or-none is also a source of much confusion: Even classical Na^+ APs are far from stereotyped in threshold, amplitude, and time course. No feature is sacred: All three change during the relative refractory period, with strength-duration and strength-length trade-offs, and all three depend on the rate of rise of the stimulus (Azouz & Gray 2000, de Polavieja et al. 2005, McCormick et al. 2007, Platkiewicz & Brette 2011, Shu et al. 2006, Tateno et al. 2004, Wilent & Contreras 2005, Yu et al. 2008). Typically, during an AP train, amplitude decreases, duration increases, and threshold

TERMS

These only partially overlap and are not synonymous.

Regenerative event: a self-driven event (waveform) involving a positive feedback loop, e.g., between voltage and current (or calcium and voltage, or calcium-induced calcium release).

Spike: a voltage transient; in the wider world, many shapes and sizes; generally agreed it must go up then down (or vice versa, i.e. be self-terminating). In neuroscience, by convention, a spike usually has a threshold of some kind across which the response jumps qualitatively/substantially in size/shape, in a biologically/computationally significant manner. Threshold can be voltage or an input variable such as glutamate, or indeed it can be multidimensional. Threshold does not need to be fixed (e.g., strength-duration trade-offs, relative refractory periods) and may need to be teased out: It may not be apparent in all input-output (I-O) relations, even though the voltage waveform is spike-like and very similar to a supra-threshold waveform from another I-O relation that does exhibit a clear threshold (see section on Sharpness of Threshold and Spike-Sigmoid Duality below).

Action potential (AP): a membrane-voltage transient that normally results in some distinct biological action (e.g., a muscle contraction or release of neurotransmitter); usually, but not necessarily, all-or-none. The term action potential has a broader meaning than many neuroscientists may assume: for example, cardiac action potential = slow spike + plateau; *Nitella* (algae) chloride-mediated action potential; negative-going action potentials in *Ascaris* (roundworm) pharyngeal muscle (Byerly & Masuda 1979).

All-or-none: an event that, once started (i.e., by some threshold being exceeded), proceeds by itself to completion; if not started (i.e., threshold not reached), a substantially different (smaller/briefer) event results. However, all-or-none does not mean completely stereotyped in size and shape (or threshold).

Electrogenic: causing an electrical change. Includes positive feedback, negative feedback (such as sag, undershoot, afterhyperpolarizations), and delayed combinations of the two (e.g., oscillations, pacemaking).

rises (Spruston et al. 1995b). This progression is very marked during high-frequency bursts: Changes in amplitude or duration can be as much as twofold, owing to Na^+ channel inactivation, Ca^{2+} accumulation, and the opening of calcium-activated potassium channels ($gK_{Ca}s$) (Storm 1990). Neuromodulation and inhibition can produce additional changes. So, the term all-or-none does not imply a completely stereotyped waveform; it implies only the existence of some instantaneous threshold below which the response is small and above which the response is clearly bigger, or substantially different (see sidebar, Terms). A similar contention arises over whether dendritic spikes are actually sharply spike-like in shape. In particular, Ca^{2+} spikes and NMDA spikes are often plateau-shaped and are referred to as such (Antic et al. 2010, Milojkovic et al. 2007, Schwindt & Crill 1999, Suzuki et al. 2008, Wei et al. 2001). Considering the biological literature as a whole, APs do not have to be fast: cardiac APs can be hundreds of milliseconds long.

The NMDA spike as a hallmark of electrogenesis in thin dendrites. A consensus is emerging that in neocortical excitatory neurons the dominant depolarization-activated conductance in thin (usually submicron-diameter) dendrites is the NMDA receptor (NMDAR) channel (Antic et al. 2010, Branco & Häusser 2011, Larkum et al. 2009, Lavzin et al. 2012, Major et al. 2008, Mel 1993, Nevian et al. 2007, Schiller et al. 2000). NMDA spikes represent an important conceptual leap because they are inherently ligand dependent, i.e. dependent on glutamate and D-serine binding, and therefore subject not just to local membrane potential but also to the spatial distribution of these transmitters along the dendrite. The clear existence of NMDA spikes in vitro, with an order-of-magnitude safety factor, demonstrates that the input-output relations of thin dendrites are neither fully predictable nor constrainable in a purely bottom-up, reductionist manner.

In vitro, NMDA spikes (or plateau potentials) have been found in all classes of thin dendrite of neocortical excitatory neurons (basals, apical obliques, apical tufts) and in all neocortical areas and layers examined to date (Branco & Häusser 2011, Gordon et al. 2006, Lavzin et al. 2012, Milojkovic et al. 2004, Nevian et al. 2007, Schiller et al. 2000). They also probably occur in hippocampal apical tufts (Wei et al. 2001). Sodium spikelets, however, are much more difficult to initiate in neocortical pyramidal neuron thin dendrites in brain slices, and are weak and variable in size, if they can be triggered at all (Larkum et al. 2009, Milojkovic et al. 2005b, Nevian et al. 2007). In vitro, calcium spikes are not evoked in basal dendrites either by direct current injection or by glutamate stimulation (Major et al. 2008, Nevian et al. 2007, Schiller et al. 2000); Ca^{2+} spikes are rarely initiated by current injection into single distal (thin) tuft dendrites (Larkum et al. 2009), although most of the depolarization from Ca^{2+} spikes initiated in the distal trunk/bifurcation zone does propagate into the distal tuft, which also then exhibits a Ca^{2+} transient (but see Xu et al. 2012).

NMDA spikes evoked focally by glutamate iontophoresis or 1-photon uncaging are associated with a local high calcium zone within $\sim$10 μm of the input site, mirroring the distribution of activated NMDA channels (Major et al. 2008). This is accompanied by a lower (but above resting) calcium zone all the way to the dendritic tip, resulting most likely from calcium channels opening as the entire dendrite distal to the input site is depolarized: It is difficult for charge to escape from this segment. Simulations suggest that $\sim$20% of the charge flow during an NMDA spike is via calcium conductances. The peak calcium in both zones increases with the spike/plateau duration (Major et al. 2008). Preliminary data suggest that brief NMDA spikes that result from distributed stimulation have correspondingly smaller, spread-out calcium transients, lacking a single obvious "hot zone" (Lavzin et al. 2012).

A key point is that at many dendritic locations it takes only a small number of synaptic inputs to evoke an NMDA spike: As few as $\sim$10 clustered single spine inputs may suffice (**Figure 2**). This is a small fraction of the synapses on a single thin dendrite, typically

100–400 dendritic spines, depending on the species (Chen et al. 2011, Larkman et al. 1992). An equally crucial point is that NMDA spikes can also be elicited by distributed inputs (**Figure 2**): Clustering is not a prerequisite for a clear threshold. A third important point is that both depolarization and glutamate prebinding from previous activation reduce the glutamate threshold, allowing cooperativity between NMDA spikes [**Figure 2*a*,*e***; **Supplemental Figure 1** (follow the **Supplemental Material link** from the Annual Reviews home page at **http://www.annualreviews.org**)] (Major et al. 2008, Polsky et al. 2009).

BIOPHYSICAL MECHANISMS OF NMDA SPIKES

A detailed discussion of the biophysical mechanisms of NMDA spikes is warranted because of their importance and prevalence in thin dendrites, which, as mentioned, receive the vast majority of inputs to pyramidal neurons.

NMDAR Channels

The properties of NMDAR channels have been reviewed extensively elsewhere (Paoletti 2011, Yuan et al. 2008). Here, we highlight important aspects relating to their regenerative nature. NMDARs are strongly electrogenic. Why would this characteristic have evolved if the sole function of NMDARs was synaptic plasticity? If their role was purely to detect and signal presynaptic-postsynaptic firing coincidences, using calcium influx to initiate a plasticity cascade, NMDARs could have been much smaller pure calcium conductances. As it is, however, typically ~90% of the charge through NMDAR channels is actually carried by Na^+ and K^+ ions (Garaschuk et al. 1996, Jahr & Stevens 1993, Schneggenburger et al. 1993). Why? The calcium permeability of NMDAR channels can also be reduced drastically by activation of metabotropic γ-aminobutyric acid B ($GABA_B$) receptors (Chalifoux & Carter 2010) and by other kinds of plasticity (Sobczyk & Svoboda 2007), without appreciably altering the total current flow.

I-V Curve Families for a Unified Understanding of NMDAR Regenerative Events: Graded versus Thresholded

A helpful approach for understanding dendritic integration and spiking is to generalize the well-known idea of current-voltage relations (I-V curves). NMDA-dependent dendritic electrogenesis has both thresholded and graded aspects, both of which can be predicted from the basic I-V curve of the NMDAR conductance and can be unified under the same conceptual framework. From this perspective, the fundamental question for any given dendritic compartment is whether its instantaneous net I-V curve is N-shaped with three zero-current crossings (**Figure 3*a***). If so, the membrane is bistable and is capable of firing a spike (of some kind) if kicked over the threshold. Following a pulse of glutamate onto a single electrical compartment of a dendrite, as the openable NMDAR conductance (g_{max}) rises, peaks, then falls, the instantaneous I-V relation progresses through a succession of different curves: an I-V movie (**Figure 3*b***) (**Supplemental Videos 1–4**). The I-V curve starts downstable (blue). The right-hand trough progressively deepens, enhancing the N-shape. Eventually, if there is sufficient NMDA conductance, and other parameters allow, the I-V curve can morph into the bistable regime (green). If the voltage is then driven (or is already) above threshold, the membrane goes into the up state, and a spike/plateau results. This switches off by itself once the NMDA conductance falls back down below the minimum required for the I-V trough to dip below the zero-current axis, and the up state ceases to exist (**Figure 3*b***: uppermost green curve → lowest blue curve; see below). A dendritic I-V relation is complex because it generally evolves over time. This is due in large part to synaptic conductances, which are major contributors and themselves rise and fall over time.

A spike is only all-or-none in the sense that it has some threshold below which it fails and above which it runs to completion.

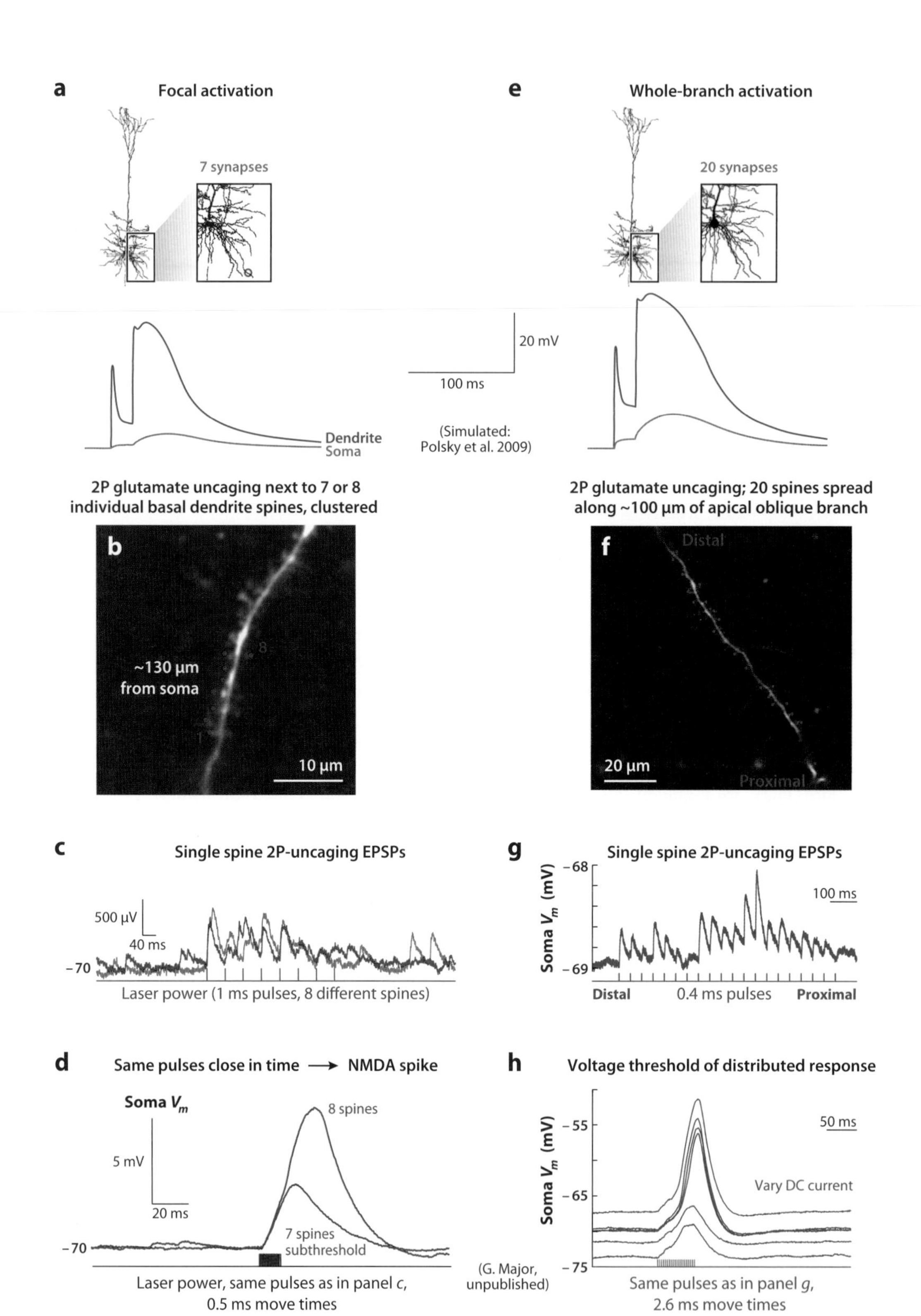
a
Focal activation
7 synapses
e
Whole-branch activation
20 synapses
20 mV
100 ms
Dendrite
Soma
(Simulated:
Polsky et al. 2009)
2P glutamate uncaging next to 7 or 8
individual basal dendrite spines, clustered
b
~130 µm
from soma
10 µm
2P glutamate uncaging; 20 spines spread
along ~100 µm of apical oblique branch
f
Distal
20 µm
Proximal
c
Single spine 2P-uncaging EPSPs
500 µV
40 ms
-70
Laser power (1 ms pulses, 8 different spines)
g
Single spine 2P-uncaging EPSPs
Soma V_m (mV)
-68
-69
100 ms
Distal
0.4 ms pulses
Proximal
d
Same pulses close in time ⟶ NMDA spike
Soma V_m
8 spines
5 mV
20 ms
-70
7 spines
subthreshold
Laser power, same pulses as in panel c,
0.5 ms move times
(G. Major,
unpublished)
h
Voltage threshold of distributed response
Soma V_m (mV)
-55
-65
-75
50 ms
Vary DC current
Same pulses as in panel g,
2.6 ms move times

Figure 2

NMDA spikes can be elicited by stimulating a relatively small number of clustered or distributed synapses. (*a*, *e*) Neuron simulations; due to NMDAR priming, the second pulse of a paired-pulse stimulus produces an NMDA spike. *Other panels*: brain slice 2-photon glutamate uncaging experiments (layer 5). (*b*, *f*) Positions of uncaging spots (*red*) next to dendritic spines. (*c*, *g*) Corresponding single-spot uncaging EPSPs; similar to single synapse quantal EPSPs. (*d*, *h*) The same stimuli in quicker succession produce either a subthreshold response or an NMDA spike. (*d*) Blue traces = first 7 spots, red = all 8 spots stimulated, demonstrating glutamate threshold. (*h*) NMDA spike fails below a distinct voltage threshold.

All-or-none implies nothing about the constancy of that threshold, the stereotypy of the ensuing suprathreshold response, or the size or shape of that response relative to the biggest just-subthreshold response. This is, of course, not to mention further confusion caused by waveform changes as potentials propagate along a dendritic tree.

In large layer 5 pyramidal neuron basal dendrites, the NMDA spike glutamate threshold increases ~5-fold from distal to proximal input sites, and the amplitude (reaching the soma) increases ~7-fold (as does the maximal just-subthreshold response). These spatial gradients mirror the local input conductance (Major et al. 2008). If the stimulus is focal, once the response is suprathreshold, further increases in the stimulus produce relatively minor increases in the amplitude of the spike/plateau; however, the duration of the spike/plateau grows almost linearly with the stimulus and may reach hundreds of milliseconds (Major et al. 2008, Milojkovic et al. 2005a). This may serve as a prolonged time window for integration.

Investigators have shown experimentally that both features of NMDA regenerativity (graded and spike/plateau) exist, often simultaneously (Branco & Häusser 2011, Major et al. 2008, Schiller et al. 2000). Sharp thresholds for NMDA spikes have been found using focal synaptic stimulation (Gordon et al. 2006, Larkum et al. 2009, Nevian et al. 2007, Polsky et al. 2004, Schiller et al. 2000), glutamate UV-laser uncaging (Gordon et al. 2006, Major et al. 2008, Polsky et al. 2004, Schiller et al. 2000), and multispot two-photon uncaging (G. Major, unpublished observations; **Figure 2*b*–*d***). In addition, it seems plausible, a priori, that whereas focal stimulation can lead to sharp thresholds, highly distributed stimulation might be expected to lead to a blurring of thresholds and more sigmoidal input-output relations. In fact, perhaps counterintuitively, both experimental data and models show that distributed stimulation is still perfectly compatible with sharp thresholds (**Figure 2*e*–*h***; **Supplemental Figure 3*f*–*g***, discussed below).

Different experimental setups have found a range of NMDA regenerative behavior, from graded boosting to full-blown spikes. This range may reflect various laboratories exploring different regions within the space of possible dendritic NMDAR-dependent I-V curves (e.g., **Figure 3*b***). Experimental differences may be responsible, including, for example, the degree of input clustering/synchrony and effective AMPA/NMDA ratios. Covert spikes may also play a role (see below).

Partially Overlapping Time Courses of AMPAR and NMDAR Channels

Dendrites contain electrical mechanisms that operate over a range of time scales. At the fast end of the spectrum are AMPA receptor conductances, with decay time constants as short as ~0.5 ms (or faster) at body temperature (Gardner et al. 2001, Postlethwaite et al. 2007, Zhang & Trussell 1994). These conductances are similar in time scale to those underlying the partially active backpropagation of APs, duration also ~0.5 ms (Stuart et al. 1997), the fast voltage switching of NMDAR conductances [dominant time constants submillisecond (Kampa et al. 2004, Spruston et al. 1995a)], and the fast inward rectifier conductance (Yamada et al. 1998).

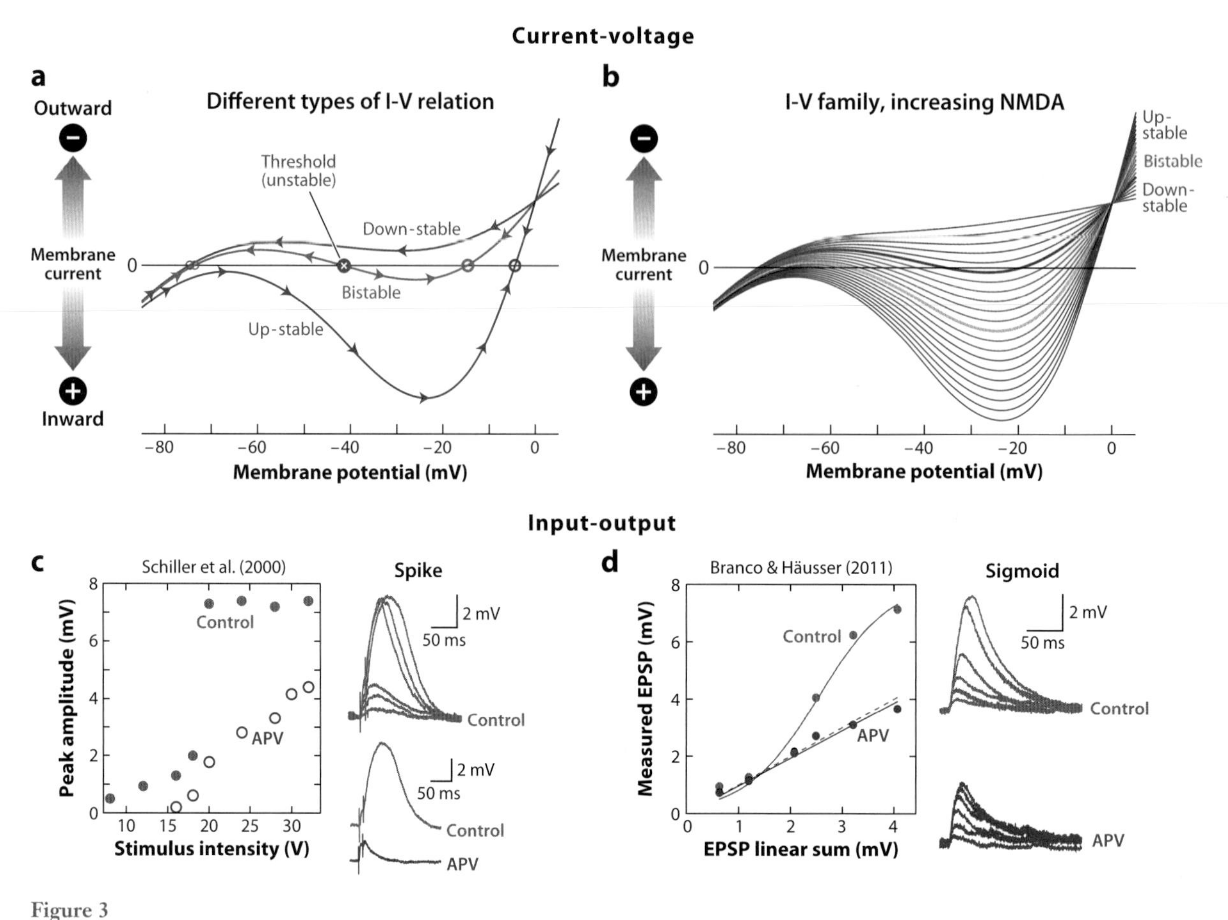

Figure 3

Understanding threshold and spike size: instantaneous current-voltage (I-V) relations and bistability. (*a*) Three types of instantaneous I-V relation. Arrows indicate system flow; open circles indicate stable states. Details in **Supplemental Figure 4** (follow the **Supplemental Material link** from the Annual Reviews home page at **http://www.annualreviews.org**). (*b*) I-V curve shape changes smoothly as maximum openable NMDAR conductance increases, but curve type switches suddenly from downstable to bistable at a critical level of NMDAR conductance (and again from bistable to upstable if NMDA conductance rises enough). *Thick dark green*: just bistable I-V curve with a high threshold, and a small jump from threshold to upstate (**Supplemental Figure 4*c***). A large enough AMPA (2-amino-3-(3-hydroxy-5-methyl-isoxazol-4-yl)propanoic acid) component could kick the dendrite over the threshold, resulting in a small spike riding on a large PSP and a sigmoid input-output relation (I-O). *Thick light green:* bistable I-V curve with more NMDA conductance, a low threshold, and a big jump (**Supplemental Figure 4*e***). An input with a low AMPA fraction may reach this threshold, yielding a big spike on a small PSP, and a step-like I-O. See **Supplemental Videos 1–4**. (*c–d*) Experiments: somatic voltage transients from basal dendritic stimulation. (*c*) NMDA spike, step-like I-O. (*d*) More graded response, sigmoidal I-O. In both cases, nonlinearity was abolished by NMDAR block.

Substantially slower is the activation time course of the NMDAR upon binding glutamate: on the order of a few milliseconds (D'Angelo et al. 1994, Dalby & Mody 2003, Dzubay & Jahr 1996, Korinek et al. 2010, Popescu et al. 2004, Spruston et al. 1995a). An absolutely key point is that following a brief synaptic pulse of glutamate, from a single presynaptic release event, most of the AMPAR conductance has died away before the NMDAR conductance is half primed or activatable, that is, capable of conducting, if depolarized (Stern et al. 1992). NMDARs must bind glutamate (and D-serine) and undergo internal conformational/state changes to become capable of magnesium unblock and conducting (Kampa et al. 2004, Popescu et al. 2004), and these state changes take longer than their counterparts in AMPARs. The NMDAR decay time constant is even longer, probably ~40–60 ms. The range may be as broad as 10 to ~100 ms at body temperature, even if one restricts discussion to the faster NR2A subtype of the receptor more prevalent in adults. There is much uncertainty about Q_{10} values, which could be in the range of 1.7 to 2.4 (Dalby & Mody 2003, Korinek et al. 2010).

The observation that the AMPA current has largely decayed by the time the NMDA conductance has half primed is absolutely crucial to understanding what dendrites might be able to do and how they might do it: specifically whether they exhibit thresholds (details in next section). Note that when one considers NMDA spikes and plateaus, which are relatively slow events (half-width or duration at half amplitude ranges from ~20 ms to several hundred milliseconds or longer), the dendrites behave in a more electrically compact fashion than they would, say, for a backpropagating AP with a half-width of only 0.5 ms.

Sharpness of Threshold and Spike-Sigmoid Duality

Experiments (**Figure 3*c*,*d***) and simulations suggest that thin dendrites can produce both sigmoid and step-like input-output (I-O) relations, depending on the input parameters. In full compartmental model simulations, sigmoidal I-O relations result from higher AMPA/NMDA ratios, slower AMPAR decay kinetics, more synchronous inputs, and more spatial spread of the inputs (**Supplemental Figures 2** and **3**). Conversely, lower AMPA/NMDA ratios, faster AMPAR decay, and more temporal but less spatial dispersion of inputs all conspire to produce more step-like I-O relations.

The simplest way to understand this process is from I-V curve movies. A large AMPA component produces a large voltage kick, which can cross the threshold at a relatively low NMDA conductance g_{max} (openable conductance), i.e., when the I-V curve is only just bistable and is still downskewed with a high barrier (downstate → threshold) and a small jump (threshold → upstate; **Figure 3*b***, **Supplemental Figure 4*c***). As a result, we see a small spike riding on a large subthreshold response (**Supplemental Videos 1** and **2**). If the input is increased further, the maximum response (upstate) grows in a graded manner. The net result is a sigmoidal I-O.

Conversely, a small AMPA component produces a small voltage kick; therefore, with a low AMPA/NMDA ratio, the NMDAR g_{max} can increase to a value at which the I-V curve is upskewed bistable before the AMPA kick is enough to cross the threshold (low barrier, big jump; **Figure 3*b***, **Supplemental Figure 4*e***). When it does cross the threshold, there is a big jump in response amplitude: a large spike riding on a small subthreshold response. The I-O relation is dominated by a sharp step. This is also true for intermediate cases in which the threshold is crossed when the I-V curve is symmetrically bistable (**Figure 3*a***, **Supplemental Figure 4*d***, **Supplemental Videos 3** and **4**).

Similar logic applies to the other input parameters. Slowing down the AMPAR decay allows the threshold to be crossed at a lower NMDAR conductance, producing a smaller spike riding on a bigger subthreshold response (**Figure 3**, **Supplemental Figures 2** and **4**). Desynchronizing the inputs is roughly equivalent to reducing the AMPA component: Time jitter has far less effect on the much slower NMDA component.

There is a further plot twist. Simulations suggest that essentially the same spike waveform can participate in both sigmoidal and stepped I-O relations (compare thick with thick, medium with medium, and dashed with dashed waveforms across panels in **Supplemental Figures 2** and **3**: Each of these is just above a clear glutamate threshold in at least one plot but is hidden within a sigmoid in at least one other plot). Thus, there is a degree of spike-sigmoid duality.

It is important that even spatially distributed inputs can produce stepped input-output relations with clear thresholds over a wide range of biologically plausible parameters (**Figure 2**, **Supplemental Figure 3**). What actually happens in real life must now be settled with experiments in awake, behaving animals: The necessary technology is gradually becoming available.

Cooperativity

Depolarization reduces the NMDA spike glutamate threshold (Major et al. 2008, Polsky et al. 2009); this depolarization can be provided by another NMDA spike/plateau, including at a more distal location in the same dendrite (**Supplemental Figure 1**), leading to a number of interesting computational capabilities such as directionally biased responses, proximal-distal synapse interplay, and classical receptive field–contextual interactions (Behabadi et al. 2012, Branco et al. 2010, Major et al. 2008). Depolarizing drive (i.e., positive current), from whatever source, shifts the I-V curve vertically downward, which makes the bistable regime start at a lower glutamate-bound NMDAR conductance g_{max} (just a small shift converts the lower blue curve in **Figure 3*b*** to bistable). Within the bistable regime, this downward shift in the I-V curve has three effects at a given NMDAR g_{max}: The downstate moves more positive, the threshold more negative, and the upstate more positive (**Supplemental Figure 5**). This lowers the barrier to threshold but increases the jump above threshold [and the net spike amplitude (**Supplemental Figures 4** and **5**)], allowing the threshold to be crossed by a smaller AMPAR conductance; for a given AMPA/NMDA ratio, the I-O relation is shifted to the left (lower glutamate threshold) and becomes more stepped, with a bigger spike riding on a smaller PSP.

Inward Rectification

Bistability of instantaneous I-V curves can be made more robust (extended over a wider range of parameters) by several other conductances also conveniently found in dendrites, such as $GABA_A$ and fast inward rectifier potassium (KIR) channels (Sanders et al. 2013).

All neocortical pyramidal neurons investigated to date exhibit substantial inward rectification (their input resistance and time constants can decrease up to ~4-fold as the membrane potential is varied from around −50 mV to −90 mV (Major 1992, p. 72; Waters & Helmchen 2006). KIR channels are major players and are, in many ways, mirror images of NMDAR channels (Kashiwagi et al. 2002, Yamada et al. 1998). They have similar structures but are inverted in the membrane. KIR channels are blocked by intracellular (as opposed to extracellular) Mg^{2+}; they open rapidly with hyperpolarization: the opposite polarity to NMDARs. The symmetry between NMDA and KIR goes further still: Some types of KIR channel are even gated by transmitters, albeit more slowly via $GABA_B$ metabotropic receptors and G proteins (Yamada et al. 1998).

The left-hand part of an N-shaped I-V curve is selectively enhanced by adding inward rectifier conductance, which makes the overall N more symmetrical, strengthening the stability of the otherwise fragile downstate by providing more corrective current on either side (Sanders et al. 2013). This in turn expands the range of baseline voltages (DC current offsets) and inputs (e.g., NMDA conductances) compatible with instantaneous bistability and, more exactingly, a sharp dendritic spike threshold. Compared with a non-voltage-dependent K^+ conductance, by closing with depolarization, inward rectifier conductance moves the unstable threshold point to the left to more hyperpolarized voltages, while moving the upstate to the right. These shifts lower the barrier to threshold and increase the jump above threshold, leading to a more pronounced step in the I-O relation. Inward rectification could thus both increase the robustness and sharpen the threshold of dendritic spikes.

Inhibition

$GABA_A$ inhibition can similarly increase the robustness and extent (in parameter space) of the bistable regime, stabilizing the downstate not only by moving the I-V curve vertically upward, but also by steepening it (Lisman et al. 1998). The price is a higher threshold (and an increase in the number of synapses required for an NMDA spike).

If negative current is injected within the same dendritic compartment, the I-V curve family shifts vertically upward (Jadi et al. 2012) (**Supplemental Figure 5**). For some parameter combinations this can destroy the bistability of weakly bistable curves, pushing the trough above the zero-current axis and rendering them downstable (**Figure 3**). In the case of more deeply bistable curves, negative current moves the threshold to the right and both down- and upstates to the left, raising the barrier but reducing the jump. If we start with a glutamate stimulus that elicits an NMDA spike and progressively hyperpolarize the cell, eventually a baseline voltage will be reached (prevention threshold), below which the same glutamate stimulus cannot push the dendrite across the threshold; thus the NMDA spike fails (**Figure 2*b***) (Jadi et al. 2012, Major et al. 2008). If, however, an N-shaped I-V curve is initially upstable (**Figure 3*a*,*b***) (e.g., relatively small leak/inward rectifier + large NMDA conductance), sufficient negative current can make it bistable by pushing the left-hand peak above the zero current axis.

Spike/Plateau Termination and Calcium-Activated Potassium Channels

However it gets there, once the dendrite is in an upstate, eventually the NMDA component starts to decay, and the level of the upstate (plateau top) declines slowly (this can be seen in experimental NMDA spike/plateau waveforms lasting more than ~40 ms, i.e., a couple of membrane time constants). After a time delay that may be several NMDAR decay time constants long, depending on the original safety factor,[2] the NMDA conductance falls to the point where the right-hand local minimum of the N-shaped I-V curve slips just above the zero-current axis, and the upstate suddenly ceases to exist. A relatively rapid downstroke ensues, and the membrane ends up back in the downstate (now the only stable state). This decay of the plateau top (upstate) can be accelerated by calcium accumulation opening gK_{Ca}s, in particular the apamin-sensitive SK (small conductance) channel, which colocalizes with NMDARs in spines (Cai et al. 2004, Ngo-Anh et al. 2005, Wei et al. 2001). On the other hand, in other brain areas, calcium-activated nonspecific cation conductances can be activated, causing the reverse effect and leading to a much longer plateau (reviewed in Major & Tank 2004). At any time during the slow decline of the upstate, providing the I-V curve is bistable (not upstable), a sufficiently strong brief hyperpolarizing pulse, for example a GABA IPSP (inhibitory postsynaptic potential), could flip the dendrite prematurely into the downstate and curtail the NMDA spike/plateau.

[2]How many times more NMDA g_{max} (openable conductance) is present, at its peak, than required for bistability.

CONCEPTUAL FRAMEWORK FOR UNDERSTANDING THE PYRAMIDAL NEURON: COMPARTMENTS, COMPUTATIONAL UNITS, CANONICAL DESCRIPTIONS

Canonical Abstractions

Taken together, the observations outlined above provide powerful constraints on models of neocortical pyramidal neurons. Actually understanding any given neuron requires the right level of abstraction (Branco & Häusser 2010, Häusser & Mel 2003, Herz et al. 2006, Koch et al. 1983, Koch & Segev 2000, Larkum et al. 2001, Mel 1994, Poirazi et al. 2003, Spruston 2008). The goal is to encapsulate the salient input/output properties of the neuron within a minimal description (Mirsky et al. 1998, Spruston 2008). Investigators have suggested several canonical abstractions for pyramidal cells (e.g., **Figure 4**).

In the three-compartment (three-computational subunit) model (**Figure 4*a*,*b***), the pyramidal neuron bauplan is split into three major dendritic arbors: basals, apical trunk/obliques, and distal trunk/tuft (Larkum et al. 2001); dendrites within each arbor are lumped together into a single functional compartment. This model also takes into account the location of the two major spike initiation zones, the axo-somatic sodium initiation zone and the distal apical calcium initiation zone, separated by the apical trunk (Amitai et al. 1993,

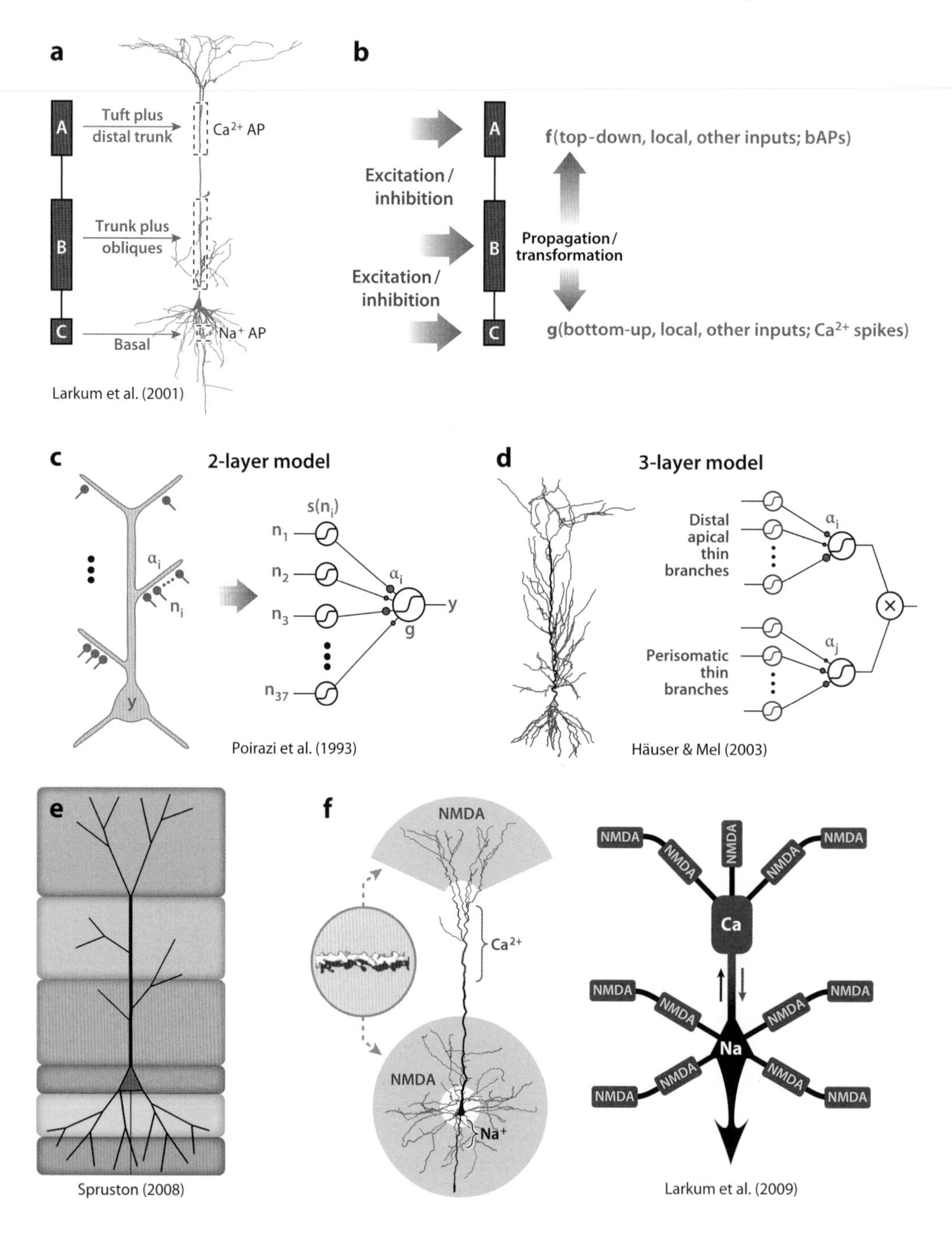

←

Figure 4

Encapsulating the properties of pyramidal neurons. (*a, b*) Three-compartment model of a layer 5 pyramidal neuron designed to include the influence of the three major dendritic arborizations: tuft + distal trunk (A), apical trunk/obliques (B), and basal dendrites (C). Each functional compartment incorporates/abstracts distinct electrical properties/channel distributions. Compartments A and C can initiate Ca^{2+} and Na^{+} spikes, respectively; f(. . .) and g(. . .) represent their input-output functions. Excitatory and inhibitory influences act together on individual compartments. The tuft integrates prominent long-range top-down feedback with local and other inputs, but the threshold for generating output is modulated by backpropagating action potentials (bAPs, i.e., output spikes of the cell). The basal compartment integrates bottom-up, local, and other inputs but is influenced heavily by Ca^{2+} spikes in the tuft. The trunk modulates the coupling between the other two compartments. (*c*) 2-layer neuron; each dendrite behaves as a neural network neuron-like unit, sending weighted output to the soma. (*d*) 3-layer neuron; similar, but two major spike initiation zones integrate input from different subtrees and interact nonlinearly. (*e*) Compartmentalization based on cortical layer. Synaptic inputs from different sources arrive on dendrites in different layers, which may lead to functional domains within the dendritic tree. (*f*) Electrogenesis based on dendrite thickness. Most synaptic inputs are onto thin dendrites, dynamically changing segments or regions of which may produce local NMDA spikes; these interact bidirectionally with one another and with other inputs to influence the nearest of the two intrinsic conductance-based spike initiation zones, which also interact via bidirectional signals along the thick apical trunk (as in the three-compartment model).

Schiller et al. 1997, Yuste et al. 1994). Distal Ca^{2+} spikes greatly enhance the influence of the tuft; inputs that may otherwise have had a negligible effect on output firing can even come to dominate it: In large layer 5 pyramidal neurons, the Ca^{2+} spike generally contributes to a burst of somatic APs (Larkum & Zhu 2002, Williams & Stuart 1999, Zhu 2000). Basals and proximal obliques can be further collapsed together functionally, depending on the layer specificity of inputs (Briggs & Callaway 2005, Lübke & Feldmeyer 2007, Petreanu et al. 2009, Schubert et al. 2001). However, the apical trunk clearly has a special layer-spanning role that signals the output of each dendritic compartment to the others (Larkum et al. 1999, Larkum 2013).

The formalization of the cell in three compartments ignores the possibility of local computations performed in thin basal, oblique, and tuft dendrites, for instance with local NMDA and sodium spikes (Larkum et al. 2009, Losonczy & Magee 2006, Mel 1993, Polsky et al. 2004, Schiller et al. 2000). These local computations are partially captured by 2- and 3-layered feedforward network models (**Figure 4*c,d***) (Häusser & Mel 2003, Mel 1993, Poirazi et al. 2003). Within this framework, local inputs are summed nonlinearly (e.g., sigmoidally) within fine dendritic branches and then integrated again at one or more main intrinsic spike initiation zones. However, these abstractions allow only unidirectional signaling, ignoring backpropagating APs and cross talk or cooperativity between different dendrites or segments of dendrites (Major et al. 2008) (**Supplemental Figure 1**). Bidirectional signaling allows more complex processing capabilities (Behabadi et al. 2012, Branco et al. 2010, London et al. 2008). For instance, a ~30-ms time window exists for the detection of coincident tuft and basal inputs (Larkum et al. 1999, Polsky et al. 2004), which can be modified by the location/activity of the apical oblique dendrites (Schaefer et al. 2003).

Pyramidal neurons are embedded in a layered structure (the cortex); the distribution of synaptic input is constrained, e.g., tuft dendrites are likely to receive long-range inputs (Binzegger et al. 2004). Inputs from different sources are not randomly distributed but tend to terminate in a layer-specific manner (Briggs & Callaway 2005, Lübke & Feldmeyer 2007, Petreanu et al. 2009, Schubert et al. 2001). Such targeting may result in different functional compartmentalization in different neurons (**Figure 4*e***). In the logical extreme, each individual thin dendrite could function as a multicompartmental unit with local NMDA-dependent computations (**Figure 4*f***). This abstraction is, in effect, an extension of the three-compartment model, with dynamic bunches of extra NMDAR-dependent compartments, depending on the input pattern. Each functional compartment can be understood in terms of the I-V analysis presented above; interactions between compartments can be understood via multidimensional I-V analysis (see **Supplemental Discussion**).

Functional Compartmentalization of Inputs in Thin Dendrites

Inclusion of the distribution of synaptic input as a key parameter complicates the issue of determining the most appropriate dendritic computational subunit (Häusser & Mel 2003, Polsky et al. 2004). The extent of functional compartmentalization becomes strongly dependent on the uniformity versus nonuniformity of both the intrinsic conductances and the pattern of inputs (extrinsic conductances). Leaving aside spatiotemporal sequences (Branco & Häusser 2011), one can start by differentiating between the case of fairly uniform versus clustered inputs (Larkum & Nevian 2008, Mel 1993). In the case of uniform input activation, it may be helpful to treat a terminal dendritic branch as a single computational unit (Poirazi et al. 2003). One can also lump several similar dendrites within a subtree into a single equivalent dendrite if they all receive comparable inputs. With very sparse activation, integration takes place in the linear/sublinear regime of input summation (see above). However, because of cooperativity between thin dendrites, we predict that increasing the density of activated inputs (<5%, see below) should be sufficient to initiate distributed regenerative boosting (Branco & Häusser 2011) or a distributed spike spread out over that group of branches (Lavzin et al. 2012).

In contrast with a random distribution of synapses, one of the leading hypotheses is that small clusters of synapses onto dendritic branchlets perform nonlinear local integration (Häusser & Mel 2003, Poirazi et al. 2003). In this framework, dendrites perform local computations such as calculating binocular rivalry (Archie & Mel 2000) and the direction of movement (Borst & Egelhaaf 1992, Branco et al. 2010, Major et al. 2008). Both modeling and experiments suggest that dendritic subunits are not fixed. They are dynamic and can change in size and location according to the input pattern (Larkum et al. 2009, Major et al. 2008, Polsky et al. 2004). Even if neocortical connectivity were totally random and firing were sparse, random inputs may still be surprisingly clustered. Thus, the arrival of ~10 inputs onto the same dendritic branch could occur reasonably often. In fact, a number of studies suggest that synaptic input is not entirely random, which would accentuate clustering (Kleindienst et al. 2011, Takahashi et al. 2012).

Finally, inhibition targeted to certain portions of the dendritic tree could play a role in compartmentalizing the neuron (Palmer et al. 2012). In vivo, GABA inhibition is an important player that would be expected to keep NMDA spikes in check (Gentet et al. 2012, Jadi et al. 2012, Liu 2004, Rhodes 2006). Neighboring branches could be decoupled by inhibitory conductances, either synaptic or intrinsic (e.g., potassium conductances; see above).

Minimum Number of Synapses Required to Initiate an NMDA Spike

How many synapses are needed to evoke an NMDA spike, and what is the spatial distribution of these synapses? The effective AMPA/NMDA ratio is crucial because it influences both the likelihood of a local NMDA spike and the sharpness of its threshold. Numerous studies have investigated AMPA/NMDA ratios at excitatory synapses in different neocortical areas, and the maximal (openable) conductances are generally comparable: If anything, there may be more available NMDAR conductance than AMPAR conductance (Myme et al. 2003, Nimchinsky et al. 2004). In addition, NMDARs have ~100-fold higher glutamate affinity than do AMPARs (Patneau & Mayer 1990, Trussell & Fischbach 1989) and are far less prone to desensitization (Korinek et al. 2010, Trussell & Fischbach 1989, Vyklicky et al. 1990). Any desynchronization between inputs will tend to favor NMDA over AMPA components (see below). Thus under many stimulus/activity patterns in vivo, NMDAR conductances may dominate AMPAR conductances, particularly if firing rates or ambient glutamate is high. Even moderate levels of glutamate are likely to desensitize AMPARs and prebind NMDARs, priming the latter to behave as purely depolarization-activated channels.

Modeling and two-photon spine uncaging data suggest that NMDA spikes in distal dendrites can be evoked by as few as ~10 clustered single spine inputs (**Figure 2*a–d***) or by ~20 inputs distributed randomly along much of the dendrite's length (**Figure 2*e–h***) (Major et al. 2008, Polsky et al. 2009; G. Major, unpublished observations). Using the two-photon uncaging method, a wide range of uncaging EPSP sizes can be dialed up at a particular dendritic spine, from zero to several times larger than the typical miniature (single synapse?) EPSP, simply by varying the intensity or location of the uncaging laser pulses (and/or the concentration of caged glutamate). Uncertainty over the correct average miniature EPSP size (and the "correct" AMPA/NMDA ratio) therefore complicates the interpretation of the number of synapses needed. Nevertheless, a typical terminal dendritic branch bears hundreds of dendritic spines (Larkman 1991), which represents a massive safety margin: Only a small fraction of the total synapses onto a branch may need to be activated, maybe as few as 5%.

The threshold number is only a rough estimate and will change as a function of baseline membrane potential. Depolarization reduces the glutamate threshold, allowing cooperativity between NMDA spikes from different locations (see above). The threshold number of synapses will also change with the NMDA/AMPA ratio, depending on the recent history of glutamate and coagonist exposure. In addition, any conductances with broadly similar I-V relations to NMDARs, e.g., depolarization-activated Na^+ or Ca^{2+} channels, will reduce the amount of NMDA conductance required to achieve a certain sharpness of threshold (Major et al. 2008, Schiller et al. 2000)—an effect similar to boosting the NMDA/AMPA ratio.

NMDA SPIKES IN VIVO

Dendritic NMDA spikes were discovered in brain slices, and their fundamental properties have been extensively characterized in vitro. NMDA spikes are very robust in thin dendrites of excitatory neurons in all neocortical layers and areas so far tested, with a large safety factor (Branco & Häusser 2011, Gordon et al. 2006, Larkum et al. 2009, Lavzin et al. 2012, Major et al. 2008, Nevian et al. 2007, Schiller et al. 2000).[3]

In vivo, investigators may already have made tantalizing preliminary sightings of events involving NMDA spike/plateaus, supporting the participation of NMDAR regenerativity across the board, from receptive fields to working memory and beyond (Coyle et al. 2003, Daw et al. 1993, de Kock et al. 2007, Major & Tank 2004, Self et al. 2012). A recent in vivo whole-cell patch recording study found a substantial NMDAR component in the angular tuning and artificial whisking responses of layer-4 barrel cortex neurons, which was blocked by an intracellular NMDAR blocker (MK801) or by membrane hyperpolarization, both of which have minimal effects on the rest of the circuit (Lavzin et al. 2012). Modeling suggests that the most likely explanation for the size and duration of the NMDAR-dependent component is dendritic NMDAR-dependent regenerative responses, potentially including multibranch/globally scattered NMDA spikes.

Large, distributed Ca^{2+} transients have been observed in the apical tufts of layer 5 neurons in the somatosensory cortex during sensorimotor behaviors involving active touch (Xu et al. 2012). A plausible explanation is that these Ca^{2+} transients are caused by distal plateau potentials, which are greatly enhanced by top-town excitatory inputs into the tuft. This is what would be expected for distributed tuft NMDA spike/plateaus triggering and cooperatively interacting with Ca^{2+} spike/plateaus in the distal apical trunk/proximal tuft Ca^{2+} zone. Equally intriguingly, voltage-dependent place fields have recently been observed in CA1 pyramidal cells, recorded whole-cell during active

[3]Diameter less than ~1.5 μm, with the possible exception of thin apical trunks (and very proximal basal segments).

maze exploration (Lee et al. 2012). Depolarizing events underlying place field firing appear above a distinct (somatic) voltage threshold in each cell and vanish into noise at more hyperpolarized membrane potentials. Again, this is exactly what one might expect if an important role was played by NMDAR-dependent dendritic regenerative excitation, conditional on the network firing pattern.

Some studies appear to indicate that, at least under some conditions in vivo (e.g., visual gratings, anesthesia), summation of inputs can be essentially linear (Jagadeesh et al. 1993). Recent reports combining two-photon imaging with whole-cell recordings in vivo from layer 2/3 pyramidal neurons in mouse visual, somatosensory, and auditory cortices also noted a lack of participation of dendritic regenerative mechanisms (Chen et al. 2011, Jia et al. 2010, Varga et al. 2011). However, these experiments were performed in unfavorable conditions for dendritic spike generation, including using anesthetized animals and keeping the resting membrane potential below AP threshold, which generally requires negative current injection [for examples of NMDAR regenerativity being blocked by hyperpolarization, see Lavzin et al. (2012)]. Despite these experimental conditions, sensory stimulation evoked scattered dendritic calcium hot spots with transients similar in size to those from several backpropagating APs. Hot spots and spine transients were substantially reduced by NMDAR blockers (including intracellular MK801), pointing to a possible role for regenerative NMDAR activation. Moreover, the number of activated spines per dendrite in vivo (up to ~30 per 100 μm) reported in these studies may well be in the appropriate range for triggering NMDA spikes, given that not all spines were visible (Chen et al. 2011, Varga et al. 2011). It is crucial to emphasize that an NMDA spike does not require neighboring spines to have the same receptive fields/stimulus preferences, nor does it require coactivation of neighboring spines (clustering), nor (if brief and distributed) will it necessarily produce large dendritic shaft Ca^{2+} transients [especially if the NMDAR Ca^{2+} permeability is reduced (Chalifoux & Carter 2010)]. The slow Ca^{2+} transients observed in vivo (Chen et al. 2011, Jia et al. 2010) are consistent with the time courses of NMDA spike/plateaus but appear far slower than what one might expect from single quantal synaptic events.

The functional consequences of NMDA spike/plateaus for cortical processing are potentially versatile and may depend on the cortical region. NMDA plateaus could help sustain network upstates and the persistent firing needed for working memory in the prefrontal cortex and other areas (Antic et al. 2010, Major & Tank 2004, Milojkovic et al. 2005a, Sanders et al. 2013). NMDA spikes could also create and sharpen receptive field selectivity in primary cortices and contribute to top-down/bottom-up interactions (Self et al. 2012), perhaps even within individual dendrites. Combined in vivo dendritic voltage and calcium recording is still in its infancy, but in the not-too-distant future, this method should begin to reveal whether regenerative events or spikes do indeed occur in thin dendrites (and if so, which type). Researchers have barely begun to explore the vast parameter space of possible brain states, neuronal types, dendritic recording sites, and stimuli: Dogma is premature, particularly in light of brain slice experiments suggesting that the biophysical machinery is present in abundance, with a massive safety factor, to support highly nonlinear input summation.

DISCLOSURE STATEMENT

The authors are not aware of any affiliations, memberships, funding, or financial holdings that might be perceived as affecting the objectivity of this review.

LITERATURE CITED

Agmon-Snir H, Segev I. 1993. Signal delay and input synchronization in passive dendritic structures. *J. Neurophysiol.* 70:2066–85

Amitai Y, Friedman A, Connors BW, Gutnick MJ. 1993. Regenerative activity in apical dendrites of pyramidal cells in neocortex. *Cereb. Cortex* 3:26–38

Antic SD, Zhou WL, Moore AR, Short SM, Ikonomu KD. 2010. The decade of the dendritic NMDA spike. *J. Neurosci. Res.* 88:2991–3001

Archie KA, Mel BW. 2000. A model for intradendritic computation of binocular disparity. *Nat. Neurosci.* 3:54–63

Ariav G, Polsky A, Schiller J. 2003. Submillisecond precision of the input-output transformation function mediated by fast sodium dendritic spikes in basal dendrites of CA1 pyramidal neurons. *J. Neurosci.* 23:7750–58

Azouz R, Gray CM. 2000. Dynamic spike threshold reveals a mechanism for synaptic coincidence detection in cortical neurons in vivo. *Proc. Natl. Acad. Sci. USA* 97:8110–15

Behabadi BF, Polsky A, Jadi M, Schiller J, Mel BW. 2012. Location-dependent excitatory synaptic interactions in pyramidal neuron dendrites. *PLoS Comput. Biol.* 8(7):e1002599

Berger T, Larkum ME, Lüscher HR. 2001. High I(h) channel density in the distal apical dendrite of layer V pyramidal cells increases bidirectional attenuation of EPSPs. *J. Neurophysiol.* 85:855–68

Berger T, Lüscher HR. 2003. Timing and precision of spike initiation in layer V pyramidal cells of the rat somatosensory cortex. *Cereb. Cortex* 13:274–81

Binzegger T, Douglas RJ, Martin KA. 2004. A quantitative map of the circuit of cat primary visual cortex. *J. Neurosci.* 24:8441–53

Borst A, Egelhaaf M. 1992. In vivo imaging of calcium accumulation in fly interneurons as elicited by visual motion stimulation. *Proc. Natl. Acad. Sci. USA* 89:4139–43

Branco T, Clark BA, Häusser M. 2010. Dendritic discrimination of temporal input sequences in cortical neurons. *Science* 329:1671–75

Branco T, Häusser M. 2010. The single dendritic branch as a fundamental functional unit in the nervous system. *Curr. Opin. Neurobiol.* 20:494–502

Branco T, Häusser M. 2011. Synaptic integration gradients in single cortical pyramidal cell dendrites. *Neuron* 69:885–92

Briggs F, Callaway EM. 2005. Laminar patterns of local excitatory input to layer 5 neurons in macaque primary visual cortex. *Cereb. Cortex* 15:479–88

Byerly L, Masuda MO. 1979. Voltage-clamp analysis of the potassium current that produces a negative-going action potential in Ascaris muscle. *J. Physiol.* 288:263–84

Cai X, Liang CW, Muralidharan S, Kao JP, Tang CM, Thompson SM. 2004. Unique roles of SK and Kv4.2 potassium channels in dendritic integration. *Neuron* 44:351–64

Cash S, Yuste R. 1999. Linear summation of excitatory inputs by CA1 pyramidal neurons. *Neuron* 22:383–94

Cauller LJ, Connors BW. 1994. Synaptic physiology of horizontal afferents to layer-I in slices of rat SI neocortex. *J. Neurosci.* 14:751–62

Chalifoux JR, Carter AG. 2010. GABAB receptors modulate NMDA receptor calcium signals in dendritic spines. *Neuron* 66:101–13

Chen X, Leischner U, Rochefort NL, Nelken I, Konnerth A, et al. 2011. Functional mapping of single spines in cortical neurons in vivo. *Nature* 475:501–5

Coyle JT, Tsai G, Goff D. 2003. Converging evidence of NMDA receptor hypofunction in the pathophysiology of schizophrenia. *Ann. N. Y. Acad. Sci.* 1003:318–27

Dalby NO, Mody I. 2003. Activation of NMDA receptors in rat dentate gyrus granule cells by spontaneous and evoked transmitter release. *J. Neurophysiol.* 90:786–97

D'Angelo E, Rossi P, Taglietti V. 1994. Voltage-dependent kinetics of N-methyl-D-aspartate synaptic currents in rat cerebellar granule cells. *Eur. J. Neurosci.* 6:640–45

Dantzker JL, Callaway EM. 2000. Laminar sources of synaptic input to cortical inhibitory interneurons and pyramidal neurons. *Nat. Neurosci.* 3:701–7

Daw NW, Stein PS, Fox K. 1993. The role of NMDA receptors in information processing. *Annu. Rev. Neurosci.* 16:207–22 de Kock CP, Bruno RM, Spors H, Sakmann B. 2007. Layer- and cell-type-specific suprathreshold stimulus representation in rat primary somatosensory cortex. *J. Physiol.* 581:139–54

de Polavieja GG, Harsch A, Kleppe I, Robinson HP, Juusola M. 2005. Stimulus history reliably shapes action potential waveforms of cortical neurons. *J. Neurosci.* 25:5657–65

Dzubay JA, Jahr CE. 1996. Kinetics of NMDA channel opening. *J. Neurosci.* 16:4129–34

Garaschuk O, Schneggenburger R, Schirra C, Tempia F, Konnerth A. 1996. Fractional Ca^{2+} currents through somatic and dendritic glutamate receptor channels of rat hippocampal CA1 pyramidal neurones. *J. Physiol.* 491:757–72

Gardner SM, Trussell LO, Oertel D. 2001. Correlation of AMPA receptor subunit composition with synaptic input in the mammalian cochlear nuclei. *J. Neurosci.* 21:7428–37

Gentet LJ, Kremer Y, Taniguchi H, Huang ZJ, Staiger JF, Petersen CC. 2012. Unique functional properties of somatostatin-expressing GABAergic neurons in mouse barrel cortex. *Nat. Neurosci.* 15:607–12

Gordon U, Polsky A, Schiller J. 2006. Plasticity compartments in basal dendrites of neocortical pyramidal neurons. *J. Neurosci.* 26:12717–26

Gulledge AT, Kampa BM, Stuart GJ. 2005. Synaptic integration in dendritic trees. *J. Neurobiol.* 64:75–90

Hardingham NR, Read JC, Trevelyan AJ, Nelson JC, Jack JJ, Bannister NJ. 2010. Quantal analysis reveals a functional correlation between presynaptic and postsynaptic efficacy in excitatory connections from rat neocortex. *J. Neurosci.* 30:1441–51

Häusser M, Mel B. 2003. Dendrites: bug or feature? *Curr. Opin. Neurobiol.* 13:372–83

Häusser M, Spruston N, Stuart GJ. 2000. Diversity and dynamics of dendritic signaling. *Science* 290:739–44

Herz AV, Gollisch T, Machens CK, Jaeger D. 2006. Modeling single-neuron dynamics and computations: a balance of detail and abstraction. *Science* 314:80–85

Jadi M, Polsky A, Schiller J, Mel BW. 2012. Location-dependent effects of inhibition on local spiking in pyramidal neuron dendrites. *PLoS Comput. Biol.* 8:e1002550

Jagadeesh B, Wheat HS, Ferster D. 1993. Linearity of summation of synaptic potentials underlying direction selectivity in simple cells of the cat visual cortex. *Science* 262:1901–4

Jahr CE, Stevens CF. 1993. Calcium permeability of the N-methyl-D-aspartate receptor channel in hippocampal neurons in culture. *Proc. Natl. Acad. Sci. USA* 90:11573–77

Jia H, Rochefort NL, Chen X, Konnerth A. 2010. Dendritic organization of sensory input to cortical neurons in vivo. *Nature* 464:1307–12

Johnston D, Magee JC, Colbert CM, Cristie BR. 1996. Active properties of neuronal dendrites. *Annu. Rev. Neurosci.* 19:165–86

Judkewitz B, Roth A, Häusser M. 2006. Dendritic enlightenment: using patterned two-photon uncaging to reveal the secrets of the brain's smallest dendrites. *Neuron* 50:180–83

Kampa BM, Clements J, Jonas P, Stuart GJ. 2004. Kinetics of Mg^{2+} unblock of NMDA receptors: implications for spike-timing dependent synaptic plasticity. *J. Physiol.* 556:337–45

Kashiwagi K, Masuko T, Nguyen CD, Kuno T, Tanaka I, et al. 2002. Channel blockers acting at N-methyl-D-aspartate receptors: differential effects of mutations in the vestibule and ion channel pore. *Mol. Pharmacol.* 61:533–45

Kleindienst T, Winnubst J, Roth-Alpermann C, Bonhoeffer T, Lohmann C. 2011. Activity-dependent clustering of functional synaptic inputs on developing hippocampal dendrites. *Neuron* 72:1012–24

Koch C, Poggio T, Torre V. 1983. Nonlinear interactions in a dendritic tree: localization, timing, and role in information processing. *Proc. Natl. Acad. Sci. USA* 80:2799–802

Koch C, Segev I. 2000. The role of single neurons in information processing. *Nat. Neurosci.* 3(Suppl.):1171–77

Kole MH, Hallermann S, Stuart GJ. 2006. Single Ih channels in pyramidal neuron dendrites: properties, distribution, and impact on action potential output. *J. Neurosci.* 26:1677–87

Kole MH, Stuart GJ. 2008. Is action potential threshold lowest in the axon? *Nat. Neurosci.* 11:1253–55

Korinek M, Sedlacek M, Cais O, Dittert I, Vyklicky L Jr. 2010. Temperature dependence of N-methyl-D-aspartate receptor channels and N-methyl-D-aspartate receptor excitatory postsynaptic currents. *Neuroscience* 165:736–48

Larkman AU. 1991. Dendritic morphology of pyramidal neurones of the visual cortex of the rat: III. Spine distributions. *J. Comp. Neurol.* 306:332–43

Larkman AU, Major G, Stratford KJ, Jack JJ. 1992. Dendritic morphology of pyramidal neurones of the visual cortex of the rat. IV: Electrical geometry. *J. Comp. Neurol.* 323:137–52
Larkum M. 2013. A cellular mechanism for cortical associations: an organizing principle for the cerebral cortex. *Trends Neurosci.* 36:141–51
Larkum ME, Nevian T. 2008. Synaptic clustering by dendritic signalling mechanisms. *Curr. Opin. Neurobiol.* 18(3):321–31
Larkum ME, Nevian T, Sandler M, Polsky A, Schiller J. 2009. Synaptic integration in tuft dendrites of layer 5 pyramidal neurons: a new unifying principle. *Science* 325:756–60
Larkum ME, Zhu JJ. 2002. Signaling of layer 1 and whisker-evoked Ca^{2+} and Na^{+} action potentials in distal and terminal dendrites of rat neocortical pyramidal neurons in vitro and in vivo. *J. Neurosci.* 22:6991–7005
Larkum ME, Zhu JJ, Sakmann B. 1999. A new cellular mechanism for coupling inputs arriving at different cortical layers. *Nature* 398:338–41
Larkum ME, Zhu JJ, Sakmann B. 2001. Dendritic mechanisms underlying the coupling of the dendritic with the axonal action potential initiation zone of adult rat layer 5 pyramidal neurons. *J. Physiol.* 533:447–66
Lavzin M, Rapoport S, Polsky A, Garion L, Schiller J. 2012. Non-linear dendritic processing determines angular tuning of barrel cortex neurons in-vivo. *Nature* 490:397–401
Lee D, Lin B-J, Lee AK. 2012. Hippocampal place fields emerge upon single-cell manipulation of excitability during behavior. *Science* 337:849–53
Lisman JE, Fellous JM, Wang XJ. 1998. A role for NMDA-receptor channels in working memory. *Nat. Neurosci.* 1:273–75
Liu G. 2004. Local structural balance and functional interaction of excitatory and inhibitory synapses in hippocampal dendrites. *Nat. Neurosci.* 7:373–79
Llinás R, Nicholson C, Freeman JA, Hillman DE. 1968. Dendritic spikes and their inhibition in alligator Purkinje cells. *Science* 160:1132–35
London M, Häusser M. 2005. Dendritic computation. *Annu. Rev. Neurosci.* 28:503–32
London M, Larkum ME, Hausser M. 2008. Predicting the synaptic information efficacy in cortical layer 5 pyramidal neurons using a minimal integrate-and-fire model. *Biol. Cybern.* 99:393–401
Losonczy A, Magee JC. 2006. Integrative properties of radial oblique dendrites in hippocampal CA1 pyramidal neurons. *Neuron* 50:291–307
Lübke J, Feldmeyer D. 2007. Excitatory signal flow and connectivity in a cortical column: focus on barrel cortex. *Brain Struct. Funct.* 212:3–17
Magee JC. 2000. Dendritic integration of excitatory synaptic input. *Nat. Rev. Neurosci.* 1:181–90
Major G. 1992. *The physiology, morphology and modelling of cortical pyramidal neurones*. PhD thesis. Oxford Univ., Oxford, UK
Major G, Larkman AU, Jonas P, Sakmann B, Jack JJ. 1994. Detailed passive cable models of whole-cell recorded CA3 pyramidal neurons in rat hippocampal slices. *J. Neurosci.* 14:4613–38
Major G, Polsky A, Denk W, Schiller J, Tank DW. 2008. Spatiotemporally graded NMDA spike/plateau potentials in basal dendrites of neocortical pyramidal neurons. *J. Neurophysiol.* 99:2584–601
Major G, Tank DW. 2004. Persistent neural activity: prevalence and mechanisms. *Curr. Opin. Neurobiol.* 14:675–84
McCormick DA, Shu Y, Yu Y. 2007. Neurophysiology: Hodgkin and Huxley model—still standing? *Nature* 445:E1–2; discussion E2–3
Mel BW. 1993. Synaptic integration in an excitable dendritic tree. *J. Neurophysiol.* 70:1086–101
Mel BW. 1994. Information processing in dendritic trees. *Neural Comput.* 6:1031–85
Migliore M, Shepherd GM. 2002. Emerging rules for the distributions of active dendritic conductances. *Nat. Rev. Neurosci.* 3:362–70
Milojkovic BA, Radojicic MS, Antic SD. 2005a. A strict correlation between dendritic and somatic plateau depolarizations in the rat prefrontal cortex pyramidal neurons. *J. Neurosci.* 25:3940–51
Milojkovic BA, Radojicic MS, Goldman-Rakic PS, Antic SD. 2004. Burst generation in rat pyramidal neurones by regenerative potentials elicited in a restricted part of the basilar dendritic tree. *J. Physiol.* 558:193–211
Milojkovic BA, Wuskell JP, Loew LM, Antic SD. 2005b. Initiation of sodium spikelets in basal dendrites of neocortical pyramidal neurons. *J. Membr. Biol.* 208:155–69

Milojkovic BA, Zhou WL, Antic SD. 2007. Voltage and calcium transients in basal dendrites of the rat prefrontal cortex. *J. Physiol.* 585(Pt. 2):447–68

Mirsky JS, Nadkarni PM, Healy MD, Miller PL, Shepherd GM. 1998. Database tools for integrating and searching membrane property data correlated with neuronal morphology. *J. Neurosci. Methods* 82:105–21

Myme CI, Sugino K, Turrigiano GG, Nelson SB. 2003. The NMDA-to-AMPA ratio at synapses onto layer 2/3 pyramidal neurons is conserved across prefrontal and visual cortices. *J. Neurophysiol.* 90:771–79

Nevian T, Larkum ME, Polsky A, Schiller J. 2007. Properties of basal dendrites of layer 5 pyramidal neurons: a direct patch-clamp recording study. *Nat. Neurosci.* 10:206–14

Ngo-Anh TJ, Bloodgood BL, Lin M, Sabatini BL, Maylie J, Adelman JP. 2005. SK channels and NMDA receptors form a Ca^{2+}-mediated feedback loop in dendritic spines. *Nat. Neurosci.* 8:642–49

Nimchinsky EA, Yasuda R, Oertner TG, Svoboda K. 2004. The number of glutamate receptors opened by synaptic stimulation in single hippocampal spines. *J. Neurosci.* 24:2054–64

Palmer LM, Schulz JM, Murphy SC, Ledergerber D, Murayama M, Larkum ME. 2012. The cellular basis of $GABA_B$-mediated interhemispheric inhibition. *Science* 335:989–93

Paoletti P. 2011. Molecular basis of NMDA receptor functional diversity. *Eur. J. Neurosci.* 33:1351–65

Patneau DK, Mayer ML. 1990. Structure-activity relationships for amino acid transmitter candidates acting at N-methyl-D-aspartate and quisqualate receptors. *J. Neurosci.* 10:2385–99

Petreanu L, Mao T, Sternson SM, Svoboda K. 2009. The subcellular organization of neocortical excitatory connections. *Nature* 457:1142–45

Platkiewicz J, Brette R. 2011. Impact of fast sodium channel inactivation on spike threshold dynamics and synaptic integration. *PLoS Comput. Biol.* 7:e1001129

Poirazi P, Brannon T, Mel BW. 2003. Pyramidal neuron as two-layer neural network. *Neuron* 37:989–99

Polsky A, Mel B, Schiller J. 2009. Encoding and decoding bursts by NMDA spikes in basal dendrites of layer 5 pyramidal neurons. *J. Neurosci.* 29:11891–903

Polsky A, Mel BW, Schiller J. 2004. Computational subunits in thin dendrites of pyramidal cells. *Nat. Neurosci.* 7:621–27

Popescu G, Robert A, Howe JR, Auerbach A. 2004. Reaction mechanism determines NMDA receptor response to repetitive stimulation. *Nature* 430:790–93

Postlethwaite M, Hennig MH, Steinert JR, Graham BP, Forsythe ID. 2007. Acceleration of AMPA receptor kinetics underlies temperature-dependent changes in synaptic strength at the rat calyx of Held. *J. Physiol.* 579:69–84

Priebe NJ, Ferster D. 2010. Neuroscience: each synapse to its own. *Nature* 464:1290–91

Rall W. 1964. Theoretical significance of dendritic trees for neuronal input-output relations. In *Neural Theory and Modeling*, ed. R Reiss, pp. 73–97. Stanford, CA: Stanford Univ. Press

Rall W. 1977. Core conductor theory and cable properties of neurons. In *Handbook of Physiology, the Nervous System, Cellular Biology of Neurons.* Sect. 1. Vol. 1, Part 1, ed. ER Kandel, pp. 39–98. Bethesda, MD: Am. Physiol. Soc.

Rapp M, Yarom Y, Segev I. 1996. Modeling back propagating action potential in weakly excitable dendrites of neocortical pyramidal cells. *Proc. Natl. Acad. Sci. USA* 93:11985–90

Rhodes P. 2006. The properties and implications of NMDA spikes in neocortical pyramidal cells. *J. Neurosci.* 26:6704–15

Rhodes PA. 1999. Functional implications of active currents in the dendrites of pyramidal neurons. In *Cerebral Cortex*, Vol. 13, ed. PS Ulinski, EG Jones, A Peters, pp. 139–200. New York: Plenum

Rhodes PA, Llinas RR. 2001. Apical tuft input efficacy in layer 5 pyramidal cells from rat visual cortex. *J. Physiol.* 536:167–87

Rinzel J, Rall W. 1974. Transient response in a dendritic neuron model for current injected at one branch. *Biophys. J.* 14:759–90

Sanders H, Berends M, Major G, Goldman MS, Lisman JE. 2013. NMDA and $GABA_B$ (KIR) conductances: the "perfect couple" for bistability. *J. Neurosci.* 33:424–29

Schaefer AT, Larkum ME, Sakmann B, Roth A. 2003. Coincidence detection in pyramidal neurons is tuned by their dendritic branching pattern. *J. Neurophysiol.* 89:3143–54

Schiller J, Major G, Koester HJ, Schiller Y. 2000. NMDA spikes in basal dendrites of cortical pyramidal neurons. *Nature* 404:285–89

Schiller J, Schiller Y. 2001. NMDA receptor-mediated dendritic spikes and coincident signal amplification. *Curr. Opin. Neurobiol.* 11:343–48

Schiller J, Schiller Y, Stuart G, Sakmann B. 1997. Calcium action potentials restricted to distal apical dendrites of rat neocortical pyramidal neurons. *J. Physiol.* 505(Pt. 3):605–16

Schneggenburger R, Zhou Z, Konnerth A, Neher E. 1993. Fractional contribution of calcium to the cation current through glutamate receptor channels. *Neuron* 11:133–43

Schubert D, Staiger JF, Cho N, Kotter R, Zilles K, Luhmann HJ. 2001. Layer-specific intracolumnar and transcolumnar functional connectivity of layer V pyramidal cells in rat barrel cortex. *J. Neurosci.* 21:3580–92

Schwindt P, Crill W. 1999. Mechanisms underlying burst and regular spiking evoked by dendritic depolarization in layer 5 cortical pyramidal neurons. *J. Neurophysiol.* 81:1341–54

Schwindt PC, Crill WE. 1997. Local and propagated dendritic action potentials evoked by glutamate iontophoresis on rat neocortical pyramidal neurons. *J. Neurophysiol.* 77:2466–83

Self MW, Kooijmans RN, Super H, Lamme VA, Roelfsema PR. 2012. Different glutamate receptors convey feedforward and recurrent processing in macaque V1. *Proc. Natl. Acad. Sci. USA* 109:11031–36

Shu Y, Hasenstaub A, Duque A, Yu Y, McCormick DA. 2006. Modulation of intracortical synaptic potentials by presynaptic somatic membrane potential. *Nature* 441:761–65

Sobczyk A, Svoboda K. 2007. Activity-dependent plasticity of the NMDA-receptor fractional Ca(2+) current. *Neuron* 53:17–24

Spruston N. 2008. Pyramidal neurons: dendritic structure and synaptic integration. *Nat. Rev. Neurosci.* 9:206–21

Spruston N, Jaffe DB, Johnston D. 1994. Dendritic attenuation of synaptic potentials and currents: the role of passive membrane properties. *Trends Neurosci.* 17:161–66

Spruston N, Jonas P, Sakmann B. 1995a. Dendritic glutamate receptor channels in rat hippocampal CA3 and CA1 pyramidal neurons. *J. Physiol.* 482(Pt. 2):325–52

Spruston N, Schiller Y, Stuart G, Sakmann B. 1995b. Activity-dependent action-potential invasion and calcium influx into hippocampal CA1 dendrites. *Science* 268:297–300

Stafstrom CE, Schwindt PC, Crill WE. 1984. Cable properties of layer V neurons from cat sensorimotor cortex in vitro. *J. Neurophysiol.* 52:278–89

Stern P, Edwards FA, Sakmann B. 1992. Fast and slow components of unitary EPSCs on stellate cells elicited by focal stimulation in slices of rat visual cortex. *J. Physiol.* 449:247–78

Storm JF. 1990. Potassium currents in hippocampal pyramidal cells. *Prog. Brain Res.* 83:161–87

Stratford K, Mason A, Larkman AU, Major G, Jack JJ. 1989. The modelling of pyramidal neurons in the visual cortex. In *The Computing Neuron*, ed. R Durbin, C Miall, G Mitchison, pp. 296–321. Wokingham, UK: Addison-Wesley

Stuart G, Spruston N. 1998. Determinants of voltage attenuation in neocortical pyramidal neuron dendrites. *J. Neurosci.* 18:3501–10

Stuart G, Spruston N, Sakmann B, Häusser M. 1997. Action potential initiation and backpropagation in neurons of the mammalian CNS. *Trends Neurosci.* 20:125–31

Stuart GJ, Sakmann B. 1994. Active propagation of somatic action potentials into neocortical pyramidal cell dendrites. *Nature* 367:69–72

Stuart GJ, Spruston N, Häusser M. 1999. *Dendrites*. New York: Oxford Univ. Press.

Suzuki T, Kodama S, Hoshino C, Izumi T, Miyakawa H. 2008. A plateau potential mediated by the activation of extrasynaptic NMDA receptors in rat hippocampal CA1 pyramidal neurons. *Eur. J. Neurosci.* 28:521–34

Takahashi N, Kitamura K, Matsuo N, Mayford M, Kano M, et al. 2012. Locally synchronized synaptic inputs. *Science* 335:353–56

Tateno T, Harsch A, Robinson HP. 2004. Threshold firing frequency-current relationships of neurons in rat somatosensory cortex: type 1 and type 2 dynamics. *J. Neurophysiol.* 92:2283–94

Trussell LO, Fischbach GD. 1989. Glutamate receptor desensitization and its role in synaptic transmission. *Neuron* 3:209–18

Varga Z, Jia H, Sakmann B, Konnerth A. 2011. Dendritic coding of multiple sensory inputs in single cortical neurons in vivo. *Proc. Natl. Acad. Sci. USA* 108:15420–25

Vyklický L Jr, Benveniste M, Mayer ML. 1990. Modulation of *N*-methyl-D-aspartic acid receptor desensitization by glycine in mouse cultured hippocampal neurones. *J. Physiol.* 428:313–31

Waters J, Helmchen F. 2006. Background synaptic activity is sparse in neocortex. *J. Neurosci.* 26:8267–77

Wei DS, Mei YA, Bagal A, Kao JP, Thompson SM, Tang CM. 2001. Compartmentalized and binary behavior of terminal dendrites in hippocampal pyramidal neurons. *Science* 293:2272–75

Wilent WB, Contreras D. 2005. Stimulus-dependent changes in spike threshold enhance feature selectivity in rat barrel cortex neurons. *J. Neurosci.* 25:2983–91

Williams SR. 2004. Spatial compartmentalization and functional impact of conductance in pyramidal neurons. *Nat. Neurosci.* 7:961–67

Williams SR, Stuart GJ. 1999. Mechanisms and consequences of action potential burst firing in rat neocortical pyramidal neurons. *J. Physiol.* 521:467–82

Williams SR, Stuart GJ. 2000. Site independence of EPSP time course is mediated by dendritic I-h in neocortical pyramidal neurons. *J. Neurophysiol.* 83:3177–82

Williams SR, Stuart GJ. 2002. Dependence of EPSP efficacy on synapse location in neocortical pyramidal neurons. *Science* 295:1907–10

Williams SR, Stuart GJ. 2003. Role of dendritic synapse location in the control of action potential output. *Trends Neurosci.* 26:147–54

Wong RKS, Prince DA, Basbaum AI. 1979. Intra-dendritic recordings from hippocampal neurons. *Proc. Natl. Acad. Sci. USA* 76:986–90

Xu NL, Harnett MT, Huber D, O'Connor DH, Svoboda K, Magee JC. 2012. Nonlinear dendritic integration of sensory and motor input during an active sensing task. *Nature* 492:247–51

Yamada M, Inanobe A, Kurachi Y. 1998. G protein regulation of potassium ion channels. *Pharmacol. Rev.* 50:723–60

Yu Y, Shu Y, McCormick DA. 2008. Cortical action potential backpropagation explains spike threshold variability and rapid-onset kinetics. *J. Neurosci.* 28:7260–72

Yuan H, Geballe MT, Hansen KB, Traynelis SF. 2008. Structure and function of the NMDA receptor. In *Structural and Functional Organization of the Synapse*, ed. JW Hell, MD Ehlers, pp. 289–316. New York: Springer Sci.+Bus. Media

Yuste R, Gutnick MJ, Saar D, Delaney KR, Tank DW. 1994. Ca^{2+} accumulations in dendrites of neocortical pyramidal neurons: an apical band and evidence for two functional compartments. *Neuron* 13:23–43

Zhang S, Trussell LO. 1994. Voltage clamp analysis of excitatory synaptic transmission in the avian nucleus magnocellularis. *J. Physiol.* 480(Pt. 1):123–36

Zhu JJ. 2000. Maturation of layer 5 neocortical pyramidal neurons: amplifying salient layer 1 and layer 4 inputs by Ca^{2+} action potentials in adult rat tuft dendrites. *J. Physiol.* 526(Pt. 3):571–87

Episodic Neurologic Disorders: Syndromes, Genes, and Mechanisms

Jonathan F. Russell,[1,2,3] Ying-Hui Fu,[1] and Louis J. Ptáček[1,2,3]

[1]Department of Neurology, [2]Medical Scientist Training Program, [3]Howard Hughes Medical Institute, School of Medicine, University of California, San Francisco, California 94158; email: jonathan.russell@ucsf.edu, ying-hui.fu@ucsf.edu, ljp@ucsf.edu

Annu. Rev. Neurosci. 2013. 36:25–50

First published online as a Review in Advance on April 29, 2013

The *Annual Review of Neuroscience* is online at neuro.annualreviews.org

This article's doi:
10.1146/annurev-neuro-062012-170300

Keywords

paroxysmal, neurogenetics, Mendelian, exome sequencing, channelopathy, synaptopathy

Abstract

Many neurologic diseases cause discrete episodic impairment in contrast with progressive deterioration. The symptoms of these episodic disorders exhibit striking variety. Herein we review what is known of the phenotypes, genetics, and pathophysiology of episodic neurologic disorders. Of these, most are genetically complex, with unknown or polygenic inheritance. In contrast, a fascinating panoply of episodic disorders exhibit Mendelian inheritance. We classify episodic Mendelian disorders according to the primary neuroanatomical location affected: skeletal muscle, cardiac muscle, neuromuscular junction, peripheral nerve, or central nervous system (CNS). Most known Mendelian mutations alter genes that encode membrane-bound ion channels. These mutations cause ion channel dysfunction, which ultimately leads to altered membrane excitability as manifested by episodic disease. Other Mendelian disease genes encode proteins essential for ion channel trafficking or stability. These observations have cemented the channelopathy paradigm, in which episodic disorders are conceptualized as disorders of ion channels. However, we expand on this paradigm to propose that dysfunction at the synaptic and neuronal circuit levels may underlie some episodic neurologic entities.

Contents

INTRODUCTION

Whereas many diseases of the nervous system cause progressive deterioration, a sizable fraction of them are predominantly episodic in nature. In this subset of disorders, a patient's neurologic function is impaired during an episode (also known as an attack). Although some patients may suffer from superimposed, chronic neurologic dysfunction, between attacks patients are usually completely normal. Episodes are often triggered by mundane stimuli, such as hunger, fatigue, emotions, stress, exercise, diet, temperature, or hormones. Why these commonplace stimuli trigger episodes of neurologic impairment in some patients but not others is poorly understood.

Many episodic neurologic disorders exist, encompassing a protean range of symptoms. These may include weakness, stiffness, paralysis, arrhythmias, pain, ataxia, migraine, involuntary movements, and seizures. Most episodic neurologic disorders exhibit complex inheritance—that is, disease seems to develop primarily owing to environmental influences rather than genetic ones. In this review, we briefly address the complex disorders that are commonly encountered in clinical practice: transient ischemic attack, syncope, epilepsy, and migraine (**Figure 1**). Aside from these four diseases, which are complex and common, there exist a myriad of symptomatically similar diseases that are complex but rare. Of these, many are in fact progressive neurologic disorders that happen to feature episodic symptoms but are not primarily episodic in nature. However, others are indeed primarily episodic. In this review, we focus on those complex, rare disorders with autoimmune etiology because they have provided substantial pathophysiological insight.

For the same reason, the remaining bulk of our review focuses on episodic neurologic disorders that are Mendelian (**Figure 1**). Each is rare. For these disorders, single gene mutations are sufficient to cause disease. However, even in these genetic diseases, environmental factors can still be important in triggering attacks. Over the past two decades, medical geneticists have extensively clarified the known phenotypes, identified many novel phenotypes, and pinpointed scores of disease genes. In many cases, disease gene discovery has directly led to pathophysiological insight and, in a few cases, even novel treatments. We organize these diverse disorders on the basis of the primary neuroanatomical location affected: skeletal muscle, cardiac muscle, neuromuscular junction (NMJ), peripheral nerve, or CNS. As much as is possible given practical constraints, for each disorder we review the clinical presentation, genetics, and pathophysiology, with particular emphasis on new discoveries and unanswered questions. Finally, in the concluding section we present our view of the field's urgent challenges.

Complex disorder: a disease that develops primarily due to environmental influences, although polygenic inheritance may increase susceptibility

NMJ: neuromuscular junction

Episodic disorder: a disease in which symptoms occur in discrete paroxysms; between paroxysms, patients appear to be normal or near normal

COMPLEX DISORDERS

Complex episodic neurologic disorders develop primarily due to environmental factors, although in most disorders some evidence indicates polygenic inheritance, which remains largely undeciphered (Poduri & Lowenstein 2011, Shyti et al. 2011, Della-Morte et al. 2012). Complex episodic disorders are very common in aggregate. This group includes a legion of causes that are each individually rare, such as autoimmune episodic disorders (see below). Also, four complex disorders are commonly encountered in clinical practice: transient ischemic attack, syncope, epilepsy, and migraine (**Figure 1**).

Figure 1

The landscape of episodic neurologic disorders. There are many episodic neurologic disorders, with a vast range in genetic contribution and prevalence. Prevalence is depicted in crude approximation by the size of each ellipse. This fascinating group includes four common complex disorders (*left*), many rare complex disorders including autoimmune (*lower left*), and many rare Mendelian forms (*right*). Some Mendelian phenotypes exhibit profound similarity to particular complex disorders, as depicted by identical coloration (e.g., *light green* for idiopathic epilepsy and for Mendelian epilepsy syndromes). Often, Mendelian phenotypes share phenotypic and/or genetic characteristics, as depicted by overlap and/or similar coloration of each corresponding ellipse. Some complex disorders seem to have a substantial genetic contribution, particularly epilepsy, but most of the complex inheritance remains unexplained (*middle*). Abbreviations: ADPEAF, autosomal dominant partial epilepsy with auditory features; AF, atrial fibrillation; ATS, Andersen-Tawil syndrome; AHC, alternating hemiplegia of childhood; CIP, congenital insensitivity to pain; CMS, congenital myasthenic syndromes; CPVT, catecholaminergic polymorphic ventricular tachycardia; EA, episodic ataxia; ES, Escobar syndrome; FADS, fetal akinesia deformation sequence; FCM, familial cortical myoclonus; FHM, familial hemiplegic migraine; GEFS+, generalized epilepsy with febrile seizures plus; HH, hereditary hyperekplexia; HyperKPP, hyperkalemic periodic paralysis; HypoKPP, hypokalemic periodic paralysis; IEM, inherited erythromelalgia; JME, juvenile myoclonic epilepsy; LE, Lambert-Eaton myasthenic syndrome; LQTS, long QT syndrome; MC, myotonia congenita; MDS, myoclonus-dystonia syndrome; MG, myasthenia gravis; PAM, potassium-aggravated myotonia; PED, paroxysmal exercise-induced dyskinesia; PEPD, paroxysmal extreme pain disorder; PKD, paroxysmal kinesigenic dyskinesia; PMC, paramyotonia congenita; PME, progressive myoclonic epilepsy; PNKD, paroxysmal nonkinesigenic dyskinesia; SeSAME, seizures, sensorineural deafness, ataxia, mental retardation, and electrolyte imbalance; SQTS, short QT syndrome; TIA, transient ischemic attack; TPP, thyrotoxic periodic paralysis.

Four Common Complex Disorders

A transient ischemic attack (TIA) results from diminished cerebral perfusion that causes abrupt, focal neurological symptoms in a pattern corresponding to the compromised vascular distribution (reviewed by Della-Morte et al. 2012). Cerebral hypoperfusion usually arises from platelet emboli or thrombi that transiently lodge in a cerebral artery but are dislodged before permanent neurologic injury develops; by definition, TIA symptoms resolve within 24 h. Despite prompt resolution, TIAs typically recur over the course of days to weeks with a stereotypic symptom cluster. These patients should be promptly evaluated and treated, usually by addressing the source of emboli or by anticoagulation, to decrease the risk of progression to ischemic stroke (Della-Morte et al. 2012).

Transient cerebral hypoperfusion also results in syncope (i.e., fainting) (reviewed

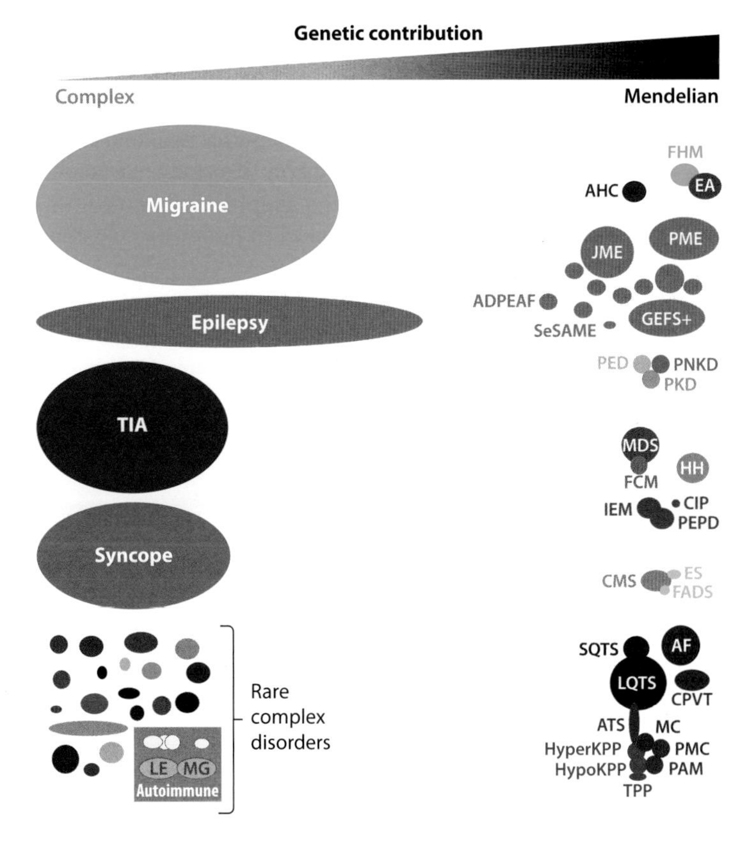

in Gauer 2011). Causes of hypoperfusion include orthostatic hypotension, neurovascular disease, decreased cardiac output (usually from arrhythmia), and neural reflexes. For example, a classic scenario is an unpleasant stimulus, such as extreme fright, triggering a vasovagal reflex of bradycardia and hypotension that leads to syncope. The clinical history and premonitory symptoms—fading vision, nausea, pallor, sweating, etc.—are typically diagnostic. Syncopal patients occasionally exhibit brief, mild myoclonic limb jerks or incontinence but are fully oriented upon awakening, distinguishing syncope from the postictal confusion of a true seizure.

Epilepsy is defined as recurrent seizures, which result from episodic cortical hyperexcitability. Generalized hyperexcitability manifests as unconsciousness, tonic-clonic convulsions, cyanosis, reactive hyperventilation, excessive salivation, and postictal confusion. In comparison, when hyperexcitability involves a focal neurologic region, the symptoms reflect the affected cortical region and vary widely depending on the particular epilepsy syndrome (reviewed by Berg et al. 2010, Berg & Scheffer 2011). Epilepsy has manifold pathophysiologies, primarily structural, metabolic, neurodegenerative, idiopathic, and genetic (including some Mendelian forms; see below and **Figure 1**). Seizures can be triggered by stressors such as infection, psychosomatic trauma, or menses.

The final common complex disorder is migraine. Migraine is characterized by episodic severe headache accompanied by nausea, photophobia, and phonophobia. Many patients experience prodromal symptoms hours to days before headache, which vary widely, and about one-quarter of patients experience aura, commonly visual, which immediately precedes the headache. The pathophysiology of migraine remains hotly disputed but probably involves both alterations in cortical excitability (i.e., cortical spreading depression) as well as vasodilatation of cerebral and meningeal vessels (reviewed in Dodick 2008). Like epilepsy, migraine is commonly triggered by stressors (Haut et al. 2006).

A Myriad of Rare Complex Disorders

Episodic neurologic symptoms also occur in a smorgasbord of complex disorders, each of which is individually rare (**Figure 1**). Many are diseases of progressive deterioration that happen to feature episodic symptoms, whereas others are primarily episodic in nature. For a given clinical finding, the differential diagnosis is typically extensive (**Table 1**). For example, myoclonus is a component of more than 200

Table 1 Diagnosis of complex episodic disorders. Episodic neurologic symptoms occur in a variety of complex disorders. The differential diagnosis for a given finding can be very broad. We provide here episodic presentations, along with references containing diagnostic approaches. We have omitted other presentations, such as weakness, stiffness, paralysis, arrhythmia, hemiplegia, pain, and ataxia.

Presentation	Reference(s) with diagnostic approach
Transient ischemic attack (TIA)	Della-Morte et al. (2012)
Syncope	Gauer (2011)
Seizures	Berg et al. (2010), Berg & Scheffer (2011)
Migraine	Haut et al. (2006)
Exaggerated startle	Dreissen et al. (2012)
Dyskinesia	Fahn (2011)
Myoclonus	Caviness & Brown (2004)
Sleep disorder	Sehgal & Mignot (2011)
Ophthalmic disorder	Sheffield & Stone (2011)

disorders that span the spectrum of neurologic disease: structural malformations, infections, storage disorders, spinocerebellar degenerations, dementias, metabolic derangements, toxin/drug exposures, posthypoxia, malabsorption (celiac disease), various epilepsy syndromes, and many more (Caviness & Brown 2004). The differential diagnosis can be just as broad for other episodic presentations (**Table 1**). Usually the diagnosis is suggested by the entire clinical history and examination rather than by episodic symptoms per se. If a diagnosis remains elusive, probability of an autoimmune or Mendelian cause is increased.

Rare complex disorders with autoimmune etiologies have provided special insight into pathophysiology (reviewed in Vincent et al. 2006, Kleopa 2011). Classically, these autoimmune episodic disorders are caused by autoantibodies against ion channels (**Figure 2*b***). For example, ion channels at the NMJ are the targets of autoantibodies that cause muscle weakness in Lambert-Eaton myasthenic syndrome (LEMS) and myasthenia gravis (MG) (Vincent et al. 2006). Clinically, LEMS is characterized by proximal muscle weakness and autonomic dysfunction, whereas MG patients exhibit striking fatigability, particularly of ocular muscles. Both LEMS and MG must be distinguished from congenital myasthenic syndromes (see below), which also present with weakness but are juvenile onset and Mendelian rather than autoimmune in etiology. LEMS is caused by autoantibodies against presynaptic voltage-gated calcium channels, whereas MG is usually caused by autoantibodies against the nicotinic acetylcholine receptor (AChR) on the motor end plate. Recent work has tied AChR-seronegative MG to autoantibodies against muscle-specific tyrosine kinase (MuSK) (Hoch et al. 2001) or low-density lipoprotein receptor-related protein 4 (Lpr4) (Higuchi et al. 2011, Pevzner et al. 2012). Neither of these targets are ion channels; instead, they promote postsynaptic clustering of the AChR channel (**Figure 2*c***).

Another example of an autoimmune episodic disorder is Isaac's syndrome (neuromyotonia). Isaac's syndrome is a disorder of motor nerve hyperexcitability that can present with hyperhidrosis and a range of muscle symptoms including fasciculations, cramps, stiffness, myokymia (quivering), and pseudomyotonia (slow relaxation). For many years, Isaac's syndrome and two related disorders, Morvan's syndrome and limbic encephalitis, were thought to result from autoantibodies against voltage-gated potassium channels (Vincent et al. 2006). However, the data were mixed. Dalmau et al. recently presented strong evidence that these disorders are instead caused by autoantibodies against the Caspr2-Lgi1 complex, which associates with voltage-gated potassium channels on the motor nerve (Lai et al. 2010, Lancaster et al. 2011, Irani et al. 2012, Loukaides et al. 2012). These findings further illustrate an emerging understanding that in addition to targeting channels directly (**Figure 2*b***), autoantibodies may target channel-associated

STIFF-MAN SYNDROME

A compelling, recently elucidated example of autoantibodies interfering with targets other than ion channels to cause episodic disease is stiff-man syndrome (SMS). SMS is characterized by extreme muscle cramps superimposed on progressive, fluctuating muscle rigidity and stiffness. Tragically, these symptoms are so severe that they often cause joint deformities, skeletal fractures, and even muscle rupture. Cramp attacks are triggered by movement, unanticipated somatosensory stimuli, stress, and strong emotions. Solimena et al. (1988) showed that 80% of patients develop autoantibodies against glutamic acid decarboxylase (GAD), but anti-GAD antibody infusion into model animals does not passively transfer SMS symptoms. Recently, Geis et al. (2010) achieved passive transfer in rats by infusing antiamphiphysin antibodies collected from human SMS patients. Furthermore, antiamphiphysin antibodies were internalized into CNS GABAergic neurons where they inhibited GABA (γ-aminobutyric acid) release. This work demonstrates that SMS is caused by autoantibodies directed against not ion channels but intracellular, presynaptic targets (**Figure 2*c***). It seems likely that other autoimmune or idiopathic disorders may be caused by autoantibodies targeting intracellular, synaptic, or even nonneuronal targets (Lennon et al. 2005).

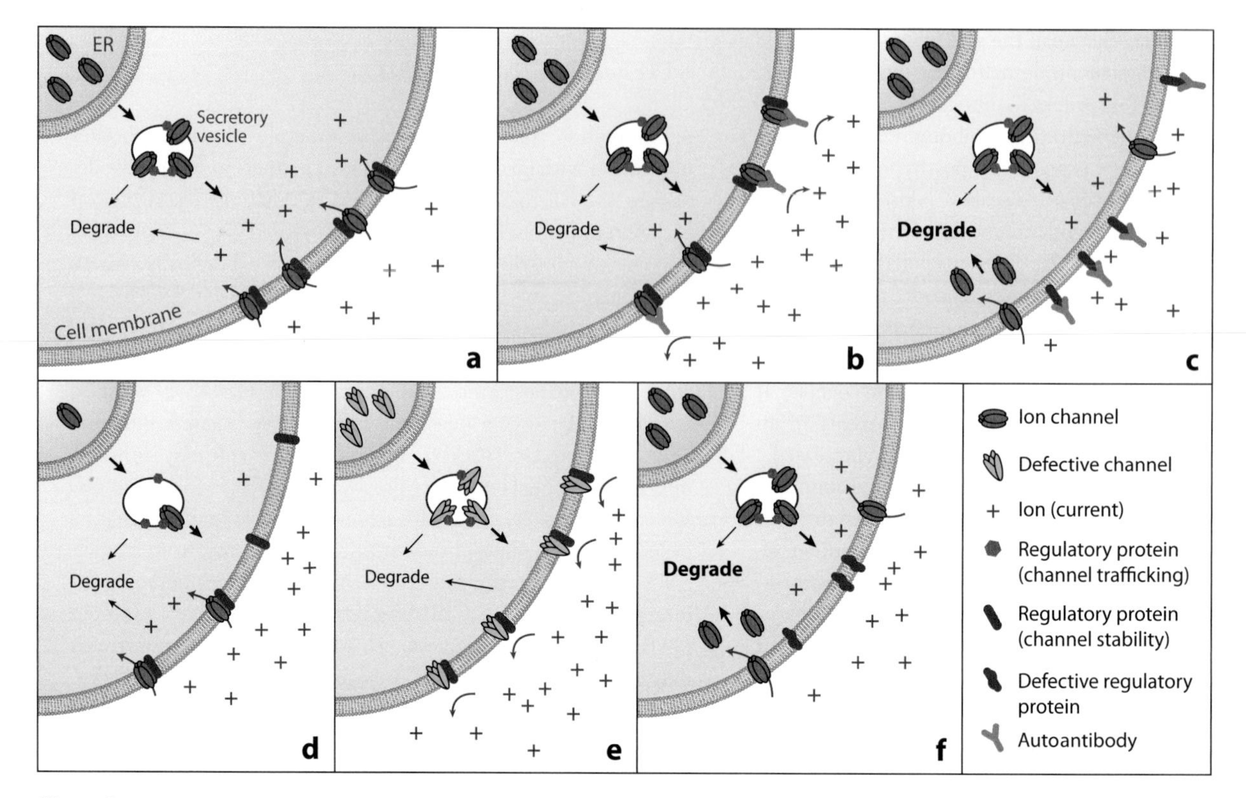

Figure 2

Channelopathy mechanisms. (*a*) Normally, ion channels are trafficked from the endoplasmic reticulum (ER) via secretory vesicles to the cell membrane, where they conduct ions across the membrane to control cellular membrane potential and hence excitability. (*b*) Many autoimmune episodic neurologic disorders are caused by autoantibodies binding to ion channels, which causes defective channel function, such as decreased ion permeability (shown here). (*c*) Investigators have recently associated numerous autoimmune episodic disorders with autoantibodies against targets other than channels. These autoantibodies bind to channel-associated regulatory proteins, thereby indirectly causing defective channel function. Shown here are autoantibodies against a regulatory protein that stabilizes the channel at the membrane. Inhibition of this regulatory protein by autoantibody binding results in decreased channel stability, so more channels are degraded and overall current is decreased. (*d–e*) Most episodic neurologic disorders that exhibit Mendelian inheritance are caused by mutations in ion channel genes. There are many different possible effects of mutations, including absent or decreased expression (*d*), defective trafficking or stability leading to premature degradation, decreased ion permeability (*e*), increased ion permeability, and altered channel kinetics (e.g., delayed inactivation). (*f*) Aside from mutations in ion channels themselves, recent work has identified mutations in genes that do not encode channels. Like the targets of some autoantibodies (*c*), these genes encode regulatory proteins that bind to channels and are critical for channel stability and localization. When mutated, defective regulatory proteins result in aberrant channel trafficking or stability, premature channel degradation, and hence decreased current. For panels *b-f*, see text for examples.

regulatory proteins to cause channel dysfunction indirectly (**Figure 2*c***).

MENDELIAN DISORDERS

General Characteristics

Although each is individually rare, many distinct episodic disorders exhibit Mendelian inheritance. Despite very strong genetic contributions, these diseases share striking similarity with the complex disorders discussed above because patients are often completely normal between attacks, and attacks are often triggered by commonplace environmental stimuli.

We have organized the Mendelian episodic disorders on the basis of the focus of pathology within the nervous system: skeletal muscle, cardiac muscle, NMJ, peripheral nerve, or CNS (**Figure 1**). Most are juvenile onset and autosomal dominant. The vast majority of known disease genes encode ion channels, which has led to use of the term channelopathies to describe this group of disorders (**Figure 2**) (Kullmann 2010, Ryan & Ptáček 2010). However, this usage is a misnomer for two reasons. First, complex episodic disorders can also result from channel dysfunction, as exemplified by the autoimmune diseases discussed above. Second, some nonneurologic Mendelian diseases are caused by mutations in ion channels (Benoit et al. 2010). Thus, the term channelopathy should be reserved for any disorder, complex or Mendelian, neurologic or nonneurologic, caused by channel dysfunction.

Most Mendelian channelopathies affect primarily a single organ system, presumably because a typical ion channel is expressed in one cell type or a limited number of cell types. The exact pathophysiology depends on the mutation severity (e.g., missense or nonsense) and on the type of channel that is mutated (Kullman 2010, Ryan & Ptáček 2010). Missense mutations are often gain-of-function, causing increased ion flux. However, missense mutations can certainly cause loss-of-function (**Figure 2*e***), and dominant-negative mechanisms are also common because some channels are composed of subunits encoded by separate genes that homo- or heteromultimerize into a functional channel. Nonsense (truncation) mutations are almost always loss-of-function or dominant-negative (**Figure 2*d***). Although exceptions abound, generally mutations in sodium channel genes cause gain-of-function, whereas potassium and chloride channel mutations cause loss-of-function. Sodium, potassium, and chloride channels usually cause myocyte or neuronal dysfunction. In contrast, AChR, $GABA_A$ receptor, glycine receptor, and calcium channel mutations typically disrupt synaptic transmission. In any case, in the PNS the ultimate pathophysiology rests on whether the mutation renders the affected cell hypoexcitable or hyperexcitable. In the CNS, pathophysiology rests on whether inhibitory or excitatory neurons are preferentially affected, thereby resulting in a net hypoexcitable or hyperexcitable network (**Figure 3*b***).

Skeletal Muscle

Primary skeletal muscle disorders were the first episodic disorders for which causative mutations were identified (Ptáček et al. 1991, Rojas et al. 1991, McClatchey et al. 1992, Ptáček et al. 1992). These entities were central to establishing the channelopathy paradigm, as all known disease genes encode ion channels. The molecular and cellular pathophysiology has been thoroughly elucidated, and in some cases this insight has led to clinical trials and successful treatments (Tawil et al. 2000). Each disorder falls on a spectrum ranging from muscle hyperexcitability to hypoexcitability. Hyperexcitable muscle presents clinically as myotonia: After contraction, the muscle is slow to relax. In contrast, hypoexcitable muscle presents clinically as weakness or paralysis.

Channelopathy: a disease caused by dysfunction of ion channels; can be inherited (Mendelian) or complex (e.g., autoimmune), neurologic or nonneurologic

Myotonia congenita (MC) constitutes the far hyperexcitable end of the spectrum. Patients suffer stiffness, particularly after prolonged inactivity, which is relieved by repetitive muscle activity (reviewed by Lossin & George 2008). Mutations in *CLCN1*, the skeletal muscle chloride channel, cause myotonia congenita in either autosomal dominant (Thomsen disease; OMIM 160800) or autosomal recessive (Becker disease; OMIM 255700) forms (Koch et al. 1992). Myotonia in Becker disease tends to be more severe and can even be accompanied by episodic weakness.

Like Becker disease, paramyotonia congenita (PMC; OMIM 168300) is characterized by both myotonia and weakness (reviewed in Jurkat-Rott et al. 2010). PMC can be distinguished from MC because PMC patients exhibit paradoxical myotonia, in which myotonia is exacerbated by exercise and can transition to weakness (whereas myotonia in MC is relieved by exercise). Also, PMC attacks are prominently triggered by cold and mostly affect the upper extremities and face. PMC is caused by mutations in *SCN4A*, a skeletal muscle voltage-gated sodium channel (Ptáček

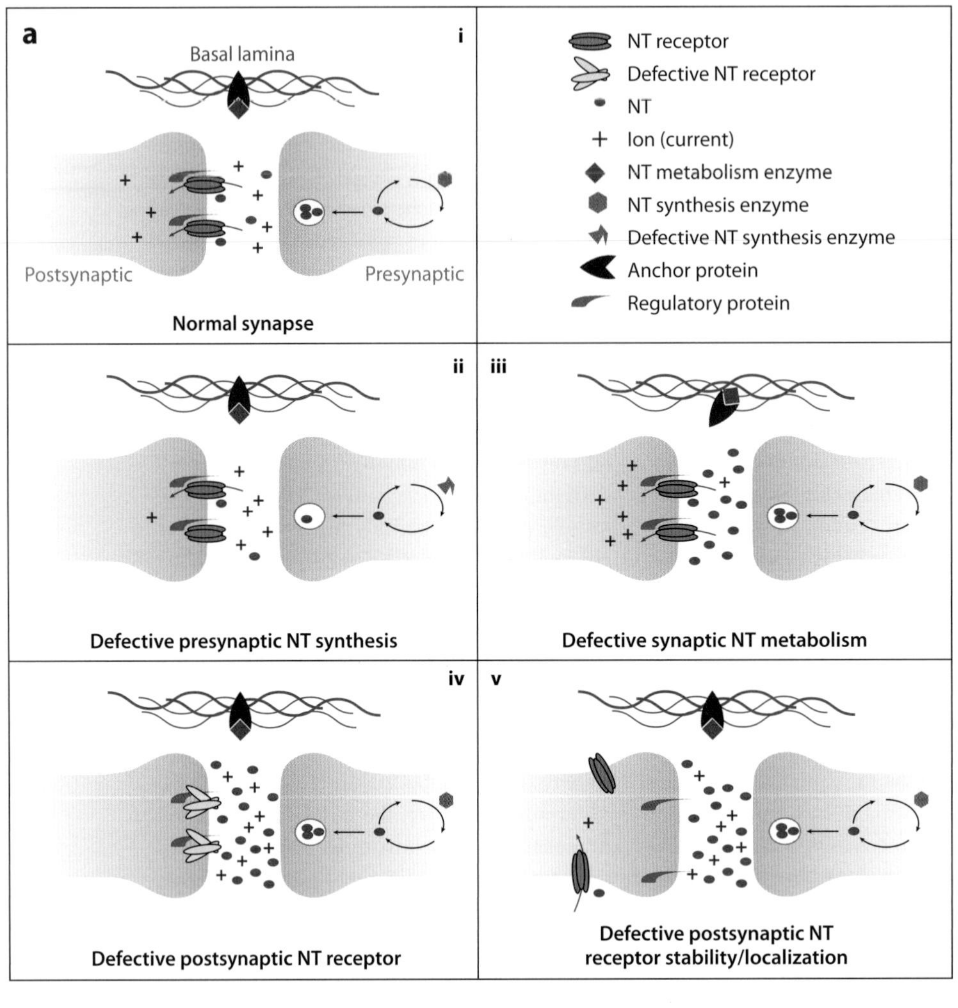
a
i
Basal lamina
Postsynaptic
Presynaptic
Normal synapse
NT receptor
Defective NT receptor
NT
Ion (current)
NT metabolism enzyme
NT synthesis enzyme
Defective NT synthesis enzyme
Anchor protein
Regulatory protein
ii
Defective presynaptic NT synthesis
iii
Defective synaptic NT metabolism
iv
Defective postsynaptic NT receptor
v
Defective postsynaptic NT receptor stability/localization

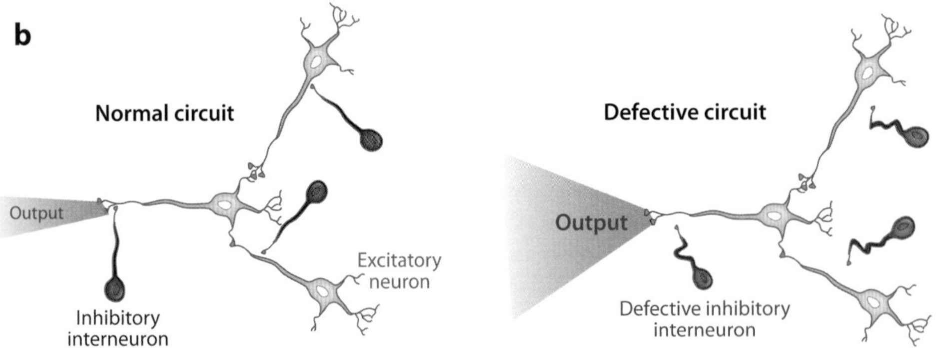
b
Normal circuit
Output
Inhibitory interneuron
Excitatory neuron
Defective circuit
Output
Defective inhibitory interneuron

et al. 1992). Different *SCN4A* mutations cause potassium-aggravated myotonia (PAM; OMIM 608390) (Ptáček et al. 1994b), in which myotonia is instead triggered by potassium. However, in PAM the myotonia never transitions to weakness (Jurkat-Rott et al. 2010).

On the opposite, hypoexcitable end of the spectrum are hyperkalemic periodic paralysis (HyperKPP; OMIM 170500) and hypokalemic periodic paralysis (HypoKPP; OMIM 170400) (Jurkat-Rott et al. 2010). Patients suffer from episodes of weakness or paralysis, triggered by exercise or stress. During attacks, HypoKPP patients are always hypokalemic, whereas HyperKPP patients are often normokalemic. However, for purposes of diagnosis HyperKPP attacks can be induced by a potassium load. Aside from serum potassium levels, the periodic paralyses are clinically distinguishable because HypoKPP never causes myotonia, whereas HyperKPP causes myotonia early in an attack before evolving to weakness/paralysis. Both HyperKPP and HypoKPP patients can develop progressive fixed weakness in those muscles prone to paralytic attacks. Like PMC and PAM, HyperKPP and HypoKPP are caused by mutations in *SCN4A* (Ptáček et al. 1991, Bulman et al. 1999), which highlights the relatedness of these disorders. Other cases of HypoKPP are caused by mutations in *CACNA1S*, which encodes a skeletal muscle voltage-gated calcium channel (Ptáček et al. 1994a). Genotype-phenotype correlations and pathophysiological mechanisms are reviewed elsewhere (Raja Rayan & Hanna 2010).

A variant of HypoKPP is thyrotoxic periodic paralysis (TPP; OMIM 613239) (Jurkat-Rott et al. 2010). TPP patients suffer weakness/paralysis in attacks triggered by thyrotoxicosis. TPP usually afflicts young adult males of Asian ancestry. Ryan et al. (2010) recently demonstrated that some TPP cases are caused by mutations in *KCNJ18*, encoding a skeletal muscle potassium channel. *KCNJ18* mutations have since been discovered in a few patients with nonfamilial HypoKPP but normal thyroid function, so-called "sporadic periodic paralysis" (Cheng et al. 2011). However, *KCNJ18* mutations account for only one-fourth to one-third of TPP cases. We have sequenced many known ion channel genes in *KCNJ18*-mutation negative TPP patients but found no mutations (L.J. Ptáček, unpublished observations), and the genetic basis underlying these cases remains to be elucidated.

Finally, periodic paralysis (either HypoKPP or HyperKPP) is observed in Andersen-Tawil syndrome (ATS; OMIM 170390) (reviewed by Tristani-Firouzi & Etheridge 2010). ATS is a pleiotropic disorder: Periodic paralysis may be accompanied by neurocognitive deficits, skeletal dysmorphisms, and, of paramount clinical importance, long QT syndrome (see below). ATS is caused by mutations in *KCNJ2*, another

Figure 3

Beyond the channelopathy paradigm: mechanisms of synaptopathy and circuitopathy. (*a*) A compelling area for future study is the role of disease genes controlling excitability at the synaptic level, i.e., synaptopathy. In a normal synapse (*i*), a neurotransmitter (NT) is enzymatically synthesized in the presynaptic cell and then released into the synaptic cleft, where it activates postsynaptic NT receptors that then pass current. Regulatory proteins modulate NT receptor stability and localization. NT is metabolized in the synaptic cleft by enzymes that can be anchored to the presynaptic cell, the postsynaptic cell, or the basal lamina (depicted). Defects in these processes can alter synaptic transmission and excitability, as exemplified by defective neuromuscular junction (NMJ) synaptic transmission in the congenital myasthenic syndromes (CMS). CMS can be caused by mutations in a NT synthesis enzyme (*ii*), mutations in proteins that anchor a NT metabolism enzyme to the basal lamina (*iii*), mutations in the NT receptor itself (*iv*), and mutations in proteins that regulate NT receptor stability/localization (*v*). Although this pathophysiology has been heretofore demonstrated only for one type of synapse—the NMJ in CMS—other episodic disorders of the CNS are likely caused by dysfunction of higher-order synapses. Additional mechanisms of synaptopathy are conceivable, such as defects in synaptic vesicle release. (*b*) Another potential mechanism of episodic disease is defective regulation at the circuit level, i.e., circuitopathy. For example, one type of defective circuit is exemplified by GEFS+, which is caused by *SCN1A* mutations that result in decreased GABAergic inhibition by interneurons. Certainly many other types of defective circuits are possible, but whether they can cause episodic neurologic disease is not yet known.

potassium channel (Plaster et al. 2001). *KCNJ2* mutations are found in only 60% of ATS families, suggesting the existence of at least one additional disease gene.

Cardiac Muscle

Numerous Mendelian diseases feature episodic dysfunction of cardiac muscle. These include atrial fibrillation and four ventricular arrhythmias: long QT syndrome, short QT syndrome, Brugada syndrome, and catecholaminergic polymorphic ventricular tachycardia. Most of the disease genes encode ion channels, but some do not. For example, atrial fibrillation (AF) can be caused by autosomal dominant mutations in five potassium channel genes (*KCNA5*, *KCNE2*, *KCNE5*, *KCNJ2*, *KCNQ1*) and three sodium channel genes (*SCN1B*, *SCN2B*, *SCN5A*) (Mahida et al. 2011). However, monogenic AF can also be caused by mutations in *GJA5*, *NPPA*, or *NUP155*, which encode a gap junction protein, an atrial natriuretic peptide, and a nucleoporin, respectively. Pathophysiological mechanisms for these non-ion channel genes remain elusive (Mahida et al. 2011).

A well-known ventricular arrhythmia, long QT syndrome (LQTS) is defined by an elongated QT interval per EKG. This electrical abnormality reflects delayed cardiomyocyte repolarization, which predisposes to torsades de pointes arrhythmia that manifests clinically as palpitations, syncope, or sudden cardiac death. LQTS presents in four clinical subtypes: Andersen-Tawil syndrome (see above), Romano-Ward syndrome (most common; OMIM 192500), Jervell and Lange-Nielsen syndrome (includes congenital deafness; OMIM 220400), and Timothy syndrome (includes cardiac malformations, syndactyly, and autism spectrum disorders; OMIM 601005) (reviewed by McBride & Garg 2010). Like AF, LQTS is genetically heterogeneous, with 13 known genes, including six potassium channel genes (*KCNE1*, *KCNE2*, *KCNH2*, *KCNJ2*, *KCNJ5*, *KCNQ1*), two sodium channel genes (*SCN4B*, *SCN5A*), one calcium channel gene (*CACNA1C*), and four genes not encoding channels: *AKAP9*, *ANK2*, *CAV3*, and *SNTA1*. Mutations in the four nonchannel genes seem to disrupt trafficking or stability of cardiomyocyte ion channels (**Figure 2*f***) (Mohler et al. 2003, Vatta et al. 2006, Chen et al. 2007, Ueda et al. 2008).

Patients with short QT syndrome (SQTS; OMIM 609620) suffer from a shortened QT interval that, like LQTS, predisposes to ventricular arrhythmia and sudden cardiac death. SQTS is caused by autosomal dominant mutations in three potassium channel genes (*KCNH2*, *KCNJ2*, *KCNQ1*), two of which are also LQTS genes (McBride & Garg 2010). Thus, SQTS and LQTS constitute a spectrum ranging from prolonged to delayed cardiomyocyte repolarization. SQTS can also present in concert with Brugada syndrome (OMIM 601144)—defined by elevation of the ST segment in select EKG leads—in patients with mutations in *CACNA1C* and *CACNB2*, which encode calcium channel subunits. Alternatively, isolated Brugada syndrome results from mutations in three sodium channel genes (*SCN1B*, *SCN3B*, *SCN5A*), one potassium channel gene (*KCNE3*), and *GPD1L*, which encodes a protein that regulates SCN5A phosphorylation and thereby modulates sodium current density (**Figure 2*f***) (Valdivia et al. 2009, McBride & Garg 2010).

Finally, another important cause of sudden cardiac death in children is catecholaminergic polymorphic ventricular tachycardia (CPVT; OMIM 604772). In CPVT, the catecholaminergic surge associated with strong emotions or exercise can trigger ventricular tachycardia. The four disease genes, *RYR2*, *CASQ2*, *TRDN*, and *CALM1*, encode essential components of cardiomyocyte calcium signaling (McBride & Garg 2010, Nyegaard et al. 2012, Roux-Buisson et al. 2012). Recently, Watanabe et al. (2009) elegantly identified flecainide as a potent inhibitor of arrhythmias in a CPVT mouse model. This work was validated in human trials (van der Werf et al. 2011), suggesting an effective treatment for this otherwise lethal disease.

Neuromuscular Junction

Mendelian disorders of the neuromuscular junction (NMJ) are known as congenital myasthenic syndromes (CMS). CMS are distinguished from the complex, autoimmune NMJ disorders LEMS and MG (see above) because CMS cannot be treated by immunosuppression. Although CMS subtypes are clinically and genetically heterogeneous, they are usually characterized by episodic ocular and respiratory weakness (reviewed in Barisic et al. 2011). Weakness results from impaired neuromuscular transmission.

Most CMS subtypes are autosomal recessive, caused by mutations in 1 of 14 known genes. The subtypes/genes are classified by the NMJ component that is primarily affected: presynaptic, synaptic, or postsynaptic (**Figure 3*a***) (Barisic et al. 2011). Presynaptic CMS (OMIM 254210) features prominent episodic apnea and is caused by mutations in *CHAT*, encoding an enzyme critical for acetylcholine synthesis (**Figure 3*a*, *part ii***). Synaptic CMS (OMIM 603034) can be caused by mutations in *COLQ* (Mihaylova et al. 2008) and *LAMB2* (Maselli et al. 2009), which encode proteins that anchor acetylcholinesterase to the basal lamina (**Figure 3*a*, *part iii***). The most common type of CMS, by far, is postsynaptic CMS (OMIM 608931), usually caused by defects in AChR subunit genes *CHRNA1*, *CHRNB1*, *CHRND*, and *CHRNE* (**Figure 3*a*, *part iv***). Mutations in another AChR subunit gene, *CHRNG*, cause Escobar syndrome (OMIM 265000), characterized by joint contractures, pterygia (webbing), and in utero CMS-like respiratory distress that resolves by birth (Hoffmann et al. 2006). Finally, rare cases of postsynaptic CMS are caused by mutations in non-AChR genes, namely *AGRN*, *DOK7*, *GFPT1*, *MUSK*, and *RAPSN*. These genes constitute a molecular pathway essential for AChR aggregation and positioning on the postsynaptic membrane (**Figure 3*a*, *part v***) (Barisic et al. 2011). Mutations in some of these genes (*CHRNA1*, *CHRNB1*, *CHRND*, *DOK7*, and *RASPN*) cause fetal akinesia deformation sequence (OMIM 208150), a perinatal lethal syndrome characterized by developmental anomalies such as pterygia as well as fetal akinesia. Given the clinical and genetic overlap, fetal akinesia deformation sequence is considered an extreme phenotype on a continuum that includes Escobar syndrome and CMS. Identifying which gene is mutated in a CMS patient is critical because certain genetic subtypes respond robustly to otherwise toxic medications (Barisic et al. 2011). About half of CMS cases await genetic diagnosis, suggesting a fruitful area for human genetics to provide further insights into synaptic physiology.

Mendelian disorder: a disease in which a mutation in a single gene causes the phenotype; cf. complex genetic disorder

Congenital myasthenic syndromes (CMS): Mendelian disorders of the neuromuscular junction

Peripheral Nerve

Recent studies have shown that mutations in *SCN9A* cause an intriguing trio of pain perception disorders. *SCN9A* encodes a sodium channel that is specifically expressed in those peripheral sensory neurons that function as nociceptors. Mutations lead to aberrant excitability of nociceptive nerves and thus alter the patient's sensitivity to painful stimuli. Autosomal dominant, gain-of-function mutations cause hypersensitivity to pain in two disorders: inherited erythromelalgia (IEM; OMIM 133020) and paroxysmal extreme pain disorder (PEPD; OMIM 167400) (Yang et al. 2004, Fertleman et al. 2006). Burning pain occurs in discrete episodes, accompanied by erythema and swelling. IEM affects the extremities and is commonly triggered by exercise, heat, or dietary components, whereas PEPD affects submandibular, ocular, and rectal areas and is triggered by perianal stimulation (e.g., bowel movements).

Autosomal recessive, loss-of-function *SCN9A* mutations cause the opposite phenotype: congenital insensitivity to pain (CIP; OMIM 243000), characterized by complete absence of pain sensation (Cox et al. 2006). Although ostensibly appealing, patients with CIP suffer substantial injuries and early deaths because of inadvertent trauma. Early studies suggested that CIP patients are otherwise normal, but Weiss et al. (2011) recently

Knockin mouse: a mouse engineered to carry a mutation, usually missense, found in humans; the mutant gene is otherwise intact

demonstrated that the patients cannot smell; furthermore, mice with olfactory sensory neuron-specific *SCN9A* knockout also exhibit anosmia (Weiss et al. 2011). Nevertheless, the specificity and degree of pain relief achieved by genetic inactivation of this channel make it a promising target for developing drugs to treat pain (Clare 2010).

Central Nervous System

A mélange of Mendelian episodic disorders afflict the CNS, with diverse symptoms depending on which region of the CNS is affected. For example, the cerebellum is the focus of pathology in episodic ataxia (EA). EA is distinguished by attacks of ataxia (imbalance and incoordination) without impaired consciousness (reviewed by Jen et al. 2007, Jen 2008). Sometimes, attacks include weakness or are superimposed on progressive ataxia. Seven subtypes (EA1–EA7) vary in associated symptoms, such as myokymia, nystagmus, tinnitus, vertigo, and hemiplegic migraine. Most subtypes share exertion, emotions, and startle as triggers. Each is autosomal dominant, with mutations in *KCNA1* (EA1; OMIM 160120), *CACNA1A* (EA2; OMIM 108500), *CACNB4* (EA5; OMIM 613855), and *SLC1A3* (EA6; OMIM 612656). Despite demonstrated linkage, the genes for EA3 (OMIM 606554), EA4 (OMIM 606552), and EA7 (OMIM 611907) have proven elusive. *KCNA1* and *CACNA1A*/*CACNB4* encode subunits of potassium and calcium channels, respectively, that are highly expressed in Purkinje cells of the cerebellum (Tomlinson et al. 2009), and indeed, mice expressing mutant channels exhibit aberrant Purkinje cell activity (Jen et al. 2007). The EA6 gene, *SLC1A3*, encodes a glutamate reuptake transporter expressed in cerebellar astrocytes (Jen et al. 2005), but how mutant *SLC1A3* alters cerebellar output remains unknown.

EA2 features migraine, so it is also termed familial hemiplegic migraine (FHM) type 1 (OMIM 141500). FHM patients suffer from attacks of headache with hemiplegia during aura. Whereas FHM1 is associated with ataxia, two other subtypes, FHM2 and FHM3, are not. For all subtypes, inheritance is autosomal dominant. FHM2 (OMIM 602481) is caused by mutations in *ATP1A2* (De Fusco et al. 2003), which encodes a sodium-potassium ATPase. FHM3 (OMIM 609634) patients carry mutations in the sodium channel *SCN1A* (Dichgans et al. 2005). Knockin mouse models for both FHM1 and FHM2 have increased susceptibility to cortical spreading depression (CSD) (Tottene et al. 2009, Leo et al. 2011), in keeping with the theory that aberrant cortical excitability is at least partially responsible for migraine pathophysiology (see above).

A related disorder is alternating hemiplegia of childhood (AHC; OMIM 104290), characterized by recurrent attacks of hemiplegia (reviewed by Neville & Ninan 2007). AHC often presents with concomitant epilepsy and developmental delay. As a very rare, sporadic disorder, the etiology of AHC has long remained a mystery, but Heinzen et al. (2012) recently showed that AHC is caused by de novo mutations in *ATP1A3*, another sodium-potassium ATPase gene. The mechanism linking sodium-potassium ATPases and hemiplegia in FHM2 and AHC is not clear. Distinct *ATP1A3* mutations cause a quite dissimilar phenotype: autosomal dominant rapid-onset dystonia-parkinsonism (OMIM 128235; de Carvalho Aguiar et al. 2004).

Some families exhibit autosomal dominant migraine without hemiplegia (OMIM 613656). Investigators have proposed two genes so far. The first, *KCNK18*, a potassium channel, was mutated in a single large affected family (Lafreniere et al. 2010). Moreover, the mutant subunit suppressed wild-type channel function in vitro through a dominant negative effect (Lafreniere et al. 2010). However, the same group (Andres-Enguix et al. 2012) later discovered *KCNK18* variants in unaffected controls, variants that also completely abrogate wild-type channel function. How to reconcile these data? One possibility is that *KCNK18* mutation alone is not sufficiently causative and that the single affected family carries additional migraine susceptibility variants. However, it is extremely unlikely that another locus would cosegregate with the phenotype in the large family (nine individuals affected), which suggests that either the *KCNK18* linkage region itself contains additional susceptibility variants or that *KCNK18* is not causally related to the phenotype. On balance, it is our view that *KCNK18* mutations are likely not causative, although we would happily recant upon identification of additional affected families with *KCNK18* mutations. A stronger case can be made for the second candidate gene, *CSNK1D*, which encodes a kinase, because two independent families carry distinct mutations (Brennan et al. 2013). These mutations alter nearby residues that reside in the highly conserved kinase domain and were shown in vitro to disrupt kinase activity (Xu et al. 2005; K.C. Brennan, E.A. Bates, R.E. Shapiro, J. Zyuzin, W.C. Hallows, H.Y. Lee, C.R. Jones, Y.H. Fui, A.C. Charles, L.J. Ptáček, forthcoming). Furthermore, a mutant mouse model exhibits increased peripheral allodynia, cortical spreading depression, and arterial dilation, all physiological markers of migraine (Brennan et al. 2013). In any case, the overlapping, well-characterized phenotypes of these three families strongly argue for the existence of Mendelian migraine that is distinct from FHM and distinct from migraine with complex inheritance (Eriksen et al. 2004). Heretofore unnamed, we propose the term autosomal dominant migraine (ADM) for this disorder.

Hereditary hyperekplexia (HH) is a disorder of the brain stem, featuring an exaggerated startle reaction (reviewed in Dreissen et al. 2012). Most patients exhibit stiffness at birth that lasts through infancy. Stiffness is exacerbated by handling and is so pronounced that the baby can be held vertically or horizontally without a change in posture. Consciousness is always preserved. Although prolonged stiffness resolves after infancy, throughout the rest of their lives patients suffer from stiffness for a few seconds after an exaggerated startle reaction to unexpected stimuli. HH inheritance can be autosomal dominant, autosomal recessive, or sporadic and is usually caused by mutations in *GLRA1* (OMIM 149400; Shiang et al. 1993). *GLRA1* encodes a subunit of the glycine receptor located in the postsynaptic membrane of glycinergic neurons (**Figure 3*a*, *part iv***) (Dreissen et al. 2012). Less commonly, patients carry mutations in *SCL6A5* (which encodes a presynaptic glycine transporter) (OMIM 614618; Rees et al. 2006) or, very rarely, mutations in *GLRB*, *GPHN*, or *ARHGEF9* (all encode postsynaptic glycinergic proteins; OMIM 138492, 149400, 300607, respectively). These mutations decrease the inhibition exerted by glycinergic neurons in the spinal cord and brain stem, resulting in excessive excitation as reflected by stiffness and exaggerated startle (Dreissen et al. 2012).

Allodynia: pain resulting from a typically innocuous stimulus; a cardinal symptom of migraine

Another fascinating group of episodic disorders are the paroxysmal dyskinesias. In these diseases, excessive excitation manifests as attacks of involuntary movements that can include dystonia (sustained contractions), athetosis (writhing), and chorea (small dance-like movements) (reviewed by Bhatia 2011). There are three Mendelian paroxysmal dyskinesias: paroxysmal exercise-induced dyskinesia (PED; OMIM 612126), paroxysmal nonkinesigenic dyskinesia (PNKD; OMIM 118800), and paroxysmal kinesigenic dyskinesia (PKD; OMIM 128200). All three are autosomal dominant with juvenile onset.

PED is usually triggered by exercise and causes dystonia in the heavily exercised muscles. The PED gene, *SLC2A1*, encodes the main glucose transporter in the brain (Suls et al. 2008, Weber et al. 2008). Mutations impair glucose import into the brain such that the increased energy demand after exercise renders the basal ganglia hypoglycemic. However, this defect must not be specific to the basal ganglia because PED often presents with concomitant neurologic illness that may include hemiplegic migraine, developmental delay, and especially epilepsy. Indeed, De Vivo disease, which is also caused by *SLC2A1* mutations, features severe, global developmental delay and epilepsy; PED may not be appreciable (De Vivo et al. 1991, Seidner et al. 1998). Diagnosis of any phenotype along this PED–De Vivo spectrum is

Pleiotropy: a mutation in a single gene causes multiple phenotypic effects, such as in multiple organ systems

critical because the ketogenic diet is a highly effective treatment (Leen et al. 2010). Ketone bodies use a different transporter to enter the CNS and thereby provide an alternative energy source.

In contrast with PED, PKD attacks are often triggered by startle or sudden movements (hence kinesigenic) (Bruno et al. 2004). PNKD attacks are, by definition, not triggered by movement. Instead, PNKD is induced by ethanol, caffeine, or stress. In both PKD and PNKD, hormones play a role: PKD attacks peak in puberty but decrease in pregnancy, and PNKD attacks increase with menses and thereafter improve with age. However, the exact role of hormones in the genesis of attacks is unclear. PNKD is caused by mutations in the gene *PNKD*, which encodes an enzyme that seems to modulate dopamine release in the striatum in response to ethanol, caffeine, and redox status (Lee et al. 2004, 2012b; Rainier et al. 2004). One hypothesis is that *PNKD* mutations, which alter protein stability and cleavage (Ghezzi et al. 2009, Shen et al. 2011), are gain-of-function, rendering a patient more susceptible to stimuli that trigger dopamine dysregulation in the basal ganglia (Lee et al. 2012b).

Numerous groups recently identified the PKD disease gene, *PRRT2* (Chen et al. 2011, Wang et al. 2011, Heron et al. 2012, Lee et al. 2012a, Li et al. 2012). Within affected families, there is remarkable pleiotropy; some patients suffer from episodic ataxia or hemiplegic migraine (Cloarec et al. 2012, Gardiner et al. 2012, Marini et al. 2012), and others from benign, afebrile infantile epilepsy prior to PKD onset [termed infantile convulsions with choreoathetosis (ICCA)] (Cloarec et al. 2012, Heron et al. 2012, Lee et al. 2012a). In fact, some patients suffer from benign familial infantile epilepsy (BFIE) that resolves in infancy and is never succeeded by PKD (Heron et al. 2012). Given the phenotypic and genetic overlap of these disorders, we have proposed the term PKD/infantile convulsions (PKD/IC) for the diagnosis of any *PRRT2* mutation-positive patient with BFIE, PKD, or both (ICCA) (Cloarec et al. 2012, Lee et al. 2012a). *PRRT2* encodes a transmembrane protein that lacks characteristic ion channel motifs, and its function is not known. Lee et al. (2012a) found that mutations disrupt in vitro binding of PRRT2 to SNAP-25, a synaptic protein integral for neurotransmitter release. However, PRRT2 predominantly localizes to axons rather than to dendritic processes (Lee et al. 2012a), and it is a widespread misconception that individual protein-protein interactions are critical to physiological function (Gillis & Pavlidis 2012). Nevertheless, one possible unifying hypothesis is that PKD and PNKD are both disorders of synaptic regulation (**Figure 3*a***), although this certainly remains unproven.

Two other Mendelian movement disorders are marked by the primary symptom of myoclonus. Myoclonus is defined as sudden, brief, involuntary movements, i.e., twitches. Myoclonus is commonly a component of epilepsy, but in these two disorders seizures do not occur. The first, myoclonus-dystonia syndrome (MDS; OMIM 159900), is characterized by juvenile-onset myoclonus and/or dystonia (Nardocci et al. 2008, Roze et al. 2008). MDS patients suffer severe psychiatric comorbidity, especially depression, although MDS symptoms are clearly ameliorated by ethanol, so depression may simply be a by-product of self-medication by intoxication. MDS is caused by autosomal dominant mutations in *SGCE* (Zimprich et al. 2001), which, like *PRRT2*, encodes a non-ion channel transmembrane protein. Although *SGCE* was cloned in 2001, there has been almost no mechanistic insight since, and its function continues to be obscure.

The second myoclonic disorder, familial cortical myoclonus (FCM), was recently described by our group (Russell et al. 2012). Several features distinguish FCM from MDS: FCM myoclonus is adult onset and slowly progressive, there is neither dystonia nor psychiatric comorbidity, ethanol does not ameliorate symptoms, and FCM exhibits cortical rather than subcortical hyperexcitability. FCM is autosomal dominant and is likely caused by mutation in the gene *NOL3* (Russell et al. 2012). Although we presented substantial genetic, bioinformatic, and biochemical evidence that *NOL3* is the FCM gene (Russell et al. 2012), we could identify only a single, albeit large, affected family, so definitive assignment of *NOL3* as the disease gene awaits discovery of independent FCM families with *NOL3* mutations and/or validation via a knockin animal model (both of which are in progress). *NOL3* encodes a well-characterized inhibitor of apoptosis (Koseki et al. 1998, Nam et al. 2004, Donath et al. 2006), but the mechanism linking *NOL3* mutations and hyperexcitability, as manifested by myoclonus in patients, remains entirely speculative.

The broadest category of inherited episodic CNS disorders is composed of the Mendelian epilepsy syndromes (reviewed by Poduri & Lowenstein 2011). An illustrative example is generalized epilepsy with febrile seizures plus (GEFS+). Whereas febrile seizures are common and typically benign in young children, their persistence after age six defines GEFS+. Most GEFS+ cases are genetically complex, but ~10% are autosomal dominant. So far, all known disease genes encode ion channels, including three voltage-gated sodium channel genes (*SCN1A*, *SCN1B*, *SCN2A*) and two $GABA_A$ receptor ($GABA_AR$) subunit genes: *GABRG2* and *GABRD* (Wallace et al. 1998, Escayg et al. 2000, Baulac et al. 2001, Sugawara et al. 2001, Dibbens et al. 2004). Thirteen additional loci have been linked to GEFS+ and await gene identification (Morar et al. 2011, Poduri & Lowenstein 2011). Mutations in *SCN1A* are most common. In fact, other *SCN1A* mutations cause more severe phenotypes along the GEFS+ continuum: severe myoclonic epilepsy of infancy (SMEI, also known as Dravet syndrome), borderline SMEI, and intractable epilepsy of childhood (IEC) (OMIM 604403; Stafstrom 2009). *SCN1A* knockout and knockin mice die young from epilepsy, and their hippocampal GABAergic interneurons are hypoexcitable, leading to a net hyperexcitable state (**Figure 3*b***) (Yu et al. 2006, Martin et al. 2010). Given that GEFS+ can also be caused by mutations in $GABA_AR$ subunits, interneuron dysfunction is likely a common mechanism underlying the entire GEFS+ continuum, although to our knowledge this hypothesis remains to be tested in *GABRG2* and *GABRD* knockout mice.

Knockout mouse: a mouse engineered to entirely lack a gene

Predictably, many other Mendelian epilepsy syndromes are caused by mutations in ion channels. These phenotypes and the associated genes are extensively reviewed elsewhere (Helbig et al. 2008, Mantegazza et al. 2010, Nicita et al. 2012). The known genes include two $GABA_AR$ subunits, two AChR subunits, the brain glucose transporter, a sodium-potassium ATPase, four potassium channels, one calcium channel subunit, one chloride channel, and one sodium channel. Two other sodium channel genes, *SCN3A* and *SCN8A*, have been associated with childhood epilepsy (Holland et al. 2008, Estacion et al. 2010, Veeramah et al. 2012), but mutations were each detected in only a single patient; therefore, definitive assignment of these genes will require the discovery of distinct mutations in additional patients.

Recent work has demonstrated that mutations in nonchannel genes can cause Mendelian epilepsy. For example, patients with a phenotype along the GEFS+ spectrum who lack an *SCN1A* mutation and are female sometimes harbor *PCDH19* mutations (OMIM 300088; Depienne et al. 2009, 2011). *PCDH19* encodes a calcium-dependent cell adhesion protein (Morishita & Yagi 2007). Two fascinating unanswered questions are, how do mutations in a cell adhesion protein cause epilepsy, and why do these mutations cause disease only in females?

Another well-characterized epilepsy syndrome caused by mutations in a nonchannel gene is autosomal dominant partial epilepsy with auditory features (ADPEAF; OMIM 600512). The disease gene is *LGI1* (Kalachikov et al. 2002). Lgi1 associates with voltage-gated potassium channels, and autoantibodies to the Lgi1-Caspr2 complex are associated with the autoimmune, peripheral nerve disorder known as Isaac's syndrome (see above). How *LGI1* mutations cause temporal lobe epilepsy without any peripheral nerve hyperexcitability is not

Exome sequencing: a method of sequencing and analyzing all protein-coding sequence from a given patient (that patient's "exome")

clear, and in fact, Lgi1 function is essentially unknown. Recent work suggests that Lgi1 inhibits seizure-induced trafficking of potassium channels in thalamocortical neurons (**Figure 2*c***) (Smith et al. 2012); however, it also seems to function in remodeling of synapses and sensory axons (Zhou et al. 2009, 2012), and it is unclear how these findings can be reconciled into a unifying hypothesis.

Yet another nonchannel epilepsy gene, *EFHC1*, is mutated in one subset of juvenile myoclonic epilepsy (JME; OMIM 254770) (Suzuki et al. 2004). *EFHC1* encodes a microtubule-associated protein that regulates cell division and neuronal migration during cortical development (de Nijs et al. 2009). In fact, many genes that function in neuronal migration are mutated in Mendelian syndromes that feature epilepsy as one symptom along with dramatic, radiologically evident malformations of cortical development (Andrade 2009, Barkovich et al. 2012). For example, severe mutations in a gene essential for interneuron migration, *ARX*, cause gross cortical malformations, but milder mutations result in less severe phenotypes such as early infantile epileptic encephalopathy or even isolated mental retardation (Kitamura et al. 2002, Stromme et al. 2002, Shoubridge et al. 2010). Likewise, severe infantile epilepsy phenotypes are caused by mutations in *CDKL5*, *STXBP1*, and *TBC1D24*, which are nonchannel genes that are clearly essential for normal brain development, although their exact function remains unknown (Weaving et al. 2004, Saitsu et al. 2008, Corbett et al. 2010, Falace et al. 2010). On the basis of these data, it seems likely that many complex cases of epilepsy—which have a substantial genetic contribution (**Figure 1**)—may result from a constellation of more subtle, genetically influenced defects in cortical development.

One last class of Mendelian epilepsies is progressive myoclonic epilepsy (PME): juvenile-onset, myoclonic epilepsy in association with neurodegeneration, dementia, and early death (reviewed by Ramachandran et al. 2009). There are many PME subtypes and causative genes, mostly encoding lysosomal proteins (Ramachandran et al. 2009). Some clinical variants also feature substantial pathology outside the CNS, such as action myoclonus-renal failure (AMRF; OMIM 254900) syndrome (Badhwar et al. 2004). A similar disorder, deemed SeSAME syndrome (seizures, sensorineural deafness, ataxia, mental retardation, and electrolyte imbalance; OMIM 612780), is caused by mutations in the potassium channel *KCNJ10* (Bockenhauer et al. 2009, Scholl et al. 2009). Epilepsy in SeSAME syndrome is less severe, does not progress, and is not accompanied by neurodegeneration, so it does not qualify as a PME subtype. We highlight it here to emphasize a somewhat unusual case in which ion channel mutations cause dramatic pleiotropy in diverse organ systems.

Finally, primary episodic sleep disorders, a few of which are Mendelian, are reviewed elsewhere (Sehgal & Mignot 2011, Zhang et al. 2011). Mendelian ophthalmic disorders are extraordinarily diverse and have been very well characterized; however, they are usually progressive rather than episodic, as reviewed by Sheffield & Stone (2011).

BEYOND THE CHANNELOPATHY PARADIGM

We have highlighted the immense progress made in characterizing the phenotypes, genetics, and pathophysiology of episodic neurologic disorders. In our view, four main objectives should be the focus of future work.

The first two objectives are broadly applicable to human genetics. First, we should identify all Mendelian phenotypes and disease genes. This goal is realistic given the advent of inexpensive, high-throughput sequencing (Gonzaga-Jauregui et al. 2012). Many sporadic or seemingly idiopathic cases of severe, stereotyped disorders are likely the result of mutations that are remarkably straightforward to detect via exome sequencing (Choi et al. 2009, Bamshad et al. 2011). On the other hand, many of the disorders described above were characterized in large families with highly significant linkage; yet, cloning of the disease genes at linked loci remained elusive for years, often because of the sheer number of candidate genes within the critical regions. This problem is now easily circumvented by high-throughput sequencing (Lee et al. 2012b, Russell et al. 2012).

However, there will be challenges. Foremost among them is evaluating whether a rare variant is truly causative. Numerous "disease genes" have been assigned on the basis of a single affected patient carrying a variant (Holland et al. 2008, Veeramah et al. 2012). Although these data certainly represent grounds for functional investigation, the gold standard should continue to be allelic heterogeneity. In fact, every human carries hundreds of rare, novel variants (Tennessen et al. 2012), so even using the identification of two rare variants in the same gene from a large collection of patients to claim causality may be unwarranted (O'Roak et al. 2012, Sanders et al. 2012). Instead, large families with the statistical power to detect linkage will remain valuable because linkage constrains the pool of rare variants that must be considered for causality. Even when the evidence includes a highly penetrant phenotype, large families, linkage, and allelic heterogeneity, some mutations are not sufficient to cause disease in unrelated patients (Klassen et al. 2011). The sobering reality is that determining the causal relationships between mutations and Mendelian diseases may take many years to unravel, particularly for genes of unknown function.

The second main objective is to identify genetic risk factors for related, genetically complex disorders (**Figure 1**). It was hoped that genome-wide association studies (GWAS) would provide an unbiased method for doing so; however, except for a few remarkable early findings (Hageman et al. 2005, Duerr et al. 2006), despite extensive patient collections the calculated effect sizes have been very small. Consequently, the overwhelming majority of GWAS associations have been insufficient to induce researchers to pursue functional biological investigation or, when investigated, are found to have no functional effect. It remains an open question whether high-throughput sequencing will prove fruitful where GWAS was not, although we remain hopeful. In our view, one possibility merits serious consideration: the null hypothesis. Perhaps the "missing heritability" (Eichler et al. 2010) is not missing after all but has instead been grossly overestimated by inherently biased measures of heritability. Only time will tell. Given this history, we are puzzled as to why more resources are not directed toward the tried-and-true approach of applying our comprehension of rare Mendelian disorders to understand pathophysiology of related complex diseases, as exemplified by Brown & Goldstein's (2009) elucidation of familial hypercholesterolemia, which sparked development of the blockbuster statin drugs. This approach has seemingly fallen out of favor. In this regard, episodic neurologic disorders are particularly tantalizing because Mendelian forms exhibit very specific symptoms and symptom clusters (e.g., congenital insensitivity to pain) that may allow for pharmacological treatments with minimal side effects.

A third goal is to understand why these disorders are episodic in nature. Typically, patients appear to be normal between attacks and yet suffer extreme dysfunction during an attack. Furthermore, attacks are triggered by precipitants that are routinely encountered by affected patients and unaffected patients alike, and even in affected patients, these precipitants do not always trigger an attack. The link between the precipitant and an attack is clear for some disorders, such as the primary skeletal muscle disorders in which altered extracellular ion concentrations affect myocyte excitability. Another well-characterized example is PED, in which exercise depletes blood glucose to cause CNS hypoglycemia and hence dyskinesia. However, other triggers remain baffling. For example, strong emotion is a common trigger, but how do psychological factors trigger neurological dysfunction? No one knows.

Finally, the fourth objective is to expand on the channelopathy paradigm (**Figure 3**). Although more mutations in ion channels will likely be found, it has become evident that many

Allelic heterogeneity: distinct mutations in the same gene cause identical/similar phenotypes; the gold standard of establishing causality in genetics

Genome-wide association study (GWAS): method using sets of cases and controls in which polymorphisms across the genome are tested for statistical association with disease

genes that do not encode channels can be mutated to cause episodic disorders. For some, the effect of gene mutation is easily tied to changes in excitability, such as when the genes encode proteins essential for ion channel trafficking, stability, or function (**Figure 2*f***). However, as we have repeatedly noted, for many genes the link to cellular excitability remains poorly understood. We propose that rather than disease genes affecting excitability in a cell-intrinsic way (e.g., ion channel expression or localization on the cell membrane), a compelling area for future study is the role of disease genes in modulating excitability at the synaptic level. This concept of a synaptopathy is certainly not new because the congenital myasthenic syndromes have long been known to be disorders of synaptic regulation (**Figure 3*a***). However, the concept of synaptopathy has, to date, been restricted to the NMJ (**Figure 3*a***), and it seems probable that higher synapses may be dysfunctional in episodic disorders of the CNS. We cannot help but speculate that the lessons learned by investigating synaptic function in Mendelian episodic disorders may apply to various complex disorders such as autism that are known synaptopathies (Grabrucker et al. 2011).

Likewise, disordered regulation of excitability at the circuit level (circuitopathy) likely contributes to episodic disorders of the CNS (**Figure 3*b***). For example, *SCN1A* mutations in GEFS+, in which sodium channel dysfunction results in aberrant interneurons, can be conceptualized as a channelopathy or a circuitopathy because ion channels and also neuronal circuits are defective. Many other types of aberrant circuits are possible, and we anticipate that some nonchannel genes, especially those implicated in brain development and/or mutated in Mendelian epilepsy syndromes, cause disease by altering circuit wiring.

In summary, the past two decades have borne witness to the description of many novel episodic neurologic phenotypes, the identification of causative mutations, and the elucidation of underlying pathophysiology. On all three fronts—syndromes, genes, and mechanisms—much work remains. With the widespread application of high-throughput genomic technology, we expect progress to continue apace. In time, we expect that these fronts will be conquered and the spoils will redound in the form of novel treatments for these tragic diseases. We owe as much to our patients.

SUMMARY POINTS

1. Episodic neurologic disorders cause symptoms in discrete attacks. Between attacks, patients appear to be normal.
2. Attacks are often triggered by commonplace stimuli such as hunger or emotional stress. For most disorders, we do not understand how these stimuli trigger attacks.
3. Episodic neurologic disorders can be caused by a mutation in a single gene (Mendelian). Alternatively, they may be genetically complex: influenced primarily by environmental factors, with some polygenic genetic contribution. The four common complex disorders are transient ischemic attack, syncope, epilepsy, and migraine.
4. Many rare complex episodic neurologic disorders exist. For example, autoimmune episodic disorders are caused mostly by autoantibodies against ion channels or channel-related proteins.
5. Many Mendelian episodic neurologic disorders exist, each of which is rare. Most affect a single anatomical location: skeletal muscle, cardiac muscle, neuromuscular junction, peripheral nerve, or CNS.

6. Most Mendelian episodic disorder genes encode ion channels. Mutant channels are dysfunctional, and ensuing alterations in membrane excitability cause disease.
7. Investigators have recently identified many causative genes that do not encode ion channels. Some alter expression, localization, or function of channels. However, for many others we do not know yet how the mutant gene leads to changes in excitability.
8. Recent progress indicates that episodic neurologic disorders may also be caused by dysfunction at the synaptic and neuronal circuit levels, suggesting an expansion of the channelopathy paradigm.

FUTURE ISSUES

1. Characterize all Mendelian phenotypes, and for each disorder, identify all causative genes.
2. Identify genetic risk factors for related, genetically complex disorders.
3. Investigate why these disorders are episodic in nature.
4. Expand on the channelopathy paradigm: Investigate dysfunction at the level of the synapse and neuronal circuit in episodic disorders of the CNS.

DISCLOSURE STATEMENT

The authors are not aware of any affiliations, memberships, funding, or financial holdings that might be perceived as affecting the objectivity of this review.

ACKNOWLEDGMENTS

We thank S.R. Russell, MD, for critical analysis of the manuscript. This work was supported by National Institutes of Health grant F31NS077533 to J.F.R.

LITERATURE CITED

Andrade DM. 2009. Genetic basis in epilepsies caused by malformations of cortical development and in those with structurally normal brain. *Hum. Genet.* 126:173–93

Andres-Enguix I, Shang L, Stansfeld PJ, Morahan JM, Sansom MSP, et al. 2012. Functional analysis of missense variants in the TRESK (KCNK18) K^+ channel. *Sci. Rep.* 2:237

Badhwar A, Berkovic S, Dowling JP, Gonzales M, Narayanan S, et al. 2004. Action myoclonus-renal failure syndrome: characterisation of a unique cerebro-renal disorder. *Brain* 127:2173–82

Bamshad MJ, Ng SB, Bigham AW, Tabor HK, Emond MJ, et al. 2011. Exome sequencing as a tool for Mendelian disease gene discovery. *Nat. Rev. Genet.* 12:745–55

Barišić N, Chaouch A, Müller JS, Lochmüller H. 2011. Genetic heterogeneity and pathophysiological mechanisms in congenital myasthenic syndromes. *Eur. J. Paediatr. Neurol.* 15:189–96

Barkovich AJ, Guerrini R, Kuzniecky RI, Jackson GD, Dobyns WB. 2012. A developmental and genetic classification for malformations of cortical development: update 2012. *Brain* 135:1348–69

Baulac S, Huberfeld G, Gourfinkel-An I, Mitropoulou G, Beranger A, et al. 2001. First genetic evidence of GABA(A) receptor dysfunction in epilepsy: a mutation in the gamma2-subunit gene. *Nat. Genet.* 28:46–48

Benoit G, Machuca E, Heidet L, Antignac C. 2010. Hereditary kidney diseases: highlighting the importance of classical Mendelian phenotypes. *Ann. N. Y. Acad. Sci.* 1214:83–93

Berg AT, Berkovic SF, Brodie MJ, Buchhalter J, Cross JH, et al. 2010. Revised terminology and concepts for organization of seizures and epilepsies: report of the ILAE Commission on Classification and Terminology, 2005–2009. *Epilepsia* 51:676–85

Berg AT, Scheffer IE. 2011. New concepts in classification of the epilepsies: entering the 21st century. *Epilepsia* 52:1058–62

Bhatia KP. 2011. Paroxysmal dyskinesias. *Mov. Disord.* 26:1157–65

Bockenhauer D, Feather S, Stanescu HC, Bandulik S, Zdebik AA, et al. 2009. Epilepsy, ataxia, sensorineural deafness, tubulopathy, and KCNJ10 mutations. *N. Engl. J. Med.* 360:1960–70

Brennan KC, Bates EA, Shapiro RE, Zyuzin J, Hallows WC, et al. 2013. Casein kinase iδ mutations in familial migraine and advanced sleep phase. *Sci. Transl. Med.* 5:183ra56

Bruno MK, Hallett M, Gwinn-Hardy K, Sorensen B, Considine E, et al. 2004. Clinical evaluation of idiopathic paroxysmal kinesigenic dyskinesia: new diagnostic criteria. *Neurology* 63:2280–87

Bulman DE, Scoggan KA, van Oene MD, Nicolle MW, Hahn AF, et al. 1999. A novel sodium channel mutation in a family with hypokalemic periodic paralysis. *Neurology* 53:1932–36

Caviness J, Brown P. 2004. Myoclonus: current concepts and recent advances. *Lancet Neurol.* 3:598–607

Chen L, Marquardt ML, Tester DJ, Sampson KJ, Ackerman MJ, Kass RS. 2007. Mutation of an A-kinase-anchoring protein causes long-QT syndrome. *Proc. Natl. Acad. Sci. USA* 104:20990–95

Chen WJ, Lin Y, Xiong ZQ, Wei W, Ni W, et al. 2011. Exome sequencing identifies truncating mutations in PRRT2 that cause paroxysmal kinesigenic dyskinesia. *Nat. Genet.* 43:1252–55

Cheng CJ, Lin SH, Lo YF, Yang SS, Hsu YJ, et al. 2011. Identification and functional characterization of Kir2.6 mutations associated with non-familial hypokalemic periodic paralysis. *J. Biol. Chem.* 286:27425–35

Choi M, Scholl UI, Ji W, Liu T, Tikhonova IR, et al. 2009. Genetic diagnosis by whole exome capture and massively parallel DNA sequencing. *Proc. Natl. Acad. Sci. USA* 106:19096–101

Clare JJ. 2010. Targeting voltage-gated sodium channels for pain therapy. *Expert Opin. Investig. Drugs* 19:45–62

Cloarec R, Bruneau N, Rudolf G, Massacrier A, Salmi M, et al. 2012. PRRT2 links infantile convulsions and paroxysmal dyskinesia with migraine. *Neurology* 79:2097–103

Corbett MA, Bahlo M, Jolly L, Afawi Z, Gardner AE, et al. 2010. A focal epilepsy and intellectual disability syndrome is due to a mutation in TBC1D24. *Am. J. Hum. Genet.* 87:371–75

Cox JJ, Reimann F, Nicholas AK, Thornton G, Roberts E, et al. 2006. An SCN9A channelopathy causes congenital inability to experience pain. *Nature* 444:894–98

de Carvalho Aguiar P, Sweadner KJ, Penniston JT, Zaremba J, Liu L, et al. 2004. Mutations in the Na^+/K^+-ATPase alpha3 gene ATP1A3 are associated with rapid-onset dystonia parkinsonism. *Neuron* 43:169–75

De Fusco M, Marconi R, Silvestri L, Atorino L, Rampoldi L, et al. 2003. Haploinsufficiency of ATP1A2 encoding the Na^+/K^+ pump alpha2 subunit associated with familial hemiplegic migraine type 2. *Nat. Genet.* 33:192–96

de Nijs L, Léon C, Nguyen L, Loturco JJ, Delgado-Escueta AV, et al. 2009. EFHC1 interacts with microtubules to regulate cell division and cortical development. *Nat. Neurosci.* 12:1266–74

De Vivo DC, Trifiletti RR, Jacobson RI, Ronen GM, Behmand RA, Harik SI. 1991. Defective glucose transport across the blood-brain barrier as a cause of persistent hypoglycorrhachia, seizures, and developmental delay. *N. Engl. J. Med.* 325:703–9

Della-Morte D, Guadagni F, Palmirotta R, Testa G, Caso V, et al. 2012. Genetics of ischemic stroke, stroke-related risk factors, stroke precursors and treatments. *Pharmacogenomics* 13:595–613

Depienne C, Bouteiller D, Keren B, Cheuret E, Poirier K, et al. 2009. Sporadic infantile epileptic encephalopathy caused by mutations in PCDH19 resembles Dravet syndrome but mainly affects females. *PLoS Genet.* 5:e1000381

Depienne C, Trouillard O, Bouteiller D, Gourfinkel-An I, Poirier K, et al. 2011. Mutations and deletions in PCDH19 account for various familial or isolated epilepsies in females. *Hum. Mutat.* 32:E1959–75

Dibbens LM, Feng HJ, Richards MC, Harkin LA, Hodgson BL, et al. 2004. GABRD encoding a protein for extra- or peri-synaptic GABAA receptors is a susceptibility locus for generalized epilepsies. *Hum. Mol. Genet.* 13:1315–19

Dichgans M, Freilinger T, Eckstein G, Babini E, Lorenz-Depiereux B, et al. 2005. Mutation in the neuronal voltage-gated sodium channel SCN1A in familial hemiplegic migraine. *Lancet* 366:371–77

Dodick DW. 2008. Examining the essence of migraine—is it the blood vessel or the brain? A debate. *Headache* 48:661–67

Donath S, Li P, Willenbockel C, Al-Saadi N, Gross V, et al. 2006. Apoptosis repressor with caspase recruitment domain is required for cardioprotection in response to biomechanical and ischemic stress. *Circulation* 113:1203–12

Dreissen YEM, Bakker MJ, Koelman JHTM, Tijssen MAJ. 2012. Exaggerated startle reactions. *Clin. Neurophysiol.* 123:34–44

Duerr RH, Taylor KD, Brant SR, Rioux JD, Silverberg MS, et al. 2006. A genome-wide association study identifies IL23R as an inflammatory bowel disease gene. *Science* 314:1461–63

Eichler EE, Flint J, Gibson G, Kong A, Leal SM, et al. 2010. Missing heritability and strategies for finding the underlying causes of complex disease. *Nat. Rev. Genet.* 11:446–50

Eriksen MK, Thomsen LL, Andersen I, Nazim F, Olesen J. 2004. Clinical characteristics of 362 patients with familial migraine with aura. *Cephalalgia* 24:564–75

Escayg A, MacDonald BT, Meisler MH, Baulac S, Huberfeld G, et al. 2000. Mutations of *SCN1A*, encoding a neuronal sodium channel, in two families with GEFS+. *Nat. Genet.* 24:343–45

Estacion M, Gasser A, Dib-Hajj SD, Waxman SG. 2010. A sodium channel mutation linked to epilepsy increases ramp and persistent current of Nav1.3 and induces hyperexcitability in hippocampal neurons. *Exp. Neurol.* 224:362–68

Fahn S. 2011. Classification of movement disorders. *Mov. Disord.* 26:947–57

Falace A, Filipello F, La Padula V, Vanni N, Madia F, et al. 2010. TBC1D24, an ARF6-interacting protein, is mutated in familial infantile myoclonic epilepsy. *Am. J. Hum. Genet.* 87:365–70

Fertleman CR, Baker MD, Parker KA, Moffatt S, Elmslie FV, et al. 2006. SCN9A mutations in paroxysmal extreme pain disorder: allelic variants underlie distinct channel defects and phenotypes. *Neuron* 52: 767–74

Gardiner AR, Bhatia KP, Stamelou M, Dale RC, Kurian MA, et al. 2012. *PRRT2* gene mutations: from paroxysmal dyskinesia to episodic ataxia and hemiplegic migraine. *Neurology* 79:2115–21

Gauer RL. 2011. Evaluation of syncope. *Am. Fam. Phys.* 84:640–50

Geis C, Weishaupt A, Hallermann S, Grünewald B, Wessig C, et al. 2010. Stiff person syndrome-associated autoantibodies to amphiphysin mediate reduced GABAergic inhibition. *Brain* 133:3166–80

Ghezzi D, Viscomi C, Ferlini A, Gualandi F, Mereghetti P, et al. 2009. Paroxysmal nonkinesigenic dyskinesia is caused by mutations of the MR-1 mitochondrial targeting sequence. *Hum. Mol. Genet.* 18:1058–64

Gillis J, Pavlidis P. 2012. "Guilt by association" is the exception rather than the rule in gene networks. *PLoS Comput. Biol.* 8:e1002444

Goldstein JL, Brown MS. 2009. The LDL receptor. *Arterioscler. Thromb. Vasc. Biol.* 29:431–38

Gonzaga-Jauregui C, Lupski JR, Gibbs RA. 2012. Human genome sequencing in health and disease. *Annu. Rev. Med.* 63:35–61

Grabrucker AM, Schmeisser MJ, Schoen M, Boeckers TM. 2011. Postsynaptic ProSAP/Shank scaffolds in the cross-hair of synaptopathies. *Trends Cell Biol.* 21:594–603

Hageman GS, Anderson DH, Johnson LV, Hancox LS, Taiber AJ, et al. 2005. A common haplotype in the complement regulatory gene factor H (HF1/CFH) predisposes individuals to age-related macular degeneration. *Proc. Natl. Acad. Sci. USA* 102:7227–32

Haut SR, Bigal ME, Lipton RB. 2006. Chronic disorders with episodic manifestations: focus on epilepsy and migraine. *Lancet Neurol.* 5:148–57

Heinzen EL, Swoboda KJ, Hitomi Y, Gurrieri F, Nicole S, et al. 2012. De novo mutations in ATP1A3 cause alternating hemiplegia of childhood. *Nat. Genet.* 44:1030–34

Helbig I, Scheffer IE, Mulley JC, Berkovic SF. 2008. Navigating the channels and beyond: unraveling the genetics of the epilepsies. *Lancet Neurol.* 7:231–45

Heron SE, Grinton BE, Kivity S, Afawi Z, Zuberi SM, et al. 2012. PRRT2 mutations cause benign familial infantile epilepsy and infantile convulsions with choreoathetosis syndrome. *Am. J. Hum. Genet.* 90:152–60

Higuchi O, Hamuro J, Motomura M, Yamanashi Y. 2011. Autoantibodies to low-density lipoprotein receptor-related protein 4 in myasthenia gravis. *Ann. Neurol.* 69:418–22

Hoch W, McConville J, Helms S, Newson-Davis J, Melms A, Vincent A. 2001. Auto-antibodies to the receptor tyrosine kinase MuSK in patients with myasthenia gravis without acetylcholine receptor antibodies. *Nat. Med.* 7:365–68

Hoffmann K, Muller JS, Stricker S, Megarbane A, Rajab A, et al. 2006. Escobar syndrome is a prenatal myasthenia caused by disruption of the acetylcholine receptor fetal gamma subunit. *Am. J. Hum. Genet.* 79:303–12

Holland KD, Kearney JA, Glauser TA, Buck G, Keddache M, et al. 2008. Mutation of sodium channel SCN3A in a patient with cryptogenic pediatric partial epilepsy. *Neurosci. Lett.* 433:65–70

Irani SR, Pettingill P, Kleopa KA, Schiza N, Waters P, et al. 2012. Morvan syndrome: clinical and serological observations in 29 cases. *Ann. Neurol.* 72:241–55

Jen JC. 2008. Hereditary episodic ataxias. *Ann. N. Y. Acad. Sci.* 1142:250–53

Jen JC, Graves TD, Hess EJ, Hanna MG, Griggs RC, et al. 2007. Primary episodic ataxias: diagnosis, pathogenesis and treatment. *Brain* 130:2484–93

Jen JC, Wan J, Palos TP, Howard BD, Baloh RW. 2005. Mutation in the glutamate transporter EAAT1 causes episodic ataxia, hemiplegia, and seizures. *Neurology* 65:529–34

Jurkat-Rott K, Lerche H, Weber Y, Lehmann-Horn F. 2010. Hereditary channelopathies in neurology. *Adv. Exp. Med. Biol.* 686:305–34

Kalachikov S, Evgrafov O, Ross B, Winawer M, Barker-Cummings C, et al. 2002. Mutations in LGI1 cause autosomal-dominant partial epilepsy with auditory features. *Nat. Genet.* 30:335–41

Kitamura K, Yanazawa M, Sugiyama N, Miura H, Iizuka-Kogo A, et al. 2002. Mutation of ARX causes abnormal development of forebrain and tests in mice and X-linked lissencephaly with abnormal genitalia in humans. *Nat. Genet.* 32:359–69

Klassen T, Davis C, Goldman A, Burgess D, Chen T, et al. 2011. Exome sequencing of ion channel genes reveals complex profiles confounding personal risk assessment in epilepsy. *Cell* 145:1036–48

Kleopa KA. 2011. Autoimmune channelopathies of the nervous system. *Curr. Neuropharmacol.* 9:458–67

Koch MC, Steinmeyer K, Lorenz C, Ricker K, Wolf F, et al. 1992. The skeletal muscle chloride channel in dominant and recessive human myotonia. *Science* 257:797–800

Koseki T, Inohara N, Chen S, Núñez G. 1998. ARC, an inhibitor of apoptosis expressed in skeletal muscle and heart that interacts selectively with caspases. *Proc. Natl. Acad. Sci. USA* 95:5156–60

Kullmann DM. 2010. Neurological channelopathies. *Annu. Rev. Neurosci.* 33:151–72

Lafrenière RG, Zameel Cader M, Poulin J-F, Andres-Enguix I, Simoneau M, et al. 2010. A dominant-negative mutation in the TRESK potassium channel is linked to familial migraine with aura. *Nat. Med.* 16:1157–60

Lai M, Huijbers MG, Lancaster E, Graus F, Bataller L, et al. 2010. Investigation of LGI1 as the antigen in limbic encephalitis previously attributed to potassium channels: a case series. *Lancet Neurol.* 9:776–85

Lancaster E, Huijbers MG, Bar V, Boronat A, Wong A, et al. 2011. Investigations of caspr2, an autoantigen of encephalitis and neuromyotonia. *Ann. Neurol.* 69:303–11

Lee HY, Huang Y, Bruneau N, Roll P, Roberson EDO, et al. 2012a. Mutations in the novel protein PRRT2 cause paroxysmal kinesigenic dyskinesia with infantile convulsions. *Cell Rep.* 1:2–12

Lee HY, Nakayama J, Xu Y, Fan X, Karouani M, et al. 2012b. Dopamine dysregulation in a mouse model of paroxysmal nonkinesigenic dyskinesia. *J. Clin. Investig.* 122:507–18

Lee HY, Xu Y, Huang Y, Ahn AH, Auburger GW, et al. 2004. The gene for paroxysmal nonkinesigenic dyskinesia encodes an enzyme in a stress response pathway. *Hum. Mol. Genet.* 13:3161–70

Leen WG, Klepper J, Verbeek MM, Leferink M, Hofste T, et al. 2010. Glucose transporter-1 deficiency syndrome: the expanding clinical and genetic spectrum of a treatable disorder. *Brain* 133:655–70

Lennon VA, Kryzer TJ, Pittock SJ, Verkman AS, Hinson SR. 2005. IgG marker of optic-spinal multiple sclerosis binds to the aquaporin-4 water channel. *J. Exp. Med.* 202:473–77

Leo L, Gherardini L, Barone V, De Fusco M, Pietrobon D, et al. 2011. Increased susceptibility to cortical spreading depression in the mouse model of familial hemiplegic migraine type 2. *PLoS Genet.* 7:e1002129

Li J, Zhu X, Wang X, Sun W, Feng B, et al. 2012. Targeted genomic sequencing identifies PRRT2 mutations as a cause of paroxysmal kinesigenic choreoathetosis. *J. Med. Genet.* 49:76–78

Lossin C, George AL Jr. 2008. Myotonia congenita. *Adv. Genet.* 63:25–55

Loukaides P, Schiza N, Pettingill P, Palazis L, Vounou E, et al. 2012. Morvan's syndrome associated with antibodies to multiple components of the voltage-gated potassium channel complex. *J. Neurol. Sci.* 312:52–56

Mahida S, Lubitz SA, Rienstra M, Milan DJ, Ellinor PT. 2011. Monogenic atrial fibrillation as pathophysiological paradigms. *Cardiovasc. Res.* 89:692–700

Mantegazza M, Rusconi R, Scalmani P, Avanzini G, Franceschetti S. 2010. Epileptogenic ion channel mutations: from bedside to bench and, hopefully, back again. *Epilepsy Res.* 92:1–29

Marini C, Conti V, Mei D, Battaglia D, Lettori D, et al. 2012. PRRT2 mutations in familial infantile seizures, paroxysmal dyskinesia, and hemiplegic migraine. *Neurology* 79:2109–14

Martin MS, Dutt K, Papale LA, Dubé CM, Dutton SB, et al. 2010. Altered function of the SCN1A voltage-gated sodium channel leads to gamma-aminobutyric acid-ergic (GABAergic) interneuron abnormalities. *J. Biol. Chem.* 285:9823–34

Maselli RA, Ng JJ, Anderson JA, Cagney O, Arredondo J, et al. 2009. Mutations in LAMB2 causing a severe form of synaptic congenital myasthenic syndrome. *J. Med. Genet.* 46:203–8

McBride KL, Garg V. 2010. Impact of Mendelian inheritance in cardiovascular disease. *Ann. N. Y. Acad. Sci.* 1214:122–37

McClatchey AI, McKenna-Yasek D, Cros D, Worthen HG, Kuncl RW, et al. 1992. Novel mutations in families with unusual and variable disorders of the skeletal muscle sodium channel. *Nat. Genet.* 2:148–52

Mihaylova V, Muller JS, Vilchez JJ, Salih MA, Kabiraj MM, et al. 2008. Clinical and molecular genetic findings in COLQ-mutant congenital myasthenic syndromes. *Brain* 131:747–59

Mohler PJ, Schott JJ, Gramolini AO, Dilly KW, Guatimosim S, et al. 2003. Ankyrin-B mutation causes type 4 long-QT cardiac arrhythmia and sudden cardiac death. *Nature* 421:634–39

Morar B, Zhelyazkova S, Azmanov DN, Radionova M, Angelicheva D, et al. 2011. A novel GEFS+ locus on 12p13.33 in a large Roma family. *Epilepsy Res.* 97:198–207

Morishita H, Yagi T. 2007. Protocadherin family: diversity, structure, and function. *Curr. Opin. Cell Biol.* 19:584–92

Nam YJ, Mani K, Ashton AW, Peng CF, Krishnamurthy B, et al. 2004. Inhibition of both the extrinsic and intrinsic death pathways through nonhomotypic death-fold interactions. *Mol. Cell* 15:901–12

Nardocci N, Zorzi G, Barzaghi C, Zibordi F, Ciano C, et al. 2008. Myoclonus-dystonia syndrome: clinical presentation, disease course, and genetic features in 11 families. *Mov. Disord.* 23:28–34

Neville BG, Ninan M. 2007. The treatment and management of alternating hemiplegia of childhood. *Dev. Med. Child Neurol.* 49:777–80

Nicita F, De Liso P, Danti FR, Papetti L, Ursitti F, et al. 2012. The genetics of monogenic idiopathic epilepsies and epileptic encephalopathies. *Seizure* 21:3–11

Nyegaard M, Overgaard MT, Sondergaard MT, Vranas M, Behr ER, et al. 2012. Mutations in calmodulin cause ventricular tachycardia and sudden cardiac death. *Am. J. Hum. Genet.* 91:703–12

O'Roak BJ, Vives L, Girirajan S, Karakoc E, Krumm N, et al. 2012. Sporadic autism exomes reveal a highly interconnected protein network of de novo mutations. *Nature* 485:246–50

Pevzner A, Schoser B, Peters K, Cosma NC, Karakatsani A, et al. 2012. Anti-LRP4 autoantibodies in AChR- and MuSK-antibody-negative myasthenia gravis. *J. Neurol.* 259:427–35

Plaster NM, Tawil R, Tristani-Firouzi M, Canun S, Bendahhou S, et al. 2001. Mutations in Kir2.1 cause the developmental and episodic electrical phenotypes of Andersen's syndrome. *Cell* 105:511–19

Poduri A, Lowenstein D. 2011. Epilepsy genetics—past, present, and future. *Curr. Opin. Genet. Dev.* 21:325–32

Ptáček LJ, George AL Jr, Barchi RL, Griggs RC, Riggs JE, et al. 1992. Mutations in an S4 segment of the adult skeletal muscle sodium channel cause paramyotonia congenita. *Neuron* 8:891–97

Ptáček LJ, George AL Jr, Griggs RC, Tawil R, Kallen RG, et al. 1991. Identification of a mutation in the gene causing hyperkalemic periodic paralysis. *Cell* 67:1021–27

Ptáček LJ, Tawil R, Griggs RC, Engel AG, Layzer RB, et al. 1994a. Dihydropyridine receptor mutations cause hypokalemic periodic paralysis. *Cell* 77:863–68

Ptáček LJ, Tawil R, Griggs RC, Meola G, McManis P, et al. 1994b. Sodium channel mutations in acetazolamide-responsive myotonia congenita, paramyotonia congenita, and hyperkalemic periodic paralysis. *Neurology* 44:1500–3

Rainier S, Thomas D, Tokarz D, Ming L, Bui M, et al. 2004. Myofibrillogenesis regulator 1 gene mutations cause paroxysmal dystonic choreoathetosis. *Arch. Neurol.* 61:1025–29

Raja Rayan DL, Hanna MG. 2010. Skeletal muscle channelopathies: nondystrophic myotonias and periodic paralysis. *Curr. Opin. Neurol.* 23:466–76

Ramachandran N, Girard JM, Turnbull J, Minassian BA. 2009. The autosomal recessively inherited progressive myoclonus epilepsies and their genes. *Epilepsia* 50(Suppl. 5):29–36

Rees MI, Harvey K, Pearce BR, Chung SK, Duguid IC, et al. 2006. Mutations in the gene encoding GlyT2 (SLC6A5) define a presynaptic component of human startle disease. *Nat. Genet.* 38:801–6

Rojas CV, Wang JZ, Schwartz LS, Hoffman EP, Powell BR, Brown RH Jr. 1991. A Met-to-Val mutation in the skeletal muscle Na+ channel alpha-subunit in hyperkalaemic periodic paralysis. *Nature* 354:387–89

Roux-Buisson N, Cacheux M, Fourest-Lieuvin A, Fauconnier J, Brocard J, et al. 2012. Absence of triadin, a protein of the calcium release complex, is responsible for cardiac arrhythmia with sudden death in humans. *Hum. Mol. Genet.* 21:2759–67

Roze E, Apartis E, Clot F, Dorison N, Thobois S, et al. 2008. Myoclonus-dystonia: clinical and electrophysiologic pattern related to SGCE mutations. *Neurology* 70:1010–16

Russell JF, Steckley JL, Coppola G, Hahn AFG, Howard MA, et al. 2012. Familial cortical myoclonus with a mutation in NOL3. *Ann. Neurol.* 72:175–83

Ryan DP, da Silva MR, Soong TW, Fontaine B, Donaldson MR, et al. 2010. Mutations in potassium channel Kir2.6 cause susceptibility to thyrotoxic hypokalemic periodic paralysis. *Cell* 140:88–98

Ryan DP, Ptáček LJ. 2010. Episodic neurological channelopathies. *Neuron* 68:282–92

Saitsu H, Kato M, Mizuguchi T, Hamada K, Osaka H, et al. 2008. De novo mutations in the gene encoding STXBP1 (MUNC18-1) cause early infantile epileptic encephalopathy. *Nat. Genet.* 40:782–88

Sanders J, Murtha MT, Gupta AR, Murdoch JD, Raubeson MJ, et al. 2012. De novo mutations revealed by whole-exome sequencing are strongly associated with autism. *Nature* 485:237–41

Scholl UI, Choi M, Liu T, Ramaekers VT, Häusler MG, et al. 2009. Seizures, sensorineural deafness, ataxia, mental retardation, and electrolyte imbalance (SeSAME syndrome) caused by mutations in KCNJ10. *Proc. Natl. Acad. Sci. USA* 106:5842–47

Sehgal A, Mignot E. 2011. Genetics of sleep and sleep disorders. *Cell* 146:194–207

Seidner G, Alvarez MG, Yeh JI, O'Driscoll KR, Klepper J, et al. 1998. GLUT-1 deficiency syndrome caused by haploinsufficiency of the blood-brain barrier hexose carrier. *Nat. Genet.* 18:188–91

Sheffield VC, Stone EM. 2011. Genomics and the eye. *N. Engl. J. Med.* 364:1932–42

Shen Y, Lee HY, Rawson J, Ojha S, Babbitt P, et al. 2011. Mutations in PNKD causing paroxysmal dyskinesia alters protein cleavage and stability. *Hum. Mol. Genet.* 20:2322–32

Shiang R, Ryan SG, Zhu YZ, Hahn AF, O'Connell P, Wasmuth JJ. 1993. Mutations in the alpha 1 subunit of the inhibitory glycine receptor cause the dominant neurologic disorder, hyperekplexia. *Nat. Genet.* 5:351–58

Shoubridge C, Fullston T, Gécz J. 2010. ARX spectrum disorders: making inroads into the molecular pathology. *Hum. Mutat.* 31:889–900

Shyti R, de Vries B, van den Maagdenberg A. 2011. Migraine genes and the relation to gender. *Headache* 51:880–90

Smith SE, Xu L, Kasten MR, Anderson MP. 2012. Mutant LGI1 inhibits seizure-induced trafficking of Kv4.2 potassium channels. *J. Neurochem.* 120:611–21

Solimena M, Folli F, Denis-Donini S, Comi GC, Pozza G, et al. 1988. Autoantibodies to glutamic acid decarboxylase in a patient with stiff-man syndrome, epilepsy, and type I diabetes mellitus. *N. Engl. J. Med.* 318:1012–20

Stafstrom CE. 2009. Severe epilepsy syndromes of early childhood: the link between genetics and pathophysiology with a focus on SCN1A mutations. *J. Child Neurol.* 24:S15–23

Strømme P, Mangelsdorf ME, Shaw MA, Lower KM, Lewis SM, et al. 2002. Mutations in the human ortholog of Aristaless cause X-linked mental retardation and epilepsy. *Nat. Genet.* 30:441–45

Sugawara T, Tsurubuchi Y, Agarwala KL, Ito M, Fukuma G, et al. 2001. A missense mutation of the Na^+ channel alpha II subunit gene Na(v)1.2 in a patient with febrile and afebrile seizures causes channel dysfunction. *Proc. Natl. Acad. Sci. USA* 98:6384–89

Suls A, Dedeken P, Goffin K, Van Esch H, Dupont P, et al. 2008. Paroxysmal exercise-induced dyskinesia and epilepsy is due to mutations in SLC2A1, encoding the glucose transporter GLUT1. *Brain* 131:1831–44

Suzuki T, Delgado-Escueta AV, Aguan K, Alonso ME, Shi J, et al. 2004. Mutations in EFHC1 cause juvenile myoclonic epilepsy. *Nat. Genet.* 36:842–49

Tawil R, McDermott MP, Brown R Jr, Shapiro BC, Ptáček LJ, et al. 2000. Randomized trials of dichlorphenamide in the periodic paralyses. Working Group on Periodic Paralysis. *Ann. Neurol.* 47:46–53

Tennessen JA, Bigham AW, O'Connor TD, Fu W, Kenny EE, et al. 2012. Evolution and functional impact of rare coding variation from deep sequencing of human exomes. *Science* 337:64–69

Tomlinson SE, Hanna MG, Kullmann DM, Tan SV, Burke D. 2009. Clinical neurophysiology of the episodic ataxias: insights into ion channel dysfunction in vivo. *Clin. Neurophysiol.* 120:1768–76

Tottene A, Conti R, Fabbro A, Vecchia D, Shapovalova M, et al. 2009. Enhanced excitatory transmission at cortical synapses as the basis for facilitated spreading depression in Ca(v)2.1 knockin migraine mice. *Neuron* 61:762–73

Tristani-Firouzi M, Etheridge SP. 2010. Kir 2.1 channelopathies: the Andersen-Tawil syndrome. *Pflugers Arch.* 460:289–94

Ueda K, Valdivia C, Medeiros-Domingo A, Tester DJ, Vatta M, et al. 2008. Syntrophin mutation associated with long QT syndrome through activation of the nNOS-SCN5A macromolecular complex. *Proc. Natl. Acad. Sci. USA* 105:9355–60

Valdivia CR, Ueda K, Ackerman MJ, Makielski JC. 2009. GPD1L links redox state to cardiac excitability by PKC-dependent phosphorylation of the sodium channel SCN5A. *Am. J. Physiol. Heart Circ. Physiol.* 297:H1446–52

van der Werf C, Kannankeril PJ, Sacher F, Krahn AD, Viskin S, et al. 2011. Flecainide therapy reduces exercise-induced ventricular arrhythmias in patients with catecholaminergic polymorphic ventricular tachycardia. *J. Am. Coll. Cardiol.* 57:2244–54

Vatta M, Ackerman MJ, Ye B, Makielski JC, Ughanze EE, et al. 2006. Mutant caveolin-3 induces persistent late sodium current and is associated with long-QT syndrome. *Circulation* 114:2104–12

Veeramah KR, O'Brien JE, Meisler MH, Cheng X, Dib-Hajj SD, et al. 2012. De novo pathogenic SCN8A mutation identified by whole-genome sequencing of a family quartet affected by infantile epileptic encephalopathy and SUDEP. *Am. J. Hum. Genet.* 90:502–10

Vincent A, Lang B, Kleopa KA. 2006. Autoimmune channelopathies and related neurological disorders. *Neuron* 52:123–38

Wallace RH, Wang DW, Singh R, Scheffer IE, George AL Jr, et al. 1998. Febrile seizures and generalized epilepsy associated with a mutation in the Na^+-channel beta1 subunit gene SCN1B. *Nat. Genet.* 19:366–70

Wang JL, Cao L, Li XH, Hu ZM, Li JD, et al. 2011. Identification of *PRRT2* as the causative gene of paroxysmal kinesigenic dyskinesias. *Brain* 134:3493–3501

Watanabe H, Chopra N, Laver D, Hwang HS, Davies SS, et al. 2009. Flecainide prevents catecholaminergic polymorphic ventricular tachycardia in mice and humans. *Nat. Med.* 15:380–83

Weaving LS, Christodoulou J, Williamson SL, Friend KL, McKenzie OL, et al. 2004. Mutations of CDKL5 cause a severe neurodevelopmental disorder with infantile spasms and mental retardation. *Am. J. Hum. Genet.* 75:1079–93

Weber YG, Storch A, Wuttke TV, Brockmann K, Kempfle J, et al. 2008. GLUT1 mutations are a cause of paroxysmal exertion-induced dyskinesias and induce hemolytic anemia by a cation leak. *J. Clin. Investig.* 118:2157–68

Weiss J, Pyrski M, Jacobi E, Bufe B, Willnecker V, et al. 2011. Loss-of-function mutations in sodium channel Nav1.7 cause anosmia. *Nature* 472:186–90

Xu Y, Padiath QS, Shapiro RE, Jones CR, Wu SC, et al. 2005. Functional consequences of a CKIdelta mutation causing familial advanced sleep phase syndrome. *Nature* 434:640–44

Yang Y, Wang Y, Li S, Xu Z, Li H, et al. 2004. Mutations in SCN9A, encoding a sodium channel alpha subunit, in patients with primary erythermalgia. *J. Med. Genet.* 41:171–74

Yu FH, Mantegazza M, Westenbroek RE, Robbins CA, Kalume F, et al. 2006. Reduced sodium current in GABAergic interneurons in a mouse model of severe myoclonic epilepsy in infancy. *Nat. Neurosci.* 9:1142–49

Zhang L, Jones CR, Ptáček LJ, Fu YH. 2011. The genetics of the human circadian clock. *Adv. Genet.* 74: 231–47

Zhou YD, Lee S, Jin Z, Wright M, Smith SE, Anderson MP. 2009. Arrested maturation of excitatory synapses in autosomal dominant lateral temporal lobe epilepsy. *Nat. Med.* 15:1208–14

Zhou YD, Zhang D, Ozkaynak E, Wang X, Kasper EM, et al. 2012. Epilepsy gene LGI1 regulates postnatal developmental remodeling of retinogeniculate synapses. *J. Neurosci.* 32:903–10

Zimprich A, Grabowski M, Asmus F, Naumann M, Berg D, et al. 2001. Mutations in the gene encoding epsilon-sarcoglycan cause myoclonus-dystonia syndrome. *Nat. Genet.* 29:66–69

RELATED RESOURCES

Online Mendelian Inheritance in Man (database created by Johns Hopkins Univ. for the Natl. Cent. Biotechnol. Inf.): **http://www.ncbi.nlm.nih.gov/omim**

Developmental Mechanisms of Topographic Map Formation and Alignment

Jianhua Cang[1] and David A. Feldheim[2]

[1]Department of Neurobiology, Northwestern University, Evanston, Illinois 60208; email: cang@northwestern.edu

[2]Department of Molecular, Cell and Developmental Biology, University of California, Santa Cruz, California 95064; email: dfeldhei@ucsc.edu

Annu. Rev. Neurosci. 2013. 36:51–77

First published online as a Review in Advance on April 29, 2013

The *Annual Review of Neuroscience* is online at neuro.annualreviews.org

This article's doi:
10.1146/annurev-neuro-062012-170341

Keywords

superior colliculus, visual cortex, ephrin, Eph, retinal waves, experience-dependent plasticity

Abstract

Brain connections are organized into topographic maps that are precisely aligned both within and across modalities. This alignment facilitates coherent integration of different categories of sensory inputs and allows for proper sensorimotor transformations. Topographic maps are established and aligned by multistep processes during development, including interactions of molecular guidance cues expressed in gradients; spontaneous activity-dependent axonal and dendritic remodeling; and sensory-evoked plasticity driven by experience. By focusing on the superior colliculus, a major site of topographic map alignment for different sensory modalities, this review summarizes current understanding of topographic map development in the mammalian visual system and highlights recent advances in map alignment studies. A major goal looking forward is to reveal the molecular and synaptic mechanisms underlying map alignment and to understand the physiological and behavioral consequences when these mechanisms are disrupted at various scales.

Contents

INTRODUCTION

The brain detects and processes sensory information using separate, parallel pathways and then combines them to create a unified percept of the outside world and to issue appropriate motor commands. Parallel pathways include those from different sensory modalities, those from bilateral sensory organs within the same modality, and those that signal different aspects of the same stimulus. The ability to integrate multiple channels of information in an organized way improves the sensitivity to detect stimuli and solves ambiguities that arise from the processing of a single attribute (Stein & Meredith 1993). For example, viewing through both eyes and hearing with both ears greatly improve our ability to perceive depth and to localize sound sources. Sensory integration is not only beneficial, but also necessary for our proper perception and interaction with the environment. Failure to integrate different sensory modalities may contribute to a number of neurological deficits, including autism spectrum disorders, synesthesia, dyslexia, and many psychoses (Bargary & Mitchell 2008, Brock et al. 2002, Geschwind & Levitt 2007).

Sensory integration: when neuronal circuits combine inputs from multiple sensory modalities to generate meaningful outputs

Thus two fundamental goals of neuroscience research are to understand the developmental mechanisms that create the brain circuitry capable of accurate multichannel integration and to understand the physiological and behavioral consequences of disrupting these mechanisms.

Topographic mapping is a feature of brain organization that may facilitate or even be required for proper sensory integration and sensorimotor transformations. In the brain, neurons are often organized into maps according to their functional properties, such as where in space they respond and the stimulus features to which they are tuned. These maps have stereotypical organization: Neighboring neurons share similar response properties, and the value of the mapped attribute changes progressively. Some topographic maps arise from preserving the spatial order of the receptor neurons in the sensory epithelium. These include retinotopic maps of visual space, tonotopic maps of sound frequency, and somatotopic maps of body surface. In these maps, neuronal neighbor-to-neighbor relationships in the projecting structure are maintained in their targets, thus preserving a continuous representation in many brain areas (Luo & Flanagan 2007). Other maps are generated after the sensory information is transformed by neuronal computations (Knudsen et al. 1987). The value of the computed parameter, instead of the input's spatial relationship, is then mapped systematically. These include maps of sound location in the auditory system and maps of orientation preference in the visual cortex (Hubel et al. 1977, Knudsen & Konishi 1978a). These computational maps can then be preserved in successive brain areas by topographically organized projections. Both types of topographic maps also exist in the motor systems: Neurons may be organized according to which body areas they control or the movement amplitude they command (Penfield & Boldrey 1937, Wurtz & Goldberg 1972).

The orderly organization of sensory and motor maps offers a structural substrate for proper multichannel integration and sensorimotor transformation. For example, to integrate the sight and sound of an object, visual and auditory neurons that respond to the same locations need to be connected or converge on the same target. Aligning the visual and auditory space maps naturally puts these neurons near each other, thus simplifying the connectivity pattern between the two maps. Theoretically, individual neurons in the two systems could still be connected according to their receptive field locations even in the absence of topographic maps, but such a connectivity pattern would be extremely difficult to establish during development and costly in terms of required wiring (Chklovskii & Koulakov 2004).

The past decade or so has seen an explosion in knowledge of the molecular and activity-dependent mechanisms that are used to form topographic maps, especially within the mammalian visual system. These studies have provided a theoretical framework, and perhaps more importantly, genetic models, that can now be used to understand how multiple maps become aligned and the resulting behavioral consequences when the alignment is disrupted. In this review, we first summarize our current understanding of topographic map development in the mammalian visual system and then review recent advances in map alignment studies. In particular, we focus most of our discussion on the superior colliculus (SC) and its nonmammalian homolog, the optic tectum (OT).

Sensorimotor transformation: the process that the brain converts sensory stimuli into motor commands

Receptive field: the portion of sensory space that can elicit neuronal responses when stimulated

SC: superior colliculus

OT: optic tectum

RGC: retinal ganglion cell

V1: primary visual cortex

dLGN: dorsal lateral geniculate nucleus

Multimodal integration: when neuronal circuits combine inputs from multiple sensory modalities to generate meaningful outputs

The Superior Colliculus: An Integrative Sensorimotor Center with Aligned Topographic Maps

In the mammalian visual system, the retinal ganglion cells (RGCs) send projections to several targets involved in image- and non-image-forming visual behaviors. Two of the main targets are the SC in the midbrain and the dorsal lateral geniculate nucleus (dLGN) in the thalamus. The dLGN, in turn, projects to V1 in the posterior cerebral cortex, which subsequently projects to several higher visual areas responsible for conscious vision and multimodal integration. Each component of the visual system

A-P: anterior-posterior

M-L: medial-lateral

N-T: nasal-temporal

D-V: dorsal-ventral

S1: primary somatosensory cortex

is mapped topographically (Andermann et al. 2011, Lewin 1994, Marshel et al. 2011, Wang & Burkhalter 2007). Because of its importance in multimodal sensorimotor transformation and experimental accessibility, the SC/OT has been a leading model to study how topographic maps form and align during development. The architecture of the SC is conserved among mammals and has two key features: lamination along the superficial to deep axis, which organizes inputs according to where they originate; and topography along the anterior-posterior (A-P) axis (also called the rostral-caudal axis) and the medial-lateral (M-L) axis, which gives spatial identity to a stimulus (**Figure 1**).

The superficial to deep axis of the SC contains seven alternating fibrous and cellular laminae, which can be divided into superficial and deep layers (Gandhi & Katnani 2011, Stein 1984). The most superficial cellular lamina of the SC, the stratum griseum superficiale (SGS), receives RGC input from both eyes and is also laminated with different RGC types that terminate in different sublaminae within the SGS (Inoue & Sanes 1997). The SC also receives visual input from V1 that terminates in the lower SGS (Wang & Burkhalter 2013). These three sources of visual inputs are mapped topographically with respect to the visual field and are aligned in retinotopic register with each other. The nasal-temporal (N-T) axis of the visual field (azimuth) is represented along the A-P axis of the SC and the dorsal-ventral (D-V) axis (elevation) represented along the M-L axis of the SC (**Figure 1**).

The deep layers of the SC receive inputs from the primary somatosensory cortex (S1), the auditory nuclei of the midbrain, the trigeminal nucleus of the brain stem, and many other sensory and motor areas (May 2005). The somatosensory and auditory inputs are organized into topographic maps that are aligned with the visual maps in the superficial layers (**Figure 1**) such that neurons in the same vertical column have corresponding receptive field locations, either in visual and auditory space or on the body surface (Drager & Hubel 1975, Stein 1984). Multisensory integration takes place in the SC as many neurons in the deep layers respond to more than one modality owing to topographically organized projections from the superficial to deep layers. Finally, these sensory maps are also aligned with motor maps in the deep layers of the SC that encode the direction and amplitude of saccadic eye and head movements (Sparks et al. 1990, Wurtz & Albano 1980). The alignment of these maps allows the SC to integrate visual, auditory, and tactile information and to initiate orienting movements to redirect attention toward a stimulus.

MOLECULAR AND ACTIVITY-DEPENDENT MECHANISMS ARE USED TO CREATE TOPOGRAPHIC MAPS

Investigators have traditionally proposed two opposing mechanisms for topographic mapping, which apply to the development of neural circuits in general (Cline 2003). One is a genetic model whereby the generation of precise neural circuitry results from specific gene expression, which gives each neuron instructions about where to project and with whom to synapse. The other is an activity-dependent model, which postulates that the firing patterns of a neuron's action potentials are used to instruct map creation. Not surprisingly, our current understanding is that the development of topographic maps uses a combination of both mechanisms. Because the mechanistic details of map formation have been extensively reviewed in recent years (Feldheim & O'Leary 2012, Flanagan 2006, McLaughlin & O'Leary 2005, Suetterlin et al. 2012, Triplett & Feldheim 2012), this section is limited to the general principles of topographic mapping that are relevant to understanding map alignment.

Genetic Mechanisms of Retinocollicular Map Formation

The initial insights that led to genetic models of map development came from studies in lower vertebrates, where retinal axons regenerate after optic nerve severing and reconnect to the

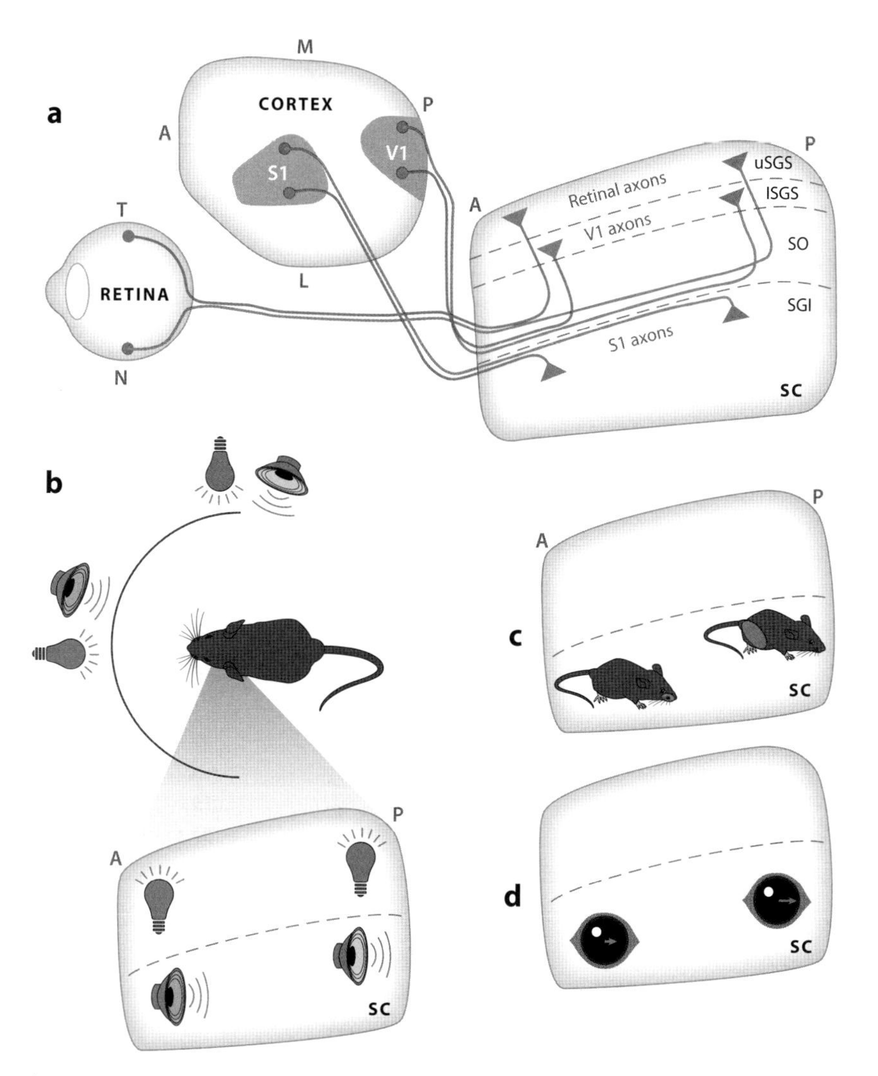

Figure 1

Alignment of multiple maps in the superior colliculus (SC). (*a*) A schematic diagram showing the laminated nature of the SC and inputs from the retina, the visual cortex (V1), and the somatosensory cortex (S1). Each projection is mapped topographically and aligned with the others. Abbreviations: N, nasal; T, temporal; A, anterior; P, posterior; L, lateral; M, medial; uSGS, upper stratum griseum superficiale; lSGS, lower stratum griseum superficiale; SO, stratum opticum; SGI, stratum griseum intermediale. (*b–c*) The orderly projections result in aligned sensory maps in the SC such that neurons in a vertical column have corresponding receptive field locations, either in visual and auditory space (*b*) or on the body surface (*c*), as illustrated by the same color in the diagrams. (*d*) Stimulation of deep layers of the anterior SC evokes small eye movements, whereas posterior SC stimulation evokes larger movements, consistent with the sensory map orientation.

OT in the same patterns as they did during development. If the eyes were rotated 180°, the regenerating axons still grew back to their original destination rather than reorienting to maintain the layout of the visual field, even though this led to maladaptive behaviors in response to visual stimuli (Sperry 1943). These results led Sperry to propose the chemoaffinity hypothesis. Sperry predicted that genetically encoded labels exist in both the retina and the OT; furthermore, if these labels are expressed in continuous gradients across both structures, each RGC axon and target neuron would have a unique concentration of labels that could be matched up to establish topography (Sperry 1963). Bonhoeffer and colleagues later demonstrated that these labels were cell surface proteins using an in vitro stripe assay that recapitulated some key aspects of retinotectal mapping along the N-T axis (Walter et al. 1987a). In this assay, temporal or nasal retinal explants were presented with alternating stripes of membranes derived from the anterior or posterior chick OT. The temporal axons avoided membranes from the posterior OT, whereas nasal axons showed no preference for either substrate. Treatment of posterior membranes with either a protease or PI-phospholipase C (PI-PLC) abolished the preference of temporal axons for anterior membranes. This result suggested that a glycosylphosphatidylinositol (GPI)-linked cell surface protein that was enriched in the posterior OT and triggered repulsion was responsible for the choice of temporal axons (Walter et al. 1987a,b).

Identification of EphAs and Ephrin-As

Two different approaches led to the discovery of the Eph and ephrin family of proteins whose properties were predicted by Sperry and Bonhoeffer (for reviews see Drescher et al. 1997, Flanagan & Vanderhaeghen 1998); we now know that Eph/ephrin signaling molecules are found in all multicellular species but have been greatly expanded in mammals (Chisholm & Tessier-Lavigne 1999, Mellott & Burke 2008). Ephs are receptor tyrosine kinases (RTKs) that bind to ephrins. Among the major RTK/ligand families, Eph/ephrin signaling is distinctive because both ephrins and Ephs are membrane bound, resulting in signals that are localized to areas of cell contact. After binding, both Eph- and ephrin-expressing cells can transduce signals that lead to changes in cellular behavior. Thus, Ephs and ephrins signal bidirectionally: ephrin induction of Eph signaling is termed forward signaling, and Eph induction of ephrin signaling is called reverse signaling. Bidirectional signaling facilitates the regulation of axon and target cell responses concomitantly after an interaction (Kullander & Klein 2002).

Ephs and ephrins can each be subdivided into two classes. Ephrin-As are GPI-linked to the membrane, whereas ephrin-Bs are transmembrane proteins that have cytoplasmic domains with C-terminal PDZ binding motifs. EphAs and EphBs are distinguished by sequence similarities that dictate their binding specificity for either ephrin-As or -Bs. Binding partners are promiscuous in vitro within each subfamily; each EphA can bind multiple ephrin-As, and each EphB can bind multiple ephrin-Bs. Some crosstalk also exists between subfamilies. For example, EphA4 can bind all ephrin-Bs, and EphB2 can bind to ephrin-A5 (Gale et al. 1996, Himanen et al. 2004). Although discussed here in the context of topographic mapping, Ephs and ephrins are important in a multitude of cellular and developmental events and can act as oncogenes and tumor suppressors when mutated or misregulated (Holmberg et al. 2005, 2006; Miao & Wang 2012; Pasquale 2010).

Both Forward Signaling and Reverse Signaling Build Topographic Maps Along the N-T Mapping Axis

In all species that researchers have examined, including humans, multiple EphA and ephrin-A family members are expressed in complementary gradients along the N-T axis of the retina and along the N-T mapping axis in the SC/OT (Lambot et al. 2005). Multiple EphAs are expressed in a high-temporal to low-nasal gradient in the retina and in a high-anterior to low-posterior gradient in the SC. Ephrin-As are expressed in a complementary manner, having a high-nasal to low-temporal expression in the retina and a low-anterior to high-posterior expression pattern in the OT (**Figure 2**). These counter gradients, together with the finding that ephrin-A-bearing axons are repelled by EphAs (Rashid et al. 2005), help explain topographic map formation. Temporal RGC axons have high levels of EphA receptors, are thus easily repelled by collicular derived ephrin-As, and therefore map to the anterior SC/OT, where ephrin-A expression is lowest (forward signaling). Nasal axons contain high levels of ephrin-As, are repelled by EphAs in the anterior SC, and therefore map to the posterior SC, where repulsion is lowest (reverse signaling). Axons will terminate in a region of the SC/OT where their sensitivity to the two repellent activities is minimized (Nakamoto et al. 1996, O'Leary & McLaughlin 2005, Rashid et al. 2005). Consistent with this idea, perturbation of EphA and ephrin-A levels in the retina or in the SC disrupts topographic mapping, which can be revealed by both anatomical and functional assays (**Figure 3**) (Cang et al. 2008b; Carreres et al. 2011; Feldheim et al. 2000, 2004; Frisén et al. 1998; Rashid et al. 2005).

One benefit of having counter gradients of Ephs and ephrins is that the same labels can be used repeatedly at successive stages within the same pathway. EphA and ephrin-A counter gradients are found in each visual area and guide the maps from the retina to the SC and the dLGN, from the dLGN to V1, and V1 projections to the dLGN and the SC (Cang et al. 2005a, Feldheim et al. 1998, Pfeiffenberger et al. 2006, Torii & Levitt 2005, Triplett et al. 2009) (**Figure 2**). In addition, EphA/ephrin-A signaling is required for the topographic mapping of many other projections, including those of the somatosensory, motor, auditory, hippocampal, and olfactory systems (Cramer 2005, Cutforth et al. 2003, Galimberti et al. 2010, Serizawa et al. 2006, Vanderhaeghen et al. 2000).

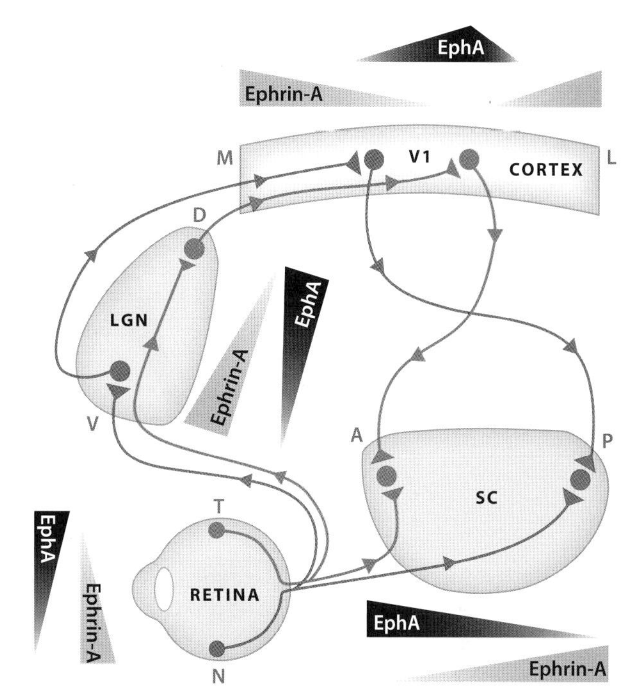

Figure 2

EphA and ephrin-A counter gradients are found in each area of the developing visual system, including the retina, the superior colliculus (SC), the lateral geniculate nucleus (LGN), and the primary visual cortex (V1). Abbreviations: N, nasal; T, temporal; A, anterior; P, posterior; D, dorsal; V, ventral; L, lateral; M, medial.

Relative Levels of EphAs Are Used to Map RGC Axons Topographically

An important insight into how EphA/ephrin-A gradients direct topographic mapping comes from experiments by Lemke and colleagues, which demonstrated that topography is established through relative, as opposed to absolute, levels of EphA signaling (Bevins et al. 2011, Brown et al. 2000, Reber et al. 2004). Taking advantage of the fact that EphA3 is not endogenously expressed in the mouse retina, Brown et al. (2000) introduced EphA3 into the 3′ untranslated region of the Islet-2 locus. Because Islet-2 is expressed in approximately half of all RGCs distributed relatively evenly across the retina, two populations of RGCs exist: (*a*) Islet-2^{-} cells, which express endogenous levels of EphA4, EphA5, and EphA6 according to their positions along the T-N axis; and (*b*) Islet-2^{+} cells, which express exogenous EphA3 in high amounts in addition to the endogenous EphAs gradients. As a result, the RGCs expressing EphA3 are compressed into the anterior half of the SC, whereas the Islet-2^{-} RGCs, although having wild-type (WT) levels of EphA, are restricted to the posterior half of the SC (Brown et al. 2000). This anatomical phenotype results in two complete and functional representations of the visual fields as revealed by intrinsic imaging (Triplett et al. 2009).

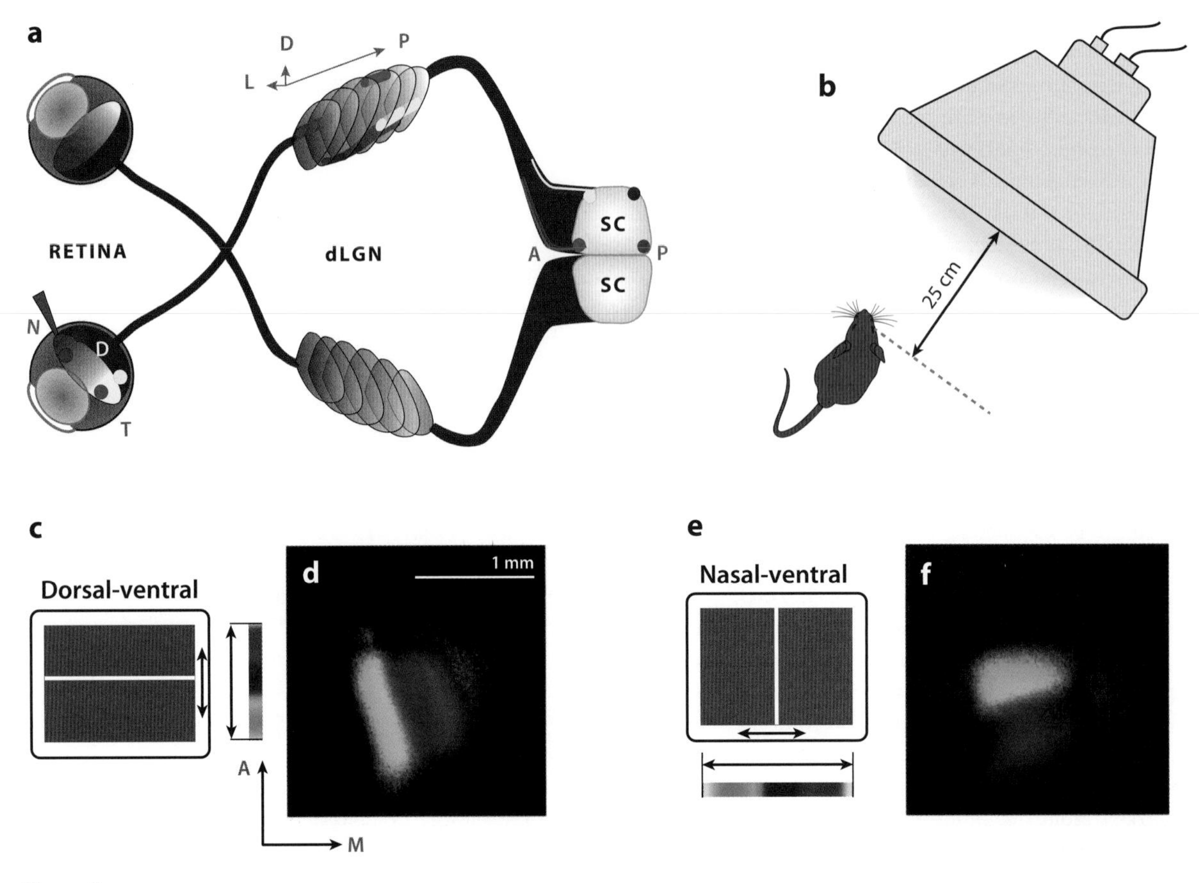

Figure 3

Determining retinotopic maps using anatomical tracing and functional imaging. (*a*) Anatomical tracing of topographic maps. Focal injections of the lipophilic fluorescent DiI in the retina label RGC axons and the termination zones in their targets. Shown is a summary of the topography of the retinogeniculate and retinocollicular projections in wild-type mice. (*b–f*) Functional maps in the SC revealed by optical imaging of intrinsic signals. An anesthetized mouse is positioned to view a computer monitor through the eye contralateral to the hemisphere under study (*b*). Stimuli are thin moving bars (*c* and *e*). The elevation map of a WT mouse is shown in panel *d*; the retinotopy is represented according to the color scale shown in panel *c*, and the response magnitude is illustrated by brightness. The azimuth map of the same SC is shown in panel *f*. Abbreviations: N, nasal; T, temporal; A, anterior; P, posterior; D, dorsal; L, lateral; M, medial; dLGN, dorsal lateral geniculate nucleus.

Competition for Target Space in Topographic Mapping

How do axons read relative rather than absolute levels of ephrins? Axon-axon competition for some limiting factor(s) in the target may be important (Brown et al. 2000, Koulakov & Tsigankov 2004, Prestige & Willshaw 1975, Triplett et al. 2011, Tsigankov & Koulakov 2006). This limiting factor would attract axons, which could provide a force that counterbalances the repellent gradient of ephrin-As. Competition for this factor would tend to drive axons with lower EphA expression in a posterior direction. Competition models also explain a large body of work involving ablations of parts of the retina or OT/SC. These experiments indicated that, at least under some experimental conditions, retinal axons tend to spread out and fill the available space in the target (Finlay et al. 1979, Fraser & Hunt 1980, Simon et al. 1994).

Experiments in fish and mice have tested the role of axonal competition in topographic mapping. Gosse et al. (2008) created transgenic zebrafish that contained only one RGC and assayed its projection patterns in the OT. Competition-based models predicted that the RGC's termination would shift anteriorly because positive factors were available in the anterior SC/OT. Instead, the most distal branch of the axon projected to a position similar to what it would have under the crowded conditions in a WT retina (Gosse et al. 2008). Experiments in mice gave an opposite result. In Math5 (Atoh7) mutant mice, which have severely reduced numbers of RGCs (Brown et al. 2001), the RGC axons do not fill the entire SC but instead are enriched in the antero-medial portion. These data suggest that in the absence of competition all RGCs prefer to project to where repulsion from ephrin-A ligands is minimized. Lower vertebrates may use competition-based mechanisms to a lesser degree than mammals, which could be due to the difference in how RGCs find their appropriate termination zone (TZ) position in fish and mammals (O'Leary & McLaughlin 2005). In fish, the RGC axon extends directly to the correct location. In contrast, mouse RGCs overshoot their final locations and then, over a period of a few days, form TZs by interstitial branching, which may be influenced more easily by competition.

The competition in mice could be mediated by comparing axonal EphA levels, which would result in the spatial segregation of retinal axons in the SC (Reber et al. 2004). Alternatively, a recent computational study (Grimbert & Cang 2012) suggested that the competition could instead rely on the optimization of collicular resources mediated by neurotrophic receptors such as p75, in addition to p75's role in reverse signaling as a transmembrane signaling partner for ephrin-As (Lim et al. 2008).

Mapping the Dorsal-Ventral Axis of the Retina onto the Superior Colliculus

An important result from the gain- and loss-of-function experiments that manipulate EphA and ephrin-As is that the topographic mapping defects are restricted predominantly to the N-T mapping axis, leaving the D-V mapping axis intact. This result strongly indicates that each axis uses independent mechanisms for mapping (Cang et al. 2008a). Although the mechanisms of D-V mapping are not as well understood, investigators have proposed that a combination of pretarget sorting and graded signaling by EphB/ephrin-Bs is used to sort axons along the D-V mapping axis of the SC.

Optic chiasm: located at the bottom of the brain; where the optic nerves partially cross

As RGC axons exit the optic disc at the back of the eye, they tightly fasciculate to form the optic nerve and join axons from the opposite eye once they reach the optic chiasm at the base of the brain. They subsequently defasciculate prior to entering their targets, including the SC. Experiments in which the dorsal or ventral retina is labeled with DiI reveal that D-V order is established within the optic tract before RGC axons enter the SC (Plas et al. 2005, Simon & O'Leary 1991). Mutations that affect D-V patterning in the retina also affect D-V sorting and mapping in the SC, suggesting that optic tract sorting is important for topographic mapping (Plas et al. 2005). Although the molecular mechanisms regulating RGC axon sorting are unknown, similar tract organization in the olfactory system relies on Sema3a and its receptor, neuropilin-1 (Bozza et al. 2009, Imai et al. 2009). On the other hand, RGC axons also refine their M-L position within the SC by interstitial branching off their primary axon toward their appropriate position. These branches can form either medial or lateral to the primary axon, suggesting that the exact entrance location of the axon into the SC is not critical (Hindges et al. 2002, McLaughlin et al. 2003a, Nakamura & O'Leary 1989).

The expression patterns and in vivo loss- and gain-of-function experiments have led researchers to propose that both EphB/ephrin-B forward and reverse signaling regulate D-V topographic map formation in the SC/OT (Hindges et al. 2002, Mann et al. 2002, Thakar et al. 2011). Dual gradients of EphBs and ephrin-Bs are expressed within the retina; Ephrin-Bs are expressed in a high-dorsal to low-ventral gradient, whereas EphBs are expressed in the opposite, high-ventral to low-dorsal gradient (Braisted et al. 1997, Hindges et al. 2002, Holash & Pasquale 1995, Mann et al. 2002, Marcus et al. 1996). However, the expression patterns of EphBs and ephrin-Bs are not as obviously graded in the M-L axis of the SC/OT. In the mouse, mRNA in situ hybridization reveals ephrin-B1 in a high medial to low lateral gradient restricted to the ventricular zone of the SC, far from the superficial retinorecipient layers. A similar pattern of expression exists in the chick OT, where ephrin-B1 protein expression is associated with radial glia that span the depth of the OT, suggesting expression on these cells could guide incoming retinal axons (Braisted et al. 1997, Hindges et al. 2002). EphBs are widely expressed in the vertebrate tectum, but not in an obvious medial–lateral gradient, making it hard to understand how EphB/epbrin-B interactions could generate positional identity on their own (Higenell et al. 2011, Scalia & Feldheim 2005, Thakar et al. 2011). Despite the lack of obvious gradients in the SC/OT, EphB/ephrin-B mutants do display topographic mapping defects. The retinocollicular maps have been analyzed in mouse lines that remove all or partial functions of EphB/ephrin-B signaling (Hindges et al. 2002, Thakar et al. 2011). In each case, many RGCs make M-L and A-P mapping errors, but most RGCs do map correctly, indicating that other D-V mapping molecules are still to be identified.

Other Molecules in Mapping Retinocollicular Projection

Along the N-T mapping axis, the GPI-linked protein repulsive guidance molecule (RGM) and its receptor neogenin are expressed in gradients along the N-T mapping axis. RGM can repel temporal but not nasal axons in vitro (Monnier et al. 2002, Rajagopalan et al. 2004). Perturbation of RGM levels by overexpression or by using RNAi disturbs the retinal-tectal projection in chicks (Monnier et al. 2002, Rajagopalan et al. 2004). Multiple semaphorins and their receptors plexins and neuropilins are expressed in the retina and the OT/SC during development, and Sema3A can repel *Xenopus* retinal axons in vitro (Campbell et al. 2001).

Molecules that have been implicated in D-V mapping include Wnts and their receptors and the cell adhesion molecules L1 and AL-CAM. Wnt3 is expressed in a high to low M-L gradient in chick OT and mouse SC, and members of the Ryk and Frizzled families of Wnt receptors are expressed in an overall high to low V-D gradient by RGCs. In vitro assays show that Ryk mediates RGC axon repulsion by higher levels of Wnt3, whereas Frizzled receptors mediate an attractant effect of lower Wnt3 levels on dorsal RGC axons (Schmitt et al. 2006).

Mice mutant for the cell adhesion molecules L1 or ALCAM have topographic mapping defects, and L1 has shown to be phosphorylated by EphBs, suggesting that cellular adhesion regulation is also important for D-V mapping (Buhusi et al. 2009, Demyanenko & Maness 2003). Other molecules that may be involved in topographic mapping are a secreted form of the transcription factor engrailed 2 (Brunet et al. 2005), the haywire homolog Phr1, and the type-2 membrane protein Ten_M3 (also called ODZ and teneurin) (Leamey et al. 2007).

Neural Activity Is Required to Generate Fine Structured Topography

Activity-dependent mapping models postulate that neurons with similar activity patterns are more likely to synapse on the same or adjacent target cells than are neurons with different activity patterns (Butts 2002, Debski & Cline 2002). Because topographic mapping in the mammalian visual system occurs before eye opening, this mechanism is thought to rely on spontaneous patterned retinal activity. Throughout the first postnatal week in mice, RGCs discharge highly correlated bursts of action potentials that propagate across the retina as waves; these retinal waves have been theorized to drive synaptic refinement through activity-dependent mechanisms (Butts 2002, Butts et al. 2007, Meister et al. 1991, Wong et al. 1993).

Perhaps the most compelling evidence for this idea came from studies of mice that lack the β2 subunit (β2 knockout) of the nicotinic acetylcholine receptor (Picciotto et al. 1995, Xu et al. 1999). Retinal waves during the first postnatal week require cholinergic neurotransmission, and β2 knockout retinas have defects in the normal patterns of retinal activity (Bansal et al. 2000, Feller et al. 1996, McLaughlin et al. 2003b, Stafford et al. 2009). In these β2 knockout mice, multiple aspects of refinement are altered in the dLGN and SC, including retinotopy and eye segregation, demonstrating a correlation between retinal waves as measured in vitro and mapping and behavioral defects in vivo (Cang et al. 2005b, Chandrasekaran et al. 2005, Grubb et al. 2003, McLaughlin et al. 2003b, Mrsic-Flogel et al. 2005, Muir-Robinson et al. 2002, Wang et al. 2009). However, the correlative nature of these studies led to rigorous debates about whether normal retinal waves are required or "instructive" for visual system development (Chalupa 2009, Feller 2009). This uncertainty came about because the in vitro activity patterns of RGCs in β2 mutants differed among the labs (Feller 2009, Stafford et al. 2009, Sun et al. 2008), and the in vivo analysis was performed in a global knockout mouse, suggesting that the lack of β2 outside the retina could be responsible for the mapping errors.

Xu et al. (2011) recently reintroduced the β2 gene into only the RGCs of β2 mutant mice and restored many aspects of retinal waves. Although the waves were smaller than those of WT mice, these small waves could rescue topography in monocular regions of the SC. These results suggest that the exact spatiotemporal profile of retinal waves is important for refining topographic connections, but the exact information included in waves, and how they execute refinement, still needs to be determined, especially in vivo. Similarly, the exact molecular mechanisms downstream of retinal waves are not well understood, although cAMP generation via adenylyl cyclase 1 (AC1) is involved (Dhande et al. 2012).

Ephrin-As and Patterned Activity Act Together to Guide Map Formation in the Visual System

The above findings suggest a mechanism whereby EphA/ephrin-A signaling acts to form a rough topography and neural activity acts to cluster correlated inputs to promote its fine structure. In support of this two-step model, intrinsic optical imaging of the ephrin-A2/-A3/-A5 triple mutant mouse finds that their functional SC maps are discontinuous with patches of the SC responding to visual signals from topographically incorrect locations. In other words, large groups of neurons project together to topographically inappropriate locations but can still refine and become integrated into visual circuits. These clusters disappear in mice that are deficient in both ephrin-As and structured neural activity in the retina, leading to a nearly complete loss of topography along the N-T axis of each visual area (Cang et al. 2008a,b; Pfeiffenberger et al. 2006).

Topographic Mapping of Different Retinal Ganglion Cell Types

In the above section, we have treated the retinocollicular projection as a single continuous topographic map that is composed of RGC axons as a single, functionally identical group. In reality, the retinocollicular map is the superimposition of many individual maps that originate from different types of RGCs, each of which is aligned with all the others. Each type of RGC is tuned to a specific visual feature such as motion, color, or changes in luminance. Different RGC types project to specific sets of visual areas and terminate within stereotypic sublaminae within these targets. For example, an ON-OFF direction-selective RGC type projects to the superficial lamina of the SGS, whereas an Off RGC type projects to the deeper SGS (Huberman et al. 2008b, 2009; Kay et al. 2011). Little is known about how different RGC types that project to different sublaminae align with each other. One possibility is that each RGC type sorts topographically using

EphA/ephrin-A signaling and uses a different set of molecules to find the correct lamina. A second possibility is that one RGC type uses EphA/ephrin-A signaling, whereas the other types use activity-dependent (e.g., retinal waves) cues to align with the initial template. Some evidence shows that not all RGCs require Eph/ephrin signaling to map (Cang et al. 2005a, Pfeiffenberger et al. 2006), but we still do not know if the RGCs that map normally in the absence of ephrin-As represent a specific type(s).

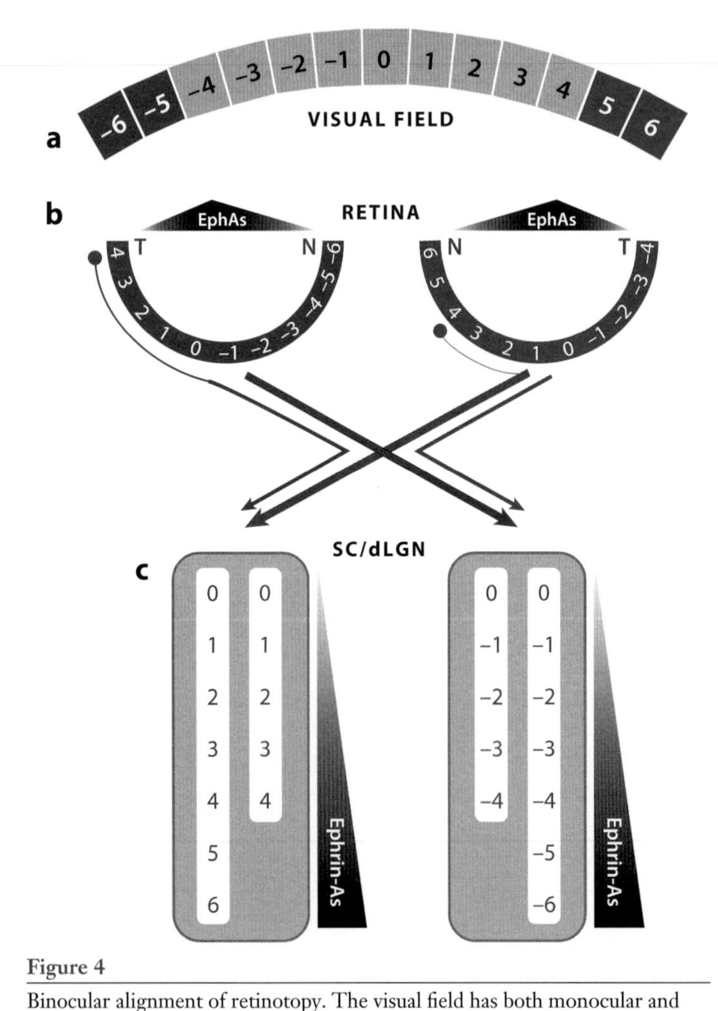

Figure 4

Binocular alignment of retinotopy. The visual field has both monocular and binocular areas (represented here as regions from −6 to 6, *panel a*) that project onto the retina of each eye (*panel b*); the central part is projected onto both eyes. The retinal projections of each eye are segregated in the SC/dLGN and are topographically aligned (*panel c*). The alignment requires that RGCs of different N-T coordinates of each eye project to nearby areas. For example, position 4 in the left SC/LGN receives inputs that originate from the most temporal retina of the left eye (*red*) and from the nasal retina of the right eye (*blue*). A center-to-periphery gradient of EphAs in the retina gives these axons similar levels of mapping labels to align on the basis of their similar sensitivities to a gradient of ephrin-As in the target.

ALIGNMENT OF MULTIPLE VISUAL MAPS

Binocular Alignment of Retinotopy

In most mammals, central visual space is seen by both eyes. The extent of this binocular field is determined by eye position and varies across species (Leamey et al. 2008). In primates and humans, RGCs in the temporal retina that view the contralateral half of the binocular visual field project to the SC and dLGN in the same side (**Figure 4**), where they meet the axons of the nasal RGCs from the other eye that see the same hemifield. Although mice have a smaller binocular field because their eyes are positioned more laterally, similar projection patterns are seen except that there is a much smaller percentage of ipsilaterally projecting RGCs (∼3–5%, Petros et al. 2008) that originate from the ventral-temporal crescent of the retina. The crossed and uncrossed axons are segregated into eye-specific layers in the SC and dLGN. Within each layer, the inputs are organized retinotopically, and the two maps are aligned between the layers. Relay neurons from different layers of the dLGN project to V1, where they converge onto the same postsynaptic neurons, resulting in the first stage of binocular processing.

For proper binocular alignment in the brain, individual RGC axons have to determine correctly whether to cross the midline. This decision is facilitated by the interaction between molecules expressed on the RGC growth cones and those at the optic chiasm (for a review, see Petros et al. 2008). Once the crossed and uncrossed axons reach their primary targets, their retinotopic maps align. This alignment requires the topography of ipsilateral axons to be reversed along the N-T axis relative to the contralateral axons. In the SC, for example, the temporal RGC axons of the contralateral retina terminate more anteriorly and nasal axons more posteriorly, and the order is reversed for axons from the ipsilateral retina (**Figure 4**). How can the same set of guidance cues, such as ephrin-As/EphAs, mediate the formation of two maps that have opposite polarity? Sperry predicted that, in binocular species, the N-T gradient of the mapping labels would become a radial gradient (center to periphery) to give nasal axons from one eye and temporal axons from the other eye similar levels of mapping labels (Sperry 1963). Remarkably, EphA receptors are indeed expressed in a center high and peripheral low gradient along the N-T axis in the developing retinas of both humans and ferrets (Huberman et al. 2005, Lambot et al. 2005), consistent with their potential roles in binocular retinotopic mapping.

A similar center-to-peripheral gradient is more difficult to demonstrate in mice because only a very small number of RGCs project ipsilaterally in this species. Although not noted in the original studies, such a gradient may indeed exist in mice (Feldheim et al. 2000, figure *6c*), suggesting a remarkable conservation in evolution. Consistent with their expression patterns, EphA/ephrin-A signaling may indeed be involved in aligning the contralateral and ipsilateral maps in mice. In ephrin-A2/A5 double knockout mice, the binocular alignment of functional maps in the SC is disrupted (Haustead et al. 2008), though the interpretation is complicated because these mice have mapping errors in both contralateral and ipsilateral projections (Feldheim et al. 2000, Haustead et al. 2008). In contrast, a transmembrane protein, Ten_m3, is required only for mapping the ipsilateral projections to the dLGN (Leamey et al. 2007). The exact mechanism of this process is still unknown, as Ten_m3 expression is not restricted to the ipsilaterally projecting RGCs (Leamey et al. 2007). The binocular misalignment is associated with visual behavioral deficits in both ephrin-A2/A5 knockouts and Ten_m3 knockouts, and the behavior deficits could be reversed by occluding or silencing one of the two eyes (Haustead et al. 2008, Leamey et al. 2007).

In the developing dLGN and SC, the RGC axons from the two eyes are initially intermingled and then retract to segregate into eye-specific layers (Godement et al. 1984, Jaubert-Miazza et al. 2005, Jeffery 1985, Lund et al. 1974). These data suggest that interactions between the crossed and uncrossed retinal axons may be important for binocular alignment. Indeed, when the contralateral projection is removed by neonatal enucleation in rats, the ipsilateral projections from the remaining eye fail to establish normal topography along the N-T axis in the SC (but not in the dLGN; Reese 1986), suggesting that the ipsilateral projections may use the contralateral ones as a template to align the two maps. The interactions between the eye-specific axons may be mediated by molecular cues, such as EphAs/ephrin-As (Haustead et al. 2008) and/or spontaneous neuronal activity. Xu et al. generated a line of transgenic mice that express the β2 subunit of the acetylcholine receptor only in the RGCs (see above) and showed that they have smaller retinal waves during development. Retinotopic refinement in these mice is normal except in the binocular zone of the dLGN and SC (Xu et al. 2011), suggesting that the activity-dependent segregation of contralateral and ipsilateral axons is mechanistically linked with retinotopic refinement and potentially with their alignment as well.

The first stage of binocular visual processing takes place in V1. The converging eye-specific geniculocortical inputs are aligned retinotopically, giving rise to binocular neurons that represent similar points in the visual space through both eyes. The development of cortical retinotopy is understood better for the azimuth map through the contralateral eye, which is guided by a combination of EphA/ephrin-A signaling and spontaneous neural activity driven by retinal waves (Cang et al. 2005a,b, 2008a). In mice, the retinotopic order of geniculocortical projections appears before P8 when RGC axons from the two eyes are still largely intermingled in the dLGN (Pfeiffenberger et al. 2005). During this period, graded EphA expression is seen along the M-L axis of V1 (Cang et al.

Orientation selectivity: the property that visual cortical neurons are more responsive to bars presented over a narrow range of orientation

Orientation maps: cortical organization where neurons within a vertical column prefer similar orientations

2005a), and ephrin-As display a single gradient in the dLGN along the lines of retinotopy regardless of eye-specific layers (Feldheim et al. 1998). Consequently, the same developmental processes that guide single map formation will likely map both inputs together to achieve binocular alignment. Consistent with this notion, the binocular misalignment seen in the dLGN of Ten_m3 knockout mice propagates to V1 and results in binocular cells that have mismatched receptive fields through the two eyes (Leamey et al. 2007, Merlin et al. 2012).

Binocular Matching of Orientation Tuning

Two major transformations occur when visual information reaches the cortex. First, cortical cells are binocular as discussed above; and second, cortical cells are selective for stimulus orientation (Hubel & Wiesel 1962). As a result, binocular cells in the cortex are tuned to similar orientations through the two eyes (Bridge & Cumming 2001, Ferster 1981, Hubel & Wiesel 1962, Nelson et al. 1977, Wang et al. 2010), a property presumably necessary for normal binocular vision. In carnivores and primates, cortical neurons with similar orientation preferences are organized into columns (Blasdel & Salama 1986, Grinvald et al. 1986, Hubel & Wiesel 1962), and the maps of orientation columns are also binocularly matched (Crair et al. 1998). According to the feedforward model proposed originally by Hubel & Wiesel (1962), orientation selectivity arises from the specific arrangement of geniculate inputs, and the preferred orientation of individual cortical cells is determined by the layout of the elongated ON (responding to light increment) and OFF (responding to light decrement) subregions in their receptive fields (Ferster & Miller 2000, Priebe & Ferster 2012). An alternative series of models, the feedback models, propose that orientation selectivity is an emergent property of intracortical circuitry (Adorjan et al. 1999, Ben-Yishai et al. 1997, Somers et al. 1995). Although researchers are still debating the exact synaptic mechanisms of orientation selectivity, many studies have investigated its development and, more recently, its binocular matching.

Orientation selectivity develops after retinotopic mapping and eye-specific segregation in the dLGN and matures after eye opening (Espinosa & Stryker 2012, Huberman et al. 2008a). The development of monocular orientation selectivity does not require visual experience, but its binocular matching, at least in mice, does. Mice do not have orientation columns, but individual V1 neurons are highly selective (Niell & Stryker 2008) and prefer similar orientations through the two eyes (Wang et al. 2010, Sarnaik et al. 2013). In young mice (P20, ~1 week after eye opening), orientation tuning is binocularly mismatched, and the mismatch decreases and reaches adult levels by P30. The binocular matching process occurs in a time window that coincides with the critical period for ocular dominance plasticity (Espinosa & Stryker 2012) and requires normal binocular vision (Wang et al. 2010).

In contrast, the matching of large-scale orientation maps appears to be experience-independent. Binocularly matched orientation maps are seen in very young kittens (Crair et al. 1998) and can develop even in the absence of coordinated binocular activity (Godecke & Bonhoeffer 1996). The degree of binocular orientation matching at the single-cell level may improve further after matching the maps in an experience-dependent manner, but this possibility has not be studied in species that have orientation maps. One study reported a small number of binocularly mismatched V1 neurons in visually deprived kittens (Blakemore & Van Sluyters 1974, figure 4), consistent with the results in mice (Wang et al. 2010).

Development of orientation selectivity requires spontaneous neuronal activity in the visual cortex (reviewed in Espinosa & Stryker 2012, Huberman et al. 2008a). In ferrets, binocularly correlated spontaneous activity exists in the LGN before eye opening because of corticogeniculate feedback (Weliky & Katz 1999). Whether such activity contributes to binocular matching of orientation is not known, but a Hebbian plasticity–based model predicts that binocularly matched orientation maps would develop if the same-polarity dLGN inputs from the two eyes (On-On, Off-Off) are temporally correlated at the same retinotopic locations (Erwin & Miller 1998). Finally, the experience-independent binocular matching of orientation maps suggests that molecular cues may be involved in the development of orientation selectivity, orientation maps, and their binocular matching. However, such molecules have not yet been discovered.

Template-Driven Alignment between Retinal and Cortical Visual Maps in the Superior Colliculus

In addition to the retinal inputs from both eyes, the SC also receives cortical projections that are aligned with the retinocollicular map (Rhoades & Chalupa 1978). This corticocollicular projection provides a link between the two major streams of visual processing: the retinocollicular pathway, used for some reflexive visual behaviors; and the retino-geniculo-cortical pathway involved in conscious vision.

Two distinct models can explain how the retinocollicular and corticocollicular maps become aligned in the SC. A gradient-matching model postulates that gradients of molecules expressed by both V1 and RGC axons match graded labels expressed in the SC to specify each map (**Figure 5*a***). In this case, corticocollicular mapping is independent of retinocollicular mapping, but because both projections use information provided by the same target molecules, they become aligned. A retinal-matching model for map alignment proposes that Hebbian-type mechanisms or direct axon-axon interactions are used to direct V1 and RGC axons that monitor the same point in space to terminate in the same region of the SC (**Figure 5*b***). In this model, the retinocollicular map is established using the mapping mechanisms described above. V1 axons would then grow in and form synapses with the SC neurons that are innervated by the RGCs that share common activity patterns or cell surface molecules. In this mechanism, the retinal axons serve as a template that orients and organizes the alignment with incoming projections.

Hebbian plasticity: form of synaptic plasticity in which synaptic efficacy increases as a result of correlated activity between pre- and postsynaptic cells

The *Islet2-EphA3* knockin (EphA3$^{ki/ki}$) mouse described above was an ideal genetic tool to distinguish between these models. Using optical imaging of intrinsic signals to visualize functional maps in the brain, Triplett et al. (2009) found that the EphA3$^{ki/ki}$ mouse has a duplicated functional retinocollicular map but has a single functional map in V1 (**Figure 5*c***). Because exogenous EphA3 expression is limited to RGCs, the gradient-matching model predicts that the V1–SC projections remain normal, thus resulting in a misalignment of the retinocollicular and corticocollicular maps. In contrast, the retinal-matching model predicts that cortical projections in the SC will align with the altered retinal projections. Triplett et al. found that the corticocollicular projections indeed bifurcate to align with the duplicated retinocollicular maps in EphA3$^{ki/ki}$ mice, supporting the retinal-matching model (**Figure 5*c***). Furthermore, disruption of the spontaneous cholinergic retinal waves in EphA3$^{ki/ki}$ mice prevented map alignment. Taken together, these data showed that the V1 projection aligns with a preexisting retinocollicular map by matching activity patterns that propagate throughout the visual system prior to eye opening (Triplett et al. 2009). EphA/ephrin-A signaling and/or other guidance cues are likely used to specify the overall polarity and perhaps rough topography of the V1–SC map, as the V1 axons of enucleated mice and Math5 mutants still project to the proper half of the SC in the absence of retinal inputs (Triplett et al. 2009).

While spontaneous retinal activity drives the topographic development of V1 inputs, visually evoked activity further consolidates these inputs to form functional circuits. Eye opening induces rapid local branching of cortical axons in the SC and causes an increase in the number and strength of the synapses that individual dendrites oriented vertically (DOV) neurons receive. Visual deprivation disrupted these changes and reduced the number of

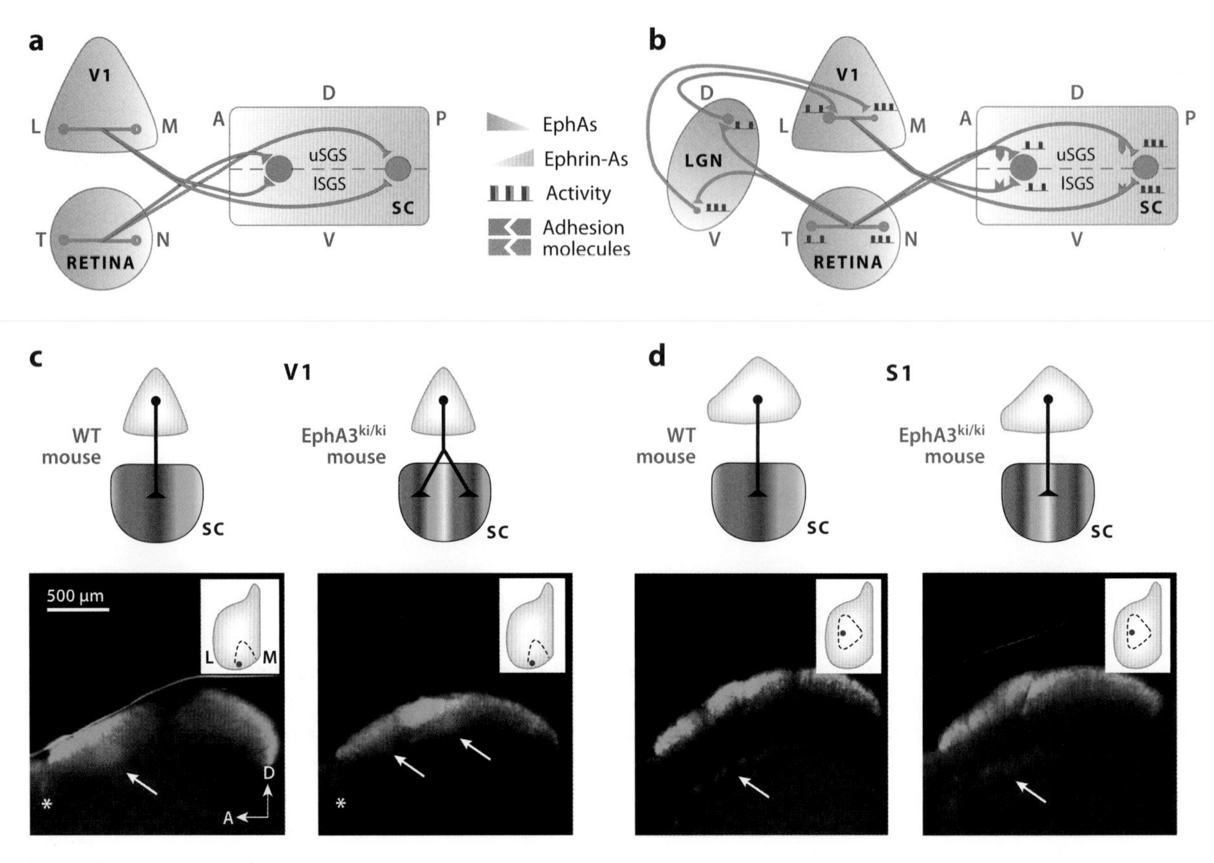

Figure 5

Alignment of the primary visual cortex (V1) and the primary somatosensory cortex (S1) inputs with retinocollicular maps in the superior colliculus (SC). (*a–b*) Models of visual map alignment in the SC. In the gradient-matching model (*a*), graded expression of EphA receptors in both the retina and V1 are used to guide topographic mapping in the SC, which expresses repulsive ephrin-A ligands in a gradient in both recipient layers. In the retinal-matching model (*b*), retinocollicular mapping is established first through the use of graded EphAs and ephrin-As. V1 projections then terminate in areas with similar activity patterns or with RGCs expressing complementary cell surface molecules. Abbreviations: N, nasal; T, temporal; A, anterior; P, posterior; D, dorsal; V, ventral; L, lateral; M, medial; uSGS, upper stratum griseum superficiale; lSGS, lower stratum griseum superficiale; WT, wild type. (*c–d*) Different outcomes of V1 and S1 inputs to the SC in EphA3$^{ki/ki}$ mice. (*c*) An injection of DiI into central V1 traces axons that project to the central SC (*white arrow*). A similar injection bifurcates and projects to each map in the EphA3$^{ki/ki}$ mouse (*white arrows*), showing that the retinal inputs to the SC direct V1 inputs; *, pretectal nucleus. (*d*) S1 axons do not alter their projection patterns in the EphA3$^{ki/ki}$ mouse, resulting in misalignment between retinal and SC maps. Diagrams and data are from Triplett et al. 2009, 2012.

spines on the V1-recipient dendrites of the DOV neurons (Phillips et al. 2011), suggesting a role for experience-driven activity in the final stages of map alignment.

MULTIMODAL MAP ALIGNMENT

Even though the above map-alignment examples are all from the visual system, the brain also integrates maps from different modalities. In the SC, aligning sensory and motor maps allows proper integration of sensory information and initiation of orienting movements to redirect attention toward the stimulus. Studies have shown that both activity-dependent and activity-independent mechanisms contribute to the establishment of map alignment between modalities.

Aligning Visual and Somatosensory Maps in the Superior Colliculus

Neurons in S1 project axons topographically to an SC lamina deeper than do V1 axons and provide a link between touch and vision. However, unlike V1 axons, S1 neurons do not receive any retinal input and do not change in the EphA3$^{ki/ki}$ mouse (**Figure 5*d***), confirming that retinal input into the SC does not influence the S1–SC map. Instead, the S1–SC topography requires ephrin-As because the ephrin-A2/A3/A5 knockout mouse has a completely disrupted S1–SC projection (Triplett et al. 2012). This observation suggests that the S1–SC projection matches EphA/ephrin-A gradients to map topographically. In combination with the above studies of V1–SC map development, a model has emerged in which distinct afferents use different mechanisms to achieve alignment.

Aligning Visual and Auditory Maps

Unlike in the visual system, where space naturally projects onto the photoreceptors across the retina, the topographic representation of the auditory space must be computed in the brain (Knudsen et al. 1987). The location of a sound source is determined by using spatial cues arising from the interactions of the head and ears with the sound stimulus, such as interaural timing differences (ITDs) and interaural level differences (ILDs). These cues are processed in frequency-specific channels in several brain stem nuclei, then combined in the posterior midbrain to generate a map of auditory space (Knudsen et al. 1987). The anatomy and computation of sound localization and its development have been studied in several species, and most notably in barn owls, where the map of auditory space is first seen in the external nucleus of the inferior colliculus (ICX) (Knudsen & Konishi 1978b). The ICX combines inputs across frequency-specific channels arriving from the central nucleus of the inferior colliculus (ICC), where ITDs and ILDs are still organized according to sound frequency (tonotopically). The auditory maps in the ICX (and brachium of the inferior colliculus in mammals; King 1999) are conveyed to the SC/OT and topographically align with maps of the visual space and of the body surface (Knudsen & Brainard 1995). As a result, many neurons in the intermediate and deep layers of the SC are bimodal or trimodal and display closely aligned auditory and visual receptive fields in space (Drager & Hubel 1975, King 1999, King & Hutchings 1987, Knudsen 1982, Knudsen & Brainard 1995, Middlebrooks & Knudsen 1984).

Interaural timing differences (ITDs): the difference in the timing of a sound reaching each ear

Interaural level differences (ILDs): the difference in the loudness of the sound entering each ear

Compared with the retinotopic map, the map of auditory space in the SC develops later and requires sensory experience. This finding is not surprising given the extensive growth and changes in head size and ear position during early life and their individual differences, which alter the values of ITD and ILD cues important for sound localization. The maturation of auditory spatial tuning and space map in the SC is a protracted process that takes place after hearing and vision onset (Vachon-Presseau et al. 2009, Wallace & Stein 1997, Withington-Wray et al. 1990a). In animals that were raised without location-specific acoustic cues or those that were visually deprived, auditory neurons in the SC/OT display disrupted spatial tuning and the auditory map is misaligned with the visual map (Efrati & Gutfreund 2011; King & Carlile 1993; Knudsen et al. 1991; Withington-Wray et al. 1990a,b). Some aspects of the SC auditory maps in the blind-reared animals are normal, suggesting that innate genetic programs may establish a crude map of space (Knudsen & Brainard 1995), although the molecules involved have not been identified.

How visual experience aligns the auditory and visual maps during development has been studied with manipulations that cause the two maps to misalign (King 1999, Knudsen 2002). For example, in barn owls, wearing prismatic spectacles that optically displace the visual field causes a misalignment between the visual and auditory spaces. During a sensitive period in early life, prism-reared owls can learn the new associations between auditory spatial cues and the shifted visual world, and the auditory and visual maps in the OT are once again in register (Brainard & Knudsen 1998, Knudsen et al. 1991). In this process, the retinotectal map does not change, but the auditory receptive fields of individual neurons and the map of the auditory space in the OT shift to match the visual displacement. As a result, turning toward the sound source again brings it to the line of sight through the prisms. Such adaptive changes are also seen in juvenile ferrets, in which surgical deviation of one eye results in a corresponding shift of the auditory space map in the contralateral SC (King et al. 1988).

Knudsen (2002) identified the ICX as the site of plasticity in barn owls, in which the ICC-to-ICX projections became broader and the spatial tuning of ICX neurons was altered adaptively in the prism-reared animals (Knudsen 2002). Because auditory map development and plasticity are under the instruction of visual experience, visual inputs must reach the ICX. The instructive visual signal originates from the OT and is topographic (Hyde & Knudsen 2001), as a small lesion in the OT eliminates auditory map plasticity only in the corresponding region in the ICX (Hyde & Knudsen 2002; see also King et al. 1998 for similar experiments in ferrets). Visually evoked responses are indeed seen in the ICX of anesthetized owls, but only after GABAergic transmission is blocked in the OT, indicating that the instructive signal from the OT to the ICX is gated by inhibition (Gutfreund et al. 2002). These visual inputs have spatially restricted receptive fields that align with the auditory receptive fields of the same sites in normal owls (Gutfreund et al. 2002). Together, these experiments support the idea that the visual map in the SC/OT is used as a topographic template to guide the transformation of spatial cues into the auditory space map, thereby achieving visual and auditory map alignment.

Aligning Visual and Motor Maps

The SC is well known for its role in initiating orienting behaviors, and the motor commands are sent out by neurons in the deep layers. These neurons project to several premotor and motor nuclei that control muscles for eye, head, ear, whisker, and body movements (Gandhi & Katnani 2011). The topographic organization of these motor functions is best studied for saccadic eye and head movements (Wurtz & Albano 1980). In primates, the motor map in the deep layers of the SC is in register with the retinotopic map in the superficial layers, such that the direction of gaze is shifted to the site of the visual stimulus after the evoked saccade (Schiller & Stryker 1972, Sparks et al. 1990, Wurtz & Goldberg 1972). In barn owls, which have very limited eye movements, the maps of head movement are similarly aligned with the visual and auditory maps in the OT (du Lac & Knudsen 1990), enabling proper visual/auditory-orienting behavior.

Both experience-dependent and -independent processes are likely involved in aligning sensory and motor maps during development. Electrical stimulation of the SC can produce eye movements in neonatal kittens, and the evoked movements show some features of topographic organization even before visual stimuli can activate neurons in the SC (Stein 1984, Stein et al. 1980). Barn owls that were reared with closed eyelids could still develop a head movement map with correct global polarity, but the movement map displayed substantial misalignment with both the visual and the auditory maps (du Lac & Knudsen 1991). Whether and how visual experience is required for visual-motor map alignment in mammals remain unknown. The lack of progress is perhaps because the mouse, a great model in developmental studies as evidenced in this review, was assumed to lack eye movements because mice do not have a fovea. However, recent studies have shown that mice do, in fact, have saccade-like rapid eye movements that can be both spontaneous and evoked by SC electrical stimulation (Sakatani & Isa 2007, 2008). The amplitude and direction of the evoked eye movements appear to be organized in a map that has the same global polarity as the retinotopic map in the SC (Wang et al. 2011). Given the powerful genetic tools and available transgenic lines that alter the visual map (e.g., the EphA3$^{ki/ki}$ mouse), mice will become a useful model to reveal the molecular and activity-dependent mechanisms of motor map development and sensorimotor alignment.

CONCLUSIONS AND FUTURE DIRECTIONS

Much progress has been made to understand how topographic maps are established; these data have provided important foundations for studying how multiple maps are aligned during development. Three main factors in map alignment have been revealed: molecular guidance cues, activity-dependent processes driven by patterned spontaneous activity, and experience-driven developmental plasticity. Although all three mechanisms are required in map alignment, their relevant contributions differ and depend on the logic of the particular alignment in question. For example, as EphA/ephrin-A gradients map both retinotopic projections in the SC and somatotopic projections in S1, the same sets of gradients can naturally be used to guide S1 inputs to the SC to align with visual maps. A similar process is probably used to map V1 projections to the SC, but because retinal and V1 inputs share patterned spontaneous activity, their alignment can be further driven by a template-matching process instructed by retinal waves. However, to align auditory space maps and eye-movement motor maps with the visual map, molecular gradients and spontaneous activity are not enough. This is because simple Eph/ephrin gradients seen in the developing visual system cannot guide the formation of the auditory space maps and eye-movement maps, where the represented values are computed and transformed from their inputs. Furthermore, these systems and the visual system do not share spontaneous activity patterns that carry spatial information. As a result, for these maps to align, experience is needed to provide the necessary instructive signal from one system to the other. The instructive signal originates from the visual system in both visual-auditory and visual-motor alignment, presumably because vision provides the most accurate and reliable information and the main purpose of the alignment is to shift the line of sight.

Despite these advances, studies of map alignment are still very descriptive. A major goal now is to reveal the molecular and synaptic mechanisms used to mediate map formation and alignment during development and to understand how they work together. First, clearly much still needs to be learned about other, non-Eph/ephrin molecules that must be involved in establishing computation maps, such as maps of orientation columns in the visual cortex of carnivores and primates, maps of auditory space in the SC/OT, and eye/head movement maps in the SC/OT. New genomic technologies such as RNA sequencing, which allow researchers to compare transcriptomes of different cell types or to compare the same cell types under different conditions (e.g., dark-reared versus normal), can reveal genes that are involved in these processes in traditionally nongenetic model organisms.

Second, what are the patterns of spontaneous activity in vivo when the maps are being formed and aligned, not only in the retina but in all visual areas? This unknown can now be addressed with recent technical advances in multiphoton imaging, which allows investigators to monitor neural activity at single-cell resolution of a large number of cells (Siegel et al. 2012, Zhang et al. 2012). Indeed, the recent report of retinal waves imaged in vivo is an exciting development along this direction (Ackman et al. 2012). Future experiments that combine these imaging techniques with transgenic mice and optogenetics will provide important information about how retinal waves instruct topographic map formation and alignment.

Furthermore, what are the synaptic and circuit mechanisms for activity-/experience-dependent map formation and alignment? Hebbian plasticity seems to be the most likely candidate, but direct evidence for this hypothesis is still lacking. It will be critical to follow connectivity patterns and functional properties of individual neurons during map formation rather than studying only the consequences of activity manipulation and visual deprivation after development.

Finally, a long-term goal of this field of research is to understand the behavioral consequences of animals that have defects in multimodal processing. These studies will generate valuable insights to help us understand neurodevelopmental diseases such as autism spectrum disorders, psychosis, and mental retardation, each of which displays impairments in aspects of sensory integration, sensorimotor transformation, social interactions, and language development. The normal development of these skills requires proper information association and integration in the nervous system, which often occur in an experience-dependent manner during critical periods in early life. By studying map alignment, the structural substrate of multichannel integration, and how it is instructed by sensory and motor experiences, we may better understand the pathogenic mechanisms of these disorders and reveal new interventions to treat them.

DISCLOSURE STATEMENT

The authors are not aware of any affiliations, memberships, funding, or financial holdings that might be perceived as affecting the objectivity of this review.

ACKNOWLEDGMENTS

We thank Jena Yamada, Ben Stafford, and Jason Triplett for help with the figures and Jena Yamada, Andrew Huberman, and Jason Triplett for comments on the manuscript. The work in our labs is supported by National Institutes of Health Grants (EY018621 and EY020950 to J.C. and EY014689, EY022117 to D.A.F), a Sloan Research Fellowship, a Klingenstein Fellowship Award in Neurosciences, and a Brain Research Foundation Seed Grant (J.C.).

LITERATURE CITED

Ackman JB, Burbridge TJ, Crair MC. 2012. Retinal waves coordinate patterned activity throughout the developing visual system. *Nature* 490(7419):219–25

Adorjan P, Levitt JB, Lund JS, Obermayer K. 1999. A model for the intracortical origin of orientation preference and tuning in macaque striate cortex. *Vis. Neurosci.* 16:303–18

Andermann ML, Kerlin AM, Roumis DK, Glickfeld LL, Reid RC. 2011. Functional specialization of mouse higher visual cortical areas. *Neuron* 72:1025–39

Bansal A, Singer JH, Hwang BJ, Xu W, Beaudet A, Feller MB. 2000. Mice lacking specific nicotinic acetylcholine receptor subunits exhibit dramatically altered spontaneous activity patterns and reveal a limited role for retinal waves in forming ON and OFF circuits in the inner retina. *J. Neurosci.* 20:7672–81

Bargary G, Mitchell KJ. 2008. Synaesthesia and cortical connectivity. *Trends Neurosci.* 31:335–42

Ben-Yishai R, Hansel D, Sompolinsky H. 1997. Traveling waves and the processing of weakly tuned inputs in a cortical network module. *J. Comput. Neurosci.* 4:57–77

Bevins N, Lemke G, Reber M. 2011. Genetic dissection of EphA receptor signaling dynamics during retinotopic mapping. *J. Neurosci.* 31:10302–10

Blakemore C, Van Sluyters RC. 1974. Reversal of the physiological effects of monocular deprivation in kittens: further evidence for a sensitive period. *J. Physiol.* 237:195–216

Blasdel GG, Salama G. 1986. Voltage-sensitive dyes reveal a modular organization in monkey striate cortex. *Nature* 321:579–85

Bozza T, Vassalli A, Fuss S, Zhang JJ, Weiland B, et al. 2009. Mapping of class I and class II odorant receptors to glomerular domains by two distinct types of olfactory sensory neurons in the mouse. *Neuron* 61:220–33

Brainard MS, Knudsen EI. 1998. Sensitive periods for visual calibration of the auditory space map in the barn owl optic tectum. *J. Neurosci.* 18:3929–42

Braisted JE, McLaughlin T, Wang HU, Friedman GC, Anderson DJ, O'Leary DDM. 1997. Graded and lamina-specific distributions of ligands of EphB receptor tyrosine kinases in the developing retinotectal system. *Dev. Biol.* 191:14–28

Bridge H, Cumming BG. 2001. Responses of macaque V1 neurons to binocular orientation differences. *J. Neurosci.* 21:7293–302

Brock J, Brown CC, Boucher J, Rippon G. 2002. The temporal binding deficit hypothesis of autism. *Dev. Psychopathol.* 14:209–24

Brown A, Yates PA, Burrola P, Ortuno D, Vaidya A, et al. 2000. Topographic mapping from the retina to the midbrain is controlled by relative but not absolute levels of EphA receptor signaling. *Cell* 102:77–88

Brown NL, Patel S, Brzezinski J, Glaser T. 2001. Math5 is required for retinal ganglion cell and optic nerve formation. *Development* 128:2497–508

Brunet I, Weinl C, Piper M, Trembleau A, Volovitch M, et al. 2005. The transcription factor Engrailed-2 guides retinal axons. *Nature* 438:94–98

Buhusi M, Demyanenko GP, Jannie KM, Dalal J, Darnell EP, et al. 2009. ALCAM regulates mediolateral retinotopic mapping in the superior colliculus. *J. Neurosci.* 29:15630–41

Butts DA. 2002. Retinal waves: implications for synaptic learning rules during development. *Neuroscientist* 8:243–53

Butts DA, Kanold PO, Shatz CJ. 2007. A burst-based "Hebbian" learning rule at retinogeniculate synapses links retinal waves to activity-dependent refinement. *PLoS Biol.* 5:e61

Campbell DS, Regan AG, Lopez JS, Tannahill D, Harris WA, Holt CE. 2001. Semaphorin 3A elicits stage-dependent collapse, turning, and branching in *Xenopus* retinal growth cones. *J. Neurosci.* 21:8538–47

Cang J, Kaneko M, Yamada J, Woods G, Stryker MP, Feldheim DA. 2005a. Ephrin-As guide the formation of functional maps in the visual cortex. *Neuron* 48:577–89

Cang J, Niell CM, Liu X, Pfeiffenberger C, Feldheim DA, Stryker MP. 2008a. Selective disruption of one Cartesian axis of cortical maps and receptive fields by deficiency in ephrin-As and structured activity. *Neuron* 57:511–23

Cang J, Renteria RC, Kaneko M, Liu X, Copenhagen DR, Stryker MP. 2005b. Development of precise maps in visual cortex requires patterned spontaneous activity in the retina. *Neuron* 48:797–809

Cang J, Wang L, Stryker MP, Feldheim DA. 2008b. Roles of ephrin-As and structured activity in the development of functional maps in the superior colliculus. *J. Neurosci.* 28:11015–23

Carreres MI, Escalante A, Murillo B, Chauvin G, Gaspar P, et al. 2011. Transcription factor Foxd1 is required for the specification of the temporal retina in mammals. *J. Neurosci.* 31:5673–81

Chalupa LM. 2009. Retinal waves are unlikely to instruct the formation of eye-specific retinogeniculate projections. *Neural Dev.* 4:25

Chalupa LM, Williams RW, eds. 2008. *Eye, Retina and Visual System of the Mouse.* Boston: MIT Press

Chandrasekaran AR, Plas DT, Gonzalez E, Crair MC. 2005. Evidence for an instructive role of retinal activity in retinotopic map refinement in the superior colliculus of the mouse. *J. Neurosci.* 25:6929–38

Chisholm A, Tessier-Lavigne M. 1999. Conservation and divergence of axon guidance mechanisms. *Curr. Opin. Neurobiol.* 9:603–15

Chklovskii DB, Koulakov AA. 2004. Maps in the brain: What can we learn from them? *Annu. Rev. Neurosci.* 27:369–92

Cline H. 2003. Sperry and Hebb: oil and vinegar? *Trends Neurosci.* 26:655–61

Crair MC, Gillespie DC, Stryker MP. 1998. The role of visual experience in the development of columns in cat visual cortex. *Science* 279:566–70

Cramer KS. 2005. Eph proteins and the assembly of auditory circuits. *Hear. Res.* 206:42–51

Cutforth T, Moring L, Mendelsohn M, Nemes A, Shah NM, et al. 2003. Axonal ephrin-As and odorant receptors: coordinate determination of the olfactory sensory map. *Cell* 114:311–22

Debski EA, Cline HT. 2002. Activity-dependent mapping in the retinotectal projection. *Curr. Opin. Neurobiol.* 12:93–99

Demyanenko GP, Maness PF. 2003. The L1 cell adhesion molecule is essential for topographic mapping of retinal axons. *J. Neurosci.* 23:530–38

Dhande OS, Bhatt S, Anishchenko A, Elstrott J, Iwasato T, et al. 2012. Role of adenylate cyclase 1 in retinofugal map development. *J. Comp. Neurol.* 520:1562–83

Drager UC, Hubel DH. 1975. Responses to visual stimulation and relationship between visual, auditory, and somatosensory inputs in mouse superior colliculus. *J. Neurophysiol.* 38:690–713

Drescher U, Bonhoeffer F, Müller BK. 1997. The Eph family in retinal axon guidance. *Curr. Opin. Neurobiol.* 7:75–80

du Lac S, Knudsen EI. 1990. Neural maps of head movement vector and speed in the optic tectum of the barn owl. *J. Neurophysiol.* 63:131–46

du Lac S, Knudsen EI. 1991. Early visual deprivation results in a degraded motor map in the optic tectum of barn owls. *Proc. Natl. Acad. Sci. USA* 88:3426–30

Efrati A, Gutfreund Y. 2011. Early life exposure to noise alters the representation of auditory localization cues in the auditory space map of the barn owl. *J. Neurophysiol.* 105:2522–35

Erwin E, Miller KD. 1998. Correlation-based development of ocularly matched orientation and ocular dominance maps: determination of required input activities. *J. Neurosci.* 18:9870–95

Espinosa JS, Stryker MP. 2012. Development and plasticity of the primary visual cortex. *Neuron* 75:230–49

Feldheim DA, Kim YI, Bergemann AD, Frisén J, Barbacid M, Flanagan JG. 2000. Genetic analysis of ephrin-A2 and ephrin-A5 shows their requirement in multiple aspects of retinocollicular mapping. *Neuron* 25:563–74

Feldheim DA, Nakamoto M, Osterfield M, Gale NW, DeChiara TM, et al. 2004. Loss-of-function analysis of EphA receptors in retinotectal mapping. *J. Neurosci.* 24:2542–50

Feldheim DA, O'Leary DD. 2012. Visual map development: bidirectional signaling, bifunctional guidance molecules, and competition. *Cold Spring Harb. Perspect. Biol.* 2:a001768

Feldheim DA, Vanderhaeghen P, Hansen MJ, Frisén J, Lu Q, et al. 1998. Topographic guidance labels in a sensory projection to the forebrain. *Neuron* 21:1303–13

Feller MB. 2009. Retinal waves are likely to instruct the formation of eye-specific retinogeniculate projections. *Neural Dev.* 4:24

Feller MB, Wellis DP, Stellwagen D, Werblin FS, Shatz CJ. 1996. Requirement for cholinergic synaptic transmission in the propagation of spontaneous retinal waves. *Science* 272:1182–87

Ferster D. 1981. A comparison of binocular depth mechanisms in areas 17 and 18 of the cat visual cortex. *J. Physiol.* 311:623–55

Ferster D, Miller KD. 2000. Neural mechanisms of orientation selectivity in the visual cortex. *Annu. Rev. Neurosci.* 23:441–71

Finlay BL, Schneps SE, Schneider GE. 1979. Orderly compression of the retinotectal projection following partial tectal ablation in the newborn hamster. *Nature* 280:153–55

Flanagan JG. 2006. Neural map specification by gradients. *Curr. Opin. Neurobiol.* 16:59–66

Flanagan JG, Vanderhaeghen P. 1998. The ephrins and Eph receptors in neural development. *Annu. Rev. Neurosci.* 21:309–45

Fraser SE, Hunt RK. 1980. Retinotectal specificity: models and experiments in search of a mapping function. *Annu. Rev. Neurosci.* 3:319–52

Frisén J, Yates PA, McLaughlin T, Friedman GC, O'Leary DDM, Barbacid M. 1998. Ephrin-A5 (AL-1/RAGS) is essential for proper retinal axon guidance and topographic mapping in the mammalian visual system. *Neuron* 20:235–43

Gale NW, Holland SJ, Valenzuela DM, Flenniken A, Pan L, et al. 1996. Eph receptors and ligands comprise two major specificity subclasses and are reciprocally compartmentalized during embryogenesis. *Neuron* 17:9–19

Galimberti I, Bednarek E, Donato F, Caroni P. 2010. EphA4 signaling in juveniles establishes topographic specificity of structural plasticity in the hippocampus. *Neuron* 65:627–42

Gandhi NJ, Katnani HA. 2011. Motor functions of the superior colliculus. *Annu. Rev. Neurosci.* 34:205–31

Geschwind DH, Levitt P. 2007. Autism spectrum disorders: developmental disconnection syndromes. *Curr. Opin. Neurobiol.* 17:103–11

Godecke I, Bonhoeffer T. 1996. Development of identical orientation maps for two eyes without common visual experience. *Nature* 379:251–54

Godement P, Salaun J, Imbert M. 1984. Prenatal and postnatal development of retinogeniculate and retinocollicular projections in the mouse. *J. Comp. Neurol.* 230:552–75

Gosse NJ, Nevin LM, Baier H. 2008. Retinotopic order in the absence of axon competition. *Nature* 452:892–95

Grimbert F, Cang J. 2012. New model of retinocollicular mapping predicts the mechanisms of axonal competition and explains the role of reverse molecular signaling during development. *J. Neurosci.* 32:9755–68

Grinvald A, Lieke E, Frostig RD, Gilbert CD, Wiesel TN. 1986. Functional architecture of cortex revealed by optical imaging of intrinsic signals. *Nature* 324:361–64

Grubb MS, Rossi FM, Changeux JP, Thompson ID. 2003. Abnormal functional organization in the dorsal lateral geniculate nucleus of mice lacking the beta 2 subunit of the nicotinic acetylcholine receptor. *Neuron* 40:1161–72

Gutfreund Y, Zheng W, Knudsen EI. 2002. Gated visual input to the central auditory system. *Science* 297:1556–59

Haustead DJ, Lukehurst SS, Clutton GT, Bartlett CA, Dunlop SA, et al. 2008. Functional topography and integration of the contralateral and ipsilateral retinocollicular projections of ephrin-A-/- mice. *J. Neurosci.* 28:7376–86

Higenell V, Han SM, Feldheim DA, Scalia F, Ruthazer ES. 2011. Expression patterns of Ephs and ephrins throughout retinotectal development in *Xenopus laevis*. *Dev. Neurobiol.* 72:547–63

Himanen JP, Chumley MJ, Lackmann M, Li C, Barton WA, et al. 2004. Repelling class discrimination: ephrin-A5 binds to and activates EphB2 receptor signaling. *Nat. Neurosci.* 7:501–9

Hindges R, McLaughlin T, Genoud N, Henkemeyer M, O'Leary DD. 2002. EphB forward signaling controls directional branch extension and arborization required for dorsal-ventral retinotopic mapping. *Neuron* 35:475–87

Holash JA, Pasquale EB. 1995. Polarized expression of the receptor protein tyrosine kinase Cek5 in the developing avian visual system. *Dev. Biol.* 172:683–93

Holmberg J, Armulik A, Senti KA, Edoff K, Spalding K, et al. 2005. Ephrin-A2 reverse signaling negatively regulates neural progenitor proliferation and neurogenesis. *Genes Dev.* 19:462–71

Holmberg J, Genander M, Halford MM, Anneren C, Sondell M, et al. 2006. EphB receptors coordinate migration and proliferation in the intestinal stem cell niche. *Cell* 125:1151–63

Hubel DH, Wiesel TN. 1962. Receptive fields, binocular interaction and functional architecture in the cat's visual cortex. *J. Physiol.* 160:106–54

Hubel DH, Wiesel TN, Stryker MP. 1977. Orientation columns in macaque monkey visual cortex demonstrated by the 2-deoxyglucose autoradiographic technique. *Nature* 269:328–30

Huberman AD, Feller MB, Chapman B. 2008a. Mechanisms underlying development of visual maps and receptive fields. *Annu. Rev. Neurosci.* 31:479–509

Huberman AD, Manu M, Koch SM, Susman MW, Lutz AB, et al. 2008b. Architecture and activity-mediated refinement of axonal projections from a mosaic of genetically identified retinal ganglion cells. *Neuron* 59:425–38

Huberman AD, Murray KD, Warland DK, Feldheim DA, Chapman B. 2005. Ephrin-As mediate targeting of eye-specific projections to the lateral geniculate nucleus. *Nat. Neurosci.* 8:1013–21

Huberman AD, Wei W, Elstrott J, Stafford BK, Feller MB, Barres BA. 2009. Genetic identification of an On-Off direction-selective retinal ganglion cell subtype reveals a layer-specific subcortical map of posterior motion. *Neuron* 62:327–34

Hyde PS, Knudsen EI. 2001. A topographic instructive signal guides the adjustment of the auditory space map in the optic tectum. *J. Neurosci.* 21:8586–93

Hyde PS, Knudsen EI. 2002. The optic tectum controls visually guided adaptive plasticity in the owl's auditory space map. *Nature* 415:73–76

Imai T, Yamazaki T, Kobayakawa R, Kobayakawa K, Abe T, et al. 2009. Pre-target axon sorting establishes the neural map topography. *Science* 325:585–90

Inoue A, Sanes JR. 1997. Lamina-specific connectivity in the brain: regulation by N-cadherin, neurotrophins, and glycoconjugates. *Science* 276:1428–31

Jaubert-Miazza L, Green E, Lo FS, Bui K, Mills J, Guido W. 2005. Structural and functional composition of the developing retinogeniculate pathway in the mouse. *Vis. Neurosci.* 22:661–76

Jeffery G. 1985. Retinotopic order appears before ocular separation in developing visual pathways. *Nature* 313:575–76

Kay JN, De la Huerta I, Kim IJ, Zhang Y, Yamagata M, et al. 2011. Retinal ganglion cells with distinct directional preferences differ in molecular identity, structure, and central projections. *J. Neurosci.* 31:7753–62

King AJ. 1999. Sensory experience and the formation of a computational map of auditory space in the brain. *BioEssays* 21:900–11

King AJ, Carlile S. 1993. Changes induced in the representation of auditory space in the superior colliculus by rearing ferrets with binocular eyelid suture. *Exp. Brain Res.* 94:444–55

King AJ, Hutchings ME. 1987. Spatial response properties of acoustically responsive neurons in the superior colliculus of the ferret: a map of auditory space. *J. Neurophysiol.* 57:596–624

King AJ, Hutchings ME, Moore DR, Blakemore C. 1988. Developmental plasticity in the visual and auditory representations in the mammalian superior colliculus. *Nature* 332:73–76

King AJ, Schnupp JW, Thompson ID. 1998. Signals from the superficial layers of the superior colliculus enable the development of the auditory space map in the deeper layers. *J. Neurosci.* 18:9394–408

Knudsen EI. 1982. Auditory and visual maps of space in the optic tectum of the owl. *J. Neurosci.* 2:1177–94

Knudsen EI. 2002. Instructed learning in the auditory localization pathway of the barn owl. *Nature* 417:322–28

Knudsen EI, Brainard MS. 1995. Creating a unified representation of visual and auditory space in the brain. *Annu. Rev. Neurosci.* 18:19–43

Knudsen EI, du Lac S, Esterly SD. 1987. Computational maps in the brain. *Annu. Rev. Neurosci.* 10:41–65

Knudsen EI, Esterly SD, du Lac S. 1991. Stretched and upside-down maps of auditory space in the optic tectum of blind-reared owls; acoustic basis and behavioral correlates. *J. Neurosci.* 11:1727–47

Knudsen EI, Konishi M. 1978a. A neural map of auditory space in the owl. *Science* 200:795–97

Knudsen EI, Konishi M. 1978b. Center-surround organization of auditory receptive fields in the owl. *Science* 202:778–80

Koulakov AA, Tsigankov DN. 2004. A stochastic model for retinocollicular map development. *BMC Neurosci.* 5:30

Kullander K, Klein R. 2002. Mechanisms and functions of Eph and ephrin signalling. *Nat. Rev. Mol. Cell Biol.* 3:475–86

Lambot MA, Depasse F, Noel JC, Vanderhaeghen P. 2005. Mapping labels in the human developing visual system and the evolution of binocular vision. *J. Neurosci.* 25:7232–37

Leamey CA, Merlin S, Lattouf P, Sawatari A, Zhou X, et al. 2007. Ten_m3 regulates eye-specific patterning in the mammalian visual pathway and is required for binocular vision. *PLoS Biol.* 5:e241

Leamey CA, Protti DA, Dreher B. 2008. Comparative survey of the mammalian visual system with reference to the mouse. See Chalupa & Williams 2008, pp. 35–60

Lewin B. 1994. On neuronal specificity and the molecular basis of perception. *Cell* 79:935–43

Lim YS, McLaughlin T, Sung TC, Santiago A, Lee KF, O'Leary DD. 2008. p75(NTR) mediates ephrin-A reverse signaling required for axon repulsion and mapping. *Neuron* 59:746–58

Lund RD, Lund JS, Wise RP. 1974. The organization of the retinal projection to the dorsal lateral geniculate nucleus in pigmented and albino rats. *J. Comp. Neurol.* 158:383–404

Luo L, Flanagan JG. 2007. Development of continuous and discrete neural maps. *Neuron* 56:284–300

Mann F, Ray S, Harris W, Holt C. 2002. Topographic mapping in dorsoventral axis of the *Xenopus* retinotectal system depends on signaling through ephrin-B ligands. *Neuron* 35:461–73

Marcus RC, Gale NW, Morrisson ME, Mason CA, Yancopoulos GD. 1996. Eph family receptors and their ligands distribute in opposing gradients in the developing mouse retina. *Dev. Biol.* 180:786–89

Marshel JH, Garrett ME, Nauhaus I, Callaway EM. 2011. Functional specialization of seven mouse visual cortical areas. *Neuron* 72:1040–54

May PJ. 2005. The mammalian superior colliculus: laminar structure and connections. *Prog. Brain Res.* 151:321–78

McLaughlin T, Hindges R, Yates PA, O'Leary DD. 2003a. Bifunctional action of ephrin-B1 as a repellent and attractant to control bidirectional branch extension in dorsal-ventral retinotopic mapping. *Development* 130:2407–18

McLaughlin T, O'Leary DD. 2005. Molecular gradients and development of retinotopic maps. *Annu. Rev. Neurosci.* 28:327–55

McLaughlin T, Torborg CL, Feller MB, O'Leary DD. 2003b. Retinotopic map refinement requires spontaneous retinal waves during a brief critical period of development. *Neuron* 40:1147–60

Meister M, Wong RO, Baylor DA, Shatz CJ. 1991. Synchronous bursts of action potentials in ganglion cells of the developing mammalian retina. *Science* 252:939–43

Mellott DO, Burke RD. 2008. The molecular phylogeny of eph receptors and ephrin ligands. *BMC Cell Biol.* 9:27

Merlin S, Horng S, Marotte LR, Sur M, Sawatari A, Leamey CA. 2013. Deletion of Ten-m3 induces the formation of eye dominance domains in mouse visual cortex. *Cereb. Cortex* 23:763–74

Miao H, Wang B. 2012. EphA receptor signaling—complexity and emerging themes. *Semin. Cell Dev. Biol.* 23:16–25

Middlebrooks JC, Knudsen EI. 1984. A neural code for auditory space in the cat's superior colliculus. *J. Neurosci.* 4:2621–34

Monnier PP, Sierra A, Macchi P, Deitinghoff L, Andersen JS, et al. 2002. RGM is a repulsive guidance molecule for retinal axons. *Nature* 419:392–95

Mrsic-Flogel TD, Hofer SB, Creutzfeldt C, Cloez-Tayarani I, Changeux JP, et al. 2005. Altered map of visual space in the superior colliculus of mice lacking early retinal waves. *J. Neurosci.* 25:6921–28

Muir-Robinson G, Hwang BJ, Feller MB. 2002. Retinogeniculate axons undergo eye-specific segregation in the absence of eye-specific layers. *J. Neurosci.* 22:5259–64

Nakamoto M, Cheng HJ, Friedman GC, McLaughlin T, Hansen MJ, et al. 1996. Topographically specific effects of ELF-1 on retinal axon guidance in vitro and retinal axon mapping in vivo. *Cell* 86:755–66

Nakamura H, O'Leary DDM. 1989. Inaccuracies in initial growth and arborization of chick retinotectal axons followed by course corrections and axon remodeling to develop topographic order. *J. Neurosci.* 9:3776–95

Nelson JI, Kato H, Bishop PO. 1977. Discrimination of orientation and position disparities by binocularly activated neurons in cat straite cortex. *J. Neurophysiol.* 40:260–83

Niell CM, Stryker MP. 2008. Highly selective receptive fields in mouse visual cortex. *J. Neurosci.* 28:7520–36

O'Leary DD, McLaughlin T. 2005. Mechanisms of retinotopic map development: Ephs, ephrins, and spontaneous correlated retinal activity. *Prog. Brain Res.* 147:43–65

Pasquale EB. 2010. Eph receptors and ephrins in cancer: bidirectional signalling and beyond. *Nat. Rev. Cancer* 10:165–80

Penfield W, Boldrey E. 1937. Somatic motor and sensory representation in the cerebral cortex of man as studied by electrical stimulation. *Brain* 60:389–443

Petros TJ, Rebsam A, Mason CA. 2008. Retinal axon growth at the optic chiasm: to cross or not to cross. *Annu. Rev. Neurosci.* 31:295–315

Pfeiffenberger C, Cutforth T, Woods G, Yamada J, Renteria RC, et al. 2005. Ephrin-As and neural activity are required for eye-specific patterning during retinogeniculate mapping. *Nat. Neurosci.* 8:1022–27

Pfeiffenberger C, Yamada J, Feldheim DA. 2006. Ephrin-As and patterned retinal activity act together in the development of topographic maps in the primary visual system. *J. Neurosci.* 26:12873–84

Phillips MA, Colonnese MT, Goldberg J, Lewis LD, Brown EN, Constantine-Paton M. 2011. A synaptic strategy for consolidation of convergent visuotopic maps. *Neuron* 71:710–24

Picciotto MR, Zoli M, Lena C, Bessis A, Lallemand Y, et al. 1995. Abnormal avoidance learning in mice lacking functional high-affinity nicotine receptor in the brain. *Nature* 374:65–67

Plas DT, Lopez JE, Crair MC. 2005. Pretarget sorting of retinocollicular axons in the mouse. *J. Comp. Neurol.* 491:305–19

Prestige MC, Willshaw DJ. 1975. On a role for competition in the formation of patterned neural connexions. *Proc. R. Soc. Lond. Ser. B* 190:77–98

Priebe NJ, Ferster D. 2012. Mechanisms of neuronal computation in mammalian visual cortex. *Neuron* 75:194–208

Rajagopalan S, Deitinghoff L, Davis D, Conrad S, Skutella T, et al. 2004. Neogenin mediates the action of repulsive guidance molecule. *Nat. Cell Biol.* 6:756–62

Rashid T, Upton AL, Blentic A, Ciossek T, Knöll B, et al. 2005. Opposing gradients of ephrin-As and EphA7 in the superior colliculus are essential for topographic mapping in the mammalian visual system. *Neuron* 47:57–69

Reber M, Burrola P, Lemke G. 2004. A relative signalling model for the formation of a topographic neural map. *Nature* 431:847–53

Reese BE. 1986. The topography of expanded uncrossed retinal projections following neonatal enucleation of one eye: differing effects in dorsal lateral geniculate nucleus and superior colliculus. *J. Comp. Neurol.* 250:8–32

Rhoades RW, Chalupa LM. 1978. Functional properties of the corticotectal projection in the golden hamster. *J. Comp. Neurol.* 180:617–34

Sakatani T, Isa T. 2007. Quantitative analysis of spontaneous saccade-like rapid eye movements in C57BL/6 mice. *Neurosci. Res.* 58:324–31

Sakatani T, Isa T. 2008. Superior colliculus and saccade generation in mice. See Chalupa & Williams 2008, pp. 233–44

Sarnaik R, Wang BS, Cang J. 2013. Experience-dependent and independent binocular correspondence of receptive field subregions in mouse visual cortex. *Cereb. Cortex* doi: 10.1093/cercor/bht027. In press

Scalia F, Feldheim DA. 2005. Eph/ephrin A- and B-family expression patterns in the leopard frog (*Rana utricularia*). *Dev. Brain Res.* 158:102–6

Schiller PH, Stryker M. 1972. Single-unit recording and stimulation in superior colliculus of the alert rhesus monkey. *J. Neurophysiol.* 35:915–24

Schmitt AM, Shi J, Wolf AM, Lu CC, King LA, Zou Y. 2006. Wnt-Ryk signalling mediates medial-lateral retinotectal topographic mapping. *Nature* 439:31–37

Serizawa S, Miyamichi K, Takeuchi H, Yamagishi Y, Suzuki M, Sakano H. 2006. A neuronal identity code for the odorant receptor-specific and activity-dependent axon sorting. *Cell* 127(5):1057–69

Siegel F, Heimel JA, Peters J, Lohmann C. 2012. Peripheral and central inputs shape network dynamics in the developing visual cortex in vivo. *Curr. Biol.* 22(3):253–58

Simon DK, O'Leary DDM. 1991. Relationship of retinotopic ordering of axons in the optic pathway to the formation of visual maps in central targets. *J. Comp. Neurol.* 307:393–404

Simon DK, Roskies AL, O'Leary DDM. 1994. Plasticity in the development of topographic order in the mammalian retinocollicular projection. *Dev. Biol.* 162:384–93

Somers DC, Nelson SB, Sur M. 1995. An emergent model of orientation selectivity in cat visual cortical simple cells. *J. Neurosci.* 15:5448–65

Sparks DL, Lee C, Rohrer WH. 1990. Population coding of the direction, amplitude, and velocity of saccadic eye movements by neurons in the superior colliculus. *Cold Spring Harb. Symp. Quant. Biol.* 55:805–11

Sperry RW. 1943. Visuomotor coordination in the newt (*Triturus viridescens*) after regeneration of the optic nerve. *J. Comp. Neurol.* 79:33–55

Sperry RW. 1963. Chemoaffinity in the orderly growth of nerve fiber patterns and connections. *Proc. Natl. Acad. Sci. USA* 50:703–10

Stafford BK, Sher A, Litke AM, Feldheim DA. 2009. Spatial-temporal patterns of retinal waves underlying activity-dependent refinement of retinofugal projections. *Neuron* 64:200–12

Stein BE. 1984. Development of the superior colliculus. *Annu. Rev. Neurosci.* 7:95–125

Stein BE, Clamann HP, Goldberg SJ. 1980. Superior colliculus: control of eye movements in neonatal kittens. *Science* 210:78–80

Stein BE, Meredith MA. 1993. *The Merging of the Senses*. Cambridge, MA: MIT Press

Suetterlin P, Marler KM, Drescher U. 2012. Axonal ephrinA/EphA interactions, and the emergence of order in topographic projections. *Semin. Cell Dev. Biol.* 23:1–6

Sun C, Warland DK, Ballesteros JM, van der List D, Chalupa LM. 2008. Retinal waves in mice lacking the beta2 subunit of the nicotinic acetylcholine receptor. *Proc. Natl. Acad. Sci. USA* 105:13638–43

Thakar S, Chenaux G, Henkemeyer M. 2011. Critical roles for EphB and ephrin-B bidirectional signalling in retinocollicular mapping. *Nat. Commun.* 2:431

Torii M, Levitt P. 2005. Dissociation of corticothalamic and thalamocortical axon targeting by an EphA7-mediated mechanism. *Neuron* 48:563–75

Triplett JW, Feldheim DA. 2012. Eph and ephrin signaling in the formation of topographic maps. *Semin. Cell Dev. Biol.* 23:7–15

Triplett JW, Owens MT, Yamada J, Lemke G, Cang J, et al. 2009. Retinal input instructs alignment of visual topographic maps. *Cell* 139:175–85

Triplett JW, Pfeiffenberger C, Yamada J, Stafford BK, Sweeney NT, et al. 2011. Competition is a driving force in topographic mapping. *Proc. Natl. Acad. Sci. USA* 108:19060–65

Triplett JW, Phan A, Yamada J, Feldheim DA. 2012. Alignment of multimodal sensory input in the superior colliculus through a gradient-matching mechanism. *J. Neurosci.* 32(15):5264–71

Tsigankov DN, Koulakov AA. 2006. A unifying model for activity-dependent and activity-independent mechanisms predicts complete structure of topographic maps in ephrin-A deficient mice. *J. Comput. Neurosci.* 21:101–14

Vachon-Presseau E, Martin A, Lepore F, Guillemot JP. 2009. Development of the representation of auditory space in the superior colliculus of the rat. *Eur. J. Neurosci.* 29:652–60

Vanderhaeghen P, Lu Q, Prakash N, Frisen J, Walsh CA, et al. 2000. A mapping label required for normal scale of body representation in the cortex. *Nat. Neurosci.* 3:358–65

Wallace MT, Stein BE. 1997. Development of multisensory neurons and multisensory integration in cat superior colliculus. *J. Neurosci.* 17:2429–44

Walter J, Henke-Fahle S, Bonhoeffer F. 1987a. Avoidance of posterior tectal membranes by temporal retinal axons. *Development* 101:909–13

Walter J, Kern-Veits B, Huf J, Stolze B, Bonhoeffer F. 1987b. Recognition of position-specific properties of tectal cell membranes by retinal axons in vitro. *Development* 101:685–96

Wang BS, Sarnaik R, Cang J. 2010. Critical period plasticity matches binocular orientation preference in the visual cortex. *Neuron* 65:246–56

Wang L, Rangarajan KV, Lawhn-Heath CA, Sarnaik R, Wang BS, et al. 2009. Direction-specific disruption of subcortical visual behavior and receptive fields in mice lacking the beta2 subunit of nicotinic acetylcholine receptor. *J. Neurosci.* 29:12909–18

Wang L, Segraves M, Cang J. 2011. A normal retinotopic map is required for the development of an eye movement map in mouse superior colliculus. *Soc. Neurosci. Neurosci. Meet. Plan.* Progr. No. 71.10

Wang Q, Burkhalter A. 2007. Area map of mouse visual cortex. *J. Comp. Neurol.* 502:339–57

Wang Q, Burkhalter A. 2013. Stream-related preferences of inputs to the superior colliculus from areas of dorsal and ventral streams of mouse visual cortex. *J. Neurosci.* 33:1696–705

Weliky M, Katz LC. 1999. Correlational structure of spontaneous neuronal activity in the developing lateral geniculate nucleus in vivo. *Science* 285:599–604

Withington-Wray DJ, Binns KE, Dhanjal SS, Brickley SG, Keating MJ. 1990a. The maturation of the superior collicular map of auditory space in the guinea pig is disrupted by developmental auditory deprivation. *Eur. J. Neurosci.* 2:693–703

Withington-Wray DJ, Binns KE, Keating MJ. 1990b. The maturation of the superior collicular map of auditory space in the guinea pig is disrupted by developmental visual deprivation. *Eur. J. Neurosci.* 2:682–92

Wong ROL, Meister M, Shatz CJ. 1993. Transient period of correlated bursting activity during development of the mammalian retina. *Neuron* 11:923–38

Wurtz RH, Albano JE. 1980. Visual-motor function of the primate superior colliculus. *Annu. Rev. Neurosci.* 3:189–226

Wurtz RH, Goldberg ME. 1972. The role of the superior colliculus in visually-evoked eye movements. *Bibl. Ophthalmol.* 82:149–58

Xu HP, Furman M, Mineur YS, Chen H, King SL, et al. 2011. An instructive role for patterned spontaneous retinal activity in mouse visual map development. *Neuron* 70:1115–27

Xu W, Orr-Urtreger A, Nigro F, Gelber S, Sutcliffe CB, et al. 1999. Multiorgan autonomic dysfunction in mice lacking the beta2 and the beta4 subunits of neuronal nicotinic acetylcholine receptors. *J. Neurosci.* 19:9298–305

Zhang J, Ackman JB, Dhande OS, Crair MC. 2012. Visualization and manipulation of neural activity in the developing vertebrate nervous system. *Front. Mol. Neurosci.* 4:43

Sleep for Preserving and Transforming Episodic Memory

Marion Inostroza[1,2] and Jan Born[1]

[1]Department of Medical Psychology and Behavioral Neurobiology and Centre for Integrative Neuroscience (CIN), University of Tübingen, 72076 Tübingen, Germany; email: jan.born@uni-tuebingen.de, marion.inostroza@uni-tuebingen.de

[2]Departamento de Psicología, Universidad de Chile, Santiago, Chile

Annu. Rev. Neurosci. 2013. 36:79–102

First published online as a Review in Advance on April 29, 2013

The *Annual Review of Neuroscience* is online at neuro.annualreviews.org

This article's doi:
10.1146/annurev-neuro-062012-170429

Keywords

slow-wave sleep, REM sleep, system consolidation, memory reactivation, schema

Abstract

Sleep is known to support memory consolidation. Here we review evidence for an active system consolidation occurring during sleep. At the beginning of this process is sleep's ability to preserve episodic experiences preferentially encoded in hippocampal networks. Repeated neuronal reactivation of these representations during slow-wave sleep transforms episodic representations into long-term memories, redistributes them toward extrahippocampal networks, and qualitatively changes them to decontextualized schema-like representations. Electroencephalographic (EEG) oscillations regulate the underlying communication: Hippocampal sharp-wave ripples coalescing with thalamic spindles mediate the bottom-up transfer of reactivated memory information to extrahippocampal regions. Neocortical slow oscillations exert a supraordinate top-down control to synchronize hippocampal reactivations of specific memories to their excitable up-phase, thus allowing plastic changes in extrahippocampal regions. We propose that reactivations during sleep are a general mechanism underlying the abstraction of temporally stable invariants from a flow of input that is solely structured in time, thus representing a basic mechanism of memory formation.

Contents

INTRODUCTION

In our mind, the continuous flow of incoming information is organized into representations covering the dimensions of time and space. Memory refers to the maintenance of these representations, or parts of them, over time. The consolidation of memory denotes a hypothetical process that transforms newly encoded representations from an initially labile into a more stable form that allows individuals to use the acquired information for future behavior and plans (Müller & Pilzecker 1900, Dudai 2012). A fundamental issue of consolidation theory refers to the stability–plasticity dilemma: With the accumulation of incoming information in a network of limited storage capacity, the storage of new information tends to overwrite older memories (Marr 1971, McClelland et al. 1995). How can the network keep stable older representations and, simultaneously, provide sufficient plasticity to incorporate new representations? As a solution, the standard consolidation theory proposes a two-stage memory system by which the flow of incoming information is acutely fed into a temporary store. From these temporarily stored memories, some information is selected to be gradually integrated with preexisting knowledge into a long-term store, thereby leaving these older memories intact. Once a representation has been redistributed to the long-term store, it is resistant to immediate interference from information continuously fed into the temporary store.

Consolidation: transforms a temporary memory representation into a more persistent long-term representation; unlike encoding and retrieval, it occurs offline

SWS: slow-wave sleep

In the past two decades, major breakthroughs in memory research indicate that sleep plays a pivotal role in this consolidation process (Stickgold 2005, Diekelmann & Born 2010). Here, we do not comprehensively review the vast amount of studies on the link between memory and sleep but, instead, focus on evidence for a specific involvement of sleep in system consolidation. We argue, basically, that episodic memory during waking hours is continuously encoded into temporary representations and is transformed by neuronal reactivations during succeeding slow-wave sleep (SWS) such that only the gist of this information becomes integrated with preexisting memories without erasing them.

A CONCEPT OF MEMORY FORMATION IN THE ADULT BRAIN

Episodic memory combines in a unique personal experience what happened, where, and when: Specific to episodic memory is that upon its one-time occurrence, the experienced event (item) becomes bound to the particular spatial and temporal context in which it took place (Tulving 2002). The episodic memory concept arose from human research with particular reference to autonoetic consciousness during recollection. Therefore, in animal studies, the term episodic-like memory is preferably used. In contrast with episodic memory, semantic memory stores knowledge about the world in the broadest sense in the absence of contextual information (e.g., objects, concepts, facts). Semantic memories can arise from the repeated encoding or activation of overlapping episodic memories.

What kind of memory is processed during sleep? Standard consolidation theory was proposed with reference to declarative memories whose retrieval is explicit (i.e., "conscious") and initially relies on the hippocampus as a temporary store with fast encoding capabilities. Over time, system consolidation is assumed to stimulate a gradual redistribution of the representation to extrahippocampal, preferentially neocortical and striatal structures that are slower to encode and serve as long-term stores so that these memories eventually become independent of hippocampal circuitry (McClelland et al. 1995, Frankland & Bontempi 2005). Importantly, this theory assumes that the mechanism underlying consolidation is equivalent for the two subtypes of declarative memory, i.e., episodic and semantic memory. Standard consolidation theory has been challenged by evidence that some memories, particularly those with a strong autobiographical component, never become completely independent from hippocampal function (Nadel & Moscovitch 1997).

Unlike standard consolidation theory, the trace transformation theory (TTT) and its harbinger, i.e., the multiple trace theory, provide an account for the specific nature of episodic memory (Nadel & Moscovitch 1997, Winocur et al. 2010). These theories posit that initially the hippocampus rapidly and sparsely encodes key features of an experienced episode, whereby hippocampal neurons serve as an index for extrahippocampal circuits that encode semantic features of the episode. When these traces are later reactivated in an altered context, further new hippocampal traces are encoded, which in turn bind new traces in extrahippocampal circuits. Consequently, based on invariant overlapping activation and statistical regularity, the gist from multiple episodes is extracted to form a semantic representation that is independent of any specific context. The TTT emphasizes the transformation that an episodic representation undergoes with repeated reactivations, resulting in abstract semantic and schema-like representations, as well as the dynamic interplay between episodic and semantic memories during the reactivation process. Importantly, if a memory is retained as a context-dependent episode, it will still require the hippocampus, but the hippocampus is basically dispensable when retrieving semantic memories. Which specific events are encoded into the hippocampal episodic memory system, as well as retrieval within this system, is essentially controlled by prefrontal cortical attention systems (Battaglia et al. 2011).

System consolidation: entails the redistribution of the (hippocampal) representation toward different (extrahippocampal) neuronal networks and qualitative changes in memory content and is sleep-dependent

Although research has experimentally described the abstraction of semantic knowledge from episodic memories, mainly with reference to perceived stimulus categories and schemata, some of the mechanisms underlying this extraction process may likewise contribute to procedural skill learning. An increasing number of studies shows that in the adult brain, at least at an initial stage of training, the acquisition of sequenced motor skills recruits hippocampal circuitry, interacting with striatal and motor cortical areas (Schendan et al. 2003, Albouy et al. 2008, Henke 2010). The conscious practice of a motor skill represents an episode featuring the repeated activation of overlapping motor-related representations, which eventually helps to shape a dominant skill representation that enables high performance on motor routines independent of contextual stimuli and effector conditions. Motor skills can also be acquired, however, independent of the hippocampal system, although at a slower pace.

PROCESSING OF HIPPOCAMPAL MEMORY DURING SLEEP

Sleep has been known for a long time to support memory retention (Jenkins & Dallenbach

SO: slow oscillation

SW-R: sharp-wave ripples

Synaptic consolidation: stabilization of a representation by synaptic mechanisms whereby the representation remains in the same neuronal networks; it occurs during sleep and wakefulness

REM sleep: rapid eye movement sleep

1924, Stickgold 2005). Initial studies ascribe the effect to sleep's passive protection of newly encoded memories from retroactive interference, i.e., from being overwritten by new information, because the encoding of new information is obviously hampered during sleep (Ellenbogen et al. 2006). Robust evidence from recent research, however, supports an active role of sleep in memory consolidation, in addition to the protection from interference, which led us and others to posit an active system consolidation view of memory formation during sleep (Born et al. 2006, Diekelmann & Born 2010, Payne & Kensinger 2010, Lewis & Durrant 2011). This concept of a sleep-dependent active system consolidation process, which gradually integrates new memories relevant for an individual into preexisting knowledge networks, is basically rooted in the standard consolidation theory. Recent developments connect the active system consolidation view with the transformation of memory representations as proposed by the TTT.

Key to the active system consolidation view on memory processing during sleep are findings in rats of neuronal replay of activity occurring in hippocampal place cell assemblies during sleep, specifically during SWS (see **Figure 1** for the basic phenomenology of sleep), in the same temporal order as during the encoding phase before the sleep period (Wilson & McNaughton 1994, Skaggs & McNaughton 1996, O'Neill et al. 2010). The active system consolidation account proposes a dialogue between the hippocampus and extrahippocampal, mainly neocortical and striatal, networks that regulates the formation of long-term memory during waking and sleep (Buzsáki 1996, Diekelmann & Born 2010). During the wake phase, episodic information to be stored is initially encoded in both extrahippocampal and hippocampal networks; the hippocampal network encodes the binding aspects of the episodes, thus placing the experienced events into a spatiotemporal context. During subsequent periods of SWS, the newly encoded hippocampal representations are repeatedly reactivated. These reactivations transiently strengthen the hippocampal episodic aspects of the representation, but they also, via efferent CA1 entorhinal pathways, simultaneously feed reactivated memory information from the hippocampus into neocortical and striatal networks. In these structures, the reactivations spreading from the hippocampus initiate plastic changes that mediate the formation of a transformed representation that preferentially resides in extrahippocampal networks and preserves the decontextualized gist of the episode. The communication between hippocampal and extrahippocampal circuitry is regulated by electroencephalographic (EEG) oscillations, primarily by the neocortical slow oscillations (SOs) that hallmark SWS and drive memory reactivations occurring conjointly with sharp-wave ripples (SW-R) in the hippocampus. As the SOs concurrently drive spindles originating from thalamocortical circuits, they allow spindle-ripple events to form as a mechanism that transfers reactivated hippocampal memory information to respective neocortical and striatal sites. The information arrives at these sites during the highly excitable SO up-state and, by triggering intracellular Ca^{2+} influx, can tag respective networks for long-term synaptic changes (Sejnowski & Destexhe 2000) (**Figure 2**). Whereas reactivations during SWS aid episodic memory transformation, synaptic consolidation processes occurring during subsequent REM sleep may help stabilize the newly transformed representation (Diekelmann & Born 2010); this idea derived from the sequential hypothesis assuming that the succession of SWS and REM sleep epochs serves a complementary function in memory processing during sleep (Giuditta et al. 1995). Synaptic consolidation mechanisms thus represent locally acting subroutines that support the system consolidation process specifically during REM sleep (Dudai 2012), i.e., a period in which both external stimulus inputs as well as communication between brain regions are reduced to a minimum (Achermann & Borbely 1998, Axmacher et al. 2008, Montgomery et al. 2008). Below, we discuss important experimental data relevant to this concept.

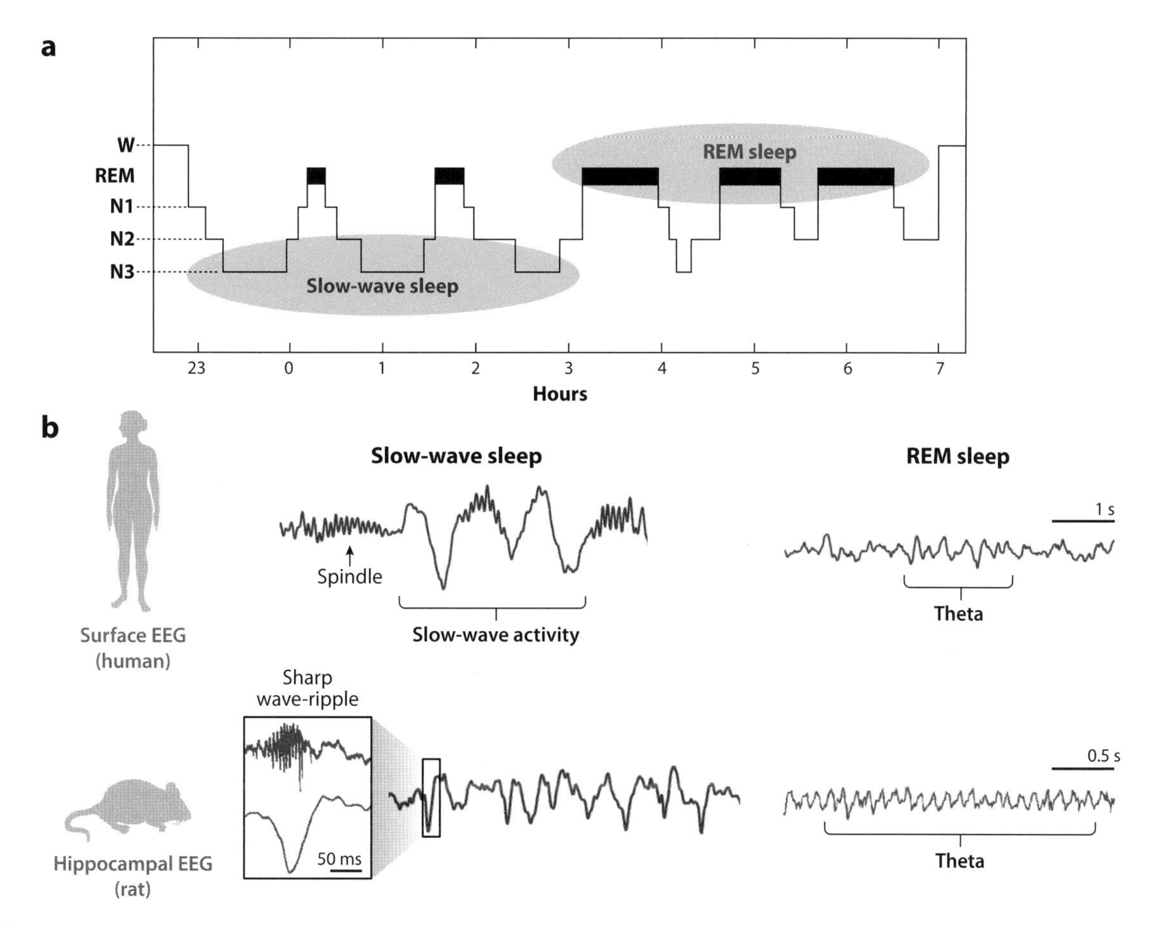

Figure 1

Physiological sleep. (*a*) Human nocturnal sleep profile. Sleep is determined mainly by continuous recordings of electroencephalogram (EEG) complemented by electrooculogram and electromyogram, subsequent 30-s epochs of which are classified into wakefulness (W), the non–rapid eye movement (non-REM) sleep stages N1, N2, N3, and REM sleep, according to standard criteria. The deepest non-REM sleep stage, N3, defines slow-wave sleep (SWS). In rodents, classification of sleep stages concentrates on SWS and REM sleep. An equivalent of N2 sleep is not discriminated. Non-REM and REM sleep periods alternate in cycles of ~90 min, which, during early nocturnal sleep, are dominated by SWS with little REM sleep and, during late sleep, by REM sleep with little SWS. (*b*) EEG characteristics of SWS and REM sleep. During SWS, the surface EEG (*upper trace*) is hallmarked by high-amplitude slow-wave activity in the frequency range between 0.5 and 4 Hz, whereby the <1 Hz range refers to the slow oscillation and the 1–4 Hz range to the delta waves. SWS, and even more so N2 sleep, is also characterized by spindle activity (12–15 Hz). The rat hippocampal EEG (*lower trace*) reveals sharp wave-ripples (SW-R); the sharp waves represent fast depolarizing events generated in CA3 that are superimposed on (100–250 Hz) ripple activity originating in CA1. The enlarged detail (*left*) shows the ripple (*upper trace*) and sharp wave (*lower trace*) components of a SW-R in separately filtered recordings at higher temporal resolution. During REM sleep, the surface EEG is characterized by a low-amplitude EEG of mixed fast frequencies. Especially the rodent hippocampal EEG shows high (4–8 Hz) theta activity.

DOES SLEEP PREFERENTIALLY CONSOLIDATE EPISODIC MEMORY?

Assuming that reactivations during SWS originate from hippocampal representations, the sleep-associated consolidation process is expected to affect episodic features more strongly than aspects of memories represented in extrahippocampal networks. Nevertheless, the proposed transformation of the memory established during sleep implicates secondary effects on extrahippocampal representations,

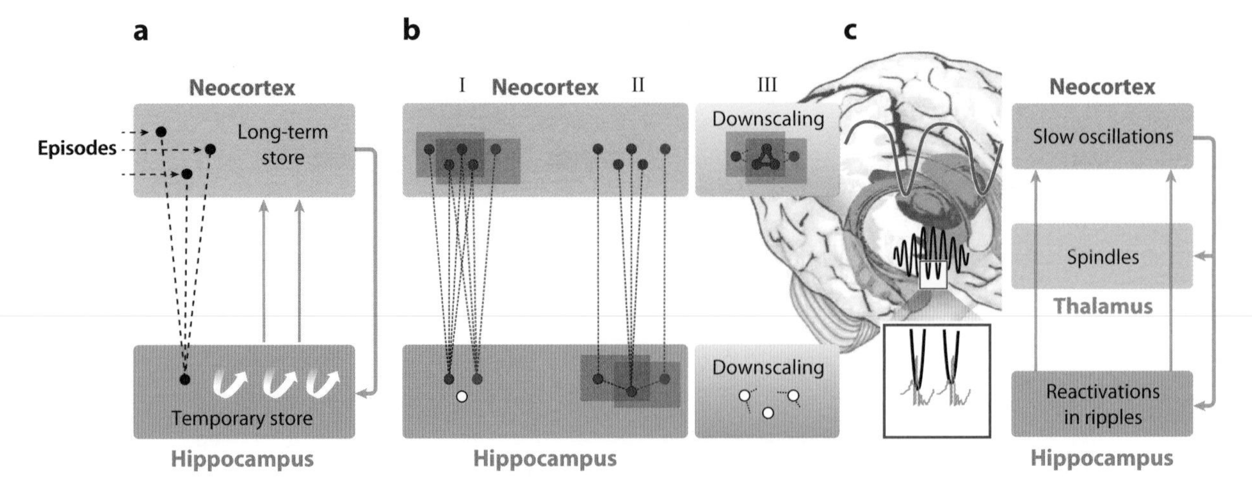

Figure 2

Active system consolidation during sleep. (*a*) System consolidation relies on a dialogue between the hippocampus serving, for part of the information, as temporary store, and extrahippocampal, mainly neocortical networks serving as long-term store. During the wake phase, an episode is encoded in both hippocampal and extrahippocampal networks (*dashed purple arrows*), whereby the hippocampus encodes aspects of the representation binding an experienced event into its unique spatiotemporal context. During subsequent slow-wave sleep (SWS), repeated reactivations of the hippocampal representation (*circular white arrows*) support its redistribution such that parts of it become preponderantly represented in extrahippocampal regions (*dark yellow arrows*). The redistribution of episodic memory representations is accompanied by a transformation toward more decontextualized schema-like representations. (*b*) The abstraction of a schema-like representation resulting from reactivations of episodic representations that overlap, i.e., share neuron assemblies either (*I*) in the neocortex or (*II*) in the hippocampus. Reactivations originate in the hippocampus and spread to neocortical networks. (*III*) Shared assemblies undergo stronger reactivation that, based on Hebbian learning, potentiates and strengthens (*thick gray lines*) the connections between neurons contributing to the overlap. Concurrent global processes of synaptic downscaling generally weaken and eventually erase connections (*truncated dashed red, blue, and black lines*) of the nonoverlapping portion of the representations as these undergo less reactivation. Downscaling may be more effective in hippocampal networks, where it also counteracts the persistence of potentiation in overlapping portions of a representation. Nonoverlapping areas preferentially represent idiosyncratic details and specific context of an episode (modified from Lewis & Durrant 2011). (*c*) The communication between the hippocampus and extrahippocampal circuitry during consolidation is regulated by EEG oscillations. During SWS, the depolarizing up-phases of slow oscillations (SO, *orange*) originating preferentially from the prefrontal cortex act top-down repeatedly to drive neuronal reactivations of representations in the hippocampus, where these reactivations occur simultaneously with sharp wave-ripples (SW-R, *green*). The concurrent drive of the SO up-phase on thalamocortical spindles (*blue*) allows spindle-ripple events to form; ripples and enwrapped reactivated memory information become nested into the excitatory phases, i.e., troughs, of the spindle oscillation (*shown enlarged in insert*). Spindle-ripple events represent a mechanism mediating the bottom-up transfer of reactivated hippocampal memory information to extrahippocampal (mainly neocortical) regions. Reactivated memory information nested in spindle troughs arrives at these regions still during the excitable SO up-phase, which initiates Ca^{2+}-dependent intracellular processes of synaptic plasticity that mediate the long-term storage of the information in these extrahippocampal regions.

i.e., semantic and procedural aspects of a memory, which might appear only with longer periods of intervening sleep. Overall there is a paucity of studies directly dissociating effects of sleep on hippocampal and extrahippocampal representations.

Human Studies

Numerous studies have demonstrated that sleep after learning of declarative tasks improves the retention of these memories, which rely on hippocampal function (reviewed in Marshall & Born 2007). In fact, the first experimental work in this field relied exclusively on the use of declarative tasks such as the learning of nonsense syllables, words, and paired associate words, among others (e.g., Jenkins & Dallenbach 1924), and the beneficial effects of postlearning sleep on memory retention seen in these early studies were confirmed in more recent studies (e.g., Yaroush et al. 1971, Fowler et al. 1973). Compared with a postlearning wake interval, sleep not only distinctly slowed the trajectory of forgetting over time, but also made the memory more resistant to interference (Ellenbogen et al. 2006). Comparisons of sleep periods rich in SWS, as they occur during the early night, with those rich in REM sleep, as occurring during the late night, indicated that SWS is much more relevant for strengthening declarative memory than is REM sleep (Plihal & Born 1997). REM-rich sleep appeared to strengthen preferentially procedural memories considered less dependent on hippocampal function, although these findings were questioned by later studies using pharmacological REM-sleep suppression (Rasch et al. 2009). However, although they differentiated declarative from procedural memories, few of these studies provided clues about which type of declarative information is influenced by sleep, i.e., semantic or episodic memory or item versus spatiotemporal context aspects of episodic memory.

Emerging evidence from recent studies supports preferential strengthening of context over item (what) memory during sleep, if context is considered to refer more generally to relational features of an episode, comprising where (spatial) and when (temporal) an event happened, as well as to background features and details of the event that are deemed more or less irrelevant in the context of learning (Aly & Moscovitch 2010). Indeed, "context" often lacks a clear conceptual definition in relevant studies. Neuroimaging and lesion studies indicate that contextual features of episodic memory critically depend on hippocampal function, whereas item memory is supported by extrahippocampal structures, mainly the perirhinal cortex (Davachi 2006, Eichenbaum et al. 2007). Napping (maximally 120 min) was indeed revealed to selectively enhance signs of context memory, leaving item memory unaffected (van der Helm et al. 2011). In this study, participants learned two lists of words (defining item memory) while facing two different posters (contexts). At a delayed retrieval test, the nap group, compared with a no-nap control, showed significantly better memory for the poster associated with a word, whereas recognition of the words per se did not differ. The enhanced context memory was correlated with the amount of non-REM sleep stage 2 and EEG spindles during the nap. However, while learning, subjects were instructed to form associations between the words and the poster images, which may question whether the posters represented context in the strict sense. Sleep-dependent enhancements in context memory in the absence of changes in item memory were likewise revealed with the what-where-when task, during which participants learned two lists of nouns (items) one after the other (when), with the words written either on the top or on the bottom of a page (where) (Rauchs et al. 2004). The forgetting rate for temporal context was lowest across SWS-rich sleep, whereas memory for spatial context appeared to be relatively enhanced following late REM-rich sleep. Fittingly, sleep also enhanced temporal order, as a principle underlying episodic memory, when memory for temporal sequences of word-triplets was studied in isolation by comparing retrieval of forward and backward associations in these triplets (Drosopoulos et al. 2007b). Sleep selectively strengthened the forward associations of the learned word triplets, matching the observation that memory replay in hippocampal neuron assemblies during SWS exclusively occurs in a forward direction (Foster & Wilson 2006). Sleep also strengthened memory for temporal order in picture sequences (Griessenberger et al. 2012), and such effects were blocked by administration of the stress hormone cortisol, which affects limbic-hippocampal circuitry in particular (Wilhelm et al. 2011b). Benefits of sleep on context memory were enhanced by varying the emotionality of the context (Lewis et al. 2011).

Interestingly, contextual associations also strengthened sleep-dependent benefits for procedural tasks that may not require hippocampal involvement per se. Comparing the effects of sleep on implicitly trained noncontextual and contextual versions of a serial reaction time task (SRTT), Spencer et al. (2006)

revealed a sleep-dependent gain in response time only for the contextual but not for the noncontextual SRTT version. Rather than responding directly to spatial cues, participants in the contextual version responded to the color of the cues (again following a sequence of spatial targets). Because the formation of such contextual associations depends on the hippocampus, these observations strongly support the view that gains in procedural skill from sleep also represent a hippocampus-mediated process. Presumably via amygdalar-hippocampal interactions, this process is subject to emotional influences because sleep-dependent gains in mirror tracing, a task in which hippocampal involvement is also not mandatory, were greater when the skill had been practiced with aversive than with neutral stimulus materials before sleep (Javadi et al. 2011).

Clues about sleep's specific influence on the episodic features of a memory can be also obtained from the remember/know paradigm (Yonelinas 2001, Rugg & Yonelinas 2003). Remembering is associated with the conscious recollection of the encoded event and thus reflects episodic memory features, whereas knowing to have seen something but being unable to recollect the specific features of the encoded event merely invokes a sense of familiarity that is not considered episodic in nature. Explicit recollection critically relies on hippocampal function, whereas familiarity-based implicit processes of recognition can be achieved by extrahippocampal regions alone. In direct comparisons of explicit recollection and familiarity-based judgments on memory for words and pictures, postlearning sleep consistently enhanced explicit recollection of memories, whereas familiarity-based judgments remained unaffected (Rauchs et al. 2004, Drosopoulos et al. 2005, Daurat et al. 2007, Atienza & Cantero 2008). Two of these studies (Rauchs et al. 2004, Daurat et al. 2007) also revealed a significant link between enhanced recollection and the occurrence of SWS during postlearning sleep.

Overall the picture emerging from these human studies suggests that sleep, in particular SWS, exerts an immediate strengthening effect on episodic memory. The effect is preponderantly seen on contextual aspects of episodic memory, where most consistent benefits of sleep were revealed for temporal contexts.

Animal Studies

Tasks examining the binding of item memory to a spatiotemporal context also demonstrated the presence of episodic-like memory and its genuine dependence on the hippocampus in animals (Kart-Teke et al. 2006, Devito & Eichenbaum 2010). Rats with lesions to dorsal CA3 and rats that, after kainate-induced epilepsy, showed some minor cell loss restricted to CA3, CA1, and ventral dentate gyrus performed well on tasks separately testing what, where, and when aspects but failed on a task requiring the binding of these three aspects into an episodic-like representation, a result that may be related to a disturbance in the coherent processing between these hippocampal areas of item and contextual information conveyed from lateral and medial entorhinal inputs, respectively (Li & Chao 2008, Inostroza et al. 2010). Notably, this binding of an event into the spatiotemporal context required sleep if rats needed to retain the episodic memory for more than 80 min (Inostroza et al. 2013). This study compared memory retention across 80-min postencoding intervals during which the rats either had normal morning sleep or were sleep deprived. A third condition took place during the evening hours when the rats were spontaneously awake during the retention interval. At retrieval testing, only in the sleep condition the rats displayed significant memory for the episode, i.e., signs that the events experienced during encoding were remembered in correct temporal order and correct place. As an indicator of memory, the study used the rat's natural novelty preference, i.e., the tendency to explore more intensely (*a*) objects that were placed at a novel rather than at the same location as that used during the encoding phase and (*b*) objects that were encountered earlier than objects encountered later in the encoding phase (Mitchell & Laiacona 1998, Kart-Teke et al. 2006). The study also separately assessed memory for the spatial and temporal contexts using an object-place recognition task and a temporal memory task, respectively. Retention of both spatial and temporal memory critically relied on the occurrence of sleep during the 80-min retention interval. By contrast, item memory tested in a novel-object recognition task did not require sleep but was retained in the wake and sleep-deprivation conditions to virtually the same extent as was found in the sleep condition. Overall, this pattern corroborates the view that sleep specifically supports the maintenance of hippocampus-dependent memories because novel-object recognition was the only task employed in the study which does not require the integrity of this structure (Bussey et al. 2000, Mumby et al. 2002). Of note, the tasks of temporal and object-place memory used in this study to dissect the effects of sleep on memory for respective spatiotemporal context aspects, per se, also comprise clear episodic features; i.e., context-item binding, inasmuch as recall of the contexts (i.e., the location and temporal order of the objects) implies that the rat also correctly remembers the objects encountered during learning. Because as in a figure-ground relationship context is, by definition, uniquely bound to an event, an isolated assessment of context memory, in the absence of any item memory formation, is basically impossible.

That the retention of spatial context over a short period of 2 h requires sleep has been confirmed in a further study using the object-place recognition task (Binder et al. 2012). The effect was associated with high amounts of both SWS and spindles. However, in mice, different from rats, sleep also enhanced item memory tested in a novel-object recognition task (Palchykova et al. 2006, Rolls et al. 2011). The reason for this divergence is not clear. Periods of sleep deprivation (>4 h) and retention intervals (24 h) in these studies were longer than those used in the rat studies, which may increase the likelihood for secondary memory effects on extrahippocampal representations. The relational features of the tasks may also have been more complex and demanding, forcing the mice into hippocampus-based strategies. In another study, mice trained on a spatial maze shifted from a hippocampus-dependent spatial strategy to a striatum-dependent response strategy when deprived from sleep (for 5 h) after each daily training (Hagewoud et al. 2010).

Fear conditioning, although stressful, bears features of episodic memory because learning is achieved typically in a single trial, thus avoiding confounding effects resulting from repetitive stimulus presentations (Maren 2001). Contextual fear conditioning, a task that depends on hippocampal function, was revealed to be sensitive to sleep deprivation, whereas cued fear conditioning, which does not require hippocampal involvement, did not profit from sleep (Graves et al. 2003). Only sleep in the 5-h interval immediately following conditioning was effective; sleep occurring 5–10 h after conditioning was not effective. Later studies confirmed the sensitivity of context fear memory to sleep (Cai et al. 2009b) and identified cyclic adenosine 3′,5′-monophosphate (cAMP)-protein kinase A (PKA)-dependent synaptic plasticity as a mechanism mediating this effect in hippocampal circuitry (Vecsey et al. 2009, Hagewoud et al. 2011), but studies were not clear about the contributions of specific sleep stages. In addition to SWS, contributions of REM sleep are likely due to the stressful nature of the task. This notion is suggested by studies using the Morris water maze, in which sleep (particularly REM sleep) benefited spatial memory only in the hidden-platform version of the task, which requires hippocampal functioning, but not in the visible-platform version, which does not require hippocampal functioning (Smith & Rose 1996, 1997).

Altogether the reviewed studies in rodents rather consistently point to a preferential consolidation of episodic memory during sleep, although mostly they compared hippocampus-dependent tasks with tasks that do not essentially rely on hippocampal function. By contrast, memories thought to be represented primarily in extrahippocampal circuitry do not require sleep to be maintained in the short term, i.e., within the first couple

IS THERE SYSTEM CONSOLIDATION DURING WAKING SIMILAR TO THAT DURING SWS?

Assembly reactivations together with hippocampal sharp wave-ripples (SW-R) occur during quiet wakefulness as well, which has stimulated the view that system consolidation would take place during waking in much the same way as it does during slow-wave sleep (SWS) (Mednick et al. 2011). However, this view neglects that reactivations during SWS and waking occur in entirely different neurochemical milieus. Acetylcholinergic activity is minimal during SWS but maximal in the wake state in which it mediates a tonic suppression of hippocampal CA1 outputs (Hasselmo & McGaughy 2004). Owing to minimal cholinergic activity, this inhibition is released during SWS, thus permitting reactivated memory information to spread to extrahippocampal areas in this state. Other signals, e.g., low corticosteroid levels during SWS, add to this effect. In humans, consonant with reconsolidation theory, wake reactivations transiently destabilized representations, whereas reactivations during SWS immediately strengthened them (Diekelmann et al. 2011). These findings suggest distinct functions of reactivations depending on the brain state: Wake reactivations and subsequent memory destabilization may provide an acute opportunity to update a representation with new information, whereas reactivations in SWS may essentially serve to incorporate hippocampal memory information within preexisting extrahippocampal representations.

of hours after learning, which does not exclude secondary effects on such representations occurring more gradually over time. Benefits for hippocampal representations appear to be linked to SWS, but REM sleep may also contribute in certain conditions, in particular when memories are emotional. Note, despite the evidence for preferential consolidation of hippocampus-dependent memory during sleep, sleep, perhaps through different mechanisms, can also enhance memories that do not involve the hippocampus (e.g., Frank 2011).

Neuronal reactivation: temporally sequenced replay of neuronal firing patterns that accompany episodic memory encoded during subsequent offline periods of sleep or wakefulness

NEURONAL REACTIVATION OF REPRESENTATIONS DURING SLEEP

The active system consolidation view assumes as a key mechanism the neuronal reactivation of memory representations encoded during waking during subsequent periods of sleep (Pavlides & Winson 1989; for reviews, see Sutherland & McNaughton 2000, O'Neill et al. 2010). Recordings of spike activity in hippocampal CA1 place cells revealed patterns of correlated activity in these cell assemblies while the rat was running along a track for food reward, and these correlation patterns were reactivated during subsequent SWS (Wilson & McNaughton 1994). Importantly, assembly reactivation during SWS occurs, although at a faster speed, in the same temporal order of place cell spiking and in the same forward direction as that observed during wake encoding of the spatial task (Skaggs & McNaughton 1996, Nadasdy et al. 1999, Ji & Wilson 2007). Initially, investigators questioned the link between assembly reactivations during SWS and a one-time episodic experience because in the first studies the rats were highly overtrained on the tasks; thus, the tasks lacked any new learning. However, in later studies, reactivations during SWS were similarly revealed after rats had engaged in exploring new environments (Ribeiro et al. 2004, O'Neill et al. 2008). In these studies, neuron assemblies linked to place fields that were longer or more frequently explored showed stronger reactivation during succeeding sleep.

Assembly reactivations have been observed almost exclusively during SWS and rarely during REM sleep. In rats highly familiar with a track, REM sleep–associated reactivations in the hippocampus were shifted in phases toward the troughs of the EEG theta rhythm, suggesting that such reactivations help erase superfluous episodic memory information from hippocampal circuitry once a task becomes familiar (Poe et al. 2000, Booth & Poe 2006). However, others failed to identify patterned reactivations during REM sleep, questioning their existence (e.g., Kudrimoti et al. 1999). During SWS, reactivations in the hippocampus typically co-occur with SW-R events (O'Neill et al. 2010) that also occur during quiet wakefulness but rarely occur during REM sleep (see sidebar, Is There System Consolidation During Waking Similar to that During SWS?).

Neuronal reactivations during SWS are also observed in extrahippocampal regions, including frontal, parietal, and visual cortical areas as well as the ventral striatum (Pennartz et al. 2004, Euston et al. 2007, Ji & Wilson 2007, Lansink et al. 2009, Peyrache et al. 2009). Assembly reactivations in these extrahippocampal areas slightly followed (by less than 50 ms) signs of reactivation in hippocampal circuitry, which is consistent with the notion that hippocampal reactivations play a leading role for reactivations occurring in distributed representational networks outside the hippocampus. Reactivations originating from newly encoded hippocampal memories and spreading to neocortical and striatal sites may help strengthen these extrahippocampal parts of the representation, thereby also fitting them into preexisting knowledge networks. Brain-imaging studies in humans confirmed that reactivations occur during non-REM sleep and SWS in hippocampal as well as extrahippocampal, i.e., neocortical, regions after learning of hippocampus-dependent declarative (spatial navigation, face-scene associations; Peigneux et al. 2004, Bergmann et al. 2012) and procedural tasks (visual texture discrimination; Yotsumoto et al. 2009).

The causal role reactivations play for memory consolidation has been demonstrated in both humans and rats. In humans, olfactory and auditory cuing was used to experimentally induce reactivations of newly encoded memories during sleep (Rasch et al. 2007, Rudoy et al. 2009). An odor presented while subjects learned place-object associations enhanced memories for the learned associations if it was presented again during subsequent SWS but not when reexposed during REM sleep. Reexposure of the odor during SWS reactivated the hippocampus; this reactivation was distinctly stronger than that observed during wakefulness, suggesting that the hippocampus during SWS is particularly sensitive to stimuli capable of reactivating memories. In rats, conditions of diminished pattern reactivation and accompanying SW-R were consistently associated with impaired retention of hippocampus-dependent memories (Girardeau et al. 2009, Bendor & Wilson 2012). Diminished formation of spatial memory in old compared with young rats was associated primarily with a disturbed temporal sequence in the reactivation patterns during sleep, while patterns of mere coactivation persisted (Gerrard et al. 2008). Indeed, considering the importance of temporal sequence in neuronal replay activity during SWS and its origin from hippocampal circuitry, it is tempting to conclude that these reactivations reflect primarily reprocessing of episodic memory aspects, although reactivations during sleep after strictly episodic tasks have not been assessed so far.

TRANSFORMATION OF MEMORY REPRESENTATIONS DURING SLEEP

The active system consolidation theory assumes that, in addition to immediately enhancing the memory representation per se, reactivations during sleep induce the transformation of memory representations. Specifically, reactivations originating from the hippocampal circuitry during SWS are expected to decontextualize episodic memory information, whereby semantic and skill representations are generated that can be applied independently from their specific spatiotemporal context during acquisition. Such transformation implies that memories change in quality during sleep, an aspect that has so far been less intensely studied, probably because persistence, rather than its dynamic nature, is traditionally considered the hallmark of memory (Dudai 2012).

Human functional magnetic imaging studies have consistently indicated that post-encoding sleep, aside from increased hippocampal activation, favors the redistribution of declarative memory representations (for word pairs, pictures, etc.) toward enhanced neocortical, mainly medial prefrontal cortical involvement as well as increased functional connectivity between medial frontal cortical and hippocampal areas at delayed retrieval (e.g., Gais et al. 2007, Sterpenich et al. 2007, Takashima et al. 2009, Payne &

Assimilation: process in which newly encoded memory information is integrated into preexisting knowledge and schema-like representations

Abstraction: process in which invariant patterns are distilled from multiple overlapping episodic memory representations to eventually form a more generalized schema-like representation

Kensinger 2011). For tasks with a strong procedural motor component, postencoding sleep appears to enhance the involvement of striatal regions and their functional connectivity with the hippocampus (Fischer et al. 2005, Orban et al. 2006). Yet, despite the obvious system-level reorganization that memory representations undergo during sleep, the qualitative changes in memory associated with such reorganization are presently not well characterized.

Available evidence points to two major processes that might be supported by system consolidation during sleep: assimilation and abstraction (Lewis & Durrant 2011). Assimilation refers to the integration of newly encoded memory information into preexisting knowledge networks and schemas. In rats, once an associative spatial schema representation (of a large event arena) is established in neocortical long-term memory networks, newly learned additional locations in this arena can be incorporated into this schema within 48 h (Tse et al. 2007). As expected, memory for the new locations was impaired by lesions to the hippocampus made 3 h after learning. However, it remained intact with lesions made 48 h after learning. Although not systematically explored in these experiments, sleep may have been critical for this relatively fast assimilation to extrahippocampal schemas, since only the 48-h but not the 3-h interval contained sleep. In humans, sleep promoted the assimilation into existing knowledge networks of learned spoken novel words that phonologically overlapped with familiar words, as measured by a lexical competition task (Dumay & Gaskell 2007), and this integration was associated with increased spindle activity during postlearning sleep (Tamminen et al. 2010).

Abstraction refers to a process in which rules and regularities are distilled from multiple episodic memory representations to eventually form a more generalized schema in long-term memory. Thus, in different tasks, such as the number reduction task, the SRTT, and statistical learning tasks, sleep consistently promoted the abstraction of explicit knowledge from hidden structures embedded in the implicitly learned materials (Wagner et al. 2004, Fischer et al. 2006, Drosopoulos et al. 2011, Durrant et al. 2011). When asked to generate deliberately the sequence underlying an SRTT trained under implicit conditions before a retention interval of sleep or wakefulness, only after sleep had subjects developed a significant amount of explicit sequence knowledge (Fischer et al. 2006, Drosopoulos et al. 2011). Postencoding sleep also improved performance on transitive inference tasks (Ellenbogen et al. 2007) and enhanced production of false memories in the Deese-Roediger-McDermott (DRM) paradigm (Payne et al. 2009, Diekelmann et al. 2010). In infants, sleep promoted grammar-related abstraction processes in a language-learning task (Gomez et al. 2006, Hupbach et al. 2009). In the infants who had napped, memory for the words per se was worse than in the wake control infants, suggesting that sleep acts primarily to transform rather than to enhance representations directly.

Hints at abstraction processes supported by sleep were likewise obtained in studies of procedural memory. On a finger-sequence tapping task, posttraining sleep favored the development of an effector-independent representation, i.e., sleep benefited tapping the sequence of target keys independent of whether tapping was performed with the same hand as during training or with the other hand (Cohen et al. 2005, Witt et al. 2010). Sleep enhanced sequence finger-tapping performance when learning occurred by observation (van der Werf et al. 2009b). Thus, sleep appears to transform skill representations such that they become less dependent on a specific stimulus context or effector system. Beyond this decontextualization of memory by sleep, the studies showed for the discussed declarative and procedural tasks that the transformation of representations is linked to non-REM sleep and SWS and associated EEG oscillations (e.g., Yordanova et al. 2008, Tamminen et al. 2010, Durrant et al. 2011, Yordanova et al. 2012); there were no consistent clues for additional contributions of REM sleep, although the possibility cannot be excluded (Cai et al. 2009a, Walker & Stickgold 2010).

Transformation Toward Increased Executive Control

Findings that postencoding sleep facilitates insight and awareness of rules hidden in implicitly learned materials suggest that the abstraction process during sleep goes along with an increased explicitness of memory in the sense that various effector systems (verbal, motor) can be flexibly used to express the memory in different stimulus conditions (Marshall & Born 2007). Sleep reorganizes representations such that they become not only decontextualized but also more accessible via the prefrontal-hippocampal system, i.e., the executive control system mediating explicit recollection. This view could also explain that enhancing effects of sleep on memory are typically more robust when delayed retrieval is tested by free recall rather than by a recognition procedure (see, for example, regarding the production of critical lures in the DRM paradigm; Diekelmann et al. 2008, 2010; Payne et al. 2009). Against this backdrop, it is tempting to speculate that the transformation of memory representations during SWS serves primarily to enhance their accessibility to executive behavioral control by prefrontal cortex structures rather than to enhance the content per se. Yet, such a distinction is difficult to make on the basis of behavioral retrieval measures alone.

MECHANISMS UNDERLYING MEMORY TRANSFORMATION

The transformation of episodic memory that leads to the formation of decontextualized schema-like representations is thought to be a gradual process originating from the repeated activation of multiple overlapping episodes that share common items, so that spatiotemporal context and idiosyncratic details of the episodic representation attenuate and eventually completely disappear (Nadel & Moscovitch 1997). Accordingly, overlap between episodic representations determines how memories are transformed during sleep-associated reactivations. However, it is not clear how hippocampus and extrahippocampal areas in representing episodic memories contribute to such overlapping reactivation (**Figure 2*b***). Overlap in spatial representations can already emerge during encoding in hippocampal CA3 and CA1 as a result of experience on two tasks performed in environments with graded dissimilarities (Guzowski et al. 2004, Lee et al. 2004, Leutgeb et al. 2004). Reactivation of such generalized mapping across episodes in CA1, as the major hippocampal output region, may then be considered a principal mechanism underlying the induction of abstract representations in extrahippocampal networks. However, the exact (spatial) conditions under which CA1, in the course of encoding different episodes, forms overlapping generalized or separated maps are not well understood. Alternatively, Buzsáki (2005) speculated that SW-R covering wider areas of hippocampal circuitry create opportunities to cross multiple separate episodes at the timescale of synaptic plasticity, the resulting concordant reactivation of these representations allowing abstract representations to form in neocortical areas. In this case, overlap would be acutely generated only in the course of redistributing representations to extrahippocampal networks in a process that may also integrate top-down inputs to the hippocampus originating from preexisting knowledge networks.

The "information overlap to abstract" concept proposed by Lewis & Durrant (2011) provides a more theoretical account of how neocortical overlap in representations could contribute to the formation of abstract schema memories. The concept assumes that repeated reactivations during SWS of newly encoded representations that show overlapping cortical areas, based on Hebbian learning rules, lead to the gradual strengthening, i.e., abstraction, of conceptual schemas representing those areas that, owing to the overlap, undergo the strongest reactivation. Concurrently, global synaptic downscaling processes gradually erode the more idiosyncratic aspects of each single representation. The model receives preliminary support from a study of interference learning (Drosopoulos et al. 2007a), which compared the effects of retention sleep and wakefulness after subjects learned two-word pair lists that, according to an A-B, A-C paradigm, overlapped and thus produced retroactive interference, or, according to an A-B, C-D paradigm, did not overlap and thus did not produce interference. Consistent with the information overlap to abstract concept, the sleep-dependent improvement in recall of A-B associations was greater when these overlapped during learning with A-C associations, compared with the learning of nonoverlapping word lists. Related findings were reported by others for declarative (Ekstrand 1967) and also for procedural types of memories (Fenn et al. 2003, Walker et al. 2003). However, the superior benefit of sleep for A-B associations when learned in the overlapping interference condition may be due to their generally weaker strength when entering sleep rather than resulting from reactivations of overlapping representations. In fact, some evidence indicates that profits from sleep are greatest for memories with an intermediate presleep encoding strength (e.g., Tucker & Fishbein 2008, Wilhelm et al. 2012a).

EEG coherence: phase synchronization of field potential oscillations, coordinates the timing of neuronal spiking and resulting synaptic plasticity across distributed brain regions

Control of Communication During Sleep-Dependent System Consolidation

The transformation of episodic representations during sleep implicates the redistribution of reactivated hippocampal memory toward preponderant representations of the memory information in extrahippocampal regions, a process that entails a fine-tuned communication between these regions. System communication between brain regions is considered to be basically controlled by electrical field potential rhythms (Buzsáki & Draguhn 2004). The EEG coherence, specifically the phase coherence in these rhythms, provides a mechanism whereby neuronal activity sent from one region can activate another network in a temporally coordinate manner, allowing also for the induction of Hebbian and spike time-dependent synaptic plasticity underlying the formation of representations in the receiving network (Benchenane et al. 2011). The spread of reactivated hippocampal memory information during SWS is thought to be orchestrated by three different oscillatory rhythms: the ~0.75 Hz SO, the classical 12–15 Hz spindles, and the SW-R (the ripples oscillate between 100 and 300 Hz) (**Figures 1** and **2*c***; Diekelmann & Born 2010).

SOs comprise highly synchronous alterations of virtually every cortical neuron between periods of membrane depolarization accompanied by sustained firing (up-state) and periods of hyperpolarization associated with neuronal silence (down-state). The SO thus provides a global time frame whereby the network is reset by the hyperpolarizing phase and processing is limited to the subsequent depolarizing up-phase (Steriade 2006, Mölle & Born 2011). The SO is generated primarily in neocortical networks; the depolarizing up-states are presumably triggered by summation of miniature excitatory postsynaptic potentials as a residual synaptic activity in local synaptic circuits, which is increased after potentiation of respective synapses, e.g., when information encodes in these circuits during prior waking (Bazhenov et al. 2002). Spindles originate from GABAergic thalamic networks; glutamatergic thalamocortical projections mediate their widespread propagation to cortical regions (Gennaro & Ferrara 2003). In the neocortex, spindles, independent of their synchronization with the central thalamic spindle generator, typically emerge as local phenomena that are restricted to specific circuitry (Nir et al. 2011, Ayoub et al. 2012). SW-R accompany memory reactivations in the hippocampus where they impact firing of discrete local circuits (Csicsvari et al. 1999). Evidence indicates that all three types of oscillations, SOs, spindles, and ripples, preferentially occur in previously potentiated synaptic networks (e.g., Behrens et al. 2005, Tononi & Cirelli 2006, Bergmann et al. 2008) and, conversely, can support plastic synaptic processes such as long-term potentiation (King 1999, Rosanova & Ulrich 2005). For SOs and ripples, studies have demonstrated a causal involvement in consolidation of hippocampus-dependent memories during sleep by directly suppressing or enhancing them through electrical stimulation (Marshall et al. 2006, Girardeau et al. 2009, Marshall et al. 2011). For spindle activity, studies have shown robust increases during postlearning sleep that predicted the retention of the acquired memories, including specifically episodic aspects in these memories (e.g., Gais et al. 2002, Fogel & Smith 2011, van der Helm et al. 2011).

The dialogue between the hippocampus and extrahippocampal regions underlying memory transformation during SWS appears to be controlled mainly by the SOs that arise most powerfully from prefrontal circuitry engaged in information encoding during prior waking. SOs globally entrain neuronal activity not only in the neocortex but also in many other structures, including the thalamus and the hippocampus, where spindle activity and SW-R together with reactivated memory information likewise become synchronized with the SO up-state (Clemens et al. 2007, Csercsa et al. 2010). Prior learning strengthens the top-down control of SOs on spindles and ripples and also promotes the occurrence of trains of several succeeding SOs (Mölle & Born 2011, Ruch et al. 2012). In these SO trains, spindles appeared not only to be driven by the SO up-state but, conversely, to enforce also the succeeding SO, suggesting a key role for spindles in maintaining memory processing (Mölle et al. 2011).

The synchronous drive of the SO up-state on the thalamus and the hippocampus allows spindle-ripple events to form during this depolarizing period, when ripples and reactivated memory information enwrapped in these ripples are nested into the succeeding troughs of a spindle (Siapas & Wilson 1998, Clemens et al. 2011). Mutually stimulating influences between spindles and ripples may add to the formation of spindle-ripple events (Mölle et al. 2009, Wierzynski et al. 2009). [Note, these relationships do not apply to the slow 10–12 Hz frontal spindles, which represent a separate type of spindles that occur later in the SO cycle (Mölle et al. 2011, Peyrache et al. 2011)]. Spindle-ripple events are a strong candidate mechanism for the bottom-up transfer of memory information to neocortical and striatal regions (Sirota & Buzsáki 2005, Mölle & Born 2011), where they might induce plastic processes to support specifically the storage of semantic and procedural features in episodic memories. Indeed, in humans signs of conjoint reactivations in relevant neocortical and hippocampal regions occurred in temporal synchrony with spindles after subjects learned face-scene associations (Bergmann et al. 2012). Also, spindles, in addition to ripples, phase-lock EEG gamma-band activity as an indicator of coherent information processing in local neocortical networks (Ayoub et al. 2012). In rats, spindle activity following exposure to a novel spatiotactile experience predicted immediate early gene (*arc*) expression in the somatosensory cortex during subsequent REM sleep, suggesting that, by facilitating Ca^{2+} influx into pyramidal cells, spindles help tag the newly formed neocortical representations for persistent synaptic strengthening during subsequent REM sleep (Ribeiro et al. 2007).

CONSOLIDATION OF HIPPOCAMPAL MEMORIES IS SELECTIVE

The brain's capacity to store memories is limited. Even if the enhancing effect of sleep on memory was restricted to the episodic memory system, overflow would inevitably occur. Such considerations have stimulated concepts that sleep entails processes counteracting any imminent overload resulting from encoding of information during wakefulness. In this vein, the synaptic homeostasis hypothesis proposed that SWS via slow-wave activity induces a global and proportional downscaling of synapses that were potentiated during wakeful information encoding (Tononi & Cirelli 2006). However, proportional synaptic downscaling alone can explain neither why memories are enhanced by sleep nor why this enhancing effect is selective. It is obvious that some episodic memory content is strengthened by sleep, whereas some other content is not (Diekelmann & Born 2010).

Selectivity of memory enhancement can be adequately explained in the context of memory transformation occurring during sleep, specifically in the context of abstracting and assimilating episodic memory information into schemas. Notwithstanding this theoretical account, experimental work has revealed that sleep preferentially strengthens those memories that are relevant for the individual's future plans. Sleep strongly supported the maintenance and delayed execution of plans in prospective memory paradigms (Scullin & McDaniel 2010, Diekelmann et al. 2012). The enhancing effect of sleep on declarative word-pair memories was distinctly greater in subjects who were informed (or merely suspected) that recall would be tested later as compared with subjects who had no such expectations; only in the informed subjects did later recall performance correlate with slow-wave activity during postlearning sleep (Wilhelm et al. 2011a). Comparable results were revealed for procedural skill memories (Cohen et al. 2005, Fischer & Born 2009).

Processing of goal-directed anticipatory aspects of behavior and retrieval is strongly associated with prefrontal cortex executive functions (Miller & Cohen 2001, Polyn & Kahana 2008). In allocating relevance (for future actions), the prefrontal structures may tag hippocampal memories to facilitate their access to system consolidation during sleep. Indeed, growing experimental evidence supports the notion that offline system consolidation is not only achieved in a bottom-up process, in which reactivation of hippocampal representations promotes the incorporation of some of this information into preexisting neocortical knowledge networks, but also entails significant top-down processing (Morris 2006). Thus during spatial learning, medial and limbic prefrontal regions can quickly (in a single trial) encode associations that overlap with a preexisting spatial schema (Lesburgueres et al. 2011, Tse et al. 2011), and such rapid prefrontal encoding may also enable a top-down tagging of hippocampal memories as a prerequisite for subsequent offline consolidation.

A candidate mechanism conveying prefrontal tagging of memories for consolidation during sleep is the EEG theta rhythm (Benchenane et al. 2011). In a Y-maze task theta coherence between the prefrontal cortex and the hippocampus was increased at the choice point of the maze as soon as the rats had learned to choose the correct arm (Benchenane et al. 2010). Notably, assembly pattern firing in the prefrontal cortex present during increased theta coherence was likely reactivated during subsequent SWS. In addition to prefrontal-hippocampal circuitry, the theta spanning network includes regions such as the ventral tegmental area and the amygdala (Fujisawa & Buzsáki 2011, Lesting et al. 2011), which may help integrate reward-related and emotional aspects into the tagging process. Together with studies of rhythmic electrical stimulation (Marshall et al. 2006, Kirov et al. 2009), these observations converge on the idea that the same prefrontal-hippocampal network producing theta during wake encoding to tag memories changes into the SO rhythm during subsequent SWS to consolidate the tagged memories.

PERSPECTIVE

A wealth of evidence now indicates that sleep contributes to the formation of long-term memories in an active system consolidation process. Although the main features of this concept appear to be firmly anchored in human and rodent research, a number of questions are left unanswered. What exactly is the role of prefrontal top-down control in this consolidation process? Do neocortical SOs transfer schema-related information that supports the selection of representations to be reactivated in the hippocampus? Do they moreover contribute to downscaling and erasing of superfluous memory information from hippocampal circuits (van der Werf et al. 2009a, Grosmark et al. 2012)? How exactly do hippocampal reactivations contribute in a bottom-up fashion to transforming representations toward more abstract schema-like representations? What happens when schemas are not readily available in long-term memory, i.e., during early development? Sleep appears to play a pivotal role for memory formation during development (Frank 2011, Wilhelm et al. 2012b). Thus, sleep-dependent formation of song representations in young birds has been conceptualized as a bottom-up feedforward process in which sensory inputs that are too complex and dynamic to be integrated into song production online are reactivated during sleep to program motor song areas in the absence of potentially disturbing acute feedback (Konishi 2004, Margoliash & Schmidt 2010). Indeed, offline neuronal reactivation during sleep may work as a principal mechanism to form any kind of memory, i.e., as a mechanism that serves to abstract temporally stable invariants from a complex stream of inputs that is dynamic and only structured in time. Regardless of what answers we find for these questions, the past research described here has firmly established a picture of sleep as a brain state most essential to the genuine understanding of memory.

DISCLOSURE STATEMENT

The authors are not aware of any affiliations, memberships, funding, or financial holdings that might be perceived as affecting the objectivity of this review.

ACKNOWLEDGMENTS

We are grateful to Gordon Feld, Manfred Hallschmid, Edo Kelemen, and Hannes Noack for comments and discussions. This work is supported by a grant from the Deutsche Forschungsgemeinschaft, SFB 654 "Plasticity and Sleep."

LITERATURE CITED

Achermann P, Borbely AA. 1998. Temporal evolution of coherence and power in the human sleep electroencephalogram. *J. Sleep Res.* 7(Suppl. 1):36–41

Albouy G, Sterpenich V, Balteau E, Vandewalle G, Desseilles M, et al. 2008. Both the hippocampus and striatum are involved in consolidation of motor sequence memory. *Neuron* 58(2):261–72

Aly M, Moscovitch M. 2010. The effects of sleep on episodic memory in older and younger adults. *Memory* 18(3):327–34

Atienza M, Cantero JL. 2008. Modulatory effects of emotion and sleep on recollection and familiarity. *J. Sleep Res.* 17(3):285–94

Axmacher N, Helmstaedter C, Elger CE, Fell J. 2008. Enhancement of neocortical-medial temporal EEG correlations during non-REM sleep. *Neural Plast.* 2008: doi: 10.1155/2008/563028

Ayoub A, Molle M, Preissl H, Born J. 2012. Grouping of MEG gamma oscillations by EEG sleep spindles. *Neuroimage* 59(2):1491–500

Battaglia FP, Benchenane K, Sirota A, Pennartz CM, Wiener SI. 2011. The hippocampus: hub of brain network communication for memory. *Trends Cogn. Sci.* 15(7):310–18

Bazhenov M, Timofeev I, Steriade M, Sejnowski TJ. 2002. Model of thalamocortical slow-wave sleep oscillations and transitions to activated states. *J. Neurosci.* 22(19):8691–704

Behrens CJ, van den Boom LP, de Hoz L, Friedman A, Heinemann U. 2005. Induction of sharp wave-ripple complexes in vitro and reorganization of hippocampal networks. *Nat. Neurosci.* 8(11):1560–67

Benchenane K, Peyrache A, Khamassi M, Tierney PL, Gioanni Y, et al. 2010. Coherent theta oscillations and reorganization of spike timing in the hippocampal-prefrontal network upon learning. *Neuron* 66(6):921–36

Benchenane K, Tiesinga PH, Battaglia FP. 2011. Oscillations in the prefrontal cortex: a gateway to memory and attention. *Curr. Opin. Neurobiol.* 21(3):475–85

Bendor B, Wilson MA. 2012. Biasing the content of hippocampal replay during sleep. *Nat. Neurosci.* 15:1439–44

Bergmann TO, Molle M, Diedrichs J, Born J, Siebner HR. 2012. Sleep spindle-related reactivation of category-specific cortical regions after learning face-scene associations. *Neuroimage* 59(3):2733–42

Bergmann TO, Molle M, Marshall L, Kaya-Yildiz L, Born J, Roman SH. 2008. A local signature of LTP- and LTD-like plasticity in human NREM sleep. *Eur. J. Neurosci.* 27(9):2241–49

Binder S, Baier PC, Molle M, Inostroza M, Born J, Marshall L. 2012. Sleep enhances memory consolidation in the hippocampus-dependent object-place recognition task in rats. *Neurobiol. Learn. Mem.* 97(2):213–19

Booth V, Poe GR. 2006. Input source and strength influences overall firing phase of model hippocampal CA1 pyramidal cells during theta: relevance to REM sleep reactivation and memory consolidation. *Hippocampus* 16(2):161–73

Born J, Rasch B, Gais S. 2006. Sleep to remember. *Neuroscientist* 12(5):410–24

Bussey TJ, Duck J, Muir JL, Aggleton JP. 2000. Distinct patterns of behavioural impairments resulting from fornix transection or neurotoxic lesions of the perirhinal and postrhinal cortices in the rat. *Behav. Brain Res.* 111(1–2):187–202

Buzsáki G. 1996. The hippocampo-neocortical dialogue. *Cereb. Cortex* 6(2):81–92

Buzsáki G. 2005. Theta rhythm of navigation: link between path integration and landmark navigation, episodic and semantic memory. *Hippocampus* 15(7):827–40

Buzsáki G, Draguhn A. 2004. Neuronal oscillations in cortical networks. *Science* 304:1926–29

Cai DJ, Mednick SA, Harrison EM, Kanady JC, Mednick SC. 2009a. REM, not incubation, improves creativity by priming associative networks. *Proc. Natl. Acad. Sci. USA* 106:10130–34

Cai DJ, Shuman T, Gorman MR, Sage JR, Anagnostaras SG. 2009b. Sleep selectively enhances hippocampus-dependent memory in mice. *Behav. Neurosci.* 123(4):713–19

Clemens Z, Mölle M, Eross L, Barsi P, Halasz P, Born J. 2007. Temporal coupling of parahippocampal ripples, sleep spindles and slow oscillations in humans. *Brain* 130:2868–78

Clemens Z, Mölle M, Eross L, Jakus R, Rasonyi G, et al. 2011. Fine-tuned coupling between human parahippocampal ripples and sleep spindles. *Eur. J. Neurosci.* 33(3):511–20

Cohen DA, Pascual-Leone A, Press DZ, Robertson EM. 2005. Off-line learning of motor skill memory: a double dissociation of goal and movement. *Proc. Natl. Acad. Sci. USA* 102(50):18237–41

Csercsa R, Dombovari B, Fabo D, Wittner L, Eross L, et al. 2010. Laminar analysis of slow wave activity in humans. *Brain* 133(9):2814–29

Csicsvari J, Hirase H, Czurko A, Mamiya A, Buzsáki G. 1999. Oscillatory coupling of hippocampal pyramidal cells and interneurons in the behaving rat. *J. Neurosci.* 19(1):274–87

Daurat A, Terrier P, Foret J, Tiberge M. 2007. Slow wave sleep and recollection in recognition memory. *Conscious. Cogn.* 16(2):445–55

Davachi L. 2006. Item, context and relational episodic encoding in humans. *Curr. Opin. Neurobiol.* 16(6):693–700

Devito LM, Eichenbaum H. 2010. Distinct contributions of the hippocampus and medial prefrontal cortex to the "what-where-when" components of episodic-like memory in mice. *Behav. Brain Res.* 215(2):318–25

Diekelmann S, Born J. 2010. The memory function of sleep. *Nat. Rev. Neurosci.* 11(2):114–26

Diekelmann S, Born J, Wagner U. 2010. Sleep enhances false memories depending on general memory performance. *Behav. Brain Res.* 208(2):425–29

Diekelmann S, Büchel C, Born J, Rasch B. 2011. Labile or stable: opposing consequences for memory when reactivated during waking and sleep. *Nat. Neurosci.* 14(3):381–86

Diekelmann S, Landolt HP, Lahl O, Born J, Wagner U. 2008. Sleep loss produces false memories. *PLoS One* 3(10):e3512

Diekelmann S, Wilhelm I, Wagner U, Born J. 2012. Sleep to implement an intention. *Sleep* 36:149–53

Drosopoulos S, Harrer D, Born J. 2011. Sleep and awareness about presence of regularity speed the transition from implicit to explicit knowledge. *Biol. Psychol.* 86(3):168–73

Drosopoulos S, Schulze C, Fischer S, Born J. 2007a. Sleep's function in the spontaneous recovery and consolidation of memories. *J. Exp. Psychol. Gen.* 136:169–83
Drosopoulos S, Wagner U, Born J. 2005. Sleep enhances explicit recollection in recognition memory. *Learn. Mem.* 12(1):44–51
Drosopoulos S, Windau E, Wagner U, Born J. 2007b. Sleep enforces the temporal order in memory. *PLoS One* 2(4):e376
Dudai Y. 2012. The restless engram: Consolidations never end. *Annu. Rev. Neurosci.* 35:227–47
Dumay N, Gaskell MG. 2007. Sleep-associated changes in the mental representation of spoken words. *Psychol. Sci.* 18(1):35–39
Durrant SJ, Taylor C, Cairney S, Lewis PA. 2011. Sleep-dependent consolidation of statistical learning. *Neuropsychologia* 49(5):1322–31
Eichenbaum H, Yonelinas AP, Ranganath C. 2007. The medial temporal lobe and recognition memory. *Annu. Rev. Neurosci.* 30:123–52
Ekstrand BR. 1967. Effect of sleep on memory. *J. Exp. Psychol.* 75(1):64–72
Ellenbogen JM, Hu PT, Payne JD, Titone D, Walker MP. 2007. Human relational memory requires time and sleep. *Proc. Natl. Acad. Sci. USA* 104(18):7723–28
Ellenbogen JM, Hulbert JC, Stickgold R, Dinges DF, Thompson-Schill SL. 2006. Interfering with theories of sleep and memory: sleep, declarative memory, and associative interference. *Curr. Biol.* 16(13):1290–94
Euston DR, Tatsuno M, McNaughton BL. 2007. Fast-forward playback of recent memory sequences in prefrontal cortex during sleep. *Science* 318(5853):1147–50
Fenn KM, Nusbaum HC, Margoliash D. 2003. Consolidation during sleep of perceptual learning of spoken language. *Nature* 425(6958):614–16
Fischer S, Born J. 2009. Anticipated reward enhances offline learning during sleep. *J. Exp. Psychol. Learn. Mem. Cogn.* 35(6):1586–93
Fischer S, Drosopoulos S, Tsen J, Born J. 2006. Implicit learning—explicit knowing: a role for sleep in memory system interaction. *J. Cogn. Neurosci.* 18(3):311–19
Fischer S, Nitschke MF, Melchert UH, Erdmann C, Born J. 2005. Motor memory consolidation in sleep shapes more effective neuronal representations. *J. Neurosci.* 25(49):11248–55
Fogel SM, Smith CT. 2011. The function of the sleep spindle: a physiological index of intelligence and a mechanism for sleep-dependent memory consolidation. *Neurosci. Biobehav. Rev.* 35(5):1154–65
Foster DJ, Wilson MA. 2006. Reverse replay of behavioural sequences in hippocampal place cells during the awake state. *Nature* 440(7084):680–83
Fowler MJ, Sullivan MJ, Ekstrand BR. 1973. Sleep and memory. *Science* 179(4070):302–4
Frank MG. 2011. Sleep and developmental plasticity not just for kids. *Prog. Brain Res.* 193:221–32
Frankland PW, Bontempi B. 2005. The organization of recent and remote memories. *Nat. Rev. Neurosci.* 6(2):119–30
Fujisawa S, Buzsáki G. 2011. A 4 Hz oscillation adaptively synchronizes prefrontal, VTA, and hippocampal activities. *Neuron* 72(1):153–65
Gais S, Albouy G, Boly M, Dang-Vu TT, Darsaud A, et al. 2007. Sleep transforms the cerebral trace of declarative memories. *Proc. Natl. Acad. Sci. USA* 104(47):18778–83
Gais S, Mölle M, Helms K, Born J. 2002. Learning-dependent increases in sleep spindle density. *J. Neurosci.* 22(15):6830–34
Gennaro L, Ferrara M. 2003. Sleep spindles: an overview. *Sleep Med. Rev.* 7(5):423–40
Gerrard JL, Burke SN, McNaughton BL, Barnes CA. 2008. Sequence reactivation in the hippocampus is impaired in aged rats. *J. Neurosci.* 28(31):7883–90
Girardeau G, Benchenane K, Wiener S. 2009. Selective suppression of hippocampal ripples impairs spatial memory. *Nat. Neurosci.* 12(10):1222–23
Giuditta A, Ambrosini MV, Montagnese P, Mandile P, Cotugno M, et al. 1995. The sequential hypothesis of the function of sleep. *Behav. Brain Res.* 69(1–2):157–66
Gomez RL, Bootzin RR, Nadel L. 2006. Naps promote abstraction in language-learning infants. *Psychol. Sci.* 17(8):670–74
Graves LA, Heller EA, Pack AI, Abel T. 2003. Sleep deprivation selectively impairs memory consolidation for contextual fear conditioning. *Learn. Mem.* 10(3):168–76

Griessenberger H, Hoedlmoser K, Heib DP, Lechinger J, Klimesch W, Schabus M. 2012. Consolidation of temporal order in episodic memories. *Biol. Psychol.* 91(1):150–55

Grosmark AD, Mizuseki K, Pastalkova E, Diba K, Buzsáki G. 2012. REM sleep reorganizes hippocampal excitability. *Neuron* 75:1001–7

Guzowski JF, Knierim JJ, Moser EI. 2004. Ensemble dynamics of hippocampal regions CA3 and CA1. *Neuron* 44(4):581–84

Hagewoud R, Bultsma LJ, Barf RP, Koolhaas JM, Meerlo P. 2011. Sleep deprivation impairs contextual fear conditioning and attenuates subsequent behavioural, endocrine and neuronal responses. *J. Sleep Res.* 20(2):259–66

Hagewoud R, Havekes R, Tiba P. 2010. Coping with sleep deprivation: shifts in regional brain activity and learning strategy. *Sleep* 33(11):1465–73

Hasselmo ME, McGaughy J. 2004. High acetylcholine levels set circuit dynamics for attention and encoding and low acetylcholine levels set dynamics for consolidation. *Prog. Brain Res.* 145:207–31

Henke K. 2010. A model for memory systems based on processing modes rather than consciousness. *Nat. Rev. Neurosci.* 11(7):523–32

Hupbach A, Gomez RL, Bootzin RR, Nadel L. 2009. Nap-dependent learning in infants. *Dev. Sci.* 12(6):1007–12

Inostroza M, Binder S, Born J. 2013. Sleep-dependency of episodic-like memory consolidation in rats. *Behav. Brain Res.* 237:15–22

Inostroza M, Cid E, Gal B, Uzcateaui Y, Sandi C, Menendez de la Prida L. 2010. *Deficits of the "what-where-when" and preservation of the individual components of episodic-like memory in kainate epileptic rats.* Presented at Annu. Meet. FENS Forum Eur. Neurosci., 7th, July 3–7, Amsterdam. Poster 176.20

Javadi AH, Walsh V, Lewis P. 2011. Offline consolidation of procedural skill learning is enhanced by negative emotional content. *Exp. Brain Res.* 208(4):507–17

Jenkins JG, Dallenbach KM. 1924. Obliviscence during sleep and waking. *Am. J. Psychol.* 35:605–12

Ji D, Wilson MA. 2007. Coordinated memory replay in the visual cortex and hippocampus during sleep. *Nat. Neurosci.* 10(1):100–7

Kart-Teke E, De Souza Silva MA, Huston JP, Dere E. 2006. Wistar rats show episodic-like memory for unique experiences. *Neurobiol. Learn. Mem.* 85(2):173–82

King C, Henze DA, Leinekugel X, Buzsáki G. 1999. Hebbian modification of a hippocampal population pattern in the rat. *J. Physiol.* 521:159–67

Kirov R, Weiss C, Siebner H. 2009. Slow oscillation electrical brain stimulation during waking promotes EEG theta activity and memory encoding. *Proc. Natl. Acad. Sci. USA* 106(36):15460–65

Konishi M. 2004. The role of auditory feedback in birdsong. *Ann. N. Y. Acad. Sci.* 1016:463–75

Kudrimoti HS, Barnes CA, McNaughton BL. 1999. Reactivation of hippocampal cell assemblies: effects of behavioral state, experience, and EEG dynamics. *J. Neurosci.* 19(10):4090–101

Lansink CS, Goltstein PM, McNaughton BL, Pennartz CMA. 2009. Hippocampus leads ventral striatum in replay of place-reward information. *PLoS Biol.* 7(8):e1000173

Lee I, Rao G, Knierim JJ. 2004. A double dissociation between hippocampal subfields: differential time course of CA3 and CA1 place cells for processing changed environments. *Neuron* 42(5):803–15

Lesburgueres E, Gobbo OL, Alaux-Cantin S, Hambucken A, Trifilieff P, Bontempi B. 2011. Early tagging of cortical networks is required for the formation of enduring associative memory. *Science* 331(6019):924–28

Lesting J, Narayanan RT, Kluge C, Sangha S, Seidenbecher T, Pape H. 2011. Patterns of coupled theta activity in amygdala-hippocampal-prefrontal cortical circuits during fear extinction. *PLoS One* 6(6):e21714

Leutgeb S, Leutgeb JK, Treves A, Moser MB, Moser EI. 2004. Distinct ensemble codes in hippocampal areas CA3 and CA1. *Science* 305(5688):1295–98

Lewis PA, Cairney S, Manning L, Critchley HD. 2011. The impact of overnight consolidation upon memory for emotional and neutral encoding contexts. *Neuropsychologia* 49(9):2619–29

Lewis PA, Durrant SJ. 2011. Overlapping memory replay during sleep builds cognitive schemata. *Trends Cogn. Sci.* 15(8):343–51

Li JS, Chao YS. 2008. Electrolytic lesions of dorsal CA3 impair episodic-like memory in rats. *Neurobiol. Learn. Mem.* 89(2):192–98

Maren S. 2001. Neurobiology of Pavlovian fear conditioning. *Annu. Rev. Neurosci.* 24:897–931

Margoliash D, Schmidt MF. 2010. Sleep, off-line processing, and vocal learning. *Brain Lang.* 115(1):45–58

Marr D. 1971. Simple memory: a theory for archicortex. *Philos. Trans. R. Soc. Lond. B Biol. Sci.* 262(841):23–81

Marshall L, Born J. 2007. The contribution of sleep to hippocampus-dependent memory consolidation. *Trends Cogn. Sci. (Regul. Ed.)* 11(10):442–50

Marshall L, Helgadottir H, Mölle M, Born J. 2006. Boosting slow oscillations during sleep potentiates memory. *Nature* 444(7119):610–13

Marshall L, Kirov R, Brade J, Mölle M, Born J. 2011. Transcranial electrical currents to probe EEG brain rhythms and memory consolidation during sleep in humans. *PLoS One* 6(2):e16905

McClelland JL, McNaughton BL, O'Reilly RC. 1995. Why there are complementary learning systems in the hippocampus and neocortex: insights from the successes and failures of connectionist models of learning and memory. *Psychol. Rev.* 102(3):419–57

Mednick SC, Cai DJ, Shuman T, Anagnostaras S, Wixted JT. 2011. An opportunistic theory of cellular and systems consolidation. *Trends Neurosci.* 34(10):504–14

Miller EK, Cohen JD. 2001. An integrative theory of prefrontal cortex function. *Annu. Rev. Neurosci.* 24:167–202

Mitchell JB, Laiacona J. 1998. The medial frontal cortex and temporal memory: tests using spontaneous exploratory behaviour in the rat. *Behav. Brain Res.* 97(1–2):107–13

Mölle M, Bergmann TO, Marshall L, Born J. 2011. Fast and slow spindles during the sleep slow oscillation: disparate coalescence and engagement in memory processing. *Sleep* 34(10):1411–21

Mölle M, Born J. 2011. Slow oscillations orchestrating fast oscillations and memory consolidation. *Prog. Brain Res.* 193:93–110

Mölle M, Eschenko O, Gais S, Sara SJ, Born J. 2009. The influence of learning on sleep slow oscillations and associated spindles and ripples in humans and rats. *Eur. J. Neurosci.* 29(5):1071–81

Montgomery SM, Sirota A, Buzsáki G. 2008. Theta and gamma coordination of hippocampal networks during waking and rapid eye movement sleep. *J. Neurosci.* 28(26):6731–41

Morris RG. 2006. Elements of a neurobiological theory of hippocampal function: the role of synaptic plasticity, synaptic tagging and schemas. *Eur. J. Neurosci.* 23(11):2829–46

Müller GE, Pilzecker A. 1900. Experimentelle Beiträge zur Lehre vom Gedächtnis. *Z. Psychol. Ergänzungsband* 1:1–300f

Mumby DG, Gaskin S, Glenn MJ, Schramek TE, Lehmann H. 2002. Hippocampal damage and exploratory preferences in rats: memory for objects, places, and contexts. *Learn. Mem.* 9(2):49–57

Nádasdy Z, Hirase H, Czurko A, Csicsvari J, Buzsáki G. 1999. Replay and time compression of recurring spike sequences in the hippocampus. *J. Neurosci.* 19(21):9497–507

Nadel L, Moscovitch M. 1997. Memory consolidation, retrograde amnesia and the hippocampal complex. *Curr. Opin. Neurobiol.* 7(2):217–27

Nir Y, Staba RJ, Andrillon T, Vyazovskiy VV, Cirelli C, et al. 2011. Regional slow waves and spindles in human sleep. *Neuron* 70(1):153–69

O'Neill J, Pleydell-Bouverie B, Dupret D, Csicsvari J. 2010. Play it again: reactivation of waking experience and memory. *Trends Neurosci.* 33(5):220–29

O'Neill J, Senior TJAK, Huxter JR, Csicsvari J. 2008. Reactivation of experience-dependent cell assembly patterns in the hippocampus. *Nat. Neurosci.* 11(2):209–15

Orban P, Rauchs G, Balteau E, Degueldre C, Luxen A, et al. 2006. Sleep after spatial learning promotes covert reorganization of brain activity. *Proc. Natl. Acad. Sci. USA* 103(18):7124–29

Palchykova S, Winsky-Sommerer R, Meerlo P, Dürr R, Tobler I. 2006. Sleep deprivation impairs object recognition in mice. *Neurobiol. Learn. Mem.* 85(3):263–71

Pavlides C, Winson J. 1989. Influences of hippocampal place cell firing in the awake state on the activity of these cells during subsequent sleep episodes. *J. Neurosci.* 9(8):2907–18

Payne JD, Kensinger EA. 2010. Sleep's role in the consolidation of emotional episodic memories. *Curr. Dir. Psychol. Sci.* 19(5):290–95

Payne JD, Kensinger EA. 2011. Sleep leads to changes in the emotional memory trace: evidence from FMRI. *J. Cogn. Neurosci.* 23(6):1285–97

Payne JD, Schacter DL, Propper RE, Huang LW, Wamsley EJ, et al. 2009. The role of sleep in false memory formation. *Neurobiol. Learn. Mem.* 92(3):327–34

Peigneux P, Laureys S, Fuchs S, Collette F, Perrin F, et al. 2004. Are spatial memories strengthened in the human hippocampus during slow wave sleep? *Neuron* 44(3):535–45

Pennartz CM, Lee E, Verheul J, Lipa P, Barnes CA, McNaughton BL. 2004. The ventral striatum in off-line processing: ensemble reactivation during sleep and modulation by hippocampal ripples. *J. Neurosci.* 24(29):6446–56

Peyrache A, Battaglia FP, Destexhe A. 2011. Inhibition recruitment in prefrontal cortex during sleep spindles and gating of hippocampal inputs. *Proc. Natl. Acad. Sci. USA* 108(41):17207–12

Peyrache A, Khamassi M, Benchenane K, Wiener S, Battaglia FP. 2009. Replay of rule-learning related neural patterns in the prefrontal cortex during sleep. *Nat. Neurosci.* 12(7):919–26

Plihal W, Born J. 1997. Effects of early and late nocturnal sleep on declarative and procedural memory. *J. Cogn. Neurosci.* 9(4):534–47

Poe GR, Nitz DA, McNaughton BL, Barnes CA. 2000. Experience-dependent phase-reversal of hippocampal neuron firing during REM sleep. *Brain Res.* 855(1):176–80

Polyn SM, Kahana MJ. 2008. Memory search and the neural representation of context. *Trends Cogn. Sci. (Regul. Ed.)* 12(1):24–30

Rasch B, Büchel C, Gais S, Born J. 2007. Odor cues during slow-wave sleep prompt declarative memory consolidation. *Science* 315(5817):1426–29

Rasch B, Pommer J, Diekelmann S, Born J. 2009. Pharmacological REM sleep suppression paradoxically improves rather than impairs skill memory. *Nat. Neurosci.* 12(4):396–97

Rauchs G, Bertran F, Guillery-Girard B, Desgranges B, Kerrouche N, et al. 2004. Consolidation of strictly episodic memories mainly requires rapid eye movement sleep. *Sleep* 27(3):395–401

Ribeiro S, Gervasoni D, Soares ES, Zhou Y, Lin SC, et al. 2004. Long-lasting novelty-induced neuronal reverberation during slow-wave sleep in multiple forebrain areas. *PLoS Biol.* 2(1):E24

Ribeiro S, Shi X, Engelhard M, Zhou Y, Zhang H, et al. 2007. Novel experience induces persistent sleep-dependent plasticity in the cortex but not in the hippocampus. *Front. Neurosci.* 1(1):43–55

Rolls A, Colas D, Adamantidis A, Carter M, Lanre-Amos T, et al. 2011. Optogenetic disruption of sleep continuity impairs memory consolidation. *Proc. Natl. Acad. Sci. USA* 108(32):13305–10

Rosanova M, Ulrich D. 2005. Pattern-specific associative long-term potentiation induced by a sleep spindle-related spike train. *J. Neurosci.* 25(41):9398–405

Ruch S, Markes O, Duss SB, Oppliger D, Reber TP, et al. 2012. Sleep stage II contributes to the consolidation of declarative memories. *Neuropsychologia* 50:2389–96

Rudoy JD, Voss JL, Westerberg CE, Paller KA. 2009. Strengthening individual memories by reactivating them during sleep. *Science* 326(5956):1079

Rugg MD, Yonelinas AP. 2003. Human recognition memory: a cognitive neuroscience perspective. *Trends Cogn. Sci.* 7(7):313–19

Schendan HE, Searl MM, Melrose RJ, Stern CE. 2003. An FMRI study of the role of the medial temporal lobe in implicit and explicit sequence learning. *Neuron* 37(6):1013–25

Scullin MK, McDaniel MA. 2010. Remembering to execute a goal: Sleep on it! *Psychol. Sci.* 21(7):1028–35

Sejnowski TJ, Destexhe A. 2000. Why do we sleep? *Brain Res.* 886(1–2):208–23

Siapas AG, Wilson M. 1998. Coordinated interactions between hippocampal ripples and cortical spindles during slow-wave sleep. *Neuron* (21):1123–28

Sirota A, Buzsáki G. 2005. Interaction between neocortical and hippocampal networks via slow oscillations. *Thalamus Relat. Syst.* 3(4):245–59

Skaggs WE, McNaughton BL. 1996. Replay of neuronal firing sequences in rat hippocampus during sleep following spatial experience. *Science* 271(5257):1870–73

Smith C, Rose GM. 1996. Evidence for a paradoxical sleep window for place learning in the Morris water maze. *Physiol. Behav.* 59(1):93–97

Smith C, Rose GM. 1997. Posttraining paradoxical sleep in rats is increased after spatial learning in the Morris water maze. *Behav. Neurosci.* 111(6):1197–204

Spencer RMC, Sunm M, Ivry RB. 2006. Sleep-dependent consolidation of contextual learning. *Curr. Biol.* 16(10):1001–5

Steriade M. 2006. Grouping of brain rhythms in corticothalamic systems. *Neuroscience* 137(4):1087–106
Sterpenich V, Albouy G, Boly M, Vandewalle G, Darsaud A, et al. 2007. Sleep-related hippocampo-cortical interplay during emotional memory recollection. *PLoS Biol.* 5(11):e282
Stickgold R. 2005. Sleep-dependent memory consolidation. *Nature* 437(7063):1272–78
Sutherland GR, McNaughton BL. 2000. Memory trace reactivation in hippocampal and neocortical neuronal ensembles. *Curr. Opin. Neurobiol.* 10(2):180–86
Takashima A, Nieuwenhuis IL, Jensen O, Talamini LM, Rijpkema M, Fernandez G. 2009. Shift from hippocampal to neocortical centered retrieval network with consolidation. *J. Neurosci.* 29(32):10087–93
Tamminen J, Payne JD, Stickgold R, Wamsley EJ, Gaskell MG. 2010. Sleep spindle activity is associated with the integration of new memories and existing knowledge. *J. Neurosci.* 30(43):14356–60
Tononi G, Cirelli C. 2006. Sleep function and synaptic homeostasis. *Sleep Med. Rev.* 10:49–62
Tse D, Langston RF, Kakeyama M, Bethus I, Spooner PA, et al. 2007. Schemas and memory consolidation. *Science* 316(5821):76–82
Tse D, Takeuchi T, Kakeyama M, Kajii Y, Okuno H, et al. 2011. Schema-dependent gene activation and memory encoding in neocortex. *Science* 333(6044):891–95
Tucker MA, Fishbein W. 2008. Enhancement of declarative memory performance following a daytime nap is contingent on strength of initial task acquisition. *Sleep* 31(2):197–203
Tulving E. 2002. Episodic memory: from mind to brain. *Annu. Rev. Psychol.* 53:1–25
van der Helm E, Gujar N, Nishida M, Walker M. 2011. Sleep-dependent facilitation of episodic memory details. *PLoS One* 6(11):e27421
van der Werf YD, Altena E, Schoonheim MM, Sanz-Arigita EJ, Vis JC, et al. 2009a. Sleep benefits subsequent hippocampal functioning. *Nat. Neurosci.* 12(2):122–23
van der Werf YD, van der Helm E, Schoonheim MM, Ridderikhoff A, van Sommeren EJW. 2009b. Learning by observation requires an early sleep window. *Proc. Natl. Acad. Sci. USA* 106(45):18926–30
Vecsey CG, Baillie GS, Jaganath D, Havekes R, Daniels A, et al. 2009. Sleep deprivation impairs cAMP signalling in the hippocampus. *Nature* 461(7267):1122–25
Wagner U, Gais S, Haider H, Verleger R, Born J. 2004. Sleep inspires insight. *Nature* 427(6972):352–55
Walker MP, Brakefield T, Hobson JA, Stickgold R. 2003. Dissociable stages of human memory consolidation and reconsolidation. *Nature* 425(6958):616–20
Walker MP, Stickgold R. 2010. Overnight alchemy: sleep-dependent memory evolution. *Nat. Rev. Neurosci.* 11(3):218
Wierzynski CM, Lubenov EV, Gu M, Siapas AG. 2009. State-dependent spike-timing relationships between hippocampal and prefrontal circuits during sleep. *Neuron* 61(4):587–96
Wilhelm I, Diekelmann S, Molzow I, Ayoub A, Molle M, Born J. 2011a. Sleep selectively enhances memory expected to be of future relevance. *J. Neurosci.* 31(5):1563–69
Wilhelm I, Metzkow-Meszaros M, Knapp S, Born J. 2012a. Sleep-dependent consolidation of procedural motor memories in children and adults: the pre-sleep level of performance matters. *Dev. Sci.* 15(4):506–15
Wilhelm I, Prehn-Kristensen A, Born J. 2012b. Sleep-dependent memory consolidation—what can be learnt from children? *Neurosci. Biobehav. Rev.* 36:1718–28
Wilhelm I, Wagner U, Born J. 2011b. Opposite effects of cortisol on consolidation of temporal sequence memory during waking and sleep. *J. Cogn. Neurosci.* 23(12):3703–12
Wilson MA, McNaughton BL. 1994. Reactivation of hippocampal ensemble memories during sleep. *Science* 265(5172):676–79
Winocur G, Moscovitch M, Bontempi B. 2010. Memory formation and long-term retention in humans and animals: convergence towards a transformation account of hippocampal-neocortical interactions. *Neuropsychologia* 48(8):2339–56
Witt K, Margraf N, Bieber K, Born J, Deuschl D. 2010. Sleep consolidates the effector-independent representation of a motor skill. *Neuroscience* 171(1):227–34
Yaroush R, Sullivan MJ, Ekstrand BR. 1971. Effect of sleep on memory. II. Differential effect of the first and second half of the night. *J. Exp. Psychol.* 88(3):361–66
Yonelinas AP. 2001. Components of episodic memory: the contribution of recollection and familiarity. *Philos. Trans. R. Soc. Lond. B Biol. Sci.* 356(1413):1363–74

Yordanova J, Kolev V, Verleger R, Bataghva Z, Born J, Wagner U. 2008. Shifting from implicit to explicit knowledge: different roles of early- and late-night sleep. *Learn. Mem.* 15(7):508–15

Yordanova J, Kolev V, Wagner U, Born J, Verleger R. 2012. Increased alpha (8–12 Hz) activity during slow wave sleep as a marker for the transition from implicit knowledge to explicit insight. *J. Cogn. Neurosci.* 24(1):119–32

Yotsumoto Y, Sasaki Y, Chan P, Vasios C. 2009. Location-specific cortical activation changes during sleep after training for perceptual learning. *Curr. Biol.* 19(15):1278–82

Computational Identification of Receptive Fields

Tatyana O. Sharpee[1,2]

[1]Computational Neurobiology Laboratories, Salk Institute for Biological Studies, La Jolla, California 92037; email: sharpee@salk.edu

[2]Center for Theoretical Biological Physics, University of California at San Diego, La Jolla, California 92093

Annu. Rev. Neurosci. 2013. 36:103–20

The *Annual Review of Neuroscience* is online at neuro.annualreviews.org

This article's doi:
10.1146/annurev-neuro-062012-170253

Keywords

natural stimuli, spike-triggered average, maximum likelihood, mutual information, neural networks

Abstract

Natural stimuli elicit robust responses of neurons throughout sensory pathways, and therefore their use provides unique opportunities for understanding sensory coding. This review describes statistical methods that can be used to characterize neural feature selectivity, focusing on the case of natural stimuli. First, we discuss how such classic methods as reverse correlation/spike-triggered average and spike-triggered covariance can be generalized for use with natural stimuli to find the multiple relevant stimulus features that affect the responses of a given neuron. Second, ways to characterize neural feature selectivity while assuming that the neural responses exhibit a certain type of invariance, such as position invariance for visual neurons, are discussed. Finally, we discuss methods that do not require one to make an assumption of invariance and instead can determine the type of invariance by analyzing relationships between the multiple stimulus features that affect the neural responses.

Contents

INTRODUCTION

One way to understand how the brain works is to describe the function of each of its neurons. In sensory systems, to describe a neuron's function means to create (either explicitly or implicitly) a model that can predict the neural responses to novel stimuli. Ultimately, the goal is to predict a neuron's responses to "natural stimuli," i.e., stimuli that are taken from an animal's environment (or to approximate such stimuli) (Felsen & Dan 2005). However, one can gain significant understanding of the function of neural pathways by using simplified stimuli (Rust & Movshon 2005). In fact, much of our current understanding about the function of visual pathways has been obtained using reduced parametric stimuli, such as spots of light, edges and bars, curved contours, and elements of three-dimensional shapes. Studies using parametric stimuli have led to such fundamental insights as the establishment of orientation selectivity in the primary visual cortex (V1) as well as tuning for curvature and orientation in three dimensions at subsequent stages of visual processing (Anzai et al. 2007, Bakin et al. 2000, Desimone & Schein 1987, Hubel & Wiesel 1968, Kuffler 1953, Kunsberg & Zucker 2012, Li & Zaidi 2004, McManus et al. 2011, Pasupathy & Connor 1999). The use of parametric stimuli has many advantages. Mainly, if the stimuli can be parameterized with a small number of parameters, then the corresponding stimulus set can be probed well experimentally, resulting in models with high predictive power, at least with the parameterized stimulus set. However, the use of parametric stimuli in some sensory areas, especially those beyond V1, presents some difficulty, namely that the relevant set of parameters is either unknown or of such high dimensionality that fully sampling it is no longer feasible. As an example, one can think of the many parameters that are needed to describe facial features and expressions. In such cases, computational approaches to characterize neural feature selectivity become indispensable.

In parallel to work in vision, research in the auditory modality since the 1950s has relied much more heavily on the use of computational approaches to describe neuronal function. The basic idea is to use stimulus sets that are defined by their statistical properties, such as the mean, variance, or correlation structure, but are otherwise unconstrained. Thus, instead of using a few parameters, such as the orientation or length of a bar, to specify each particular stimulus, investigators use a few parameters to specify the properties of an entire stimulus distribution. Examples of the resulting stimulus ensembles include "white-noise" stimuli for which the mean and variance are specified but responses at different times or across different spatial locations are uncorrelated. Adding correlations between stimulus values at different times, frequencies, or locations yields ensembles of correlated Gaussian stimuli. Again, the correlations between different stimulus values can be described by a small number of parameters, such as how fast the correlations decrease with the increasing distance between pixels.

The use of statistical stimulus sets has its own advantages. First, because the stimulus set is not optimized for a given neuron, such stimuli are ideally suited for multielectrode recordings that are becoming increasingly common. Second, the use of such stimuli enables researchers to uncover types of neural feature selectivity that were not part of the original hypothesis. The use of statistical stimulus ensembles also permits investigators to proceed without a good starting hypothesis about the types of stimulus features that are relevant in a given sensory region. As a result, compared with models obtained using parametric stimuli, models derived using statistical stimulus sets usually have better predictive power when they are applied to predict the neural responses in another stimulus context.

How do statistical approaches work in general? The basic idea common to all these methods is as follows. Prior to conducting an experiment, researchers do not know which stimulus features will modulate the responses of the neuron under consideration. The goal is to find these features, referred to as the "relevant stimulus features," because they may either increase or decrease the neural spike probability relative to its average value. To find the relevant stimulus features, one can present a large number of stimuli ($\sim$20,000 to $\sim$50,000 different images or sound patterns, which is roughly equivalent to a typical sensory episode of $\sim$1 h). Although it is possible that none of these patterns exactly matches a particular neuron's relevant feature(s), many (or at least some) of the stimuli will be sufficiently close to optimal to elicit some neural responses. Once the neuronal responses to a large number of different stimuli have been recorded, the relevant stimulus features for a given neuron, as well as its preferred optimal stimulus, can be deduced by analyzing how potentially subtle changes in the neuronal firing rate are related to changes in the corresponding stimuli. Typically, if a stimulus sequence contains $\sim 10^4$ stimuli, then $\sim 10^3$ spikes will be collected, and this will be sufficient to map out the profile of the relevant stimulus features on a grid of $\sim 10^2$ points. Notably, stimuli do not need to be repeated multiple times to enable analysis of the potentially subtle changes in the neuronal firing rate with small changes in the stimulus. In fact, for most of the techniques described below, presenting many similar (but not identical) stimuli just once is preferable to having any chosen stimulus presented multiple times, even though multiple presentations of the same stimuli allow us to average out the neuronal noise. This is because presenting many similar stimuli allows researchers not only to average out the neuronal noise (assuming some continuity in the stimulus/response function of a given neuron), but also to better probe the stimulus/response function at intermediate points.

This article discusses the advantages and limitations of statistical techniques that are currently available for characterizing neural feature selectivity and that use noise-like and natural stimuli. Particular attention is given to the methods used to characterize neural responses to natural stimuli, because such stimuli are often the only type that can drive robust responses in high-level sensory areas. Largely omitted from this discussion are techniques that can lead to a more effective use of experimental time. The reader is directed to a number of recent and excellent reviews (Huys & Paninski 2009, Lewi et al. 2009, Paninski et al. 2007) on how to optimize the order in which stimuli should be presented to maximize the accuracy of derived models given the limitations of the length of the recording.

RECEPTIVE FIELD

The concept of a receptive field (RF) was first introduced in somatosensation to describe a part of the body surface where the reflex can be elicited (Sherrington 1906). In sensory systems, the term became much more widely known after Hartline (1938) used it to describe the firing properties of the retinal ganglion cells (RGCs). One way to measure the RF of an RGC in its original formulation is to plot the neuronal firing rate as a function of light position. Another way is to plot the pattern of light

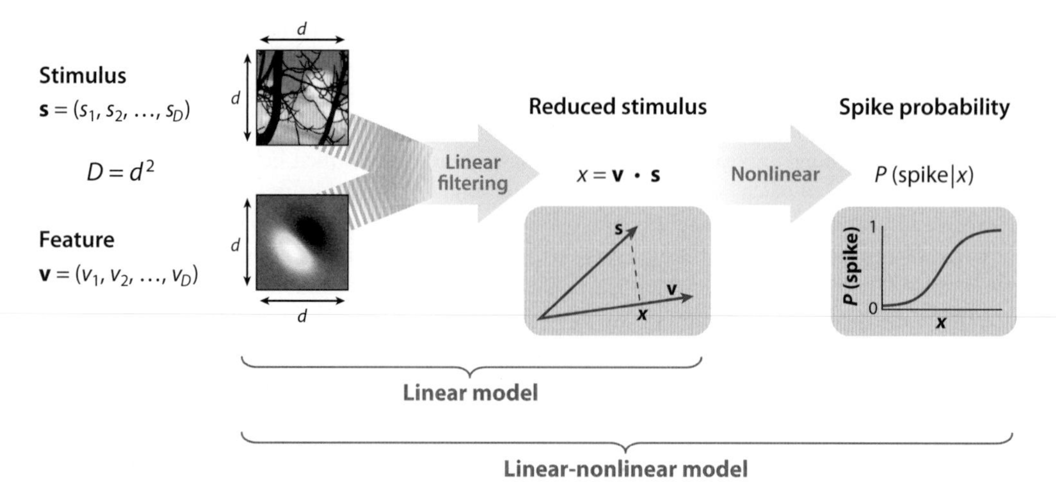

Figure 1

Geometric interpretation of the receptive field (RF) in the context of the linear and linear-nonlinear (LN) models. An example stimulus is a natural image taken from a van Hateren data set (van Hateren & van der Schaaf 1998). The stimulus has d pixels in the horizontal and vertical dimensions, which yields a stimulus of $D = d^2$ dimensions. The RF taken to mimic properties of V1 neurons is also defined in this space. The linear model predicts the spike probability as taking a projection between the stimulus and the RF. The LN model adds a nonlinear gain function to account for such properties as rectification and saturation in the neural response.

intensities that, when presented on a screen, would elicit the maximal firing rate from this neuron. If the neuron is modeled as a linear system, then the two ways of measuring the RF are equivalent. However, because the neuronal firing rate inevitably exhibits at least some nonlinear effects, for example, because it cannot be negative, the two interpretations of the RF concept will differ. Research has shown that the second formulation, whereby the RF is interpreted as the optimal stimulus for the neuron, is much more amenable to the generalizations necessary to capture a rich variety of nonlinear and contextual effects observed for sensory neurons.

LINEAR MODEL

To predict the firing rate of a neuron to a novel stimulus using a linear model, one can compare how similar that stimulus is to the optimal pattern, i.e., the RF. Mathematically, this corresponds to multiplying stimulus values pixel by pixel by the RF values and summing across all pixels. In this interpretation, the RF becomes the weighting function according to which stimulus values are combined to obtain the firing rate (**Figure 1**). The linear model also includes the coefficient of proportionality between the stimulus similarity to the RF and the neural firing rate. This coefficient of proportionality, referred to as the "gain," is the same for all stimuli.

Before discussing various ways for building nonlinear models of neural responses, it is useful to explore other ways of thinking about the RF concept. If the RF has D pixels (which could include temporal profiles), then it can also be represented as a vector in a D-dimensional space. To compare each new stimulus to the RF, stimuli should be defined on the same grid of pixel values as the RF. Then, each stimulus can be considered as a vector in the same D-dimensional space. The mathematical procedure described above of weighting each stimulus value by the RF profile corresponds to the computation of a dot product between the RF and the stimulus for which we would like to obtain the firing rate prediction. In geometrical terms, this corresponds to taking a projection of a vector that describes the stimulus onto the vector that describes the RF (**Figure 1**). In other words, this procedure corresponds to either finding a stimulus component along the RF or filtering stimuli by the RF.

Spike-Triggered Average

These geometrical interpretations suggest ways to find the RF from neural recordings. When only the stimulus component along the RF affects the neural firing rate, averaging all stimuli that have elicited a spike will result in an averaging out of all the stimulus components along directions in the stimulus space other than that of the RF. This spike-triggered average (STA) should then yield a vector that is proportional to the RF (**Figure 2**). The STA could also contain the average stimulus that is not associated with spiking, which in many cases equals zero. If the mean of all stimuli (both those that elicited and those that did not elicit a spike) is nonzero, then this term should be subtracted from the STA to yield an estimate of the neuron's RF.

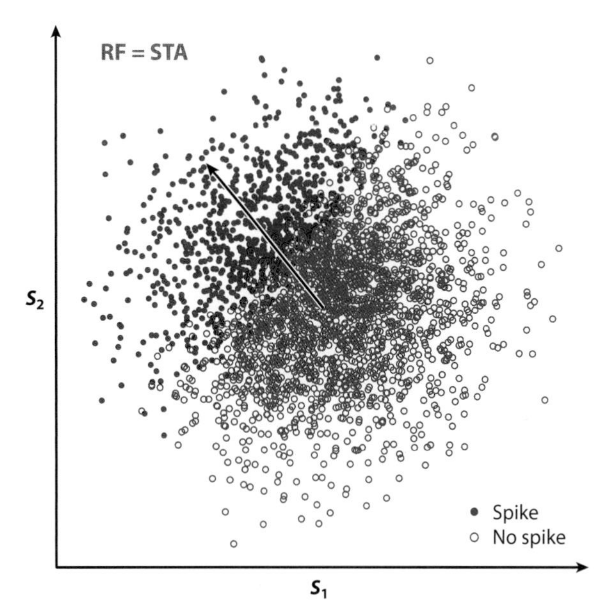

Figure 2

Illustration of how the receptive field (RF) of a neuron can be estimated from its responses to white-noise stimuli by computing the spike-triggered average (STA). Each dot represents a high-dimensional stimulus projected onto a plane. Each stimulus was taken from an uncorrelated, white-noise Gaussian distribution. Stimuli that elicited a spike are marked with red filled dots. Their average yields a vector, which is the relevant stimulus dimension for generating spikes from this neuron.

One can prove that this intuition for computing the STA is mathematically rigorous if the stimulus ensemble is "circularly symmetric," meaning that the stimulus ensemble probes the neuron's responses in all directions equally (Chichilnisky 2001). If so, the STA yields an unbiased estimate of the neuron's RF that will converge to the true RF as longer recordings are obtained. One example of a circularly symmetric stimulus distribution is the so-called white-noise stimulus ensemble, which has zero mean and independent variations along all stimulus dimensions that follow a Gaussian distribution. More generally, the probability distribution of observing a stimulus with a certain amplitude can be non-Gaussian. Such distributions are also of relevance. For example, analyses of image intensities in the natural environment may follow a Laplace distribution (Ruderman & Bialek 1994, Simoncelli & Olshausen 2001). However, as long as the distribution is circularly symmetric, the STA will correspond to the RF asymptotically.

What happens if the stimulus ensemble is not circularly symmetric? For example, in the natural environment, image intensities at nearby locations are often positively correlated with each other (Field 1987). As a consequence, the sum of intensities at the two locations will cover a much broader range of values than their difference. Stimulus ensembles with such correlations are not circularly symmetric. In other words, variance is not equal along different dimensions in the stimulus space (**Figure 3*b***). How does this affect our ability to estimate the relevant stimulus features by computing the STA? Consider the following hypothetical example: The neuron's spikes are triggered when the light intensity at one location exceeds a certain threshold value. The light intensities at this location are correlated with light intensities at nearby locations. The STA will, therefore, show a peak in the light intensity at the location that is relevant for eliciting neuronal spikes. However, the nearby locations will also show significant deviations from zero (**Figure 3**). In technical terms, the STA provides a biased

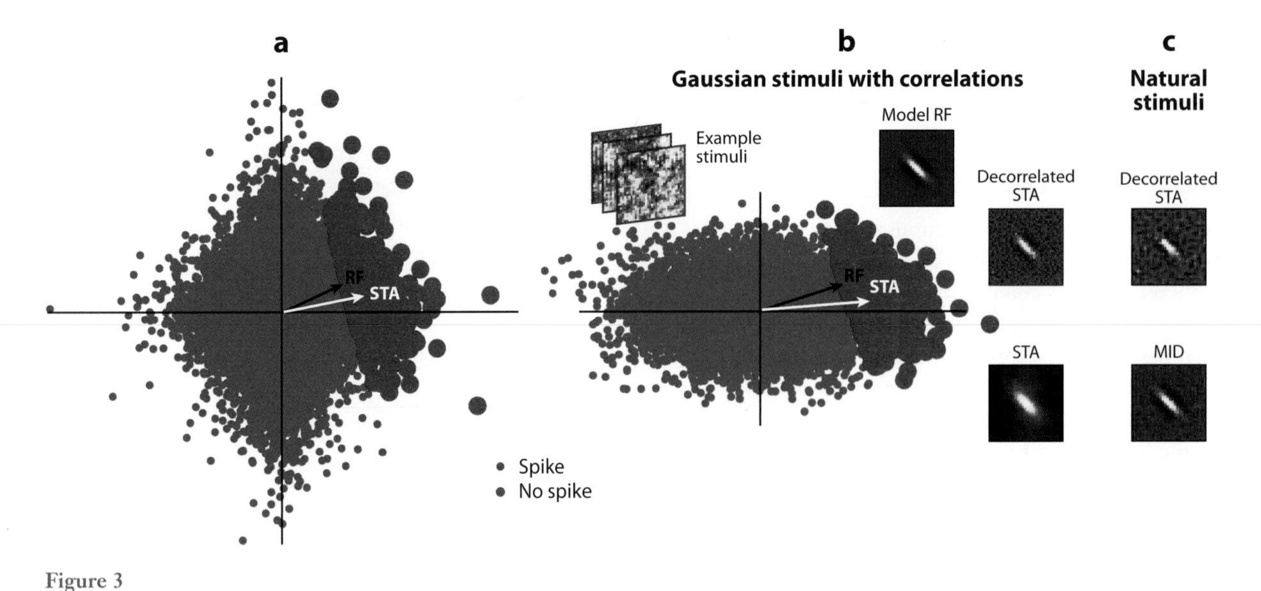

Figure 3

Differences between the spike-triggered average (STA) and the receptive field (RF) for noncircular probability distributions. (*a*) Example of a non-Gaussian distribution of stimuli without correlations. (*b*) Example of a Gaussian distribution with correlations. In both examples, the STA does not always yield a correct estimate of the RF. However, for the Gaussian distribution, it can be corrected according to Equation 1 to yield an accurate estimate. (*c*) In the case of natural stimuli, which are non-Gaussian and have strong correlations, even with corrections for the second-order correlations, the STA does not yield a correct RF estimate. However, the RF can be estimated as a maximally informative dimension (MID).

estimate of the relevant stimulus feature. Even if we collect an infinite amount of data, the STA will not provide a correct estimate of the relevant stimulus feature in cases where stimulus values are correlated across different dimensions. However, it is possible to compensate for the presence of stimulus correlations with one relatively simple step. In the linear model, the effect of stimulus correlations is limited to pairwise correlations between different stimulus values, which are specified by the stimulus covariance matrix. To obtain an unbiased estimate of the relevant stimulus dimension from the STA, one simply needs to multiply it by the inverse of the stimulus covariance matrix (Rieke et al. 1997; Theunissen et al. 2000, 2001):

$$v_i = C_{ij}^{-1}(\mathrm{STA})_j, \qquad 1.$$

where v_i are components of the relevant stimulus feature, matrix C_{ij} is the stimulus covariance matrix, $(\mathrm{STA})_j$ represents components of the STA, and summation over the repeated index j is implied. The covariance matrix is obtained by averaging stimulus deviations from the mean. This procedure is known as decorrelation or deconvolution. The resultant vector is termed the decorrelated STA (Rieke et al. 1997; Theunissen et al. 2000, 2001) and is analogous to deconvolution that is sometimes done in microscopy to obtain a deblurred image by compensating for the microscope's point spread function (Press et al. 1992). The stimulus covariance matrix is the analogue of the point spread function in the context of neural coding. It captures the expected stimulus values at nearby locations given a "point source" at the chosen location.

Although correcting the STA for stimulus correlations according to Equation 1 is a simple linear operation, practical implementations can be difficult because a very strong asymmetry is often found in the range of stimulus values contained within the input ensemble. This is especially true for stimuli derived from the natural environment. Here, the intensities averaged across space (the image) or time have a much wider range of values compared with the range of values explored by their differences. More generally, components of natural stimuli at high spatial and temporal frequencies are much less well sampled than are components corresponding to low spatial and temporal frequencies (Field 1987). Because of this asymmetry, some stimulus dimensions are much less well probed than others. The process of compensating for the difference in sampling along different stimulus dimensions according to Equation 1 amplifies noise at the stimulus dimensions that were less well probed. In mathematical terms, the covariance matrix C is ill-defined, and its eigenvalues have widely different amplitudes. In the context of natural stimuli, small eigenvalues of the covariance matrix correspond to high temporal and spatial frequencies and reflect the small power found at these frequencies.

The division of the STA by the stimulus covariance matrix amplifies noise at the components that are underrepresented in the stimulus. To overcome these issues, one can perform "regularization." This process is based on the observation that at some frequencies the amount of noise will exceed the measured signal; for those frequencies, it is better to assume that the signal is equal to zero. Deciding which stimulus components are sufficiently well sampled to be included often has to be determined on a neuron by neuron basis and in some cases can introduce additional biases that are not present in the decorrelated STA computed without regularization according to Equation 1 (Sharpee et al. 2008). Several advanced statistical techniques have been developed to tackle this issue by incorporating priors such as STA smoothness and sparsity (Ahrens et al. 2008, Christianson et al. 2008, Paninski et al. 2007, Park & Pillow 2011b, Sahani & Linden 2003).

LINEAR-NONLINEAR MODEL

The linear model can often well characterize neural feature selectivity by specifying the neuron's RF (Theunissen et al. 2000, 2001). However, the linear model cannot account for many nonlinear and contextual effects in the neural response (Gilbert & Wiesel 1990, Nothdurft et al. 1999, Schwartz et al. 2007, Series et al. 2003, Sharpee & Victor 2008, Sillito & Jones 1996, Zipser et al. 1996). One way to generalize the linear model of neural feature selectivity to capture some of the nonlinear aspects of the neural computation is to assume that these nonlinear effects are in some sense weak. Then, one can write the nonlinear transformation as a power series containing linear, quadratic, and higher-order terms. This expansion corresponds to the Volterra/Wiener series approximation (Marmarelis & Marmarelis 1978, Victor & Purpura 1998, Victor & Shapley 1979). Given enough terms in the expansion, the series is guaranteed to approximate any arbitrary function well. However, in practice, this approximation can extend only to linear and quadratic terms. Thus, in the Wiener approach, only quadratic functions of the stimulus can be modeled. Unfortunately, neural responses often contain much sharper nonlinearities than can be described by a quadratic function. For example, in a simple threshold model where the neuron produces a spike only when the stimulus value exceeds a certain value, the quadratic function is too smooth to approximate the nonlinearity well.

A very elegant, simple, yet agile way to capture sharp nonlinearities in the neural response is provided by the so-called linear-nonlinear (LN) model (de Boer & Kuyper 1968, Meister & Berry 1999, Victor & Shapley 1980). In statistics, this model is also known as the generalized linear model (Weisberg & Welsh 1994). Whereas the linear model requires that the firing rate of a neuron depend on the stimulus similarity to the RF (computed as the stimulus projection onto the RF, a purely linear operation), the LN model allows the firing rate to be an arbitrary nonlinear function of the stimulus projection on the RF (**Figure 1**). This function is often referred to as the nonlinear gain function to emphasize that the gain between the firing rate and the stimulus projection onto the RF now depends on this projection. In the LN model, the RF is often referred to as the filter or the relevant stimulus dimension.

The LN model has many computational advantages. For example, its linear component—the relevant dimension representing the RF—can be found using the linear techniques described above if the stimulus ensemble is circularly symmetric (Chichilnisky 2001, de Boer & Kuyper 1968) or, for certain nonlinearities, in the more general case of finite energy band-limited (Lazar & Slutskiy 2012, Victor et al. 2006, Victor & Knight 2003). The circularly symmetric stimulus includes the case of the white-noise Gaussian stimulus without correlations. The linear techniques also work for determining the linear part of the LN model if stimuli are correlated but are Gaussian (Ringach et al. 2002, Sharpee et al. 2004) or for models with specific nonlinearities. In this case, the RF is computed as the decorrelated STA according to Equation 1 (Theunissen et al. 2000, 2001). If necessary, regularization may also be used. However, as in the case of the linear model, doing so may introduce biases into the RF estimates (Sharpee et al. 2008). The mathematical proof that the linear component of the LN model can be estimated with linear techniques in the case of Gaussian stimuli relies on the special property of Gaussian stimuli: Here, the average of any number of Gaussian variables is equal to the sum of products of pairwise averages between the variables (de Boer & Kuyper 1968). In other words, by specifying the stimulus covariance matrix, researchers can fully explain the statistics of stimulus correlations in terms of pairwise correlations between different stimulus values.

In a more general case, where correlations of higher than second order cannot be predicted from the knowledge of pairwise stimulus correlations, the linear techniques do not provide unbiased estimates of the RFs of the LN model. In effect, the procedure for estimating the linear and the nonlinear parts can no longer be done independently of one another, as was done for the Gaussian stimuli. Importantly, for sensory neuroscience, natural stimuli in visual, auditory, or olfactory modalities exhibit strong non-Gaussian correlations that extend beyond the second order (Ruderman 1997, Simoncelli 2003, Singh & Theunissen 2003, Vickers et al. 2001). The presence of strong higher-order correlations in the stimulus ensemble is thought to be driven by the fact that natural stimuli are composed of objects. By contrast, the presence of Gaussian second-order correlations typically yields "cloud-like" stimulus patterns that are devoid of edges and object boundaries (Field 1987). Examples of such stimuli are provided in **Figure 3*b***.

Maximally Informative Dimensions

To characterize the feature selectivity with natural stimuli and other stimuli with non-Gaussian correlations, the presence of higher-order stimulus correlations in such stimulus ensembles must be taken into account. One approach is to evaluate the relevance of different stimulus dimensions for eliciting the neural response using measures that do not rely exclusively on the first- and second-order moments of the stimulus distributions. The Kullback-Leibler (KL) distance between probability distributions provides such a measurement (Cover & Thomas 1991). When the two distributions are the same, this distance is zero. In addition, when the KL distance is computed between two probability distributions, one of which reflects all stimuli $P(\mathbf{s})$ and the other stimuli that elicited a spike $P(\mathbf{s} \mid \text{spike})$, it corresponds to the mutual information between stimuli and the neural spike rate (Brenner et al. 2000):

$$I(\mathbf{v}) = \int d\mathbf{s} P(\mathbf{s} \mid \text{spike}) \log_2 \left[\frac{P(\mathbf{s} \mid \text{spike})}{P(\mathbf{s})} \right].$$

If we look only at the probability distributions along a single dimension or a set of stimulus dimensions, we obtain the mutual Shannon information between these dimensions and the neural spike rate (Adelman et al. 2003, Sharpee et al. 2004). If the spike probability does not depend on this stimulus dimension, then these two probability distributions (one computed across all stimuli and the other for stimuli that elicited a spike) will be the same, indicating that these dimensions carry zero information about the neural spikes. By contrast, the dimensions that capture all the information in the neural response correspond to those upon which the decision to spike was based. Therefore, the RF of the LN model may be found by scarching for the maximally informative dimension about the neural response (Sharpee et al. 2004). Because the KL distance upon which the mutual information is based is sensitive to any deviations between the probability distributions regardless of whether these deviations are described by differences in the first-, second-, or higher-order moments of the distribution, this procedure can be used with different kinds of stimuli, including natural stimuli (**Figure 3*c***).

The proof that the RF of the LN model corresponds to the maximally informative dimension about the neural response is based on the so-called data-processing inequality (Cover & Thomas 1991). This inequality states that adding any extra processing of inputs can only decrease the amount of information about the output of a system. Thus, if we take stimulus components along the same dimensions as were considered in the process of generating the neural spikes, then no extra processing steps are added. Otherwise, if we take stimulus components along dimensions that do not exactly correspond to the dimensions that elicited the neural spikes, then the output information will be reduced. Finding the RF of the LN model is equivalent to the maximum likelihood fitting of the LN model (Kouh & Sharpee 2009). Thus, although other distance measures can be used (Paninski 2003, Sharpee 2007) to compare changes in the probability distributions between all presented stimuli and stimuli that elicited a spike, information maximization yields the smallest variance in RF estimates versus other unbiased methods.

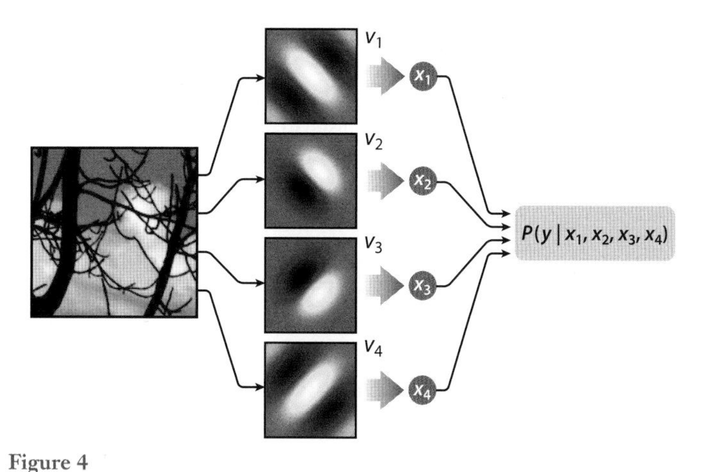

Figure 4

Schematic of a linear-nonlinear model with multiple relevant stimulus features.

Multidimensional Feature Selectivity

The LN model discussed above describes the neural responses as being triggered according to the degree to which a single stimulus feature is present in the stimulus. However, responses of many types of sensory neurons exhibit a variety of contextual effects wherein their responses to the primary stimulus feature are modulated by the presence of other stimulus features in the stimulus. Examples include cross-orientation suppression (Carandini et al. 1998, Priebe & Ferster 2006) and contrast-invariant orientation tuning for V1 cells (Troyer et al. 1998). Marr (1982) has argued that to distinguish a faint edge of a given orientation from a brighter edge of a nearby orientation requires that an orientation-selective neuron be suppressed by the presence of edges orthogonal to its preferred orientation. Another example is the feature selectivity of complex cells in the primary visual cortex whose responses indicate the presence of an edge while allowing for some degree of position invariance. This property can be modeled with several visual features that correspond to Gabor patterns with different spatial phases (Adelson & Bergen 1985). To account for these aspects of neural coding, the traditional LN model is generalized such that the spike probability is a nonlinear function of several stimulus components (Rust et al. 2005). Similar to the case of a one-dimensional LN model, the nonlinearity can take an arbitrary shape, but now with respect to several stimulus components (**Figure 4**).

Spike-Triggered Covariance

How can we determine these relevant stimulus features from the neural responses? The STA is one of these features, and it is computed by analyzing the change in the mean between the stimulus distribution conditional on a spike and the distribution of all stimuli that were presented in the experiment (**Figure 2**). Analyzing the change in the variance between these stimulus distributions allows one to find all the relevant dimensions in cases where the stimuli are described by a Gaussian distribution (Bialek & de Ruyter van Steveninck 2005, de Ruyter van Steveninck & Bialek 1988, Schwartz et al. 2006) or by a circularly symmetric distribution (Samengo & Gollisch 2013). The intuition behind this procedure is that any dimension along which the variance is different from those expected a priori is associated with neural spikes. Unlike the change in the mean that defines a single dimension, the variance can differ along many dimensions.

In mathematical terms, the spike-triggered covariance method consists of two steps. The first step is to compute a difference between the covariance matrix of all stimuli and that of the stimuli that elicited a spike:

$$\Delta C = C - C^{\text{spike}}.$$

The second step is to diagonalize this matrix to find dimensions along which the variance is significantly different from zero. The variance is encoded in the eigenvalues of matrix ΔC. The eigenvectors that correspond to the significant eigenvalues describe the dimensions along which the variance is significantly different between the ensemble of stimuli that elicited a spike and that of all stimuli. To determine the significance, these two steps are repeated using shuffled spike trains that contain as many spikes as the recorded spike trains. Another way of breaking the correlations between stimuli and spikes that preserves all the structure in the neural spike trains is to shift forward these spike trains relative to the stimuli (Bialek & de Ruyter van Steveninck 2005). The spike-triggered covariance can, in principle, be used in the presence of stimulus correlations (Bialek & de Ruyter van Steveninck 2005, Schwartz et al. 2006). To obtain the relevant dimensions in this case, the eigenvectors of matrix ΔC must be divided by the stimulus covariance C_{ij} in a manner analogous to Equation 1. However, preliminary evidence indicates that this procedure can present more difficulties to execute in practice than arise when removing the effect of correlations from the STA (Aljadeff et al. 2013).

Maximally Informative Subspaces

Multiple relevant stimulus dimensions can also be found by maximizing the mutual information (Sharpee et al. 2004). Several dimensions that together account for a maximal amount of information in the neural response may be obtained by computing the KL distance between the probability distribution along these dimensions for all presented stimuli and the distribution for the stimuli that elicited a spike. For example, a pair of maximally informative dimensions may be found by maximizing

$$I(\mathbf{v}_1, \mathbf{v}_2) = \int dx_1 dx_2 P_{\mathbf{v}_1,\mathbf{v}_2}(x_1, x_2 \mid \text{spike}) \log_2 \times \left[\frac{P_{\mathbf{v}_1,\mathbf{v}_2}(x_1, x_2 \mid \text{spike})}{P_{\mathbf{v}_1,\mathbf{v}_2}(x_1, x_2)}\right]. \quad 2.$$

This equation yields the KL distance between the probability distribution $P_{\mathbf{v}_1,\mathbf{v}_2}(x_1, x_2)$ of stimulus projections x_1 and x_2, respectively, along dimensions $\mathbf{v}_1$ and $\mathbf{v}_2$, and the probability distribution $P_{\mathbf{v}_1,\mathbf{v}_2}(x_1, x_2 \mid \text{spike})$ of these projections across the stimuli that elicited a spike. Accordingly, a two-dimensional probability distribution must be sampled to jointly characterize two dimensions in terms of the amount of information about the neural responses that they provide.

In principle, one can find N maximally informative dimensions by sampling the stimulus probability distribution across N dimensions, although doing so is difficult to achieve in practice for more than three or four dimensions. Qualitatively, the number of samples that are available to map out the spike-conditional distribution $P_{\mathbf{v}_1,\mathbf{v}_2}(x_1, x_2 \mid \text{spike})$ is related to the number of recorded spikes. This is often the limiting factor in the computation of information. Note that the stimulus distribution

$P_{v_1,v_2}(x_1, x_2)$ is easier to sample, and in some cases, an analytic expression may be available. These constraints, known as the curse of dimensionality (Belleman 1961), limit the number of dimensions that can be simultaneously characterized according to the mutual information (Rowekamp & Sharpee 2011) as well as other divergence measures (Paninski 2003), which include the percentage of explained variance.

In some cases, it is possible to bypass the curse of dimensionality by searching for the relevant dimensions in an iterative fashion, for example, by first finding the relevant dimension, then finding the second relevant dimension in the subspace orthogonal to the first, etc. (Rapela et al. 2010, Rapela et al. 2006, Rowekamp & Sharpee 2011). Typically, this procedure works well when stimuli are not correlated. However, with uncorrelated stimuli, the relevant dimensions may also be found using spike-triggered covariance, which is a much simpler procedure. When stimuli are correlated, sequential searching can find the first relevant dimension accurately if stimulus correlations are Gaussian. The sequential search for secondary relevant dimensions is complicated by the presence of stimulus correlations: The sequential search can return a dimension that accounts for a large amount of information, not because it is relevant to the neural response, but because stimulus components along this dimension are strongly correlated with the primary stimulus dimension (Rowekamp & Sharpee 2011).

QUADRATIC NONLINEAR MODELS

Maximally Informative Quadratic Models

A different approach for finding multiple relevant stimulus dimensions from the neural responses to natural stimuli is to modify the structure of the LN model. The model considered above allows for an arbitrary nonlinear function of a few stimulus components. A recently proposed alternative is to describe the spike probability as an arbitrary nonlinear function of a quadratic form of stimuli (Fitzgerald et al. 2011a, Rajan & Bialek 2012):

$$P(\text{spike} \mid \mathbf{s}) = F(s_i v_i + s_i J_{ij} s_j), \qquad 3.$$

where the sum over repeated indices is implied and F is an arbitrary nonlinear function. The parameters of the model are v_i, which is analogous to the RF, and J_{ij}. By analogy with the LN model, this model may be termed the quadratic-nonlinear (QN) model. The structure of the QN model is motivated by the need to capture such properties of sensory neurons as divisive normalization and contrast gain control (Carandini et al. 1997) where responses of one neuron are normalized by a squared output of the responses of other neurons in the circuit. As discussed by Rajan & Bialek (2012), the QN model is also well matched both to the properties of complex cells (Adelson & Bergen 1985) in the primary visual cortex and to nonphase locked auditory neurons (Hudspeth & Corey 1977).

The QN model is also congruent with some types of the LN model. For example, the matrix J can have a low-dimensional structure. In such cases, the neuronal response will be described as a quadratic function of a small number of stimulus components, as in the standard LN model. However, the LN model can, in principle, describe arbitrary interactions between the relevant dimensions, whereas these interactions are limited in the QN model (Equation 3) to sums and differences between the squares of relevant stimulus components. Nevertheless, the quadratic model provides a way forward to determine multiple relevant stimulus dimensions from the neural responses to natural stimuli. At the same time, by incorporating an arbitrary nonlinearity, the QN model represents an advance over the Wiener approach where in practice only a quadratic form of the stimulus can be estimated.

How can we estimate the parameters of the QN model from the neural responses to natural stimuli? As in the LN model, this can be done by finding its parameters—the values of the linear term v and the quadratic matrix J—that account for the maximal amount of information in the neural response (Fitzgerald et al. 2011a, Rajan & Bialek 2012). All the arguments for self-consistency of estimators carry over from the LN model because the QN model can be reformulated as an LN model with respect to the expanded stimulus $\{s_i; s_i s_i\}$, where indices i and j go from 1 to D (the dimensionality of the original stimulus). Even though the resulting expanded stimulus will have a very large dimensionality, analysis of model neurons suggests that the procedure remains feasible (Fitzgerald et al. 2011a). As in the spike-triggered covariance methods, diagonalizing the matrix J of the QN model will yield the stimulus dimensions that are relevant for spiking. Finding the parameters of the QN or LN models by maximizing information is equivalent to a maximum likelihood estimation, at least for a Poisson model of spiking (Kouh & Sharpee 2009). If the nonlinearity F in the QN model is constrained to be an exponential, then prior assumptions can be incorporated via maximum likelihood optimization onto the smoothness of the relevant stimulus features or the sparseness of their values (Park & Pillow 2011a).

Minimal Quadratic Models: A Maximum Noise Entropy Approach

The approaches described above rely on finding a suitable model structure to describe the neural circuit and then fitting parameters of these models according to the chosen criterion, such as information maximization, maximum likelihood estimation, or percent explained variance. For complex and hierarchical circuits, the appropriate model structure can be difficult to determine. An alternative to finding a suitable model structure to fit the neural responses is to construct what is known as a minimal model. The goal is to construct a model that is consistent with a given set of measurements of the neural responses and stimuli, but that is otherwise as unconstrained as possible. This approach is theoretically similar to the maximum entropy principle (Jaynes 1957) and in machine learning is known as the conditional Markov random fields (Lafferty et al. 2001).

According to the maximum entropy principle, when many distributions may be consistent with a given set of measurements, the distribution that has the maximal entropy (least constrained) should be chosen to obtain the least-biased model. Such a choice often yields models with the best predictive power on a novel set of data (Jaynes 2003). Recent studies show that this approach is fruitful in characterizing the responses of neural populations (Schneidman et al. 2006, Shlens et al. 2006). In the current discussion, the aim is to adopt this principle to build minimal models of input/output functions. Thus, the focus is on input/output functions for single neurons, although extensions to multiple neurons are certainly possible (Globerson et al. 2009, Granot-Atedge et al. 2012).

To build a minimal model of the neural input/output function, the chosen model should yield the highest entropy of the neural response for a given stimulus, averaged over all stimuli. The corresponding quantity is known as the noise entropy (Brenner et al. 2000, Strong et al. 1998). For binary responses (at sufficiently small time resolution, all neural responses are binary if patterns of spikes in time are not considered), the maximum noise entropy model has three appealing properties. First, this model is analytically solvable and the response function has a simple structure: It is a logistic function whose argument is a sum of stimulus parameters whose correlations with the neural response represent the constraints that the model needs to satisfy (Fitzgerald et al. 2011b). For example, the minimal model that is consistent with the measurements of the STA and spike-triggered covariance is

$$P(\text{spike} \mid \mathbf{s}) = 1/(1 + \exp(c + s_i v_i + s_i J_{ij} s_j)).$$

Here, parameters of the model are c, v_i, and J_{ij}. These parameters are adjusted such that the STA and spike-triggered covariance predicted by the model match those measured experimentally. The relevant stimulus dimensions can be found by diagonalizing the matrix J, just as in the QN model.

Second, although these models are similar, the main difference between the minimal quadratic model and the QN model is that the nonlinearity is a logistic function in the former but can take an arbitrary form in the latter. Thus, the parameters of a minimal model can be found through a convex optimization that is not plagued by local minima. This practical advantage of minimal models makes it possible to find their parameters even with high-dimensional stimuli. For example, in the analysis of simulated neurons with six relevant dimensions, the relevant stimulus dimensions from a minimal quadratic model were a slightly better match to the model dimensions than the dimensions from a maximally informative QN model (Fitzgerald et al. 2011a). However, in cases where only one or two relevant dimensions are needed, the maximally informative LN model yields a better match to model dimensions than either the maximally informative QN model or the minimal quadratic model.

Third, minimal models often make it possible to quantify, in information-theoretic terms, the relative importance of different constraints (**Figure 5**). For example, matching the minimal model to the experimental data in terms of the mean spike rate (which can be considered a zeroth order constraint because it does not involve stimuli) fully determines the overall entropy of the neural responses if they are binary or Poisson. With the entropy of the neural response fixed, maximizing the noise entropy is equivalent to seeking the model that provides the least amount of information while satisfying the necessary constraints (Globerson et al. 2009). As pointed out by Globerson et al. (2009), the information captured by the model that provides the minimal amount of information while satisfying a given set of constraints is a direct measure of the information content of these constraints (**Figure 3**). In general, minimally informative models are not analytically solvable, but in cases where they coincide with models computed by maximum noise entropy, the analytical solution is provided by the logistic function mentioned above. In addition to the binary and Poisson responses, it is

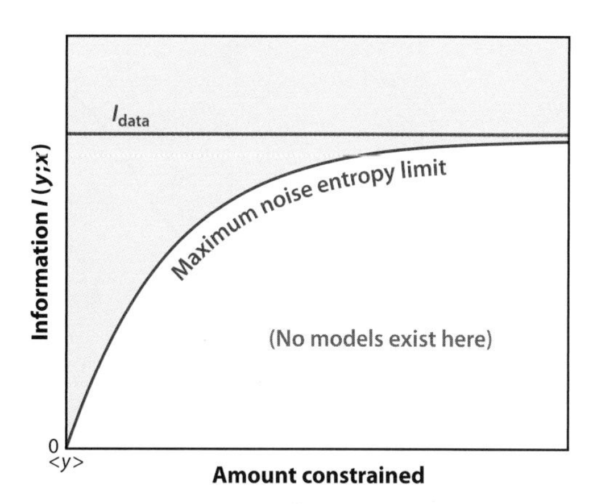

Figure 5

Geometric intuition of information transmission in minimal models. For a given set of constraints, the maximal noise entropy model provides the smallest amount of information between the neural response $y = 0,1$ and the stimulus x. The point of origin with zero information corresponds to the model where only the mean spike rate $<y>$ is constrained; this model carries no information about the stimulus. As more constraints are added, information captured by the minimal model will approach the true information in the data I_{data}. As a result, the information content of any constraint can be quantified. In contrast, a maximally informative model is used to find the value of parameters that account for the greatest amount of I_{data} in one step. Although optimization to find the maximally informative set of parameters is nonconvex, the corresponding optimization for the minimal models is convex. Given the right set of constraints, both models will converge to the same input/output function of the neuron.

possible to find a minimally informative model by maximizing noise entropy for Gaussian neural responses through the addition of the mean and variance of the spike rate to the set of constraints. Again, this is because the mean and variance are sufficient to specify the entropy of the neural response across a set of stimuli. Finally, the emergence of the logistic function as the least-constrained model that represents the necessary constraints between inputs and outputs suggests a possible functional explanation for the ubiquity of logistic input/output functions in systems biology, ranging from transcription control to neural gain functions (Clemens et al. 2012, Fairhall et al. 2006, Sharpee et al. 2011, Tyson et al. 2003). It can also explain nonmonotonic gain functions, which are especially common in the auditory system wherein the neurons encode not just the mean of the relevant stimulus feature, but also its variance (**Figure 6*b***).

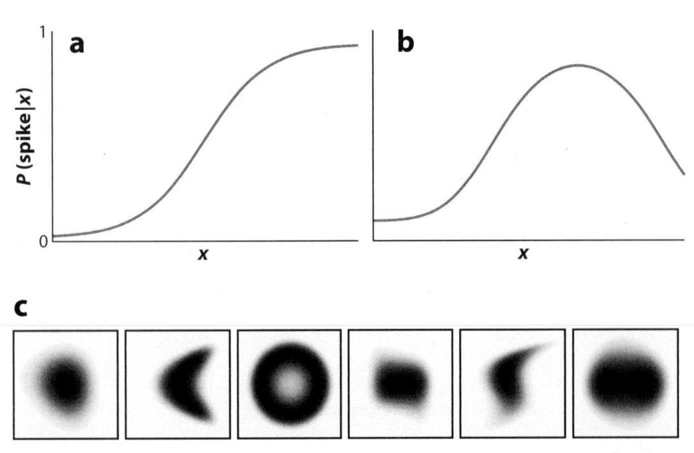

Figure 6

Maximum noise entropy models can account for variety of nonlinear effects. (*a*) The logistic function arises when the neural response encodes stimuli linearly. (*b*) The logistic function of a quadratic argument can be observed in models where the neural response encodes both the mean and the variance of stimulus components along the relevant stimulus features. In this case, the nonlinear gain function is nonmonotonic. (*c*) Maximum noise entropy models based on two relevant stimulus features can characterize a variety of nonlinear computations when extended to moments higher than two. These include "ring," "bimodal," and "crescent-shaped" nonlinearities previously observed in the retina (Fairhall et al. 2006).

OUTLOOK

Ultimately, computational methods for characterizing neural feature selectivity will be able to relate the neural responses to the underlying neural circuitry even in cases where many stages of nonlinear processing separate stimuli from the recorded neural responses. The progress that has been made in the field over the past ten years goes a long way toward this goal by recovering the multiple stimulus features that are relevant to the responses of high-level sensory neurons. However, much remains to be done. In particular, the stimulus features upon which the responses of a given neuron are triggered usually have specific relationships to each other, as indicated by their co-occurrence in the natural environment. However, orthogonal representations of these features often make it difficult to deduce these relationships. For example, when the neural responses are triggered by the same image feature that is centered at different positions in the visual field, the features obtained by spike-triggered covariance make it difficult to guess the computational relationship between the features (Rust et al. 2005). Recent methods are beginning to characterize the neural feature selectivity by taking invariance directly into account (Eickenberg et al. 2012, Vintch et al. 2012). The resulting models can economically account for the observed neural responses as triggered by a small number of image features, while allowing for their different positioning within the visual field. In other sensory systems, relevant invariance properties are either more difficult to parameterize than position invariance or are not known. One possible way to discover types of invariance present in the responses of a particular neuron is to find such linear combinations of the relevant stimulus features that make it possible to describe the observed responses as logical OR operations with respect to a set of inputs (Kaardal et al. 2013). Given a set of relevant stimulus features, finding the appropriate linear combinations of features is numerically and computationally easy, although the methods so far have not been tested in higher-level sensory areas. The hope is that, by building on combinations of the available computational tools to identify receptive fields, future works will be able to characterize the neural feature selectivity in the presence of complex and, in some cases, still unknown types of invariance of high-level sensory neurons in vision and other sensory modalities.

DISCLOSURE STATEMENT

The author is not aware of any affiliations, memberships, funding, or financial holdings that might be perceived as affecting the objectivity of this review.

ACKNOWLEDGMENTS

The author thanks Johnatan Aljadeff, William Bialek, Michael J. Berry II, Michael Eickenberg, Jeffrey D. Fitzgerald, Joel Kaardal, Minjoon Kouh, Kenneth D. Miller, Michael P. Stryker, Ryan Rowekamp, and Adrian Wanner for many helpful discussions. In addition, I thank Johnatan Aljadeff and Brian Sharpee for comments on the manuscript and Jeffrey Fitzgerald for help with some of the figures. This research was supported by grant R01EY019493 from the National Institutes of Health, grant 0712852 from the National Science Foundation, the Alfred P. Sloan Research Fellowship, Searle Funds, the McKnight Scholarship, the Ray Thomas Edwards Career Development Award in Biomedical Sciences, and the W.M. Keck Foundation Research Excellence Award. Additional resources were provided by the Center for Theoretical Biological Physics (NSF PHY-0822283).

LITERATURE CITED

Adelman TL, Bialek W, Olberg RM. 2003. The information content of receptive fields. *Neuron* 40:823–33

Adelson EH, Bergen JR. 1985. Spatiotemporal energy models for the perception of motion. *J. Opt. Soc. Am. A* 2:284–99

Ahrens MB, Linden JF, Sahani M. 2008. Nonlinearities and contextual influences in auditory cortical responses modeled with multilinear spectrotemporal methods. *J. Neurosci.* 28:1929–42

Aljadeff J, Segev R, Berry MJ II, Sharpee TO. 2013. Singular dimensions in spike triggered ensembles of correlated stimuli. *COSYNE.* Abstr.

Anzai A, Peng X, Van Essen DC. 2007. Neurons in monkey visual area V2 encode combinations of orientations. *Nat. Neurosci.* 10:1313–21

Bakin JS, Nakayama K, Gilbert CD. 2000. Visual responses in monkey areas V1 and V2 to three-dimensional surface configurations. *J. Neurosci.* 20:8188–98

Belleman R. 1961. *Adaptive Processes: A Guided Tour.* Princeton, NJ: Princeton Univ. Press

Bialek W, de Ruyter van Steveninck RR. 2005. *Features and dimensions: motion estimation in fly vision.* **http://arxiv.org/abs/q-bio/0505003**

Brenner N, Strong SP, Koberle R, Bialek W, de Ruyter van Steveninck RR. 2000. Synergy in a neural code. *Neural. Comput.* 12:1531–52

Carandini M, Heeger DJ, Movshon JA. 1997. Linearity and normalization in simple cells of the macaque primary visual cortex. *J. Neurosci.* 17:8621–44

Carandini M, Movshon JA, Ferster D. 1998. Pattern adaptation and cross-orientation interactions in the primary visual cortex. *Neuropharmacology* 37:501–11

Chichilnisky EJ. 2001. A simple white noise analysis of neuronal light responses. *Network* 12:199–213

Christianson GB, Sahani M, Linden JF. 2008. The consequences of response nonlinearities for interpretation of spectrotemporal receptive fields. *J. Neurosci.* 28:446–55

Clemens J, Wohlgemuth S, Ronacher B. 2012. Nonlinear computations underlying temporal and population sparseness in the auditory system of the grasshopper. *J. Neurosci.* 32:10053–62

Cover TM, Thomas JA. 1991. *Elements of Information Theory.* New York: Wiley-Interscience

de Boer E, Kuyper P. 1968. Triggered correlation. *IEEE Trans. Biomed. Eng.* 15:169–79

de Ruyter van Steveninck RR, Bialek W. 1988. Real-time performance of a movement-sensitive neuron in the blowfly visual system: coding and information transfer in short spike sequences. *Proc. R. Soc. Lond. Ser. B* 234:379–414

Desimone R, Schein SJ. 1987. Visual properties of neurons in area V4 of the macaque: sensitivity to stimulus form. *J. Neurophysiol.* 57:835–68

Eickenberg M, Rowekamp RJ, Kouh M, Sharpee TO. 2012. Characterizing responses of translation-invariant neurons: maximally informative invariant dimensions. *Neural Comput.* 24:2384–421

Fairhall AL, Burlingame CA, Narasimhan R, Harris RA, Puchalla JL, Berry MJ 2nd. 2006. Selectivity for multiple stimulus features in retinal ganglion cells. *J. Neurophysiol.* 96:2724–38

Felsen G, Dan Y. 2005. A natural approach to studying vision. *Nat. Neurosci.* 8:1643–46

Field DJ. 1987. Relations between the statistics of natural images and the response properties of cortical cells. *J. Opt. Soc. Am. A* 4:2379–94

Fitzgerald JD, Rowekamp RJ, Sincich LC, Sharpee TO. 2011a. Second order dimensionality reduction using minimum and maximum mutual information models. *PLoS Comput. Biol.* 7:e1002249

Fitzgerald JD, Sincich LC, Sharpee TO. 2011b. Minimal models of multidimensional computations. *PLoS Comput. Biol.* 7:e1001111

Gilbert CD, Wiesel TN. 1990. The influence of contextual stimuli on the orientation selectivity of cells in primary visual cortex of the cat. *Vis. Res.* 30:1689–701

Globerson A, Stark E, Vaadia E, Tishby N. 2009. The minimum information principle and its application to neural code analysis. *Proc. Natl. Acad. Sci. USA* 106:3490–95

Granot-Atedge E, Tkacik G, Segev R, Schneidman E. 2012. Stimulus-dependent maximum entropy models of neural population codes. *PLoS Comput. Biol.* 9:e1002922

Hartline HK. 1938. The response of single optic nerve fibers of the vertebrate eye to illumination of the retina. *Am. J. Physiol.* 121:400–15

Hubel DH, Wiesel TN. 1968. Receptive fields and functional architecture of monkey striate cortex. *J. Physiol.* 195:215–43

Hudspeth AJ, Corey DP. 1977. Sensitivity, polarity, and conductance change in the response of vertebrate hair cells to controlled mechanical stimuli. *Proc. Natl. Acad. Sci. USA* 74:2407–11

Huys QJ, Paninski L. 2009. Smoothing of, and parameter estimation from, noisy biophysical recordings. *PLoS Comput. Biol.* 5:e1000379

Jaynes ET. 1957. Information theory and statistical mechanics. *Phys. Rev.* 106:620–30

Jaynes ET. 2003. *Probability Theory: The Logic of Science*. Cambridge, UK: Cambridge Univ. Press

Kaardal J, Fitzgerald JD, Berry MJ II, Sharpee TO. 2013. Identifying functional bases for multidimensional neural computations. *Neural Comput.* In press

Kouh M, Sharpee TO. 2009. Estimating linear-nonlinear models using Renyi divergences. *Network* 20:49–68

Kuffler SW. 1953. Discharge patterns and functional organization of mammalian retina. *J. Neurophysiol.* 16:37–68

Kunsberg B, Zucker SW. 2012. Shape-from-shading and cortical computation: a new formulation. *J. Vis.* 12:233

Lafferty J, McCallum A, Pereira FCN. 2001. Conditional random fields: probabilistic models for segmenting and labeling sequence data. *Proc. Int. Conf. Mach. Learn., 18th, Williamstown*, June 28–July 1, pp. 282–89. San Francisco, CA: Morgan Kaufmann

Lazar AA, Slutskiy YB. 2012. Channel identification machines. *Comput. Intell. Neurosci.* 2012: Art. 209590

Lewi J, Butera R, Paninski L. 2009. Sequential optimal design of neurophysiology experiments. *Neural Comput.* 21:619–87

Li A, Zaidi Q. 2004. Three-dimensional shape from non-homogeneous textures: carved and stretched surfaces. *J. Vis.* 4:860–78

Marmarelis VZ, Marmarelis PZ. 1978. *Analysis of Physiological Systems. The White Noise Approach.* Chacon: Lavoisier

Marr D. 1982. *Vision: A Computational Investigation into the Human Representation and Processing of Visual Information*. New York: WH Freeman

McManus JN, Li W, Gilbert CD. 2011. Adaptive shape processing in primary visual cortex. *Proc. Natl. Acad. Sci. USA* 108:9739–46

Meister M, Berry MJ. 1999. The neural code of the retina. *Neuron* 22:435–50

Nothdurft HC, Gallant JL, Van Essen DC. 1999. Response modulation by texture surround in primate area V1: correlates of "popout" under anesthesia. *Vis. Neurosci.* 16:15–34

Paninski L. 2003. Convergence properties of three spike-triggered analysis techniques. *Network* 14:437–64

Paninski L, Pillow J, Lewi J. 2007. Statistical models for neural encoding, decoding, and optimal stimulus design. *Prog. Brain Res.* 165:493–507

Park IMM, Pillow JW. 2011a. Bayesian spike-triggered covariance analysis. *Adv. Neural Inf. Process. Syst.* 24:1692–700

Park M, Pillow JW. 2011b. Receptive field inference with localized priors. *PLoS Comput. Biol.* 7:e1002219

Pasupathy A, Connor CE. 1999. Responses to contour features in macaque area V4. *J. Neurophysiol.* 82:2490–502
Press WH, Teukolsky SA, Vetterling WT, Flannery BP. 1992. *Numerical Recipes in C: The Art of Scientific Computing*. Cambridge, UK: Cambridge Univ. Press
Priebe NJ, Ferster D. 2006. Mechanisms underlying cross-orientation suppression in cat visual cortex. *Nat. Neurosci.* 9:552–61
Rajan K, Bialek W. 2012. *Maximally informative "stimulus energies" in the analysis of neural responses to natural signals*. **http://arxiv.org/abs/1201.0321**
Rapela J, Felsen G, Touryan J, Mendel JM, Grzywacz NM. 2010. ePPR: a new strategy for the characterization of sensory cells from input/output data. *Network* 21:35–90
Rapela J, Mendel JM, Grzywacz NM. 2006. Estimating nonlinear receptive fields from natural images. *J. Vis.* 6:441–74
Rieke F, Warland D, de Ruyter van Steveninck R, Bialek WB. 1997. *Spikes: Exploring the Neural Code*. Cambridge, MA: MIT Press
Ringach DL, Hawken MJ, Shapley R. 2002. Receptive field structure of neurons in monkey primary visual cortex revealed by stimulation with natural image sequences. *J. Vis.* 2:12–24
Rowekamp RJ, Sharpee TO. 2011. Analyzing multicomponent receptive fields from neural responses to natural stimuli. *Network* 22:1–29
Ruderman DL. 1997. Origins of scaling in natural images. *Vis. Res.* 37:3385–98
Ruderman DL, Bialek W. 1994. Statistics of natural images: scaling in the woods. *Phys. Rev. Lett.* 73:814–7
Rust NC, Movshon JA. 2005. In praise of artifice. *Nat. Neurosci.* 8:1647–50
Rust NC, Schwartz O, Movshon JA, Simoncelli EP. 2005. Spatiotemporal elements of macaque V1 receptive fields. *Neuron* 46:945–56
Sahani M, Linden JF. 2003. Evidence optimization techniques for estimating stimulus-response functions. In *Advances in Neural Information Processing Systems 15*, ed. S Becker, S Thrun, K Obermayer, pp. 301–8. Cambridge, MA: MIT Press
Samengo I, Gollisch T. 2013. Spike-triggered covariance: geometric proof, symmetry properties, and extension beyond Gaussian stimuli. *J. Comput. Neurosci.* 34:137–61
Schneidman E, Berry MJ 2nd, Segev R, Bialek W. 2006. Weak pairwise correlations imply strongly correlated network states in a neural population. *Nature* 440:1007–12
Schwartz O, Hsu A, Dayan P. 2007. Space and time in visual context. *Nat. Rev. Neurosci.* 8:522–35
Schwartz O, Pillow JW, Rust NC, Simoncelli EP. 2006. Spike-triggered neural characterization. *J. Vis.* 6:484–507
Series P, Lorenceau J, Fregnac Y. 2003. The "silent" surround of V1 receptive fields: theory and experiments. *J. Physiol. Paris* 97:453–74
Sharpee T, Rust NC, Bialek W. 2004. Analyzing neural responses to natural signals: maximally informative dimensions. *Neural Comput.* 16:223–50
Sharpee TO. 2007. Comparison of information and variance maximization strategies for characterizing neural feature selectivity. *Stat. Med.* 26:4009–31
Sharpee TO, Miller KD, Stryker MP. 2008. On the importance of static nonlinearity in estimating spatiotemporal neural filters with natural stimuli. *J. Neurophysiol.* 99:2496–509
Sharpee TO, Nagel KI, Doupe AJ. 2011. Two-dimensional adaptation in the auditory forebrain. *J. Neurophysiol.* 106:1841–61
Sharpee TO, Victor JD. 2008. Contextual modulation of V1 receptive fields depends on their spatial symmetry. *J. Comput. Neurosci.* 26:203–18
Sherrington CS. 1906. *The Integrative Action Of The Nervous System*. New York: Schribner
Shlens J, Field GD, Gauthier JL, Grivich MI, Petrusca D, et al. 2006. The structure of multi-neuron firing patterns in primate retina. *J. Neurosci.* 26:8254–66
Sillito AM, Jones HE. 1996. Context-dependent interactions and visual processing in V1. *J. Physiol. Paris* 90:205–9
Simoncelli EP. 2003. Vision and the statistics of the visual environment. *Curr. Opin. Neurobiol.* 13:144–49
Simoncelli EP, Olshausen BA. 2001. Natural image statistics and neural representation. *Annu. Rev. Neurosci.* 24:1193–216

Singh NC, Theunissen FE. 2003. Modulation spectra of natural sounds and ethological theories of auditory processing. *J. Acoust. Soc. Am.* 114:3394–411

Strong SP, de Ruyter van Steveninck RR, Bialek W, Koberle R. 1998. On the application of information theory to neural spike trains. *Pac. Symp. Biocomput.* 1998:621–32

Theunissen FE, David SV, Singh NC, Hsu A, Vinje WE, Gallant JL. 2001. Estimating spatio-temporal receptive fields of auditory and visual neurons from their responses to natural stimuli. *Network* 12:289–316

Theunissen FE, Sen K, Doupe AJ. 2000. Spectral-temporal receptive fields of nonlinear auditory neurons obtained using natural sounds. *J. Neurosci.* 20:2315–31

Troyer TW, Krukowski AE, Priebe NJ, Miller KD. 1998. Contrast-invariant orientation tuning in cat visual cortex: feedforward tuning and correlation-based intracortical connectivity. *J. Neurosci.* 18:5908–27

Tyson JJ, Chen KC, Novak B. 2003. Sniffers, buzzers, toggles and blinkers: dynamics of regulatory and signaling pathways in the cell. *Curr. Opin. Cell Biol.* 15:221–31

van Hateren JH, van der Schaaf A. 1998. Independent component filters of natural images compared with simple cells in primary visual cortex. *Proc. R. Soc. Lond. Ser. B* 265:359–66

Vickers NJ, Christensen TA, Baker TC, Hildebrand JG. 2001. Odour-plume dynamics influence the brain's olfactory code. *Nature* 410:466–70

Victor JD, Knight BW. 2003. Simultaneously band and space limited functions in two dimensions, and receptive fields of visual neurons. In *Perspectives and Problems in Nolinear Science*, ed. E Kaplan, J Marsden, KR Sreenivasan, pp. 375–420. New York: Springer-Verlag. Springer Appl. Math. Sci. Ser.

Victor JD, Mechler F, Repucci MA, Purpura KP, Sharpee TO. 2006. Responses of V1 neurons to two-dimensional Hermite functions. *J. Neurophysiol.* 95:379–400

Victor JD, Shapley R. 1980. A method of nonlinear analysis in the frequency domain. *Biophys. J.* 29:459–83

Victor JD, Purpura KP. 1998. Spatial phase and the temporal structure of the response to gratings in V1. *J. Neurophysiol.* 80:554–71

Victor JD, Shapley RM. 1979. The nonlinear pathway of Y ganglion cells in the cat retina. *J. Gen. Physiol.* 74:671–89

Vintch B, Zaharia A, Movshon JA, Simoncelli EP. 2012. Efficient and direct estimation of a neural subunit model for sensory coding. In *Advances in Neural Information Processing Systems 25*, ed. P Bartlett, FCN Pereira, CJC Burges, L Bottou, KQ Weinberger. NIPS. La Jolla, CA: NIPS Found. **http://books.nips.cc/papers/files/nips25/NIPS2012_1432.pdf**

Weisberg S, Welsh AH. 1994. Adapting for the missing link. *Ann. Stat.* 22:1674–700

Zipser K, Lamme VA, Schiller PH. 1996. Contextual modulation in primary visual cortex. *J. Neurosci.* 16:7376–89

The Evolution of *Drosophila melanogaster* as a Model for Alcohol Research

Anita V. Devineni[1,2] and Ulrike Heberlein[1,3]

[1]Program in Neuroscience and Department of Anatomy, University of California, San Francisco, California 94158;

[2]Current address: Department of Neuroscience, Columbia University, New York, NY 10032; email: ad3030@columbia.edu

[3]Howard Hughes Medical Center, Janelia Farm Research Campus, Ashburn, Virginia 20147; email: heberleinu@janelia.hhmi.org

Annu. Rev. Neurosci. 2013. 36:121–38

First published online as a Review in Advance on April 29, 2013

The *Annual Review of Neuroscience* is online at neuro.annualreviews.org

This article's doi:
10.1146/annurev-neuro-062012-170256

Keywords

ethanol, addiction, fruit flies

Abstract

Animal models have been widely used to gain insight into the mechanisms underlying the acute and long-term effects of alcohol exposure. The fruit fly *Drosophila melanogaster* encounters ethanol in its natural habitat and possesses many adaptations that allow it to survive and thrive in ethanol-rich environments. Several assays to study ethanol-related behaviors in flies, ranging from acute intoxication to self-administration and reward, have been developed in the past 20 years. These assays have provided the basis for studying the physiological and behavioral effects of ethanol and for identifying genes mediating these effects. In this review we describe the ecological relationship between flies and ethanol, the effects of ethanol on fly development and behavior, the use of flies as a model for alcohol addiction, and the interaction between ethanol and social behavior. We discuss these advances in the context of their utility to help decipher the mechanisms underlying the diverse effects of ethanol, including those that mediate ethanol dependence and addiction in humans.

Contents

INTRODUCTION

Alcohol is one of the most widely used and abused drugs in the world. Acute ethanol exposure produces dramatic short-term changes in behavior. These changes include both stimulant effects, such as disinhibition and euphoria, which occur at lower ethanol doses, and depressant effects, such as fatigue, cognitive and motor impairment, and loss of consciousness, which occur at higher ethanol doses (Babor et al. 1983, Pohorecky 1977). In addition to these short-term effects, long-term consumption of ethanol can lead to alcohol use disorders (AUDs), which include alcohol abuse and dependence. Alcohol abuse is characterized by continued alcohol consumption despite physically hazardous situations, legal or social problems, and a failure to fulfill obligations (APA 1994). Alcohol dependence, a more severe condition, is characterized by physical symptoms, such as tolerance and withdrawal, as well as signs of uncontrolled use, such as unsuccessful attempts to curtail alcohol consumption (APA 1994). The estimated prevalence of AUDs is 8.5% in the United States, thus affecting more than 17 million Americans (Grant et al. 2004).

Several studies have suggested a relationship between acute sensitivity to ethanol and propensity for developing an AUD. Specifically, risk of AUD has been associated with increased sensitivity to the stimulant effects of ethanol and decreased sensitivity to its depressant effects (Morean & Corbin 2010, Newlin & Thomson 1990, Schuckit 1994). Thus, studying the mechanisms underlying the short-term effects of ethanol, which are generally easier to measure, may also provide insight into the process of alcohol addiction. Genetic factors influence both acute ethanol sensitivity and susceptibility to AUDs; familial, twin, and adoption studies have estimated that the genetic contribution to AUDs is ~40–60% (Prescott & Kendler 1999, Schuckit et al. 2001). Despite these findings, investigators have identified very few genes that influence susceptibility to AUDs. Gene-expression profiling and linkage studies have identified many candidate genes (Mayfield et al. 2008), but much work remains to be done to link these genes conclusively to AUDs and to develop effective therapeutic strategies.

The difficulty and cost of human studies have led to the development of animal models to investigate the genetic, molecular, and neural mechanisms underlying both the short- and long-term effects of ethanol. Rodent models have been most widely used and have provided important insights into these mechanisms (Crabbe et al. 2006). More recently, the fruit fly *Drosophila melanogaster* has been developed as a model to study the effects of ethanol. Unless noted otherwise, the terms *Drosophila*, flies, or fruit flies are used here to refer to the species *D. melanogaster*. The experimental and genetic accessibility of *Drosophila* makes it an ideal organism to study molecular, cellular, developmental, and neurobiological processes, most of which are conserved between flies and mammals. Flies are easy and inexpensive to maintain and breed, they have a rapid life cycle (~10 days at 25°C), and their small size allows for hundreds or thousands of genotypes to be maintained in a typical laboratory. More importantly, useful genetic tools are being developed continuously in *Drosophila*. These tools include publicly available collections of mutants and RNA interference (RNAi) lines to inactivate nearly any fly gene of interest, lines to target specific tissues or cell types, and transgenes to manipulate neuronal activity (Kaun et al. 2012, Venken et al. 2011).

In this review, we describe the use of *D. melanogaster* to study the physiological and behavioral effects of ethanol, and we examine these effects in the context of human conditions associated with excessive alcohol use. Specifically, we discuss (*a*) the ecological relationship between flies and ethanol, (*b*) the effects of acute and developmental ethanol exposure on fly behavior, (*c*) ethanol preference behaviors that may model aspects of alcohol addiction, and (*d*) the relationship between ethanol responses and social behavior. Because recent reviews have comprehensively described the molecular and neural mechanisms that mediate ethanol-related behaviors in flies, we do not examine these subjects here (Kaun et al. 2012, Rodan & Rothenfluh 2010).

ETHANOL IN THE NATURAL HABITAT OF *DROSOPHILA*

Adaptations of *D. Melanogaster* to Ethanol-Containing Environments

The relationship between *D. melanogaster* and ethanol was initially investigated from an ecological and evolutionary perspective. Fruit flies encounter ethanol in their natural environment: They inhabit and breed in ripe fruits and other plant materials that can contain up to 5% ethanol, produced by fermentation of sugars by various yeasts (Gibson et al. 1981). A long evolutionary history of ethanol exposure has allowed flies to expand to new ecological niches. For example, they frequently reside in wineries, in which wine seepages can contain up to 11% ethanol (Gibson et al. 1981).

ADH: alcohol dehydrogenase

ALDH: acetaldehyde dehydrogenase

Flies efficiently metabolize ethanol and can use it as an energy source or a substrate for lipid synthesis (Daly & Clarke 1981, Geer et al. 1985). Ethanol is first converted to acetaldehyde by the enzyme alcohol dehydrogenase (ADH), and acetaldehyde is metabolized to acetate by acetaldehyde dehydrogenase (ALDH). Both ADH and ALDH functions are essential to promote resistance to ethanol toxicity (David et al. 1976, Fry & Saweikis 2006). Natural polymorphisms in ADH have been well studied. The two major alleles that exist in natural populations are Adh^F and Adh^S. ADH^F shows higher enzymatic activity than does ADH^S, and Adh^F flies are more resistant to ethanol toxicity than are Adh^S flies (Briscoe et al. 1975, Cavener 1979, van Delden et al. 1978). Some studies have reported that Adh^F flies also show greater larval preference and adult oviposition (egg-laying) preference for ethanol compared with Adh^S flies, but other studies have disputed this relationship (Cavener 1979, Gelfand & McDonald 1983, Siegal & Hartl 1999, van Delden & Kamping 1990).

Drosophila ADH is a short-chain dehydrogenase, and its structure and enzymatic mechanism differ from those of vertebrate ADH, a medium-chain dehydrogenase (Hernández-Tobías et al. 2011, Jornvall et al. 1981). Some researchers have proposed that *Drosophila* ADH evolved as an adaptive response to detoxify environmental ethanol and colonize ethanol-containing environments (Ashburner 1998, Hernández-Tobías et al. 2011). Several findings support this hypothesis: (*a*) Flies carrying short-chain ADH evolved shortly after ethanol became available in fermenting fruits ~130–140 mya (Hernández-Tobías et al. 2011), (*b*) larval ADH is upregulated by the presence of ethanol in food (McKechnie & Geer 1984), and (*c*) ADH is critical for survival of flies on ethanol-containing food (see above). Additionally, flies are much more resistant to ethanol toxicity compared with toxicity induced by other primary alcohols, and this selective resistance is due to ADH (David et al. 1976). Thus, flies likely evolved greater resistance to ethanol to inhabit ethanol-containing environments, allowing flies to utilize ethanol as a food source as well as to consume other nutrients available in fermenting fruits.

In addition to providing nutritional benefits, a recent study suggests that ethanol-containing environments may also be favorable for *Drosophila* to fight infection by parasitic wasps, which lay eggs inside *Drosophila* larvae (Milan et al. 2012). In this study, environmental ethanol reduced wasp infection of *Drosophila* larvae, and ethanol consumption by infected larvae also reduced the survival of the wasp progeny. Moreover, infected larvae showed a stronger preference for ethanol-containing food compared with uninfected larvae, suggesting that flies in the wild may actively seek out ethanol-containing substrates not only for nutritional purposes, but also to combat parasitic infection.

Ethanol as a Food Source

Early studies showed that flies can use ethanol as a source of calories because providing ethanol as the sole caloric source increased survival relative to complete starvation (Van Herrewege & David 1980). Ethanol is converted to acetate by ADH and ALDH and is then converted to acetyl-CoA, which can enter the tricarboxylic acid cycle. However, some studies have suggested that the calories present in ethanol are poorly utilized by flies. For example, although providing ethanol as the sole caloric source prolonged survival compared with complete starvation, it was much less effective to prolong survival than was an isocaloric concentration of sucrose (Xu et al. 2012). Although this result may be due to differences in consumption, the large disparity in survival times suggests that flies fail to utilize some or most of the calories present in ethanol. In support of this hypothesis, flies modulate their consumption of different sucrose concentrations to maintain their caloric intake at a remarkably constant level; they do not, however, modulate their consumption of different ethanol concentrations, suggesting that these solutions provide few calories (Xu et al. 2012). A previous study also observed only a modest compensatory feeding reduction of ethanol-containing food, leading to the proposal that calories in ethanol may not be utilized effectively by flies (Ja et al. 2007).

Both the detoxification of ethanol and its utilization as a caloric source require metabolism by ADH and ALDH, suggesting that the efficiency of these two processes may be correlated. Ethanol detoxification can be quantified as the survival time upon exposure to high, toxic doses of ethanol; in contrast, caloric utilization has been measured as the increase in survival time when low ethanol concentrations (up to 5%) are provided as the sole caloric source (as compared with completely starved flies). As expected, Adh^F flies show greater utilization of the calories in ethanol compared with Adh^S flies, indicating that both ethanol detoxification and utilization are regulated by the rate of ethanol conversion to acetaldehyde (Daly & Clarke 1981). However, selective breeding studies showed that although raising fly strains on ethanol for many generations increased their resistance to ethanol toxicity, these selected lines did not always show improved caloric utilization of ethanol (Van Herrewege & David 1980). This dissociation is supported by a study that surveyed a variety of alcohols and found that detoxification and caloric utilization of these alcohols were not well correlated (Karan et al. 1999). Thus, although the detoxification and caloric utilization of ethanol both represent important adaptations that may have allowed *Drosophila* to survive in ethanol-containing environments, these processes are dissociable and may reflect partially distinct biological pathways.

Species and Population Differences

The physiological and behavioral effects of ethanol have been compared among various

Drosophila species as well as among different *D. melanogaster* populations. Overall, these studies support the view that flies inhabiting ethanol-containing environments are more resistant to ethanol toxicity and may show greater preference for ethanol.

Several studies have compared *D. melanogaster* to *D. simulans*, a closely related species that appears to lack the adaptations to ethanol that *D. melanogaster* has evolved. *D. melanogaster* is found more frequently in ethanol-containing environments, such as wine cellars, compared with *D. simulans* (Gibson & Wilks 1988, McKenzie & McKechnie 1979, McKenzie & Parsons 1972). *D. melanogaster* also shows greater resistance to ethanol toxicity than does *D. simulans*, which is likely due to the higher ADH activity of *D. melanogaster* (David et al. 1976, McKenzie & Parsons 1972, Pipkin & Hewitt 1972). Some studies suggest that *D. melanogaster* shows greater larval preference and adult oviposition preference for ethanol compared with *D. simulans*, although conflicting results have been reported (Gelfand & McDonald 1983, McKenzie & Parsons 1972, Parsons & King 1977, Richmond & Gerking 1979).

Several studies have reported population differences within *D. melanogaster* associated with latitude as well as microenvironment. For example, resistance to ethanol toxicity in both adults and larvae tends to increase with increasing latitude (David et al. 1986, David & Bocquet 1975). Larval ethanol preference similarly increases further from the equator (Parsons 1980). These latitudinal clines correlate with a gradient of Adh^F versus Adh^S distribution: Adh^F is more frequent at higher latitudes, which generally correspond with lower temperatures (David et al. 1986, Oakeshott et al. 1982, Parkash et al. 1999). Researchers have sought to explain the correlation between the allelic and phenotypic distributions. One study found that ADH^S is more heat-resistant than ADH^F, suggesting that in warmer climates ADH^S may be superior to ADH^F in detoxifying ethanol (Sampsell & Sims 1982). However, others have failed to observe this biochemical difference and instead found that the activity of both ADH^S and ADH^F was similarly increased at higher temperatures (McElfresh & McDonald 1986). These authors proposed an alternative explanation for the genotypic and phenotypic correlation: At high temperatures, increased activity of the already more active ADH^F isoform may actually be detrimental to ethanol detoxification because the rapid conversion of ethanol to acetaldehyde, a toxic by-product, may surpass the rate at which acetaldehyde can be detoxified by ALDH. Corroborating the importance of acetaldehyde metabolism, Fry et al. (2008) showed that latitude is correlated with a polymorphism in ALDH that affects enzymatic activity. Thus, polymorphisms in ADH and ALDH likely provide a molecular explanation for the latitudinal variation in ethanol toxicity, although the evolutionary reason for this gradient is still unknown.

In addition to latitude, the microenvironment of flies strongly influences ethanol resistance. Many studies have compared populations of *D. melanogaster* found within high-ethanol-containing environments, such as wine cellars, to populations found elsewhere. The populations collected from ethanol-containing environments show greater resistance to ethanol toxicity, but the genetic factors mediating this resistance are unknown (Gibson & Wilks 1988, McKenzie & McKechnie 1978, McKenzie & Parsons 1974, Vouidibio et al. 1989). Some studies have observed a higher frequency of Adh^F in these ethanol-adapted populations (Briscoe et al. 1975, Vouidibio et al. 1989), but most studies found no difference in Adh^F or Adh^S allele frequencies (Gibson & Wilks 1988, McKenzie & McKechnie 1978, McKenzie & Parsons 1974).

Laboratory selection studies have been conducted to mimic natural selection that may have occurred as flies encountered ethanol-containing environments. In accordance with observations from natural populations, raising fly strains for many generations on ethanol-containing food typically increases their resistance to ethanol toxicity (Barbancho et al. 1987, Kerver et al. 1992, Van Herrewege

& David 1980). One study additionally observed increased oviposition preference for ethanol in the ethanol-selected lines (van Delden & Kamping 1990). In some studies, the ethanol-selected lines showed an increase in *Adh*^{F} frequency (Cavener & Clegg 1981, van Delden et al. 1978), but other studies using lines homozygous for *Adh*^{S} or *Adh*^{F} saw no effect of selection on ADH activity (Barbancho et al. 1987, Kerver et al. 1992). Finally, a behavioral selection study selecting for greater resistance to the motor-impairing effect of ethanol also generated lines that were more resistant to ethanol toxicity, which suggests that ethanol-induced toxicity and motor impairment are mediated by overlapping sets of genes (Cohan & Hoffmann 1986).

Thus, *Drosophila*'s colonization of ethanol-containing environments is linked to the evolution of greater resistance to ethanol toxicity, allowing flies to derive multiple benefits from ethanol ingestion. Some investigators have hypothesized that the evolutionary origins of human alcoholism derive from a strong historical association between primate frugivory and ethanol consumption. Specifically, if ethanol consumption during frugivory was advantageous for our ancestors, then alcoholism could be viewed as a maladaptive consequence of modern environments in which alcohol is readily available (Dudley 2000, 2002). Given this potential shared history of ethanol consumption in flies and humans, it is interesting that many of the effects observed in flies upon acute and chronic ethanol exposure in the laboratory are remarkably similar to those observed in vertebrates, including humans. The remainder of the review focuses on these similarities and the efforts to develop *Drosophila* as a model for alcohol addiction.

DEVELOPMENTAL ETHANOL EXPOSURE

Exposure to ethanol during developmental stages of the *Drosophila* life cycle leads to profound effects on development and adult behavior. Early studies showed that rearing larvae on food containing 4.5% ethanol delays development and decreases survival compared with larvae reared on control food (Geer et al. 1991, Ranganathan et al. 1987). However, the surviving adults were heavier than those grown on regular food (Geer et al. 1991), which may be due to less crowded growth conditions or utilization of ethanol as a food source. A more recent study, in which flies were exposed to at least 5% ethanol throughout embryonic, larval, and pupal development, also reported delayed development and decreased survival (McClure et al. 2011). The developmental delay was due to ethanol exposure during the larval stages (a period of extensive growth), whereas the decreased survival was due to exposure during both the larval stages as well as metamorphosis, a period of complex structural and neural remodeling. In this study, flies reared continuously on food containing 5% ethanol (or higher) were smaller, showed decreased adult weight, and had reduced larval brain size owing to decreased cell proliferation (McClure et al. 2011). The conflicting findings regarding the effect of developmental ethanol on adult weight may be explained by differences in the ethanol exposure protocol: Both Geer et al. (1991) and McClure et al. (2011) fed ethanol-containing food to larvae, but only the latter study additionally maintained the animals in an environment containing ethanol vapor, thereby avoiding ethanol evaporation from the food (which otherwise occurs within a few days) and ensuring ethanol exposure during metamorphosis, when flies do not feed.

How do we reconcile the negative consequences of developmental ethanol exposure in the laboratory with the fact that, in nature, flies frequently reside in ethanol-containing environments? First, the ethanol concentrations used in laboratory experiments may be higher than those most commonly occurring in nature (Gibson et al. 1981). Second, in their natural environment larvae can move away from highly concentrated ethanol sources and may thus experience only intermittent ethanol exposure.

Developmental ethanol exposure also affects behavioral responses to ethanol in adult flies.

In adult flies, acute exposure to ethanol vapor causes locomotor hyperactivity (observed at low to moderate doses) and sedation (observed at higher doses; see below for detailed description of the assays). Flies reared on 5% ethanol show greater ethanol-induced locomotor hyperactivity but reduced sensitivity to ethanol sedation compared with unexposed control flies, even though ethanol-reared flies showed normal ethanol pharmacokinetics (McClure et al. 2011). The decrease in sedation sensitivity may represent a long-lasting form of tolerance (acquired resistance to ethanol caused by prior exposure; see below). In support of this model, McClure et al. (2011) showed that when ethanol-reared flies were tested for their ability to develop further tolerance after exposure to ethanol vapor during adulthood, they showed reduced tolerance compared with control flies that had been reared on normal food. These data are consistent with the idea that developmental ethanol exposure induces a form of tolerance and that the mechanisms underlying developmental ethanol tolerance and adult tolerance likely overlap.

The phenotypes of ethanol-reared flies, such as developmental delay and reduced size, are similar to phenotypes observed upon impairing insulin/insulin-like growth factor (IGF) signaling in flies. A potential connection between *Drosophila* insulin-like peptides (dILPs) and developmental ethanol exposure was therefore examined. Indeed, developmental ethanol exposure leads to reduced expression of both dILPs and the insulin/IGF receptor (InR), and mutants with reduced expression of InR are more strongly affected by developmental ethanol exposure than are wild-type flies. Moreover, the behavioral abnormalities observed in ethanol-reared flies can be reversed by increasing expression of one dILP (dILP2) in the brain (McClure et al. 2011).

Prenatal alcohol exposure in humans can cause fetal alcohol syndrome (FAS), a complex disorder characterized by multiple developmental, morphological, and neurological deficits (Jones & Smith 1973). The fly model recapitulates many FAS-related phenotypes observed in humans and rodent models, including persistent growth deficits, deficient brain growth, and altered responses to ethanol as adults (Becker et al. 1993, 1996). Perhaps most intriguing is the evolutionarily conserved effect of insulin/IGF: Developmental ethanol exposure inhibits the expression of insulin and IGF receptors in the rodent brain, and IGFs reduce ethanol toxicity in cultured neurons as well as motor coordination defects caused by developmental ethanol exposure in rats (Barclay et al. 2005, de la Monte et al. 2005, McGough et al. 2009). Thus, an unbiased genetic and molecular analysis of developmental ethanol effects in *Drosophila* should shed light on the mechanisms underlying the teratogenic effects of developmental ethanol exposure in mammals, including humans.

BEHAVIORAL EFFECTS OF ETHANOL EXPOSURE

Intoxication

Behavioral studies of acute ethanol intoxication in *Drosophila* began in the late 1980s. The inebriometer, a vertical column containing mesh baffles through which ethanol vapor is circulated, was the first apparatus used to measure ethanol-induced motor impairment, primarily in the context of selective breeding experiments (Cohan & Graf 1985, Weber 1988). Flies naturally exhibit negative geotaxis and remain at the top of the column, but ethanol (which they inhale) causes them to lose postural control and gradually fall from one baffle to the next. Ethanol sensitivity can therefore be quantified as the time required for flies to reach the bottom of the column. Another early assay for ethanol sensitivity was the line-crossing assay, in which flies were exposed to ethanol vapor in a chamber containing a grid, and locomotor activity was quantified by recording the number and pattern of gridlines crossed as a function of time (Singh & Heberlein 2000). This assay revealed that in contrast with the motor impairment induced by high doses of ethanol, lower concentrations of ethanol stimulate locomotor activity and cause flies to turn more frequently. Upon flies' continued ethanol exposure, their locomotor activity decreases as flies show motor incoordination and eventually sedation. Thus, even these early studies indicated that flies exhibit a biphasic response to acute ethanol exposure, similar to that in humans and other mammals (Pohorecky 1977): At lower concentrations, ethanol acts as a stimulant, inducing greater locomotor activity, whereas at higher concentrations, ethanol acts as a depressant, causing motor impairment and sedation.

Investigators have subsequently developed more sophisticated automated systems to measure the effects of ethanol on locomotor activity (Parr et al. 2001, Wolf et al. 2002). For example, one method relies on filming the flies during ethanol exposure and using image-processing software to calculate their locomotor velocity (Wolf et al. 2002). This approach enabled precise quantification of the kinetics of ethanol-induced hyperactivity and revealed that ethanol increases the length, but not the frequency, of activity bouts to produce an overall increase in activity (Wolf et al. 2002). Several groups of flies can be tested and analyzed simultaneously using these automated methods, thereby making it possible to conduct forward genetic screens to identify molecules that regulate ethanol responses (Kaun et al. 2012, Rodan & Rothenfluh 2010).

New assays to quantify the sedative phase of ethanol intoxication have also been developed. The ethanol rapid iterative negative geotaxis (eRING) assay measures the loss of negative geotaxis during ethanol exposure. In this assay, the experimenter knocks the flies to the bottom of a vial and then measures the vertical distance that they climb (Bhandari et al. 2009). The loss-of-righting reflex (LORR) assay measures ethanol-induced sedation (Corl et al. 2009, Rothenfluh et al. 2006). To quantify LORR, the experimenter knocks over the flies by mechanically agitating the horizontally positioned vials and then counts the fraction of flies that are unable to regain upright posture. Flies' recovery from ethanol sedation has also been used as a measure of ethanol sensitivity (Berger et al. 2004). The eRING, LORR, and sedation recovery assays all have the advantage that they generally require a smaller number of flies per experiment than does the inebriometer. These assays may also be more specific in the behavioral effect that they measure because sensitivity in the inebriometer reflects a combination of several factors, including propensity for negative geotaxis, loss of postural control, and ability to cling to the baffles. Overall, these assays have demonstrated that acute ethanol exposure produces specific behavioral alterations in *Drosophila*, including both stimulant and depressant effects, that bear strong resemblance to those observed in mammals, including humans.

The relationship between the stimulant and depressant effects of ethanol has been analyzed in flies by measuring both types of responses in various sets of mutants. One study observed a small correlation between ethanol-induced hyperactivity and sedation (Kong et al. 2010). However, in this study the two measures were not entirely independent because ethanol hyperactivity was quantified as the total distance traveled during ethanol exposure, including the end of the assay when many flies are sedated. In contrast, a different study quantified peak hyperactivity during exposure to a nonsedating concentration of ethanol and found no significant correlation between ethanol hyperactivity and sedation (Devineni et al. 2011). These results suggest that ethanol-induced hyperactivity and sedation reflect ethanol's distinct effects on the nervous system and may be largely influenced by different genes.

Acute ethanol responses in larvae have also been studied, although not as extensively as in adults. Low doses of ethanol (producing an internal ethanol concentration of ~7 mM) impair associative olfactory learning in larvae without affecting general olfactory or locomotor behavior (Robinson et al. 2012b). This effect may model the ethanol-induced impairment observed in humans and rodents on various short-term memory tasks (Acheson et al. 1998, Gibson 1985, Givens & McMahon 1997, Lister et al. 1991). It remains to be investigated whether higher ethanol levels also produce stimulant and/or sedative effects in larvae, as they do in adult flies.

Tolerance, Sensitization, and Withdrawal

Ethanol tolerance is defined as acquired resistance to the behavioral effects of ethanol due to repeated ethanol exposure. Tolerance to the sedative effects of ethanol is generally quantified as the decrease in ethanol sensitivity after ethanol preexposure and can be measured using any of the ethanol sensitivity assays described above (Berger et al. 2004, Bhandari et al. 2009, Scholz et al. 2000). Although tolerance studies often vary in the timing and concentration of ethanol exposure, two broad classes of ethanol tolerance have been characterized in flies, termed rapid and chronic (Berger et al. 2004). Rapid tolerance is induced by relatively brief exposure (~30 min) to a high, sedating concentration of ethanol, whereas chronic tolerance is induced by a longer exposure (~24 h) to a low ethanol concentration that does not produce overt signs of intoxication. These two types of tolerance are mediated by partially distinct molecular mechanisms (Berger et al. 2004), although some genes appear to affect both (Berger et al. 2008). A third type of tolerance present in mammals, acute functional tolerance, develops within a single intoxicating session (Kalant et al. 1971). This form of tolerance has not yet been characterized in flies because it is difficult to distinguish from naive ethanol sensitivity using existing assays.

Most studies examining the effect of multiple ethanol exposures have focused on the sedative effects of ethanol. However, one study tested the effect of repeated ethanol exposure on ethanol hyperactivity. Flies that had been previously exposed to ethanol showed an earlier onset and overall increase in ethanol hyperactivity compared with naive flies (Kong et al. 2010). Thus, flies appear to show sensitization (increased response after preexposure) to the stimulant effect of ethanol, which contrasts with the tolerance that they show to its sedative effects. These findings closely match observations in rodents, which also tend to show sensitization to ethanol's stimulant effects and tolerance to its sedative effects (Crabbe et al. 1994, Phillips et al. 1997).

Ethanol withdrawal, like tolerance and sensitization, reflects long-term neural changes that occur in response to ethanol exposure. In humans, the presence of withdrawal symptoms upon alcohol abstinence, which can include dysphoria, anxiety, cognitive impairment, and seizures, is considered a sign of alcohol dependence (Kosten & O'Connor 2003, Stavro et al. 2012). Withdrawal seizures occur because the depressant effects of ethanol in the central nervous system induce homeostatic adaptations promoting neural excitability, which can lead to seizures when ethanol is removed (Littleton 1998). A recent study described a model for ethanol withdrawal seizures in flies (Ghezzi et al. 2012). In this study, seizure susceptibility was measured in an electrophysiological preparation by electrically stimulating the brain and recording seizure-like activity in a flight muscle. The stimulation threshold for inducing seizures was lowered in flies that were preexposed to a sedating dose of ethanol. The mechanisms underlying withdrawal and tolerance in flies at least partially overlap because both processes rely on the large conductance (BK) calcium-activated potassium channel encoded by the gene *slo* (Cowmeadow et al. 2005, Ghezzi et al. 2012).

In addition to the physiological model of ethanol withdrawal described above, a behavioral model for withdrawal has been recently characterized in *Drosophila* larvae (Robinson et al. 2012a). This study first showed that although acute ethanol exposure impairs associative learning in larvae, larvae that were chronically exposed to ethanol for six days showed normal learning, likely owing to neural adaptations related to tolerance. However, chronically exposed larvae that were removed from ethanol for 6 h showed reduced learning, representing a withdrawal-induced learning deficit, and this effect was reversed upon reexposure to ethanol for 1 h. Ethanol-withdrawn larvae also showed enhanced nervous system excitability, similar to the findings in adult flies: Ethanol-withdrawn larvae were more sensitive than control larvae to seizures induced by picrotoxin, a GABA receptor antagonist that increases neural excitability, and this effect was partially reversed by ethanol reinstatement. Thus, flies have now been established as a model for both physiological and behavioral signs of withdrawal.

In summary, *Drosophila* shows both acute and chronic behavioral responses to ethanol that are similar to those observed in rodents as well as in humans. During acute ethanol exposure, adult flies exhibit locomotor hyperactivity followed by motor incoordination and sedation, whereas larvae show learning deficits. Repeated ethanol exposures can induce either tolerance or sensitization in adult flies, depending on the behavioral response being measured, and chronic ethanol exposure induces alcohol dependence in larvae, which show withdrawal signs when ethanol is removed. All these behaviors are amenable to genetic analyses to help decipher the underlying molecular and neural mechanisms.

ETHANOL SELF-ADMINISTRATION AND REWARD

Ethanol Consumption

As discussed above, initial studies investigated larval positional and adult ovipositional preference for ethanol, but they did not assess whether flies choose to consume ethanol. Assays to measure ethanol consumption (or feeding in general) were fairly limited until recently. One early study measured the duration of proboscis (the feeding organ of the fly) extension on ethanol-containing food as a proxy for ethanol intake (Cadieu et al. 1999) because this action is necessary for food consumption. When given a choice between ethanol-containing food and regular food, naive flies showed a slight preference for the ethanol-containing food, and ethanol preexposure increased this preference.

The development of the capillary feeder (CAFE) assay has allowed for more precise quantification of ethanol consumption (Ja et al. 2007). In this assay, flies consume liquid food from microcapillary tubes placed vertically through the top of a vial, and the volume consumed is quantified by measuring the descent of the liquid column. When flies are offered a choice between food with or without ethanol, they show a robust, dose-dependent preference for food containing 5–25% ethanol, and this preference increases over several days (Devineni & Heberlein 2009). The increase in ethanol preference over time may represent a positional learning effect because altering the position of the ethanol-containing capillary on a daily basis resulted in a constant, rather than increasing, level of preference (Xu et al. 2012). Preference for consuming ethanol is not dependent entirely on chemosensory attraction to ethanol because it persists in flies with olfactory or gustatory defects (Devineni & Heberlein 2009, Xu et al. 2012).

Role of Calories in Ethanol Preference

The question of whether flies consume ethanol for its caloric value or its pharmacological properties is a topic of ongoing debate. Although the calories in ethanol may be poorly utilized by flies, as discussed above, the caloric value of ethanol may nevertheless represent an attractive stimulus. Initial studies suggested that preference for consuming ethanol was not strongly dependent on its caloric value because it was not significantly affected by varying the caloric ratio between the ethanol and nonethanol food (Devineni & Heberlein 2009). In support of this idea, ethanol preference can be dissociated from general food consumption, which likely reflects attraction to calories. For example, (*a*) sexual experience modulates ethanol preference (see below) without affecting general food consumption (Shohat-Ophir et al. 2012), and (*b*) mutation of the gene *rutabaga* decreased both general feeding and ethanol preference; however, restoring *rutabaga* function in adult neurons reinstated normal ethanol preference but not normal overall feeding (Xu et al. 2012). Thus, manipulations that alter attraction to calories do not necessarily affect ethanol preference, and vice versa.

The role of calories in ethanol preference was more directly tested by using a variety of nutritive carbohydrates to equalize the caloric content of the ethanol-containing and nonethanol food (Pohl et al. 2012). Of several carbohydrates tested, three were believed to have neither an attractive nor aversive taste based on the proboscis extension assay. Equalizing the caloric content of food with and without ethanol by using two of these three carbohydrates caused flies to lose their preference for ethanol; in the third case, flies showed relatively normal ethanol preference. These results suggest that the caloric value of ethanol likely represents a significant attractive force driving ethanol consumption in flies but that other factors may also contribute. One caveat of these experiments is that the published caloric values of each compound are estimated for humans, not flies, so the number of calories available to a fly is not precisely known. In fact, if flies cannot utilize all the calories in ethanol, as discussed above, then these caloric-balancing experiments do not actually equalize the number of utilizable calories but instead add excess calories to the nonethanol food. Overall, calories likely contribute to ethanol preference in *Drosophila*, but their relative importance has yet to be conclusively determined.

Role of Hedonic Effects in Ethanol Preference

In addition to the caloric value of ethanol, the hedonic effects of ethanol likely contribute to ethanol consumption and preference. In support of this notion, flies will consume ethanol and develop ethanol preference even when it is laced with the aversive, bitter-tasting compound quinine (Devineni & Heberlein 2009). In addition, flies rapidly return to peak levels of ethanol consumption after a period of ethanol abstinence, suggesting a relapse-like effect (Devineni & Heberlein 2009). These properties are at least outwardly reminiscent of addiction-like behavior, as both humans and rodent models consume ethanol despite its initially aversive taste and other negative consequences, and they rapidly return to high levels of ethanol consumption after a period of abstinence (Morse & Flavin 1992, Rodd et al. 2004, Wolffgramm & Heyne 1991).

Some have argued that because flies contain a low internal ethanol concentration after each drinking bout ($\leq$4 mM), they are not likely to reach intoxication during the ethanol preference assay, and the pharmacological effects of ethanol are therefore unlikely to contribute to ethanol consumption (Pohl et al. 2012). However, ethanol concentration in flies may reach higher levels after a cluster of multiple drinking bouts, which is important because the frequency of drinking bouts increases over time (Devineni & Heberlein 2009). Even if this is not the case, low internal ethanol levels (in the range of 5 mM) may still have behavioral relevance in flies; they can induce locomotor hyperactivity (Kaun et al. 2011), tolerance (Berger et al. 2004), and conditioned preference (see below). Further experiments will be required to determine whether the ethanol consumed during the preference assay induces either immediate or long-term behavioral changes related to its intoxicating or hedonic effects.

Conditioned Ethanol Preference

A recently developed assay for conditioned ethanol preference directly addresses the question of whether flies find intoxicating doses of ethanol to be rewarding or aversive (Kaun et al. 2011). Similar to conditioned place preference in mammals, in this assay flies are exposed to two neutral odor cues, one of which is paired with ethanol vapor exposure. Flies are later offered a choice between the two odors. Preference for the ethanol-associated odor is thought to represent the rewarding effect of ethanol.

At early time points following ethanol exposure (~30 min), flies show aversion to the ethanol-associated cue; however, within

24 h this aversion transforms into long-lasting preference (Kaun et al. 2011). Flies show conditioned preference when exposed to moderate concentrations of ethanol that induce locomotor hyperactivity (Kaun et al. 2011), but not when exposed to lower concentrations that fail to produce behavioral effects or higher concentrations that induce sedation (K. Kaun and U. Heberlein, unpublished data). Flies therefore form a positive association with the stimulant effects of ethanol but not with its sedative effects (although it is possible that sedation impairs flies' ability to remember the association). These findings bear resemblance to humans: The stimulant effects of ethanol are typically perceived as positive, whereas its sedative effects are perceived negatively (Babor et al. 1983). These results are also supported by an earlier finding showing that although ethanol preexposure increases subsequent attraction to ethanol-containing food, ethanol preexposure along with an ADH inhibitor instead induces aversion (Cadieu et al. 1999). Thus, the high internal levels of ethanol that accumulate in the absence of ADH function may be aversive to flies, whereas flies form a positive association with lower internal ethanol levels. In the conditioned preference assay, some flies will endure electric shock to attain the ethanol-associated cue, representing a strong attraction to ethanol reminiscent of addiction-like behavior.

In summary, flies show not only strong attraction to ethanol in multiple types of assays, but also some traits associated with addictive behavior. Flies prefer to consume ethanol-containing solutions, they return to peak levels of ethanol consumption after a period of abstinence, and they form a positive association with an intoxicating dose of ethanol. Furthermore, flies continue to show both ethanol consumption and conditioned ethanol preference in the presence of aversive cues (quinine or electric shock, respectively). Taken together with the recent results describing ethanol withdrawal in flies (see above), *Drosophila* has now become a fairly comprehensive model to study multiple facets of alcohol addiction.

ETHANOL AND SOCIAL BEHAVIOR

For the remainder of this review, we focus on an emerging area of study in *Drosophila*: the relationship between ethanol and social behavior. Recent studies have shown not only that ethanol can alter social behavior, such as courtship, but also that social experience can affect ethanol responses.

One study examined the acute and long-term effects of ethanol on one type of social behavior, courtship, which males typically display toward virgin females but not toward other males (Lee et al. 2008). When males are exposed to an intoxicating dose of ethanol for the first time, they rarely court either females or other males while intoxicated. However, repeated daily ethanol exposures altered their behavior dramatically: Males that had been previously exposed to ethanol showed a strong increase in courtship toward both females and other males while intoxicated. This increase is not likely explained by tolerance to the motor-impairing effects of ethanol, which developed from the first to second ethanol exposure and was not enhanced by subsequent exposures, whereas the level of male-male courtship continued to increase from the second to the fourth ethanol exposure. In addition, the effect of ethanol preexposure on male-male courtship could not be explained by an increase in the attractiveness of ethanol-exposed males because preexposed males also courted ethanol-naive males. Ethanol-preexposed males did not show male-male courtship in the absence of ethanol. Thus, although an acute effect of ethanol is necessary to stimulate male-male courtship, it does so only in males that have been repeatedly exposed to ethanol and have likely experienced some form of molecular or neural plasticity. The mechanisms underlying this effect are largely unknown, although both dopamine and the sex-determination gene *fruitless* seem to be required (Lee et al. 2008).

Whereas this courtship study characterized the effect of ethanol on social behavior, two other studies have demonstrated an effect of social experience on ethanol responses. The first study examined the effect of adult social isolation on sensitivity to ethanol sedation. Male flies that were isolated for 6 days showed a strong decrease in ethanol sensitivity compared with males that were housed in groups of 20 (Eddison et al. 2011). Adult isolation was previously shown to reduce the number of synapses of a subset of brain neurons (Donlea et al. 2009), suggesting that decreased synapse number may account for decreased ethanol sensitivity observed in isolated flies. In support of this model, the mutant *arouser* shows both increased ethanol sensitivity and increased synapse number, and adult isolation rescues both of these phenotypes (Eddison et al. 2011). Evidence from rodent models also indicates that social isolation reduces sensitivity to ethanol sedation (Jones et al. 1990). In addition, isolation increases ethanol consumption and operant responding for ethanol in rodents (McCool & Chappell 2009, Sanna et al. 2011). Social isolation is thought to represent a form of stress in rodents and induces many behavioral deficits (Fone & Porkess 2008). It will be interesting to determine whether isolation is also a stressor in flies and whether other social stressors may similarly affect ethanol sensitivity.

A second study examined the relationship between social experience and ethanol consumption (Shohat-Ophir et al. 2012). In this study, the sexual experience of male flies was varied, taking advantage of the fact that virgin females will readily mate with males, but previously mated females will actively reject the courtship advances of males. This study found that mated males show decreased ethanol consumption and preference compared with rejected males. This effect is due primarily to successful mating, rather than the experience of rejection, because mated males also showed lower ethanol preference than did naive males that had not interacted with females. In addition, mated males showed lower ethanol preference than did males allowed to interact with decapitated virgin females, which retain chemosensory cues typical of normal virgin females and elicit male courtship but not copulation. Successful mating therefore appears to be the key factor in reducing ethanol preference.

Several lines of evidence suggest that the effect of sexual experience on ethanol preference is mediated by neuropeptide F (NPF), the fly homolog of neuropeptide Y (NPY) in mammals (Shohat-Ophir et al. 2012). First, NPF levels in males are increased by mating and reduced by rejection. Second, downregulation of NPF signaling increases ethanol preference in mated males, mimicking the effect of rejection. Third, activation of NPF neurons to stimulate NPF release decreases ethanol preference in virgin males, mimicking the effect of mating. Activation of NPF neurons not only decreased ethanol preference, but also diminished the rewarding properties of ethanol, as assessed by the conditioned ethanol preference assay.

One possible explanation for the relationship between sexual experience, ethanol consumption, and NPF is a reward homeostat model (Shohat-Ophir et al. 2012). In this model, NPF signaling may serve as an internal representation of the state of the reward system: Experiences that decrease NPF signaling (such as sexual rejection) stimulate reward-seeking behavior (such as increased ethanol consumption), and vice versa. In support of this model, the conditioned preference assay demonstrates that both mating and ethanol intoxication are perceived as rewarding, as is artificial activation of the NPF pathway (Kaun et al. 2011, Shohat-Ophir et al. 2012). In rodents, NPY regulates ethanol intake (Thiele et al. 1998) but has not been shown to link social experience to ethanol consumption, so it will be interesting to determine whether this function is conserved from flies to mammals.

CONCLUSIONS

Drosophila and its ancestors have reproduced and grown on ethanol-containing substrates for many millions of years (Ashburner 1998, Dudley 2002). Flies have evolved mechanisms to metabolize ethanol for the purposes of both detoxification and nutrition (although it is uncertain exactly how many of the available calories in ethanol flies utilize), and flies likely use their ability to smell ethanol to locate ripe fruit in the wild. It is therefore not surprising that in a laboratory setting flies develop tolerance when chronically exposed to low levels of ethanol, show preference for ethanol-containing food, and find exposure to low levels of ethanol to be rewarding. What at first glance appears surprising, however, is the observation that behaviors induced by exposure to moderate-to-high ethanol doses (which flies would never have encountered in their natural habitat) are remarkably similar in flies and humans. As mentioned in this review, with increasing ethanol doses flies become hyperactive, no longer walk in straight lines, display incoordination and loss of postural control, and eventually succumb to sedation. Because the majority of the known targets of ethanol are present in flies (Heberlein et al. 2004), these similarities are perhaps not that surprising after all; the molecular and cellular actions of ethanol and the ensuing physiological changes may in fact be the same, whether in a fly, rodent, or human. We would argue, then, that using the powerful behavioral and molecular genetic approaches that are feasible in *Drosophila* should continue to provide novel insights into the mechanisms by which ethanol affects behavior. In particular, investigators can exploit recently developed assays that model aspects of addiction, such as withdrawal, voluntary consumption, relapse, and use despite adverse consequences, although more laborious than the simpler assays for acute intoxication, to reveal the underlying molecular mechanisms.

DISCLOSURE STATEMENT

The authors are not aware of any affiliations, memberships, funding, or financial holdings that might be perceived as affecting the objectivity of this review.

LITERATURE CITED

Acheson SK, Stein RM, Swartzwelder HS. 1998. Impairment of semantic and figural memory by acute ethanol: age-dependent effects. *Alcohol Clin. Exp. Res.* 22:1437–42

Am. Psychiatr. Assoc. (APA). 1994. *Diagnostic and Statistical Manual of Mental Disorders.* Washington, DC: APA. 4th ed.

Ashburner M. 1998. Speculations on the subject of alcohol dehydrogenase and its properties in *Drosophila* and other flies. *BioEssays* 20:949–54

Babor TF, Berglas S, Mendelson JH, Ellingboe J, Miller K. 1983. Alcohol, affect, and the disinhibition of verbal behavior. *Psychopharmacology* 80:53–60

Barbancho M, Sánchez-Cañete FJ, Dorado G, Pineda M. 1987. Relation between tolerance to ethanol and alcohol dehydrogenase (ADH) activity in *Drosophila melanogaster*: selection, genotype and sex effects. *Heredity* 58:443–50

Barclay DC, Hallbergson AF, Montague JR, Mudd LM. 2005. Reversal of ethanol toxicity in embryonic neurons with growth factors and estrogen. *Brain Res. Bull.* 67:459–65

Becker HC, Diaz-Granados JL, Randall CL. 1996. Teratogenic actions of ethanol in the mouse: a minireview. *Pharmacol. Biochem. Behav.* 55:501–13

Becker HC, Hale RL, Boggan WO, Randall CL. 1993. Effects of prenatal ethanol exposure on later sensitivity to the low-dose stimulant actions of ethanol in mouse offspring: possible role of catecholamines. *Alcohol Clin. Exp. Res.* 17:1325–36

Berger KH, Heberlein U, Moore MS. 2004. Rapid and chronic: two distinct forms of ethanol tolerance in *Drosophila*. *Alcohol Clin. Exp. Res.* 28:1469–80

Berger KH, Kong EC, Dubnau J, Tully T, Moore MS, Heberlein U. 2008. Ethanol sensitivity and tolerance in long-term memory mutants of *Drosophila melanogaster*. *Alcohol Clin. Exp. Res.* 32: 895–908

Bhandari P, Kendler KS, Bettinger JC, Davies AG, Grotewiel M. 2009. An assay for evoked locomotor behavior in *Drosophila* reveals a role for integrins in ethanol sensitivity and rapid ethanol tolerance. *Alcohol Clin. Exp. Res.* 33:1794–805

Briscoe DA, Robertson A, Malpica JM. 1975. Dominance at Adh locus in response of adult *Drosophila melanogaster* to environmental alcohol. *Nature* 255:148–49

Cadieu N, Cadieu J-C, El Ghadraoui L, Grimal A, Lamboeuf Y. 1999. Conditioning to ethanol in the fruit fly—a study using an inhibitor of ADH. *J. Insect Physiol.* 45:579–86

Cavener D. 1979. Preference for ethanol in *Drosophila melanogaster* associated with the alcohol dehydrogenase polymorphism. *Behav. Genet.* 9:359–65

Cavener DR, Clegg MT. 1981. Multigenic response to ethanol in *Drosophila melanogaster*. *Evolution* 35:1–10

Cohan FM, Graf J-D. 1985. Latitudinal cline in *Drosophila melanogaster* for knockdown resistance to ethanol fumes and for rates of response to selection for further resistance. *Evolution* 39:278–93

Cohan FM, Hoffmann AA. 1986. Genetic divergence under uniform selection. II. Different responses to selection for knockdown resistance to ethanol among *Drosophila melanogaster* populations and their replicate lines. *Genetics* 114:145–64

Corl AB, Berger KH, Ophir-Shohat G, Gesch J, Simms JA, et al. 2009. Happyhour, a Ste20 family kinase, implicates EGFR signaling in ethanol-induced behaviors. *Cell* 137:949–60

Cowmeadow RB, Krishnan HR, Atkinson NS. 2005. The slowpoke gene is necessary for rapid ethanol tolerance in *Drosophila*. *Alcohol Clin. Exp. Res.* 29:1777–86

Crabbe JC, Belknap JK, Buck KJ. 1994. Genetic animal models of alcohol and drug abuse. *Science* 264:1715–23

Crabbe JC, Phillips TJ, Harris RA, Arends MA, Koob GF. 2006. Alcohol-related genes: contributions from studies with genetically engineered mice. *Addict. Biol.* 11:195–269

Daly K, Clarke B. 1981. Selection associated with the alcohol dehydrogenase locus in *Drosophila melanogaster*: differential survival of adults maintained on low concentrations of ethanol. *Heredity* 46:219–26

David JR, Bocquet C. 1975. Similarities and differences in latitudinal adaptation of two *Drosophila* sibling species. *Nature* 257:588–90

David JR, Bocquet C, Arens MF, Fouillet P. 1976. Biological role of alcohol dehydrogenase in the tolerance of *Drosophila melanogaster* to aliphatic alcohols: utilization of an ADH-null mutant. *Biochem. Genet.* 14:989–97

David JR, Mercot H, Capy P, McEvey SF, Van Herrewege J. 1986. Alcohol tolerance and Adh gene frequencies in European and African populations of *Drosophila melanogaster*. *Genet. Sel. Evol.* 18:405–16

de la Monte SM, Xu XJ, Wands JR. 2005. Ethanol inhibits insulin expression and actions in the developing brain. *Cell. Mol. Life Sci.* 62:1131–45

Devineni AV, Heberlein U. 2009. Preferential ethanol consumption in *Drosophila* models features of addiction. *Curr. Biol.* 19:2126–32

Devineni AV, McClure KD, Guarnieri DJ, Corl AB, Wolf FW, et al. 2011. The genetic relationships between ethanol preference, acute ethanol sensitivity, and ethanol tolerance in *Drosophila melanogaster*. *Fly* 5:191–99

Donlea JM, Ramanan N, Shaw PJ. 2009. Use-dependent plasticity in clock neurons regulates sleep need in *Drosophila*. *Science* 324:105–8

Dudley R. 2000. Evolutionary origins of human alcoholism in primate frugivory. *Q. Rev. Biol.* 75:3–15

Dudley R. 2002. Fermenting fruit and the historical ecology of ethanol ingestion: Is alcoholism in modern humans an evolutionary hangover? *Addiction* 97:381–88

Eddison M, Guarnieri DJ, Cheng L, Liu C-H, Moffat KG, et al. 2011. *arouser* reveals a role for synapse number in the regulation of ethanol sensitivity. *Neuron* 70:979–90

Fone KCF, Porkess MV. 2008. Behavioural and neurochemical effects of post-weaning social isolation in rodents—relevance to developmental neuropsychiatric disorders. *Neurosci. Biobehav. Rev.* 32:1087–102

Fry JD, Donlon K, Saweikis M. 2008. A worldwide polymorphism in aldehyde dehydrogenase in *Drosophila melanogaster*: evidence for selection mediated by dietary ethanol. *Evolution* 62:66–75

Fry JD, Saweikis M. 2006. Aldehyde dehydrogenase is essential for both adult and larval ethanol resistance in *Drosophila melanogaster*. *Genet. Res.* 87:87–92

Geer BW, Langevin ML, McKechnie SW. 1985. Dietary ethanol and lipid synthesis in *Drosophila melanogaster*. *Biochem. Genet.* 23:607–22

Geer BW, McKechnie SW, Heinstra PWH, Pyka MJ. 1991. Heritable variation in ethanol tolerance and its association with biochemical traits in *Drosophila melanogaster*. *Evolution* 45:1107–19

Gelfand LJ, McDonald JF. 1983. Relationship between alcohol dehydrogenase (ADH) activity and behavioral response to environmental alcohol in five *Drosophila* species. *Behav. Genet.* 13:281–93

Ghezzi A, Krishnan HR, Atkinson NS. 2012. Susceptibility to ethanol withdrawal seizures is produced by BK channel gene expression. *Addict. Biol.* doi: 10.1111/j.1369-1600.2012.00465.x. In press

Gibson JB, May TW, Wilks AV. 1981. Genetic variation at the alcohol dehydrogenase locus in *Drosophila melanogaster* in relation to environmental variation: ethanol levels in breeding sites and allozyme frequencies. *Oecologia* 51:191–98

Gibson JB, Wilks AV. 1988. The alcohol dehydrogenase polymorphism of *Drosophila melanogaster* in relation to environmental ethanol, ethanol tolerance and alcohol dehydrogenase activity. *Heredity* 60:403–14

Gibson WE. 1985. Effects of alcohol on radial maze performance in rats. *Physiol. Behav.* 35:1003–5

Givens B, McMahon K. 1997. Effects of ethanol on nonspatial working memory and attention in rats. *Behav. Neurosci.* 111:275–82

Grant BF, Dawson DA, Stinson FS, Chou SP, Dufour MC, Pickering RP. 2004. The 12-month prevalence and trends in DSM-IV alcohol abuse and dependence: United States, 1991–1992 and 2001–2002. *Drug Alcohol Depend.* 74:223–34

Heberlein U, Wolf FW, Rothenfluh A, Guarnieri DJ. 2004. Molecular genetic analysis of ethanol intoxication in *Drosophila melanogaster*. *Integr. Comp. Biol.* 44:269–74

Hernández-Tobías A, Julián-Sánchez A, Piña E, Riveros-Rosas H. 2011. Natural alcohol exposure: Is ethanol the main substrate for alcohol dehydrogenases in animals? *Chem. Biol. Interact.* 191:14–25

Ja WW, Carvalho GB, Mak EM, de la Rosa NN, Fang AY, et al. 2007. Prandiology of *Drosophila* and the CAFE assay. *Proc. Natl. Acad. Sci. USA* 104:8253–56

Jones BC, Connell JM, Erwin G. 1990. Isolate housing alters ethanol sensitivity in long-sleep and short-sleep mice. *Pharmacol. Biochem. Behav.* 35:469–72

Jones KL, Smith DW. 1973. Recognition of the fetal alcohol syndrome in early infancy. *Lancet* 302:999–1001

Jornvall H, Persson M, Jeffery J. 1981. Alcohol and polyol dehydrogenases are both divided into two protein types, and structural properties cross-relate the different enzyme activities within each type. *Proc. Natl. Acad. Sci. USA* 78:4226–30

Kalant H, LeBlanc AE, Gibbins RJ. 1971. Tolerance to, and dependence on, some non-opiate psychotropic drugs. *Pharmacol. Rev.* 23:135–91

Karan D, Parkash R, David JR. 1999. Microspatial genetic differentiation for tolerance and utilization of various alcohols and acetic acid in *Drosophila* species from India. *Genetica* 105:249–58

Kaun KR, Azanchi R, Maung Z, Hirsh J, Heberlein U. 2011. A *Drosophila* model for alcohol reward. *Nat. Neurosci.* 14:612–19

Kaun KR, Devineni AV, Heberlein U. 2012. *Drosophila melanogaster* as a model to study drug addiction. *Hum. Genet.* 131:959–75

Kerver JW, Wolf W, Kamping A, van Delden W. 1992. Effects on ADH activity and distribution, following selection for tolerance to ethanol in *Drosophila melanogaster*. *Genetica* 87:175–83

Kong EC, Allouche L, Chapot PA, Vranizan K, Moore MS, et al. 2010. Ethanol-regulated genes that contribute to ethanol sensitivity and rapid tolerance in *Drosophila*. *Alcohol Clin. Exp. Res.* 34:302–16

Kosten TR, O'Connor PG. 2003. Management of drug and alcohol withdrawal. *N. Engl. J. Med.* 348:1786–95

Lee HG, Kim YC, Dunning JS, Han KA. 2008. Recurring ethanol exposure induces disinhibited courtship in *Drosophila*. *PLoS One* 3:e1391

Lister RG, Gorenstein C, Fisher-Flowers D, Weingartner HJ, Eckardt MJ. 1991. Dissociation of the acute effects of alcohol on implicit and explicit memory processes. *Neuropsychologia* 29:1205–12

Littleton J. 1998. Neurochemical mechanisms underlying alcohol withdrawal. *Alcohol Health Res. World* 22:13–24

Mayfield RD, Harris RA, Schuckit MA. 2008. Genetic factors influencing alcohol dependence. *Br. J. Pharmacol.* 154:275–87

McClure KD, French RL, Heberlein U. 2011. A *Drosophila* model for fetal alcohol syndrome disorders: role for the insulin pathway. *Dis. Model. Mech.* 4:335–46

McCool BA, Chappell AM. 2009. Early social isolation in male long-evans rats alters both appetitive and consummatory behaviors expressed during operant ethanol self-administration. *Alcohol Clin. Exp. Res.* 33:273–82

McElfresh KC, McDonald JF. 1986. The effect of temperature on biochemical and molecular properties of *Drosophila* alcohol dehydrogenase. *Biochem. Genet.* 24:873–89

McGough NN, Thomas JD, Dominguez HD, Riley EP. 2009. Insulin-like growth factor-I mitigates motor coordination deficits associated with neonatal alcohol exposure in rats. *Neurotoxicol. Teratol.* 31:40–48

McKechnie SW, Geer BW. 1984. Regulation of alcohol dehydrogenase in *Drosophila melanogaster* by dietary alcohol and carbohydrate. *Insect Biochem.* 14:231–42

McKenzie JA, McKechnie SW. 1978. Ethanol tolerance and the Adh polymorphism in a natural population of *Drosophila melanogaster*. *Nature* 272:75–76

McKenzie JA, McKechnie SW. 1979. A comparative study of resource utilization in natural populations of *Drosophila melanogaster* and *D. simulans*. *Oecologia* 40:299–309

McKenzie JA, Parsons PA. 1972. Alcohol tolerance: an ecological parameter in the relative success of *Drosophila melanogaster* and *Drosophila simulans*. *Oecologia* 10:373–88

McKenzie JA, Parsons PA. 1974. Microdifferentiation in a natural population of *Drosophila melanogaster* to alcohol in the environment. *Genetics* 77:385–94

Milan NF, Kacsoh BZ, Schlenke TA. 2012. Alcohol consumption as self-medication against blood-borne parasites in the fruit fly. *Curr. Biol.* 22:488–93

Morean ME, Corbin WR. 2010. Subjective response to alcohol: a critical review of the literature. *Alcohol Clin. Exp. Res.* 34:385–95

Morse RM, Flavin DK. 1992. The definition of alcoholism. *JAMA* 268:1012–14

Newlin DB, Thomson JB. 1990. Alcohol challenge with sons of alcoholics: a critical review and analysis. *Psychol. Bull.* 108:383–402

Oakeshott JG, Gibson JB, Anderson PR, Knibb WR, Anderson DG, Chambers GK. 1982. Alcohol dehydrogenase and glycerol-3-phosphate dehydrogenase clines in *Drosophila melanogaster* on different continents. *Evolution* 36:86–96

Parkash R, Karan D, Munjal AK. 1999. Geographical variation in ADH^F and alcoholic resource utilization in Indian populations of *D. melanogaster*. *Biol. J. Linn. Soc.* 66:205–14

Parr J, Large A, Wang X, Fowler SC, Ratzlaff KL, Ruden DM. 2001. The inebri-actometer: a device for measuring the locomotor activity of *Drosophila* exposed to ethanol vapor. *J. Neurosci. Methods* 107:93–99

Parsons PA. 1980. Larval responses to environmental ethanol in *Drosophila melanogaster*: variation within and among populations. *Behav. Genet.* 10:183–90

Parsons PA, King SB. 1977. Ethanol: larval discrimination between two *Drosophila* sibling species. *Experientia* 33:898–99

Phillips TJ, Roberts AJ, Lessov CN. 1997. Behavioral sensitization to ethanol: genetics and the effects of stress. *Pharmacol. Biochem. Behav.* 57:487–93

Pipkin SB, Hewitt NE. 1972. Variation of alcohol dehydrogenase levels in *Drosophila* species hybrids. *J. Hered.* 63:267–70

Pohl JB, Baldwin BA, Dinh BL, Rahman P, Smerek D, et al. 2012. Ethanol preference in *Drosophila melanogaster* is driven by its caloric value. *Alcohol Clin. Exp. Res.* 36:1903–12

Pohorecky LA. 1977. Biphasic action of ethanol. *Biobehav. Rev.* 1:231–40

Prescott CA, Kendler KS. 1999. Genetic and environmental contributions to alcohol abuse and dependence in a population-based sample of male twins. *Am. J. Psychiatry* 156:34–40

Ranganathan S, Davis DG, Hood RD. 1987. Developmental toxicity of ethanol in *Drosophila melanogaster*. *Teratology* 36:45–49

Richmond RC, Gerking JL. 1979. Oviposition site preference in *Drosophila*. *Behav. Genet.* 9:233–41

Robinson BG, Khurana S, Kuperman A, Atkinson NS. 2012a. Neural adaptation leads to cognitive ethanol dependence. *Curr. Biol.* 22:2338–41

Robinson BG, Khurana S, Pohl JB, Li WK, Ghezzi A, et al. 2012b. A low concentration of ethanol impairs learning but not motor and sensory behavior in *Drosophila* larvae. *PLoS One* 7:e37394

Rodan AR, Rothenfluh A. 2010. The genetics of behavioral alcohol responses in *Drosophila*. *Int. Rev. Neurobiol.* 91:25–51

Rodd ZA, Bell RL, Sable HJ, Murphy JM, McBride WJ. 2004. Recent advances in animal models of alcohol craving and relapse. *Pharmacol. Biochem. Behav.* 79:439–50

Rothenfluh A, Threlkeld RJ, Bainton RJ, Tsai LT-Y, Lasek AW, Heberlein U. 2006. Distinct behavioral responses to ethanol are regulated by alternate RhoGAP18B isoforms. *Cell* 127:199–211

Sampsell B, Sims S. 1982. Effect of adh genotype and heat stress on alcohol tolerance in *Drosophila melanogaster*. *Nature* 296:853–55

Sanna E, Talani G, Obili N, Mascia MP, Mostallino MC, et al. 2011. Voluntary ethanol consumption induced by social isolation reverses the increase of α(4)/δ GABA(A) receptor gene expression and function in the hippocampus of C57BL/6J mice. *Front. Neurosci.* 5:15

Scholz H, Ramond J, Singh CM, Heberlein U. 2000. Functional ethanol tolerance in *Drosophila*. *Neuron* 28:261–71

Schuckit MA. 1994. Low level of response to alcohol as a predictor of future alcoholism. *Am. J. Psychiatry* 151:184–89

Schuckit MA, Edenberg HJ, Kalmijn J, Flury L, Smith TL, et al. 2001. A genome-wide search for genes that relate to a low level of response to alcohol. *Alcohol Clin. Exp. Res.* 25:323–29

Shohat-Ophir G, Kaun KR, Azanchi R, Heberlein U. 2012. Sexual deprivation increases ethanol intake in *Drosophila*. *Science* 335:1351–55

Siegal ML, Hartl DL. 1999. Oviposition-site preference in *Drosophila* following interspecific gene transfer of the alcohol dehydrogenase locus. *Behav. Genet.* 29:199–204

Singh CM, Heberlein U. 2000. Genetic control of acute ethanol-induced behaviors in *Drosophila*. *Alcohol Clin. Exp. Res.* 24:1127–36

Stavro K, Pelletier J, Potvin S. 2012. Widespread and sustained cognitive deficits in alcoholism: a meta-analysis. *Addict. Biol.* 18:203–13

Thiele TE, Marsh DJ, Ste Marie L, Bernstein IL, Palmiter RD. 1998. Ethanol consumption and resistance are inversely related to neuropeptide Y levels. *Nature* 396:366–69

van Delden W, Boerema AC, Kamping A. 1978. The alcohol dehydrogenase polymorphism in populations of *Drosophila melanogaster*. I. Selection in different environments. *Genetics* 90:161–91

van Delden W, Kamping A. 1990. Genetic variation for oviposition behavior in *Drosophila melanogaster*. II. Oviposition preferences and differential survival. *Behav. Genet.* 20:661–73

Van Herrewege J, David JR. 1980. Alcohol tolerance and alcohol utilisation in *Drosophila*: partial independence of two adaptive traits. *Heredity* 44:229–35

Venken KJ, Simpson JH, Bellen HJ. 2011. Genetic manipulation of genes and cells in the nervous system of the fruit fly. *Neuron* 72:202–30

Vouidibio J, Capy P, Defaye D, Pla E, Sandrin J, et al. 1989. Short-range genetic structure of *Drosophila melanogaster* populations in an Afrotropical urban area and its significance. *Proc. Natl. Acad. Sci. USA* 86:8442–46

Weber KE. 1988. An apparatus for measurement of resistance to gas-phase agents. *Drosoph. Inf. Serv.* 67:90–92

Wolf FW, Rodan AR, Tsai LT-Y, Heberlein U. 2002. High-resolution analysis of ethanol-induced locomotor stimulation in *Drosophila*. *J. Neurosci.* 22:11035–44

Wolffgramm J, Heyne A. 1991. Social behavior, dominance, and social deprivation of rats determine drug choice. *Pharmacol. Biochem. Behav.* 38:389–99

Xu S, Chan T, Shah V, Zhang S, Pletcher SD, Roman G. 2012. The propensity for consuming ethanol in *Drosophila* requires rutabaga adenylyl cyclase expression within mushroom body neurons. *Genes Brain Behav.* 11:727–39

From Atomic Structures to Neuronal Functions of G Protein–Coupled Receptors

Krzysztof Palczewski and Tivadar Orban

Department of Pharmacology, School of Medicine, Case Western Reserve University, Cleveland, Ohio 44106-4965; email: kxp65@case.edu

Annu. Rev. Neurosci. 2013. 36:139–64

First published online as a Review in Advance on May 15, 2013

The *Annual Review of Neuroscience* is online at neuro.annualreviews.org

This article's doi:
10.1146/annurev-neuro-062012-170313

Keywords

GPCRs, signal transduction, membrane biology, transmembrane receptors, receptor pharmacology, allosteric regulation, rhodopsin, crystal structure

Abstract

G protein–coupled receptors (GPCRs) are essential mediators of signal transduction, neurotransmission, ion channel regulation, and other cellular events. GPCRs are activated by diverse stimuli, including light, enzymatic processing of their N-termini, and binding of proteins, peptides, or small molecules such as neurotransmitters. GPCR dysfunction caused by receptor mutations and environmental challenges contributes to many neurological diseases. Moreover, modern genetic technology has helped identify a rich array of mono- and multigenic defects in humans and animal models that connect such receptor dysfunction with disease affecting neuronal function. The visual system is especially suited to investigate GPCR structure and function because advanced imaging techniques permit structural studies of photoreceptor neurons at both macro and molecular levels that, together with biochemical and physiological assessment in animal models, provide a more complete understanding of GPCR signaling.

Contents

G PROTEIN–MEDIATED SIGNALING: A BRIEF HISTORICAL PERSPECTIVE

GPCR: G protein–coupled receptor

7-TM: 7-transmembrane

Although it is difficult to identify a single event that initiates a new scientific field, the discovery of cyclic adenosine monophosphate (cAMP) is generally perceived to have initiated the study of G protein–mediated signaling in earnest. In the late 1950s, Earl W. Sutherland, Jr., Chair of Pharmacology at Western Reserve University (later renamed Case Western Reserve University), and his postdoctoral fellow, Theodore W. Rall (a subsequent Chair of the same department), discovered this distinctive nucleotide (Rall & Sutherland 1958, Sutherland & Rall 1958). Trained by Carl F. Cori (Washington University in St. Louis), a Czech emigrant biochemist/pharmacologist and 1947 Nobel laureate, Sutherland linked the formation and destruction of cyclic AMP (the secondary messenger) with the effects of epinephrine, glucagon, and insulin upon glycogen metabolism (Robison et al. 1968). Sutherland was the sole winner of the 1971 Nobel Prize in Physiology or Medicine "for his discoveries concerning the mechanisms of the action of hormones" (**http://www.Nobelprize.org**). Not only was Sutherland an incredibly insightful scientist, but he also initiated one of the first MD/PhD programs in the country and personally recruited Alfred G. Gilman to Cleveland to join that program. Gilman, a doctoral student of Rall (Gilman & Rall 1968a,b) who was intrigued by the role of cAMP, later discovered and characterized G proteins, molecular switches that link cellular receptors to various responses (Gilman 1987).

That hormone action is mediated via receptors had been known since the beginning of the twentieth century, but it was Rodbell and colleagues who showed that these receptors and associated proteins are organized at the cellular plasma membrane as large multiprotein complexes composed of receptors, G proteins, and effector enzymes (Schlegel et al. 1979). Both Gilman and Martin Rodbell were awarded the Nobel Prize in Physiology or Medicine in 1994 for these seminal contributions.

Today we know that GPCRs [also known as guanine-nucleotide-binding protein-coupled membrane receptors, seven-transmembrane (7-TM) receptors, or heptahelical receptors] are evolutionarily highly conserved (Strotmann et al. 2011) and are widely expressed in eukaryotic organisms where they control a vast

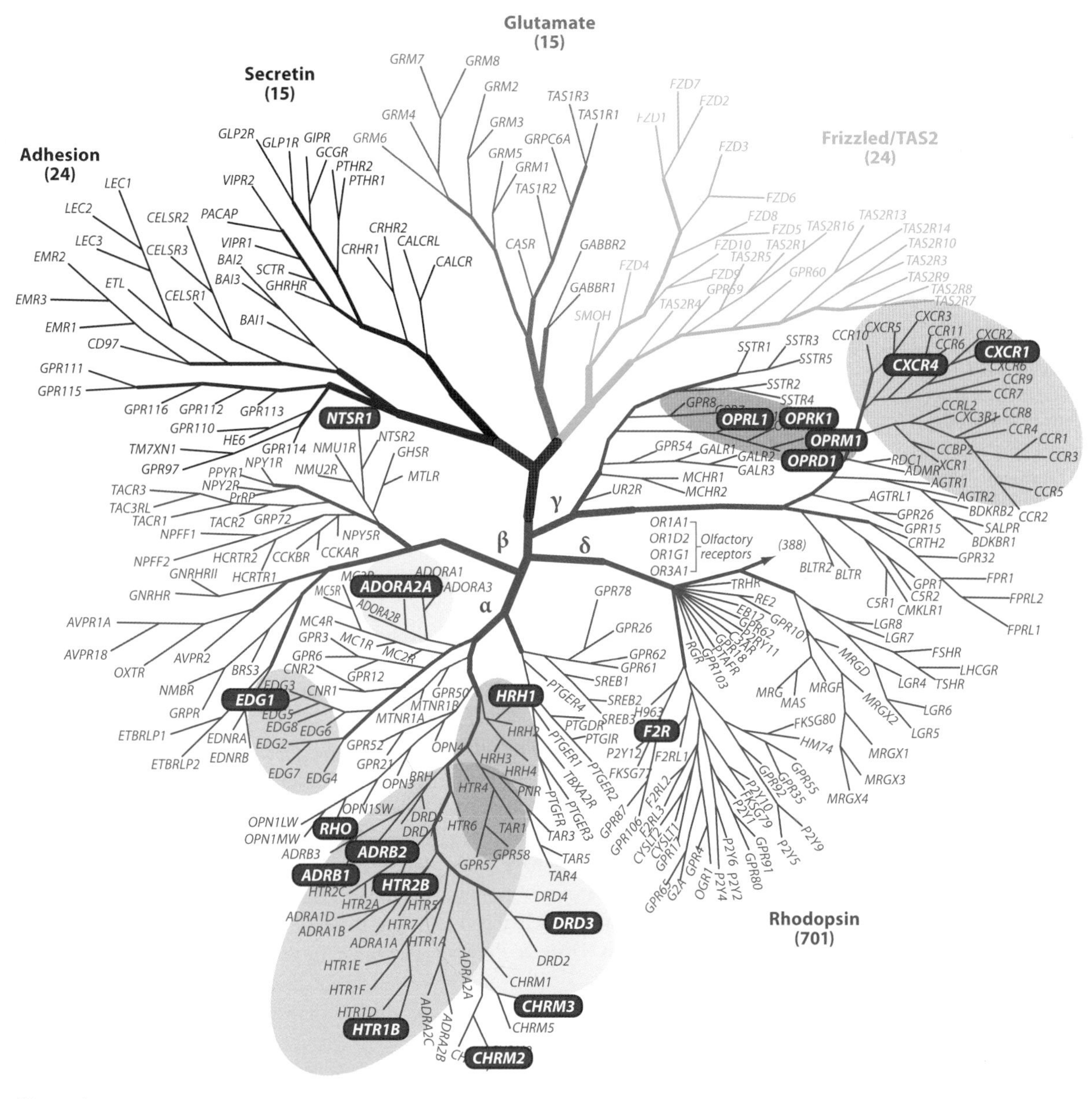

Figure 1

Current phylogenetic tree for GPCRs. GPCR gene names are located on branches on the basis of their primary sequence similarity. Solved structures are represented using gene names on blue backgrounds. Colored backgrounds indicate groups of GPCRs that share sequence homology with determined X-ray structures that are 35% or higher. Major branches such as the adhesion, secretin, glutamate, Frizzled/TAS2, and rhodopsin are labeled, as well. Figure adapted from Katritch et al. (2012).

range of biological processes. These receptors are classified into various subtypes on the basis of their extracellular domain topology, sequence similarity, activating ligands, type of G protein they activate, function, and other criteria (see, e.g., Park et al. 2008b) (**Figure 1**). All GPCRs are predicted to share a common 7-TM α-helical structure and are

Rhodopsin: the red-colored GPCR composed of an opsin covalently linked to 11-*cis*-retinal. Responsible for dim-light vision

Heterotrimeric G proteins: consist of three polypeptides: a GPCR-binding α-subunit with GTP to GDP hydrolytic activity and tightly bound βγ-subunits

Transmembrane domain: any segment of a protein that passes completely through a cell membrane

ECL: extracellular loop

ICL: intracellular loop

Orthosteric ligand-binding region: the primary ligand-binding site in a GPCR that is not subtype-selective for endogenous agonists

localized within cellular membranes. These receptors respond to a diverse array of physical and chemical entities that include photons, proteolytic events, soluble small molecules and ions, peptides, and large proteins. More than a decade ago, the first crystal structure of a native GPCR (rhodopsin) was solved (Palczewski et al. 2000), and on the basis of a vast amount of biochemical data, investigators predicted that the general features in the TM α-helical region are conserved across all GPCRs (Filipek et al. 2003b).

In neuronal systems, GPCRs fine-tune cellular functions by activating a heterotrimeric G protein–dependent cascade of signals (Gainetdinov et al. 2004). These fine-tuning events further regulate a variety of cellular events including synaptic modulation and neurotransmission, Ca^{2+} and cyclic nucleotide signaling, cellular plasticity, electrical activity, neurite outgrowth, exocytosis, gene expression, the unfolded protein response, and neuronal development/differentiation, among others. For example, a putative octopamine GPCR in sensory neurons actively suppresses innate immune responses by downregulating the expression of noncanonical unfolded-protein response genes to control stress-response pathways (Sun et al. 2011). GPCRs expressed by adult neural stem cells have also emerged as potential modulators of adult neurogenesis that are critical for self-repair of neuropathologic conditions (Doze & Perez 2012).

GPCR activities may influence higher functions such as behavior but are also involved in pathologies such as neurodegeneration, excitotoxic injury, and motor dysfunction. Here, we summarize recent progress in determining the structural biology of GPCRs and understanding GPCR signaling by using visual phototransduction as a specific example. Invertebrate and vertebrate visual systems involving phototransduction are model systems that have provided molecular and functional information about GPCR signaling for decades (Yau & Hardie 2009).

ADVANCES IN STRUCTURAL STUDIES OF GPCRs AND OTHER INTERACTING PROTEINS

GPCRs Exhibit Similar Overall Topology with Diversity in Their Ligand-Binding Sites

Among all available GPCR structures solved to date (**Figure 1**), the large structural diversity of the extracellular half, denoted by an average root-mean-square deviation of ~2.7 Å, contrasts sharply with the relatively similar structural features of the intracellular half, with an average root-mean-square deviation of ~1.5 Å. The molecular topology shared by all GPCRs consists of the seven transmembrane (TM) domain composed of α-helices spanning the biological membrane (**Figure 2*a–t***). These TM helices are connected by three extracellular (ECL) and three intracellular (ICL) loops denoted as ECL-I to -III and ICL-I to -III, respectively (Salon et al. 2011). The ECLs, N-termini, and/or part of the TM α-helical region facing the extracellular space are responsible for ligand recognition, whereas the ICL and C-termini recognize the specific protein partner required for signal transmission within the cell. The orthosteric ligand-binding region is frequently located within a cavity formed by the TM helices, but in some GPCRs, such as the metabotropic glutamate receptors, the ligands are bound by a separate amino terminal discrete domain (Pin et al. 2003). Although the extracellular region has evolved to recognize a multitude of diverse ligands, the intracellular region needs to recognize and discriminate between a relatively small number of G proteins. Because the amino terminal ligand-binding domains of some GPCRs fold properly and bind ligand independently of the TM α-helical region, researchers have elucidated several structures of these domains (reviewed in Park et al. 2008b, Pin et al. 2003). Common features of the TM helices have long been recognized (**Figure 2*u***), and efficient signal transduction through the membrane is achieved by a similar mechanism for all rhodopsin-like class A GPCRs

(Mirzadegan et al. 2003) and perhaps all GPCR classes. The mechanism by which the signal is propagated involves a network of water molecules, specific conserved amino acids, and small movements of the TM helices (**Figure 3*a***). In addition to these features, other conserved structural elements such as the disulfide bond linking ECL-II to TM-III, the (D/E)RY motif in helix III (which participates in the ionic lock between helix III and VI), the

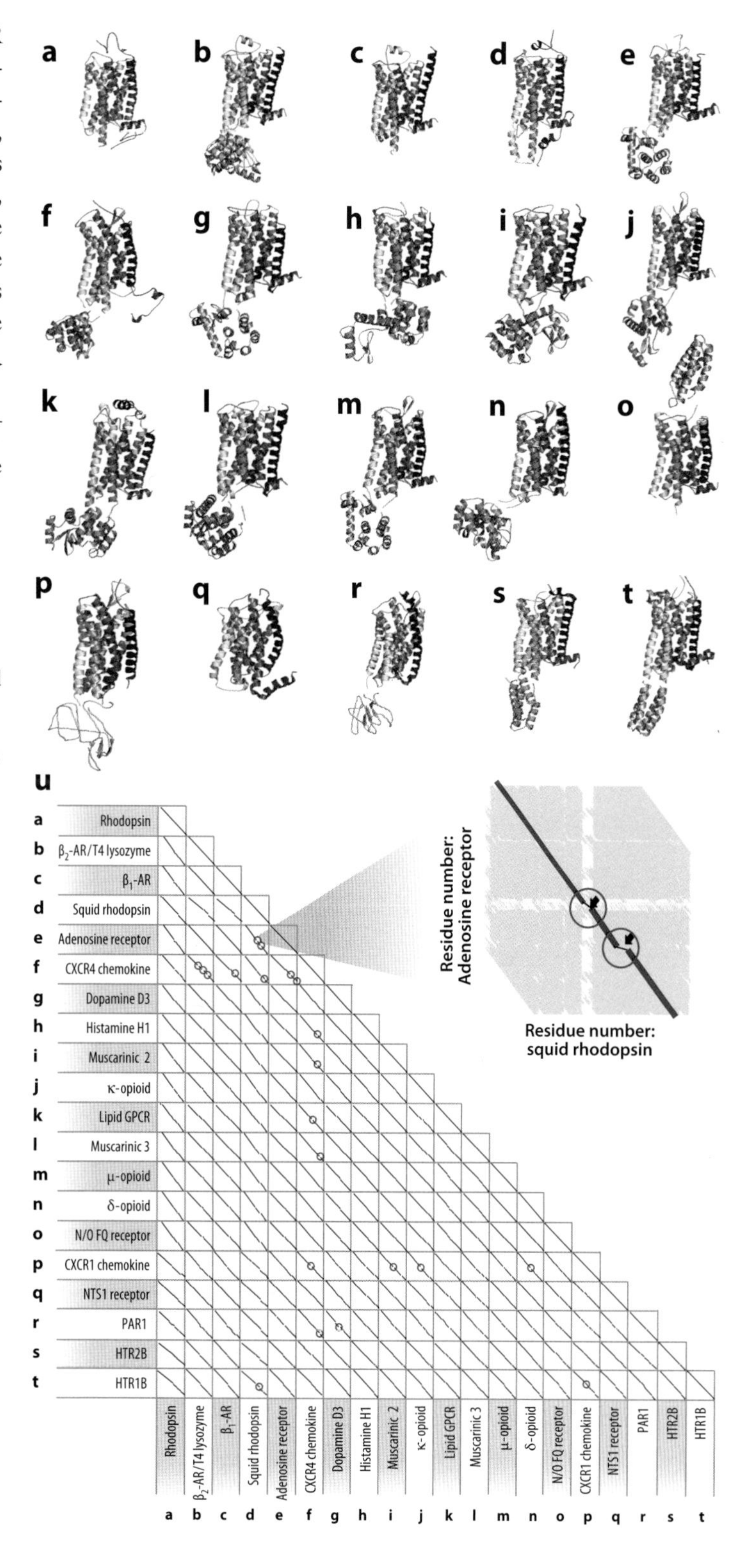

Figure 2

Gallery of selected GPCR structures determined by X-ray crystallography. Available X-ray structures are shown in the same orientation for all GPCRs together with an alignment graph of each GPCR pair obtained using flexible structure alignment by chaining aligned fragments (Ye & Godzik 2003). (*a*) bovine rhodopsin (PDB code: 1U19); (*b*) β_2-adrenergic receptor/T4 lysozyme chimera (PDB code: 2RH1); (*c*) thermostabilized β_1-adrenergic receptor (PDB code: 2Y01); (*d*) squid rhodopsin (PDB code: 3AYN); (*e*) adenosine receptor/T4 lysozyme chimera (PDB code: 3EML); (*f*) CXCR4 chemokine/T4 lysozyme chimera (PDB code: 3ODU); (*g*) dopamine D3 receptor/T4 lysozyme chimera (PDB code: 3PBL); (*h*) human histamine H1 receptor/T4 lysozyme chimera (PDB code: 3RZE); (*i*) human M2 muscarinic acetylcholine receptor/T4 lysozyme chimera (PDB code: 3UON); (*j*) κ-opioid receptor/T4 lysozyme chimera (PDB code: 4DJH); (*k*) lipid GPCR/T4 chimera (PDB code: 3V2W); (*l*) M3 muscarinic receptor/T4 lysozyme chimera (PDB code: 4DAJ); (*m*) μ-opioid receptor/T4 lysozyme chimera (PDB code: 4DKL); (*n*) δ-opioid receptor/T4 lysozyme chimera (PDB code: 4EJ4); (*o*) thermostabilized apocytochrome b_{562} (Bril)-nociceptin/orphanin FQ receptor fusion (PDB code: 4EA3); (*p*) chemokine receptor CXCR1 (PDB code: 2LNL); (*q*) neurotensin receptor NTS1 (PDB code: 4GRV); (*r*) human protease-activated receptor 1 (PAR1) (PDB code: 3VW7); (*s*) chimeric protein of 5-HT2B-BRIL (HTR2B) (PDB code: 4IB4); (*t*) chimeric protein of 5-HT1B-BRIL (HTR1B) (PDB code: 4IAQ). GPCRs are shown in a rainbow-colored scheme: N-terminal (*dark blue*), C-terminal (*red*), and nonnative fusion protein components (*gray*). (*u*) Alignment of fragment pairs derived for each combination of two GPCR structures presented in panels *a–t*. Blue circles indicate the largest differences resulting from insertion/deletions when comparing the aligned structures of the GPCRs under study.

WxP motif in helix VI, and the NPxxY motif in helix VII are also thought to be involved in efficient signal transduction. Several different analyses of the activation mechanism have been published but are beyond the scope of this review (Hofmann et al. 2009, Katritch et al. 2012, Park et al. 2008b, Salon et al. 2011).

Palczewski et al. (2000) revealed the first atomic resolution details of a GPCR with the crystallographic structure determination of rhodopsin. Ground state rhodopsin crystals prepared under different conditions yielded structures of rhodopsin in three different conformations with significant changes in ICL-III

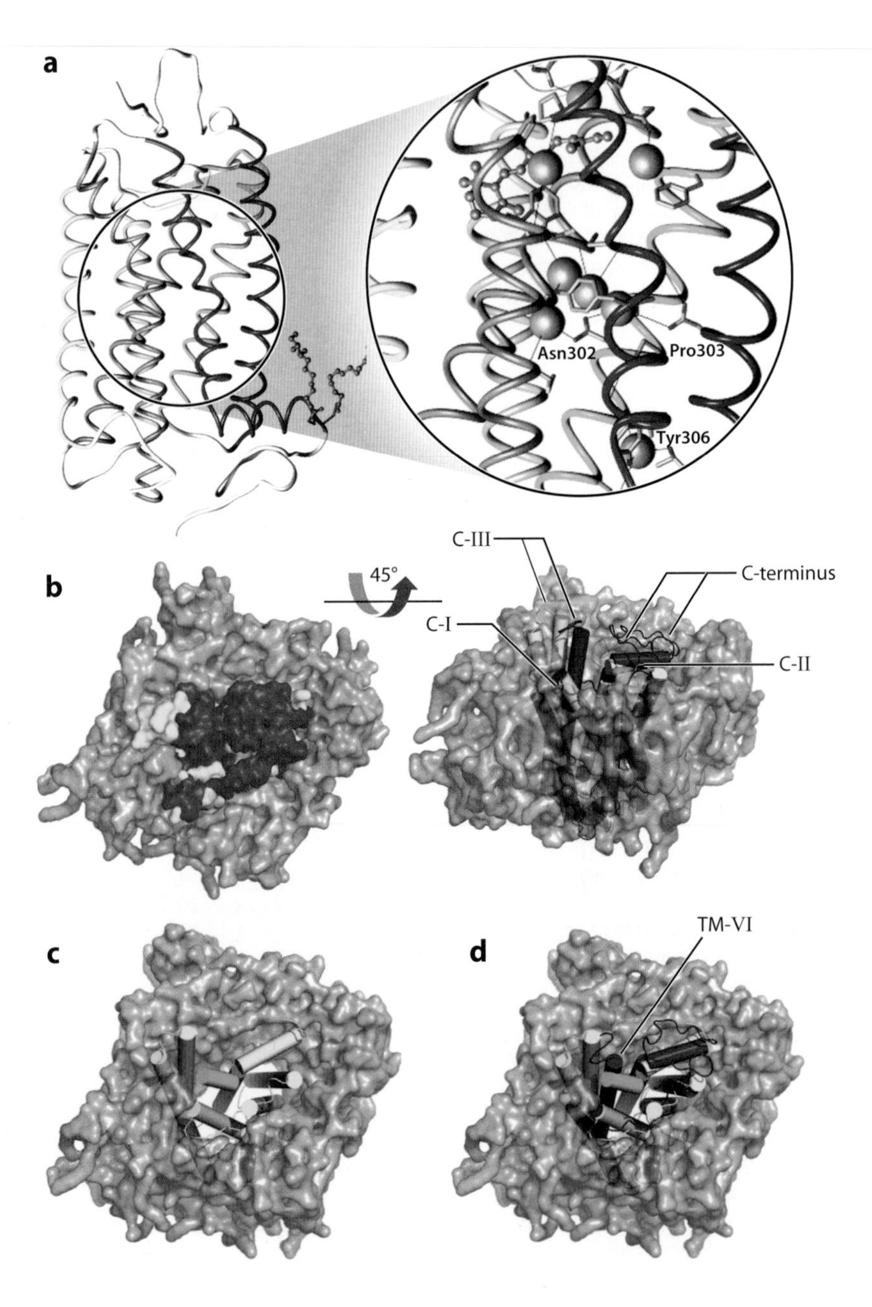

(Li et al. 2004, Okada et al. 2002, Salom et al. 2006). Subsequent collaborative efforts led to the discovery of rhodopsin dimerization in native membranes (Fotiadis et al. 2003), and the first reported photoactivated deprotonated rhodopsin structure (Salom et al. 2006).

A new era in the GPCR field began in 2007 including studies that produced the first structures of GPCRs bound to various diffusible and covalently bound ligands (Cherezov et al. 2007, Rasmussen et al. 2007, Rosenbaum et al. 2007), the structure of squid rhodopsin (Murakami & Kouyama 2008, Shimamura et al. 2008), and a glimpse into the GPCR activation process provided by the structure of bovine opsin (Park et al. 2008a). Some of these advances were facilitated by using protein engineering techniques to increase GPCR stability—a much sought after property for crystallization trials (Tate & Schertler 2009). Often N-termini, ICL-III, and/or C-termini were shortened to improve expression and crystallization (Mustafi & Palczewski 2009). Another strategy employed the fusion of GPCRs with fast-folding proteins such as T4 lysozyme (Cherezov et al. 2007, Rosenbaum et al. 2007) or thermostabilized apocytochrome b_{562} (Chun et al. 2012, Thompson et al. 2012). GPCR crystal formation has also been successfully facilitated by the formation of complexes between GPCRs and antibody Fab fragments or nanobodies (Day et al. 2007). **Figure 2*a–t*** shows all currently known three-dimensional structures of GPCRs together with **Figure 2*u***, a one-to-one comparison half matrix that illustrates the main differences between members. Inspection of **Figure 2*a–t*** indicates that GPCRs, as described above, share a high degree of structural homology except for the CXCR4 chemokine receptor, the most distant member of the cluster of known structures from the homology tree (**Figure 1**; **Figure 2*u***, column f and row f). The structural differences observed when comparing GPCR members with the CXCR4 chemokine receptor stem from the disordered nature of helix VIII in the CXCR4 chemokine receptor.

GPCR ligands: agonists, antagonists/inverse agonists, or neutral ligands that bind to orthosteric GPCR sites, often causing GPCR activation or inhibition

Investigators have observed specific diversity in the GPCR ligand-binding sites, which indicates that these receptors can respond to different groups of chemicals, peptides, or proteins (Granier & Kobilka 2012). It is an incredible evolutionary triumph that hundreds of different genes have produced proteins that fold into similar structures that diversify by fine-tuning their ligand-binding sites. Thus, it was not suprising that the 2012 Nobel Prize in Chemistry recognized these accomplishments in the field. Robert J. Lefkowitz of Duke University (Durham, North Carolina), and Brian K. Kobilka of Stanford University School of Medicine (Palo

Figure 3

Signal transduction in GPCRs. (*a*) Diagrams of the seven transmembrane (TM) helices and hydrogen-bonding network in rhodopsin (PDB ID:1U19). (*Left*) Helices are colored according to their primary sequence: helix-I (*blue*); helix-II (*blue-green*), helix-III (*green*); helix-IV (*light green*); helix-V (*yellow*); helix-VI (*orange*); helix-VII (*red*); helix-8 (*purple*). (*Right*) The chromophore is shown with balls and sticks (*pink*). Water molecules are displayed as spheres (*light blue*). The receptor is oriented such that the extracellular space is above and the G protein–interacting cytoplasmic face is below. (*b*) The superposed surface representations of ground-state rhodopsin (PDB code: 1u19 in *red*) and photoactivated-like rhodopsin (PDB code: 4a4m in *yellow*) viewed from the intracellular side of a plasma membrane. Phospholipids representing the plasma membrane are shown as a gray layer. On the right, 45° rotation reveals major secondary structure elements of rhodopsin and photoactivated rhodopsin visible outside the phospholipid membrane. Locations of intracellular loops (ICLs) and the C-terminal segment are indicated by arrows. (*c*) An intracellular view of metarhodopsin II (*yellow*) with a bound peptide (*green*) derived from the primary sequence of Gt. (*d*) Superposition of the structure of ground-state rhodopsin (*red*) on the structure of photoactivated rhodopsin (*panel c*). The overlap of the peptide derived from Gt with ground-state rhodopsin suggests that interaction of the two would be unlikely.

Alto, California) were recipients of the award for studies of G protein–coupled receptors (**http://www.Nobelprize.org**).

Gt (transducin): the heterotrimeric G protein localized to the outer segment of rod photoreceptors that couples specifically to rhodopsin

Bound water molecules: integral components of a biological macromolecule that do not readily exchange with external water and can act as cofactors

Membrane proteins: water-insoluble proteins that reside at least partially in lipid membrane bilayers

Water Molecules in the Transmembrane Segments of GPCRs

Upon photoactivation, rhodopsin, with its covalently bound ligand 11-*cis*-retinal, undergoes changes that involve both deprotonation of the Schiff base and protonation/deprotonation at the cytoplasmic surface, which intuitively suggests a role for internal water molecules in proton transfer (Hofmann et al. 2009). Crystallography, particularly of GPCR structures at high resolution, has unequivocally revealed the presence of water molecules bound in specific sites within the hydrophobic TM α-helical domain (Angel et al. 2009a). More recently, radiolytic footprinting analysis of bound waters revealed that internal water was redistributed when rhodopsin was activated and the rod photoreceptor–specific G protein (transducin or Gt) was subsequently bound (Angel et al. 2009b, Orban et al. 2012b). In addition to bound water molecules, sodium ions can play an important role in GPCR function as recently proposed for the highest resolution X-ray structure (1.8 Å) of a GPCR solved to date (Liu et al. 2012). Reorganization of water molecules identified in different states of photoactivated rhodopsin (**Figure 3*a***) and in dynamic regions noted in both photoactivated rhodopsin and Gt revealed that the receptor undergoes relaxation upon activation and then rigidifies once the G protein binds (**Figure 4**). Thus, water appears to mediate the conformational changes between the ligand-binding site and the G protein–coupling domain on the cytoplasmic surface of these receptors. A general characteristic of all transmembrane proteins, even those that are not water channels, could be their use of internal water to stabilize secondary elements and allow flexibility in assembly of these elements. This underappreciated strategy perhaps allows membrane proteins to behave like highly hydrated sponges, requiring only minimal energy from ligand binding to be sensed through the receptor. This hydration possibility should also caution researchers who interpret GPCR mutagenesis studies because changes in side chains within TM α-helical segments will inevitably reorganize the internal water molecules. This hydration phenomenon could cause serious problems for rational drug (ligand) design in general if investigators do not consider the precise localization of water molecules. The number of water molecules within the TM α-helical domain is highly conserved among GPCRs, suggesting that functional roles for water-mediated contacts can also be conserved across members of this receptor family (Angel et al. 2009a, Pardo et al. 2007).

GPCR ACTIVATION AND RECRUITMENT OF G PROTEINS AND OTHER PROTEINS

Mechanism of GPCR Activation: A Brief Account

The crystal structures of several GPCRs help researchers understand how GPCRs activate, stabilize their inactive conformations, and interact with partner proteins. These receptors likely share a common mode of activation (Smith 2012). One idea is that a rearrangement of TM-V and TM-VI opens a crevice at the cytoplasmic side of the receptor into which the C-terminus of the cognate G protein can bind (Hofmann et al. 2009). Another idea, derived from a nuclear magnetic resonance study of rhodopsin in membranes, proposes that the retinal ligand initiates collective helix fluctuations on a microsecond to millisecond timescale between activated GPCR conformers at equilibrium (Struts et al. 2011). Because membranes are integral to GPCR signaling (Inagaki et al. 2012, Jastrzebska et al. 2011a, Kaya et al. 2011), this idea deserves serious consideration.

Crystallography of GPCRs has accelerated over the past three years with the determination of a large number of antagonist-bound GPCR structures as well as with more recent work that has produced agonist-bound (activated) structures (**Figure 2**). For example, Doré et al. (2011) determined the structures of a

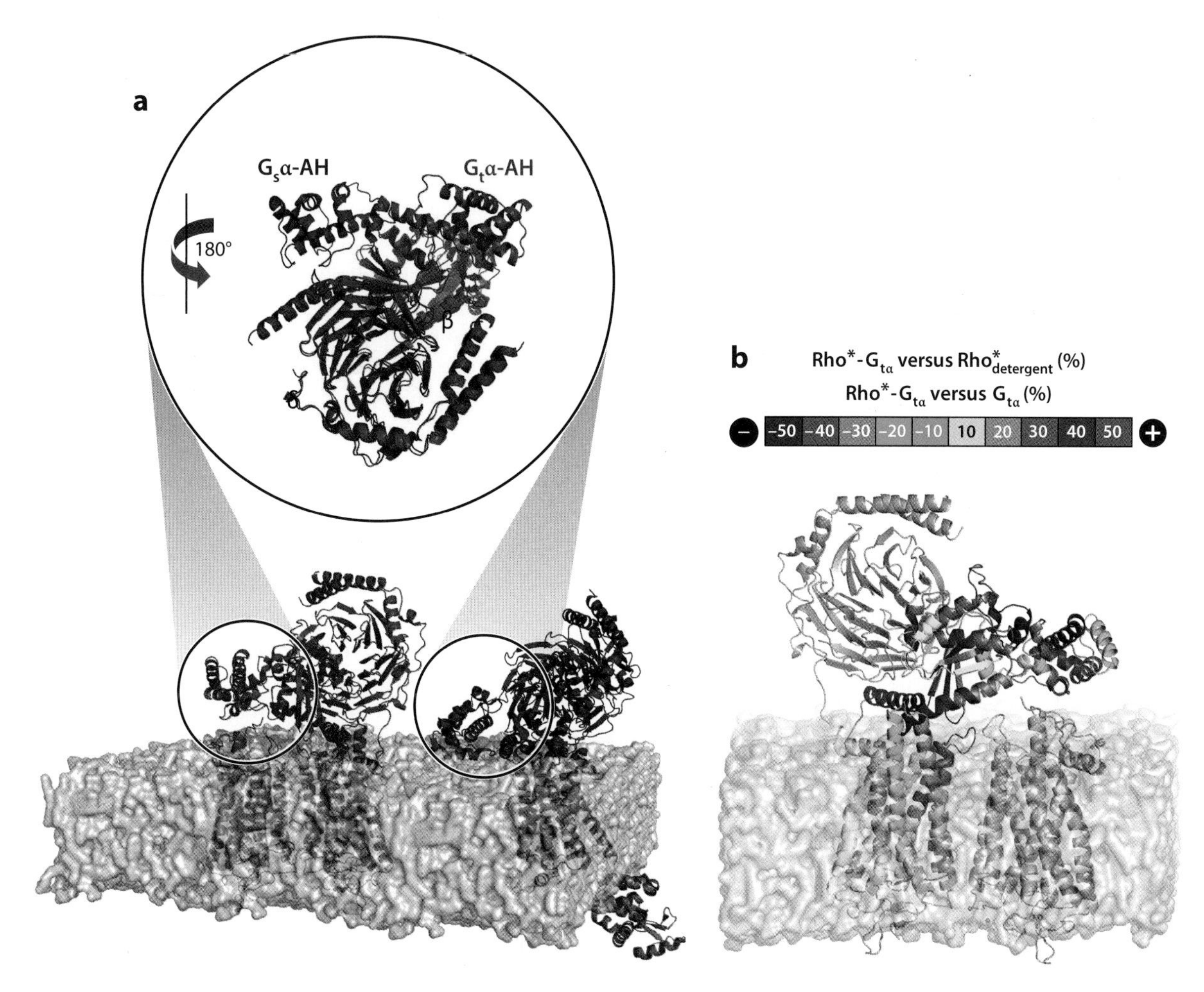

Figure 4

GPCR: G protein complex. (*a*) Superposed structures of photoactivated rhodopsin in complex with Gt (*red*) (Jastrzebska et al. 2011b) and the β_2-adrenergic receptor in complex with Gs (*blue*) (Chung et al. 2011) (*dark gray*, T4 lysozyme present in the β_2-adrenergic receptor-Gs structure; *gray transparent layer*, the phospholipid membrane). The section highlighted in detail (*large black circle*) shows superposed Gt (PDB code: 1GOT, *red*) and Gs (PDB code: 4A4M, *blue*). The averaged angle between the Gsα-AH and Gtα-AH domains evaluated from several measurements was ~95°, whereas distances between corresponding atoms of the two domains had an average value of ~40 Å (Jastrzebska et al. 2011b). Superposition of the Gsα-AH subdomain with the Gtα-AH subdomain is achieved only after a full 180° rigid body rotation around the axis shown on the left. (*b*) Differences in normalized hydrogen-deuterium exchange evaluated for photoactivated rhodopsin, photoactivated rhodopsin-Gt, and Gt. Heat maps (*dark blue to rose*) were used to evaluate differences in normalized hydrogen-deuterium exchange for free Gt and Gt in complex with photoactivated rhodopsin and for free photoactivated rhodopsin and photoactivated rhodopsin in complex with Gt (Orban et al. 2012b). Heat maps were then placed on the three-dimensional structure of the photoactivated rhodopsin-Gt complex model (Jastrzebska et al. 2011b). Negative differences in hydrogen-deuterium exchange are shown as 0–9% (*green*), 10–19% (*cyan*), 20–29% (*light blue*), 30–39% (*blue*), and 40–50% (*darker blues*). Positive differences are displayed as 0–9% (*yellow*), 10–19% (*light orange*), 20–29% (*orange*), 30–39% (*red*), and 40–50% (*magenta to purple*). The lipid bilayer where photoactivated rhodopsin is embedded is presented as a transparent gray layer. Abbreviation: AH, α-helical.

All-*trans*-retinal: vitamin A (retinol) aldehyde is metabolically produced from β, β-carotene but also results from photoisomerization of the visual pigment chromophore, 11-*cis*-retinal

Hydrogen-deuterium exchange: a chemical labeling technique used to track dynamic changes in protein structural conformation

thermostabilized adenosine A2A receptor in complex with xanthine, a xanthine amine congener, and caffeine, as well as the A2A selective inverse agonist ZM241385. This receptor was crystallized in its inactive conformation because the ionic lock in the conserved three-amino-acid DRY motif region was observed. Endorphins and opioid alkaloids bind to three members of the GPCR family: the μ-, δ-, and κ-opioid receptors. Together the structures of the μ- and κ-, the δ-opioid receptors, and the nociceptin/orphanin FQ (N/OFQ) peptide receptor structure with bound antagonists provide insight into the conserved elements of opioid ligand recognition and also reveal those features associated with ligand-subtype selectivity (Granier et al. 2012, Manglik et al. 2012, Thompson et al. 2012, Wu et al. 2012). A more complete list of crystallized GPCRs with bound agonists, antagonists, or antibody fragments is shown in **Figure 2**.

Constitutively active GPCR mutants with increased basal activity have been used to elucidate the structures of activated forms of rhodopsin. Two different constitutively active mutants of rhodopsin (E113Q and M257Y) with a bound C-terminal fragment of a G protein α-subunit were used to trap the activated state of rhodopsin with the agonist all-*trans*-retinal present in the native-binding pocket (Deupi et al. 2012, Standfuss et al. 2011). Within crystal structure uncertainty, this conformation is virtually identical to that of opsin-all-*trans*-retinal bound to a similar peptide (Choe et al. 2011). Investigators have also successfully probed the dynamics of GPCRs during the activation process using hydrogen-deuterium exchange (Lodowski et al. 2010, Orban et al. 2012b, West et al. 2011) and long-time scale computational simulations (Dror et al. 2011).

Conformational changes on the cytoplasmic side of the β_2-adrenergic receptor following binding of the agonist BI-167107 can be described by the rigid body movement of TM-VI. This alteration of the TM-VI position probably results from the disrupted hydrogen-bonding network within the core of this GPCR, as shown in a detailed snapshot of this region in rhodopsin. The hydrogen-bonding network starting in the immediate vicinity of the ECL region maps continuously throughout the entire core of the protein all the way to the ICL. This network involves internal water molecules and amino acid residues such as Asn302, Tyr306, and even the side chain of Pro303 in one instance (**Figure 3*a***). The network's stability was perturbed after activation of rhodopsin and binding of Gt (Orban et al. 2012b). However, whether disruption of the hydrogen-bonding network is the cause or the effect or works in synergy with the rigid body movement of the TM helices during agonist binding/receptor activation remains uncertain.

Structure of Light-Activated Rhodopsin in Membranes

Park and colleagues (Park et al. 2008b) proposed that GPCR action can be analogous in some respects to regulatory enzymes where the ligand-binding site is discrete from the active site (Changeux & Edelstein 2005). Transmission of the signal from the ligand-binding site of a GPCR to the nucleotide-binding site in the G protein is proposed to occur by relaxing the former, allowing transient association between the two proteins. The energy of their binding then is used to open the nucleotide-binding site and release guanosine diphosphate (GDP) from the G protein, whereas the subsequent binding of guanosine-5′-triphosphate (GTP) provides the energy to break this complex and release the G protein.

Activated rhodopsin can form a stable complex with Gt when the guanyl nucleotide-binding site in the G protein α-subunit is unoccupied (Bornancin et al. 1989). This finding was expanded in more recent studies (Jastrzebska et al. 2006, 2011a), showing that communication between the chromophore and the nucleotide-binding site on a G protein can be uncoupled once the complex is formed. The chromophore can be added in either all-*trans*- or 11-*cis*- configurations, but dissociation of the G protein restores the receptor back to the inactive intermediate (Jastrzebska et al. 2009).

As noted above, changes in the ligand-binding site are transmitted to the GPCR's cytoplasmic surface. But only the cytoplasmic surface is recognized by the G protein to initiate signal transduction (**Figure 3*b***). The deprotonated photoactivated form of rhodopsin was successfully crystallized by Salom and colleagues (2006). Insights into rhodopsin activation were obtained from studies of opsin at low pH that was assumed to have achieved an activated conformation (Park et al. 2008a, Scheerer et al. 2008), even though its catalytic properties were known to be extremely low compared with those of photoactivated rhodopsin (Buczylko et al. 1996, Jäger et al. 1996). Opsin crystals can bind all-*trans*-retinal without significant conformational changes (Choe et al. 2011, Scheerer et al. 2008) and can also bind the Gtα C-terminal peptide (Choe et al. 2011) (**Figure 3*c,d***). The 2–5 Å changes in helix V at the cytoplasmic surface end are needed to dock G protein because, in rhodopsin, this helix blocks full opening of the binding site. This simplistic docking of just the Gtα C-terminal peptide cannot be the only binding site of Gt to rhodopsin. Gt binds to dark-state rhodopsin prior to photoactivation (Hamm et al. 1987), suggesting that this nonproductive complex does not achieve the necessary conformational changes to allow nucleotide exchange.

Oligomeric Forms of GPCRs

In biochemical assays, largely of detergent-solubilized preparations, either monomeric or multimeric forms of GPCRs can activate their cognate G-proteins (Jastrzebska et al. 2006), but they assemble into homo- and heterodimers or larger oligomers in native tissues (Palczewski 2010, Park et al. 2004, Prezeau et al. 2010, Terrillon & Bouvier 2004). In the dimeric form, GPCR monomers are not functionally equivalent because only one monomer will bind to the C-terminal helix of the cognate G protein, and ligand affinities could differ between the monomer and the dimer (Fuxe et al. 2012, Jastrzebska et al. 2013, Maurice et al. 2011). Cholesterol and palmitoyl moieties present on many GPCR C-terminal regions could play a critical role in forming the dimer interface (Maeda et al. 2010, Oates & Watts 2011, Thompson et al. 2011, Zheng et al. 2012). Moreover, each monomeric unit, when occupied by a ligand, does not transmit an equivalent signal to its cognate G protein. The efficiency of activation can depend on whether one or two ligands are present in the dimer (Pellissier et al. 2011). A mathematical model was recently created to reflect this asymmetric organization and its impact on different signaling pathways (Rovira et al. 2010). On the basis of theoretical considerations, Neri et al. (2010) proposed that one monomer of rhodopsin is responsible for light detection, whereas the other serves as the G protein activator. Moreover, GPCR homodimer activity can be controlled by an allosteric ligand-binding site when different ligands bind to each receptor monomer (Zylbergold & Hébert 2009).

Visualization of GPCRs in native tissue is a key to understanding the physiological settings of these receptors. Atomic force microscopy (AFM) of rhodopsin in native rod disc membranes provides the clearest structural picture to date of the oligomeric arrangement of a GPCR (Fotiadis et al. 2003, Liang et al. 2003) (see sidebar, Imaging GPCRs in Native Settings). AFM measurements yielded a density of 30,000–55,000 rhodopsin molecules/μm^2 and $\sim 10^8$ rhodopsin molecules per rod, partially organized in para-crystalline arrays composed of rows of rhodopsin (Fotiadis et al. 2003, Liang et al. 2003). These rows of rhodopsin dimers were used to derive spatial constraints that were used to construct a molecular model of the oligomeric structure of rhodopsin in the rod outer segment (ROS) membrane (Fotiadis et al. 2004). This model indicates that the rhodopsin dimer offers a platform complementary to the binding of a single Gt or arrestin molecule (**Figure 5**).

Only limited information is available about the dynamics and stability of GPCR dimers (Lambert 2010). To this end, development of a single fluorescent-molecule imaging method

Allosteric ligand-binding site: a separate ligand-binding site on a receptor that either positively or negatively modulates the receptor's affinity for its primary ligand(s)

Atomic force microscopy (AFM): a high-resolution technique for imaging flat materials by probing them with a contacting cantilever

Rod disc membranes: Internal membranes shaped like flattened sacks that occupy the outer segments of rod photoreceptor cells

IMAGING GPCRs IN NATIVE SETTINGS

High-resolution methods are needed for two types of GPCR imaging. First is the static self-clustering and complexing of GPCRs with other proteins in native membranes. The second is real-time imaging to understand and quantify GPCR dynamics in response to cellular changes. For static imaging, atomic force microscopy, used to resolve metal orbitals on hard surfaces, has provided the highest resolution images of several membrane-embedded or reconstituted proteins in various activated states (Engel & Gaub 2008, Müller et al. 2008). Near-field scanning optical microscopy, fluorescence correlation spectroscopy, and other advanced confocal and 2-photon microscopy techniques also permit static visualization of GPCRs in membranes (Herrick-Davis et al. 2012, Ianoul et al. 2005). Dynamic changes of GPCRs are monitored predominantly by fluorescence methods. Resonance energy transfer methods together with complementation and transactivation of functional receptors provide evidence for receptor colocalization and oligomerization. These methods are also used for imaging applications (Harrison & van der Graaf 2006). Quantum dot (QD) probes can monitor movement of a single GPCR during its life cycle (Fichter et al. 2010) or follow a ligand that binds tightly to these receptors (Zhou et al. 2007).

allowed investigators to observe a dynamic equilibrium for the *N*-formyl peptide receptor in live cells, indicating that dimers rapidly fall apart and reassemble (Kasai et al. 2011). It appears that the M3 muscarinic receptor also exists in multiple dimeric/oligomeric arrangements (McMillin et al. 2011). GPCR oligomerization could be responsible for fine-tuning receptor function and signaling (Tadagaki et al. 2012).

Great advances were made using noninvasive fluorescence- and luminescence-based techniques to observe bimolecular fluorescence complementation that can result from binding fluorescent protein fragments to nonfluorescent proteins. This innovative approach allowed researchers to visualize and measure protein interactions in neuronal cells (Ciruela et al. 2010, Vidi et al. 2010). Adapted to native tissues, the same strategy also demonstrated the presence of oxytocin receptor dimers and/or oligomers in the mammary gland (Albizu et al. 2010). An elegant application of fluorescence correlation spectroscopy with photon counting from histograms led to the recent conclusion that GPCR dimers represent the basic signaling unit (Herrick-Davis et al. 2006). Some other evidence for the functional significance of heterodimerization, such as variants of the serotonin receptor or the metabotropic glutamate receptor, is questionable on the basis of pharmacological assays using the heterologous expression system (Delille et al. 2012) (see sidebar, Heterologous Expression of GPCRs). Effects of oligomerization itself must be distinguished from the integration of inputs from different receptors. Data derived from native tissues in a physiologically relevant setting are most critical. For example, Park et al. (2012) demonstrated a functional heterodimer for monomeric GPCRs known as DAF-37 and Daf-38 in live *C. elegans* (Park et al. 2012).

With such a large number of GPCRs, some mechanisms of oligomerization and signaling will apply to some receptors, whereas others will not. Oligomerization or parallel signaling could have dramatic effects on medically relevant treatments. For disorders such as schizophrenia and dementia, metabotropic glutamate 2 receptor/serotonin 2A receptor–mediated changes in Gi and Gq activity could affect the psychoactive behavioral effects of a variety of pharmacological compounds (Fribourg et al. 2011).

Complexes Between GPCRs and G Proteins

The only available structural model for the photoactivated rhodopsin-Gt complex is a recently described low-resolution electron microscopic reconstruction by single particle analysis (Jastrzebska et al. 2011b). After purifying and analyzing photoactivated rhodopsin and Gt, investigators determined a 2:1 molar ratio of photoactivated rhodopsin to Gt and calculated a 22 Å structure from projections of the negatively stained complexes. The molecular envelope accommodated two rhodopsin molecules together with one Gt heterotrimer, consistent

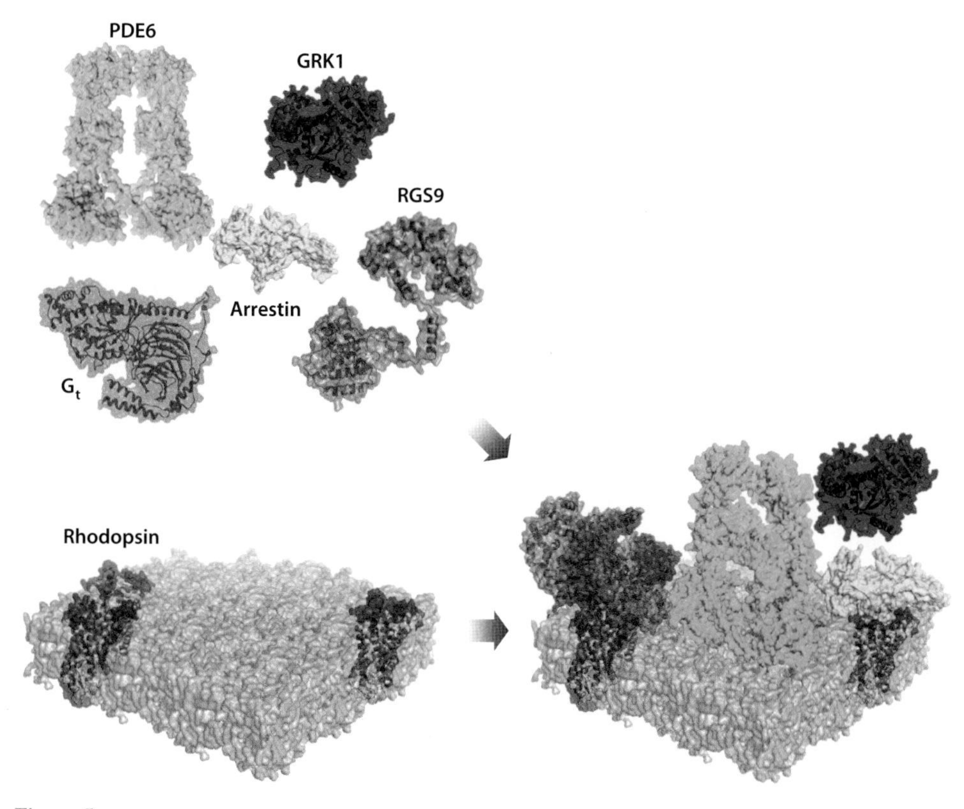

Figure 5

Hypothetical model for assembly on a cell surface of major proteins required for effective visual signal transduction. Visual signal transduction is carried out by a multitude of proteins and messengers. (*Top left*) Structures of proteins involved in phototransduction. (*Bottom left*) A phospholipid cell membrane represented as a gray transparent layer containing a dimer and monomer of rhodopsin (*red*). Various complexes of these proteins assembled on the membranes are modeled on the right. Following activation, heterotrimeric Gt (*pink*) is recruited to the cytoplasmic cell surface where it binds to photoactivated rhodopsin, forming the photoactivated rhodopsin-Gt complex. Gt-bound guanosine diphosphate (GDP) is then exchanged for guanosine-5′-triphosphate (GTP). Next, phosphodiesterase 6 holoenzyme (PDE6, *green*) is recruited to the membrane and carries out hydrolysis of cyclic guanosine monophosphate (cGMP). The activity of Gt is, in turn, suppressed by RGS9 (*orange*), which promotes its inactivation. Finally, arrestin (*yellow*) and GRK1 (*blue*) both help deactivate photoactivated rhodopsin, a step required for resetting the dark state of rhodopsin prior to another light stimulus.

with a heteropentameric structure for this complex (Jastrzebska et al. 2011b) (**Figure 4*a***).

Another breakthrough in the understanding of how an activated GPCR binds to G proteins came with the recent crystal structure of an active state complex composed of (*a*) agonist-occupied monomeric β_2-adrenergic receptor-T4L-lyzosome fusion, (*b*) nucleotide-free Gs heterotrimer, and (*c*) a nanobody (Rasmussen et al. 2011) (**Figure 4*a***). In this complex, the cytoplasmic end of TM-VI and an α-helical extension of the cytoplasmic end of TM-V were displaced by several Å compared with the inactive receptor. As predicted from biochemical studies (Oldham & Hamm 2008), the amino- and carboxy-terminal α-helices of the G protein exhibited conformational changes that propagated to its nucleotide-binding pocket. It was surprising that the GPCR was in a monomeric state because single-particle analyses of a similar preparation indicated that the receptor exists to some

HETEROLOGOUS EXPRESSION OF GPCRs

Most GPCRs are expressed at such low levels in native tissues that investigators must overexpress them to investigate their structure and function. Rhodopsin constitutes a notable exception because it is expressed at 1 mg per bovine retina and 5 mM in rod outer segments (Palczewski 2012). GPCRs in taste buds and olfactory receptor neurons are also expressed at high concentrations, but receptors in these tissues are scarce (DeMaria & Ngai 2010), making purification difficult. For these reasons, we developed an in vivo system to express heterologous GPCRs in rod photoreceptors of *Xenopus* (Zhang et al. 2005) and mice (Li et al. 2007, Salom et al. 2008). More recently, researchers achieved overexpression of heterologous GPCRs in *C. elegans* (Cao et al. 2012, Salom et al. 2012). However, simpler cultured insect (Schneider & Seifert 2010) and mammalian cells (Li et al. 2005) are most often employed for expressing heterologous GPCRs because they can perform the appropriate complex posttranslational modifications and their membranes mimic native membranes. GPCRs have been successfully expressed in yeast and cell-free systems, whereas expression in bacteria typically requires refolding (Lundstrom 2006, McCusker et al. 2007). To be useful, heterologously expressed GPCRs must also display the same pharmacology as native receptors.

degree in a dimeric form (Westfield et al. 2011). But the most unexpected observation was a major displacement of the entire α-helical (AH) domain of Gα relative to the Ras-like GTPase domain (Rasmussen et al. 2011).

Recent models of photoactivated rhodopsin-Gt and β_2-adrenergic receptor-Gs complexes provide important details about the interaction of a GPCR molecule with its corresponding G protein. Both photoactivated rhodopsin and the β_2-adrenergic receptor interact with their corresponding G proteins through the C-terminal end of the α-domain (**Figure 4*a***). Superposition of the Gt and Gs structures reveals a relatively small root-mean-square deviation (~0.5–1.0 Å) for the β and γ subunits and the Ras subdomains of the α-domains. However, the positions of the AH subdomains of the α-domains differ markedly in Gt and Gs. This difference is due to the rigid body–type displacement of these subdomains (**Figure 4*a***). (The averaged angle between these subdomains was ~130°, whereas the distance between the AH subdomains of Gt and Gs was found to be ~40 Å.) For the two domains to superpose, the Gsα-AH subdomain alone requires a full ~180° rotation (**Figure 4*a***). This displacement could result from the nanobody that is stabilizing the β_2-adrenergic receptor-Gs complex in the crystal.

AH: α-helical domain

As do rhodopsin and Gt, the M3 muscarinic receptor forms inactive-state complexes with Gq heterotrimers in intact cells (Qin et al. 2011). Using ~250 combinations of cysteine-substituted M3 muscarinic receptors and Gαq proteins, investigators tested which of these mutants undergo cross-link formation. This strategy identified M3 muscarinic receptor–Gαq contact sites that help build low- and high-resolution models of the complex (Hu et al. 2010).

In summary, allosteric communication between the ligand-binding site in photoactivated rhodopsin or β_2-adrenergic receptors and the nucleotide-binding site of a G protein (Gs or Gt) at a distance of ~50–60 Å is achieved by precise conformational changes that include helical movements and a reorganization of internal water molecules. The complex between the GPCR and G protein suggests that binding of a G protein confers a stabilizing effect on the receptor whose rigidity resembles that of ground-state rhodopsin (**Figure 4*b***). This observation sharply contrasts with the highly dynamic structure of photoactivated rhodopsin, which forms a broad constellation of conformers.

Other Proteins Interacting with GPCRs

GPCRs interact with receptor kinases, arrestins, and a variety of other proteins, in addition to their G proteins. Together they form large functional complexes that directly mediate receptor signaling or play key roles in trafficking and cellular localization. These multipart protein assemblages could influence pharmacological properties of these receptors (Ritter & Hall 2009).

Substrates favoring the activated ghrelin state initiated by agonist-induced arrestin recruitment cause ghrelin to adopt a different conformation from that observed in the absence of arrestin (Mary et al. 2012). Thus, G proteins affect the balance between active and inactive ghrelin receptors. Another study also showed that the ghrelin conformation stabilized by a G protein/agonist is distinct from that stabilized by arrestin-based agonists (Rahmeh et al. 2012). Molecular understanding of the interplay between GPCRs and their kinases has advanced significantly with more precise elucidation of receptor recognition and phospholipid binding that leads to kinase activation (Boguth et al. 2010). Huang & Tesmer (2011) have proposed that the interaction of receptor kinases and arrestins, despite differences in their folding, occurs via a common molecular mechanism. In the presence of ATP, the influence of photoactivated rhodopsin versus ground-state rhodopsin on rhodopsin kinase dynamics was negligible (Orban et al. 2012a). Few experimental techniques can assess the orientation of peripheral membrane proteins in their native environment. Sum-frequency generation vibrational spectroscopy was used to determine the membrane orientation of the GPCR kinase 2-G-$\beta\gamma$ complex (Boughton et al. 2011). Rhodopsin kinase phosphorylates rhodopsin at different subsets of Ser and Thr residues at the C-terminal tail, producing phosphorylated receptor variants with divergent functional properties (Maeda et al. 2003). Moreover, researchers noted a similar phenomenon for the β_2-adrenergic receptor in cell cultures (Liggett 2011, Nobles et al. 2011). Phosphorylation of GPCRs and the subsequent capping of these receptors by arrestins are essential because recent studies show that internalized GPCRs can continue either to stimulate or to inhibit sustained cAMP production by remaining associated with their cognate G protein subunits and adenylyl cyclase in endosomal compartments. Once internalized, these GPCRs can produce cellular responses distinct from those elicited at the cell surface (Jalink & Moolenaar 2010).

Large complexes with other proteins are also necessary to target GPCRs to subcellular compartments and their trafficking to and from the plasma membrane during internalization (Bockaert et al. 2010). For example, activation of Arf6 leads to β_2-adrenergic receptor degradation and accumulation and also negatively controls Rab4-dependent fast recycling to prevent the resensitization of β_2-adrenergic receptors (Macia et al. 2012). Another important physiological example is the interaction of GPCRs with voltage-gated calcium channels that can synergize neurotransmitter release (Altier & Zamponi 2011). In addition to Ca^{2+} influx and direct interaction with Ca^{2+}-binding proteins (Navarro et al. 2012), rapid depolarization-induced charge movement in GPCRs is needed for neurotransmitter control (Kupchik et al. 2011). Opioid receptors form complexes with transient receptor potential (TRP) Ca^{2+} channels as well (Yekkirala 2013).

Retinal pigment epithelium (RPE): a cell layer just outside the neurosensory retina critical for metabolite transport and photoreceptor cell maintenance

CELLULAR CONTEXT OF GPCR SIGNALING

Structural studies of GPCRs and other integral membrane proteins, such as those involving X-ray crystallography or nuclear magnetic resonance, require high concentrations of well-purified membrane proteins in detergent solution. These artificial conditions obviously limit the relevance of these studies to GPCR signaling in vivo, where these transmembrane proteins interact mostly with other integral membrane proteins and membrane-associated proteins. Although high-resolution methods are required to obtain atomic details of biochemical molecules, the resulting structures must be pieced together in both structural and functional contexts (**Figure 5**).

Rod Photoreceptor Cells

The visual system and its signaling pathways currently provide a good molecular framework—a GPCR signaling system with physiological relevance. Over the years, many methods were developed to investigate the visual system. It is the only system in which

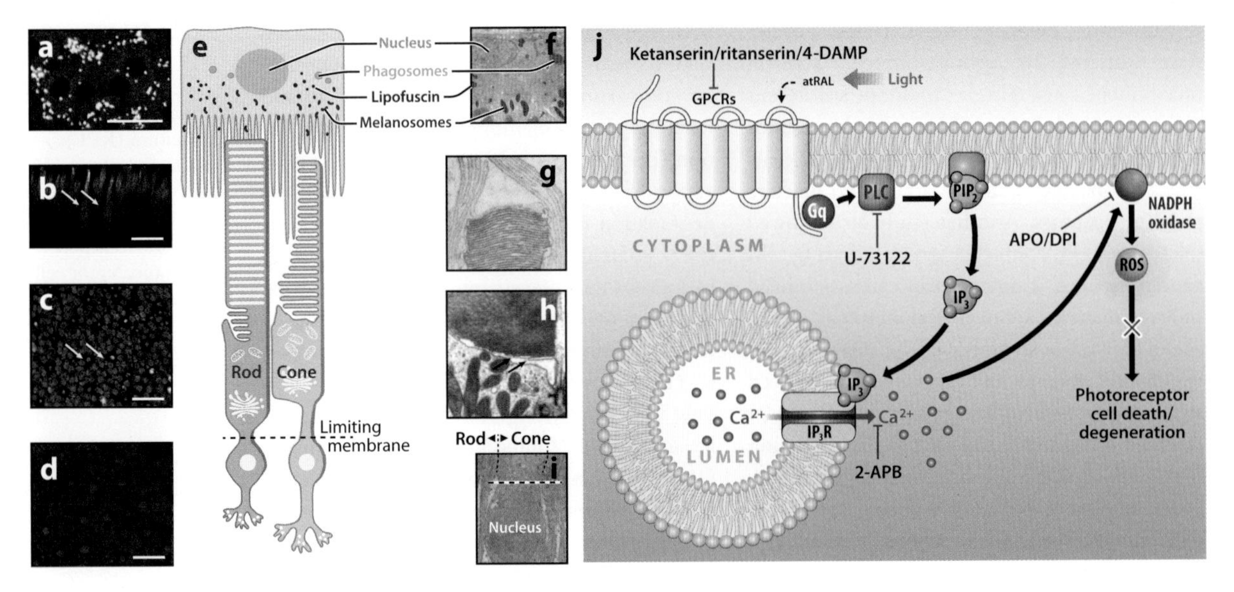

Figure 6

High-resolution images and pictorial representation of mammalian photoreceptors and the retinal pigment epithelium (RPE). (*a–d*) Fluorescence images obtained by two-photon excitation of endogenous fluorophores. Scale bars represent 20 μm. (*a*) Double-nucleated RPE cells obtained from an ex vivo unfixed C57BL/6J-*Tyr*$^{C\text{-}2J}$ mouse eye. Highly fluorescent retinosomes are visible as bright spots located along the plasma membranes. (*b*) A transverse (*xz*) image of cones and rods in cynomolgus monkey peripheral retina assembled from a series of *en face* images. Ellipsoids of two cones directly on the *xz* axis are indicated by yellow arrows. (*c*) An *en face* (*xy*) image of the rod and cone outer segment mosaic. Outer segments of two cones are indicated with yellow arrows. (*d*) *En face* (*xy*) image of the rod and cone inner segment mosaic. Two-photon images courtesy of G. Palczewska. (*e*) Pictorial representation of the RPE-photoreceptor interface. Melanosomes (*purple*) are located in the RPE cell and processes. Other elements: lipofuscin (*smaller dark red circles*), phagosomes (*larger blue circles*), cone photoreceptor cell (*green*), and rod cell (*blue*). The dashed line represents the external limiting membrane separating the nuclei and the ellipsoid with its abundant mitochondria. (*f–i*) Electron microscopic images. (*f*) Image of the RPE and a photoreceptor cell in extrafoveal rhesus monkey retina (from Anderson et al. 1980). (*g*) The tip of the outer segment of a foveal cone engulfed by RPE apical processes in rhesus monkey retina (from Anderson & Fisher 1979). Some distal discs appear separate from the cell membrane. (*h*) The cone outer segment base from a rhesus monkey retina (from Anderson et al. 1978). Black arrows indicate a new evagination of the membrane. (*i*) The nucleus of a Müller cell and the external limiting membrane. Müller cell villous processes extend beyond the external limiting membrane between rods and cones. (*j*) The cytotoxic effect of all-*trans*-retinal in light-induced photoreceptor degeneration. The illustrated signaling cascade implicates GPCRs, PLC/IP_3/Ca^{2+} signaling, and NADPH oxidase in this process. Elevated signaling of G_q-coupled GPCRs is involved in mediating all-*trans*-retinal toxicity during light-induced photoreceptor degeneration, but the precise mechanism has yet to be clarified (*black arrow with dotted line*). Activation of G_q-coupled GPCRs stimulates PLC/IP_3/Ca^{2+} pathways, which then lead to NADPH oxidase-mediated ROS production and photoreceptor degeneration (*black arrows*). Pharmacological interventions targeting G_q-coupled GPCRs, PLC/IP_3/Ca^{2+}, and NADPH oxidase protect photoreceptors from light-induced, all-*trans*-retinal-mediated degeneration (*red bars*). Modified illustration based on Chen et al. (2012, figure 8). APO/DPI, Apocynin/diphenylleneiodonium; ER, endoplasmic reticulum; NADPH, nicotinamide adenine dinucleotide phosphate; PLC, Phospholipase-C; ROS, reactive oxygen species

we can directly access the intact tissue and monitor its signal in vivo (**Figure 6**). In addition to the neural retina, the retinal pigment epithelium (RPE) layer is critically important in the eye. The RPE is responsible for (*a*) movement of metabolites between the choroidocapillaris and avascular photoreceptor cell layers; (*b*) metabolic transformation of spent all-*trans*-retinal back to its 11-*cis*-retinal light-sensitive form by a process called the visual (or retinoid) cycle; (*c*) daily phagocytosis of ~10% of photoreceptor outer segments; and (*d*) other homeostatic functions. Photoreceptor outer segments are critical for phototransduction, namely transformation of single photons of light into a biochemical cascade of events that culminate in neurotransmission. Mouse ROS lengths were estimated to be ~24 μm with a diameter of ~1.2 μm (Liang et al. 2004). Particularly fascinating are structural studies of internal structures called disc membranes. For example, a mouse ROS contains ~800 membranous discs stacked on top of each other, increasing the total membrane surface area of the cell ~1,500-fold (Mayhew & Astle 1997). The cytoplasmic space used for phototransduction represents only ~30% of the space inside a ROS [see a recent review for a summary of this structure (Palczewski 2012)].

The main protein of ROS internal discs is rhodopsin, which constitutes ~50% of the membrane volume (Filipek et al. 2003a). In addition to being a photon catcher, rhodopsin is also a critical structural protein that establishes and preserves ROS morphology. The size of ROS discs is dictated by the level of rhodopsin expression (Liang et al. 2004), and absence of rhodopsin prevents ROS formation (Humphries et al. 1997).

Network Pharmacology in the Retina

The retinal transcriptome has now been analyzed, and most of the ROS proteome components have been identified by mass spectrometry (e.g., Wong et al. 2009). The retina contains a large number of GPCRs, including the visual pigments rhodopsin, short-wavelength cone opsin, melanopsin, and other receptors (**Table 1**). However, the intersections of multiple signaling cascades and the precise interplay between them are presently unknown. Localization of different GPCRs within the retina, their associated physiological responses, and genetic alterations in their expression will provide clues about their functions and the interdependence of these functions. These GPCRs also provide targets for therapeutics that can be used to prevent or ameliorate retinal degeneration.

Interrelated oxidative stress and Ca^{2+} imbalance both contribute to the pathology of retinal degeneration (Chen et al. 2012). NADPH oxidase inhibitors and PLC/IP_3/Ca^{2+} signaling antagonists similarly protect retinas from light-mediated degeneration, suggesting that these activities are involved in the same signaling pathway (**Figure 6*j***). Indeed, one could imagine a single signaling cascadc/pathway for light-induced retinal degeneration involving GPCRs, PLC/IP_3/Ca^{2+} signaling, and NADPH oxidase (**Figure 6*j***). Pharmacological interventions targeting specific reactions in this pathway could provide novel therapeutic strategies for treating blinding retinal disorders such as Stargardt's disease and age-related macular degeneration (Chen et al. 2012, Maeda et al. 2012). The serotonin 2A receptor is an excellent candidate for activating PLC, although data regarding its involvement in light-induced retinal degeneration are limited. Serotonin 2A receptor expression is readily detectible in the retina, and its activation leads mainly to elevations in cytosolic Ca^{2+} through PLC activation (Hoyer et al. 1994). This intersection of signaling pathways could be targeted to correct or prevent aberrant signaling leading to neurodegeneration.

Oxidative stress: a condition that results when the production rate exceeds the destruction rate of toxic reactive oxygen intermediates

NADPH oxidases: enzymes catalyzing formation of reactive oxygen species implicated in redox signaling. These constitute important therapeutic targets

Age-related macular degeneration: includes both geographic atrophic (dry) and neovascular (wet) types of age-related degeneration of retinal photoreceptor cells with progressive visual loss

Phospholipase C (PLC): an enzyme that hydrolyzes phospholipids before the phosphate group

CONCLUDING REMARKS

GPCRs have evolved to play an essential role in both sensory and endocrine control systems. The complexity of GPCR signaling continues to be revealed as investigators use new-generation high-throughput sequencing to identify the full repertoire of GPCRs in many organisms. Research has made steady progress to elucidate further the role of GPCRs in normal and pathological states. The past few years have provided much more detailed information about the assembly of GPCRs at the atomic level and their mode of binding to many antagonists and some agonists. This improved level of detail has led to testable hypotheses about GPCR activation. More than 50% of all publications in the past three years that discuss GPCRs explore the exciting possibility of expanding the GPCR pool by homo- and hetero-oligomerization (Park & Palczewski 2005). The next frontier in our understanding of GPCR signaling should uncover additional

Table 1 Listed are genes detected in the C57BL/6J mouse eye with an average fragments per kilobase of exon per million fragments mapped (FPKM) value of 2 or greater that are categorized as GPCRs by gene ontology (see Mustafi et al. 2011 for details)

Gene	B6 Eye Average FPKM	Gene	B6 Eye Average FPKM	Gene	B6 Eye Average FPKM
Rho	6162.04	*Adra2c*	8.96	*Gpr158*	3.72
Rgr	355.74	*Gpr146*	8.91	*Tacr1*	3.72
Opn1sw	125.13	*Vipr2*	8.79	*Fzd8*	3.56
Drd4	93.84	*Fzd5*	8.69	*Opn4*	3.35
Opn1mw	62.97	*Gpr110*	8.59	*Tshr*	3.24
Nisch	52.38	*Adrb1*	8.43	*S1pr2*	3.2
Gprc5b	29.82	*S1pr3*	8.42	*Mrgprf*	3.18
Gpr162	29.37	*Gabbr2*	7.8	*Oprl1*	3.15
Darc	28.91	*Lphn2*	7.66	*F2rl1*	3.13
Gpr37	28.47	*Lpar1*	7.47	*S1pr5*	3.12
Ednrb	22.27	*P2ry2*	7.2	*Gpr135*	3.07
Crcp	21.88	*Adrb2*	7.13	*Crhr1*	3.02
Gpr153	20.42	*Hrh3*	7.11	*Eltd1*	3.0
Gabbr1	19.78	*Gpr143*	6.8	*Mrgpre*	2.96
Rrh	19.29	*Celsr2*	6.53	*Gpr27*	2.92
Gpr152	18.55	*Fzd7*	6.34	*Ednra*	2.87
Adora1	16.2	*Drd1a*	6.15	*Opn3*	2.55
Lphn1	15.98	*Adora2b*	6.09	*Gpr165*	2.45
Cxcr7	14.3	*Celsr3*	5.82	*Cnr1*	2.41
Cd97	12.93	*Fzd4*	5.39	*Gpr37l1*	2.39
Gpr19	12.21	*Gprc5c*	5.26	*Crhr2*	2.24
Fzd1	11.99	*Gpr179*	5.12	*P2ry14*	2.23
Fzd6	11.34	*Gpr56*	5.12	*Gpr176*	2.21
Gpr87	11.34	*Tacr3*	4.95	*Celsr1*	2.18
Lgr4	11.09	*Ramp1*	4.68	*Gpr22*	2.17
Drd2	10.82	*Adra2a*	4.6	*Lgr5*	2.1
Smo	10.75	*Gpr85*	4.56	*Gpr26*	2.06
S1pr1	10.66	*Lphn3*	4.22	*Agtr2*	2.02
Glp2r	9.94	*Htr3a*	4.14	*Calcrl*	2.02
Ptger1	9.59	*Fzd2*	3.89	*Gpr68*	2.02
Gpr124	9.56	*Fzd10*	3.86	*Cckbr*	2.0
F2r	9.31	*Gpr98*	3.79		

GPCR structures and their partner proteins. In addition, new research must further define the cellular context of these various interacting complexes, such as neurons, that pertain to the indistinct function and physiology and contribute to the physiology of the whole organism. Pharmacological manipulation of G protein signaling continues to be extremely important in defining many interacting signaling pathways. Most GPCRs are not essential to establish life, but they are critical for sustaining it. The next decade should witness exciting research breakthroughs in the GPCR field both in generation of new and improved structures and in our interpretation of these structures and their implications for signaling. Even now, we can elucidate the genetics of complex G protein–signaling pathways between different individuals of the same species owing to next-generation DNA sequencing. A rich array of experimental approaches and techniques are currently available; thus, various combinations of biochemical and structural methods should greatly enhance our understanding of GPCR structure/function and its applications to modern medicine.

SUMMARY POINTS

1. GPCRs exist in many different structural forms to accommodate a broad range of external stimuli. Interest in their function and structure has been at the center of biological science for decades.
2. The past 15 years have witnessed tremendous advances in understanding of GPCR structures and those of a subset of complexes as determined by high-resolution methods such as electron microscopy, X-ray crystallography, and nuclear magnetic resonance. These studies show that highly divergent sequences yield GPCRs with similar overall topology but with high specificity in ligand binding. The mechanism of GPCR activation and GPCR interactions with partner proteins are likely to be highly evolutionally preserved with ligand and cognate G protein specificities among subgroups of receptors.
3. GPCRs are highly dynamic molecules, and water has emerged as an important element in their assembly and activation. Internal water molecules provide the elasticity essential for the ligand-binding signal to be transmitted 40 Å through the cell membrane to the cytoplasmic surface to enable interactions with G proteins and other partner proteins. This explanation is compatible with the multiple conformational changes elicited by different ligands and the thermodynamics governing the activation process.
4. Structural information derived from high-resolution methods must be aligned with physiological information about assemblies of GPCRs within cells such as neurons to avoid misinterpretation of potential artifacts arising from the study of highly purified, detergent-solubilized systems.
5. GPCRs were shown to exist in oligomeric forms that are critically important for their stability, signaling, and other cell-biological processes. Specific allosteric regulators (ligands of the allosteric-binding sites) could also target a specific subset of these structures. Because different signaling pathways intersect within a given cell, cultures of specific cell types can be used to identify pathology and devise preventive therapies that target GPCRs.

FUTURE ISSUES

1. Determine high-resolution structures of native GPCRs and receptor–G protein and receptor-arrestin complexes with all relevant posttranslational modifications.
2. Elucidate the precise rearrangements of water molecules during GPCR activation.
3. Understand the structural complementarities of GPCR homo- and hetero-oligomerization.
4. Determine the intersections of different relevant GPCR signaling pathways in tissues of interest.

5. Obtain high-resolution structures of signaling complexes in native tissues. Determine how GPCR signaling complexes are compartmentalized within cells.
6. Understand the energy landscape of GPCR activation and folding. Identify key residues for folding and membrane insertion.
7. Use next-generation DNA and RNA sequencing to identify the impact of mutations and polymorphisms of GPCR expression levels.

DISCLOSURE STATEMENT

The authors are not aware of any affiliations, memberships, funding, or financial holdings that might be perceived as affecting the objectivity of this review.

ACKNOWLEDGMENTS

We thank Drs. Leslie T. Webster, Jr., Phoebe L. Stewart, David Salom, David Lodowski, Paul Park, Philip Kiser, and members of Dr. Palczewski's laboratory for helpful comments on the manuscript. We also express our gratitude to Grazyna Palczewska for generating **Figure *6a–i***, Dr. Ray Stevens for generating **Figure 1**, and Debarshi Mustafi for data presented in **Table 1**. This work was supported by National Institutes of Health grants EY008061 and EY019478.

LITERATURE CITED

Albizu L, Cottet M, Kralikova M, Stoev S, Seyer R, et al. 2010. Time-resolved FRET between GPCR ligands reveals oligomers in native tissues. *Nat. Chem. Biol.* 6:587–94

Altier C, Zamponi GW. 2011. Analysis of GPCR/ion channel interactions. *Methods Mol. Biol.* 756:215–25

Anderson DH, Fisher SK. 1979. The relationship of primate foveal cones to the pigment epithelium. *J. Ultrastruct. Res.* 67:23–32

Anderson DH, Fisher SK, Erickson PA, Tabor GA. 1980. Rod and cone disc shedding in the rhesus monkey retina: a quantitative study. *Exp. Eye Res.* 30:559–74

Anderson DH, Fisher SK, Steinberg RH. 1978. Mammalian cones: disc shedding, phagocytosis, and renewal. *Investig. Ophthalmol. Vis. Sci.* 17:117–33

Angel TE, Chance MR, Palczewski K. 2009a. Conserved waters mediate structural and functional activation of family A (rhodopsin-like) G protein-coupled receptors. *Proc. Natl. Acad. Sci. USA* 106:8555–60

Angel TE, Gupta S, Jastrzebska B, Palczewski K, Chance MR. 2009b. Structural waters define a functional channel mediating activation of the GPCR, rhodopsin. *Proc. Natl. Acad. Sci. USA* 106:14367–72

Bockaert J, Perroy J, Bécamel C, Marin P, Fagni L. 2010. GPCR interacting proteins (GIPs) in the nervous system: roles in physiology and pathologies. *Annu. Rev. Pharmacol. Toxicol.* 50:89–109

Boguth CA, Singh P, Huang CC, Tesmer JJG. 2010. Molecular basis for activation of G protein-coupled receptor kinases. *EMBO J.* 29:3249–59

Bornancin F, Pfister C, Chabre M. 1989. The transitory complex between photoexcited rhodopsin and transducin. Reciprocal interaction between the retinal site in rhodopsin and the nucleotide site in transducin. *Eur. J. Biochem.* 184:687–98

Boughton AP, Yang P, Tesmer VM, Ding B, Tesmer JJG, Chen Z. 2011. Heterotrimeric G protein $\beta 1\gamma 2$ subunits change orientation upon complex formation with G protein-coupled receptor kinase 2 (GRK2) on a model membrane. *Proc. Natl. Acad. Sci. USA* 108:E667–73

Buczylko J, Saari JC, Crouch RK, Palczewski K. 1996. Mechanisms of opsin activation. *J. Biol. Chem.* 271:20621–30

Cao P, Sun W, Kramp K, Zheng M, Salom D, et al. 2012. Light-sensitive coupling of rhodopsin and melanopsin to G(i/o) and G(q) signal transduction in *Caenorhabditis elegans*. *FASEB J.* 26:480–91

Changeux JP, Edelstein SJ. 2005. Allosteric mechanisms of signal transduction. *Science* 308:1424–28
Chen Y, Okano K, Maeda T, Chauhan V, Golczak M, et al. 2012. Mechanism of all-*trans*-retinal toxicity with implications for Stargardt disease and age-related macular degeneration. *J. Biol. Chem.* 287:5059–69
Cherezov V, Rosenbaum DM, Hanson MA, Rasmussen SG, Thian FS, et al. 2007. High-resolution crystal structure of an engineered human β2-adrenergic G protein-coupled receptor. *Science* 318:1258–65
Choe HW, Kim YJ, Park JH, Morizumi T, Pai EF, et al. 2011. Crystal structure of metarhodopsin II. *Nature* 471:651–55
Chun E, Thompson AA, Liu W, Roth CB, Griffith MT, et al. 2012. Fusion partner toolchest for the stabilization and crystallization of G protein-coupled receptors. *Structure* 20:967–76
Chung KY, Rasmussen SG, Liu T, Li S, DeVree BT, et al. 2011. Conformational changes in the G protein Gs induced by the β2 adrenergic receptor. *Nature* 477:611–15
Ciruela F, Vilardaga JP, Fernández-Dueñas V. 2010. Lighting up multiprotein complexes: lessons from GPCR oligomerization. *Trends Biotechnol.* 28:407–15
Day PW, Rasmussen SG, Parnot C, Fung JJ, Masood A, et al. 2007. A monoclonal antibody for G protein–coupled receptor crystallography. *Nat. Methods* 4:927–29
Delille HK, Becker JM, Burkhardt S, Bleher B, Terstappen GC, et al. 2012. Heterocomplex formation of 5-HT2A-mGlu2 and its relevance for cellular signaling cascades. *Neuropharmacology* 62:2184–91
DeMaria S, Ngai J. 2010. The cell biology of smell. *J. Cell Biol.* 191:443–52
Deupi X, Edwards P, Singhal A, Nickle B, Oprian D, et al. 2012. Stabilized G protein binding site in the structure of constitutively active metarhodopsin-II. *Proc. Natl. Acad. Sci. USA* 109:119–24
Doré AS, Robertson N, Errey JC, Ng I, Hollenstein K, et al. 2011. Structure of the adenosine A(2A) receptor in complex with ZM241385 and the xanthines XAC and caffeine. *Structure* 19:1283–93
Doze VA, Perez DM. 2012. G-protein-coupled receptors in adult neurogenesis. *Pharmacol. Rev.* 64:645–75
Dror RO, Arlow DH, Maragakis P, Mildorf TJ, Pan AC, et al. 2011. Activation mechanism of the β2-adrenergic receptor. *Proc. Natl. Acad. Sci. USA* 108:18684–89
Engel A, Gaub HE. 2008. Structure and mechanics of membrane proteins. *Annu. Rev. Biochem.* 77:127–48
Fichter KM, Flajolet M, Greengard P, Vu TQ. 2010. Kinetics of G-protein-coupled receptor endosomal trafficking pathways revealed by single quantum dots. *Proc. Natl. Acad. Sci. USA* 107:18658–63
Filipek S, Stenkamp RE, Teller DC, Palczewski K. 2003a. G protein-coupled receptor rhodopsin: a prospectus. *Annu. Rev. Physiol.* 65:851–79
Filipek S, Teller DC, Palczewski K, Stenkamp R. 2003b. The crystallographic model of rhodopsin and its use in studies of other G protein-coupled receptors. *Annu. Rev. Biophys. Biomol. Struct.* 32:375–97
Fotiadis D, Liang Y, Filipek S, Saperstein DA, Engel A, Palczewski K. 2003. Atomic-force microscopy: rhodopsin dimers in native disc membranes. *Nature* 421:127–28
Fotiadis D, Liang Y, Filipek S, Saperstein DA, Engel A, Palczewski K. 2004. The G protein-coupled receptor rhodopsin in the native membrane. *FEBS Lett.* 564:281–88
Fribourg M, Moreno JL, Holloway T, Provasi D, Baki L, et al. 2011. Decoding the signaling of a GPCR heteromeric complex reveals a unifying mechanism of action of antipsychotic drugs. *Cell* 147:1011–23
Fuxe K, Borroto-Escuela DO, Marcellino D, Romero-Fernandez W, Frankowska M, et al. 2012. GPCR heteromers and their allosteric receptor-receptor interactions. *Curr. Med. Chem.* 19:356–63
Gainetdinov RR, Premont RT, Bohn LM, Lefkowitz RJ, Caron MG. 2004. Desensitization of G protein-coupled receptors and neuronal functions. *Annu. Rev. Neurosci.* 27:107–44
Gilman AG. 1987. G proteins: transducers of receptor-generated signals. *Annu. Rev. Biochem.* 56:615–49
Gilman AG, Rall TW. 1968a. Factors influencing adenosine 3′,5′-phosphate accumulation in bovine thyroid slices. *J. Biol. Chem.* 243:5867–71
Gilman AG, Rall TW. 1968b. The role of adenosine 3′,5′-phosphate in mediating effects of thyroid-stimulating hormone on carbohydrate metabolism of bovine thyroid slices. *J. Biol. Chem.* 243:5872–81
Granier S, Kobilka B. 2012. A new era of GPCR structural and chemical biology. *Nat. Chem. Biol.* 8:670–73
Granier S, Manglik A, Kruse AC, Kobilka TS, Thian FS, et al. 2012. Structure of the δ-opioid receptor bound to naltrindole. *Nature* 485:400–4
Hamm HE, Deretic D, Hofmann KP, Schleicher A, Kohl B. 1987. Mechanism of action of monoclonal antibodies that block the light activation of the guanyl nucleotide-binding protein, transducin. *J. Biol. Chem.* 262:10831–38

Harrison C, van der Graaf PH. 2006. Current methods used to investigate G protein coupled receptor oligomerisation. *J. Pharmacol. Toxicol. Methods* 54:26–35

Herrick-Davis K, Grinde E, Lindsley T, Cowan A, Mazurkiewicz JE. 2012. Oligomer size of the serotonin 5-hydroxytryptamine 2C (5-HT2C) receptor revealed by fluorescence correlation spectroscopy with photon counting histogram analysis: evidence for homodimers without monomers or tetramers. *J. Biol. Chem.* 287:23604–14

Herrick-Davis K, Weaver BA, Grinde E, Mazurkiewicz JE. 2006. Serotonin 5-HT2C receptor homodimer biogenesis in the endoplasmic reticulum: real-time visualization with confocal fluorescence resonance energy transfer. *J. Biol. Chem.* 281:27109–16

Hofmann KP, Scheerer P, Hildebrand PW, Choe HW, Park JH, et al. 2009. A G protein-coupled receptor at work: the rhodopsin model. *Trends Biochem. Sci.* 34:540–52

Hoyer D, Clarke DE, Fozard JR, Hartig PR, Martin GR, et al. 1994. International Union of Pharmacology classification of receptors for 5-hydroxytryptamine (Serotonin). *Pharmacol. Rev.* 46:157–203

Hu J, Wang Y, Zhang X, Lloyd JR, Li JH, et al. 2010. Structural basis of G protein-coupled receptor/G protein interactions. *Nat. Chem. Biol.* 6:541–48

Huang CC, Tesmer JJG. 2011. Recognition in the face of diversity: interactions of heterotrimeric G proteins and G protein-coupled receptor (GPCR) kinases with activated GPCRs. *J. Biol. Chem.* 286:7715–21

Humphries MM, Rancourt D, Farrar GJ, Kenna P, Hazel M, et al. 1997. Retinopathy induced in mice by targeted disruption of the rhodopsin gene. *Nat. Genet.* 15:216–19

Ianoul A, Grant DD, Rouleau Y, Bani-Yaghoub M, Johnston LJ, Pezacki JP. 2005. Imaging nanometer domains of beta-adrenergic receptor complexes on the surface of cardiac myocytes. *Nat. Chem. Biol.* 1:196–202

Inagaki S, Ghirlando R, White JF, Gvozdenovic-Jeremic J, Northup JK, Grisshammer R. 2012. Modulation of the interaction between neurotensin receptor NTS1 and Gq protein by lipid. *J. Mol. Biol.* 417:95–111

Jäger S, Palczewski K, Hofmann KP. 1996. Opsin/all-*trans*-retinal complex activates transducin by different mechanisms than photolyzed rhodopsin. *Biochemistry* 35:2901–8

Jalink K, Moolenaar WH. 2010. G protein-coupled receptors: the inside story. *Bioessays* 32:13–16

Jastrzebska B, Debinski A, Filipek S, Palczewski K. 2011a. Role of membrane integrity on G protein-coupled receptors: rhodopsin stability and function. *Prog. Lipid. Res.* 50:267–77

Jastrzebska B, Fotiadis D, Jang GF, Stenkamp RE, Engel A, Palczewski K. 2006. Functional and structural characterization of rhodopsin oligomers. *J. Biol. Chem.* 281:11917–22

Jastrzebska B, Golczak M, Fotiadis D, Engel A, Palczewski K. 2009. Isolation and functional characterization of a stable complex between photoactivated rhodopsin and the G protein, transducin. *FASEB J.* 23:371–81

Jastrzebska B, Orban T, Golczak M, Engel A, Palczewski K. 2013. Asymmetry of the rhodopsin dimer in complex with transducin. *FASEB J.* 27:1–13

Jastrzebska B, Ringler P, Lodowski DT, Moiseenkova-Bell V, Golczak M, et al. 2011b. Rhodopsin-transducin heteropentamer: three-dimensional structure and biochemical characterization. *J. Struct. Biol.* 176:387–94

Kasai RS, Suzuki KG, Prossnitz ER, Koyama-Honda I, Nakada C, et al. 2011. Full characterization of GPCR monomer-dimer dynamic equilibrium by single molecule imaging. *J. Cell Biol.* 192:463–80

Katritch V, Cherezov V, Stevens RC. 2012. Diversity and modularity of G protein-coupled receptor structures. *Trends Pharmacol. Sci.* 33:17–27

Kaya AI, Thaker TM, Preininger AM, Iverson TM, Hamm HE. 2011. Coupling efficiency of rhodopsin and transducin in bicelles. *Biochemistry* 50:3193–203

Kupchik YM, Barchad-Avitzur O, Wess J, Ben-Chaim Y, Parnas I, Parnas H. 2011. A novel fast mechanism for GPCR-mediated signal transduction–control of neurotransmitter release. *J. Cell Biol.* 192:137–51

Lambert NA. 2010. GPCR dimers fall apart. *Sci. Signal.* 3:pe12

Li J, Edwards PC, Burghammer M, Villa C, Schertler GF. 2004. Structure of bovine rhodopsin in a trigonal crystal form. *J. Mol. Biol.* 343:1409–38

Li N, Salom D, Zhang L, Harris T, Ballesteros JA, et al. 2007. Heterologous expression of the adenosine A1 receptor in transgenic mouse retina. *Biochemistry* 46:8350–59

Li S, Huang S, Peng SB. 2005. Overexpression of G protein-coupled receptors in cancer cells: involvement in tumor progression. *Int. J. Oncol.* 27:1329–39

Liang Y, Fotiadis D, Filipek S, Saperstein DA, Palczewski K, Engel A. 2003. Organization of the G protein-coupled receptors rhodopsin and opsin in native membranes. *J. Biol. Chem.* 278:21655–62

Liang Y, Fotiadis D, Maeda T, Maeda A, Modzelewska A, et al. 2004. Rhodopsin signaling and organization in heterozygote rhodopsin knockout mice. *J. Biol. Chem.* 279:48189–96

Liggett SB. 2011. Phosphorylation barcoding as a mechanism of directing GPCR signaling. *Sci. Signal.* 4:pe36

Liu W, Chun E, Thompson AA, Chubukov P, Xu F, et al. 2012. Structural basis for allosteric regulation of GPCRs by sodium ions. *Science* 337:232–36

Lodowski DT, Palczewski K, Miyagi M. 2010. Conformational changes in the g protein-coupled receptor rhodopsin revealed by histidine hydrogen-deuterium exchange. *Biochemistry* 49:9425–27

Lundstrom KH. 2006. *Structural Genomics on Membrane Proteins*. Boca Raton, FL: CRC/Taylor & Francis. 383 pp.

Macia E, Partisani M, Paleotti O, Luton F, Franco M. 2012. Arf6 negatively controls the rapid recycling of the β2 adrenergic receptor. *J. Cell Sci.* 125:4026–35

Maeda A, Golczak M, Chen Y, Okano K, Kohno H, et al. 2012. Primary amines protect against retinal degeneration in mouse models of retinopathies. *Nat. Chem. Biol.* 8:170–78

Maeda A, Okano K, Park PS, Lem J, Crouch RK, et al. 2010. Palmitoylation stabilizes unliganded rod opsin. *Proc. Natl. Acad. Sci. USA* 107:8428–33

Maeda T, Imanishi Y, Palczewski K. 2003. Rhodopsin phosphorylation: 30 years later. *Prog. Retin. Eye Res.* 22:417–34

Manglik A, Kruse AC, Kobilka TS, Thian FS, Mathiesen JM, et al. 2012. Crystal structure of the μ-opioid receptor bound to a morphinan antagonist. *Nature* 485:321–26

Mary S, Damian M, Louet M, Floquet N, Fehrentz JA, et al. 2012. Ligands and signaling proteins govern the conformational landscape explored by a G protein-coupled receptor. *Proc. Natl. Acad. Sci. USA* 109:8304–309

Maurice P, Kamal M, Jockers R. 2011. Asymmetry of GPCR oligomers supports their functional relevance. *Trends Pharmacol. Sci.* 32:514–20

Mayhew TM, Astle D. 1997. Photoreceptor number and outer segment disk membrane surface area in the retina of the rat: stereological data for whole organ and average photoreceptor cell. *J. Neurocytol.* 26:53–61

McCusker EC, Bane SE, O'Malley MA, Robinson AS. 2007. Heterologous GPCR expression: a bottleneck to obtaining crystal structures. *Biotechnol. Prog.* 23:540–47

McMillin SM, Heusel M, Liu T, Costanzi S, Wess J. 2011. Structural basis of M3 muscarinic receptor dimer/oligomer formation. *J. Biol. Chem.* 286:28584–98

Mirzadegan T, Benko G, Filipek S, Palczewski K. 2003. Sequence analyses of G-protein-coupled receptors: similarities to rhodopsin. *Biochemistry* 42:2759–67

Müller DJ, Wu N, Palczewski K. 2008. Vertebrate membrane proteins: structure, function, and insights from biophysical approaches. *Pharmacol. Rev.* 60:43–78

Murakami M, Kouyama T. 2008. Crystal structure of squid rhodopsin. *Nature* 453:363–67

Mustafi D, Kevany BM, Genoud C, Okano K, Cideciyan AV, et al. 2011. Defective photoreceptor phagocytosis in a mouse model of enhanced S-cone syndrome causes progressive retinal degeneration. *FASEB J.* 25:3157–76

Mustafi D, Palczewski K. 2009. Topology of class A G protein-coupled receptors: insights gained from crystal structures of rhodopsins, adrenergic and adenosine receptors. *Mol. Pharmacol.* 75:1–12

Navarro G, Hradsky J, Lluis C, Casadó V, McCormick PJ, et al. 2012. NCS-1 associates with adenosine A(2A) receptors and modulates receptor function. *Front. Mol. Neurosci.* 5:53

Neri M, Vanni S, Tavernelli I, Rothlisberger U. 2010. Role of aggregation in rhodopsin signal transduction. *Biochemistry* 49:4827–32

Nobles KN, Xiao K, Ahn S, Shukla AK, Lam CM, et al. 2011. Distinct phosphorylation sites on the β(2)-adrenergic receptor establish a barcode that encodes differential functions of β-arrestin. *Sci. Signal.* 4:ra51

Oates J, Watts A. 2011. Uncovering the intimate relationship between lipids, cholesterol and GPCR activation. *Curr. Opin. Struct. Biol.* 21:802–7

Okada T, Fujiyoshi Y, Silow M, Navarro J, Landau EM, Shichida Y. 2002. Functional role of internal water molecules in rhodopsin revealed by X-ray crystallography. *Proc. Natl. Acad. Sci. USA* 99:5982–87

Oldham WM, Hamm HE. 2008. Heterotrimeric G protein activation by G-protein-coupled receptors. *Nat. Rev. Mol. Cell Biol.* 9:60–71

Orban T, Huang CC, Homan KT, Jastrzebska B, Tesmer JJ, Palczewski K. 2012a. Substrate-induced changes in the dynamics of rhodopsin kinase (G protein-coupled receptor kinase 1). *Biochemistry*. 51:3404–11

Orban T, Jastrzebska B, Gupta S, Wang B, Miyagi M, et al. 2012b. Conformational dynamics of activation for the pentameric complex of dimeric G protein-coupled receptor and heterotrimeric G protein. *Structure* 20:826–40

Palczewski K. 2010. Oligomeric forms of G protein-coupled receptors (GPCRs). *Trends Biochem. Sci.* 35:595–600

Palczewski K. 2012. Chemistry and biology of vision. *J. Biol. Chem.* 287:1612–19

Palczewski K, Kumasaka T, Hori T, Behnke CA, Motoshima H, et al. 2000. Crystal structure of rhodopsin: a G protein-coupled receptor. *Science* 289:739–45

Pardo L, Deupi X, Dolker N, López-Rodríguez ML, Campillo M. 2007. The role of internal water molecules in the structure and function of the rhodopsin family of G protein-coupled receptors. *Chembiochem* 8:19–24

Park D, O'Doherty I, Somvanshi RK, Bethke A, Schroeder FC, et al. 2012. Interaction of structure-specific and promiscuous G-protein-coupled receptors mediates small-molecule signaling in *Caenorhabditis elegans*. *Proc. Natl. Acad. Sci. USA* 109:9917–22

Park JH, Scheerer P, Hofmann KP, Choe H-W, Ernst OP. 2008a. Crystal structure of the ligand-free G-protein-coupled receptor opsin. *Nature* 454:183–87

Park PS, Filipek S, Wells JW, Palczewski K. 2004. Oligomerization of G protein-coupled receptors: past, present, and future. *Biochemistry* 43:15643–56

Park PS, Lodowski DT, Palczewski K. 2008b. Activation of G protein-coupled receptors: beyond two-state models and tertiary conformational changes. *Annu. Rev. Pharmacol. Toxicol.* 48:107–41

Park PS, Palczewski K. 2005. Diversifying the repertoire of G protein-coupled receptors through oligomerization. *Proc. Natl. Acad. Sci. USA* 102:8793–94

Pellissier LP, Barthet G, Gaven F, Cassier E, Trinquet E, et al. 2011. G protein activation by serotonin type 4 receptor dimers: evidence that turning on two protomers is more efficient. *J. Biol. Chem.* 286:9985–97

Pin JP, Galvez T, Prézeau L. 2003. Evolution, structure, and activation mechanism of family 3/C G-protein-coupled receptors. *Pharmacol. Ther.* 98:325–54

Prezeau L, Rives ML, Comps-Agrar L, Maurel D, Kniazeff J, Pin JP. 2010. Functional crosstalk between GPCRs: with or without oligomerization. *Curr. Opin. Pharmacol.* 10:6–13

Qin K, Dong C, Wu G, Lambert NA. 2011. Inactive-state preassembly of G(q)-coupled receptors and G(q) heterotrimers. *Nat. Chem. Biol.* 7:740–47

Rahmeh R, Damian M, Cottet M, Orcel H, Mendre C, et al. 2012. Structural insights into biased G protein-coupled receptor signaling revealed by fluorescence spectroscopy. *Proc. Natl. Acad. Sci. USA* 109:6733–38

Rall TW, Sutherland EW. 1958. Formation of a cyclic adenine ribonucleotide by tissue particles. *J. Biol. Chem.* 232:1065–76

Rasmussen SG, Choi HJ, Rosenbaum DM, Kobilka TS, Thian FS, et al. 2007. Crystal structure of the human beta2 adrenergic G-protein-coupled receptor. *Nature* 450:383–87

Rasmussen SG, DeVree BT, Zou Y, Kruse AC, Chung KY, et al. 2011. Crystal structure of the β2 adrenergic receptor-Gs protein complex. *Nature* 477:549–55

Ritter SL, Hall RA. 2009. Fine-tuning of GPCR activity by receptor-interacting proteins. *Nat. Rev. Mol. Cell Biol.* 10:819–30

Robison GA, Butcher RW, Sutherland EW. 1968. Cyclic AMP. *Annu. Rev. Biochem.* 37:149–74

Rosenbaum DM, Cherezov V, Hanson MA, Rasmussen SG, Thian FS, et al. 2007. GPCR engineering yields high-resolution structural insights into beta2-adrenergic receptor function. *Science* 318:1266–73

Rovira X, Pin JP, Giraldo J. 2010. The asymmetric/symmetric activation of GPCR dimers as a possible mechanistic rationale for multiple signalling pathways. *Trends Pharmacol. Sci.* 31:15–21

Salom D, Cao P, Sun W, Kramp K, Jastrzebska B, et al. 2012. Heterologous expression of functional G-protein-coupled receptors in *Caenorhabditis elegans*. *FASEB J.* 26:492–502

Salom D, Lodowski DT, Stenkamp RE, Le Trong I, Golczak M, et al. 2006. Crystal structure of a photoactivated deprotonated intermediate of rhodopsin. *Proc. Natl. Acad. Sci. USA* 103:16123–28

Salom D, Wu N, Sun W, Dong Z, Palczewski K, et al. 2008. Heterologous expression and purification of the serotonin type 4 receptor from transgenic mouse retina. *Biochemistry* 47:13296–307

Salon JA, Lodowski DT, Palczewski K. 2011. The significance of G protein-coupled receptor crystallography for drug discovery. *Pharmacol. Rev.* 63:901–37

Scheerer P, Park JH, Hildebrand PW, Kim YJ, Krauss N, et al. 2008. Crystal structure of opsin in its G-protein-interacting conformation. *Nature* 455:497–502

Schlegel W, Kempner ES, Rodbell M. 1979. Activation of adenylate cyclase in hepatic membranes involves interactions of the catalytic unit with multimeric complexes of regulatory proteins. *J. Biol. Chem.* 254:5168–76

Schneider EH, Seifert R. 2010. Sf9 cells: a versatile model system to investigate the pharmacological properties of G protein-coupled receptors. *Pharmacol. Ther.* 128:387–418

Shimamura T, Hiraki K, Takahashi N, Hori T, Ago H, et al. 2008. Crystal structure of squid rhodopsin with intracellularly extended cytoplasmic region. *J. Biol. Chem.* 283:17753–56

Smith SO. 2012. Insights into the activation mechanism of the visual receptor rhodopsin. *Biochem. Soc. Trans.* 40:389–93

Standfuss J, Edwards PC, D'Antona A, Fransen M, Xie G, et al. 2011. The structural basis of agonist-induced activation in constitutively active rhodopsin. *Nature* 471:656–60

Strotmann R, Schröck K, Böselt I, Stäubert C, Russ A, Schöneberg T. 2011. Evolution of GPCR: change and continuity. *Mol. Cell Endocrinol.* 331:170–78

Struts AV, Salgado GFJ, Martinez-Mayorga K, Brown MF. 2011. Retinal dynamics underlie its switch from inverse agonist to agonist during rhodopsin activation. *Nat. Struct. Mol. Biol.* 18:392–94

Sun J, Singh V, Kajino-Sakamoto R, Aballay A. 2011. Neuronal GPCR controls innate immunity by regulating non-canonical unfolded protein response genes. *Science* 332:729–32

Sutherland EW, Rall TW. 1958. Fractionation and characterization of a cyclic adenine ribonucleotide formed by tissue particles. *J. Biol. Chem.* 232:1077–91

Tadagaki K, Jockers R, Kamal M. 2012. History and biological significance of GPCR heteromerization in the neuroendocrine system. *Neuroendocrinology* 95:223–31

Tate CG, Schertler GF. 2009. Engineering G protein-coupled receptors to facilitate their structure determination. *Curr. Opin. Struct. Biol.* 19:386–95

Terrillon S, Bouvier M. 2004. Roles of G-protein-coupled receptor dimerization. *EMBO Rep.* 5:30–34

Thompson AA, Liu JJ, Chun E, Wacker D, Wu H, et al. 2011. GPCR stabilization using the bicelle-like architecture of mixed sterol-detergent micelles. *Methods* 55:310–17

Thompson AA, Liu W, Chun E, Katritch V, Wu H, et al. 2012. Structure of the nociceptin/orphanin FQ receptor in complex with a peptide mimetic. *Nature* 485:395–99

Vidi PA, Przybyla JA, Hu CD, Watts VJ. 2010. Visualization of G protein-coupled receptor (GPCR) interactions in living cells using bimolecular fluorescence complementation (BiFC). *Curr. Protoc. Neurosci.* Chapter 5: Unit 5.29

West GM, Chien EY, Katritch V, Gatchalian J, Chalmers MJ, et al. 2011. Ligand-dependent perturbation of the conformational ensemble for the GPCR β2 adrenergic receptor revealed by HDX. *Structure* 19:1424–32

Westfield GH, Rasmussen SG, Su M, Dutta S, DeVree BT, et al. 2011. Structural flexibility of the G alpha s alpha-helical domain in the beta2-adrenoceptor Gs complex. *Proc. Natl. Acad. Sci. USA* 108:16086–91

Wong JP, Reboul E, Molday RS, Kast J. 2009. A carboxy-terminal affinity tag for the purification and mass spectrometric characterization of integral membrane proteins. *J. Proteome Res.* 8:2388–96

Wu H, Wacker D, Mileni M, Katritch V, Han GW, et al. 2012. Structure of the human κ-opioid receptor in complex with JDTic. *Nature* 485:327–32

Yau KW, Hardie RC. 2009. Phototransduction motifs and variations. *Cell* 139:246–64

Ye Y, Godzik A. 2003. Flexible structure alignment by chaining aligned fragment pairs allowing twists. *Bioinformatics* 19(Suppl. 2):ii246–55

Yekkirala AS. 2013. Two to tango: GPCR oligomers and GPCR-TRP channel interactions in nociception. *Life Sci.* 92:438–45

Zhang L, Salom D, He J, Okun A, Ballesteros J, et al. 2005. Expression of functional G protein-coupled receptors in photoreceptors of transgenic *Xenopus laevis*. *Biochemistry* 44:14509–18

Zheng H, Pearsall EA, Hurst DP, Zhang Y, Chu J, et al. 2012. Palmitoylation and membrane cholesterol stabilize μ-opioid receptor homodimerization and G protein coupling. *BMC Cell Biol.* 13:6

Zhou M, Nakatani E, Gronenberg LS, Tokimoto T, Wirth MJ, et al. 2007. Peptide-labeled quantum dots for imaging GPCRs in whole cells and as single molecules. *Bioconjug. Chem.* 18:323–32

Zylbergold P, Hébert TE. 2009. A division of labor: asymmetric roles for GPCR subunits in receptor dimers. *Nat. Chem. Biol.* 5:608–9

RELATED RESOURCES

Munger SD, Leinders-Zufall T, Zufall F. 2009. Subsystem organization of the mammalian sense of smell. *Annu. Rev. Physiol.* 71:115–40

Rogan SC, Roth BL. 2011. Remote control of neuronal signaling. *Pharmacol. Rev.* 63:291–315

Rosenbaum DM, Rasmussen SG, Kobilka BK. 2009. The structure and function of G protein-coupled receptors. *Nature* 459:356–63

Smith SO. 2010. Structure and activation of the visual pigment rhodopsin. *Annu. Rev. Biophys.* 39:309–28

Spiegel AM, Weinstein LS. 2004. Inherited diseases involving G proteins and G protein–coupled receptors. *Annu. Rev. Med.* 55:27–39

Stevens RC. 2013. GPCR structure and activation. *Annu. Rev. Pharmacol. Toxicol.* 53:531–56

Superior Colliculus and Visual Spatial Attention

Richard J. Krauzlis,[1,2] Lee P. Lovejoy,[3] and Alexandre Zénon[4]

[1]Laboratory of Sensorimotor Research, National Eye Institute, National Institutes of Health, Bethesda, Maryland 20892, USA; email: richard.krauzlis@nih.gov

[2]Systems Neurobiology Laboratory, Salk Institute for Biological Studies, La Jolla, California 92037, USA

[3]Department of Psychiatry, Columbia University and the New York State Psychiatric Institute, New York, NY 10032, USA

[4]Institute of Neuroscience, Université Catholique de Louvain, 1200 Brussels, Belgium

Annu. Rev. Neurosci. 2013. 36:165–82

First published online as a Review in Advance on May 15, 2013

The *Annual Review of Neuroscience* is online at neuro.annualreviews.org

This article's doi:
10.1146/annurev-neuro-062012-170249

Keywords

perception, pursuit, saccade, selection, neglect, eye movement

Abstract

The superior colliculus (SC) has long been known to be part of the network of brain areas involved in spatial attention, but recent findings have dramatically refined our understanding of its functional role. The SC both implements the motor consequences of attention and plays a crucial role in the process of target selection that precedes movement. Moreover, even in the absence of overt orienting movements, SC activity is related to shifts of covert attention and is necessary for the normal control of spatial attention during perceptual judgments. The neuronal circuits that link the SC to spatial attention may include attention-related areas of the cerebral cortex, but recent results show that the SC's contribution involves mechanisms that operate independently of the established signatures of attention in visual cortex. These findings raise new issues and suggest novel possibilities for understanding the brain mechanisms that enable spatial attention.

Contents

INTRODUCTION

Visual spatial attention allows animals to base decisions on relevant environmental stimuli and suppress irrelevant signals. It is associated with a variety of changes in neural processing and is controlled by a network of cortical brain areas, especially the parietal, prefrontal, and extrastriate visual cortices. The importance of this cortical network has been established using a range of techniques, including lesions, electrophysiology, and fMRI, and this work has been thoroughly summarized in several reviews (Desimone & Duncan 1995, Petersen & Posner 2012, Reynolds & Chelazzi 2004).

SC: superior colliculus

Investigators have long recognized that, beyond these cortical areas, the superior colliculus (SC), an evolutionarily conserved structure located on the roof of the vertebrate midbrain, also plays a central role in visual spatial attention (Goldberg & Wurtz 1972). A key aspect of this role is controlling orienting movements of the eyes and head that are tightly linked to shifts of spatial attention; this motor function of the SC has been reviewed elsewhere (Gandhi & Katnani 2011, Sparks 1999, Wurtz & Albano 1980). However, several lines of evidence indicate that the SC does more than simply implement the motor consequences of attention shifts prescribed by the cerebral cortex. In this article, we first review evidence that the SC is necessary for the normal control of spatial attention. We then consider several candidate circuits that could mediate this role, illustrating that key components of the brain mechanisms of attention may lie outside the scope of current models.

BASIC FEATURES OF THE SC

We first very briefly review some of the basic features of the primate SC; a more comprehensive review can be found elsewhere (May 2006).

The SC is a laminar structure delineated by alternating strata of fibers and soma. In primates, these are usually divided broadly into superficial, intermediate, and deep layers. The superficial layers receive direct projections from both retinal ganglion cells and striate cortex and contain neurons that exhibit a variety of responses to salient visual stimuli. Neurons in the intermediate and deep layers receive input from the extrastriate cortex and also respond to visual stimuli; however, consistent with the diverse anatomical connections of these layers, these neurons also respond to stimulus modalities in addition to vision, and most exhibit activity patterns related to the planning and execution of orienting movements. This diversity belies the view of the SC as simply a node in a descending motor pathway; instead, it contains multiple classes of neurons that provide points of interconnection between many circuits serving a range of sensory, motor, and cognitive functions.

In addition to their diverse features and properties, SC neurons are organized in a topographic map of visual space. The SC on each side of the brain contains a representation of the contralateral visual field, with an enlarged representation of the central visual field. In the superficial layers, this is a map of stimulus position. In the intermediate and deep layers, a corresponding map organizes activity related to orienting movements and selecting visual stimuli. These maps provide the anatomical basis for the role of the SC in visual spatial attention.

EVIDENCE FOR A ROLE IN SPATIAL ATTENTION

The Sprague Effect

As first described by Sprague (1966), lesions of the SC in cats can lead to an unexpected recovery from deficits in spatial orienting caused by damage to other parts of the attention network. After the occipital or parietal cortex is lesioned on one side of the brain, cats exhibit visual neglect—the tendency to ignore visual objects presented in the affected part of the visual field. When activity in the SC on the opposite side of the brain is then suppressed, the symptoms of visual neglect are typically relieved. In a remarkable clinical case, the symptoms of visual neglect experienced after damage to the frontal cortex were relieved by additional subsequent damage to the contralateral SC, suggesting that the Sprague effect also extends to humans (Weddell 2004).

The Sprague effect was initially thought to be a recovery from cortical blindness, in which the transmission of visual information from the superficial layers of the SC to the cortex (via pulvinar) was disinhibited by ablating the SC on the other side (Diamond & Hall 1969, Robson & Hall 1977, Trojanowski & Jacobson 1975). Subsequent work found that the effect is not caused by interactions between the two sides of the SC but instead depends on inhibitory input from the substantia nigra pars reticulata or pedunculopontine region (Ciaramitaro et al. 1997; Durmer & Rosenquist 2001; Wallace et al. 1989, 1990). The anatomy of these connections implicates the intermediate and deeper SC layers, rather than the superficial layers, in these manipulations of spatial orienting.

Selecting Visual Targets

In addition to the SC's function in the motor implementation of orienting movements, multiple lines of evidence indicate that the SC plays a central role in the process of selecting which stimuli will guide behavior.

Neuronal activity in the intermediate and deeper layers of the SC is related to evaluating possible saccade targets. SC neurons show elevated activity for visual stimuli that will be selected as the end point of saccades, relative to those that are ignored (Glimcher & Sparks 1992, Krauzlis & Dill 2002, McPeek & Keller 2002), and this activity is proportional to the probability that a saccade will be made into their response fields (Basso & Wurtz 1998, Dorris & Munoz 1998, Horwitz & Newsome 2001). For many neurons this modulation predicts the upcoming saccade, but for others it is related to the selected visual stimulus rather than to the movement itself (Horwitz et al. 2004, Horwitz & Newsome 1999, McPeek & Keller 2002).

Some aspect of SC activity is necessary for saccade target selection because chemically blocking activity disrupts saccade choices: When the target is placed in the affected part of the visual field, saccades tend to be erroneously directed to distracter stimuli located elsewhere (McPeek & Keller 2004, Nummela & Krauzlis 2010). Conversely, when SC activity is artificially raised by electrical microstimulation, saccade selection favors the stimulus location matching the activated SC site (Carello & Krauzlis 2004, Dorris et al. 2007).

Experiments using pursuit eye movements, rather than saccades, show that SC activity is involved not just in saccade selection but also in the process of target selection itself. In contrast with saccades, which rapidly redirect the line of sight, pursuit slowly rotates the eyes to follow moving targets (Krauzlis 2005, Lisberger 2010). Because pursuit depends on the motion

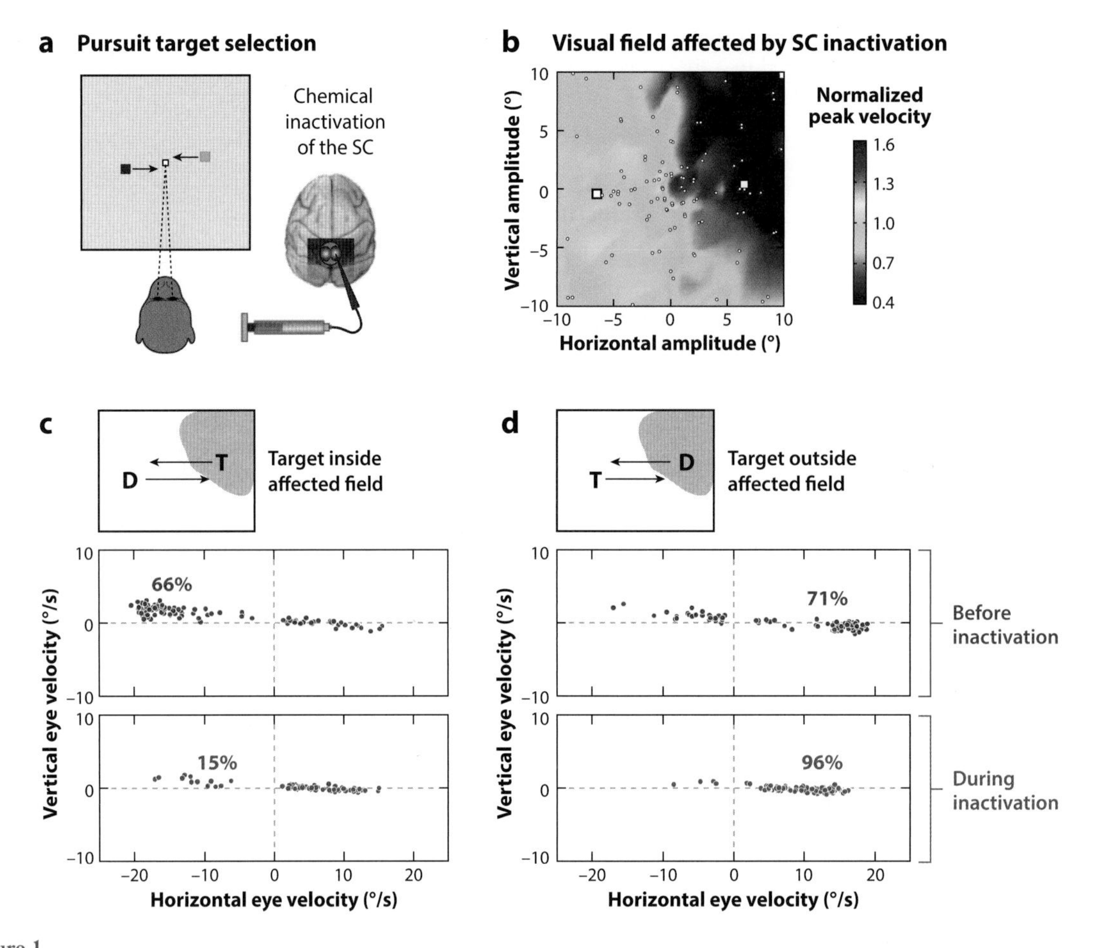

Figure 1

Effects of superior colliculus (SC) inactivation on pursuit target selection. (*a*) The task was to smoothly track the target defined by a precue. Performance on this two-alternative task was measured before and after SC inactivation. (*b*) The area affected by SC inactivation was estimated by measuring the change in peak velocity of saccades to visual targets. (*c*) Changes in performance when the target appeared inside the affected part of the visual field. The distributions of target choices are illustrated by plotting the mean horizontal eye velocity against mean vertical eye velocity over the first 100 ms of the pursuit eye movement response. SC inactivation reduced correct choices from 66% to 15% correct. (*d*) When the distracter appeared in the affected part of the visual field, correct pursuit choices improved from 71% correct to 96% correct. Overall, SC inactivation biased choices in favor of the stimulus appearing outside the affected region, even though these required movements toward the affected field. Adapted from Nummela & Krauzlis (2010).

of the target, rather than the target's location, it can dissociate target selection from movement direction; for example, if a target appears on the left and moves rightward, the subject should select the stimulus on the left even though this requires a rightward movement.

SC neurons change their activity during pursuit as expected from the retinotopic map—they increase their discharge when the image of the target falls inside their response field (Krauzlis 2003; Krauzlis et al. 1997, 2000). When there are multiple moving stimuli, SC neurons show enhanced activity for the selected target, even if doing so requires a pursuit movement directed away from the target's starting location (Krauzlis & Dill 2002). Chemically inhibiting SC activity causes a neglect-like impairment (see **Figure 1**). Choices are strongly biased against the stimulus initially located inside the affected part of the visual field, in favor of whichever stimulus is located outside the affected region (Nummela & Krauzlis 2010, 2011); similar pursuit impairments occur in patients with hemineglect (Rizzo & Hurtig 1992). Conversely, artificially activating the SC biases target choice in favor of the stimulus placed at the matching location in the visual field (Carello & Krauzlis 2004).

Moreover, inhibiting SC activity impairs target selection even when the eyes remain fixed and the response involves moving the hand. As in the eye movement tasks, chemical inactivation biases choices away from the stimulus placed in the affected part of the visual field. The effects are large when animals reach directly to targets on a touch screen (Song et al. 2011) and are also present when subjects press buttons on a response pad positioned outside the field of view (Nummela & Krauzlis 2010).

Overt and Covert Attention

The SC's role in target selection appears to be related to a broader role in spatial attention, including both overt shifts during saccades and covert allocation in the absence of orienting movements.

Signals related to spatial attention and saccade commands are intermingled in the primate SC. For example, the end points of saccades evoked by stimulating the SC are systematically deviated in the direction of spatial cues, suggesting that shifts of attention prompted by spatial cues automatically lead to saccade preparation (Kustov & Robinson 1996). The time course of these deviations reflects the origin of the shift: Peripheral cues cause deviations at short delays, consistent with a stimulus-driven effect, whereas central cues cause deviations at longer delays, consistent with a top-down effect.

Neuronal activity in the SC is modulated by peripheral cues in ways that parallel the behavioral effects on saccades. When a peripheral cue is flashed just before the visual target for a saccade, it reduces saccade latency and causes a larger neuronal response. These effects are larger when the cue is predictive of the target location, indicating that top-down, as well as stimulus-driven, factors are involved (Bell et al. 2004, Fecteau et al. 2004). These results support the view that activity in the SC reflects both the stimulus-driven and top-down factors that regulate spatial attention and saccade selection (Fecteau & Munoz 2006).

Although SC activity may have been restricted to spatial attention associated with eye movements, other studies have shown that the SC may play a role in covert attention. In one study, animals discriminated the orientation of a Landolt "C," and spatial attention was manipulated using spatial and symbolic cues (Ignashchenkova et al. 2004). The pattern of effects depended on the type of neuron (see **Figure 2**). "Visual" neurons (with visual activity but lacking saccade-related activity) had larger responses to the appearance of the "C" when the location was cued, for both spatial and symbolic cues. In contrast, "visual-motor" neurons (with both visual and saccade-related activity) showed a similar effect, but only for spatial cues and not for symbolic cues. These neurons also showed higher activity during the delay period after the spatial cue was presented but before the appearance of the "C" stimulus, suggesting that visual-motor SC neurons may be especially important for stimulus-driven shifts of attention.

Some studies have applied subthreshold microstimulation (too weak to evoke saccades directly) to test whether SC activity plays a causal role in spatial attention. One set of studies (Cavanaugh et al. 2006, Cavanaugh & Wurtz 2004) tested spatial attention using change-blindness, the inability to detect changes in a visual scene when those changes are accompanied by a full-field transient such as a blank screen (Rensink 2002). Animals were shown three patches of random-dot motion and were required to detect if any patch changed its motion direction after the blank. On some trials, the animal was given a spatial cue to indicate which motion patch might change, and on other trials the SC was microstimulated at a location matching one of the three patches. SC microstimulation caused effects nearly identical to those obtained with spatial cues: Detection performance improved

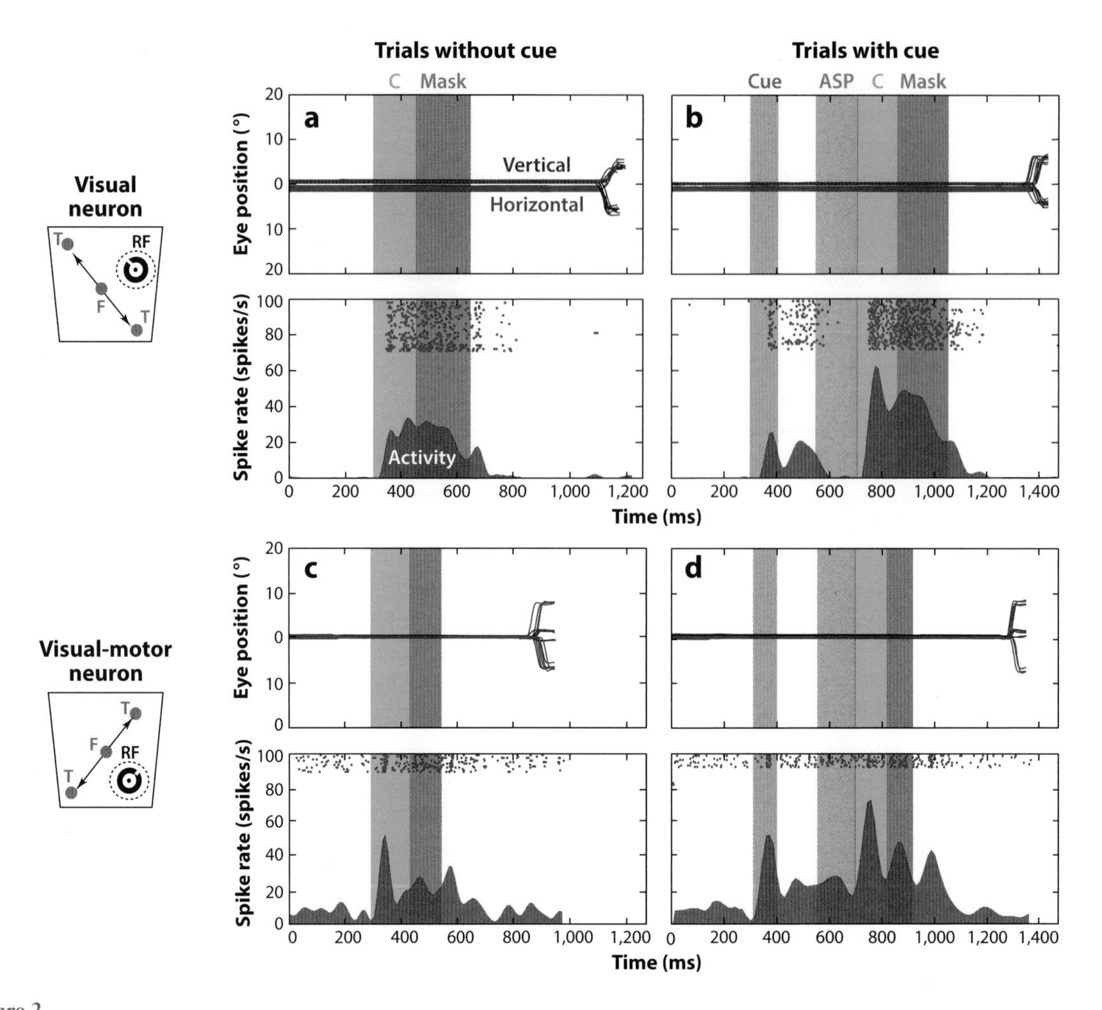

Figure 2

Changes in superior colliculus (SC) activity during covert shifts of spatial attention. The sample visual neuron showed enhanced activity (*shaded area*) when the animal was shown a spatially precise cue (*b*) compared to when the animal was shown no cue (*a*). (*c–d*) Sample visual-motor neuron under the same task conditions. Both the visual and visual-motor neurons responded to the presentation of the "C" (*blue*), but this activity was higher when the location of the "C" was cued than when it was not cued. Also, the visual-motor neuron showed activity during the time epoch before the presentation of the "C" [attention shift period (ASP), *orange*], but the visual neuron did not. Abbreviations: F, fixation spot; RF, receptive field of neuron; T, choice target. Adapted from Ignashchenkova et al. (2004).

and reaction times decreased. These effects cannot be explained by a response bias because the animals were not simply more likely to report a change but rather became better at detecting when changes occurred.

In another study, SC microstimulation was applied as animals judged the direction of motion in a random-dot motion patch, which was made more challenging by including irrelevant flickering dots elsewhere in the display (Müller et al. 2005). The performance of the animal was summarized with psychometric curves showing how the percentage of correct answers increased with signal strength. Microstimulation in the intermediate SC improved discrimination performance; the psychometric curves shifted leftward, meaning that less visual motion was needed to achieve a particular level of performance. This improvement in performance was observed only if the SC site and motion patch were at spatially coincident locations.

These microstimulation studies show that altering SC activity is sufficient to evoke attention-like changes in performance. To test whether SC activity is necessary, behavioral performance was tested before and during reversible chemical inactivation of the SC (Lovejoy & Krauzlis 2010). The task in these experiments was to discriminate the direction of motion in a random-dot motion patch that appeared at a previously cued location (see **Figure 3**). To ensure that spatial attention was necessary to perform the task, the display included an irrelevant "foil" stimulus that also contained random-dot motion but appeared at an uncued location and should therefore have been ignored.

The ability of animals to select the appropriate stimulus for basing perceptual judgments was profoundly impaired during SC inactivation. When the cued stimulus was placed in the affected part of the visual field, performance was severely impaired, but the errors were not random. They tended to be based on the irrelevant foil stimulus placed outside the affected region. This pattern of errors shows that the animal still attempted to discriminate the direction of motion after the SC was inactivated but erroneously based his choices on the stimulus at the wrong spatial location. In control experiments,

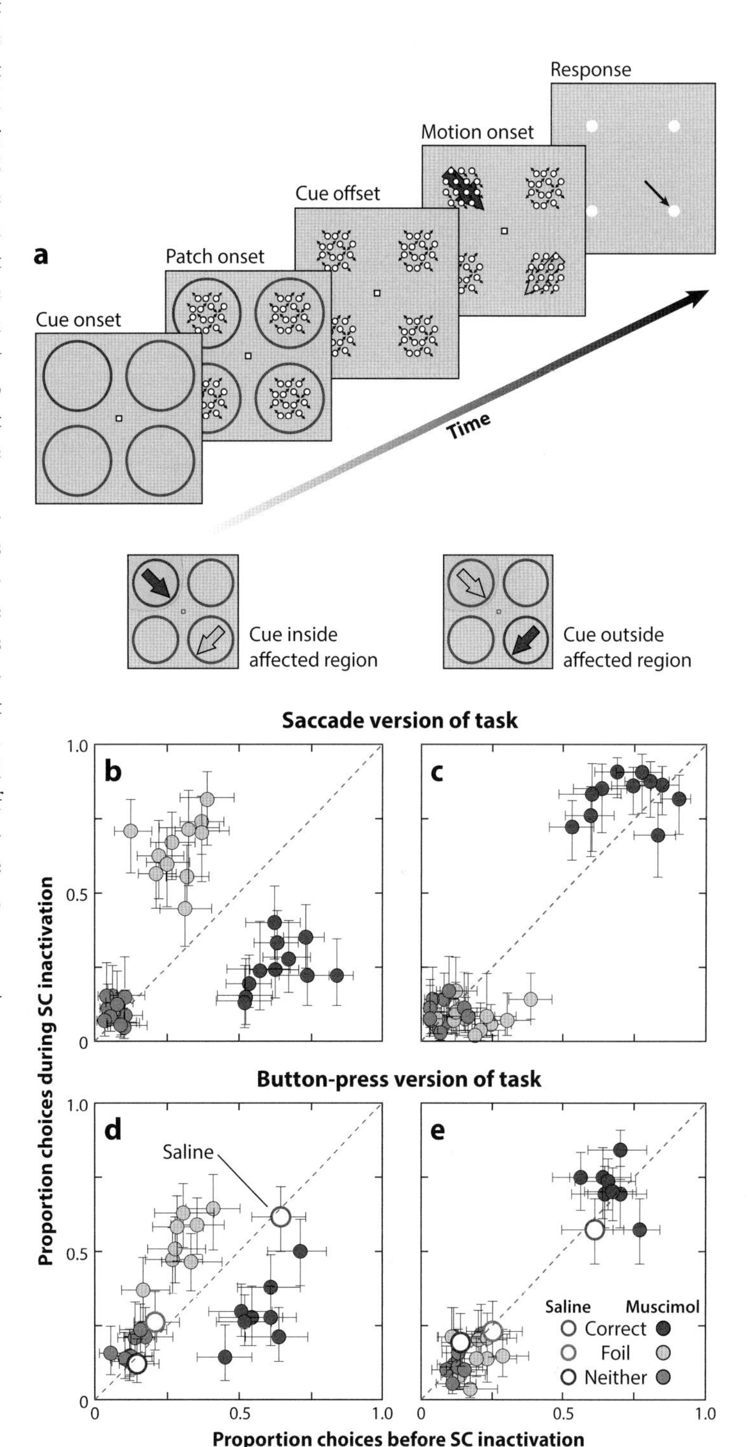

Figure 3

Impairments in covert selection of signals for perceptual judgments during superior colliculus (SC) inactivation. (*a*) Task design. A color cue indicated the relevant motion patch, which contained a brief pulse of motion in one of four possible directions. The animal indicated its choice either by making a saccade in the discriminated direction or by pressing a corresponding button. (*b*) When the cued signal was in the affected region, animals ignored this signal and instead based their choices on the foil. (*c*) Conversely, when the foil signal appeared in the affected region, subjects tended to ignore the foil. (*d–e*) Similar results were obtained in the absence of saccades during a button-press version of the task. Filled symbols show data from muscimol injection experiments; open symbols, from saline control injections. Adapted from Lovejoy & Krauzlis (2010).

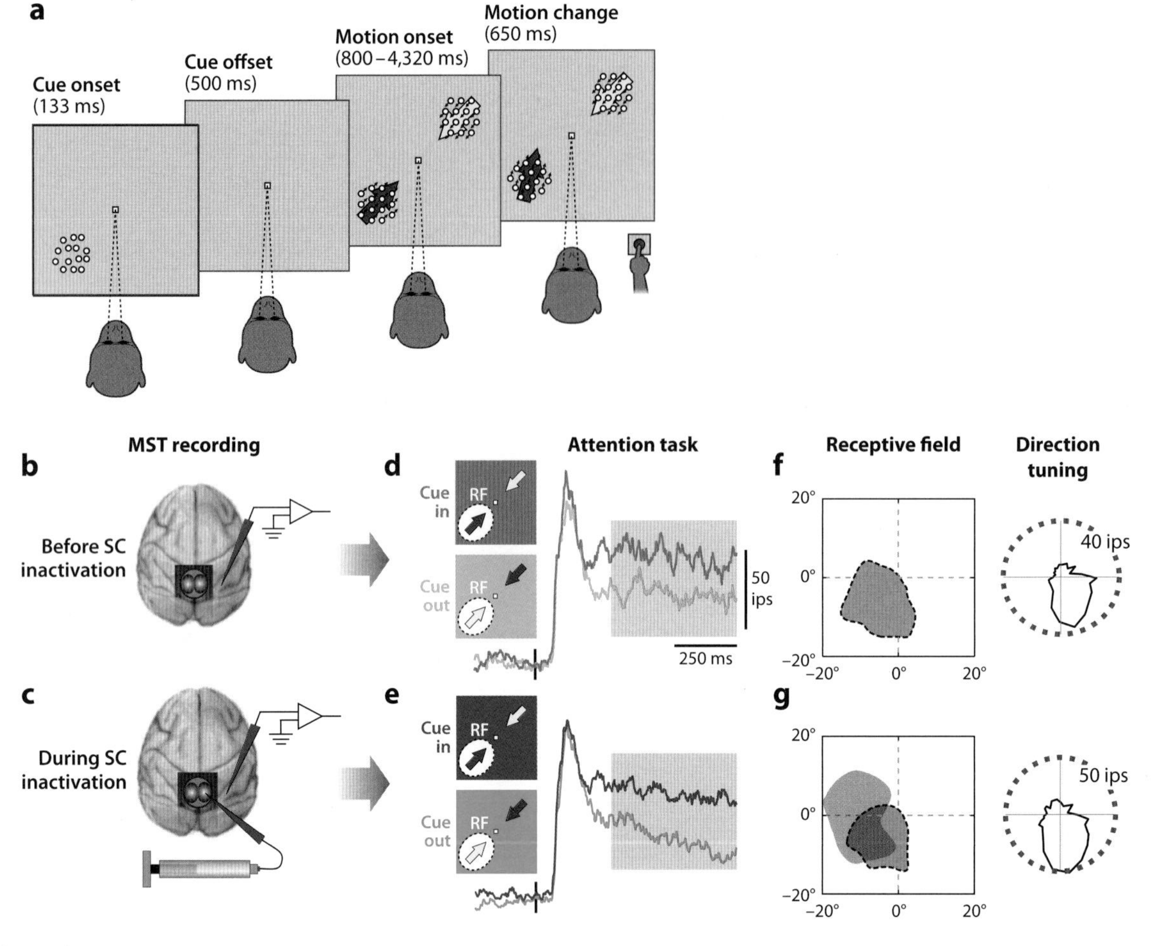

Figure 4

Neuronal correlates of spatial attention remain intact in visual cortex during superior colliculus (SC) inactivation despite behavioral deficits in attention. (*a*) Task design. A cue indicated the relevant motion patch, and later either this cued patch or the foil patch changed its direction. If the cued patch changed, the animals reported this detection by pressing a button. (*b–c*) Neuronal activity was recorded in the medial superior temporal area (MST) before (*b*) and during (*c*) SC inactivation. (*d–e*) Activity of a sample neuron during the attention task before (*d*) and during (*e*) SC inactivation. Despite large deficits in task performance caused by SC inactivation, the modulation of neuronal activity by spatial cues remained intact. (*f–g*) Receptive field and tuning properties before (*f*) and during (*g*) SC inactivation. The blue shading shows the affected region. Abbreviations: RF, receptive field. Adapted from Zénon & Krauzlis (2012).

when investigators presented only a single motion patch in the affected part of the visual field, the animal exhibited only a minor impairment in performance, showing that the impairment in the attention task cannot be explained by a deficit in visual-motion processing. This study demonstrates that the SC is not just linked to the neuronal circuit for spatial attention but in fact is necessary for its normal operation.

Interaction with the Visual Cortex

Researchers have generally assumed that if the SC plays a role in the control of spatial attention, it uses the same mechanisms responsible for the well-known effects of spatial attention in the visual cortex (Desimone & Duncan 1995, Petersen & Posner 2012, Reynolds & Chelazzi 2004). However, recent direct tests cast doubt on this assumption (see **Figure 4**). In these experiments (Zénon & Krauzlis 2012), activity in the SC was chemically inactivated during a motion-change detection task while neuronal activity was simultaneously recorded in two cortical visual areas well known to be modulated by spatial attention: the middle temporal area (MT) and the medial superior temporal area (MST). As in earlier work (Lovejoy & Krauzlis 2010), SC inactivation caused large and spatially specific deficits in spatial attention—the animals' ability to detect changes in the cued stimulus was markedly impaired when it was placed in the affected part of the visual field. These effects cannot be explained by a motor deficit because the operant response was a single-button press that was unimpaired when the attended stimulus was placed outside the affected part of the visual field. If these behavioral deficits were caused by changes in activity ascending from the SC to the visual cortex, then inactivating the SC should also change attention-related effects in the visual cortex. Contrary to this prediction, the attention-related effects in visual cortex—including each of the major signatures of spatial attention—did not change during SC inactivation. Moreover, the ability of neurons to detect the stimulus change, overall firing rates, and other aspects of their activity also did not change, despite the behavioral deficit.

These results are strong evidence against the idea that SC activity contributes to spatial attention through the same mechanisms responsible for the well-known correlates of attention in the visual cortex. The crucial changes may have taken place in other cortical areas—for example, the frontal eye fields (FEF)—but in that case one would still have expected to detect neuronal changes because of the feedback from the FEF to the visual cortex (Moore & Armstrong 2003, Moore & Fallah 2001). Instead, these results demonstrate that the SC's control of spatial attention depends on additional processes that operate independently of the known modulations of activity in the visual cortex.

POSSIBLE NEURONAL CIRCUITS

Although the SC has now been demonstrated to be a key part of the mechanisms for spatial attention, the neuronal circuits involved are unknown. In addition to establishing well-known connections with motor structures in the brain stem and cerebellum, which is important for the control of orienting movements, the SC also forms circuits with many other brain regions that could serve nonmotoric functions, including spatial attention.

FEF: frontal eye fields

Pathways to the Cortex: Inferior Pulvinar

A circuit from the SC through the inferior pulvinar might explain the possible changes in sensory processing with spatial attention (**Figure 5**). The properties of this circuit are poorly understood in primates, but in birds it plays a striking role in gating the transmission of visual signals. The superficial layers of the bird optic tectum (homologous to the primate SC) contain a distinctive class of neurons called tectal ganglion cells, which respond to visual motion across large parts of the visual field and contribute to the even larger receptive fields of neurons in the thalamic nucleus rotundus (homologous to the primate pulvinar). Cholinergic inputs to the superficial SC modulate the efficacy of retinal inputs onto these tectal ganglion cells and are involved in resolving competition among multiple stimuli (Lai et al. 2011, Marín et al. 2005, Wang et al. 2006). When this cholinergic input is locally blocked, neurons in the nucleus rotundus no longer respond to visual stimulation in the affected subregion of the visual field (Marín et al. 2007).

Whether a similar functional circuit operates in the primate is not clear, but the inferior pulvinar receives efferents from the superficial layers of the SC and projects to areas in the extrastriate cortex that are specialized for processing visual motion. Investigators identified this pathway using transsynaptic anatomical tracers (Lyon et al. 2010), and they also used electrical stimulation to identify neurons that both project to the cortex and receive inputs from the SC (Berman & Wurtz 2010). Because it provides a direct route to motion-processing areas in the cortex, this pathway has been implicated

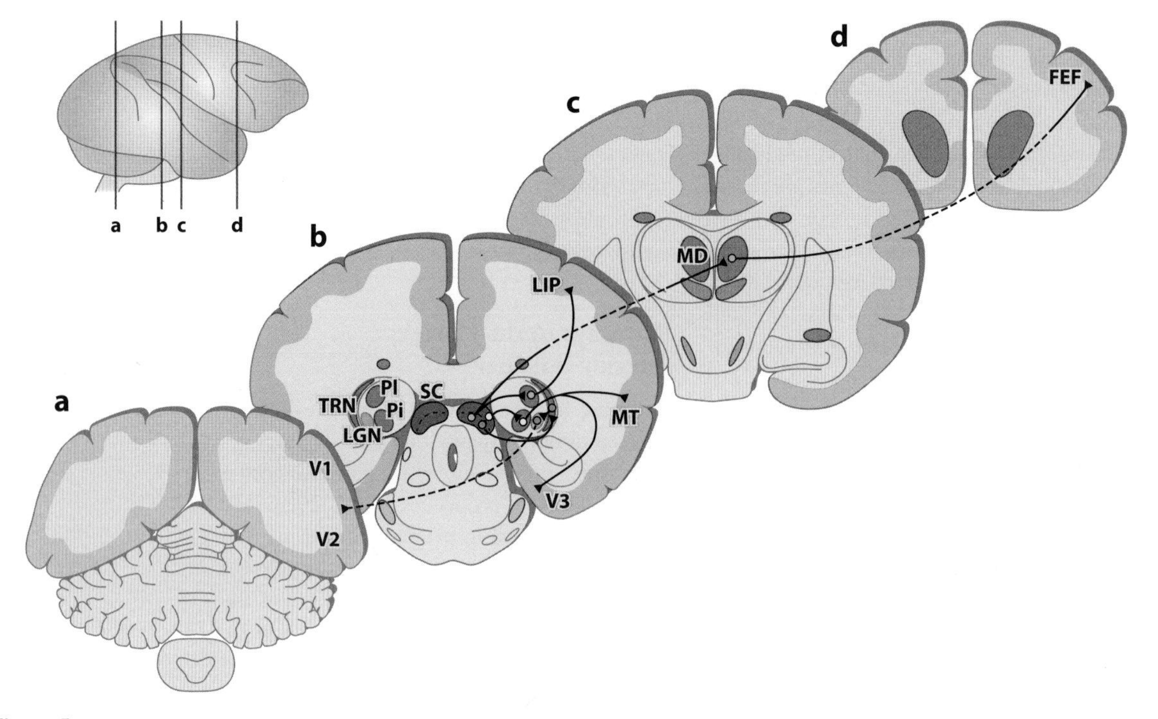

Figure 5

Possible ascending neuronal circuits accounting for the superior colliculus (SC) role in spatial attention. Three circuits are outlined in these schematic sections of the monkey brain. One pathway (*yellow*) passes through the inferior pulvinar (Pi) to the middle temporal area (MT). A second pathway (*pink*) involves the thalamic reticular nucleus (TRN), which connects to the lateral geniculate nucleus (LGN), which, in turn, projects to the visual cortex (V1 and V2). A third pathway (*light blue*) passes through the lateral pulvinar (Pl) to the lateral intraparietal area (LIP) and through the mediodorsal thalamus (MD) to the frontal eye fields (FEF).

TRN: thalamic reticular nucleus

in the changes in performance on visual-motion tasks caused by SC stimulation (Cavanaugh & Wurtz 2004, Müller et al. 2005).

However, because this circuit links the SC to the visual cortex, its role in the control of attention is not compatible with the observation that the disruption of attention following SC inactivation leaves visual cortical activity unchanged. Moreover, this putative explanation is complicated by the fact that it relies on a circuit involving the superficial layers of the SC. The strongest evidence for a role of the SC in attention implicates the intermediate layers, and the functional role of connections between the intermediate and superficial layers of the primate SC remains an unresolved issue. Therefore, we do not know if such a circuit would be sufficient to explain the role of the SC in the control of attention. Instead, recent evidence suggests that it may suppress the transmission of self-induced visual signals during saccades (Berman & Wurtz 2011).

Pathways to the Cortex: Thalamic Reticular Nucleus

Another possible circuit involves the thalamic reticular nucleus (TRN). As fibers from other thalamic nuclei pass through the TRN en route to the sensory cortex, they give off collateral branches that contact inhibitory neurons in the TRN that project back to sensory thalamic nuclei (**Figure 5**). This thalamic feedback loop through the TRN could provide a gating mechanism that regulates signal transmission to the cortex, and the topographic ordering of the connections in the TRN provides a possible anatomical correlate of the putative searchlight of attention (Crick 1984, Guillery et al. 1998). Consistent with this hypothesis, neuronal activity in the visual TRN is modulated during attention tasks, and these changes precede complementary changes in the lateral geniculate nucleus, the primary visual thalamic nucleus (McAlonan et al. 2006, 2008). Because the intermediate and deep layers of the SC project to the TRN, SC activity may help determine how the TRN gates the transmission of visual signals to the cortex.

Although this putative circuit places the intermediate SC in a central role, it still depends on the notion that the SC contributes to spatial attention by changing sensory processing, which is not consistent with the finding that attention-related changes in the visual cortex remain intact during SC inactivation (Zénon & Krauzlis 2012).

Pathways to the Cortex: Lateral Pulvinar and Mediodorsal Thalamus

The SC is also a member of circuits that could act at stages downstream of sensory processing. Signals from the intermediate and deeper layers of the SC reach areas of the prefrontal and parietal cortex through several thalamic nuclei (**Figure 5**), including the lateral pulvinar and mediodorsal nucleus (Harting et al. 1980, Robinson & Petersen 1992). The ascending input from the SC through the lateral pulvinar reaches cortical areas involved in spatial attention, most notably, the lateral intraparietal area (LIP) in the parietal lobe (Romanski et al. 1997). Area LIP contains a salience map that can guide visual attention as well as saccadic eye movements (Bisley & Goldberg 2010). The functional circuits through the lateral pulvinar are poorly understood; however, neuronal activity in the pulvinar is modulated by manipulations of spatial attention (Benevento & Port 1995, Petersen et al. 1985), and lesions of the pulvinar cause deficits in visual attention tasks (Desimone et al. 1990). Pulvinar lesions tend to produce neglect-like deficits similar to those found during SC inactivation (Lovejoy & Krauzlis 2010)—performance is especially impaired in the presence of distracters.

SPATIAL INDEXING?

The SC contains maps of spatial locations but not stimulus features. Consequently, signals in the SC cannot directly determine the content of spatial attention but must interact with signals from other brain areas that process and represent the qualities of the attended object. One possibility is that the SC is part of a spatial indexing system that identifies which signals elsewhere in the brain should be pooled or otherwise put together to determine the content of perception. In this view, when SC activity is suppressed, signals from the affected part of space can no longer be properly retrieved or shared by other brain regions, resulting in deficits in spatial attention and visual neglect.

Another major component of the ascending input to the cortex from the SC passes through the mediodorsal thalamic nucleus to the prefrontal cortex, including the FEF. One functional role of this pathway is to convey corollary discharge signals about saccadic eye movements (Sommer & Wurtz 2008). In addition to transmitting motor signals about saccades, this pathway also transmits visual and other signals from the SC (Sommer & Wurtz 2004) that may be related to the role of the FEF in the control of attention. In particular, signals from the FEF can cause attention-like changes in the processing of visual signals in the extrastriate cortex (Moore & Armstrong 2003, Moore & Fallah 2001). By modifying signal processing in FEF, the input from the SC could regulate this cortico-cortical feedback mechanism for spatial attention.

These putative circuits place the intermediate SC in a central role and act downstream of sensory processing. However, they still assume that the control of spatial attention occurs through modulation of signal processing in the visual cortex, albeit indirectly through LIP or FEF. This assumption is undermined by the finding that attention-related changes in the visual cortex remain intact during SC inactivation, despite the presence of large deficits in attention (Zénon & Krauzlis 2012).

LIP: lateral intraparietal area

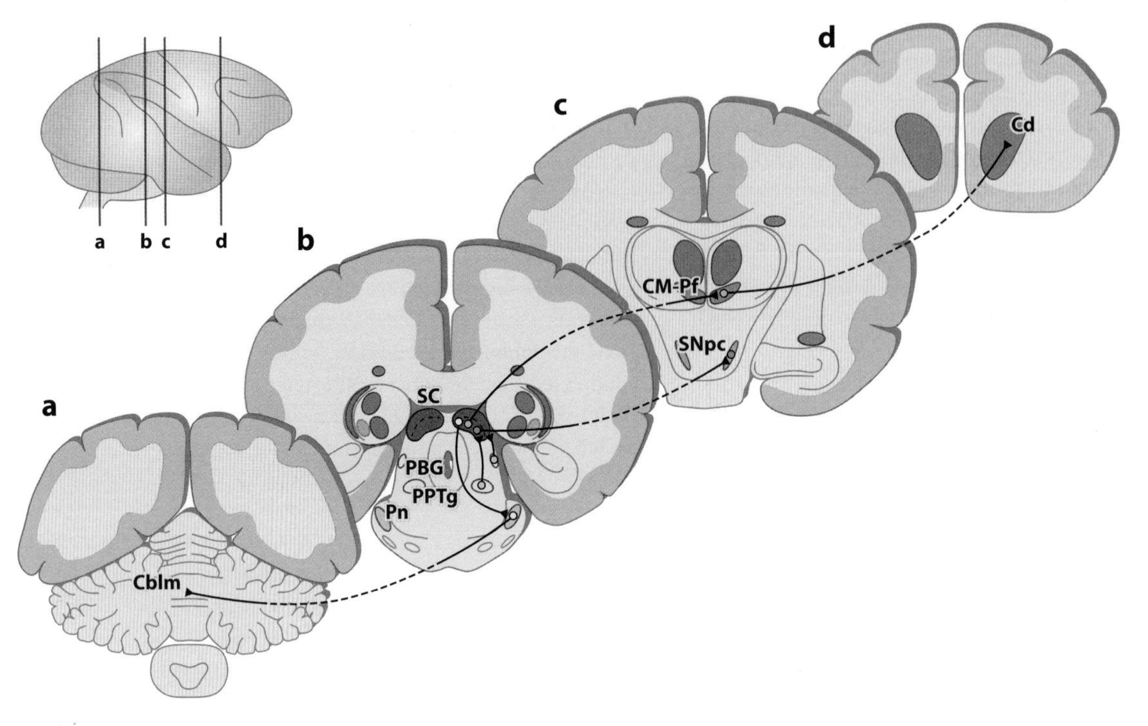

Figure 6

Possible subcortical neuronal circuits accounting for the superior colliculus (SC) role in spatial attention. One pathway (*light blue*) passes through the centromedian parafascicular complex (CM-Pf) to the caudate nucleus (Cd) in the basal ganglia. A second pathway (*pink*) projects to the substantia nigra pars compacta (SNpc). A third circuit (*green*) involves cholinergic inputs from the pedunculopontine tegmental nucleus (PPTg) and parabigeminal nucleus (PBG). A fourth pathway (*yellow*) reaches the pontine nuclei (Pn), which relay cortical signals to the cerebellum (Cblm).

Subcortical Loops Through the Basal Ganglia

Not all the ascending circuits from the SC through the thalamus have the cortex as their primary target. Instead, some SC-thalamic projections are part of subcortical circuits linking the SC to the basal ganglia (see **Figure 6**). One such circuit involves projections to the intralaminar nuclei in the thalamus, which forms part of a loop through the caudate and putamen, to the substantia nigra pars reticulata, and back to the SC. As part of the ascending reticular activating system, the intralaminar nuclei play some role in general arousal. However, portions of the intralaminar nuclei receive inputs from the intermediate and deeper layers of the SC and appear to play a more specific role in the processing of salient events (Sadikot & Rymar 2009, Smith et al. 2004, Van der Werf et al. 2002). For example, neuronal activity in the centromedian-parafascicular (CM-Pf) complex is modulated by spatial cues during attention tasks, and chemical inactivation of the CM-Pf causes spatially specific increases in reaction times (Minamimoto & Kimura 2002). Spatially selective activity has also been found in these structures in humans during covert attention tasks (Hulme et al. 2010).

The anatomical loops formed by these thalamic nuclei with the SC and basal ganglia may act independently of, or competitively with, similar loops originating from the cerebral cortex. The exact role of these loops remains controversial, but one compelling theory is that they implement value-based selection mechanisms at different stages of sensorimotor and cognitive processing (McHaffie et al. 2005; Redgrave et al. 1999, 2010). By providing access to these selection mechanisms in the basal ganglia, these circuits provide a plausible explanation for how SC inactivation could cause deficits in spatial attention without altering visual cortical responses.

Substantia Nigra Pars Compacta

In addition to a pathway through the thalamus to the basal ganglia, a direct projection from the intermediate and deeper layers of the SC to the substantia nigra pars compacta (SNpc) has been demonstrated in several species, including primates (**Figure 6**; Comoli et al. 2003, May et al. 2009, McHaffie et al. 2006). The SNpc contains dopamine neurons that play a prominent role in reinforcement learning by providing a prediction error signal (Niv & Schoenbaum 2008, Schultz 2010). However, SNpc neurons also respond to unexpected and salient sensory events (Redgrave et al. 2008, Schultz & Romo 1990), which could be driven at least in part by inputs from the SC and linked to shifts of spatial attention.

Modulatory Inputs from the Brain Stem

As described earlier, the SC receives cholinergic inputs from tegmental cholinergic nuclei that can modulate the transmission of visual signals to the thalamus (**Figure 6**). The parabigeminal nucleus [nucleus isthmi pars parvocellularis in birds (PBG/Ipc)] provides cholinergic inputs mostly to the superficial layer (Wang et al. 2006), whereas the intermediate and deep layers receive cholinergic inputs bilaterally from the pedunculopontine tegmental nuclei (Harting & Van Lieshout 1991).

The PBG/Ipc is part of a circuit involving another tegmental structure, the periparabigeminal tegmental nucleus [nucleus isthmi pars magnocellularis in birds (PPBN/Imc)], which sends inhibitory inputs to the intermediate and deep layers of the SC. The interaction between the facilitatory influence of PBG/Ipc cholinergic connections and the global inhibitory effects of the γ-aminobutyric acid (GABA)ergic PPBN/Imc inputs in the SC could mediate selection between competing visual stimuli (Lai et al. 2011, Mysore et al. 2010, Wang et al. 2004) and generate the switch-like behavior of some neurons in the bird SC that seem to respond only to the most salient stimulus (Mysore & Knudsen 2011). This mechanism has been studied prominently in birds, but if similar mechanisms operate in the primate SC, they could contribute to how SC activity selects visual targets; however, investigators would still need to define the circuits involved in relaying this target selection within the SC to structures that can influence perceptual decisions.

The other cholinergic nucleus projecting to the SC, the pedunculopontine tegmental nucleus (PPTg), is sometimes considered one of the output nuclei of the basal ganglia, exhibiting reciprocal connections with many basal ganglia structures and sending efferent projections to the SC, the thalamus, and other brain stem, medullary, and spinal nuclei (Inglis & Winn 1995). Evidence points to a role in saccade control, especially in relation to reward processing (Kobayashi et al. 2002), and this function may depend on connections with the SC. Indeed, injections of cholinergic agonists in the intermediate layers of the SC, which receive cholinergic inputs from the PPTg, affect saccadic control (Aizawa et al. 1999, Watanabe et al. 2005). However, the specific role of PPTg in spatial attention has not yet been directly investigated.

Pontine Nuclei

The pontine nuclei generally act as a relay that takes signals from the cerebral cortex and provides mossy fiber inputs to the cerebellum. The portions of the pontine nuclei receiving inputs from dorsal visual areas have received special attention for their role in providing visual signals for guiding pursuit and saccadic eye movements (Thier & Möck 2006). In addition to receiving cortical inputs, the pontine nuclei also receive inputs from multiple layers of the SC, and these terminate in dorsolateral regions overlapping with inputs from the visual cortex (Glickstein et al. 1990, Harting 1977). This overlap suggests that SC activity could modify the visual signals transmitted through the pontine nuclei (**Figure 6**). This circuit mechanism may seem unlikely because researchers do not typically consider the cerebellum to be a central player in spatial attention, and studies of this question have produced mixed results (Ignashchenkova et al. 2009, Townsend et al. 1999).

In summary, there are several possible neuronal circuits for the SC's role in attention. Some of these circuits act through established cortical mechanisms, whereas others are subcortical and could operate largely independently of the known mechanisms of attention involving modulation of visual processing in the cerebral cortex. Clarifying the functional role of these circuits is a necessary step toward understanding the complete network of brain regions that control spatial attention.

CONCLUSIONS AND SPECULATIONS

In addition to its well-established role in orienting movements of the eyes and head, the primate SC also plays a crucial role in spatial attention. Part of this role is to mediate the effects of attention during orienting movements, but SC activity is also necessary for the normal operation of spatial attention during perceptual tasks even in the absence of orienting movements. Surprisingly, the SC's role in covert spatial attention involves mechanisms that appear to operate independently of the established signatures of attention in the visual cortex, raising new questions about the organization of the neuronal circuits that underlie spatial attention.

Our working hypothesis is that the control of attention involves at least two dissociable functional modules. First, the SC is a central node in an evolutionarily older circuit with the basal ganglia; these brain areas operate together to represent the location of important objects in space. This idea contrasts with the view of attention as a unified system: the SC is usually grouped together functionally with other cortical brain regions involved in both overt and covert orienting, such as the FEF or area LIP. Unlike those cortical areas, the SC is a highly conserved component of the vertebrate brain that plays an important role in spatial orienting, even in animals that lack a well-developed neocortex; the finding that blocking SC activity causes attention deficits without altering visual cortical activity suggests that it acts downstream.

We speculate that the second module is a younger system of cortical circuits that refines the operation of the SC–basal ganglia system and that the capabilities of the mature attention system in primates are acquired through learning. Indeed, in a visually cluttered environment, developing robust stimulus-response associations is impossible without first selecting the appropriate stimulus object to associate with the reinforced action. With the expansion of the neocortex, the number of stimulus features available for classifying objects and assigning meaning has expanded far beyond the capacity of a retinotopic map of visual space; however, when it comes to selecting the content of action or perception, the central components of this evolutionarily ancient selection mechanism remain intact.

DISCLOSURE STATEMENT

The authors are not aware of any affiliations, memberships, funding, or financial holdings that might be perceived as affecting the objectivity of this review.

LITERATURE CITED

Aizawa H, Kobayashi Y, Yamamoto M, Isa T. 1999. Injection of nicotine into the superior colliculus facilitates occurrence of express saccades in monkeys. *J. Neurophysiol.* 82(3):1642–46

Basso MA, Wurtz RH. 1998. Modulation of neuronal activity in superior colliculus by changes in target probability. *J. Neurosci.* 18(18):7519–34

Bell AH, Fecteau JH, Munoz DP. 2004. Using auditory and visual stimuli to investigate the behavioral and neuronal consequences of reflexive covert orienting. *J. Neurophysiol.* 91(5):2172–84

Benevento LA, Port JD. 1995. Single neurons with both form/color differential responses and saccade-related responses in the nonretinotopic pulvinar of the behaving macaque monkey. *Vis. Neurosci.* 12(3):523–44

Berman RA, Wurtz RH. 2010. Functional identification of a pulvinar path from superior colliculus to cortical area MT. *J. Neurosci.* 30(18):6342–54

Berman RA, Wurtz RH. 2011. Signals conveyed in the pulvinar pathway from superior colliculus to cortical area MT. *J. Neurosci.* 31(2):373–84

Bisley JW, Goldberg ME. 2010. Attention, intention, and priority in the parietal lobe. *Annu. Rev. Neurosci.* 33:1–21

Carello CD, Krauzlis RJ. 2004. Manipulating intent: evidence for a causal role of the superior colliculus in target selection. *Neuron* 43(4):575–83

Cavanaugh J, Alvarez BD, Wurtz RH. 2006. Enhanced performance with brain stimulation: attentional shift or visual cue? *J. Neurosci.* 26(44):11347–58

Cavanaugh J, Wurtz RH. 2004. Subcortical modulation of attention counters change blindness. *J. Neurosci.* 24(50):11236–43

Ciaramitaro VM, Todd WE, Rosenquist AC. 1997. Disinhibition of the superior colliculus restores orienting to visual stimuli in the hemianopic field of the cat. *J. Comp. Neurol.* 387(4):568–87

Comoli E, Coizet V, Boyes J, Bolam JP, Canteras NS, et al. 2003. A direct projection from superior colliculus to substantia nigra for detecting salient visual events. *Nat. Neurosci.* 6(9):974–80

Crick F. 1984. Function of the thalamic reticular complex: the searchlight hypothesis. *Proc. Natl. Acad. Sci. USA* 81(14):4586–90

Desimone R, Duncan J. 1995. Neural mechanisms of selective visual attention. *Annu. Rev. Neurosci.* 18:193–222

Desimone R, Wessinger M, Thomas L, Schneider W. 1990. Attentional control of visual perception: cortical and subcortical mechanisms. *Cold Spring Harb. Symp. Quant. Biol.* 55:963–71

Diamond IT, Hall WC. 1969. Evolution of neocortex. *Science* 164:251–62

Dorris MC, Munoz DP. 1998. Saccadic probability influences motor preparation signals and time to saccadic initiation. *J. Neurosci.* 18(17):7015–26

Dorris MC, Olivier E, Munoz DP. 2007. Competitive integration of visual and preparatory signals in the superior colliculus during saccadic programming. *J. Neurosci.* 27(19):5053–62

Durmer JS, Rosenquist AC. 2001. Ibotenic acid lesions in the pedunculopontine region result in recovery of visual orienting in the hemianopic cat. *Neuroscience* 106(4):765–81

Fecteau JH, Bell AH, Munoz DP. 2004. Neural correlates of the automatic and goal-driven biases in orienting spatial attention. *J. Neurophysiol.* 92(3):1728–37

Fecteau JH, Munoz DP. 2006. Salience, relevance, and firing: a priority map for target selection. *Trends Cogn. Sci.* 10(8):382–90

Gandhi NJ, Katnani HA. 2011. Motor functions of the superior colliculus. *Annu. Rev. Neurosci.* 34:205–31

Glickstein M, May J, Mercier B. 1990. Visual corticopontine and tectopontine projections in the macaque. *Arch. Ital. Biol.* 128(2–4):273–93

Glimcher PW, Sparks DL. 1992. Movement selection in advance of action in the superior colliculus. *Nature* 355:542–45

Goldberg ME, Wurtz RH. 1972. Activity of superior colliculus in behaving monkey. II. Effect of attention on neuronal responses. *J. Neurophysiol.* 35(4):560–74

Guillery RW, Feig SL, Lozsádi DA. 1998. Paying attention to the thalamic reticular nucleus. *Trends Neurosci.* 21(1):28–32

Harting JK. 1977. Descending pathways from the superior collicullus: an autoradiographic analysis in the rhesus monkey (*Macaca mulatta*). *J. Comp. Neurol.* 173(3):583–612

Harting JK, Huerta MF, Frankfurter AJ, Strominger NL, Royce GJ. 1980. Ascending pathways from the monkey superior colliculus: an autoradiographic analysis. *J. Comp. Neurol.* 192(4):853–82

Harting JK, Van Lieshout DP. 1991. Spatial relationships of axons arising from the substantia nigra, spinal trigeminal nucleus, and pedunculopontine tegmental nucleus within the intermediate gray of the cat superior colliculus. *J. Comp. Neurol.* 305(4):543–58

Horwitz GD, Batista AP, Newsome WT. 2004. Representation of an abstract perceptual decision in macaque superior colliculus. *J. Neurophysiol.* 91(5):2281–96

Horwitz GD, Newsome WT. 1999. Separate signals for target selection and movement specification in the superior colliculus. *Science* 284:1158–61

Horwitz GD, Newsome WT. 2001. Target selection for saccadic eye movements: prelude activity in the superior colliculus during a direction-discrimination task. *J. Neurophysiol.* 86(5):2543–58

Hulme OJ, Whiteley L, Shipp S. 2010. Spatially distributed encoding of covert attentional shifts in human thalamus. *J. Neurophysiol.* 104(6):3644–56

Ignashchenkova A, Dash S, Dicke PW, Haarmeier T, Glickstein M, Thier P. 2009. Normal spatial attention but impaired saccades and visual motion perception after lesions of the monkey cerebellum. *J. Neurophysiol.* 102(6):3156–68

Ignashchenkova A, Dicke PW, Haarmeier T, Thier P. 2004. Neuron-specific contribution of the superior colliculus to overt and covert shifts of attention. *Nat. Neurosci.* 7(1):56–64

Inglis WL, Winn P. 1995. The pedunculopontine tegmental nucleus: where the striatum meets the reticular formation. *Prog. Neurobiol.* 47(1):1–29

Kobayashi Y, Inoue Y, Yamamoto M, Isa T, Aizawa H. 2002. Contribution of pedunculopontine tegmental nucleus neurons to performance of visually guided saccade tasks in monkeys. *J. Neurophysiol.* 88(2):715–31

Krauzlis RJ. 2003. Neuronal activity in the rostral superior colliculus related to the initiation of pursuit and saccadic eye movements. *J. Neurosci.* 23(10):4333–44

Krauzlis RJ. 2005. The control of voluntary eye movements: new perspectives. *Neuroscientist* 11(2):124–37

Krauzlis RJ, Basso MA, Wurtz RH. 1997. Shared motor error for multiple eye movements. *Science* 276:1693–95

Krauzlis RJ, Basso MA, Wurtz RH. 2000. Discharge properties of neurons in the rostral superior colliculus of the monkey during smooth-pursuit eye movements. *J. Neurophysiol.* 84(2):876–91

Krauzlis R, Dill N. 2002. Neural correlates of target choice for pursuit and saccades in the primate superior colliculus. *Neuron* 35(2):355–63

Kustov AA, Robinson DL. 1996. Shared neural control of attentional shifts and eye movements. *Nature* 384(6604):74–77

Lai D, Brandt S, Luksch H, Wessel R. 2011. Recurrent antitopographic inhibition mediates competitive stimulus selection in an attention network. *J. Neurophysiol.* 105(2):793–805

Lisberger SG. 2010. Visual guidance of smooth-pursuit eye movements: sensation, action, and what happens in between. *Neuron* 66(4):477–91

Lovejoy LP, Krauzlis RJ. 2010. Inactivation of primate superior colliculus impairs covert selection of signals for perceptual judgments. *Nat. Neurosci.* 13(2):261–66

Lyon DC, Nassi JJ, Callaway EM. 2010. A disynaptic relay from superior colliculus to dorsal stream visual cortex in macaque monkey. *Neuron* 65(2):270–79

Marín G, Mpodozis J, Sentis E, Ossandón T, Letelier JC. 2005. Oscillatory bursts in the optic tectum of birds represent re-entrant signals from the nucleus isthmi pars parvocellularis. *J. Neurosci.* 25(30):7081–89

Marín G, Salas C, Sentis E, Rojas X, Letelier JC, Mpodozis J. 2007. A cholinergic gating mechanism controlled by competitive interactions in the optic tectum of the pigeon. *J. Neurosci.* 27(30):8112–21

May PJ. 2006. The mammalian superior colliculus: laminar structure and connections. *Prog. Brain Res.* 151:321–78

May PJ, McHaffie JG, Stanford TR, Jiang H, Costello MG, et al. 2009. Tectonigral projections in the primate: a pathway for pre-attentive sensory input to midbrain dopaminergic neurons. *Eur. J. Neurosci.* 29(3):575–87

McAlonan K, Cavanaugh J, Wurtz RH. 2006. Attentional modulation of thalamic reticular neurons. *J. Neurosci.* 26(16):4444–50

McAlonan K, Cavanaugh J, Wurtz RH. 2008. Guarding the gateway to cortex with attention in visual thalamus. *Nature* 456(7220):391–94

McHaffie JG, Jiang H, May PJ, Coizet V, Overton PG, et al. 2006. A direct projection from superior colliculus to substantia nigra pars compacta in the cat. *Neuroscience* 138(1):221–34

McHaffie JG, Stanford TR, Stein BE, Coizet V, Redgrave P. 2005. Subcortical loops through the basal ganglia. *Trends Neurosci.* 28(8):401–7

McPeek RM, Keller EL. 2002. Saccade target selection in the superior colliculus during a visual search task. *J. Neurophysiol.* 88(4):2019–34

McPeek RM, Keller EL. 2004. Deficits in saccade target selection after inactivation of superior colliculus. *Nat. Neurosci.* 7(7):757–63

Minamimoto T, Kimura M. 2002. Participation of the thalamic CM-Pf complex in attentional orienting. *J. Neurophysiol.* 87(6):3090–101

Moore T, Armstrong KM. 2003. Selective gating of visual signals by microstimulation of frontal cortex. *Nature* 421:370–73

Moore T, Fallah M. 2001. Control of eye movements and spatial attention. *Proc. Natl. Acad. Sci. USA* 98(3):1273–76

Müller JR, Philiastides MG, Newsome WT. 2005. Microstimulation of the superior colliculus focuses attention without moving the eyes. *Proc. Natl. Acad. Sci. USA* 102(3):524–29

Mysore SP, Asadollahi A, Knudsen EI. 2010. Global inhibition and stimulus competition in the owl optic tectum. *J. Neurosci.* 30(5):1727–38

Mysore SP, Knudsen EI. 2011. Flexible categorization of relative stimulus strength by the optic tectum. *J. Neurosci.* 31(21):7745–52

Niv Y, Schoenbaum G. 2008. Dialogues on prediction errors. *Trends Cogn. Sci.* 12(7):265–72

Nummela SU, Krauzlis RJ. 2010. Inactivation of primate superior colliculus biases target choice for smooth pursuit, saccades, and button press responses. *J. Neurophysiol.* 104(3):1538–48

Nummela SU, Krauzlis RJ. 2011. Superior colliculus inactivation alters the weighted integration of visual stimuli. *J. Neurosci.* 31(22):8059–66

Petersen SE, Posner MI. 2012. The attention system of the human brain: 20 years after. *Annu. Rev. Neurosci.* 35:73–89

Petersen SE, Robinson DL, Keys W. 1985. Pulvinar nuclei of the behaving rhesus monkey: visual responses and their modulation. *J. Neurophysiol.* 54(4):867–86

Redgrave P, Coizet V, Comoli E, McHaffie JG, Leriche M, et al. 2010. Interactions between the midbrain superior colliculus and the basal ganglia. *Front. Neuroanat.* 4:132

Redgrave P, Gurney K, Reynolds J. 2008. What is reinforced by phasic dopamine signals? *Brain Res. Rev.* 58(2):322–39

Redgrave P, Prescott TJ, Gurney K. 1999. The basal ganglia: a vertebrate solution to the selection problem? *Neuroscience* 89(4):1009–23

Rensink RA. 2002. Change detection. *Annu. Rev. Psychol.* 53:245–77

Reynolds JH, Chelazzi L. 2004. Attentional modulation of visual processing. *Annu. Rev. Neurosci.* 27:611–47

Rizzo M, Hurtig R. 1992. Visual search in hemineglect: What stirs idle eyes? *Clin. Vis. Sci.* 7(1):39–52

Robinson DL, Petersen SE. 1992. The pulvinar and visual salience. *Trends Neurosci.* 15(4):127–32

Robson JA, Hall WC. 1977. The organization of the pulvinar in the grey squirrel (*Sciurus carolinensis*). I. Cytoarchitecture and connections. *J. Comp. Neurol.* 173(2):355–88

Romanski LM, Giguere M, Bates JF, Goldman-Rakic PS. 1997. Topographic organization of medial pulvinar connections with the prefrontal cortex in the rhesus monkey. *J. Comp. Neurol.* 379(3):313–32

Sadikot AF, Rymar VV. 2009. The primate centromedian-parafascicular complex: anatomical organization with a note on neuromodulation. *Brain Res. Bull.* 78(2–3):122–30

Schultz W. 2010. Dopamine signals for reward value and risk: basic and recent data. *Behav. Brain Funct.* 6:24

Schultz W, Romo R. 1990. Dopamine neurons of the monkey midbrain: contingencies of responses to stimuli eliciting immediate behavioral reactions. *J. Neurophysiol.* 63(3):607–24

Smith Y, Raju DV, Pare JF, Sidibe M. 2004. The thalamostriatal system: a highly specific network of the basal ganglia circuitry. *Trends Neurosci.* 27(9):520–27

Sommer MA, Wurtz RH. 2004. What the brain stem tells the frontal cortex. I. Oculomotor signals sent from superior colliculus to frontal eye field via mediodorsal thalamus. *J. Neurophysiol.* 91(3):1381–402

Sommer MA, Wurtz RH. 2008. Brain circuits for the internal monitoring of movements. *Annu. Rev. Neurosci.* 31:317–38

Song J-H, Rafal RD, McPeek RM. 2011. Deficits in reach target selection during inactivation of the midbrain superior colliculus. *Proc. Natl. Acad. Sci. USA* 108(51):E1433–40

Sparks DL. 1999. Conceptual issues related to the role of the superior colliculus in the control of gaze. *Curr. Opin. Neurobiol.* 9(6):698–707

Sprague JMJ. 1966. Interaction of cortex and superior colliculus in mediation of visually guided behavior in the cat. *Science* 153:1544–47

Thier P, Möck M. 2006. The oculomotor role of the pontine nuclei and the nucleus reticularis tegmenti pontis. *Prog. Brain Res.* 151:293–320

Townsend J, Courchesne E, Covington J, Westerfield M, Harris N, et al. 1999. Spatial attention deficits in patients with acquired or developmental cerebellar abnormality. *J. Neurosci.* 19(13):5632–43

Trojanowski JO, Jacobson S. 1975. Peroxidase labeled subcortical afferents to pulvinar in rhesus monkey. *Brain Res.* 97(1):144–50

Van der Werf YD, Witter MP, Groenewegen HJ. 2002. The intralaminar and midline nuclei of the thalamus. Anatomical and functional evidence for participation in processes of arousal and awareness. *Brain Res. Brain Res. Rev.* 39(2–3):107–40

Wallace SF, Rosenquist AC, Sprague JM. 1989. Recovery from cortical blindness mediated by destruction of nontectotectal fibers in the commissure of the superior colliculus in the cat. *J. Comp. Neurol.* 284(3):429–50

Wallace SF, Rosenquist AC, Sprague JM. 1990. Ibotenic acid lesions of the lateral substantia nigra restore visual orientation behavior in the hemianopic cat. *J. Comp. Neurol.* 296(2):222–52

Wang Y, Luksch H, Brecha NC, Karten HJ. 2006. Columnar projections from the cholinergic nucleus isthmi to the optic tectum in chicks (*Gallus gallus*): a possible substrate for synchronizing tectal channels. *J. Comp. Neurol.* 494(1):7–35

Wang Y, Major DE, Karten HJ. 2004. Morphology and connections of nucleus isthmi pars magnocellularis in chicks (*Gallus gallus*). *J. Comp. Neurol.* 469(2):275–97

Watanabe M, Kobayashi Y, Inoue Y, Isa T. 2005. Effects of local nicotinic activation of the superior colliculus on saccades in monkeys. *J. Neurophysiol.* 93(1):519–34

Weddell RA. 2004. Subcortical modulation of spatial attention including evidence that the Sprague effect extends to man. *Brain Cogn.* 55(3):497–506

Wurtz RH, Albano JE. 1980. Visual-motor function of the primate superior colliculus. *Annu. Rev. Neurosci.* 3:189–226

Zénon A, Krauzlis RJ. 2012. Attention deficits without cortical neuronal deficits. *Nature* 489:434–37

Genetic Approaches to Neural Circuits in the Mouse

Z. Josh Huang[1] and Hongkui Zeng[2]

[1]Cold Spring Harbor Laboratory, Cold Spring Harbor, New York 11724;
email: huangj@cshl.edu

[2]Allen Institute for Brain Science, Seattle, Washington 98103;
email: hongkuiz@alleninstitute.org

Annu. Rev. Neurosci. 2013. 36:183–215

First published online as a Review in Advance on May 17, 2013

The *Annual Review of Neuroscience* is online at neuro.annualreviews.org

This article's doi:
10.1146/annurev-neuro-062012-170307

Keywords

cell type, genetic targeting, genetic manipulation, transgenic mice, viral vector, driver, reporter

Abstract

To understand the organization and assembly of mammalian brain circuits, we need a comprehensive tool set that can address the challenges of cellular diversity, spatial complexity at synapse resolution, dynamic complexity of circuit operations, and multifaceted developmental processes rooted in the genome. Complementary to physics- and chemistry-based methods, genetic tools tap into intrinsic cellular and developmental mechanisms. Thus, they have the potential to achieve appropriate spatiotemporal resolution and the cellular-molecular specificity necessary for observing and probing the makings and inner workings of neurons and neuronal circuits. Furthermore, genetic analysis will be key to unraveling the intricate link from genes to circuits to systems, in part through systematic targeting and tracking of individual cellular components of neural circuits. Here we review recent progress in genetic tool development and advances in genetic analysis of neural circuits in the mouse. We also discuss future directions and implications for understanding brain disorders.

Contents

[S]ince the essence of all biological systems is that they are encoded as molecular descriptions in their genes and since genes... exert their functions through other molecules, the molecular explanation must constitute the core of understanding biological systems.

Sydney Brenner (2010)

INTRODUCTION

A major goal in neuroscience is to understand how the cellular constituents of the brain are organized into neural circuits that process information, guide behavior, and give rise to mental activities. Over a century ago, Ramón y Cajal's (1899) discovery that individual nerve cells are the building blocks of the nervous system laid a solid foundation for the fertile exploration pursued ever since. However, it also unveiled some daunting scientific challenges, many of which we still face today.

The mammalian brain consists of a vast number of highly diverse neurons that are extensively intermingled and interconnected. We are still far from generating a comprehensive catalog of neuron types in most brain regions, and a wiring diagram of their connectivity. Such are formidable challenges for contemporary neuroanatomists, requiring that neurons and circuits be imaged and reconstructed with single-synapse resolution in large volumes of highly heterogeneous tissues. The functional operation of neural circuits is similarly complex and involves dynamic, coordinated firing of neural ensembles occurring on a millisecond timescale. To understand the principles of information processing in neural networks, we must monitor and manipulate individual and populations of identified neurons at precise spatiotemporal scales within the context of behavior. Studying the assembly of brain circuits demands a similar level of experimental precision. It requires tracking different cellular building blocks through a multifaceted developmental process. Until recently, the available experimental tools made the study of neural circuits at this level nearly impossible.

The state of affairs has been changing rapidly and dramatically in the past two decades, fueled by innovations in microscopy and imaging, molecular genetic engineering, genomics, and computer science. Today, equipped with tools such as genetically encoded contrast agents, two-photon microscopes, and optrodes, contemporary neuroscientists are increasingly capable of monitoring and perturbing the activity of specific neurons as well as of mapping neuronal connectivity. At the core of this powerful "circuit-breaking" arsenal are genetic tools that deliver custom-tailored markers, sensors, and transducers to specific circuit components for structural imaging, reporting, and alteration of neuronal activity. Thus, genetic tools not only provide the entry point to unleash the

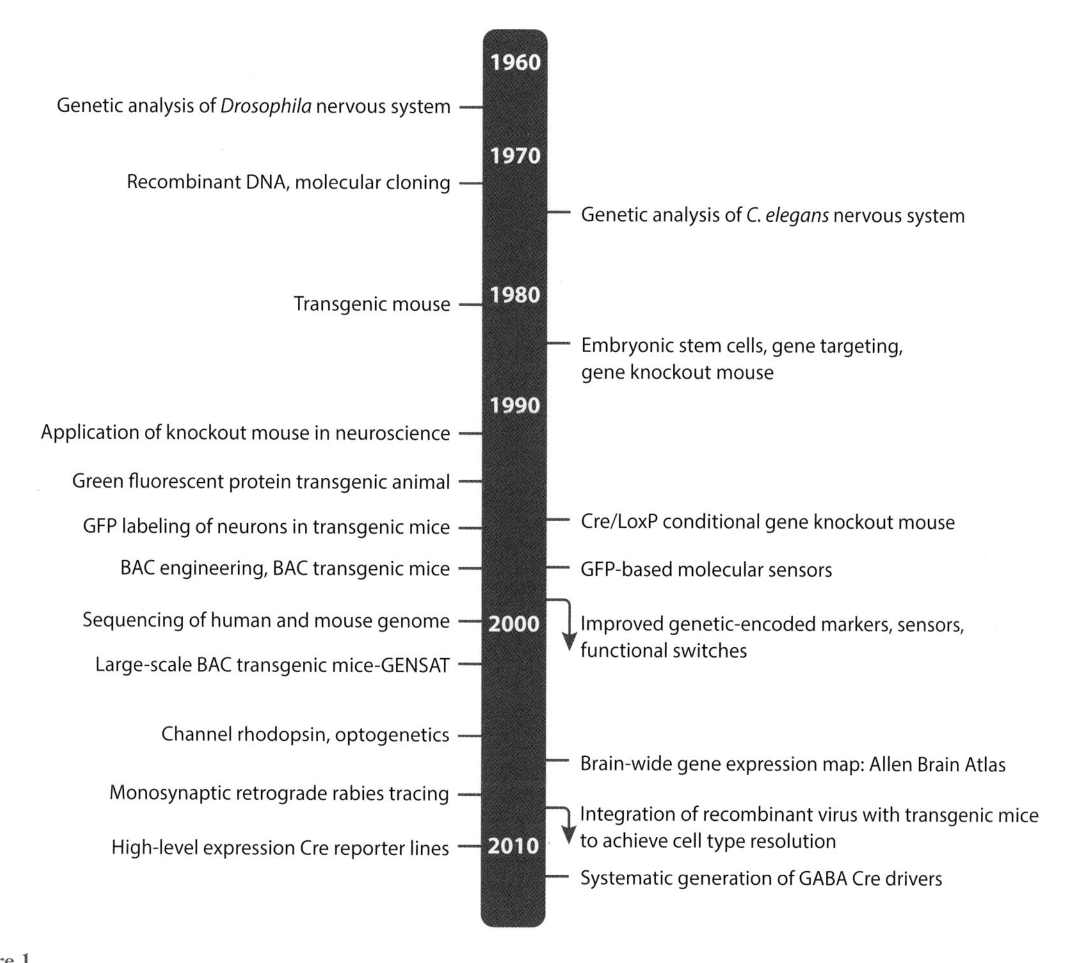

Figure 1

Significant events in the advancement of genetic approaches to neuroscience in the mouse. Downward arrow indicates continued progress. Abbreviations: BAC, bacterial artificial chromosome; GFP, green fluorescent protein.

full power of imaging and physiology, but also further integrate different approaches at the level of defined cellular components in circuit analysis.

In this article, we review genetic approaches and tools now available to neuroscientists working in the mouse. The mouse has a nervous system with much resemblance to that of humans, and it is a mammalian species in which precise and systematic genetic engineering is feasible. Beginning with a brief historical account (**Figure 1**) and discussion of the unique significance of genetic approaches, we then discuss recent advances in genetic targeting of neuronal cell types; genetic tools for genomic, anatomical, physiological, functional, and developmental studies; and strategies for exploring the circuit pathogenesis of neuropsychiatric disorders. An excellent review has covered several general aspects of the genetic dissection of neural circuits (Luo et al. 2008), and several others have reviewed optogenetic (Fenno et al. 2011, Yizhar et al. 2011), imaging (Lichtman & Denk 2011), and physiological (Scanziani & Hausser 2009) approaches.

A Brief History

Historically, studies of brain structure, function, and development belong to such classic

HR: homologous recombination

BAC: bacterial artificial chromosome

disciplines as anatomy, physiology, and embryology. Until the latter half of the twentieth century, anatomists studied brain structures using dyes of unknown chemical nature that serendipitously stained nerve cells. These classical methods began to reveal the rich diversity of neuronal morphologies. With the advent of immunohistochemistry and mRNA in situ hybridization, the distribution of specific molecules in the nervous system could be visualized, albeit often without the underlying cellular morphology. As recently as two decades ago, the possibilities of visualizing the complete morphology of specific cell types in a systematic way and of tracing the connectivity patterns of cells were difficult to fathom. In the meanwhile, physiologists were able to record electrical signals in neural populations, single neurons, subcellular compartments, and even single ion channels. Although they used pharmacology and electrical stimulation to manipulate neuronal activities, it was largely impractical to target these recordings and manipulations to specific neurons in vivo, especially in behaving animals.

The potential of harnessing the power of genetic tools to study a wide range of biological processes, including the nervous system, was recognized early on. Seymour Benzer (1967) pioneered the genetic analysis of behavior in *Drosophila*; he suggested that genes and genetic mosaicism could be used in a way similar to the scalpel in anatomy to create "composite individuals" and, thus, to "genetically dissect" nervous system development and function (Hotta & Benzer 1970). Sydney Brenner (1974) subsequently launched the study of the genetics and biology of *Caenorhabditis elegans* with a goal to "determine the complete structure of the nervous system." This led to the elucidation of a complete cell lineage and circuit wiring diagram of this organism. In the early 1970s, the advent of recombinant DNA and molecular cloning gave rise to genetic engineering, making it possible to transfer genetically defined molecular functionality between species.

The success of mouse transgenesis in the early 1980s (Brinster et al. 1981, Costantini & Lacy 1981, Gordon & Ruddle 1981) marked the beginning of genetic engineering in mammals. The invention of gene targeting through homologous recombination (HR) in embryonic stem cells (Doetschman et al. 1987, Thomas & Capecchi 1987) allowed for modification of the mouse genome with single-nucleotide precision and provided a powerful method for systematic and highly reliable genetic engineering. Initially, knockout mice were used to link gene function to cellular physiology and behavior (Silva et al. 1992). Cre-recombinase-based conditional expression methods were later developed to restrict genetic manipulation in specific brain regions and/or cell types (Tsien et al. 1996). These approaches have become widely applicable with the completion of both mouse and human genomes, the rapid expansion of genomic tools, e.g., BAC (bacterial artificial chromosome) libraries, and expression databases, e.g., Allen Brain Atlas (Lein et al. 2007). Collectively, these approaches have made mouse genetic engineering more efficient.

The generation of green fluorescent protein (GFP) transgenic animals (Chalfie et al. 1994) is of singular importance, transforming the way biological structures and processes are visualized in model organisms. There has been rapid progress in the invention and improvement of genetically encoded markers, sensors, and transducers, including optogenetic tools. These tools are driving a renaissance in neuroscience research and are changing the way scientists approach the field.

With the generation of knockin and transgenic mice that brightly label specific cell types (Mombaerts et al. 1996) or neuronal subsets (Feng et al. 2000), the genetic approach began to truly impact neural circuit analysis in the mammalian brain. GENSAT, the large-scale generation of BAC transgenic GFP reporter mice, was the first systematic effort to use gene expression pattern to visualize specific cell populations in the brain (Gong et al. 2003). Systematic generation of Cre driver lines (Gong et al. 2007, Taniguchi et al. 2011) substantially expands the number of neuronal populations accessible to genetic manipulation;

systematic generation of Cre reporter lines with sufficient expression levels facilitates the labeling, monitoring, and manipulation of these cell populations (Madisen et al. 2010). In parallel, continued innovation in viral vectors and their combination with transgenic mice substantially increase the ease and power of genetic tools. Today, most neuroscientists working with rodents readily take advantage of this wide array of genetic tools to study nearly all aspects of nervous system development and function.

Why Is a Genetic Approach Useful in Neuroscience?

Two attributes of genetic tools make them uniquely powerful in studies of the nervous system. First, physics- and chemistry-based methods (e.g., microscopes and synthetic dyes), though essential and powerful, do not have the inherent capacity to distinguish specific biological entities (e.g., different cells or proteins) or processes (e.g., different steps of neural development). In contrast, genetic approaches directly engage intrinsic molecular and cellular mechanisms that build and operate neural circuits. Thus, genetic tools have the inherent potential to achieve appropriate, and sometimes ultimate, spatiotemporal resolution and cellular-molecular specificity, making it possible to observe and perturb the inner workings of cells and their interactions within circuits. Second, the universality of the genetic code means that molecular talents in one species can be readily transferred and harnessed as markers (e.g., GFP), sensors (e.g., GCaMPs), or transducers [e.g., channelrhodopsin-2 (ChR2)] in another species. These attributes allow genetic tools to be systematically designed and tailored to specific problems on the basis of our understanding of the underlying biological mechanisms.

The usefulness of the genetic approach in neuroscience goes beyond the development of experimental tools. Genetic analysis has been the driving force for elucidating the mechanisms of many biological processes, such as the cell cycle (Hartwell 1991) and embryonic patterning (Nusslein-Volhard & Wieschaus 1980). Despite their cellular complexity and sophisticated operations that guide behavior, the fundamental plans of brain circuits, like all other biological systems, are ultimately encoded by the genome. A unique strength of genetic analysis is the use of a "genetic screen" to identify major components and their functional relationships. By using genes to target and "screen through" individual cell types, genetic analysis will help in deciphering the links from genes to neural circuits to behavior.

ChR: channelrhodopsin

TARGETING SPECIFIC CELL TYPES

Cell-type identification is central to the study of neural circuits. The debate over how to define neuronal cell types remains unresolved. A partial list of the criteria that have been used to classify neurons includes developmental genetic origin, precise location, morphology, connectivity, neurotransmitter content(s), physiological properties, and functional role in circuits and behavior. Key questions regard whether and to what extent these criteria correlate or converge and whether it is possible or necessary to adopt overarching cell-type definitions for the purpose of circuit analysis.

In invertebrate nervous systems, neurons often show stereotyped and tightly correlated lineage, location, connectivity, and function, making it easier to classify them. This is also true in certain highly organized and better-understood regions of the vertebrate nervous system, such as the olfactory bulb, retina, spinal cord, and cerebellum. Defining cell types becomes increasingly challenging in more complex circuits, especially those of the cerebral cortex. For example, although the basic division between glutamatergic excitatory neurons and GABAergic inhibitory neurons is well recognized, there are likely dozens of subclasses of excitatory and inhibitory cortical neurons. Much of the current confusion may stem from our incomplete knowledge of the comprehensive properties of different cell populations. Clarifying this situation requires, in part, that researchers reliably identify the same neuronal populations as well as compare and integrate their results across different laboratories.

Genetic targeting is arguably the best strategy to establish reliable experimental access to specific cell populations. With this approach, we can operationally define cell types and then explore their biological basis within the context of the neural circuit. We now understand that unique gene expression profiles likely reflect and contribute to cell phenotypes, which can be used to distinguish different cell populations. Although this provides the rationale for gene-based targeting of cell types, it also immediately sets the limits of this approach, because there is often no simple correlation between a single gene and an anatomically and physiologically defined cell type. Therefore, a combination of different genetic and viral strategies is often necessary to achieve specific targeting.

Genetic Targeting

A wide range of strategies have been employed to generate genetically engineered mouse lines, utilizing endogenous promoters/enhancers or synthetic/inducible promoters, or their combinations, to drive the expression of transgenes in specific cell populations.

Conventional transgenic and BAC transgenic approaches. Transgenic approaches introduce exogenous genes to a host genome via the use of promoter/enhancer regions added to a transgene cassette (~5–15 kb) or contained in a BAC (~200 kb). This drives the expression of a gene of interest and mimics endogenous expression patterns, even when the promoter/enhancer region is inserted at an ectopic genomic location (**Figure 2**). Generating transgenic mice is relatively easy, and there is potential for high levels of transgene expression if multiple copies are incorporated. However, this approach can be limiting, as transgene expression may not fully recapitulate endogenous expression and may vary among transgenic founder lines. Indeed, given that *cis*-regulatory elements (enhancers, repressors, insulators) can reside far away from the transcription start site (Kapranov et al. 2007) and can act over long distances (Bulger & Groudine 2011, Heintzman & Ren 2009), even BAC constructs may not contain the full complement of regulatory elements for the genes contained therein. Irrelevant enhancers and repressors near the integration site can influence the transgene, leading to ectopic or suppressed transcription. Transgenes inserted into a foreign chromatin environment can also be epigenetically silenced or altered in unpredictable ways. There have been many notable successes in transgenic targeting of cell types (Gong et al. 2007). For example, using a JAM-B-CreER BAC transgenic line, researchers labeled and uncovered a novel retinal ganglion cell subtype that specifically detects upward motion (Kim et al. 2008). Still, many attempts fail to achieve the intended expression. Nevertheless, ectopic expression unrelated to the targeted endogenous pattern can sometimes be useful. For example, the BAC transgenic line Rbp4-Cre labels layer-5 cortical neurons, and the Ntsr1-Cre line labels layer-6 cortical neurons fairly specifically (see **http://www.gensat.org**). Thus, the transgenic approach can be useful in identifying genetically defined cell types, but extensive screening and careful evaluation of expression patterns are critical.

Gene targeting and knockin. Gene targeting takes advantage of the full transcription mechanism of an endogenous gene to drive the expression of an exogenous gene of interest. Unlike transgenic approaches, gene targeting relies on HR to insert an exogenous gene at the endogenous gene locus embedded in its native chromatin environment with largely intact regulatory elements (**Figure 2**). The main advantage of this strategy is that expression of the inserted gene often reliably and precisely recapitulates the targeted locus, making this method particularly useful in targeting cell types or populations defined by a gene expression pattern. A compelling example is the recent report of ~20 knockin Cre driver lines that target major subpopulations of GABAergic neurons (Taniguchi

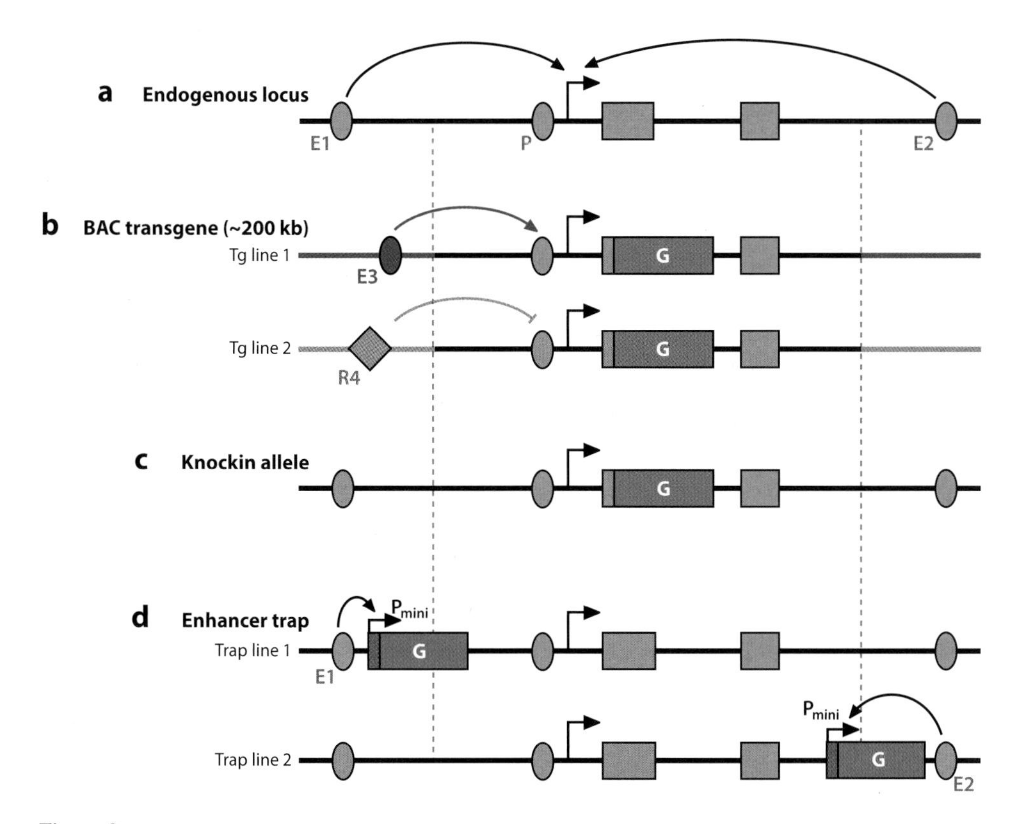

Figure 2

Mouse genetic strategies to target cell types. (*a*) Transcription of an endogenous gene (*gray boxes*) at its promoter (P) is regulated by two distant enhancers (E1 and E2). (*b*) A bacterial artificial chromosome (BAC) transgenic construct expressing a gene of interest (G) (*green boxes*) integrates at two ectopic genomic loci (*blue*, *orange lines*) and is differentially influenced by a nearby enhancer (E3) or repressor (R4) in two transgenic founder lines. (*c*) A knockin allele in which the G is targeted to the endogenous locus and thus is regulated by the endogenous elements. (*d*) In an enhancer trap, when an expression cassette driving G by a minimal promoter (P_{mini}) randomly integrates into the genome, its expression is influenced by nearby enhancers that may represent a specific component (e.g., E1 or E2) of the endogenous gene expression.

et al. 2011). Extensive characterization demonstrates that recombination patterns in almost all these Cre drivers largely match the spatial and temporal patterns of endogenous gene expression. However, there are disadvantages to this approach. For example, the expression of the targeted gene may be altered, even when a bicistronic cassette is inserted after the targeted gene via IRES or 2A. In addition, the full expression pattern of a gene may include multiple cell types or brain regions, making a partial expression pattern more desirable in some cases.

Recent developments in engineering sequence-specific designer nucleases, including transcription activator-like effector nucleases (TALENs) and zinc-finger nucleases (ZFNs), have the potential to increase dramatically the efficiency of gene targeting and HR, not just in mice, but also in many other species (Mussolino & Cathomen 2012). TALENs and ZFNs introduce DNA double-strand breaks (DSBs) at specifically targeted genomic sequences, and DSBs enhance the frequency of HR-mediated repair at the targeted locus. A more frequently occurring DSB-repair mechanism, nonhomologous end-joining (NHEJ), will lead to a small deletion or insertion at the break point, thus knocking out or mutating the target gene. On the other hand, when a transgene-containing replacement vector is co-introduced along with TALENs or ZFNs, HR-mediated repair can insert the transgene into the targeted locus. With increased HR efficiency, TALENs and ZFNs, as well as the CRISPR (clustered regularly interspaced short palindromic repeats) system (Cong et al. 2013, Mali et al. 2013), may enable multigene targeting. A major concern with this approach is the potential off-target activity of the designer nucleases. Further systematic investigation is essential to provide guidance to advance engineering toward improved specificity.

Strategies utilizing enhancer elements. It is often useful to dissect complex gene expression patterns into components that define more restricted cell populations. Enhancers are short DNA sequence elements that are recognized by transcription factors and can enhance transcription at promoter regions from both proximal and distant locations. The activity of enhancers can be modular: Multiple enhancers of the same gene can act independently or cooperatively, each contributing to the overall spatiotemporal transcription pattern. Various strategies such as "trapping" the endogenous enhancers or using identified enhancers to generate transgenic lines have been developed to direct transgene expression in particular cell types.

In enhancer trapping, a transgene driven by a minimal promoter is randomly inserted in the genome. Its expression is then determined by enhancers near the integration site (**Figure 2**). Combining enhancer trapping with the use of lentiviral vectors that promote efficient genomic integration (i.e., higher rate of transgenesis) as a single copy (i.e., reduced rate of epigenetic silencing) has yielded encouraging results in generating lacZ and Cre lines (Kelsch et al. 2012). A major hurdle with this approach is the necessity of screening through a large number of randomly integrated lines to find useful ones and to ascertain the identity of the labeled neurons.

Recently, epigenetic signatures of mammalian enhancers have been discovered (e.g., histone modification profile H3K4 trimethylation), and large-scale efforts are under way to annotate systematically these *cis*-regulatory units in the genome via methods such as multispecies sequence comparisons and ChIP-seq analyses (Bernstein et al. 2012, Sanyal et al. 2012, Shen et al. 2012, Visel et al. 2009b). Such efforts could transform our understanding of gene expression regulation in the genome. In an extensive transgenic screening effort, many of these putative enhancer elements drove consistent lacZ expression across multiple E11.5 transgenic founder embryos in neurons of specific brain regions, demonstrating the potential validity of these elements in driving cell-specific expression and a rapid transgenic approach to screen them (Pennacchio et al. 2006, Visel et al. 2009a).

Binary expression systems. Successful cell targeting needs to achieve two deceptively simple technical goals that are often difficult to reach simultaneously: cell specificity and high-level expression. In the binary system (**Figure 3*a***), gene expression is controlled by two independently engineered transgenes brought together through mouse breeding. This has several major advantages. First, one transgene (driver) can be harnessed for cell specificity, while the other (reporter) can be optimized for expression levels. Second, the combinatorial power among different drivers and reporters is accomplished simply by breeding. Third, a binary system can convert developmentally transient gene expression to stable transgene expression, thus allowing for fate mapping and lineage tracing. Currently, there are two main binary systems: one based on DNA recombination and the other on transactivation.

Recombination-based systems. The bacterial phage–derived Cre recombinase and its loxP recognition site compose the best-established binary system, allowing for versatile and efficient activation and inactivation of gene expression. In particular, a reporter allele can be driven by a strong and ubiquitous promoter

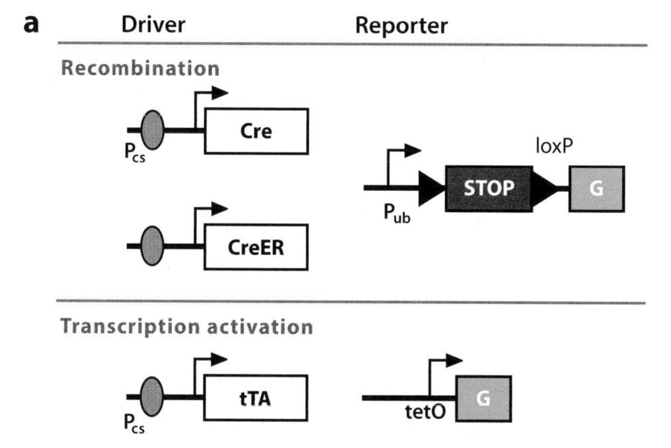

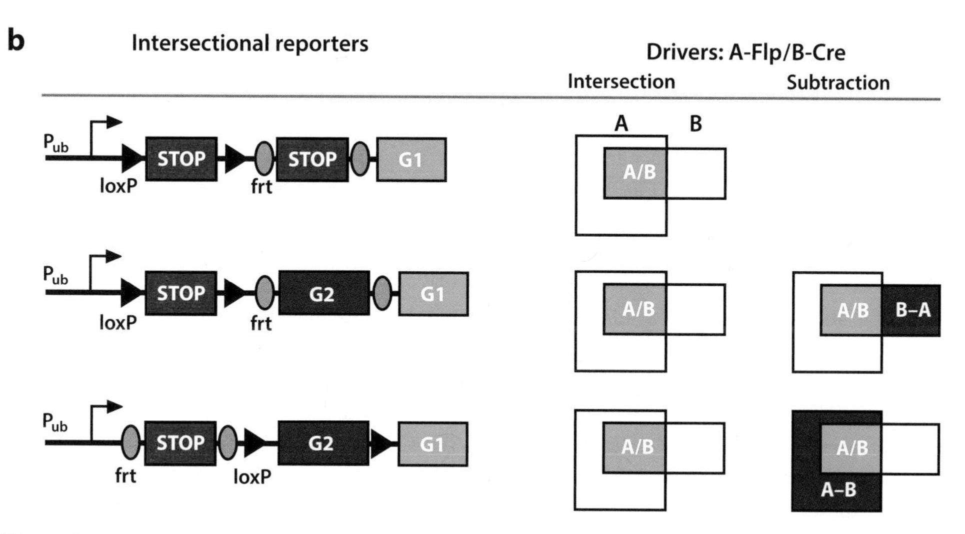

Figure 3

Binary expression systems and intersection/subtraction strategies. (*a*) Recombination-based (*upper panel*) and transactivation-based (*lower panel*) binary expression systems. Abbreviations: G, gene of interest; P_{cs}, cell -specific promoter; P_{ub}, ubiquitous promoter; STOP, transcription STOP cassette. (*b*) Intersection and subtraction strategies. Three configurations of Cre-Flp dual reporter alleles are depicted on the left. A and B represent two genes with overlapping expression patterns (*boxes*). Combined with A-Flp/B-Cre drivers, expected reporter expression patterns are shown on the right. A/B, A and B; A-B, A-not-B; B-A, B-not-A.

but is interrupted by a transcription STOP cassette flanked by loxP sites. A selective Cre driver can activate the reporter in the desired cell types. An increasing number of reporter lines, many at the Rosa26 locus, have been generated that conditionally express markers, sensors, and transducers. An important advance was the invention of the Ai9 reporter and its siblings, generating high-level expression of tool genes sufficient for in vivo imaging and physiology studies (Madisen et al. 2010, 2012). In order to combine different reporters in the same animal, we need to identify and validate other strong and ubiquitous loci for inserting reporter alleles, e.g., the HPRT (Jasin et al. 1996), Tau (Tucker et al. 2001), TIGRE (Zeng et al. 2008), H11 (Tasic et al. 2011), and β-actin (Tanaka et al. 2012) loci.

The mouse central nervous system likely contains hundreds of neuron types.

The number of currently available high-quality Cre drivers are far from providing a sufficient account of all these cell types. Most driver lines to date have been generated by individual laboratories according to their specific research interests. The NIH Blueprint for Neuroscience Research (**http://neuroscienceblueprint.nih.gov/**) and GENSAT (**http://www.gensat.org/cre.jsp**) projects as well as the Allen Institute (**http://connectivity.brain-map.org/transgenic/search/basic**) have begun to support a more systematic effort. In particular, the generation of ~20 knockin GABA Cre drivers (Taniguchi et al. 2011) has targeted major populations of GABAergic neurons and is beginning to facilitate a more systematic analysis of inhibitory circuits in many brain regions. These groups have also begun to characterize the expression patterns of the Cre driver lines and are presenting more complete data sets in their online databases. These freely available resources provide valuable information for assessing the specificity of available genetic tools.

The Flp/frt system is derived from yeast and is conceptually identical to the Cre/loxP system described above. Currently, only a few Flp drivers have been reported (Miyoshi et al. 2010). Additional site-specific recombinases from phage or yeast, including PhiC31, Dre (Anastassiadis et al. 2009), B3, and KD (Nern et al. 2011), also show promise as gene regulatory tools in the mouse, pending necessary validation regarding their efficiency, specificity, and nontoxicity. Multiple recombination systems can be used jointly to achieve flexibility and increase specificity in genetic targeting (see below).

Transactivation-based systems. Transactivation systems use a transcription activator as a driver and use its target sequence to drive the expression of a gene of interest in a reporter line. The most commonly used binary transactivation system in mouse is the tetracycline-regulated transactivator tTA, which activates its target sequence, tetO (or TRE, for tetracycline-responsive element), thus driving reporter gene expression. The yeast transcription factor Gal4 together with its target sequence, UAS, is also a binary transactivation system. Although widely used in *Drosophila*, Gal4/UAS has had very limited application in the mouse, likely owing to its potential toxicity (H. Zeng, unpublished results).

Intersection and subtraction strategies. Because a single gene often has expression in multiple cell populations, an intersectional method using expression overlap of two or more genes is often necessary to target cells with increased specificity. These approaches combine two or more strategies (e.g., Cre-Flp or Cre-tTA) to further restrict tool gene expression to the cells of interest. For example, in the Cre-Flp intersection, there are three components: a Flp driver directed by gene A; a Cre driver directed by gene B; and a dual reporter that can be activated only when both Cre and Flp are present, thus indicating the overlap of A and B (**Figure 3*b***). This method has proven successful in multiple studies (Kim & Dymecki 2009, Taniguchi et al. 2011).

Another intersectional strategy is to split the Cre protein into N- and C-terminal half proteins (NCre and CCre); each half is inactive, but when they are expressed together, recombinase activity is restored (Casanova et al. 2003). This split-Cre method is appealing in light of the expanding repertoire of well-characterized Cre-dependent reporter lines and all the existing floxed alleles for conditional manipulation of genes. There are two ways to reconstitute Cre: One method links NCre and CCre with a pair of protein-dimerization domains (Beckervordersandforth et al. 2010, Hirrlinger et al. 2009); another uses split-intein to connect NCre and CCre through protein splicing (Wang et al. 2012). Although earlier versions of the split-Cre approach showed significantly lower recombination efficiency than resulted from using Cre alone, improvement has been made with newer versions.

The Cre-Flp system can also be used to subtract one expression pattern from another to improve specificity. For example, by designing reporter alleles with different configurations of the loxP and frt cassettes, the same A-Flp and B-Cre drivers noted above can be used to achieve an expression pattern that is either A-not-B or B-not-A (**Figure 3*b***). This method has been used to improve the precision of genetic fate mapping (Dymecki & Kim 2007, Jensen et al. 2008).

Inducible systems. Inducible systems provide additional temporal control over transgene expression to target and manipulate cells. This is particularly useful when the expression pattern of the endogenous gene changes over time and/or is cell-type specific only during a certain temporal window (e.g., during particular periods of development). To achieve this level of control, an on-off switch is added to a binary expression system to confer temporal control ranging from hours to days. One widely used method is CreER, a fusion between Cre and the ligand-binding domain of the estrogen receptor, in particular the CreERT2 variant (Feil et al. 1997). Although CreER is normally retained in the cytoplasm, it translocates into the nucleus to mediate recombination upon administration of tamoxifen, an estrogen analog. FlpER is also effective (Hunter et al. 2005). However, two technical drawbacks should be noted. First, high recombination efficiency and tight inducible control in CreER systems often cannot be achieved simultaneously. Second, tamoxifen, mimicking estrogen, has strong side effects, such as an increased prenatal abortion rate, postnatal mortality, and behavioral consequences. Nonetheless, many CreER drivers have proven instrumental in addressing a wide range of questions (Dymecki & Kim 2007).

Another inducible binary system is tetracycline-dependent transcription (Berens & Hillen 2004). The expression of bacterial tetracycline-regulated transactivator (tTA) is driven by a gene promoter that then activates the expression of a reporter gene under the control of the tetracycline-responsive element, tetO, only in the absence of tetracycline or doxycycline. The system can be reversed via the expression of reverse tTA (rtTA), which is only active in the presence of doxycycline, although rtTA generally exhibits much lower efficiency than tTA in the brain. This approach has proven useful in revealing the mechanisms of memory storage and retrieval (Garner et al. 2012, Liu et al. 2012, Reijmers et al. 2007). Significant issues to be improved include the high susceptibility of this system to epigenetic silencing as well as the mosaic expression and variability in inducibility that often result.

GIFM: genetic inducible fate mapping

IEG: immediate early gene

Cell-type targeting based on lineage and birth timing. During the assembly of neural circuits, neuronal cell types or subpopulations are specified through a process of temporally regulated neurogenesis from progenitors of different lineages. This general mechanism provides the basis for genetic inducible fate mapping (GIFM) and a unique opportunity to target neuronal subpopulations precisely (Joyner & Zervas 2006). CreER expression driven by key transcription factors (TF-CreER) has been used to discover how lineage and birth timing contribute to neuronal specification and the subsequent migration and circuit integration of postmitotic neurons in the spinal cord (Arber 2012), retina (Livesey & Cepko 2001), cerebellum (Sudarov et al. 2011), and other brain circuits. Using the TF-CreER strategy, researchers can target exquisite cell types, such as chandelier cells in the neocortex, and track their development with a single transcription factor (Taniguchi et al. 2013).

Activity-dependent targeting. Activity-dependent targeting takes advantage of promoters of immediate early genes (IEGs), such as *c-fos* and *Arc*, which are activated by strong neuronal activity, to drive the expression of a marker gene such as *LacZ* or *GFP* that has been introduced via a transgenic or knockin approach. Therefore, activity-dependent targeting is unlike the genetic targeting strategies discussed above, as it does not follow a genomically predefined gene expression pattern. Instead, this strategy identifies cells based on their involvement, i.e., activity-driven IEG expression, in a particular sensory stimulation or behavioral state. In this way, an ensemble of neurons that may be functionally related to each other can be identified, regardless of whether they belong to the same "type" as defined by intrinsic genetic expression patterns.

Four activity-dependent reporter lines have been generated to label neurons activated by specific behavioral or pharmacological manipulations: a fos-tau-lacZ transgenic line (Wilson et al. 2002), a fos-GFP transgenic line (Barth et al. 2004), an Arc-GFP knockin line (Wang et al. 2006), and a fos-tTA transgenic line (Reijmers et al. 2007). The activation of IEG transcripts usually lasts several hours, extending beyond the cessation of neuronal activity, and the translation of reporter proteins and their subsequent degradation lag further behind. Thus, transiently activated neurons can be labeled for an extended period of time to allow follow up in vivo or in vitro physiological and histological studies. This property may be further exploited to compare neurons involved in past experience versus those involved in subsequent experience (Reijmers et al. 2007). In situations where long-lasting effects may confound results, a destabilized GFP protein could be used to track endogenous IEG expression more closely and increase temporal specificity (Wang et al. 2006).

Future studies will aim to generate activity-dependent driver lines expressing inducible Cre and other transgenes. The promoters of a variety of activity-activated genes other than *c-fos* and *Arc* should also be explored. Strategies alternative to the IEG expression-based approach will also be quite useful. For example, the Tango assay was designed to record a transient receptor activation event into a stable and amplifiable reporter gene signal (Barnea et al. 2008). Efforts should be made to expand the capability of tracking functionally relevant neuronal populations and investigating their properties, interconnections, and functions in the underlying circuits.

Viral Vector Targeting

Diverse neurotropic viruses have evolved ingenious mechanisms that hijack cellular machineries for their transduction, rapid genome amplification, high-level gene expression, and efficient transneuronal spread. Viral vectors from numerous systems have been engineered to harness these properties for the purposes of neuronal labeling and manipulation. Until a few years ago, the most important limitation of viral vectors was the lack of sufficient cell-type selectivity for infection and gene expression. Recent advances employing engineered tropism, engineered gene expression, and axon transduction at projection sites have begun to overcome this limitation in several widely used viral systems (**Figure 4** and **Table 1**). The delivery of viral vectors can be both temporally controlled and region specific. Thus, the combination of mouse genetic systems and viral vector technology has the potential to yield unprecedented cell-type specificity.

Engineered tropism. The specificity of a virus for a particular host tissue (its tropism) is determined by coat protein(s). The virus will preferentially infect cells that express the receptor(s) for its coat protein(s). Enveloped viruses, which include rabies virus, lentivirus, and retrovirus, can be pseudotyped with a foreign coat protein that determines which cells the virus will infect. Cell-type-specific expression of the coat protein receptor can be accomplished through targeted genetic expression systems like those detailed in the previous sections. For example, a recombination-activated reporter line expressing TVA, the receptor for an avian coat protein, EnvA, in neurons defined by a Cre driver line can direct specific infection by EnvA-pseudotyped rabies, lenti-, and retroviruses (Seidler et al. 2008).

Engineered gene expression. Engineered gene expression within viral vectors can also target transgene expression in specific cell types. Gene expression driven by specific promoters is difficult to obtain owing to the size restriction in AAV and lentiviral vectors (~5 kb and ~8 kb, respectively). However, much shorter Cre-dependent expression cassettes have been successfully incorporated.

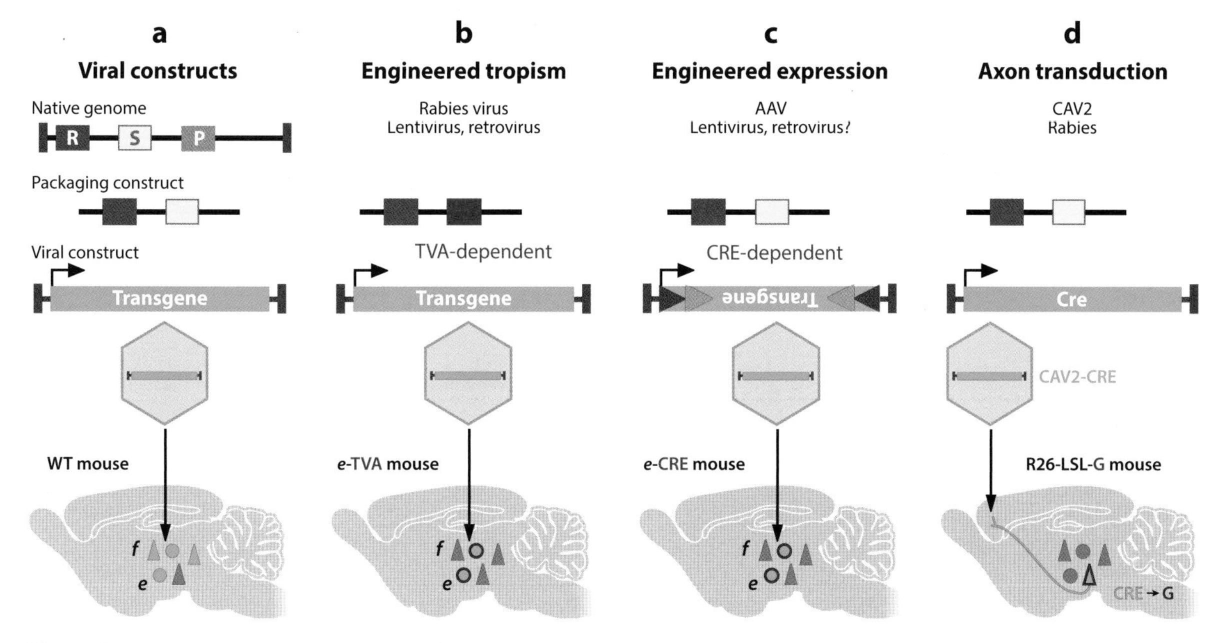

Figure 4

Integrated viral and mouse genetic strategies to target cell types. (*a*) Schematics of native viral genome, packaging construct, and vector construct. *Cis*-acting element (*magenta*), pathogenic gene (P) (*pink*), replication gene (R) (*dark blue*), and structural gene (S) (*yellow*) are shown. In vector constructs, transgenes are driven by a strong promoter (*arrow*). When viral particles are injected into brain tissue, a nonspecific subset of cells at the injection site (e.g., *e* and *f* cells) expresses the transgene. (*b*) The viral vector is pseudotyped to express a foreign coat protein (*red*) to alter its tropism. When injected into a mouse expressing the receptor (TVA) for the coat protein in *e* cells (*red outline*), only *e* cells will be selectively infected to express the transgene. (*c*) The viral vector is engineered to express the transgene in a Cre-dependent manner. When injected into a mouse expressing Cre in *e* cells (*red outline*), a transgene is active only in *e* cells, even though the virus infects nonspecifically. (*d*) CAV2 and rabies viruses transduce neuronal axons and are then transported retrogradely to the cell soma. Thus, a Cre-expressing virus injected to the axon projection site of a neuron can transport Cre to its soma and further activate a Cre-dependent transgene (G) in a Rosa26-loxP-STOP-loxP-G (R26-LSL-G) reporter mouse. Abbreviations: AAV, adeno-associated virus; WT, wild type.

Both the loxPSTOPloxP cassette and the inversion-based FLEX cassette have been used to render viral transgene expression recombination dependent (Atasoy et al. 2008, Kuhlman & Huang 2008). This approach combines the high levels of gene expression conferred by the easy-to-engineer and cost-efficient AAV vector system with the proven cell-type targeting of Cre driver lines. Thus, there has been wide application of this approach in recent years and rapid accumulation of related vectors. Engineered gene expression in viral vectors is still limited by intrinsic capacity, preventing the expression of long sequences or multiple genes. In addition, lenti- and retroviral vectors require transcription of their full genome for replication and, thus, cannot tolerate the interruption caused by a loxP-STOP-loxP cassette. In these cases, the FLEX strategy should be able to bypass this complication and confer Cre dependence to infected cells.

Axon transduction at the projection site. Where axons project is a defining feature of cell types, and viral transduction with retrograde transport from one area can be used to label neurons projecting to that area. By combining such viruses with other regional or cell-type specific targeting tools, highly specific projection neuronal types can be targeted. Recombinant rabies virus deficient in glycoprotein can efficiently infect axon terminals and restrict

Table 1 Recombinant viral vectors that confer cell-specific transduction and/or gene expression[a]

	AAV	Lentivirus	Retrovirus	CAV2	Rabies virus
Genome	Single-stranded DNA	RNA	RNA	DNA	RNA
Capacity	~5 Kb	~8 Kb	~8 Kb	~30 Kb	~? Kb
Expression onset	Weeks	Weeks	Weeks	Weeks	Days
Expression duration	Years	Years	Years	Years	~10 days until cell death
Envelope	No	Yes	Yes	No	Yes
Specificity through engineered tropism	Capsid protein, serotypes	Envelope protein; pseudotyping	Envelope protein; pseudotyping	Capsid protein?	Envelope protein; pseudotyping
Specificity through engineered expression	Cre, Flp, tTA-dependent	Cre-dependent (FLEX)?	Cre-dependent (FLEX)	N/A	N/A
Specificity through axon transduction		Yes, if pseudotyped with rabies G		Yes	Yes; ΔG prevents trans-synaptic spread
Predominant applications	Long-term transgene expression; axonal projection mapping	Long-term transgene expression	Fate mapping	Retrograde labeling	Retrograde; trans-synaptic labeling

[a]Abbreviations: AAV, adeno-associated virus; CAV2, canine adenovirus type 2; tTA, tetracycline-regulated transactivator.

expression within the infected neuron (Callaway 2008). One drawback is the cytotoxicity that kills the neuron within 2 weeks. The canine adenovirus type 2 (CAV2) also shows preferential axon transduction and a high level of retrograde transport, with low cytotoxicity and long-term expression (Soudais et al. 2001, 2004). A CAV2-Cre virus was used for cell-type-specific gene manipulation (Hnasko et al. 2006). Cre expression in retrogradely labeled cells can further activate a Cre-dependent AAV vector delivered to the projection neuron soma (**Figure 4*d***).

GENETIC TOOLS FOR NEURAL CIRCUIT STUDIES

Cell-Based Genomics

As discussed, gene expression profiles may provide a foothold for cell-type identification in the brain. To the extent that gene expression determines a cell's phenotypic properties, a genetic approach may also reveal information about the physiological states of neurons within a circuit. For example, differential expression of ion channel genes is associated with the differential electrical properties of neocortical neuronal types (Toledo-Rodriguez et al. 2004).

A key barrier to genomic analysis in brain tissue is cellular heterogeneity: Distinct cell types are the basic units of gene expression, but they are highly intermingled, making their transcriptomes difficult to isolate and purify. Two broad strategies, both based on genetic labeling, have been used to harvest genetic materials from mixed cell populations: cell isolation and molecular tagging. Fluorescence-activated cell sorting (FACS) and manual sorting can readily purify a sufficient number of labeled cells for gene expression analysis (Arlotta et al. 2005, Sugino et al. 2006), but the physical stress inherent to the sorting process may alter the physiological state of cells and their expression profiles. Laser capture microdissection (LCM) is used to collect neurons from their native circuit niche but is of low yield and requires a method suited for obtaining trace amounts of nucleic acids.

The TRAP (translating ribosome affinity purification) method uses BAC transgenic mice to express a GFP-tagged ribosomal protein, L10a, in a genetically defined cell population whose polysomal mRNA contents can be affinity purified directly from brain tissue using a GFP antibody for expression analysis (Heiman et al. 2008). In a similar approach, RiboTag uses a hemagglutinin-tagged ribosomal protein (Rpl22) for mRNA purification and delivers this tag to specific cell populations through a Cre-dependent reporter line (Sanz et al. 2009). This molecular tagging strategy has been extended to microRNAs: The miRAP (microRNA affinity purification) method can harvest cell-specific microRNAs using a GFP and MYC-tagged Argonaute2 protein expressed in specific populations using the Cre/loxP binary system (He et al. 2012). Although their efficiency and specificity need to be further validated, these models have the potential to be integrated with cell-type-targeting techniques applied in more native environments. This would represent major progress toward systematic, cell-type-based gene expression analysis in brain tissues.

Rapid advances in sequencing technology will continue to improve the sensitivity and lower the cost of genomic studies. Such may enable the analysis of single neurons or a small number of identified neurons in their native tissue environment. Furthermore, these techniques will allow researchers to interpret comprehensive genomics data from highly distinct cellular components within the context of circuit organization and function.

Genetic Neuroanatomy

The challenges in neuroanatomy have remained largely the same for the past century: how to reveal the complex and often spatially extensive dendritic and axonal morphology of a neuron at nanometer (i.e., synapse) resolution, how to elucidate the highly convergent and divergent connectivity in a heterogeneous tissue, and how to distinguish cell types or members of the same cell type embedded in neural circuits. Genetic tools have been combined with modern imaging and computational methods to make progress toward meeting some of these challenges.

Morphology. Ideally, studies of neuronal morphology would involve high-intensity labeling capable of visualizing even the most distal axonal projections while also identifying presynaptic boutons and postsynaptic densities. A few methods have been developed that begin to approach this goal. Marker expression detectable by sensitive histological methods, such as LacZ and AP (alkaline phosphatase), can visualize fine processes in exquisite detail (Mombaerts et al. 1996, Rotolo et al. 2008). On the other hand, fluorescent labeling, when coupled with two-photon imaging and other optimization techniques such as tissue clearing (Hama et al. 2011), can confer high-sensitivity and deep-tissue visualization without laborious sectioning and staining. This approach also enables long-term monitoring of morphological and synaptic dynamics in vivo (Holtmaat & Svoboda 2009, Lichtman & Sanes 2003). Furthermore, GFP fusion proteins allow for cell-specific labeling of subcellular structures such as presynaptic terminals (e.g., synaptophysin-GFP) (De Paola et al. 2003), postsynaptic density (e.g., PSD95-GFP) (Gray et al. 2006), or even gap junctions (e.g., connexin36-GFP) (Helbig et al. 2010).

Although seemingly straightforward, genetic labeling with both cell-type specificity and high intensity is technically difficult to attain. In transgenic mice, most cell-restricted promoters are weak, and strong promoters tend to have broad expression patterns. One exception is the Thy1 promoter, which can drive YFP transgene expression to very high levels while exhibiting various degrees of neuronal coverage in different founder lines as a result of multiple integrated copies and the position of random integration sites (Feng et al. 2000). Binary expression strategies can harness cell specificity and expression levels in separate mouse lines and deliver a more general solution. Although the number of Cre driver lines with cell specificity is accumulating, until recently few reporter lines provided enough high-level expression for morphological studies. By incorporating a strong and ubiquitous synthetic promoter, CAG, into the ubiquitously permissive Rosa26 locus along with the use of WPRE, a set of single-copy knockin alleles of Cre reporter transgenes were created with strong expression of fluorescent proteins (Madisen et al. 2010). Among these, the tdTomato-expressing Cre reporter line (Ai9/Ai14) is particularly robust and has become a widely used standard mouse line to evaluate various Cre lines and the cell structures they label.

Using a cleverly designed strategy of Cre-dependent stochastic activation, Brainbow mice express red, green, and blue (RGB) fluorescent proteins at different ratios in different cells, generating a unique color mix in each cell. More than 90 different RGB colors can be detected by laser-scanning confocal microscopy, enabling random mosaic labeling of a large number of cells whose processes can then be tracked over long distances (Livet et al. 2007). A limitation is that this strategy relies on many tandem copies of the transgenes all under the very strong Thy1 promoter to generate the plethora of random combinations of colors. Whether this strategy can be applied to other promoters and/or combined with more refined knockin approaches remains to be determined.

Recombinant viruses [e.g., AAV, rabies, herpes simplex virus (HSV), Sindbis, and vesicular stomatitis virus] can express fluorescent markers at very high levels. In particular, low-titer Sindbis virus produces Golgi-like fluorescent labeling of distant axonal fibers, making a full reconstruction of single neurons possible (Ghosh et al. 2011, Kuramoto et al. 2009). Fluorescent labeling with cell-type specificity can be achieved using Cre-dependent AAV and pseudotyped rabies virus when applied to appropriately engineered mouse lines (see above). Though powerful, these approaches carry the caveat that some viral strains are particularly virulent, leaving only a short window for visualization before the infected cells become unhealthy.

Connectivity. The mammalian brain connectivity spans a vast (10^7-fold linear and 10^{21}-fold volumetric) scale from nanometer-level synaptic contacts to centimeter-level long-range axonal projections. To obtain a comprehensive wiring diagram of the mouse brain, neuroanatomical approaches at macro-, meso-, and microscopic scales are essential. Genetic labeling not only enables connectivity mapping at defined cell-type and single-cell levels, but also can serve as a common thread to link and integrate these multiple scales and further engage electrophysiological and optogenetic connectivity mapping approaches.

A variety of recombinant viruses are highly useful tools for connectivity mapping. Because of its excellent intracellular filling, long-term stable expression, nontoxicity and ability to express GFP under Cre control, AAV has been used as an anterograde tracer to map axonal projections (Chamberlin et al. 1998, Harris et al. 2012). Regular and Cre-dependent AAVs are also used in large-scale efforts (e.g., Allen Mouse Brain Connectivity Atlas: **http://connectivity.brain-map.org/**) for regional and cell-type-specific projectional mapping that utilizes a large number of Cre lines and automated high-throughput two-photon imaging (Ragan et al. 2012).

Both rabies and pseudorabies viruses are capable of retrograde transneuronal spread. By creating a replication-deficient (ΔG) viral strain and pseudotyping it with the avian glycoprotein EnvA, which binds only to a specific receptor TVA, a rabies virus capable of monosynaptic retrograde traversing from defined cell populations was created (Wickersham et al. 2007). When the expression of TVA and the rabies glycoprotein are put under Cre control, Cre-dependent cell-type specific, monosynaptic, retrograde trans-synaptic tracing is achieved (Miyamichi et al. 2011, Wall et al. 2010). An array of rabies viruses expressing GCaMP3, ChR2, CreER, Flp, etc., have been generated (Osakada et al. 2011), offering ample opportunities for integrated circuit mapping and circuit analysis (see below). A pseudorabies virus vector carrying a Cre-dependent Brainbow reporter cassette has also been developed and used to trace polysynaptic connections within larger networks from a defined projection target (Card et al. 2011).

Pseudotyping with various glycoproteins can also convert other viruses into retrograde or anterograde trans-synaptic tracers. For example, pseudotyping the vesicular stomatitis virus with rabies G protein (RABV-G) enabled retrograde trans-synaptic spread, whereas psuedotyping with LCMV-G enabled anterograde trans-synaptic spread (Beier et al. 2011). Taking advantage of its capacity to effect anterograde trans-synaptic spread, the HSV-1 H129 strain was engineered to put its *HTK* gene under Cre control. As a result, Cre-dependent anterograde trans-synaptic tracing was enabled (Lo & Anderson 2011).

Several other molecular tools facilitate visualization in connectivity studies. For example, GRASP (GFP reconstitution across synaptic partners) utilizes split-GFP reconstitution across synaptic transmembrane protein partners to visualize synaptic contacts between two cells (Feinberg et al. 2008). This technique has been adopted in mice: mGRASP uses carefully designed pre- and postsynaptic protein pairs that can bind to each other at synaptic clefts specifically, but does not interfere with normal synaptic function (Kim et al. 2012). Similar to fluorescent approaches, cell-type-specific targeting of electron-dense molecular tools will open doors for cell-type electron microscopy. For example, genetically encoded electron-microscopic tags, miniSOG (Shu et al. 2011) or APEX (Martell et al. 2012), can be fused to various proteins for ultrastructural subcellular visualization.

Another genetic approach to map connectivity among cell types or individual cells involves optogenetic activation by expressing ChR2 in input neurons. This can be done at the macroscopic scale, looking at global regional connections that are inferred from correlated firing patterns (Kahn et al. 2011, Lee et al. 2010), or in a much more refined manner with cellular resolution. While a target neuron is being recorded, researchers can use light to activate input sources (neurons or axon fiber bundles) systematically to map the location and strength of the synaptic inputs to the recorded cell (Katzel et al. 2011; Petreanu et al. 2007, 2009). Two-photon activation of ChRs to induce spiking in single presynaptic cells further improves this approach to single-cell-level precision in circuit mapping (Andrasfalvy et al. 2010, Packer et al. 2012, Papagiakoumou et al. 2010, Prakash et al. 2012).

Genetic Neurophysiology

The physiological activities of neurons can be monitored via electrophysiological recordings and/or optical imaging. Although electrophysiological methods record currents and membrane potential fluctuations with high sensitivity and high temporal resolution wherever the electrodes can reach (including deep in the brain), not all cells in the targeted areas are accessible and the identity of recorded neurons is usually unknown. By contrast, optical imaging using fluorescent chemical or biological sensors can simultaneously track the activity of many labeled cells with high spatial resolution to provide reports of changes in calcium concentration, voltage, etc. Furthermore, the identity of the imaged neurons can often be determined through genetic targeting. However, optical imaging's temporal resolution and sensitivity to subthreshold activities need to be improved. Genetic approaches with the potential to strengthen and integrate these methods include the use of optogenetics to identify recorded neurons in electrophysiology and the incorporation of more sensitive and fast-acting biosensors in optical imaging.

The high-level XFP expression in specific promoter-driven or Cre-dependent transgenic

GECI: genetically encoded calcium indicator

FRET: Förster resonance energy transfer

GEVI: genetically encoded voltage indicator

mice is useful for visualizing many types of neurons in vivo through two-photon imaging, and it enables targeted intracellular recordings (Gentet et al. 2010, 2012) as well as cell identification when imaging with synthetic calcium dyes (Hofer et al. 2011, Zariwala et al. 2011).

In contrast to synthetic dyes, genetically encoded biosensors provide ways to monitor neuronal activities at cellular and subcellular resolution and over long periods of time. They can also be readily delivered to specific cell types. The most well-developed and widely used genetically encoded biosensors are calcium indicators; other biosensors include voltage indicators, pH sensors, and transmitter sensors (**Table 2**).

Genetically encoded calcium indicators. There are generally two types of genetically encoded calcium indicators (GECIs) (Mank & Griesbeck 2008, Tian et al. 2012): One is based on FRET (Förster resonance energy transfer), in which calcium binding induces a conformational change of the calcium binding protein that brings two XFP domains close enough to allow FRET to occur; the other is based on a calcium-induced change in the fluorescence level of a single XFP. Since the first FRET-based GECI, Cameleon, was invented, multiple variants have been developed, including the calmodulin (CaM)-based YC series (e.g., YC3.60 and D3cpv) and the troponin C (TnC)-based TN series (e.g., TN-XXL). Similarly, the first single-fluorphore-based GECI, Camgaroo, led to the development of the GCaMP series and the Pericam series. In addition, GECIs have now been expanded into the blue and red spectrum, including GECOs (Zhao et al. 2011) and R-CaMP (Ohkura et al. 2012), with the potential for combinatorial imaging applications in the future.

Several GECIs have been used in in vivo studies, where they have been introduced via transgenic mice or recombinant viruses (Zeng & Madisen 2012). In such studies, the long-term expression of GECIs may reduce the dynamic range of the response of GECIs to calcium, and overexpression of the calcium binding protein may lead to intracellular calcium imbalance. Despite these concerns, there has been an explosion of newly engineered GECI molecules with improved performance, particularly in vivo. For example, GCaMP3 (Tian et al. 2009) has been widely used to image in vivo neuronal activities via viral expression (Dombeck et al. 2010, Mittmann et al. 2011, O'Connor et al. 2010) and in Cre-dependent transgenic mice (Zariwala et al. 2012). The latest GCaMP5 (Akerboom et al. 2012) and GCaMP6 series (Chen et al. 2012) exhibit further improved properties in vivo, with GCaMP6 reaching single-spike detectability. Through viral expression, the Yellow Cameleon series, YC2.60, YC3.60, YC-nano, etc., also reliably reports neuronal spiking in vivo. This series can detect single spikes and has the added advantage of ratiometric imaging that is unaffected by motion (Yamada et al. 2011).

Genetically encoded voltage indicators. Since the first genetically encoded voltage indicator (GEVI) was developed in 1997, the performance of GEVIs has lagged behind that of GECIs, owing to the lower levels of signal, making these sensors more useful for wide-field imaging than for single-cell tracking. The most advanced version of GEVIs is the VSFP series, in which a voltage-sensitive domain is linked to a fluorescent protein or pair of proteins (FRET-based). VSFP2.3 and VSFP-Butterfly detect electrical responses in both brain slices and in vivo (Akemann et al. 2010, 2012). In addition, alternative types of voltage-sensing fluorescent proteins have been recently identified, including PROPS (Kralj et al. 2011), Arch (Kralj et al. 2012), and ArcLight (Jin et al. 2012). However, the functionality of these new voltage-sensing proteins has yet to be demonstrated in vivo.

Other genetically encoded indicators. Other genetically encoded fluorescent indictors include those that can monitor neurotransmitter content or vesicular release and those that can sense intracellular signaling and metabolism (Tantama et al. 2012). In particular, the pH

Table 2 **Representative genetic tools for monitoring and manipulating neuronal activities**[a]

Name	Type of molecules	Transgenic expression	Viral expression	Comments
GCaMP3, GCaMP5, GCaMP6	Calcium indicator (single FP-based)	Cre-dependent or specific promoter-driven	AAV, rabies	Widely used in vivo; GCaMP6 has single-spike detectability
R-CaMP1.07	Calcium indicator (single FP-based, red-shifted)	N/A	N/A	Tested in vitro only; possible combination with ChR2
YC2.60, YC3.60, YC-nano	Calcium indicator (FRET-based)	CAG promoter-driven (ubiquitous)	AAV, adenovirus	Single-spike detectability in vivo using virus
VSFP2.3,VSFP-Butterfly	Voltage indicator (FRET-based)	N/A	AAV	Best for wide-field fluorescent imaging; tested in vivo
ArcLight	Voltage indicator (single FP-based)	N/A	N/A	Tested in vitro only; stronger signals than other VSFPs; best for wide-field imaging
synaptopHluorin	pH sensor (single FP-based)	Specific promoter-driven	N/A	Most suited for areas with converging synaptic inputs
pHTomato	pH sensor (red-shifted)	N/A	N/A	Tested in vitro only; possible combination with GCaMP3
iGluSnFR	Glutamate sensor (single FP-based)	N/A	AAV	Suitable for one- and two-photon in vivo imaging
DTR	Inducible cell ablation	Specific promoter-driven, or Cre-dependent	N/A	Activated by DT
TeTxLc	Synaptic transmission inhibitor	tTA-dependent	N/A	Regulated by doxycycline
hM3D	DREADD (activating)	tTA-dependent	N/A	Inducible by CNO
hM4D	DREADD (inhibiting)	Cre/Flp-dependent	N/A	Inducible by CNO
ChR2, $ChR2_R$	Light-induced excitation	Specific promoter-driven or Cre-dependent	AAV, rabies, lentivirus	Widely used in vivo
$ChETA_T$, $ChETA_{TC}$	Light-induced excitation (faster kinetics)	N/A	AAV	Facilitates high-frequency stimulation
CatCh	Light-induced excitation (larger response)	N/A	N/A	Increased calcium permeability; slower kinetics; tested in vitro only
$C1V1_T$, $C1V1_{TT}$	Light-induced excitation (red-shifted)	N/A	AAV	Red-shifted; large current; slower kinetics; suitable for two-photon activation
eNpHR3.0	Light-induced silencing	Cre-dependent	AAV	Chloride pump
Arch, ArchT	Light-induced silencing	Cre-dependent	AAV, lentivirus	Proton pump

[a]Abbreviations: AAV, adeno-associated virus; ChR2, channelrhodopsin-2; CNO, clozapine-N-oxide; DREADD, designer receptors exclusively activated by designer drugs; DT, diphtheria toxin; DTR, diphtheria toxin receptor; FP, fluorescent protein; N/A, not applicable; tTA, tetracycline-regulated transactivator; TeTxLc, tetanus toxin light chain.

sensor (Miesenböck et al. 1998), synaptopHluorin, has been used successfully in olfactory neural circuit studies (Bozza et al. 2004, Li et al. 2005, McGann et al. 2005). Recently a red, pH-sensitive fluorescent protein, pHTomato, was also used in parallel with green probes (e.g., GCaMP3) to monitor neuronal activity (Li & Tsien 2012). Another recently developed glutamate sensor, iGluSnFR, has demonstrated a high signal-to-noise ratio and amenability to fast in vivo imaging (Marvin et al. 2013).

Targeted recording in vivo: channelrhodopsin tagging of single units. Single or multichannel extracellular recordings remain indispensable for studying neuronal activities in vivo. Knowing the cell-type identity of recorded neurons is crucial for understanding the role of individual neurons in a circuit. Although excitatory neurons can be distinguished from inhibitory interneurons on the basis of the spike shapes in the hippocampus and cortex, subtypes within these classes cannot be identified by conventional physiology. Coupling local optogenetic activation with electrode recording using optrodes makes it possible to assign spikes to any ChR2-expressing cell type (Lima et al. 2009, Madisen et al. 2012, Royer et al. 2012). This approach holds great promise to facilitate cell-type-specific in vivo electrophysiology.

Genetic Manipulation

To understand neural circuit function, one needs to manipulate the circuit under physiologically relevant conditions. Doing so is extremely challenging owing to the complexity of dynamic circuit operations, which include convergent, divergent, parallel, and feedback pathways as well as the rapid adaptability of the network via dynamic plasticity and homeostatic mechanisms. Unlike other approaches, such as lesion studies, microstimulation, and pharmacological manipulation, recently developed tools based on genetic manipulation make it possible to target specific cell populations with temporally precise and often reversible control (**Table 2**).

Toxins for genetic ablation. One efficient way to study the function of a circuit is to remove one of its components and observe the consequences. This form of precise, cell-type-specific ablation is possible via cytotoxic gene expression. The diphtheria toxin (DT) A chain (DTA) is an effective cell-autonomous toxin that has been used in lineage-ablation and functional-mapping studies (Gogos et al. 2000, Sakamoto et al. 2011). Owing to DTA's high toxicity, inducible and reversible expression is preferable. In another approach, because mice are naturally resistant to exogenous DT, transgenic mice were developed to express a simian or human diphtheria toxin receptor (DTR); administration of DT, which can cross the blood-brain barrier (BBB), to these transgenic mice results in selective killing of only the cells expressing human or simian DTR (Buch et al. 2005). This approach was used to ablate the Agrp-expressing neurons and demonstrate their essential role in energy homeostasis (Gropp et al. 2005, Luquet et al. 2005).

Synaptic transmission inhibitors. Agents that disrupt neurotransmitter release at selective synapses can be effective circuit breakers, making it possible to investigate the role of specific cells. Inducible and reversible transgenic expression of tetanus toxin light chain (TeTxLc) has been used to block specific synaptic transmission in the cerebellar or hippocampal circuit of behaving mice (Nakashiba et al. 2008, Yamamoto et al. 2003). Because the effect is solely dependent on the presence of the TeTxLc protein, the inactivation time course is slow (hours to days). Thus, it is suitable mainly for studies that require long-term manipulation.

DREADDs. G protein–coupled receptors (GPCRs) are the most common targets for developing therapeutic drugs. Thus, they are appealing targets for genetic engineering efforts aimed at developing designer GPCRs for precise spatiotemporal control in vivo. These engineered receptors, termed RASSLs (receptors activated solely by synthetic ligands) and DREADDs (designer receptors exclusively activated by designer drugs), are unresponsive to their endogenous ligands, but they can be activated by synthetic drug-like compounds that easily cross the BBB (Conklin et al. 2008). Of these receptors, the main type is the muscarinic ACh receptor (mAChR) family of DREADDs. At low nanomolar concentrations, the synthetic compound clozapine-N-oxide (CNO) can activate both the Gq-coupled human M3 DREADD (hM3D) and the Gi-coupled human M4 DREADD (hM4D), thereby activating or silencing the neuron, respectively (Armbruster et al. 2007). Application of CNO in transgenic mice expressing hM4D in serotonergic neurons caused silencing of these neurons and altered respiratory and body temperature control (Ray et al. 2011). CNO-mediated activation of hM3D-expressing neurons driven by the c-fos promoter could also activate specific memory traces (Garner et al. 2012).

Optogenetics. Since the discovery of the directly light-gated cation-selective channel channelrhodopsin-2 (ChR2) (Nagel et al. 2003) and the subsequent report that application of ChR2 directly activates neurons (Boyden et al. 2005), the optogenetics field has developed a large arsenal of genetic and optical tools with the exquisite capacity to control neuronal activity at the millisecond timescale in response to light. Here we briefly summarize the optogenetic tools available to date. Several reviews contain detailed descriptions of the history of optogenetics, the variety of available tools, and their applications (Fenno et al. 2011, Mattis et al. 2012, Yizhar et al. 2011).

Microbial opsins that conduct cations depolarize neurons in response to light stimulation. They can be grouped into three main types based on their light-responsive properties: (*a*) blue-light-responsive opsins (e.g., ChR2, ChIEF, ChETA), (*b*) red-shifted opsins (e.g., VChR1, C1V1), and (*c*) step-function opsins (Yizhar et al. 2011). Although validation has been demonstrated in neurons for most, if not all, these ChR variants, the vast majority of the functional studies so far have used the original ChR2 or its modestly enhanced version ChR2(H134R). The applicability of other variants to functional studies will depend on their proper protein expression, nontoxicity, sensitivity, and tightness in their responses to light.

DREADD: designer receptor exclusively activated by designer drugs

There are also microbial opsins that generate hyperpolarizing currents upon light stimulation and can inhibit electrophysiological activity when expressed in neurons (Chow et al. 2012). Halorhodopsin that has been isolated from *Natronomonas pharaonis* (NpHR) is a yellow-light-activated chloride pump. A membrane-targeting enhanced version of the protein, eNpHR3.0, functions well in the mammalian brain in vivo (Gradinaru et al. 2010). Another type of inhibitory opsins includes light-activated proton pumps, such as Arch, MAC, and eBR. In particular, Arch (archaerhodopsin-3) and its variant ArchT enable robust neuronal silencing upon yellow-light stimulation (Chow et al. 2010, Han et al. 2011).

Cre-dependent AAVs as well as a range of specific promoter-driven or Cre reporter transgenic mice expressing ChR2(H134R), eNpHR3.0, and Arch have all been generated over the past few years and proven to be powerful tools for investigating the roles of specific groups of neurons in circuit dynamics and in behavior (Madisen et al. 2012, Yizhar et al. 2011, Zeng & Madisen 2012, Zhao et al. 2011).

Optogenetic approaches can also be used to modulate intracellular pathways. For instance, photoactivated adenylyl cyclase (PAC) can be expressed in mammalian cells to manipulate cAMP levels (Schroder-Lang et al. 2007), and chimeric molecules formed from ChR and several types of GPCRs, the so-called optoXRs, allow light-mediated regulation of various G protein–coupled signaling pathways (Airan et al. 2009). There are also light-dependent protein-protein interacting molecules, although their applicability to mammalian systems has yet to be demonstrated (Tucker 2012).

Other activators, silencers, and modulators. Other ligand/receptor pair systems that have been used to inducibly and reversibly modulate neuronal activities in mice include the capsaicin/TRPV1 system for ligand-mediated neuronal activation (Arenkiel et al. 2008) as well as the allatostatin/AlstR (Tan et al. 2006) and the IVM/GluCl (Lerchner et al. 2007) systems for ligand-mediated neuronal silencing. An efficient way to regulate protein function rapidly is by controlling protein degradation. One such approach takes advantage of destabilizing domains that can direct protein degradation when fused to any protein of interest. A recent system was developed in which fusion proteins containing the destabilizing domain of a bacterial protein DHFR can be inducibly stabilized by an antibiotic, trimethoprim (TMP) (Iwamoto et al. 2010). The ability of TMP to cross the BBB easily makes it an ideal tool to regulate protein expression in the mouse brain.

In general, several technical issues need to be resolved to achieve effective genetic manipulation in vivo. These include the requirement of high-level expression of the transgenes, the effective translation and correct subcellular and membrane targeting of nonmammalian genes, and the BBB permeability of small-molecule ligand inducers of genetic-encoded transducers. Currently, the most successful genetic manipulation approaches applied to the mouse brain include TeTxLc, DREADDs, and optogenetics, whose respective scales are in the range of hours to days, minutes to hours, and milliseconds to seconds.

Tracking Circuit Assembly in Neurodevelopment

Studying neural circuit assembly entails gaining a comprehensive understanding of how diverse cells originate, navigate to appropriate and distant brain locations, differentiate into elaborate morphology, and connect with specific partners. In past decades, developmental neurobiology has been divided into multiple subfields specializing in neurogenesis, cell migration, axon guidance, or synapse formation. With few exceptions, it has been exceedingly difficult to integrate studies of different developmental episodes in the unifying context of circuit assembly. This is in part due to the difficulty of tracking the "life history" of any specific cellular component as it emerges through the multifaceted and convoluted "circuitgenic" process in its native environment and assembles into its destined functional circuit. Fluorescent labeling–based lineage and fate mapping has the potential to visualize the life history of specific circuit elements (i.e., cell types), from their origin to their circuit integration, and to report key developmental steps that are otherwise invisible.

Using CreER driven by a key transcription factor, GIFM can permanently label cells on the basis of their genetically defined lineage. This method is extremely powerful in tracking the developmental trajectory of cellular components during circuit formation. For example, studies using the Ascl1-CreER line revealed the sequential origin of distinct cell types in cerebellum and provided insight into their assembly into cerebellar local circuits (Sudarov et al. 2011). Combined with a MADM reporter, GIFM can also be used to examine whether and to what extent a clonal relationship contributes to circuit formation (Hippenmeyer et al. 2010).

Patterns of cell division and regulation of neuroepithelial cells along the ventricle wall are crucial to the process of neuronal cell fate specification. The antiproliferative gene *Tis21* is selectively expressed in neuron-generating but not proliferative neuroepithelial cells. Notably, a Tis21-GFP knockin mouse line selectively revealed neurogenic cell divisions. This has led to the finding that the basal neuroepithelial cells of the telencephalon are a major site of neurogenesis (Haubensak et al. 2004) and produce pyramidal neurons for all layers of the cortex (Kowalczyk et al. 2009). Furthermore, cell cycle–dependent ubiquitination machinery was harnessed to elucidate the spatiotemporal dynamics of cell cycle progression in transgenic mice (Sakaue-Sawano et al. 2008).

In addition to transgenic tools, retroviruses selectively infect dividing neural stem cells and progenitors. This provides a powerful method for lineage and clonal analysis. By imaging virally labeled live radial glial cells, researchers discovered intermediate progenitors that undergo symmetric cell divisions in the production of cortical pyramidal neurons (Noctor et al. 2004). Importantly, labeling of ontogenetic radial clones of excitatory pyramidal neurons followed by in vitro physiological recording revealed that clonally related sister cells preferentially establish synaptic connections. In vivo imaging further demonstrated that sister cells show similar receptive field properties, suggesting the formation of functional microcircuits (Li et al. 2012, Yu et al. 2009). Furthermore, pseudotyped retrovirus was used to label transgenically defined progenitors of the medial ganglionic eminence. These studies provide evidence that GABAergic interneurons are produced as spatially organized clonal units and suggest that lineage relationship also plays a role in the assembly of cortical inhibitory circuits (Brown et al. 2011).

MODELING NEUROPSYCHIATRIC DISORDERS

Aided by increasingly powerful genomics technology, human genetic studies are identifying increasingly more genetic mutations that predispose individuals to various neurological and psychiatric disorders. Researchers now know that a large number of rare and specific mutations individually contribute as major etiological factors to a set of overlapping behavioral and cognitive traits that clinically define categories of neurological and mental illnesses (Mitchell et al. 2011). These mutations, which are increasingly identified at the molecular level, represent concrete entry points to understanding the underlying neurobiology and pathogenic mechanisms. Importantly, investigators can now generate mouse models that recapitulate human genetic etiology (Mitchell et al. 2011). Combined with the increasingly powerful circuit analysis tools described in the sections above, these advances make studying abnormal brain development and function in the mouse feasible.

Modeling Mutant Alleles as well as Genes

Current understanding of the genetic architecture of brain development suggests that individual genes often play pleiotropic roles at different stages of neurodevelopment and contribute to the assembly and function of multiple brain circuits. Although loss-of-function mutations that cause certain brain disorders [e.g., MeCP2 and Rett syndrome (Tate et al. 1996)] have been identified, most mutations appear to alter "more subtly" the activity or expression of a gene. These alterations impact a subset of gene functions or alter a subset of brain circuit developmental processes and operations. Indeed, different alleles of the same gene can have distinct phenotypes in neuroanatomy and behavior. In this context, most studies using germline gene knockout mice to "model" human mutations are not ideal—it is critical to recapitulate more precisely the specific human mutant alleles, whether they are copy-number variations (CNVs), missense point mutations, or alterations in gene regulatory elements. Current mouse engineering technologies allow for the precise recapitulation of human genetic etiology (Horev et al. 2011, Stark et al. 2008). Such bona fide mouse models are a critical first step toward exploring the pathogenic mechanisms.

Identifying Behavior and Circuit Deficits

Even with a bona fide mouse model, it is often enormously challenging to identify the aberrant brain circuits underlying behavioral deficits. This effort will require brain-wide analyses of the altered function and assembly of neural circuits. Recent studies based on genetic models of schizophrenia (22q11) (Sigurdsson et al. 2010) and autism (CNTNAP2) (Penagarikano et al. 2011) have begun to provide insight into the pathology, pathophysiology, and behavioral deficits involved in these disorders, but the data do not reach the level of specific circuits and their cellular constituents. To achieve this, one needs to incorporate the full range of genetic circuit analysis tools into etiological mouse models.

Several unique challenges arise when studying models of brain disorders. Because of the complex genetic architecture of neuropsychiatric diseases and the convoluted relationship between single-gene mutations and behavior changes, it is often unclear where (brain regions), when (developmental stages), and how (functional and behavior contexts) one should look for circuit deficits. Although in some cases an overt behavioral deficit may guide circuit studies (Peca et al. 2011), it will usually be necessary to establish experimental paradigms and pipelines that allow high-throughput screens with cellular resolution to detect circuit and behavioral deficits (Bohland et al. 2009, Ragan et al. 2012).

Exploring Pathogenic Mechanisms

A mutation's direct impact is on the expression and/or molecular function of the gene product. However, the cumulative effects of genetic mutations can further alter the developmental trajectory of circuit formation and function. This is likely why understanding altered neural circuit development is equally, if not more, challenging than understanding normal circuit formation. Even when they start with the best genetic etiology models, few studies to date have been able to provide a coherent and integrated explanation of the process that leads from genetic mutation to aberrant circuit formation and function and ultimately to behavioral deficits (Mitchell et al. 2011). Indeed, understanding how mutations and adverse environments alter neural circuit formation and function, i.e., connecting the dots from genes to social-cognitive functions, presents formidable challenges even in light of spectacular advances in modern neuroscience. Guided by identified circuit and behavior deficits, genetic tools that allow scientists to track the developmental trajectory of cellular components during circuit assembly may reveal specific alterations and provide wedges for exploring pathogenic processes.

OUTLOOK

We envision a time when mouse genetic tools and resources will allow neuroscientists to systematically study neural circuits throughout the central nervous system. Distinct cell types or populations will be systematically tagged with genetic markers that allow researchers to monitor their structure and activity while manipulating their function with the goal of determining their role in behavior. The knowledge gained, along with data-based theoretical modeling, may reveal the principles underlying brain information processing. Beyond experimental access, genetic analysis of circuit assembly promises an opportunity to trace the complex link from genes to circuit function and to behavior, which may even reveal the genetic architecture of mental functions and disorders. Recent progress toward targeted transgenesis through TALENs and ZFNs in several species raises the hope that what we learn in the mouse brain can be extended to other mammals.

Despite significant progress in the development and application of mouse genetic tools, we are still far from reaching a "critical mass" that begins to enable systematic analysis of functional components of the circuitry throughout the nervous system. Here we present a "wish list" that, if fulfilled, we believe will further transform mammalian neuroscience from traditional anatomy and physiology to contemporary circuit analysis, with reproducible experimental access that bridges different levels of explanation.

1. Many more driver lines that target a sizable proportion of mouse neurons are needed. Given their success for precise recapitulation of gene expression, more knockin driver lines expressing Cre, Flp, and other novel recombinases should be generated. Complementing the knockin approach, BAC transgenic and enhancer trapping approaches should be coupled with rigorous screening and characterization efforts. To reduce variability and unpredictable silencing effects resulting from random integration,

site-specific transgenesis should be further developed.
2. We need improved inducible systems that have low toxicity, low background, high inducibility, and a fast time course, and equal efficacy at all developmental and adult stages.
3. Novel approaches for activity-dependent genetic targeting are highly desirable. IEG-based strategies should continue to be improved. Novel strategies that can track faster and more transient neuronal activities that are insufficient to drive IEG activation, such as fewer action potentials, calcium rises, and specific receptor activation, etc., should be explored. These strategies will be particularly useful in labeling neurons with specific response properties or those involved in specific phases of a behavior sequence.
4. In parallel, more high-level expression reporter lines should be designed at multiple genomic loci (in addition to Rosa26), including those suited for intersection and subtraction cell targeting. It will be very useful to develop a method to insert multiple copies of a transgene into a defined genomic locus to facilitate high-level expression and Brainbow-like strategies. Also useful is the development of mouse lines that target molecular tools to specific subcellular compartments or that tag specific endogenous proteins (such as ion channels) to track their subcellular distributions, all in a cell-type-specific manner.
5. More efforts should be devoted to harness the unusual properties of viruses from more diverse and exotic sources. New viral strains should be engineered that can be combined with transgenic mice for cell-type-specific infection or gene expression, that are capable of more efficient and specific monosynaptic retrograde or anterograde trans-synaptic tracing, and/or that have increased payload and reduced toxicity.
6. Finally, strategies should be developed to enable combinatorial labeling, activity sensing, and manipulation of multiple cell types in the same mouse. This may involve the combination of different intersectional strategies, with both transgenic and viral methods. By monitoring the activities of multiple cell types simultaneously and by differentially manipulating their activities, these tools will further enhance our ability to observe and probe the dynamic operation of brain circuits.

DISCLOSURE STATEMENT

The authors are not aware of any affiliations, memberships, funding, or financial holdings that might be perceived as affecting the objectivity of this review.

LITERATURE CITED

Airan RD, Thompson KR, Fenno LE, Bernstein H, Deisseroth K. 2009. Temporally precise in vivo control of intracellular signalling. *Nature* 458:1025–29

Akemann W, Mutoh H, Perron A, Park YK, Iwamoto Y, Knopfel T. 2012. Imaging neural circuit dynamics with a voltage-sensitive fluorescent protein. *J. Neurophysiol.* 108:2323–37

Akemann W, Mutoh H, Perron A, Rossier J, Knopfel T. 2010. Imaging brain electric signals with genetically targeted voltage-sensitive fluorescent proteins. *Nat. Methods* 7:643–49

Akerboom J, Chen TW, Wardill TJ, Tian L, Marvin JS, et al. 2012. Optimization of a GCaMP calcium indicator for neural activity imaging. *J. Neurosci.* 32:13819–40

Anastassiadis K, Fu J, Patsch C, Hu S, Weidlich S, et al. 2009. Dre recombinase, like Cre, is a highly efficient site-specific recombinase in *E. coli*, mammalian cells and mice. *Dis. Models Mech.* 2:508–15

Andrasfalvy BK, Zemelman BV, Tang J, Vaziri A. 2010. Two-photon single-cell optogenetic control of neuronal activity by sculpted light. *Proc. Natl. Acad. Sci. USA* 107:11981–86

Arber S. 2012. Motor circuits in action: specification, connectivity, and function. *Neuron* 74:975–89

Arenkiel BR, Klein ME, Davison IG, Katz LC, Ehlers MD. 2008. Genetic control of neuronal activity in mice conditionally expressing *TRPV1*. *Nat. Methods* 5:299–302

Arlotta P, Molyneaux BJ, Chen J, Inoue J, Kominami R, Macklis JD. 2005. Neuronal subtype-specific genes that control corticospinal motor neuron development in vivo. *Neuron* 45:207–21

Armbruster BN, Li X, Pausch MH, Herlitze S, Roth BL. 2007. Evolving the lock to fit the key to create a family of G protein–coupled receptors potently activated by an inert ligand. *Proc. Natl. Acad. Sci. USA* 104:5163–68

Atasoy D, Aponte Y, Su HH, Sternson SM. 2008. A FLEX switch targets channelrhodopsin-2 to multiple cell types for imaging and long-range circuit mapping. *J. Neurosci.* 28:7025–30

Barnea G, Strapps W, Herrada G, Berman Y, Ong J, et al. 2008. The genetic design of signaling cascades to record receptor activation. *Proc. Natl. Acad. Sci. USA* 105:64–69

Barth AL, Gerkin RC, Dean KL. 2004. Alteration of neuronal firing properties after in vivo experience in a FosGFP transgenic mouse. *J. Neurosci.* 24:6466–75

Beckervordersandforth R, Tripathi P, Ninkovic J, Bayam E, Lepier A, et al. 2010. In vivo fate mapping and expression analysis reveals molecular hallmarks of prospectively isolated adult neural stem cells. *Cell Stem Cell* 7:744–58

Beier KT, Saunders A, Oldenburg IA, Miyamichi K, Akhtar N, et al. 2011. Anterograde or retrograde transsynaptic labeling of CNS neurons with vesicular stomatitis virus vectors. *Proc. Natl. Acad. Sci. USA* 108:15414–19

Benzer S. 1967. Behavioral mutants of *Drosophila* isolated by countercurrent distribution. *Proc. Natl. Acad. Sci. USA* 58:1112–19

Berens C, Hillen W. 2004. Gene regulation by tetracyclines. *Genet. Eng.* 26:255–77

Bernstein BE, Birney E, Dunham I, Green ED, Gunter C, Snyder M. 2012. An integrated encyclopedia of DNA elements in the human genome. *Nature* 489:57–74

Bohland JW, Wu C, Barbas H, Bokil H, Bota M, et al. 2009. A proposal for a coordinated effort for the determination of brainwide neuroanatomical connectivity in model organisms at a mesoscopic scale. *PLoS Comput. Biol.* 5:e1000334

Boyden ES, Zhang F, Bamberg E, Nagel G, Deisseroth K. 2005. Millisecond-timescale, genetically targeted optical control of neural activity. *Nat. Neurosci.* 8:1263–68

Bozza T, McGann JP, Mombaerts P, Wachowiak M. 2004. In vivo imaging of neuronal activity by targeted expression of a genetically encoded probe in the mouse. *Neuron* 42:9–21

Brenner S. 1974. The genetics of *Caenorhabditis elegans*. *Genetics* 77:71–94

Brenner S. 2010. Sequences and consequences. *Philos. Trans. R. Soc. B* 365:207–12

Brinster RL, Chen HY, Trumbauer M, Senear AW, Warren R, Palmiter RD. 1981. Somatic expression of herpes thymidine kinase in mice following injection of a fusion gene into eggs. *Cell* 27:223–31

Brown KN, Chen S, Han Z, Lu CH, Tan X, et al. 2011. Clonal production and organization of inhibitory interneurons in the neocortex. *Science* 334:480–86

Buch T, Heppner FL, Tertilt C, Heinen TJ, Kremer M, et al. 2005. A Cre-inducible diphtheria toxin receptor mediates cell lineage ablation after toxin administration. *Nat. Methods* 2:419–26

Bulger M, Groudine M. 2011. Functional and mechanistic diversity of distal transcription enhancers. *Cell* 144:327–39

Callaway EM. 2008. Transneuronal circuit tracing with neurotropic viruses. *Curr. Opin. Neurobiol.* 18:617–23

Card JP, Kobiler O, McCambridge J, Ebdlahad S, Shan Z, et al. 2011. Microdissection of neural networks by conditional reporter expression from a Brainbow herpesvirus. *Proc. Natl. Acad. Sci. USA* 108:3377–82

Casanova E, Lemberger T, Fehsenfeld S, Mantamadiotis T, Schutz G. 2003. α Complementation in the Cre recombinase enzyme. *Genesis* 37:25–29

Chalfie M, Tu Y, Euskirchen G, Ward WW, Prasher DC. 1994. Green fluorescent protein as a marker for gene expression. *Science* 263:802–5

Chamberlin NL, Du B, de Lacalle S, Saper CB. 1998. Recombinant adeno-associated virus vector: use for transgene expression and anterograde tract tracing in the CNS. *Brain Res.* 793:169–75

Chen T-W, Wardill TJ, Hasseman JP, Tsegaye G, Fosque BF, et al. 2012. Engineering next generation GCaMP calcium indicators using neuron-based screening. *Soc. Neurosci. Meet. Plann.* Abstr. 207.14DDD34

Chow BY, Han X, Boyden ES. 2012. Genetically encoded molecular tools for light-driven silencing of targeted neurons. *Prog. Brain Res.* 196:49–61

Chow BY, Han X, Dobry AS, Qian X, Chuong AS, et al. 2010. High-performance genetically targetable optical neural silencing by light-driven proton pumps. *Nature* 463:98–102

Cong L, Ran FA, Cox D, Lin S, Barretto R, et al. 2013. Multiplex genome engineering using CRISPR/Cas systems. *Science* 339:819–23

Conklin BR, Hsiao EC, Claeysen S, Dumuis A, Srinivasan S, et al. 2008. Engineering GPCR signaling pathways with RASSLs. *Nat. Methods* 5:673–78

Costantini F, Lacy E. 1981. Introduction of a rabbit β-globin gene into the mouse germ line. *Nature* 294:92–94

De Paola V, Arber S, Caroni P. 2003. AMPA receptors regulate dynamic equilibrium of presynaptic terminals in mature hippocampal networks. *Nat. Neurosci.* 6:491–500

Doetschman T, Gregg RG, Maeda N, Hooper ML, Melton DW, et al. 1987. Targetted correction of a mutant *HPRT* gene in mouse embryonic stem cells. *Nature* 330:576–78

Dombeck DA, Harvey CD, Tian L, Looger LL, Tank DW. 2010. Functional imaging of hippocampal place cells at cellular resolution during virtual navigation. *Nat. Neurosci.* 13:1433–40

Dymecki SM, Kim JC. 2007. Molecular neuroanatomy's "three Gs": a primer. *Neuron* 54:17–34

Feil R, Wagner J, Metzger D, Chambon P. 1997. Regulation of Cre recombinase activity by mutated estrogen receptor ligand-binding domains. *Biochem. Biophys. Res. Commun.* 237:752–57

Feinberg EH, Vanhoven MK, Bendesky A, Wang G, Fetter RD, et al. 2008. GFP reconstitution across synaptic partners (GRASP) defines cell contacts and synapses in living nervous systems. *Neuron* 57:353–63

Feng G, Mellor RH, Bernstein M, Keller-Peck C, Nguyen QT, et al. 2000. Imaging neuronal subsets in transgenic mice expressing multiple spectral variants of GFP. *Neuron* 28:41–51

Fenno L, Yizhar O, Deisseroth K. 2011. The development and application of optogenetics. *Annu. Rev. Neurosci.* 34:389–412

Garner AR, Rowland DC, Hwang SY, Baumgaertel K, Roth BL, et al. 2012. Generation of a synthetic memory trace. *Science* 335:1513–16

Gentet LJ, Avermann M, Matyas F, Staiger JF, Petersen CC. 2010. Membrane potential dynamics of GABAergic neurons in the barrel cortex of behaving mice. *Neuron* 65:422–35

Gentet LJ, Kremer Y, Taniguchi H, Huang ZJ, Staiger JF, Petersen CC. 2012. Unique functional properties of somatostatin-expressing GABAergic neurons in mouse barrel cortex. *Nat. Neurosci.* 15:607–12

Ghosh S, Larson SD, Hefzi H, Marnoy Z, Cutforth T, et al. 2011. Sensory maps in the olfactory cortex defined by long-range viral tracing of single neurons. *Nature* 472:217–20

Gogos JA, Osborne J, Nemes A, Mendelsohn M, Axel R. 2000. Genetic ablation and restoration of the olfactory topographic map. *Cell* 103:609–20

Gong S, Doughty M, Harbaugh CR, Cummins A, Hatten ME, et al. 2007. Targeting Cre recombinase to specific neuron populations with bacterial artificial chromosome constructs. *J. Neurosci.* 27:9817–23

Gong S, Zheng C, Doughty ML, Losos K, Didkovsky N, et al. 2003. A gene expression atlas of the central nervous system based on bacterial artificial chromosomes. *Nature* 425:917–25

Gordon JW, Ruddle FH. 1981. Integration and stable germ line transmission of genes injected into mouse pronuclei. *Science* 214:1244–46

Gradinaru V, Zhang F, Ramakrishnan C, Mattis J, Prakash R, et al. 2010. Molecular and cellular approaches for diversifying and extending optogenetics. *Cell* 141:154–65

Gray NW, Weimer RM, Bureau I, Svoboda K. 2006. Rapid redistribution of synaptic PSD-95 in the neocortex in vivo. *PLoS Biol.* 4:e370

Gropp E, Shanabrough M, Borok E, Xu AW, Janoschek R, et al. 2005. Agouti-related peptide–expressing neurons are mandatory for feeding. *Nat. Neurosci.* 8:1289–91

Hama H, Kurokawa H, Kawano H, Ando R, Shimogori T, et al. 2011. Scale: a chemical approach for fluorescence imaging and reconstruction of transparent mouse brain. *Nat. Neurosci.* 14:1481–88

Han X, Chow BY, Zhou HH, Klapoetke NC, Chuong A, et al. 2011. A high-light sensitivity optical neural silencer: development and application to optogenetic control of non-human primate cortex. *Front. Syst. Neurosci.* 5:1

Harris JA, Oh SW, Zeng H. 2012. Adeno-associated viral vectors for anterograde axonal tracing with fluorescent proteins in nontransgenic and Cre driver mice. *Curr. Protoc. Neurosci.* Suppl. 59:Unit 1.20.1–18; doi: 10.1002/0471142301.ns0120s59

Hartwell LH. 1991. Twenty-five years of cell cycle genetics. *Genetics* 129:975–80

Haubensak W, Attardo A, Denk W, Huttner WB. 2004. Neurons arise in the basal neuroepithelium of the early mammalian telencephalon: a major site of neurogenesis. *Proc. Natl. Acad. Sci. USA* 101:3196–201

He M, Liu Y, Wang X, Zhang MQ, Hannon GJ, Huang ZJ. 2012. Cell-type-based analysis of microRNA profiles in the mouse brain. *Neuron* 73:35–48

Heiman M, Schaefer A, Gong S, Peterson JD, Day M, et al. 2008. A translational profiling approach for the molecular characterization of CNS cell types. *Cell* 135:738–48

Heintzman ND, Ren B. 2009. Finding distal regulatory elements in the human genome. *Curr. Opin. Genet. Dev.* 19:541–49

Helbig I, Sammler E, Eliava M, Bolshakov AP, Rozov A, et al. 2010. In vivo evidence for the involvement of the carboxy terminal domain in assembling connexin 36 at the electrical synapse. *Mol. Cell Neurosci.* 45:47–58

Hippenmeyer S, Youn YH, Moon HM, Miyamichi K, Zong H, et al. 2010. Genetic mosaic dissection of *Lis1* and *Ndel1* in neuronal migration. *Neuron* 68:695–709

Hirrlinger J, Scheller A, Hirrlinger PG, Kellert B, Tang W, et al. 2009. Split-Cre complementation indicates coincident activity of different genes in vivo. *PLoS ONE* 4:e4286

Hnasko TS, Perez FA, Scouras AD, Stoll EA, Gale SD, et al. 2006. Cre recombinase-mediated restoration of nigrostriatal dopamine in dopamine-deficient mice reverses hypophagia and bradykinesia. *Proc. Natl. Acad. Sci. USA* 103:8858–63

Hofer SB, Ko H, Pichler B, Vogelstein J, Ros H, et al. 2011. Differential connectivity and response dynamics of excitatory and inhibitory neurons in visual cortex. *Nat. Neurosci.* 14:1045–52

Holtmaat A, Svoboda K. 2009. Experience-dependent structural synaptic plasticity in the mammalian brain. *Nat. Rev. Neurosci.* 10:647–58

Horev G, Ellegood J, Lerch JP, Son YE, Muthuswamy L, et al. 2011. Dosage-dependent phenotypes in models of 16p11.2 lesions found in autism. *Proc. Natl. Acad. Sci. USA* 108:17076–81

Hotta Y, Benzer S. 1970. Genetic dissection of the *Drosophila* nervous system by means of mosaics. *Proc. Natl. Acad. Sci. USA* 67:1156–63

Hunter NL, Awatramani RB, Farley FW, Dymecki SM. 2005. Ligand-activated Flpe for temporally regulated gene modifications. *Genesis* 41:99–109

Iwamoto M, Bjorklund T, Lundberg C, Kirik D, Wandless TJ. 2010. A general chemical method to regulate protein stability in the mammalian central nervous system. *Chem. Biol.* 17:981–88

Jasin M, Moynahan ME, Richardson C. 1996. Targeted transgenesis. *Proc. Natl. Acad. Sci. USA* 93:8804–8

Jensen P, Farago AF, Awatramani RB, Scott MM, Deneris ES, Dymecki SM. 2008. Redefining the serotonergic system by genetic lineage. *Nat. Neurosci.* 11:417–19

Jin L, Han Z, Platisa J, Wooltorton JR, Cohen LB, Pieribone VA. 2012. Single action potentials and subthreshold electrical events imaged in neurons with a fluorescent protein voltage probe. *Neuron* 75:779–85

Joyner AL, Zervas M. 2006. Genetic inducible fate mapping in mouse: establishing genetic lineages and defining genetic neuroanatomy in the nervous system. *Dev. Dyn.* 235:2376–85

Kahn I, Desai M, Knoblich U, Bernstein J, Henninger M, et al. 2011. Characterization of the functional MRI response temporal linearity via optical control of neocortical pyramidal neurons. *J. Neurosci.* 31:15086–91

Kapranov P, Willingham AT, Gingeras TR. 2007. Genome-wide transcription and the implications for genomic organization. *Nat. Rev. Genet.* 8:413–23

Katzel D, Zemelman BV, Buetfering C, Wolfel M, Miesenböck G. 2011. The columnar and laminar organization of inhibitory connections to neocortical excitatory cells. *Nat. Neurosci.* 14:100–7

Kelsch W, Stolfi A, Lois C. 2012. Genetic labeling of neuronal subsets through enhancer trapping in mice. *PLoS ONE* 7:e38593

Kim IJ, Zhang Y, Yamagata M, Meister M, Sanes JR. 2008. Molecular identification of a retinal cell type that responds to upward motion. *Nature* 452:478–82

Kim J, Zhao T, Petralia RS, Yu Y, Peng H, et al. 2012. mGRASP enables mapping mammalian synaptic connectivity with light microscopy. *Nat. Methods* 9:96–102

Kim JC, Dymecki SM. 2009. Genetic fate-mapping approaches: new means to explore the embryonic origins of the cochlear nucleus. *Methods Mol. Biol.* 493:65–85

Kowalczyk T, Pontious A, Englund C, Daza RA, Bedogni F, et al. 2009. Intermediate neuronal progenitors (basal progenitors) produce pyramidal-projection neurons for all layers of cerebral cortex. *Cereb. Cortex* 19:2439–50

Kralj JM, Douglass AD, Hochbaum DR, Maclaurin D, Cohen AE. 2012. Optical recording of action potentials in mammalian neurons using a microbial rhodopsin. *Nat. Methods* 9:90–95

Kralj JM, Hochbaum DR, Douglass AD, Cohen AE. 2011. Electrical spiking in *Escherichia coli* probed with a fluorescent voltage-indicating protein. *Science* 333:345–48

Kuhlman SJ, Huang ZJ. 2008. High-resolution labeling and functional manipulation of specific neuron types in mouse brain by Cre-activated viral gene expression. *PLoS ONE* 3:e2005

Kuramoto E, Furuta T, Nakamura KC, Unzai T, Hioki H, Kaneko T. 2009. Two types of thalamocortical projections from the motor thalamic nuclei of the rat: a single neuron-tracing study using viral vectors. *Cereb. Cortex* 19:2065–77

Lee JH, Durand R, Gradinaru V, Zhang F, Goshen I, et al. 2010. Global and local fMRI signals driven by neurons defined optogenetically by type and wiring. *Nature* 465:788–92

Lein ES, Hawrylycz MJ, Ao N, Ayres M, Bensinger A, et al. 2007. Genome-wide atlas of gene expression in the adult mouse brain. *Nature* 445:168–76

Lerchner W, Xiao C, Nashmi R, Slimko EM, van Trigt L, et al. 2007. Reversible silencing of neuronal excitability in behaving mice by a genetically targeted, ivermectin-gated Cl- channel. *Neuron* 54:35–49

Li Y, Lu H, Cheng PL, Ge S, Xu H, et al. 2012. Clonally related visual cortical neurons show similar stimulus feature selectivity. *Nature* 486:118–21

Li Y, Tsien RW. 2012. pHTomato, a red, genetically encoded indicator that enables multiplex interrogation of synaptic activity. *Nat. Neurosci.* 15:1047–53

Li Z, Burrone J, Tyler WJ, Hartman KN, Albeanu DF, Murthy VN. 2005. Synaptic vesicle recycling studied in transgenic mice expressing synaptopHluorin. *Proc. Natl. Acad. Sci. USA* 102:6131–36

Lichtman JW, Denk W. 2011. The big and the small: challenges of imaging the brain's circuits. *Science* 334:618–23

Lichtman JW, Sanes JR. 2003. Watching the neuromuscular junction. *J. Neurocytol.* 32:767–75

Lima SQ, Hromadka T, Znamenskiy P, Zador AM. 2009. PINP: a new method of tagging neuronal populations for identification during in vivo electrophysiological recording. *PLoS ONE* 4:e6099

Liu X, Ramirez S, Pang PT, Puryear CB, Govindarajan A, et al. 2012. Optogenetic stimulation of a hippocampal engram activates fear memory recall. *Nature* 484:381–85

Livesey FJ, Cepko CL. 2001. Vertebrate neural cell-fate determination: lessons from the retina. *Nat. Rev. Neurosci.* 2:109–18

Livet J, Weissman TA, Kang H, Draft RW, Lu J, et al. 2007. Transgenic strategies for combinatorial expression of fluorescent proteins in the nervous system. *Nature* 450:56–62

Lo L, Anderson DJ. 2011. A Cre-dependent, anterograde transsynaptic viral tracer for mapping output pathways of genetically marked neurons. *Neuron* 72:938–50

Luo L, Callaway EM, Svoboda K. 2008. Genetic dissection of neural circuits. *Neuron* 57:634–60

Luquet S, Perez FA, Hnasko TS, Palmiter RD. 2005. NPY/AgRP neurons are essential for feeding in adult mice but can be ablated in neonates. *Science* 310:683–85

Madisen L, Mao T, Koch H, Zhuo JM, Berenyi A, et al. 2012. A toolbox of Cre-dependent optogenetic transgenic mice for light-induced activation and silencing. *Nat. Neurosci.* 15:793–802

Madisen L, Zwingman TA, Sunkin SM, Oh SW, Zariwala HA, et al. 2010. A robust and high-throughput Cre reporting and characterization system for the whole mouse brain. *Nat. Neurosci.* 13:133–40

Mali P, Yang L, Esvelt KM, Aach J, Guell M, et al. 2013. RNA-guided human genome engineering via Cas9. *Science* 339:823–26

Mank M, Griesbeck O. 2008. Genetically encoded calcium indicators. *Chem. Rev.* 108:1550–64

Martell JD, Deerinck TJ, Sancak Y, Poulos TL, Mootha VK, et al. 2012. Engineered ascorbate peroxidase as a genetically encoded reporter for electron microscopy. *Nat. Biotechnol.* 30:1143–48

Marvin JS, Borghuis BG, Tian L, Cichon J, Harnett MT, et al. 2013. An optimized fluorescent probe for visualizing glutamate neurotransmission. *Nat. Methods* 10:162–70

Mattis J, Tye KM, Ferenczi EA, Ramakrishnan C, O'Shea DJ, et al. 2012. Principles for applying optogenetic tools derived from direct comparative analysis of microbial opsins. *Nat. Methods* 9:159–72

McGann JP, Pirez N, Gainey MA, Muratore C, Elias AS, Wachowiak M. 2005. Odorant representations are modulated by intra- but not interglomerular presynaptic inhibition of olfactory sensory neurons. *Neuron* 48:1039–53

Miesenböck G, De Angelis DA, Rothman JE. 1998. Visualizing secretion and synaptic transmission with pH-sensitive green fluorescent proteins. *Nature* 394:192–95

Mitchell KJ, Huang ZJ, Moghaddam B, Sawa A. 2011. Following the genes: a framework for animal modeling of psychiatric disorders. *BMC Biol.* 9:76

Mittmann W, Wallace DJ, Czubayko U, Herb JT, Schaefer AT, et al. 2011. Two-photon calcium imaging of evoked activity from L5 somatosensory neurons in vivo. *Nat. Neurosci.* 14:1089–93

Miyamichi K, Amat F, Moussavi F, Wang C, Wickersham I, et al. 2011. Cortical representations of olfactory input by trans-synaptic tracing. *Nature* 472:191–96

Miyoshi G, Hjerling-Leffler J, Karayannis T, Sousa VH, Butt SJ, et al. 2010. Genetic fate mapping reveals that the caudal ganglionic eminence produces a large and diverse population of superficial cortical interneurons. *J. Neurosci.* 30:1582–94

Mombaerts P, Wang F, Dulac C, Chao SK, Nemes A, et al. 1996. Visualizing an olfactory sensory map. *Cell* 87:675–86

Mussolino C, Cathomen T. 2012. TALE nucleases: tailored genome engineering made easy. *Curr. Opin. Biotechnol.* 23:644–50

Nagel G, Szellas T, Huhn W, Kateriya S, Adeishvili N, et al. 2003. Channelrhodopsin-2, a directly light-gated cation-selective membrane channel. *Proc. Natl. Acad. Sci. USA* 100:13940–45

Nakashiba T, Young JZ, McHugh TJ, Buhl DL, Tonegawa S. 2008. Transgenic inhibition of synaptic transmission reveals role of CA3 output in hippocampal learning. *Science* 319:1260–64

Nern A, Pfeiffer BD, Svoboda K, Rubin GM. 2011. Multiple new site-specific recombinases for use in manipulating animal genomes. *Proc. Natl. Acad. Sci. USA* 108:14198–203

Noctor SC, Martínez-Cerdeño V, Ivic L, Kriegstein AR. 2004. Cortical neurons arise in symmetric and asymmetric division zones and migrate through specific phases. *Nat. Neurosci.* 7:136–44

Nusslein-Volhard C, Wieschaus E. 1980. Mutations affecting segment number and polarity in *Drosophila*. *Nature* 287:795–801

O'Connor DH, Peron SP, Huber D, Svoboda K. 2010. Neural activity in barrel cortex underlying vibrissa-based object localization in mice. *Neuron* 67:1048–61

Ohkura M, Sasaki T, Kobayashi C, Ikegaya Y, Nakai J. 2012. An improved genetically encoded red fluorescent Ca^{2+} indicator for detecting optically evoked action potentials. *PLoS ONE* 7:e39933

Osakada F, Mori T, Cetin AH, Marshel JH, Virgen B, Callaway EM. 2011. New rabies virus variants for monitoring and manipulating activity and gene expression in defined neural circuits. *Neuron* 71:617–31

Packer AM, Peterka DS, Hirtz JJ, Prakash R, Deisseroth K, Yuste R. 2012. Two-photon optogenetics of dendritic spines and neural circuits. *Nat. Methods* 9:1202–5

Papagiakoumou E, Anselmi F, Begue A, de Sars V, Gluckstad J, et al. 2010. Scanless two-photon excitation of channelrhodopsin-2. *Nat. Methods* 7:848–54

Peca J, Feliciano C, Ting JT, Wang W, Wells MF, et al. 2011. *Shank3* mutant mice display autistic-like behaviours and striatal dysfunction. *Nature* 472:437–42

Penagarikano O, Abrahams BS, Herman EI, Winden KD, Gdalyahu A, et al. 2011. Absence of CNTNAP2 leads to epilepsy, neuronal migration abnormalities, and core autism-related deficits. *Cell* 147:235–46

Pennacchio LA, Ahituv N, Moses AM, Prabhakar S, Nobrega MA, et al. 2006. In vivo enhancer analysis of human conserved non-coding sequences. *Nature* 444:499–502

Petreanu L, Huber D, Sobczyk A, Svoboda K. 2007. Channelrhodopsin-2-assisted circuit mapping of long-range callosal projections. *Nat. Neurosci.* 10:663–68

Petreanu L, Mao T, Sternson SM, Svoboda K. 2009. The subcellular organization of neocortical excitatory connections. *Nature* 457:1142–45

Prakash R, Yizhar O, Grewe B, Ramakrishnan C, Wang N, et al. 2012. Two-photon optogenetic toolbox for fast inhibition, excitation and bistable modulation. *Nat. Methods* 9:1171–79

Ragan T, Kadiri LR, Venkataraju KU, Bahlmann K, Sutin J, et al. 2012. Serial two-photon tomography for automated ex vivo mouse brain imaging. *Nat. Methods* 9:255–58

Ramón y Cajal S. 1899. *La textura del sistema nerviosa del hombre y los vertebrados*. Madrid: Moya. 1st ed.

Ray RS, Corcoran AE, Brust RD, Kim JC, Richerson GB, et al. 2011. Impaired respiratory and body temperature control upon acute serotonergic neuron inhibition. *Science* 333:637–42

Reijmers LG, Perkins BL, Matsuo N, Mayford M. 2007. Localization of a stable neural correlate of associative memory. *Science* 317:1230–33

Rotolo T, Smallwood PM, Williams J, Nathans J. 2008. Genetically-directed, cell type-specific sparse labeling for the analysis of neuronal morphology. *PLoS ONE* 3:e4099

Royer S, Zemelman BV, Losonczy A, Kim J, Chance F, et al. 2012. Control of timing, rate and bursts of hippocampal place cells by dendritic and somatic inhibition. *Nat. Neurosci.* 15:769–75

Sakamoto M, Imayoshi I, Ohtsuka T, Yamaguchi M, Mori K, Kageyama R. 2011. Continuous neurogenesis in the adult forebrain is required for innate olfactory responses. *Proc. Natl. Acad. Sci. USA* 108:8479–84

Sakaue-Sawano A, Ohtawa K, Hama H, Kawano M, Ogawa M, Miyawaki A. 2008. Tracing the silhouette of individual cells in S/G2/M phases with fluorescence. *Chem. Biol.* 15:1243–48

Sanyal A, Lajoie BR, Jain G, Dekker J. 2012. The long-range interaction landscape of gene promoters. *Nature* 489:109–13

Sanz E, Yang L, Su T, Morris DR, McKnight GS, Amieux PS. 2009. Cell-type-specific isolation of ribosome-associated mRNA from complex tissues. *Proc. Natl. Acad. Sci. USA* 106:13939–44

Scanziani M, Hausser M. 2009. Electrophysiology in the age of light. *Nature* 461:930–39

Schroder-Lang S, Schwarzel M, Seifert R, Strunker T, Kateriya S, et al. 2007. Fast manipulation of cellular cAMP level by light in vivo. *Nat. Methods* 4:39–42

Seidler B, Schmidt A, Mayr U, Nakhai H, Schmid RM, et al. 2008. A Cre-loxP-based mouse model for conditional somatic gene expression and knockdown *in vivo* by using avian retroviral vectors. *Proc. Natl. Acad. Sci. USA* 105:10137–42

Shen Y, Yue F, McCleary DF, Ye Z, Edsall L, et al. 2012. A map of the *cis*-regulatory sequences in the mouse genome. *Nature* 488:116–20

Shu X, Lev-Ram V, Deerinck TJ, Qi Y, Ramko EB, et al. 2011. A genetically encoded tag for correlated light and electron microscopy of intact cells, tissues, and organisms. *PLoS Biol.* 9:e1001041

Sigurdsson T, Stark KL, Karayiorgou M, Gogos JA, Gordon JA. 2010. Impaired hippocampal-prefrontal synchrony in a genetic mouse model of schizophrenia. *Nature* 464:763–67

Silva AJ, Stevens CF, Tonegawa S, Wang Y. 1992. Deficient hippocampal long-term potentiation in α-calcium-calmodulin kinase II mutant mice. *Science* 257:201–6

Soudais C, Laplace-Builhe C, Kissa K, Kremer EJ. 2001. Preferential transduction of neurons by canine adenovirus vectors and their efficient retrograde transport in vivo. *FASEB J.* 15:2283–85

Soudais C, Skander N, Kremer EJ. 2004. Long-term in vivo transduction of neurons throughout the rat CNS using novel helper-dependent CAV-2 vectors. *FASEB J.* 18:391–93

Stark KL, Xu B, Bagchi A, Lai WS, Liu H, et al. 2008. Altered brain microRNA biogenesis contributes to phenotypic deficits in a 22q11-deletion mouse model. *Nat. Genet.* 40:751–60

Sudarov A, Turnbull RK, Kim EJ, Lebel-Potter M, Guillemot F, Joyner AL. 2011. Ascl1 genetics reveals insights into cerebellum local circuit assembly. *J. Neurosci.* 31:11055–69

Sugino K, Hempel CM, Miller MN, Hattox AM, Shapiro P, et al. 2006. Molecular taxonomy of major neuronal classes in the adult mouse forebrain. *Nat. Neurosci.* 9:99–107

Tan EM, Yamaguchi Y, Horwitz GD, Gosgnach S, Lein ES, et al. 2006. Selective and quickly reversible inactivation of mammalian neurons in vivo using the *Drosophila* allatostatin receptor. *Neuron* 51:157–70

Tanaka KF, Matsui K, Sasaki T, Sano H, Sugio S, et al. 2012. Expanding the repertoire of optogenetically targeted cells with an enhanced gene expression system. *Cell Rep.* 2:397–406

Taniguchi H, He M, Wu P, Kim S, Paik R, et al. 2011. A resource of Cre driver lines for genetic targeting of GABAergic neurons in cerebral cortex. *Neuron* 71:995–1013

Taniguchi H, Lu J, Huang ZJ. 2013. The spatial and temporal origin of chandelier cells in mouse neocortex. *Science* 339:70–74

Tantama M, Hung YP, Yellen G. 2012. Optogenetic reporters: fluorescent protein-based genetically encoded indicators of signaling and metabolism in the brain. *Prog. Brain Res.* 196:235–63

Tasic B, Hippenmeyer S, Wang C, Gamboa M, Zong H, et al. 2011. Site-specific integrase-mediated transgenesis in mice via pronuclear injection. *Proc. Natl. Acad. Sci. USA* 108:7902–7

Tate P, Skarnes W, Bird A. 1996. The methyl-CpG binding protein MeCP2 is essential for embryonic development in the mouse. *Nat. Genet.* 12:205–8

Thomas KR, Capecchi MR. 1987. Site-directed mutagenesis by gene targeting in mouse embryo-derived stem cells. *Cell* 51:503–12

Tian L, Akerboom J, Schreiter ER, Looger LL. 2012. Neural activity imaging with genetically encoded calcium indicators. *Prog. Brain Res.* 196:79–94

Tian L, Hires SA, Mao T, Huber D, Chiappe ME, et al. 2009. Imaging neural activity in worms, flies and mice with improved GCaMP calcium indicators. *Nat. Methods* 6:875–81

Toledo-Rodriguez M, Blumenfeld B, Wu C, Luo J, Attali B, et al. 2004. Correlation maps allow neuronal electrical properties to be predicted from single-cell gene expression profiles in rat neocortex. *Cereb. Cortex* 14:1310–27

Tsien JZ, Chen DF, Gerber D, Tom C, Mercer EH, et al. 1996. Subregion- and cell type–restricted gene knockout in mouse brain. *Cell* 87:1317–26

Tucker CL. 2012. Manipulating cellular processes using optical control of protein-protein interactions. *Prog. Brain Res.* 196:95–117

Tucker KL, Meyer M, Barde YA. 2001. Neurotrophins are required for nerve growth during development. *Nat. Neurosci.* 4:29–37

Visel A, Blow MJ, Li Z, Zhang T, Akiyama JA, et al. 2009a. ChIP-seq accurately predicts tissue-specific activity of enhancers. *Nature* 457:854–58

Visel A, Rubin EM, Pennacchio LA. 2009b. Genomic views of distant-acting enhancers. *Nature* 461:199–205

Wall NR, Wickersham IR, Cetin A, De La Parra M, Callaway EM. 2010. Monosynaptic circuit tracing in vivo through Cre-dependent targeting and complementation of modified rabies virus. *Proc. Natl. Acad. Sci. USA* 107:21848–53

Wang KH, Majewska A, Schummers J, Farley B, Hu C, et al. 2006. In vivo two-photon imaging reveals a role of arc in enhancing orientation specificity in visual cortex. *Cell* 126:389–402

Wang P, Chen T, Sakurai K, Han BX, He Z, et al. 2012. Intersectional Cre driver lines generated using split-intein mediated split-Cre reconstitution. *Sci. Rep.* 2:497

Wickersham IR, Lyon DC, Barnard RJ, Mori T, Finke S, et al. 2007. Monosynaptic restriction of transsynaptic tracing from single, genetically targeted neurons. *Neuron* 53:639–47

Wilson Y, Nag N, Davern P, Oldfield BJ, McKinley MJ, et al. 2002. Visualization of functionally activated circuitry in the brain. *Proc. Natl. Acad. Sci. USA* 99:3252–57

Yamada Y, Michikawa T, Hashimoto M, Horikawa K, Nagai T, et al. 2011. Quantitative comparison of genetically encoded Ca indicators in cortical pyramidal cells and cerebellar Purkinje cells. *Front. Cell Neurosci.* 5:18

Yamamoto M, Wada N, Kitabatake Y, Watanabe D, Anzai M, et al. 2003. Reversible suppression of glutamatergic neurotransmission of cerebellar granule cells in vivo by genetically manipulated expression of tetanus neurotoxin light chain. *J. Neurosci.* 23:6759–67

Yizhar O, Fenno LE, Davidson TJ, Mogri M, Deisseroth K. 2011. Optogenetics in neural systems. *Neuron* 71:9–34

Yu YC, Bultje RS, Wang X, Shi SH. 2009. Specific synapses develop preferentially among sister excitatory neurons in the neocortex. *Nature* 458:501–4

Zariwala HA, Borghuis BG, Hoogland TM, Madisen L, Tian L, et al. 2012. A Cre-dependent GCaMP3 reporter mouse for neuronal imaging in vivo. *J. Neurosci.* 32:3131–41

Zariwala HA, Madisen L, Ahrens KF, Bernard A, Lein ES, et al. 2011. Visual tuning properties of genetically identified layer 2/3 neuronal types in the primary visual cortex of Cre-transgenic mice. *Front. Syst. Neurosci.* 4:162

Zeng H, Horie K, Madisen L, Pavlova MN, Gragerova G, et al. 2008. An inducible and reversible mouse genetic rescue system. *PLoS Genet.* 4:e1000069

Zeng H, Madisen L. 2012. Mouse transgenic approaches in optogenetics. *Prog. Brain Res.* 196:193–213

Zhao S, Ting JT, Atallah HE, Qiu L, Tan J, et al. 2011. Cell type-specific channelrhodopsin-2 transgenic mice for optogenetic dissection of neural circuitry function. *Nat. Methods* 8:745–52

Zhao Y, Araki S, Wu J, Teramoto T, Chang YF, et al. 2011. An expanded palette of genetically encoded Ca^{2+} indicators. *Science* 333:1888–91

Early Olfactory Processing in *Drosophila*: Mechanisms and Principles

Rachel I. Wilson

Department of Neurobiology, Harvard Medical School, Boston, Massachusetts 02115;
email: rachel_wilson@hms.harvard.edu

Annu. Rev. Neurosci. 2013. 36:217–41

The *Annual Review of Neuroscience* is online at neuro.annualreviews.org

This article's doi:
10.1146/annurev-neuro-062111-150533

Keywords

olfactory receptor neurons, transduction, synapses, antennal lobe, concentration, lateral inhibition

Abstract

In the olfactory system of *Drosophila* melanogaster, it is relatively straightforward to target in vivo measurements of neural activity to specific processing channels. This, together with the numerical simplicity of the *Drosophila* olfactory system, has produced rapid gains in our understanding of *Drosophila* olfaction. This review summarizes the neurophysiology of the first two layers of this system: the peripheral olfactory receptor neurons and their postsynaptic targets in the antennal lobe. We now understand in some detail the cellular and synaptic mechanisms that shape odor representations in these neurons. Together, these mechanisms imply that interesting neural adaptations to environmental statistics have occurred. These mechanisms also place some fundamental constraints on early sensory processing that pose challenges for higher brain regions. These findings suggest some general principles with broad relevance to early sensory processing in other modalities.

Contents

INTRODUCTION

This review focuses on the physiology of the first stages of the adult olfactory system in *Drosophila melanogaster*. Recent reviews have surveyed the development of this system (Brochtrup & Hummel 2011) as well as that of homologous structures in the larvae (Stocker 2008). These topics are not covered here. The study of olfactory processing in *Drosophila* also owes an enormous debt to the study of olfactory processing in other insects—chiefly locusts, moths, and bees (Martin et al. 2011)—but that literature is not reviewed here for space reasons.

CNS: central nervous system

This review is divided into two major sections corresponding to the first two layers of the olfactory system. Each section begins with general observations of how odors are represented in one of these layers, followed by a discussion of the underlying mechanisms at play in that layer. Next, I have tried to extract some general principles and to relate them to higher olfactory processing and the challenges faced by the organism. Finally, each section closes with a summary of key open questions.

Why the fly? One can perform certain experiments in *Drosophila* that are not currently possible in any other species. In particular, one can easily monitor neural activity from individual neurons corresponding to a targeted olfactory processing channel. These neurons are "identified" in the strongest sense of the word: Not only do they have known (or knowable) connectivity to other neurons in the circuit, but their connectivity and odor responses are also relatively stereotyped across individuals.

A major reason for studying the *Drosophila* olfactory system is its strong similarity to the vertebrate olfactory system. Beyond this, there are also looser analogies between the anatomy of this structure and that of other structures that perform early sensory processing. In particular, there are appealing parallels between olfactory structures and visual processing circuits. Thus, studies of the *Drosophila* olfactory system should yield insight into fundamental principles of sensory processing that have general relevance across sensory modalities (Cleland 2010, Mu et al. 2012, Singer et al. 2009).

It is currently taken for granted that the *Drosophila* central nervous system (CNS) is a useful preparation for systems neurophysiology; however, this viewpoint is relatively recent. Until the past decade or so, the neurophysiology of the *Drosophila* CNS was a black box. This situation changed with the widespread application of the visualized "blow and seal" technique for whole-cell patch-clamp recording (Stuart et al. 1993). Starting in the late 1990s, this technique was applied to the intact larval or embryonic *Drosophila* CNS (Baines & Bate 1998, Choi et al. 2004, Rohrbough & Broadie 2002). The development of genetically encoded fluorescent sensors of neural activity, together with the development of modular transgenic systems for expressing these sensors in the fly, was another revolution (Brand & Perrimon 1993, Miesenböck 2004). The first studies to exploit these fluorescent sensors studied the adult brain in semireduced preparations (Ng et al. 2002, Wang et al. 2003). These studies were soon followed by the first field potential recordings (Nitz et al. 2002) and whole-cell patch-clamp recordings (Wilson et al. 2004) from the adult brain in vivo.

OLFACTORY PROCESSING IN RECEPTOR NEURONS

Anatomical Organization

The fly is unusual in that its olfactory receptor neurons (ORNs) are relatively accessible to in vivo electrophysiological recording. ORNs are housed in the antennae and maxillary palps, which are covered by finger-like protrusions called sensilla. These sensilla contain the dendrites of ORNs, and each sensillum typically houses exactly two ORNs (although some types of sensilla house three or four ORNs). By inserting a tungsten or glass electrode into a sensillum, the spikes of both of its ORNs can be recorded simultaneously, and each spike can typically be attributed unequivocally to one of the two ORNs in that sensillum.

ORNs can be segregated into discrete types on the basis of their odor responses (de Bruyne et al. 1999, 2001; van der Goes van Naters & Carlson 2007; Yao et al. 2005). These types turn out to map rather neatly onto patterns of odorant receptor expression (Benton et al. 2009, Hallem et al. 2004). In total, there are ~50 ORN types, corresponding roughly to the 50–60 odorant receptors expressed in the adult antennae and maxillary palps (Benton et al. 2009; Couto et al. 2005; de Bruyne et al. 1999, 2001; Elmore et al. 2003; Fishilevich et al. 2005; van der Goes van Naters & Carlson 2007; Yao et al. 2005).

ORN: olfactory receptor neuron

Phenomenology of Odor Responses

Several studies have systematically surveyed ORN responses using large and chemically diverse sets of stimuli. These studies have characterized odorant receptors either in their native context (de Bruyne et al. 1999, 2001; Silbering et al. 2011; Yao et al. 2005) or in an expression system that captures most of their native properties (Dobritsa et al. 2003, Hallem & Carlson 2006, Hallem et al. 2004). As a result, the chemical selectivities of almost all ORN types have now been described, which is an enormous asset to the field. In addition, several other studies have surveyed ORN responses by systematically varying stimuli in the time domain (A.J. Kim et al. 2011; Nagel & Wilson 2011; Schuckel & French 2008; Schuckel et al. 2008, 2009). As a group, these studies have revealed some general observations about how stimuli are encoded in *Drosophila* ORNs.

- Most individual ORN types respond to multiple ligands, and most individual ligands activate multiple ORN types. The best ligands for a neuron often do not fall into a single chemical class (de Bruyne et al. 1999, 2001; Hallem & Carlson 2006; Silbering et al. 2011; Yao et al. 2005).
- Individual ORN types can be broadly tuned, narrowly tuned, or in between (Hallem & Carlson 2006).
- ORN firing rates rise with increasing ligand concentration; they have a typical dynamic range of approximately two orders of magnitude in odor concentration. Increasing concentration tends to recruit responses in a larger number of ORN types, and ORNs become more broadly tuned at higher concentrations (Hallem & Carlson 2006).
- ORNs spike even in the absence of ligands. Some ligands are actually inhibitory, meaning they suppress the cell's

spike rate below its spontaneous rate (de Bruyne et al. 1999, 2001; Hallem & Carlson 2006; Nagel & Wilson 2011; Schuckel et al. 2009; Silbering et al. 2011; Yao et al. 2005).

- ORN responses are dynamic. Spike rates peak rapidly and subsequently relax to a tonic level of activity. After odor offset, spike rates are often suppressed below spontaneous rates. The dynamics of these responses depend on ORN type, ligand, and concentration (A.J. Kim et al. 2011; Nagel & Wilson 2011; Schuckel & French 2008; Schuckel et al. 2008, 2009).

The mechanisms underlying these observations are now understood at the molecular and cellular levels, at least to a large degree. The next several sections summarize these mechanisms and some of their proposed functional consequences.

Diverse Receptors, Generic Cells

In general, each *Drosophila* ORN expresses a single odorant receptor gene that specifies the odor tuning of that neuron (Vosshall et al. 2000), although a few types of ORNs express multiple receptors (Abuin et al. 2011, Dobritsa et al. 2003, Goldman et al. 2005). Importantly, swapping receptors between ORNs swaps their odor responses (Hallem et al. 2004). Receptor swap also recapitulates the dynamics of odor responses. Thus, all of the diversity in ORN odor responses is likely due to diversity in ORN odorant receptor expression. In other words, the different ORN types are functionally generic, except that they express different receptors. The only exception to this rule is that some ORNs also have specialized accessory protein machinery needed to traffic the transduction complex to the correct subcellular location (Abuin et al. 2011, Larsson et al. 2004).

Given that the diversity among ORNs can be attributed to diversity in odorant receptor expression, we can understand many of the principles of ORN odor coding as arising from the properties of odorant receptor proteins themselves, namely, the molecular pharmacology of these receptors (Hallem et al. 2004, Nagel & Wilson 2011). In general, each receptor binds multiple ligands, and each ligand binds multiple receptors. Some receptors evidently have high affinity for many ligands, whereas others have high affinity for only a few ligands. At high ligand concentrations, a receptor can be activated by both low- and high-affinity ligands. A receptor is less selective at high concentrations versus low concentrations because high ligand concentrations tend to saturate the receptor.

Receptors for Social Odors

Some of the most selective ORNs respond to social odors. For example, two types of ORNs respond to *cis*-vaccenyl acetate, which is produced exclusively by males (Clyne et al. 1997, Ha & Smith 2006, van der Goes van Naters & Carlson 2007, Xu et al. 2005). Other ORN types respond to other male scents or to scents produced by female virgins (van der Goes van Naters & Carlson 2007). These ORNs have not yet been characterized in detail, in part because most of the chemical constituents of social odors have not yet been identified. Notably, ORN types that respond to social odors are generally inhibited by most other odors, which is unusual. Social odors are interesting to neurobiologists because these odors trigger robust behaviors. Some of the central neurons postsynaptic to these ORNs have unusual properties or patterns of connectivity, suggesting specialization for social odor processing (Chou et al. 2010, Datta et al. 2008, Jefferis et al. 2007, Ruta et al. 2010, Schlief & Wilson 2007). Identifying the chemical constituents of social odors will be an important step in understanding the specialization of central circuits and the roles of social odors and their cognate ORNs in various social behaviors.

Spontaneous Transduction and Odor-Evoked Inhibition

All ORNs fire spontaneously, and each ORN type has a characteristic spontaneous firing rate (de Bruyne et al. 1999, 2001; van der Goes van Naters & Carlson 2007; Yao et al. 2005). Mutating the odorant receptor that an ORN normally expresses diminishes its spontaneous firing rate (Dobritsa et al. 2003, Olsen et al. 2007), implying that spontaneous firing reflects the receptor's tendency to reside in the active state even in the absence of ligand. Different odorant receptors likely have different equilibria between their active and inactive states, which would explain why swapping receptors between ORNs can swap their spontaneous firing rates (Hallem et al. 2004).

In some cases, an odor can inhibit spontaneous spiking. Most odors inhibit at least one ORN type while exciting other types (Hallem & Carlson 2006), meaning no odors are inhibitory per se. If an odorant receptor mediates inhibition in response to a particular ligand in its native ORN, it will also generate an inhibitory response to the same ligand in a different ORN whose native receptor has been removed (Hallem et al. 2004). This result argues that inhibitory responses simply reflect inverse agonism; that is, the ligand stabilizes the inactive state more than it stabilizes the active state and thereby suppresses activation below spontaneous levels (Hallem et al. 2004, Nagel & Wilson 2011). Inhibitory responses can also suppress responses to simultaneously applied excitatory odors (Turner & Ray 2009).

Spontaneous activity in ORNs is puzzling from a functional standpoint because it simply adds noise to the system. Why hasn't the fly evolved odorant receptors that are inactive when unbound? Spontaneous transduction might be useful because it depolarizes the cell's resting potential to near its spike threshold. Alternatively, it might just be difficult to evolve a receptor protein with the requisite specificity and kinetics that is never activated in the absence of a ligand.

Transduction Speed

Current evidence suggests that odorant receptors in *Drosophila* are ligand-gated ion channels, not metabotropic receptors (as they are in vertebrates). This is clear for the so-called IR family of odorant receptors, which bears structural homology to ionotropic glutamate receptors in vertebrates (Abuin et al. 2011, Benton et al. 2009), but the issue is less clear for the OR family of odorant receptors, for which most evidence favors an ionotropic mechanism (Benton et al. 2006, Sato et al. 2008, Smart et al. 2008; but see Wicher et al. 2008, Yao & Carlson 2010).

Although ionotropic transduction should be faster than metabotropic transduction, transduction in *Drosophila* is still slower than the dynamics of the odor stimuli themselves, in part because of the time required for odors to diffuse from the surface of the olfactory organ to the receptor sites. The concentration of an odor near its source can fluctuate steeply at high rates, with substantial power at frequencies >10 Hz (Dekker & Carde 2011, Nagel & Wilson 2011, Schuckel & French 2008). Transduction is slower than the fastest odor fluctuations, so responses to odor plume fluctuations are severely attenuated at frequencies greater than 1–10 Hz, and the cutoff frequency depends on the odor-receptor combination (Nagel & Wilson 2011). The onset and decay rates of transduction depend on both the odor and the receptor, implying that different ligand-receptor combinations produce different rise and decay times for receptor activation.

Adaptation in Transduction

In response to a prolonged and steady odor stimulus, ORN responses peak rapidly, then decay. A prolonged stimulus also reduces responses to subsequent stimuli (de Bruyne et al. 1999). What mechanisms produce adaptation? If an ORN is engineered to simultaneously express two different receptors that are activated independently by different ligands, these receptors can cross-adapt each other. Also, an inhibitory odor response can actually potentiate a subsequent excitatory response, suggesting that spontaneous transduction produces a basal level of adaptation and that the excitatory response has been de-adapted following inhibition of spontaneous transduction (Nagel & Wilson 2011). Together, these results argue that adaptation is mediated by a diffusible factor that accumulates in the cell as a result of transduction. This diffusible factor might be calcium, as odorant receptor activation increases the cytoplasmic calcium concentration (Sato et al. 2008). Consistent with a role for cytoplasmic calcium, adaptation is reduced by mutations in either IP_3 receptors or the TRP channel (Deshpande et al. 2000, Stortkuhl et al. 1999). Adaptation slows transduction onset rates, suggesting that it involves a decrease in ligand binding affinity and/or a decrease in the efficacy of receptor activation (Nagel & Wilson 2011). In functional terms, adaptation in sensory systems is thought to be useful because it allows neurons to use their dynamic range efficiently: neurons decrease their sensitivity when stimuli are strong and increase it when stimuli are weak (Wark et al. 2007).

After odor offset, ORN firing rates are often inhibited below spontaneous rates (de Bruyne et al. 1999, 2001). Both offset inhibition and adaptation increase with odor pulse duration (Nagel & Wilson 2011), suggesting that both processes reflect a common mechanism. Adaptation and offset inhibition may be due to a decrease in the efficacy of receptor activation. Assuming that there is some basal level of receptor activation in the absence of odor, a process that inhibits receptor activation will suppress spontaneous activity. Both adaptation and offset inhibition depend on the identity of the receptor and the identity of the odor (Hallem et al. 2004, Nagel & Wilson 2011). This makes sense because changing the ligand-receptor combination would change the rate constants governing transitions between the active and inactive states of the receptor.

From Transduction to Spiking

In some circumstances, isolated receptor potentials and spikes can be recorded simultaneously from the same ORNs (Nagel & Wilson 2011). These experiments demonstrate that spike rate in ORNs is not simply related to the magnitude of transduction. Rather, it is related to both the magnitude and rate of change of transduction. Spike rates peak when transduction is increasing rapidly, and they can be suppressed below baseline when transduction begins to rapidly decay. As a result, ORN spike rates encode both odor concentration and its rate of change (A.J. Kim et al. 2011). Consistent with theoretical models of spiking behavior, ORN spiking behavior can be altered by manipulating sodium channel expression levels in these neurons (Nagel & Wilson 2011).

Because the spike rate of an ORN depends on the rate of change in transduction, the dynamics of spiking tend to be more complex than the dynamics of transduction (Nagel & Wilson 2011). Nevertheless, the relationship between transduction and spiking is similar across ORN types. This similarity helps explain why swapping odorant receptors is sufficient to swap all of the dynamics of an ORN's response to a ligand: Because the relationship between spiking and transduction is similar, receptor swap recapitulates not only the simpler dynamics of transduction but also the more complex dynamics of spiking.

Some Fundamental Principles

The previous sections have detailed the mechanisms underlying ORN odor responses. What do these mechanisms mean for downstream neurons? The following list of fundamental principles of odor coding in *Drosophila* ORNs places special emphasis on how peripheral mechanisms shape the format of information flowing to higher brain regions. These themes are revisited in the second half of this review, which follows olfactory information into the brain.

ORNs are noisy. On average, a *Drosophila* ORN fires 8 spikes/s in the absence of an odor (de Bruyne et al. 1999, 2001). Because each antenna contains 1,200 ORNs (Stocker et al. 1990), the brain is continuously barraged by ~20,000 ORN spikes/s, even when no odor is present. Moreover, ORN odor responses are also noisy, so ORN noise likely places a fundamental limit on the ability of downstream neurons to detect dilute or transient odor stimuli.

ORNs fire most strongly at odor onset. At the onset of a rapid increase in odor concentration, transduction rises more slowly than odor concentration. As a result, ORN responses are delayed and responses to transient stimuli are attenuated. This should limit downstream neurons' abilities to detect odor rapidly and to detect transient odor filaments. However, because the spike rates of ORNs depend on the rate of change in transduction, not the absolute transduction level, spike rates peak before transduction does. This increases the speed with which rapid odor fluctuations are encoded. As discussed below, a similar process of speeding also occurs downstream.

Most odors are encoded by the combined activity of several ORN types with overlapping receptive fields. Multiple ORN types are generally coactivated by a single stimulus, which has important implications for downstream odor processing. As we shall see, the signals sent by different ORN types can influence each other at the very first stage of olfactory processing in the brain. The recruitment of multiple ORN types is also important because each type is sensitive to concentration over a restricted concentration range. Thus, the organism's ability to resolve concentration differences over a wide range likely depends on the recruitment of multiple receptors with different affinities for the same ligand (Kreher et al. 2008).

ORNs conjointly encode physical features of the stimulus that must be extracted independently. Every ORN odor response depends on (*a*) odor identity, (*b*) odor concentration, and (*c*) the rate of change in odor concentration. In the natural world, all three of these features are constantly changing. Nevertheless, behavioral experiments indicate that odor identity and concentration are encoded independently in the *Drosophila* brain. Flies can be conditioned to avoid an odor irrespective of its concentration and to discriminate between different concentrations of the same odor (Borst 1983, Dudai 1977, Masek & Heisenberg 2008, Yarali et al. 2009). The problem of encoding odor identity and concentration independently creates a challenge for downstream neurons.

ORNs have correlated odor selectivity. A stimulus that evokes a high firing rate in a given ORN type also tends to evoke a high firing rate in many other ORN types. Conversely, a stimulus that elicits unusually weak activity in a given ORN type also tends to evoke weak or little activity in most other ORNs. In other words, there is substantial redundancy in ORN odor representations (Haddad et al. 2010, Luo et al. 2010, Olsen et al. 2010). This too has important implications for downstream odor processing.

Comparisons with Vertebrates

There are many similarities between *Drosophila* and vertebrate ORNs. In vertebrates, most odors are encoded by the combined activity of several ORN types, and increasing the concentration of an odor recruits more ORNs (Reisert & Restrepo 2009). Vertebrate receptors can be narrowly tuned, broadly tuned, or anything in between (Saito et al. 2009). Vertebrate ORNs are noisy and spontaneously active, partly owing to spontaneous transduction in odorant receptors (Reisert 2010). Vertebrate ORNs also preferentially signal the onset of odor responses, owing to adaptation in transduction and spike generation (Reisert & Matthews 2000). As in *Drosophila*, adaptation in vertebrate transduction reflects, at least in part, an apparent reduction in receptor affinity, such that adapted responses resemble responses to a lower ligand concentration (Liu et al. 1994). Finally, vertebrate ORNs resemble *Drosophila* ORNs in that both have correlated odor selectivity (Haddad et al. 2010).

An important difference between vertebrate and *Drosophila* ORNs is the speed of transduction. In vertebrates, the response to a brief pulse of odor (25 ms) requires 400 ms to peak and

Glomerulus: a neuropil compartment where ORN axons form synapses with PN and LN dendrites

1,000 ms to terminate (Bhandawat et al. 2005). By comparison, *Drosophila* ORN responses can peak in 30 ms and terminate in 200 ms (Nagel & Wilson 2011, Schuckel et al. 2009). Speed may be more important for insects because they experience rapidly fluctuating, wind-borne odor filaments, and they can potentially use the information contained in these fluctuations to locate an odor source (Murlis et al. 1992, Silbering & Benton 2010). By contrast, speed is probably less important for vertebrates; terrestrial vertebrates draw air into their noses before it encounters ORNs, a process that likely disperses odor filaments and smoothes fluctuations in concentration (Schoenfeld & Cleland 2005).

Another difference is that vertebrate ORNs are reportedly sensitive to air speed (Mozell et al. 1991, Scott et al. 2006, Sobel & Tank 1993). Terrestrial vertebrates actively control air flow through their noses and thereby use air speed to modulate olfactory transduction (Johnson et al. 2003, Schoenfeld & Cleland 2005). *Drosophila* have comparatively little control over air flow across their olfactory organs, so it may be advantageous that *Drosophila* ORNs are insensitive to air speed (Zhou & Wilson 2012).

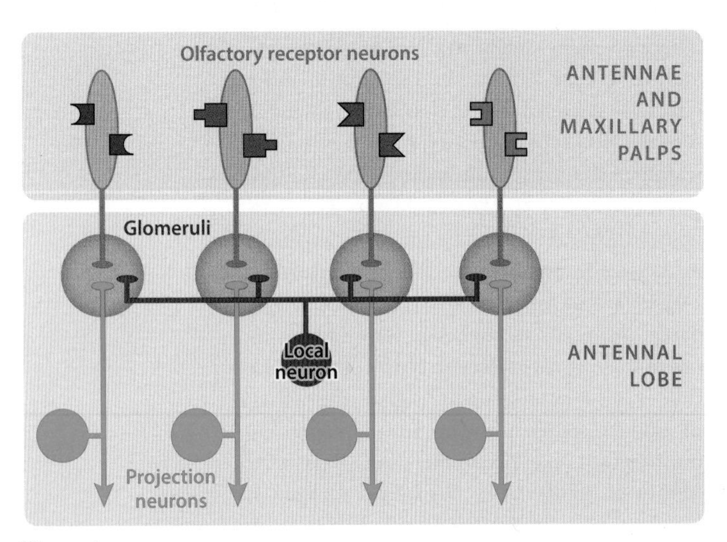

Figure 1

Anatomy of the *Drosophila* olfactory system. Olfactory receptor neuron (ORN) cell bodies and dendrites (*brown*) reside in peripheral olfactory organs. All of the ORNs that express a given odorant receptor converge onto the same glomerulus in the antennal lobe, schematized here as a single ORN per glomerulus. Each projection neuron (PN) (*blue*) sends a dendrite into a single glomerulus (*purple*), where it receives monosynaptic input from ORNs. Although each glomerulus contains the dendrites of several PNs, only one PN for each glomerulus is shown here. Glomeruli are laterally interconnected by a network of local neurons (LNs) (*gray*), which interact with PNs, ORNs, and other LNs. Many individual LNs innervate most or all glomeruli, but some are more selective.

Key Open Questions

Although olfactory processing in ORNs is arguably better understood in *Drosophila* than in any other species, several important questions remain unanswered:

- **Are odorant receptors in the OR family really ligand-gated ion channels?** If so, the structure of these receptors must be unusual, as they are predicted to have seven transmembrane domains (Vosshall et al. 1999). Neither the structure nor the function of these receptors has received much attention from structural biologists or biophysicists.
- **Might ORs also be metabotropic?** It has been suggested that ORs might be both ion channels and G protein–coupled receptors, although the latter pathway might be a minor one. This would reconcile some recent findings (Wicher et al. 2008, Yao & Carlson 2010).
- **What are the mechanisms of transduction adaptation?** Progress on this question depends on a better understanding of transduction mechanisms.
- **What molecules do *Drosophila* use for olfactory social communication?** Also, what receptors and ORN types mediate responses to each of these ligands?

OLFACTORY PROCESSING IN THE ANTENNAL LOBE

Anatomical Organization

The antennal lobe is the first brain region of the fly olfactory system. Thus, it is analogous to the vertebrate olfactory bulb, and like the bulb, it is organized into discrete neuropil compartments, called glomeruli (**Figure 1**). All of the ORNs that express a given odorant receptor converge onto the same glomerulus (Vosshall et al. 2000). There they make excitatory synapses with second-order neurons called projection neurons (PNs). Like mitral cells in the vertebrate olfactory bulb, each antennal lobe PN is postsynaptic to a single glomerulus (Stocker et al. 1990). Each glomerulus contains the dendrites of several PNs, termed sister PNs; these sister PNs have highly correlated patterns of activity (Kazama & Wilson 2009).

Glomeruli are interconnected by a network of local neurons (LNs). LNs lack axons and release the inhibitory neurotransmitter γ-aminobutyric acid (GABA) from their dendrites instead. PNs also release neurotransmitters from their dendrites (Ng et al. 2002, Wilson et al. 2004). Thus, each glomerulus is potentially the site of reciprocal interactions between these three cell types.

Phenomenology of Odor Responses

Most odorant receptors and ORN types have now been matched with their cognate glomeruli in the brain (Couto et al. 2005, Fishilevich & Vosshall 2005, Silbering et al. 2011). Several studies have made systematic comparisons between odor coding in ORNs and their cognate PNs. These studies demonstrate a coarse resemblance between the odor responses of ORNs and their postsynaptic PNs (Bhandawat et al. 2007, Ng et al. 2002, Schlief & Wilson 2007, Silbering et al. 2008, Wang et al. 2003, Wilson et al. 2004). Specifically, ligands that are unusually effective in stimulating an ORN (particularly at low concentrations) also tend to be unusually effective in stimulating postsynaptic PNs. This result is consistent with the idea that ORNs provide the major source of excitation to PNs.

That said, PN odor representations are not identical to ORN odor representations. Specifically, PN and ORN odor responses differ as follows:

- PN responses show less variability in trial-to-trial spike count than do the responses of their presynaptic ORNs to the same stimulus (Bhandawat et al. 2007).
- PN responses generally peak earlier than ORN responses and decay more quickly (Bhandawat et al. 2007, Wilson et al. 2004). This means that PNs respond most vigorously to odor onset.
- In general, PNs are more broadly tuned to odors (i.e., less selective) than their presynaptic ORNs (**Figure 2**) (Bhandawat et al. 2007, Olsen & Wilson 2008, Wilson et al. 2004).
- When only one ORN type is active, and when those ORNs are firing at a low rate, their postsynaptic PNs are disproportionately sensitive to small changes in ORN input. However, when those same ORNs are firing at a high rate, their PNs are less sensitive to small changes in presynaptic input (**Figure 2**). That is, the relationship between ORN and PN activity exhibits a compressive nonlinearity (Olsen et al. 2010).
- The odor responses of a PN can be suppressed by recruiting additional activity in other glomeruli. For example, when mixed with a second odor, an odor that elicits no response in a given PN when presented alone can inhibit that PN's response to the second odor (Olsen et al. 2010, Silbering & Galizia 2007), implying the existence of inhibitory interactions between glomerular processing channels.

The following section summarizes the mechanisms underlying these transformations and the reasons they might be useful to the organism.

PN: antennal lobe projection neuron

LN: antennal lobe local neuron

GABA: γ-aminobutyric acid

Convergence

Why are PN odor responses so sensitive to weak inputs, and why are they so reliable? Part of the answer lies in the convergence of ORNs onto PNs. Each odorant receptor is expressed in multiple ORNs, ranging from ~10 to ~100 ORNs per antenna or palp, depending on the receptor (de Bruyne et al. 2001, Shanbhag et al. 1999). Most individual ORNs project bilaterally (Stocker et al. 1990), and each PN receives input from all of the ORN axons that enter its cognate glomerulus (Kazama & Wilson 2009). Thus, each PN receives convergent bilateral input from all of the ORNs that express a given odorant receptor.

Unitary synaptic event: a synaptic event produced by a spike in a single axon synapsing onto the postsynaptic cell; this axon may form multiple vesicular release sites onto the postsynaptic neuron

The high convergence of ORNs onto PNs helps account for PN sensitivity to weak levels of presynaptic ORN input. It also helps account for why PN responses show less trial-to-trial variability than the responses of their presynaptic ORNs to the same stimulus. Recall that sister ORNs spike independently (Kazama & Wilson 2009), so pooling many ORN inputs should allow for reduced trial-to-trial variability in PN odor responses (Abbott 2008).

Olfactory Receptor Neuron Synapses

The properties of ORN-to-PN synapses also promote reliability. Each ORN spike produces a large, excitatory, unitary synaptic event in a PN (5–7 mV in amplitude; Kazama & Wilson 2008). Each ORN axon forms several dozen synaptic sites onto each postsynaptic PN, and each release site has a high vesicular release probability. Thus, each ORN spike releases several dozen vesicles onto the PN, thereby producing a highly reliable synapse. The strength of these synapses also helps explain why PNs are very sensitive to weak levels of ORN input. ORN-to-PN synapses are cholinergic and are blocked by a nicotinic acetylcholine receptor antagonist (Kazama & Wilson 2008).

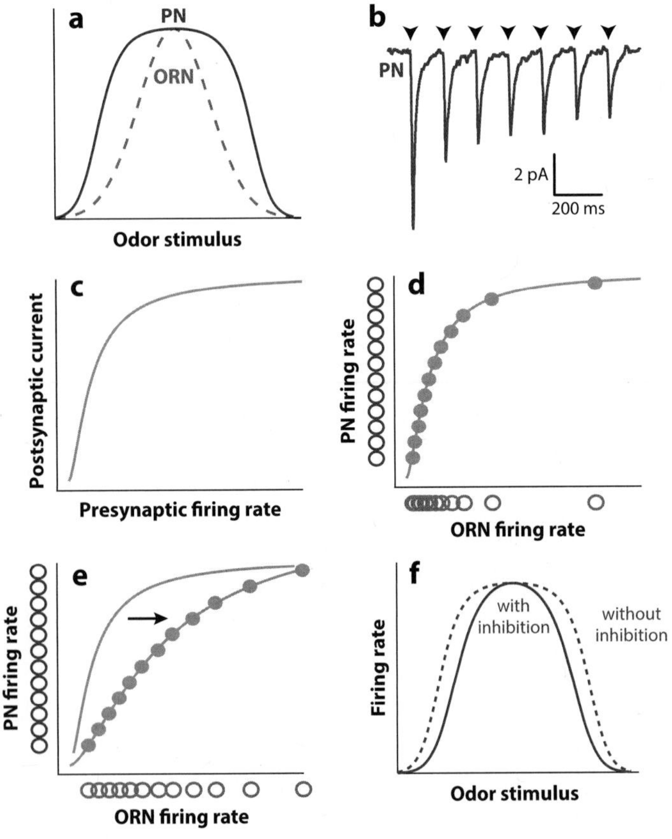

Why are PN responses more transient than ORN responses? This is partly explained by the properties of ORN-to-PN synapses. A high vesicular release probability means that synaptic vesicles should be easily depleted from this synapse. Consistent with this, ORN-to-PN synapses exhibit strong short-term depression (Kazama & Wilson 2008). Short-term synaptic depression should make PN responses more transient, meaning that PNs should respond most strongly at the onset of an odor pulse.

Why does the relationship between PN and ORN exhibit a compressive nonlinearity? Thus, it must reflect either a process at ORN-to-PN synapses or a process intrinsic to PNs. Short-term depression at ORN-to-PN synapses likely explains most of this phenomenon. Short-term synaptic depression suppresses steady-state postsynaptic responses to high presynaptic firing rates. This flattens the peak of a neuron's tuning curve (Abbott et al. 1997), and so this phenomenon can also account for why PNs are more broadly tuned than their presynaptic ORNs. Short-term synaptic depression is not the only mechanism that broadens PN tuning; lateral excitation also contributes (see below). However, contrary to early conjectures (Borst 2007, Wilson et al. 2004), lateral excitation is not strictly necessary to explain the basic phenomenon of broad PN tuning.

Interestingly, most individual ORNs arborize bilaterally (Stocker et al. 1990), which should make it difficult for the fly to lateralize odor stimuli. Nevertheless, odor lateralization behavior can be robust and rapid (Borst & Heisenberg 1982, Duistermars et al. 2009,

Gaudry et al. 2013). This is explained by a small asymmetry in ORN neurotransmitter release properties: The ORN releases ~40% more neurotransmitter per spike from its ipsilateral axon branch than from its contralateral axon branch. As a result, when an odor stimulus is lateralized, the PNs that are ipsilateral to the stimulus spike at slightly higher rates and with a slightly shorter latency than do those that are contralateral to the stimulus (Gaudry et al. 2013).

Projection Neurons

Almost all PNs send a dendritic arbor into a single glomerulus (Jefferis et al. 2001, Stocker et al. 1990), meaning that they receive direct input from a single ORN type. Analysis of a passive compartmental model suggests that approximately three synchronous unitary ORN synaptic inputs should be required to drive a PN from its resting potential to its spike initiation threshold (Gouwens & Wilson 2009). PNs express a variety of voltage-dependent conductances (Gu et al. 2009), but the contribution(s) of these conductances to PN odor responses has not been investigated. PNs spike spontaneously in the absence of odors; this behavior is mainly due to spiking input from ORNs that produces large spontaneous fluctuations in the membrane potential of the postsynaptic PN (Gouwens & Wilson 2009, Kazama & Wilson 2009).

Almost all PNs are cholinergic (Yasuyama & Salvaterra 1999). They release acetylcholine from their axonal arbors in higher brain regions and also from their dendrites in the antennal lobe (Kazama & Wilson 2008, Ng et al. 2002, Wilson et al. 2004, Yaksi & Wilson 2010). Within the antennal lobe, PNs excite other PNs in the same glomerulus; they also excite LNs.

Most individual glomeruli contain several sister PNs (Stocker 1994, Tanaka et al. 2004). Sister PNs carry highly correlated signals—i.e., they have very similar trial-averaged odor responses, particularly when sister PNs are recorded in the same fly. This finding argues that brain-to-brain variability is much larger than stochastic variability in cellular or circuit properties. In addition, sister PNs display correlated noise; i.e., trial-to-trial odor response fluctuations are similar in sister PNs, and they show correlated spiking in the absence of odors. Both correlated signals and correlated noise are consequences of the fact that sister PNs receive input from precisely the same set of ORNs (Kazama & Wilson 2009).

Figure 2

The nonlinear relationship between olfactory receptor neuron (ORN) and projection neuron (PN) firing rates. (*a*) Schematic tuning curves (i.e., plots of firing rate versus stimulus number) for an ORN (*dashed brown curve*) and a PN (*solid blue curve*). Stimuli are arbitrarily ordered so that the strongest responses are in the center of the plot because this ordering makes it easier to visually assess tuning breadth. In this example, the ORN tuning curve is shown as Gaussian, although this may not be typical. The PN tuning curve was created by transforming the ORN tuning curve using a hyperbolic ratio function (like that in panel *d*). Tuning curves are normalized to the same peak. (*b*) A recording from a PN showing synaptic currents elicited by electrical stimulation of a train of spikes in ORN axons (*arrowheads*). The synaptic currents are depressed during the train. Modified from Kazama & Wilson (2008). (*c*) Schematic illustration of how total postsynaptic current increases sublinearly as presynaptic firing rates increase, owing to synaptic depression (as in panel *b*). Modified from Kazama & Wilson (2008). (*d*) Schematic showing the typical relationship between the odor-evoked firing rates of ORNs and PNs in the same glomerulus. Each black symbol represents a different odor stimulus; odor stimuli might be different concentrations of the same chemical or different chemicals. The relationship between ORN and PN firing rates is monotonic (as shown in this schematic) in a situation in which only one ORN type is activated by the odor. The relationship is strongly sublinear, probably due to the sublinear relationship between presynaptic spiking and postsynaptic current. Projecting these points into the x- and y-axes (*brown* and *blue symbols*, respectively) makes it clear that most of the ORN responses cluster near the bottom of the cell's dynamic range; this behavior is typical of ORNs. By contrast, PN responses are more uniformly distributed throughout the cell's dynamic range. Modified from Bhandawat et al. (2007), Olsen et al. (2010). (*e*) Lateral inhibition (*black arrow*) inhibits neurotransmitter release from ORNs and thereby increases the level of ORN input required to drive the PNs to saturation. The magnitude of lateral inhibition is correlated with total ORN activity, as is the activity of each ORN type; thus, a glomerulus tends to receive strong lateral inhibition when its ORN inputs are also strong. The distribution of ORN firing rates in this schematic has been shifted to the right to represent this idea, and this shift means that a shallower curve is needed to make the PN odor responses uniformly distributed within its dynamic range (compare *brown symbols* to those in panel *d*). In this schematic, the magnitude of lateral inhibition is the same for all the odor stimuli; however, in a situation where different stimuli elicit different levels of lateral inhibition, the relationship between ORN and PN activity would not be monotonic. (*f*) Lateral inhibition makes PNs more narrowly tuned than they otherwise would be, although it does not necessarily make PNs more narrowly tuned than ORNs.

Targets of Lateral Inhibition

Lateral inhibition: inhibitory interactions between principal neurons (ORNs or PNs) in different glomeruli; this term does not necessarily imply any particular spatial organization for interglomerular interactions

The net effect of lateral input to a PN is generally inhibitory. This is clear from the fact that a PN's odor responses are typically disinhibited by silencing input to other glomeruli (Asahina et al. 2009, Olsen & Wilson 2008). Conversely, adding new odors to an odor mixture typically produces either sublinear summation or frank suppression (Olsen et al. 2010, Silbering & Galizia 2007). A PN can even be inhibited by a stimulus that actually excites its ORNs (Olsen & Wilson 2008). These sorts of mixture effects can be blocked by a combination of $GABA_A$ and $GABA_B$ receptor antagonists (Olsen et al. 2010, Silbering & Galizia 2007). Together, these results demonstrate the existence of odor-evoked lateral inhibition.

The site of lateral inhibition is predominantly presynaptic, at the ORN axon terminal. This locus of inhibition is implied by the finding that robust lateral inhibition requires active neurotransmitter release from ORN axons. When ORNs are silent, most lateral inhibition disappears (Olsen & Wilson 2008). Moreover, ORN axon terminals show immunoreactivity for GABA receptors (Root et al. 2008), and iontophoretic GABA inhibits ORN-to-PN synaptic transmission at a presynaptic locus (Olsen & Wilson 2008, Root et al. 2008). Similarly, activating LNs with odor stimuli also inhibits ORN-to-PN synaptic currents at a presynaptic locus (Olsen & Wilson 2008).

Although ORNs are perhaps the most functionally important targets of inhibition, PNs also receive synaptic inhibition. Iontophoretic GABA hyperpolarizes PNs via $GABA_A$ and $GABA_B$ receptors (Wilson & Laurent 2005). In paired recordings from GABAergic LNs and PNs, injecting depolarizing current into the LN produces a train of spikes in the LN and weak hyperpolarization of the PN (Yaksi & Wilson 2010). Interestingly, clear unitary synaptic connections are never observed in these paired recordings. Rather, a train of spikes in the LN is always required to see any measurable PN response in single trials, and the PN response grows slowly throughout the train. This suggests these connections might represent volume transmission rather than true synapses.

LNs themselves are also likely targets of inhibition. LNs are hyperpolarized by iontophoretic GABA (Wilson & Laurent 2005), and paired recordings from LN-LN pairs reveal inhibitory connections (Huang et al. 2010, Yaksi & Wilson 2010). Like LN-to-PN connections, these connections seem to be weak and slow.

Selectivity of Lateral Inhibition

In general, the overall level of inhibition in the antennal lobe rises with increasing stimulus intensity (Olsen et al. 2010, Silbering & Galizia 2007, Silbering et al. 2008). But how does the spatial pattern of inhibition depend on the odor? One study addressed this question by measuring GABA release in different glomeruli using a fluorescent sensor of vesicular release that was expressed specifically in LNs (Ng et al. 2002). That study found that the stimulus dictated the identity of the glomerulus with the largest fractional fluorescence change. For example, banana odor produced a substantial increase in fluorescence in glomerulus VA3 but hardly any change in glomerulus D; conversely, apple odor produced a fluorescence increase in glomerulus D but very little change in fluorescence in VA3. These results imply that the spatial pattern of GABA release depends on the stimulus, thereby suggesting a model where specific subsets of glomeruli are linked by inhibitory subnetworks and ORN input to a glomerulus recruits LN input to a specific subset of other glomeruli (**Figure 3**).

An alternative approach is to compare ORN and PN responses to many stimuli and to ask what determines a PN's sensitivity to its ORN inputs. Using this approach, one study found that a PN's sensitivity to its ORN inputs could be predicted on the basis of total ORN activity alone; that is, the identity of the active ORNs did not matter. Indeed, PN odor responses could be predicted with high accuracy on the basis of only two factors: the firing rate of the PN's cognate ORNs and the total firing rate

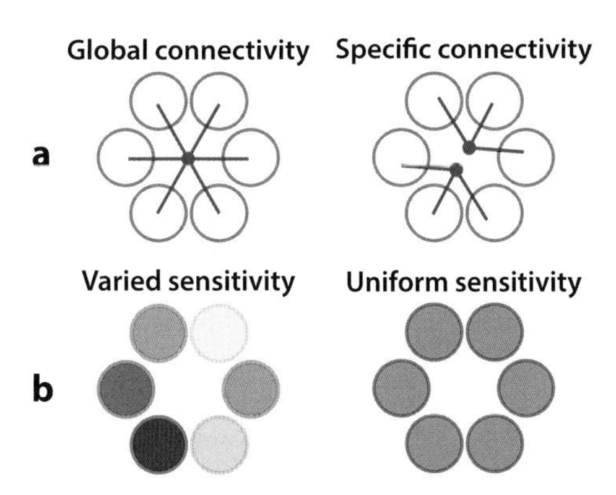

Figure 3

Possible components of specificity in lateral inhibition. (*a*) All glomeruli may be mutually interconnected, as implied by the finding that many lateral neurons (LNs) innervate most or all glomeruli. Alternatively, some glomeruli might be interconnected in specific subnetworks. These subnetworks might be created by LNs with sparse innervation patterns or by electrical compartmentalization within the arbors of broadly innervating LNs. (*b*) Glomeruli may have varied sensitivity to LN activity, possibly reflecting heterogeneous levels of GABA receptor expression or heterogeneous release properties of LN arbors. Alternatively, all glomeruli might have similar levels of sensitivity to LN activity. Note that spatial inhomogeneity would create glomerulus-specific levels of inhibition, but these spatial patterns may or may not be odor specific.

of the entire ORN population (Olsen et al. 2010). This finding suggests a model whereby inhibition is global, meaning all glomeruli inhibit each other (**Figure 3**). However, this approach is indirect, and it is impossible to exclude the idea of spatially specific inhibition using this method.

LN anatomy is consistent with either global or specific inhibition. Some LNs innervate a relatively small subset of glomeruli and could therefore permit specific interactions between glomeruli. However, most individual LNs innervate most or all glomeruli (Chou et al. 2010, Das et al. 2008, Lai et al. 2008, Okada et al. 2009, Seki et al. 2010, Shang et al. 2007, Stocker et al. 1990, Wilson & Laurent 2005). Overall, highly specific LNs represent a small fraction of all LNs. Based on the largest data set available, the individual LNs that innervate fewer than half of all glomeruli represent only 11% of all LNs (Chou et al. 2010). Most LNs are broadly tuned to odors; such tuning is consistent with broad connectivity (Chou et al. 2010, Wilson & Laurent 2005).

Notably, odor invariance in the spatial pattern of inhibition does not necessarily imply that all glomeruli receive the same level of inhibition. In principle, at least, the pattern of inhibition may be not only odor invariant but also spatially inhomogeneous (**Figure 3**). Indeed, there is evidence that the spatial pattern of inhibition may vary across glomeruli (whether or not this pattern is odor invariant). For example, a comparison of mixture suppression in two glomeruli showed that one glomerulus was systematically more sensitive to suppression than the other, although both were suppressed by the same component of the odor mix, and the suppressive component was known to act laterally (Olsen et al. 2010). The mechanistic basis for this observation is not clear. Some glomeruli are avoided by a subpopulation of LNs; this avoidance could produce unusually low levels of GABA release in those glomeruli (Chou et al. 2010, Okada et al. 2009, Seki et al. 2010). In addition, some glomeruli also have relatively low levels of GABA receptor expression (Root et al. 2008).

In summary, this important topic appears to remain an active area of debate. There are two distinct issues at hand: (*a*) whether inhibition is odor-selective and (*b*) whether sensitivity to inhibition is heterogeneous across glomeruli. Future progress on both issues will likely depend on using improved optical sensors, more selective odor stimuli, and more direct methods of measuring functional inhibition.

Functional Consequences of Lateral Inhibition

One functional consequence of inhibition is that it makes PNs less sensitive to their ORN inputs. When inhibition is absent, and when ORNs are firing at a low rate, PNs are very sensitive to small changes in ORN firing rates.

eLN: excitatory antennal lobe local neuron

In the absence of inhibition, PNs saturate only when their presynaptic ORNs fire at a high rate (**Figure 2**). In the presence of inhibition, however, PNs can be much less sensitive to small changes in ORN input, meaning that inhibition increases the ORN firing rate that is needed to drive PN firing rates to saturation (**Figure 2**). Thus, lateral inhibition allows PNs to encode changes in concentration over a broader range of concentrations.

Another functional consequence of lateral inhibition is that PN responses become more transient (Olsen et al. 2010). Because the major locus of inhibition is presynaptic rather than postsynaptic, the increased transience of PN responses is probably not dependent on any changes in the time constant of the postsynaptic membrane. Rather, it may reflect the fact that excitation is monosynaptic (ORN-to-PN), whereas the minimal pathway for inhibition is multisynaptic. As a result, inhibition is likely to be recruited later than excitation, and inhibition would have the largest effect on the later part of the PN response.

A final functional proposed consequence of inhibition is that it coordinates synchronous oscillations among PNs. Under certain conditions, odor stimuli can entrain PNs to fire oscillatory bursts of spikes. The power of these oscillations is reduced by reducing neurotransmitter release from a specific class of LNs (Tanaka et al. 2009). Oscillatory synchrony is less prominent in the *Drosophila* olfactory system than in the olfactory systems of other insects (Turner et al. 2007) and is thought to make a smaller contribution to olfactory processing in *Drosophila* than in other insects (Tanaka et al. 2009).

Lateral Excitation

Odor-induced depolarization of ORNs and PNs in a glomerulus tends to suppress activity in other glomeruli via GABAergic LNs. At the same time, however, this depolarization also tends to boost activity in other glomeruli via excitatory LNs. Thus, activity in one glomerulus elicits both excitation and inhibition in other glomeruli. This is one of the more intriguing and mysterious aspects of antennal lobe processing.

Several groups of investigators discovered lateral excitation in the *Drosophila* antennal lobe simultaneously. The basic experiment was simple: stimulate the fly with odors while recording signals from PNs directly postsynaptic to silent ORNs (ORNs silenced using either genetic tools or microdissections). As it turns out, little to no lateral inhibition was observed in these PNs, probably because the main target of lateral inhibition is the ORN axon terminal, and there is nothing to inhibit if the ORNs are essentially silent. Instead, under these conditions, an odor stimulus excites the PNs postsynaptic to the mutant ORNs, implying the existence of lateral excitation (Olsen et al. 2007, Root et al. 2007, Shang et al. 2007).

Which LNs might mediate lateral excitation? Odor-evoked lateral excitation is not blocked by GABA receptor antagonists, so it cannot be an excitatory effect of GABA. Because a minority of LNs are cholinergic, these neurons seemed like attractive candidates (Shang et al. 2007). Indeed, PNs are depolarized when cholinergic LNs are directly excited (using an optogenetic stimulus or current injection via a patch pipette). Thus, these LNs were dubbed excitatory LNs (eLNs). However, PN responses to eLNs are essentially unaffected by pharmacological blockade of nicotinic acetylcholine receptors or voltage-dependent calcium channels. Also, eLN-to-PN connections transmit both hyperpolarizing and depolarizing voltage steps (Huang et al. 2010, Yaksi & Wilson 2010). Moreover, these connections are abolished by a mutation in a gap junction subunit, and the same mutation abolishes odor-evoked lateral excitation (Yaksi & Wilson 2010), implying that lateral excitation is attributable to electrical connections formed by eLNs onto PNs. Thus, although eLNs are cholinergic, they evidently do not release acetylcholine onto PNs. eLNs themselves receive cholinergic excitation from both ORNs and PNs (Huang et al. 2010, Yaksi & Wilson 2010). Electrical connections should be fast, which helps explain why lateral excitation to a PN lags direct excitation from ORNs by less than 2 ms (Kazama & Wilson 2008).

The major source of excitatory drive to a PN is the powerful cholinergic input it receives from its cognate ORNs. That said, the contribution of eLNs to PN odor responses is not negligible. Using a gap junction mutation to remove the contribution of the eLN-PN network modestly but significantly diminishes the strength of some PN odor responses (Yaksi & Wilson 2010). Less intuitively, the same mutation actually potentiates some PN odor responses, probably because eLNs can excite GABAergic LNs, thereby recruiting PN inhibition. Consistent with this idea, after application of GABA receptor antagonists, odor responses that are potentiated by the mutation are less disinhibited. The net effect of the eLN network on a PN—either excitation or inhibition—appears to depend on both the glomerulus and the odor stimulus. Overall, the functional consequences of the eLN network are poorly understood.

Some LNs are glutamatergic (Chou et al. 2010, Das et al. 2011), and some researchers have suggested that these LNs mediate lateral excitation. However, given that lateral excitation is blocked by a gap junction mutation, it seems unlikely that glutamate plays a key role in its mediation. The synaptic actions of glutamate in the *Drosophila* brain are unknown, although they are probably widespread (Daniels et al. 2008).

Long-Term Plasticity

A variety of sensory stimuli and experimental manipulations can produce persistent changes in the output of the antennal lobe. These modulations fall into two categories: (*a*) local forms of plasticity that tend to compensate for altered overall levels of neural activity and (*b*) top-down forms of plasticity that tend to adjust the salience of sensory cues on the basis of behavioral state.

Persistent local modulations can be viewed as forms of adaptation over long timescales. These modulations are generally compensatory, meaning that they at least partially counteract changes in the overall level of neural activity. For example, rearing flies in a high concentration of carbon dioxide produces a persistent suppression of PN responses to this odor. This suppression reflects a selective increase in the density of GABAergic LN innervation in the glomerulus where these PNs reside (Sachse et al. 2007). Conversely, chronic removal of some ORN types leads to the gradual recovery of odor responsiveness in deafferented PNs, reflecting an upregulation of lateral excitation mediated by an increase in the strength of the electrical connections between eLNs and PNs (Kazama et al. 2011). Finally, decreasing PN excitability by overexpressing a potassium channel produces a compensatory increase in the strength of ORN-to-PN synaptic currents in the affected PNs. This compensatory behavior may represent a natural homeostatic mechanism for coping with the systematic differences across glomeruli in PN input resistance (Kazama & Wilson 2008). All of these phenomena appear to be local to the antennal lobe.

Persistent top-down modulations can result from changes in behavioral state. For example, hunger potentiates PN odor responses: Falling levels of circulating insulin lead to upregulation of an autocrine neuropeptide signaling pathway in ORNs, which in turn produces increased ORN neurotransmitter release. Interfering with this signaling cascade reduces searching behavior in hungry flies (Root et al. 2011). Top-down plasticity can also result from classical conditioning. Pairing an odor with an aversive electric shock to the fly's abdomen causes an increase in the odor-evoked activity of some PNs (Yu et al. 2004), but the mechanism that underlies this phenomenon is still unknown.

Some Fundamental Principles

What follows is a short list of fundamental principles of olfactory processing in the *Drosophila* antennal lobe. Creating such a list is necessarily a selective and somewhat speculative exercise; the following focuses on the relevance of olfactory processing in this circuit for downstream neurons and for the organism as a whole.

Each glomerulus pools many inputs from neurons with essentially identical odor tuning. All of the ORNs that express the same odorant receptor wire precisely to the same PNs. Why would it be useful to segregate each ORN type into a different glomerulus? Recall that even when no odor is present, ORNs as a population continuously barrage the brain with ~20,000 ORN spikes/s. If ORNs wired randomly to PNs, then the task of detecting (for example) 10 odor-evoked spikes in this barrage would seem hopeless. But, if all 10 spikes were fired nearly synchronously by ORNs that were presynaptic to the same glomerulus, then they would likely summate effectively enough to drive a PN above its spike threshold. Thus, the orderly wiring of the olfactory system represents a computational machine par excellence: an extreme and illustrative example of what has been proposed to be a generally useful strategy for organizing neural connectivity (Abbott 2008).

PNs respond most strongly at the onset of ORN spiking. Two mechanisms cause this behavior: lateral inhibition and synaptic depression at ORN-to-PN synapses. This response profile is functionally important because it predicts that PNs should respond better to fluctuating inputs than to sustained inputs. Moreover, this behavior should also speed olfactory processing. Because natural odor plumes produce large fluctuations in odor concentration (Murlis et al. 1992), onset-oriented PN responses may be an adaptation to the natural distribution of odors in the environment as well as a selective pressure for speed in olfactory behaviors. Indeed, olfactory behaviors in *Drosophila* can be observed within 100 ms of the onset of ORN activity (Bhandawat et al. 2010, Gaudry et al. 2013). Recall that ORNs spike most strongly at the onset of transduction. Here, we see that PNs spike most strongly at the onset of ORN spiking; thus, there is an iterative process of response speeding. This phenomenon is analogous to what occurs in the vertebrate retina (Field et al. 2005), where there is a similar process of response speeding that promotes rapid visual perception despite the slow dynamics of visual transduction.

PNs are most sensitive to low ORN firing rates. When sister ORNs are firing at a low rate, small increases in their firing rate cause relatively large increases in the firing rates of their postsynaptic PNs (Olsen et al. 2010). When ORNs are firing at high rates, they tend to saturate their postsynaptic PNs. Consequently, odor stimuli that elicit low ORN firing rates occupy the lion's share of a PN's dynamic range (Bhandawat et al. 2007). Because most odor-evoked ORN firing rates are low (<50 spikes/s) compared with the maximum ORN firing rate (~300 spikes/s) (Hallem & Carlson 2006), most of a PN's dynamic range may be devoted to the most common odor stimuli (**Figure 2**). Thus, this property of PN tuning should maximize rates of information transmission (termed histogram equalization; Laughlin 1981). In simulations, the compressive nonlinearity in the relationship between ORN and PN firing rates substantially improves odor discrimination by a linear encoder (Luo et al. 2010, Olsen et al. 2010).

Lateral inhibition adjusts PN sensitivity to the level of total ORN activity. LNs collectively pool input from all glomeruli, and they inhibit ORN neurotransmitter release as ORN activity increases. This behavior makes PNs less sensitive to the firing rates of their cognate ORNs (Olsen et al. 2010, Olsen & Wilson 2008, Root et al. 2008). As a consequence of inhibition, PN firing rates do not saturate as easily as they would otherwise, and their dynamic range becomes more closely matched to that of their inputs (**Figure 2**). In simulations, this type of lateral inhibition substantially improves odor discrimination by a linear decoder. In particular, it improves a decoder's ability to identify an odor in a concentration-invariant manner (Luo et al. 2010, Olsen et al. 2010), implying that lateral inhibition may help flies identify odors in spite of natural variations in odor concentration. Lateral inhibition also decorrelates the activity of different PNs.

Indeed, the need for lateral inhibition can be seen as a consequence of the highly correlated activity of different ORN types; highly correlated ORN activity could easily lead to network saturation at high firing rates in the absence of gain control (Haddad et al. 2010, Luo et al. 2010, Olsen et al. 2010). The computation implemented by this type of lateral inhibition has been called divisive normalization, and it appears to play a role in a wide variety of sensory systems (Carandini & Heeger 2012).

Comparisons with Vertebrates

The most widely noted similarity between *Drosophila* and vertebrate olfaction is that in both cases, each glomerulus pools many inputs with essentially identical odor tuning (Bargmann 2006, Su et al. 2009). The similarity between the glomerular organization of the vertebrate and *Drosophila* olfactory systems is a spectacular case of evolution hitting upon the same solution to a general problem (Eisthen 2002).

However, there are other parallels as well. For example, the properties of neurotransmitter release from ORN axon terminals are similar in *Drosophila* and vertebrates. Specifically, the probability of vesicular release from ORN axon terminals is unusually high, and synapses are strongly depressed at high presynaptic firing rates (Kazama & Wilson 2008, Murphy et al. 2004).

Another parallel is the relationship between presynaptic and postsynaptic odor-evoked firing rates within a glomerulus. ORN and mitral cell responses have been compared systematically in only one study, which focused on a single, gene-targeted glomerulus in the mouse. That study found that the firing rates of mitral cells saturate at lower odor concentrations than do those of their cognate ORNs (Tan et al. 2010). Thus, when these mitral cell firing rates are plotted against the firing rates of their presynaptic ORNs, one should see a compressive nonlinearity (**Figure 2**). This finding is exactly analogous to the situation in *Drosophila* (Olsen et al. 2010) in that it implies that, like *Drosophila* PNs, mitral cells may be more broadly tuned than their presynaptic ORNs (**Figure 2**).

Similar to the synaptic coupling of *Drosophila* PNs (Kazama & Wilson 2009), sister mitral cells are reciprocally coupled by electrochemical synapses (Christie & Westbrook 2006). However, in *Drosophila*, sister PNs have similar trial-averaged odor responses, especially when recordings are conducted in the same brain. This similarity extends to spike timing at the millisecond timescale (Kazama & Wilson 2009). Sister mitral cells in the mouse olfactory bulb are not as similar in this way: Although odor-evoked changes in their firing rates are highly correlated, spike rate modulations in sister mitral cells occur at different times within the respiration cycle (Dhawale et al. 2010). These differences in modulation timing could be due in part to differences in intrinsic properties among sister neurons (Padmanabhan & Urban 2010). When odor stimuli are not fluctuating rapidly, spike timing can become an additional dimension for encoding odor identity (Laurent 2002). Thus, diversity among sister mitral cells could expand the available coding space.

The selectivity of lateral inhibition is a major open question in vertebrates, just as it is in *Drosophila*. Although two studies have proposed the existence of highly sparse and specific interactions among olfactory bulb glomeruli (Fantana et al. 2008, D.H. Kim et al. 2011), the evidence for this phenomenon was relatively indirect. Adjacent glomeruli in the olfactory bulb can have very different odor tuning (Soucy et al. 2009), so even a small region of local connectivity could produce relatively nonselective inhibition.

Key Open Questions

Although we understand some of the fundamental principles of olfactory processing in the *Drosophila* antennal lobe, some key questions remain unanswered:

How do PNs encode rapidly fluctuating stimuli? In a natural turbulent plume, odor concentration can fluctuate rapidly (Murlis et al. 1992). No studies have examined how these sorts of stimuli are encoded at the PN level. Given that inhibition lags excitation, it is unclear whether inhibition is recruited by transient odor encounters.

Do glomeruli perform specialized computations? There are characteristic differences between identifiable glomeruli that are stereotyped across flies. For example, the number of ORNs in a given glomerulus varies across glomeruli by a factor of about four (de Bruyne et al. 2001, Shanbhag et al. 1999) and is correlated with glomerular size (Dekker et al. 2006, Kazama & Wilson 2008). Also, the number of LNs varies across glomeruli by factor of about five (Chou et al. 2010). Glomeruli differ in their levels of neuropeptide and neurotransmitter receptor expression; they may also differ in sensitivity to neurotransmitters (Nassel et al. 2008; Root et al. 2008, 2011). Finally, there are variations in the strength of lateral inhibition and lateral excitation (Olsen et al. 2007, 2010; Yaksi & Wilson 2010) as well as variations in the intrinsic properties of PNs (Kazama & Wilson 2008). These variations among glomeruli raise the following questions: Are these variables correlated or independent? Do they represent adaptations to the odors that are processed by each ORN type (Martin et al. 2011)? Are there specialized adaptations for processing social odors?

Why are LNs so diverse? Different LNs can target different portions of a glomerular compartment and can form either dense or sparse arbors within that compartment (Chou et al. 2010, Sachse et al. 2007, Seki et al. 2010). LNs also have diverse intrinsic electrophysiological (Chou et al. 2010, Seki et al. 2010) and neurochemical properties (Carlsson et al. 2010; Ignell et al. 2009; Winther et al. 2003, 2006). Do different types of LNs play different functional roles? Some evidence supports this idea (Sachse et al. 2007, Tanaka et al. 2009), but the number of characterized LN types seems to be outrunning the conceivable number of distinct functions of local interneurons. Before this idea can be tested, we need better tools for directing transgenic expression to specific LN types.

What is the function of excitatory LNs? Are these neurons actually important for boosting sensitivity near the absolute threshold for odor detection? Do they play an important role in recruiting GABAergic LNs, and if so, why (given that ORNs and PNs also provide excitatory input to GABAergic LNs)? Better genetic tools for mapping and manipulating electrical connections would help address these questions.

How is olfactory processing in the antennal lobe modulated by changes in the behavioral state of the organism? In particular, the effects of biogenic amines on antennal lobe physiology are largely uncharacterized. Serotonin reportedly inhibits ORN axon terminals while increasing PN odor responses (Dacks et al. 2009), but the mechanism of this effect is not known.

What dictates the innate hedonic valence of a particular pattern of PN activity? There is evidence that certain glomeruli are innately associated with a fixed hedonic weight. When activated individually, one glomerulus can be sufficient to elicit aversion (Suh et al. 2007), whereas the activation of a different individual glomerulus can be sufficient to elicit attraction (Semmelhack & Wang 2009). Odors that activate multiple glomeruli elicit a behavior that can be accounted for by summing the weights associated with each glomerulus (Semmelhack & Wang 2009). This summation would predict that coactivating two attractive glomeruli would always produce attraction, never aversion. Is this true? Notably, many odors are attractive at low concentrations but aversive at higher concentrations (Schlief & Wilson 2007, Wang et al. 2001). This observation can be reconciled with the sum-of-weights model, but only if the receptors for aversive glomeruli are systematically recruited only at high odor concentrations, implying low ligand-receptor affinities.

FUTURE DIRECTIONS

Ultimately, sensory systems neurophysiology has succeeded when it can account for the precision of the organism's behavioral responses (Parker & Newsome 1998). Thus, the field must develop better ways of measuring the precision of olfactory perception in *Drosophila*. What is the most dilute or transient odor stimulus that the fly can detect? What are the fastest fluctuations in odor concentration that the fly can resolve? What are the most chemically similar mixtures that the fly can discriminate? Some of the most exciting recent studies of olfaction have revealed surprising levels of behavioral performance in mammals (Smear et al. 2011, Uchida & Mainen 2003). Similar studies in the fly would be extremely useful in defining what *Drosophila* olfactory neurophysiology needs to account for.

A second important task for the field is to define the natural statistics of odors. A general principle of sensory neurophysiology is that neurons and circuits are adapted to maximize the rate of information transmission under stimulus conditions that are typical for a given organism (Simoncelli 2003, Wark et al. 2007). It is therefore important to define what constitutes a typical (or "natural") olfactory stimulus. More precisely, we would like to know the statistical distribution of olfactory stimulus parameters. What odors and what odor concentrations are typical of the natural environment? What odors naturally occur together? What are the natural temporal patterns of odor fluctuation in turbulent plumes? Answers to these questions will help us define the olfactory scenes that the nervous system might be adapted to encode.

DISCLOSURE STATEMENT

The author is not aware of any affiliations, memberships, funding, or financial holdings that might be perceived as affecting the objectivity of this review.

ACKNOWLEDGMENTS

Mehmet Fisek, Elizabeth J. Hong, Katherine I. Nagel, and Andreas Liu provided critical comments on the manuscript. Work in the author's laboratory is funded by the National Institutes of Health (R01DC008174) and a Howard Hughes Medical Institute Early Career Scientist Award.

LITERATURE CITED

Abbott LF. 2008. Theoretical neuroscience rising. *Neuron* 60:489–95

Abbott LF, Varela JA, Sen K, Nelson SB. 1997. Synaptic depression and cortical gain control. *Science* 275:220–24

Abuin L, Bargeton B, Ulbrich MH, Isacoff EY, Kellenberger S, Benton R. 2011. Functional architecture of olfactory ionotropic glutamate receptors. *Neuron* 69:44–60

Asahina K, Louis M, Piccinotti S, Vosshall LB. 2009. A circuit supporting concentration-invariant odor perception in *Drosophila*. *J. Biol.* 8:9

Baines RA, Bate M. 1998. Electrophysiological development of central neurons in the *Drosophila* embryo. *J. Neurosci.* 18:4673–83

Bargmann CI. 2006. Comparative chemosensation from receptors to ecology. *Nature* 444:295–301

Benton R, Sachse S, Michnick SW, Vosshall LB. 2006. Atypical membrane topology and heteromeric function of *Drosophila* odorant receptors in vivo. *PLoS Biol.* 4:e20

Benton R, Vannice KS, Gomez-Diaz C, Vosshall LB. 2009. Variant ionotropic glutamate receptors as chemosensory receptors in *Drosophila*. *Cell* 136:149–62

Bhandawat V, Maimon G, Dickinson MH, Wilson RI. 2010. Olfactory modulation of flight in *Drosophila* is sensitive, selective and rapid. *J. Exp. Biol.* 213:3625–35

Bhandawat V, Olsen SR, Schlief ML, Gouwens NW, Wilson RI. 2007. Sensory processing in the *Drosophila* antennal lobe increases the reliability and separability of ensemble odor representations. *Nat. Neurosci.* 10:1474–82

Bhandawat V, Reisert J, Yau KW. 2005. Elementary response of olfactory receptor neurons to odorants. *Science* 308:1931–34

Borst A. 1983. Computation of olfactory signals in *Drosophila melanogaster*. *J. Comp. Physiol. A* 152:373–83

Borst A. 2007. The broader, the better? *Drosophila* olfactory interneurons are found to respond to a wider range of odorants than their immediate sensory input. *Neuron* 54:6–8

Borst A, Heisenberg M. 1982. Osmotropotaxis in *Drosophila melanogaster*. *J. Comp. Physiol. A* 147:479–84

Brand AH, Perrimon N. 1993. Targeted gene expression as a means of altering cell fates and generating dominant phenotypes. *Development* 118:401–15

Brochtrup A, Hummel T. 2011. Olfactory map formation in the *Drosophila* brain: genetic specificity and neuronal variability. *Curr. Opin. Neurobiol.* 21:85–92

Carandini M, Heeger DJ. 2012. Normalization as a canonical neural computation. *Nat. Rev. Neurosci.* 13:51–62

Carlsson MA, Diesner M, Schachtner J, Nassel DR. 2010. Multiple neuropeptides in the *Drosophila* antennal lobe suggest complex modulatory circuits. *J. Comp. Neurol.* 518:3359–80

Choi JC, Park D, Griffith LC. 2004. Electrophysiological and morphological characterization of identified motor neurons in the *Drosophila* third instar larva central nervous system. *J. Neurophysiol.* 91:2353–65

Chou YH, Spletter ML, Yaksi E, Leong JC, Wilson RI, Luo L. 2010. Diversity and wiring variability of olfactory local interneurons in the *Drosophila* antennal lobe. *Nat. Neurosci.* 13:439–49

Christie JM, Westbrook GL. 2006. Lateral excitation within the olfactory bulb. *J. Neurosci.* 26:2269–77

Cleland TA. 2010. Early transformations in odor representation. *Trends Neurosci.* 33:130–39

Clyne P, Grant A, O'Connell R, Carlson JR. 1997. Odorant response of individual sensilla on the *Drosophila* antenna. *Invertebr. Neurosci.* 3:127–35

Couto A, Alenius M, Dickson BJ. 2005. Molecular, anatomical, and functional organization of the *Drosophila* olfactory system. *Curr. Biol.* 15:1535–47

Dacks AM, Green DS, Root CM, Nighorn AJ, Wang JW. 2009. Serotonin modulates olfactory processing in the antennal lobe of *Drosophila*. *J. Neurogenet.* 23:366–77

Daniels RW, Gelfand MV, Collins CA, DiAntonio A. 2008. Visualizing glutamatergic cell bodies and synapses in *Drosophila* larval and adult CNS. *J. Comp. Neurol.* 508:131–52

Das A, Chiang A, Davla S, Priya R, Reichert H, et al. 2011. Identification and analysis of a glutamatergic local interneuron lineage in the adult *Drosophila* olfactory system. *Neural Syst. Circuits* 1:4

Das A, Sen S, Lichtneckert R, Okada R, Ito K, et al. 2008. *Drosophila* olfactory local interneurons and projection neurons derive from a common neuroblast lineage specified by the *empty spiracles* gene. *Neural Dev.* 3:33

Datta SR, Vasconcelos ML, Ruta V, Luo S, Wong A, et al. 2008. The *Drosophila* pheromone cVA activates a sexually dimorphic neural circuit. *Nature* 452:473–77

de Bruyne M, Clyne PJ, Carlson JR. 1999. Odor coding in a model olfactory organ: the *Drosophila* maxillary palp. *J. Neurosci.* 19:4520–32

de Bruyne M, Foster K, Carlson JR. 2001. Odor coding in the *Drosophila* antenna. *Neuron* 30:537–52

Dekker T, Carde RT. 2011. Moment-to-moment flight manoeuvres of the female yellow fever mosquito (*Aedes aegypti* L.) in response to plumes of carbon dioxide and human skin odour. *J. Exp. Biol.* 214:3480–94

Dekker T, Ibba I, Siju KP, Stensmyr MC, Hansson BS. 2006. Olfactory shifts parallel superspecialism for toxic fruit in *Drosophila melanogaster* sibling, *D. sechellia*. *Curr. Biol.* 16:101–9

Deshpande M, Venkatesh K, Rodrigues V, Hasan G. 2000. The inositol 1,4,5-trisphosphate receptor is required for maintenance of olfactory adaptation in *Drosophila* antennae. *J. Neurobiol.* 43:282–88

Dhawale AK, Hagiwara A, Bhalla US, Murthy VN, Albeanu DF. 2010. Non-redundant odor coding by sister mitral cells revealed by light addressable glomeruli in the mouse. *Nat. Neurosci.* 13:1404–12

Dobritsa AA, van der Goes van Naters W, Warr CG, Steinbrecht RA, Carlson JR. 2003. Integrating the molecular and cellular basis of odor coding in the *Drosophila* antenna. *Neuron* 37:827–41

Dudai Y. 1977. Properties of learning and memory in *Drosophila melanogaster*. *J. Comp. Physiol. A* 114:69–89
Duistermars BJ, Chow DM, Frye MA. 2009. Flies require bilateral sensory input to track odor gradients in flight. *Curr. Biol.* 19:1301–7
Eisthen HL. 2002. Why are olfactory systems of different animals so similar? *Brain Behav. Evol.* 59:273–93
Elmore T, Ignell R, Carlson JR, Smith DP. 2003. Targeted mutation of a *Drosophila* odor receptor defines receptor requirement in a novel class of sensillum. *J. Neurosci.* 23:9906–12
Fantana AL, Soucy ER, Meister M. 2008. Rat olfactory bulb mitral cells receive sparse glomerular inputs. *Neuron* 59:802–14
Field GD, Sampath AP, Rieke F. 2005. Retinal processing near absolute threshold: from behavior to mechanism. *Annu. Rev. Physiol.* 67:491–514
Fishilevich E, Domingos AI, Asahina K, Naef F, Vosshall LB, Louis M. 2005. Chemotaxis behavior mediated by single larval olfactory neurons in *Drosophila*. *Curr. Biol.* 15:2086–96
Fishilevich E, Vosshall LB. 2005. Genetic and functional subdivision of the *Drosophila* antennal lobe. *Curr. Biol.* 15:1548–53
Gaudry Q, Hong EJ, Kain J, de Bivort BL, Wilson RI. 2013. Asymmetric neurotransmitter release enables rapid odour lateralization in *Drosophila*. *Nature* 493:424–28
Goldman AL, van der Goes van Naters W, Lessing D, Warr CG, Carlson JR. 2005. Coexpression of two functional odor receptors in one neuron. *Neuron* 45:661–66
Gouwens NW, Wilson RI. 2009. Signal propagation in *Drosophila* central neurons. *J. Neurosci.* 29:6239–49
Gu H, Jiang SA, Campusano JM, Iniguez J, Su H, et al. 2009. Ca_V2-type calcium channels encoded by *cac* regulate AP-independent neurotransmitter release at cholinergic synapses in adult *Drosophila* brain. *J. Neurophysiol.* 101:42–53
Ha TS, Smith DP. 2006. A pheromone receptor mediates 11-*cis*-vaccenyl acetate-induced responses in *Drosophila*. *J. Neurosci.* 26:8727–33
Haddad R, Weiss T, Khan R, Nadler B, Mandairon N, et al. 2010. Global features of neural activity in the olfactory system form a parallel code that predicts olfactory behavior and perception. *J. Neurosci.* 30:9017–26
Hallem EA, Carlson JR. 2006. Coding of odors by a receptor repertoire. *Cell* 125:143–60
Hallem EA, Ho MG, Carlson JR. 2004. The molecular basis of odor coding in the *Drosophila* antenna. *Cell* 117:965–79
Huang J, Zhang W, Qiao W, Hu A, Wang Z. 2010. Functional connectivity and selective odor responses of excitatory local interneurons in *Drosophila* antennal lobe. *Neuron* 67:1021–33
Ignell R, Root CM, Birse RT, Wang JW, Nassel DR, Winther AM. 2009. Presynaptic peptidergic modulation of olfactory receptor neurons in *Drosophila*. *Proc. Natl. Acad. Sci. USA* 106:13070–75
Jefferis GS, Marin EC, Stocker RF, Luo L. 2001. Target neuron prespecification in the olfactory map of *Drosophila*. *Nature* 414:204–8
Jefferis GS, Potter CJ, Chan AM, Marin EC, Rohlfing T, et al. 2007. Comprehensive maps of *Drosophila* higher olfactory centers: spatially segregated fruit and pheromone representation. *Cell* 128:1187–203
Johnson BN, Mainland JD, Sobel N. 2003. Rapid olfactory processing implicates subcortical control of an olfactomotor system. *J. Neurophysiol.* 90:1084–94
Kazama H, Wilson RI. 2008. Homeostatic matching and nonlinear amplification at genetically-identified central synapses. *Neuron* 58:401–13
Kazama H, Wilson RI. 2009. Origins of correlated activity in an olfactory circuit. *Nat. Neurosci.* 12:1136–44
Kazama H, Yaksi E, Wilson RI. 2011. Cell death triggers olfactory circuit plasticity via glial signaling in *Drosophila*. *J. Neurosci.* 31:7619–30
Kim AJ, Lazar AA, Slutskiy YB. 2011. System identification of *Drosophila* olfactory sensory neurons. *J. Comput. Neurosci.* 30:143–61
Kim DH, Phillips ME, Chang AY, Patel HK, Nguyen KT, Willhite DC. 2011. Lateral connectivity in the olfactory bulb is sparse and segregated. *Front. Neural Circuits* 5:5
Kreher SA, Mathew D, Kim J, Carlson JR. 2008. Translation of sensory input into behavioral output via an olfactory system. *Neuron* 59:110–24
Lai SL, Awasaki T, Ito K, Lee T. 2008. Clonal analysis of *Drosophila* antennal lobe neurons: diverse neuronal architectures in the lateral neuroblast lineage. *Development* 135:2883–93

Larsson MC, Domingos AI, Jones WD, Chiappe ME, Amrein H, Vosshall LB. 2004. *Or83b* encodes a broadly expressed odorant receptor essential for *Drosophila* olfaction. *Neuron* 43:703–14

Laughlin S. 1981. A simple coding procedure enhances a neuron's information capacity. *Z. Naturforsch. C* 36:910–12

Laurent G. 2002. Olfactory network dynamics and the coding of multidimensional signals. *Nat. Rev. Neurosci.* 3:884–95

Liu M, Chen TY, Ahamed B, Li J, Yau KW. 1994. Calcium-calmodulin modulation of the olfactory cyclic nucleotide-gated cation channel. *Science* 266:1348–54

Luo SX, Axel R, Abbott LF. 2010. Generating sparse and selective third-order responses in the olfactory system of the fly. *Proc. Natl. Acad. Sci. USA* 107:10713–18

Martin JP, Beyerlein A, Dacks AM, Reisenman CE, Riffell JA, et al. 2011. The neurobiology of insect olfaction: sensory processing in a comparative context. *Prog. Neurobiol.* 95:427–47

Masek P, Heisenberg M. 2008. Distinct memories of odor intensity and quality in *Drosophila*. *Proc. Natl. Acad. Sci. USA* 105:15985–90

Miesenböck G. 2004. Genetic methods for illuminating the function of neural circuits. *Curr. Opin. Neurobiol.* 14:395–402

Mozell MM, Kent PF, Murphy SJ. 1991. The effect of flow rate upon the magnitude of the olfactory response differs for different odorants. *Chem. Senses* 16:631–49

Mu L, Ito K, Bacon JP, Strausfeld NJ. 2012. Optic glomeruli and their inputs in *Drosophila* share an organizational ground pattern with the antennal lobes. *J. Neurosci.* 32:6061–71

Murlis J, Elkinton JS, Cardé RT. 1992. Odor plumes and how insects use them. *Annu. Rev. Entomol.* 37:505–32

Murphy GJ, Glickfeld LL, Balsen Z, Isaacson JS. 2004. Sensory neuron signaling to the brain: properties of transmitter release from olfactory nerve terminals. *J. Neurosci.* 24:3023–30

Nagel KI, Wilson RI. 2011. Biophysical mechanisms underlying olfactory receptor neuron dynamics. *Nat. Neurosci.* 14:208–16

Nassel DR, Enell LE, Santos JG, Wegener C, Johard HA. 2008. A large population of diverse neurons in the *Drosophila* central nervous system expresses short neuropeptide F, suggesting multiple distributed peptide functions. *BMC Neurosci.* 9:90

Ng M, Roorda RD, Lima SQ, Zemelman BV, Morcillo P, Miesenböck G. 2002. Transmission of olfactory information between three populations of neurons in the antennal lobe of the fly. *Neuron* 36:463–74

Nitz DA, van Swinderen B, Tononi G, Greenspan RJ. 2002. Electrophysiological correlates of rest and activity in *Drosophila melanogaster*. *Curr. Biol.* 12:1934–40

Okada R, Awasaki T, Ito K. 2009. Gamma-aminobutyric acid (GABA)-mediated neural connections in the *Drosophila* antennal lobe. *J. Comp. Neurol.* 514:74–91

Olsen SR, Bhandawat V, Wilson RI. 2007. Excitatory interactions between olfactory processing channels in the *Drosophila* antennal lobe. *Neuron* 54:89–103

Olsen SR, Bhandawat V, Wilson RI. 2010. Divisive normalization in olfactory population codes. *Neuron* 66:287–99

Olsen SR, Wilson RI. 2008. Lateral presynaptic inhibition mediates gain control in an olfactory circuit. *Nature* 452:956–60

Padmanabhan K, Urban NN. 2010. Intrinsic biophysical diversity decorrelates neuronal firing while increasing information content. *Nat. Neurosci.* 13:1276–82

Parker AJ, Newsome WT. 1998. Sense and the single neuron: probing the physiology of perception. *Annu. Rev. Neurosci.* 21:227–77

Reisert J. 2010. Origin of basal activity in mammalian olfactory receptor neurons. *J. Gen. Physiol.* 136:529–40

Reisert J, Matthews HR. 2000. Adaptation-induced changes in sensitivity in frog olfactory receptor cells. *Chem. Senses* 25:483–86

Reisert J, Restrepo D. 2009. Molecular tuning of odorant receptors and its implication for odor signal processing. *Chem. Senses* 34:535–45

Rohrbough J, Broadie K. 2002. Electrophysiological analysis of synaptic transmission in central neurons of *Drosophila* larvae. *J. Neurophysiol.* 88:847–60

Root CM, Ko KI, Jafari A, Wang JW. 2011. Presynaptic facilitation by neuropeptide signaling mediates odor-driven food search. *Cell* 145:133–44

Root CM, Masuyama K, Green DS, Enell LE, Nassel DR, et al. 2008. A presynaptic gain control mechanism fine-tunes olfactory behavior. *Neuron* 59:311–21

Root CM, Semmelhack JL, Wong AM, Flores J, Wang JW. 2007. Propagation of olfactory information in *Drosophila*. *Proc. Natl. Acad. Sci. USA* 104:11826–31

Ruta V, Datta SR, Vasconcelos ML, Freeland J, Looger LL, Axel R. 2010. A dimorphic pheromone circuit in *Drosophila* from sensory input to descending output. *Nature* 468:686–90

Sachse S, Rueckert E, Keller A, Okada R, Tanaka NK, et al. 2007. Activity-dependent plasticity in an olfactory circuit. *Neuron* 56:838–50

Saito H, Chi Q, Zhuang H, Matsunami H, Mainland JD. 2009. Odor coding by a mammalian receptor repertoire. *Sci. Signal.* 2:ra9

Sato K, Pellegrino M, Nakagawa T, Nakagawa T, Vosshall LB, Touhara K. 2008. Insect olfactory receptors are heteromeric ligand-gated ion channels. *Nature* 452:1002–6

Schlief ML, Wilson RI. 2007. Olfactory processing and behavior downstream from highly selective receptor neurons. *Nat. Neurosci.* 10:623–30

Schoenfeld TA, Cleland TA. 2005. The anatomical logic of smell. *Trends Neurosci.* 28:620–27

Schuckel J, French AS. 2008. A digital sequence method of dynamic olfactory characterization. *J. Neurosci. Methods* 171:98–103

Schuckel J, Meisner S, Torkkeli PH, French AS. 2008. Dynamic properties of *Drosophila* olfactory electroantennograms. *J. Comp. Physiol. A* 194:483–89

Schuckel J, Torkkeli PH, French AS. 2009. Two interacting olfactory transduction mechanisms have linked polarities and dynamics in *Drosophila melanogaster* antennal basiconic sensilla neurons. *J. Neurophysiol.* 102:214–23

Scott JW, Acevedo HP, Sherrill L. 2006. Effects of concentration and sniff flow rate on the rat electroolfactogram. *Chem. Senses* 31:581–93

Seki Y, Rybak J, Wicher D, Sachse S, Hansson BS. 2010. Physiological and morphological characterization of local interneurons in the *Drosophila* antennal lobe. *J. Neurophysiol.* 104:1007–19

Semmelhack JL, Wang JW. 2009. Select *Drosophila* glomeruli mediate innate olfactory attraction and aversion. *Nature* 459:218–23

Shanbhag SR, Muller B, Steinbrecht RA. 1999. Atlas of olfactory organs of *Drosophila melanogaster*. 1. Types, external organization, innervation, and distribution of olfactory sensilla. *Int. J. Insect Morphol. Embryol.* 28:377–97

Shang Y, Claridge-Chang A, Sjulson L, Pypaert M, Miesenböck G. 2007. Excitatory local circuits and their implications for olfactory processing in the fly antennal lobe. *Cell* 128:601–12

Silbering AF, Benton R. 2010. Ionotropic and metabotropic mechanisms in chemoreception: 'chance or design'? *EMBO Rep.* 11:173–79

Silbering AF, Galizia CG. 2007. Processing of odor mixtures in the *Drosophila* antennal lobe reveals both global inhibition and glomerulus-specific interactions. *J. Neurosci.* 27:11966–77

Silbering AF, Okada R, Ito K, Galizia CG. 2008. Olfactory information processing in the *Drosophila* antennal lobe: anything goes? *J. Neurosci.* 28:13075–87

Silbering AF, Rytz R, Grosjean Y, Abuin L, Ramdya P, et al. 2011. Complementary function and integrated wiring of the evolutionarily distinct *Drosophila* olfactory subsystems. *J. Neurosci.* 31:13357–75

Simoncelli EP. 2003. Vision and the statistics of the visual environment. *Curr. Opin. Neurobiol.* 13:144–49

Singer JH, Glowatzki E, Moser T, Strowbridge BW, Bhandawat V, Sampath AP. 2009. Functional properties of synaptic transmission in primary sense organs. *J. Neurosci.* 29:12802–6

Smart R, Kiely A, Beale M, Vargas E, Carraher C, et al. 2008. *Drosophila* odorant receptors are novel seven transmembrane domain proteins that can signal independently of heterotrimeric G proteins. *Insect Biochem. Mol. Biol.* 38:770–80

Smear M, Shusterman R, O'Connor R, Bozza T, Rinberg D. 2011. Perception of sniff phase in mouse olfaction. *Nature* 479:397–400

Sobel EC, Tank DW. 1993. Timing of odor stimulation does not alter patterning of olfactory bulb unit activity in freely breathing rats. *J. Neurophysiol.* 69:1331–37

Soucy ER, Albeanu DF, Fantana AL, Murthy VN, Meister M. 2009. Precision and diversity in an odor map on the olfactory bulb. *Nat. Neurosci.* 12:210–20

Stocker RF. 1994. The organization of the chemosensory system in *Drosophila melanogaster*: a review. *Cell Tissue Res.* 275:3–26

Stocker RF. 2008. Design of the larval chemosensory system. *Adv. Exp. Med. Biol.* 628:69–81

Stocker RF, Lienhard MC, Borst A, Fischbach KF. 1990. Neuronal architecture of the antennal lobe in *Drosophila melanogaster*. *Cell Tissue Res.* 262:9–34

Stortkuhl KF, Hovemann BT, Carlson JR. 1999. Olfactory adaptation depends on the Trp Ca^{2+} channel in *Drosophila*. *J. Neurosci.* 19:4839–46

Stuart GJ, Dodt HU, Sakmann B. 1993. Patch-clamp recordings from the soma and dendrites of neurons in brain slices using infrared video microscopy. *Pflugers Arch.* 423:511–18

Su CY, Menuz K, Carlson JR. 2009. Olfactory perception: receptors, cells, and circuits. *Cell* 139:45–59

Suh GS, Ben-Tabou de Leon S, Tanimoto H, Fiala A, Benzer S, Anderson DJ. 2007. Light activation of an innate olfactory avoidance response in *Drosophila*. *Curr. Biol.* 17:905–8

Tan J, Savigner A, Ma M, Luo M. 2010. Odor information processing by the olfactory bulb analyzed in gene-targeted mice. *Neuron* 65:912–26

Tanaka NK, Awasaki T, Shimada T, Ito K. 2004. Integration of chemosensory pathways in the *Drosophila* second-order olfactory centers. *Curr. Biol.* 14:449–57

Tanaka NK, Ito K, Stopfer M. 2009. Odor-evoked neural oscillations in *Drosophila* are mediated by widely branching interneurons. *J. Neurosci.* 29:8595–603

Turner GC, Bazhenov M, Laurent G. 2007. Olfactory representations by *Drosophila* mushroom body neurons. *J. Neurophysiol.* 99:734–46

Turner SL, Ray A. 2009. Modification of CO_2 avoidance behaviour in *Drosophila* by inhibitory odorants. *Nature* 461:277–81

Uchida N, Mainen ZF. 2003. Speed and accuracy of olfactory discrimination in the rat. *Nat. Neurosci.* 6:1224–29

van der Goes van Naters W, Carlson JR. 2007. Receptors and neurons for fly odors in *Drosophila*. *Curr. Biol.* 17:606–12

Vosshall LB, Amrein H, Morozov PS, Rzhetsky A, Axel R. 1999. A spatial map of olfactory receptor expression in the *Drosophila* antenna. *Cell* 96:725–36

Vosshall LB, Wong AM, Axel R. 2000. An olfactory sensory map in the fly brain. *Cell* 102:147–59

Wang JW, Wong AM, Flores J, Vosshall LB, Axel R. 2003. Two-photon calcium imaging reveals an odor-evoked map of activity in the fly brain. *Cell* 112:271–82

Wang Y, Wright NJ, Guo H, Xie Z, Svoboda K, et al. 2001. Genetic manipulation of the odor-evoked distributed neural activity in the *Drosophila* mushroom body. *Neuron* 29:267–76

Wark B, Lundstrom BN, Fairhall A. 2007. Sensory adaptation. *Curr. Opin. Neurobiol.* 17:423–29

Wicher D, Schafer R, Bauernfeind R, Stensmyr MC, Heller R, et al. 2008. *Drosophila* odorant receptors are both ligand-gated and cyclic-nucleotide-activated cation channels. *Nature* 452:1007–11

Wilson RI, Laurent G. 2005. Role of GABAergic inhibition in shaping odor-evoked spatiotemporal patterns in the *Drosophila* antennal lobe. *J. Neurosci.* 25:9069–79

Wilson RI, Turner GC, Laurent G. 2004. Transformation of olfactory representations in the *Drosophila* antennal lobe. *Science* 303:366–70

Winther AM, Acebes A, Ferrus A. 2006. Tachykinin-related peptides modulate odor perception and locomotor activity in *Drosophila*. *Mol. Cell. Neurosci.* 31:399–406

Winther AM, Siviter RJ, Isaac RE, Predel R, Nassel DR. 2003. Neuronal expression of tachykinin-related peptides and gene transcript during postembryonic development of *Drosophila*. *J. Comp. Neurol.* 464:180–96

Xu P, Atkinson R, Jones DN, Smith DP. 2005. *Drosophila* OBP LUSH is required for activity of pheromone-sensitive neurons. *Neuron* 45:193–200

Yaksi E, Wilson RI. 2010. Electrical coupling between olfactory glomeruli. *Neuron* 67:1034–47

Yao CA, Carlson JR. 2010. Role of G-proteins in odor-sensing and CO_2-sensing neurons in *Drosophila*. *J. Neurosci.* 30:4562–72

Yao CA, Ignell R, Carlson JR. 2005. Chemosensory coding by neurons in the coeloconic sensilla of the *Drosophila* antenna. *J. Neurosci.* 25:8359–67

Yarali A, Ehser S, Hapil FZ, Huang J, Gerber B. 2009. Odour intensity learning in fruit flies. *Proc. Biol. Sci.* 276:3413–20

Yasuyama K, Salvaterra PM. 1999. Localization of choline acetyltransferase-expressing neurons in *Drosophila* nervous system. *Microsc. Res. Tech.* 45:65–79

Yu D, Ponomarev A, Davis RL. 2004. Altered representation of the spatial code for odors after olfactory classical conditioning; memory trace formation by synaptic recruitment. *Neuron* 42:437–49

Zhou Y, Wilson RI. 2012. Transduction in *Drosophila* olfactory receptor neurons is invariant to air speed. *J. Neurophysiol.* 108:2051–59

RELATED RESOURCES

An atlas of the *Drosophila* brain: **http://www.virtualflybrain.org**

A database of *Drosophila* neuronal morphologies and long-range connectivity patterns: **http://www.flycircuit.tw**

A tool for visualizing odors in the space defined by the major axes of their physicochemical properties: **http://odorspace.weizmann.ac.il/content/physicochemical-space**

Masse NY, Turner GC, Jefferis GS. 2009. Olfactory information processing in *Drosophila*. *Curr. Biol.* 19:R700–13

Olsen SR, Wilson RI. 2008. Cracking neural circuits in a tiny brain: new approaches for understanding the neural circuitry of *Drosophila*. *Trends Neurosci.* 31:512–20

Simpson JH. 2009. Mapping and manipulating neural circuits in the fly brain. *Adv. Genet.* 65:79–143

Su CY, Menuz K, Carlson JR. 2009. Olfactory perception: receptors, cells, and circuits. *Cell* 139:45–59

RNA Protein Interaction in Neurons

Robert B. Darnell

Department of Molecular Neuro-Oncology, Howard Hughes Medical Institute, The Rockefeller University, New York, NY 10065; email: darnelr@rockefeller.edu

Annu. Rev. Neurosci. 2013. 36:243–70

First published online as a Review in Advance on May 20, 2013

The *Annual Review of Neuroscience* is online at neuro.annualreviews.org

This article's doi:
10.1146/annurev-neuro-062912-114322

Keywords

HITS-CLIP, Nova, Elavl, FMRP, microRNA, neuron-specific splicing factors

Abstract

Neurons have their own systems for regulating RNA. Several multigene families encode RNA binding proteins (RNABPs) that are uniquely expressed in neurons, including the well-known neuron-specific markers ELAV and NeuN and the disease antigen NOVA. New technologies have emerged in recent years to assess the function of these proteins in vivo, and the answers are yielding insights into how and why neurons may regulate RNA in special ways—to increase cellular complexity, to localize messenger RNA (mRNA) spatially, and to regulate their expression in response to synaptic stimuli. The functions of such restricted neuronal proteins are likely to be complemented by more widely expressed RNABPs that may themselves have developed specialized functions in neurons, including Argonaute/microRNAs (miRNAs). Here we review what is known about such RNABPs and explore the potential biologic and neurologic significance of neuronal RNA regulatory systems.

Contents

INTRODUCTION

The idea that neurons have developed unique systems for regulating RNA metabolism—processing, localization, and expression—has several interesting roots. First are observations on memory. A large body of evidence suggests that memory, and physiologic correlates such as long-term potentiation (LTP), require protein synthesis during a brief window after a memorable experience (Lynch 2004). Physical evidence—the presence of detectable messenger RNA (mRNA) and ribosomes, and the ability to detect local translation (Aakalu et al. 2001, Schuman et al. 2006, Steward & Schuman 2001)—and physiologic evidence—the protein synthesis requirement for axon guidance (Holt & Bullock 2009, Ming et al. 2002, Zhang & Poo 2002) or for LTP after Hebbian stimuli (Lynch 2004)—support the hypothesis that synaptic plasticity requires the regulation of some mRNA transcripts within the synapse itself.

Second, early characterization of neuronal transcripts suggested that pre-mRNAs are differentially processed in the brain relative to other tissues. This idea was raised by the work of Amara and Evans, who discovered that the transcript encoding calcitonin, normally made in the thyroid gland, had two isoforms in thyroid tumors. The alternate isoform was a poly(A)/splice variant normally found only in neurons, where it encoded an entirely different protein, the neuropeptide transmitter calcitonin-gene related peptide, CGRP (Amara et al. 1982). This observation, and many follow-up studies, underlies the idea that neurons leverage RNA processing to generate orders of magnitude increases in complexity from a limited size protein-coding genome (Licatalosi & Darnell 2010).

A third line of evidence came from the study of neuronal markers. Neurobiologists have long sought out genes that are uniquely expressed in neurons to demarcate neurons histologically and to develop genetic tools. Although many markers have been identified that are specific to neurons within the nervous system (versus glia, for example), most are robustly expressed in other tissues at some point in development. However, a few were identified that are extremely specific to neurons throughout life (one common exception being germ cells). The first such marker was the *Drosophila* ELAV (embryonic lethal abnormal visual) protein. ELAV was identified through a screen for behavioral defects in flies (Homyk et al. 1980) that were found to have abnormal electroretinograms and abnormal axonal tracts. Ultimately, RNA and protein expression studies revealed that ELAV was expressed in all neurons, but not in neuroblasts, glia, or any other tissue type (Bier et al. 1988, Campos et al. 1987, Yao et al. 1993). Subsequently, ELAV became widely used as a specific neuronal marker and as a neuron-specific driver for genetic studies.

In mammalian systems, early putative neuronal markers such as neuron-specific enolase were found to label glia (Magavi & Macklis 2008), prompting empiric searches for additional markers. These led to the mammalian homolog of ELAV [originally termed the Hu antigen, discussed below, now termed ELAV-like (ELAVL)] and NeuN (neuronal nuclei). Wolf et al. (1996) found that NeuN, identified with a monoclonal antibody generated against brain nuclei and originally thought to be a transcription factor, can reliably identify mature neurons in the adult brain. Amazingly, each of these classical neuron-specific markers, *Elav/Elavl* and *NeuN* (see below), were found to encode RNABPs. These discoveries have since been augmented with an expanding list of additional neuron-specific RNABPs and investigators have recognized that at least some more generally expressed RNABPs may act in concert with the neuronal RNABPs (**Table 1**).

A fourth and equally unexpected line of evidence was the discovery of neuron-specific RNA binding proteins in humans, which came from the work on a set of cancer-related autoimmune neurologic disorders, termed the paraneoplastic neurologic disorders (Darnell & Posner 2011). Autoimmune antisera from these patients identified neuron-specific proteins that were ectopically expressed in tumor cells, triggering an antitumor immune response that ultimately crossed into the brain, leading to the neurologic disease. In the 1980s, the Darnell laboratory established that these antisera could be used to clone complementary DNA (cDNAs) encoding these antigens using λgt11 expression vectors, and two different multigene families turned out to encode the neuron-specific RNA binding proteins Nova and Elavl (Darnell 1996, Musunuru & Darnell 2001, Darnell & Posner 2011).

The significance of the discovery that the brain expresses its own neuronal RNABPs relates to our attempt to understand what underlies complexity in brain function. To a first approximation, the genomic protein-coding capacity of the human and the worm are very similar (in number and types of protein-coding genes). This observation has shifted interest in understanding complexity as a consequence of the ways in which these genes are deconvoluted in the RNA world into understanding how pre-mRNA gene copies are alternatively spliced and polyadenylated, edited, localized throughout the neuron, and translationally regulated. This much more robust complexity, relative to the control of DNA transcription, will likely play a key role in the evolution of complex cellular function, neuronal plasticity, and brain function (Licatalosi & Darnell 2010). This review describes the approaches used to identify the functions of neuronal RNABPs and what is known about each in brain function and disease and then discusses future directions.

Neuronal nuclei (NeuN): a neuron-specific RNA binding protein

APPROACHES TO STUDYING NEURONAL RNA BINDING PROTEIN FUNCTION

To appreciate the work done by many laboratories to establish the roles of neuronal RNABPs, one must appreciate the methods used to establish their functions. Three major approaches have been established, which, when used in combination with modern bioinformatics, combine to form a powerful means to define in vivo functions for RNABPs. A prior review detailed the combined use of these approaches (Licatalosi & Darnell 2010), which are outlined below.

Traditional Biochemical Approaches

The importance of knowing whether a neuronal protein is an RNABP is underscored by the original reports that described NeuN as a transcription factor. The traditional means to define a protein as an RNABP came from the lab of Gideon Dreyfuss, who characterized a large number of heteronuclear ribonuclear proteins (hnRNPs) as RNABPs. The fundamental assay, still valid as a screen, was to bind purified proteins to ribohomopolymer

Table 1 Neuronal RNA binding proteins

RNABP	Comments	Expression	Other Paralogs	Position-dependent splicing map/other function	Expression references	Knockout mouse
Neuron-specific RNA binding proteins						
Nova1–2	Nova1: first neuronal RNABP for which a genetic-null mouse was engineered	CNS-specific Nova2: trace expression in lung unknown cell type)	None	Yes		Nova1: Jensen et al. 2000a Nova2: Huang et al. 2005
Elavl2–4	Elavl4: frequent marker neuron-specific marker	Elavl2: Most weakly expressed Elavl3: most unique expression patterns Extra-CNS expression in autonomic neurons (e.g., intestinal expression in myenteric plexus neurons)	Elavl1	Yes	Okano & Darnell 1997	Elavl3: Akamatsu et al. 2005, Ince-Dunn et al. 2012 Elavl4: Akamatsu et al. 2005
Rbfox3 (NeuN)	Frequent neuron-specific marker; Rbfox1 implicated in autism	Intestinal expression restricted to myenteric plexus neurons	Rbfox1/2	Yes	Dredge & Jensen 2011, Kim et al. 2009, Voineagu et al. 2011	Rbfox1: Gehman et al. 2011
Ptbp2		Expression in rare subsets of glial cells and in mitotic neuronal progenitors	Ptbp1	Yes	Licatalosi et al. 2012	Licatalosi et al. 2012
nSR100		Highly neural enriched by nonquantitative RT-PCR		?	Calarco et al. 2009	
RNABPs not unique to neurons, but of biologic interest						
CELF	Family includes CELF1 (CUGBP1), a splicing factor implicated in myotonic dystrophy; literature unclear regarding neuronal specificity of other family members	Celf4: Weak RNA expression by Northern blot outside nervous system, but cross-reactivity among paralogs is unclear	Celf1, -2, -4–6	?	Cooper et al. 2009, Meins et al. 2002	Wagnon et al. 2011
FMRP		Neuron-specific in the brain; widely expressed in other tissues	FXR1/2 (wide expression)	Inhibits translation by arresting ribosomal elongation	Darnell et al. 2011	Fmr1 null: Bakker et al. 1994 Fmr1 inactive: Zang et al. 2009

(*Continued*)

Table 1 (*Continued*)

Neuron-specific RNA binding proteins						
RNABP	**Comments**	**Expression**	**Other Paralogs**	**Position-dependent splicing map/other function**	**Expression references**	**Knockout mouse**
Mbnl1-3		Mbnl2: neuron-specific in the brain; robust expression in lung		Yes		Mbnl1: Kanadia et al. 2003 Mbnl2: Charizanis et al. 2012
Msi1		Expressed in mitotic neuronal progenitors in embryonic and adult brain; also expressed in some astrocytes and some neurons [e.g., deep cerebellar nuclei, stellate neurons, as well as ovaries and small intestine (unknown cell type)]	Msi2	?		Msi1: Sakakibara et al. 2002
ZBP	ZBP1 particularly well-defined functions in mRNA localization and translation	Widely expressed in the brain and other tissues	ZBP2	?	Hansen et al. 2004	

Abbreviations: mRNA, messenger RNA; NeuN, neuronal nuclei; RNABP, RNA binding proteins.

columns and to measure their retention under increasingly stringent salt washes. In this way, for example, after the gene encoding the NOVA1 protein was cloned, NOVA1 was found to bind ribohomopolymers in up to 1.0 M salt, evidence of robust RNA affinity (Buckanovich et al. 1996), whereas the Fragile X mental retardation protein, FMRP, also bound to ribohomopolymers, but with much less affinity (Siomi et al. 1993). These approaches allowed Dreyfuss and colleagues to classify RNABPs according to the presence of several canonical motifs (Burd & Dreyfuss 1994). This, in turn, accelerated the classification of many newly discovered proteins as RNABPs, although we note that new high-affinity RNA binding motifs continue to be described.

A second level of analysis was to identify preferred RNA binding motifs in vitro. This strategy relied on the development of methods to use recombinant proteins to affinity purify RNA sequences from random RNA libraries in an iterative manner, termed in vitro RNA selection (developed using affinity chromatography; Ellington & Szostak 1990, Green et al. 1991) or RNA SELEX (developed using filter-binding strategies; Tuerk & Gold 1990). Early validation of these methods included their use to identify RNAs bound to the HIV-1 Rev protein (Ellington & Szostak 1990) and to T4 DNA polymerase (Tuerk & Gold 1990) and to

NOVA: Onconeural ventral antigen; a neuron-specific RNA binding protein

Table 2 Neuron-specific RNABPs—RNA binding sites

Neuron-specific RNABP	Methods used in identification	Binding motif	Comment	Reference
NOVA1 NOVA2	In vitro RNA selection CLIP HITS-CLIP Bayesian network	Stem, loop sequence: UCAU × 3 YCAY clusters	Stem sequence variable; may play structural role to keep YCAY binding site available in sequence single loop	Buckanovich & Darnell 1997, Jensen et al. 2000b, Yang et al. 1998
nELAVL2 (Hel-N1)	In vitro RNA selection	UUUAUUU		Gao et al. 1994
nELAVL2-3, -4	In vitro RNA selection HITS-CLIP	U-rich element with single purine	G slightly preferred over A as purine	Ince-Dunn et al. 2012
RBFOX1	In vitro RNA selection HITS-CLIP	GCAUG UGCAUG		Jin et al. 2003, Yeo et al. 2009
PTBP2	HITS-CLIP	CU-repeat and UCUY-rich elements		Licatalosi et al. 2012

Abbreviation: HITS-CLIP, high-throughput sequencing, cross-linking immunoprecipitation.

confirm binding of U1 small nuclear ribonuclear protein-A (snRNP-A) to sequences in U1 RNA (Tsai et al. 1991). These approaches have been used to identify in vitro RNA ligands for many of the mammalian neuronal RNABPs discussed in this review (**Table 2**), and these have proved to be extremely valuable in cross-checking binding motifs identified by complementary methods described below.

Mammalian Genetics

For many neuronal RNABPs, in vitro biochemistry has been complemented by validation of predicted functions in the brains of RNABP-knockout mice. This approach has been important because neither cell quality (particularly the specialized neuronal cell types in the brain), cell biology (particularly the complex synaptic interactions among many cell types), nor the stoichiometry of RNA-protein interactions can be faithfully reproduced in tissue-culture cells or primary neurons.

The first neuronal RNABP for which a genetic-null mouse was engineered was the neuron-specific RNABP *Nova1* (Jensen et al. 2000a). Null mice have been generated subsequently for most of the RNABPs discussed in this review (see **Table 1**). In addition, both null mice (Bakker et al. 1994) and mice harboring an inactivating point mutant (Zang et al. 2009) have been generated for *Fmr1*, the gene encoding the Fragile X protein FMRP, providing useful overlapping means to validate biochemical predictions of in vivo biology (Darnell et al. 2011). For each of these RNABPs, as detailed below, the most robust data regarding their function in neurons relate predictions from biochemical experiments to examination of RNA variants in wild-type compared with knockout mice.

Global RNA Analyses

One significant limitation of traditional biochemical approaches is that they study one RNA-protein interaction at a time; thus, making generalizations from such data is difficult. The past decade or so saw an emergence and maturation of methods to analyze RNA complexity on a global scale. By complexity, we refer to the unique species of RNA present in a given biological sample. The initial enumeration of RNA complexity came from analysis of microarrays. As previously discussed (Blencowe et al. 2009, Mortazavi et al. 2008), the difficulties with such arrays were their inherent signal:noise problem; different probe sets had varying sensitivity and specificity in detecting a range of transcript levels, leading to burdensome and sometimes inaccurate normalization requirements. Nonetheless, such arrays were important to provide genome-wide means to approximate transcript levels in different tissues.

A subsequent generation of exon arrays allowed analysis of alternative splice variants using probe sets that spanned exon-exon junctions. These exon junction microarrays were especially informative in the analysis of neuronal RNA complexity. They initially were used to enumerate alternative splice variants present in the brain relative to other tissues (Johnson et al. 2003, Pan et al. 2004) and then more specifically to analyze splicing alterations evident when specific factors were missing. Such analyses included genome-wide descriptions of splice variants in *Nova*-null mice (Ule et al. 2005b) and, in *Drosophila*, in flies in which various splicing factors were knocked down (Blanchette et al. 2005). Subsequent studies expanded our understanding of splicing variation in different tissues (Sugnet et al. 2006) and of splicing variants related to a number of different neuronal RNABPs. These include studies of splicing dependent on PTBP2 in N2A neuroblastoma cells (Boutz et al. 2007), the neural-specific serine-arginine rich (SR) protein nSR100 in tissue culture and zebrafish (Calarco et al. 2009), and PTBP2 (Licatalosi et al. 2012), MBNL1 (Du et al. 2010), RBFOX1 (Gehman et al. 2011), and ELAVL3 (Ince-Dunn et al. 2012) in knockout mouse brain.

A more refined and accurate picture of transcript variants in the brain (and other tissues) has now been afforded by applying next-generation sequencing—RNA-seq—a method originally developed to quantify transcriptomes in yeast (Nagalakshmi et al. 2008) and mammalian tissues (Mortazavi et al. 2008, Wang et al. 2008), including the brain. Even with early-generation high-throughput sequencing, read depth was sufficient to begin delineating alternative splicing patterns unique to the brain (Mortazavi et al. 2008, Wang et al. 2008). Such studies, combined with pair-end sequencing methods and decreasing costs of sequencing, promise to become the most sensitive and specific means to enumerate RNA variants. Recent applications include analysis of RNA variants generated by alternative splicing (Calarco et al. 2011), alternative polyadenylation (Weill et al. 2012), and RNA editing (Silberberg & Ohman 2011).

HITS-CLIP: high-throughput sequencing, cross-linking immunoprecipitation. Methods for covalently cross-linking RNA-protein complexes and sequencing the bound RNA, yielding genome-wide in vivo footprints

HITS-CLIP

A critical view of data enumerating RNA variants underscores its correlative nature (Licatalosi & Darnell 2010). For example, alternative exons whose levels vary in proportion to expression of a neuronal splicing factor may be directly regulated by that factor or indirectly regulated by, for example, an action on a transcript encoding an intermediate RNABP. Moreover, to understand RNABP function, it is important to identify both the target RNAs on which they directly act as well as the actual sites of action. Although traditional biochemical experiments may demarcate potential binding interactions, even with high-level bioinformatic analysis they have not been able to predict fully and accurately the range and variability of interactions seen in living cells. Resolving these issues is crucial to understanding the mechanisms of RNABP actions in the brain.

Among several approaches taken, those involving cross-linking of RNA-protein complexes have become the gold standard for identifying sites of functional interaction in vivo, including in the brain (reviewed in Calarco et al. 2011, Darnell 2010, Konig et al. 2012, Licatalosi & Darnell 2010). These methods originated with the development of cross-linking immunoprecipitation (CLIP) (**Figure 1**) for identifying Nova-RNA interactions in the brain (Ule et al. 2003). CLIP was initially adopted slowly, although it proved versatile for a host of species and RNABPs (see Darnell 2010). However, the application of high-throughput sequencing methods to CLIP (HITS-CLIP) (Licatalosi et al. 2008; see sidebar, Discovery of Mammalian Neuron-Specific RNABPs Through Study of the Paraneoplastic Neurologic Disorders) provided a breakthrough and is now in wide use in the field.

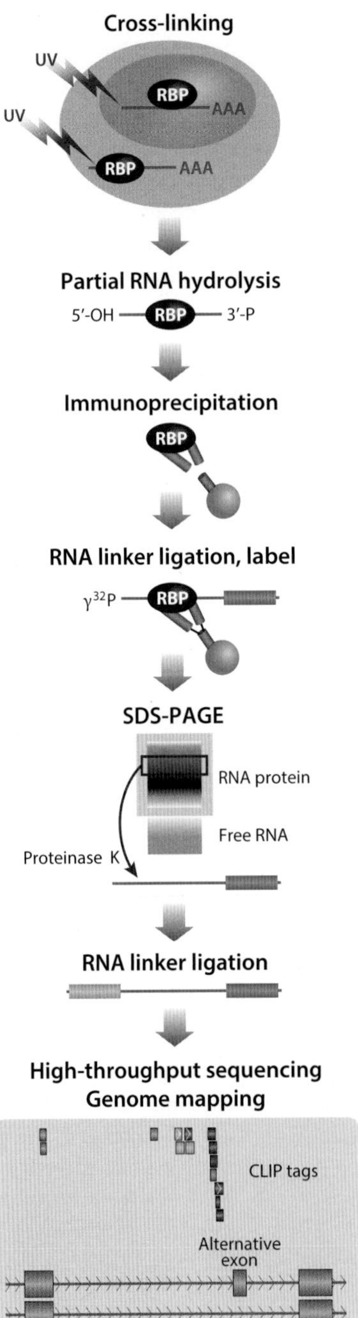

NEURON-SPECIFIC RNA BINDING PROTEINS

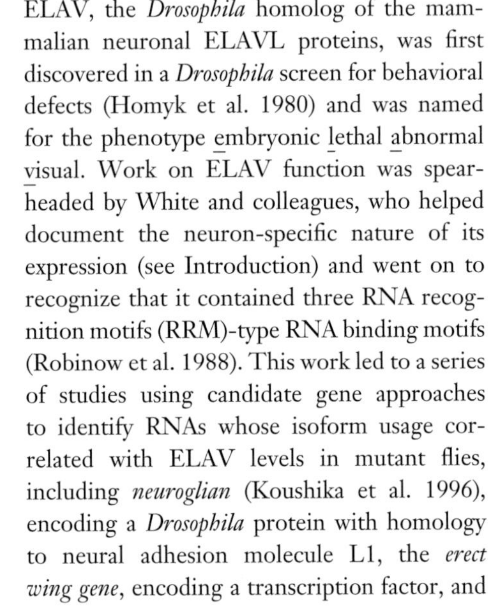

Elavl

ELAV, the *Drosophila* homolog of the mammalian neuronal ELAVL proteins, was first discovered in a *Drosophila* screen for behavioral defects (Homyk et al. 1980) and was named for the phenotype embryonic lethal abnormal visual. Work on ELAV function was spearheaded by White and colleagues, who helped document the neuron-specific nature of its expression (see Introduction) and went on to recognize that it contained three RNA recognition motifs (RRM)-type RNA binding motifs (Robinow et al. 1988). This work led to a series of studies using candidate gene approaches to identify RNAs whose isoform usage correlated with ELAV levels in mutant flies, including *neuroglian* (Koushika et al. 1996), encoding a *Drosophila* protein with homology to neural adhesion molecule L1, the *erect wing gene*, encoding a transcription factor, and

Figure 1

Schematic of the HITS-CLIP method. Schematic of major steps in HITS-CLIP (modified from Licatalosi & Darnell 2010; see sidebar, HITS-CLIP, for details about CLIP). In brief, tissues or cells to be analyzed are irradiated with ultraviolet-light, inducing crosslinking between RNA-protein complexes within the cell, in situ. RNA is partially hydrolyzed, using nucleases or chemical hydrolysis, to a desired modal size (e.g., 30–50 nucleotides), RNA-protein complexes are purified (typically by immunoprecipitation), and RNA is labeled and run on autoradiograms to separate cross-linked RNA-protein complexes by size and transferred to nitrocellulose (through which any free RNA passes). The RNABP is removed by proteinase K treatment, RNA linker ligation is completed, and RNA polymerase chain reaction (PCR) is amplified and sequenced.

armadillo, whose vertebrate homolog is β-catenin (Koushika et al. 2000, Soller & White 2003). More recent evidence suggests that ELAV bound to an intronic adenine and uridine (AU)-rich element could mediate effects on *ewg* splicing (Soller & White 2005).

Studies of paraneoplastic neurologic disorders led to the discovery of three mammalian homologs of *Elav*, each with expression that is restricted entirely to neurons, termed *Elavl2-4* (originally HuB, HuC, and HuD; there is also one nonneuronal paralog, *Elavl1*, or HuA/HuR). These neuronal ELAVL (termed nELAVL) proteins were discovered using autoantibodies to screen for the target antigen in the paraneoplastic subacute sensory neuropathy/encephalomyelopathy syndrome or in the Hu syndrome neurologic disorders (see sidebar, Discovery of Mammalian Neuron-Specific RNABPs). A careful description of their expression revealed that the three nELAVL proteins were indeed restricted to central and peripheral neurons in overlapping but unique patterns in all postmitotic neurons, with expression persisting throughout adulthood (Okano & Darnell 1997).

RNA selection experiments first performed with nELAVL2/HuB revealed a consensus sequence for U-rich elements interspersed with purines, interpreted as a preference for binding AU-rich elements (AREs) (Gao et al. 1994, Levine et al. 1993). Moreover, in elegant biochemical studies, nELAVL1/HuA copurified with the ARE in the *c-fos* gene (Myer et al. 1997). ARE elements mediate rapid decay of mRNAs through shortening of the poly(A) tail, followed by mRNA degradation (Chen & Shyu 2011) (see Meisner & Filipowicz 2011 for review of nELAVL1 function). A number of studies demonstrated that overexpression of mammalian ELAVL proteins could stabilize ARE-containing messages [including studies of ELAVL1 (Dean et al. 2001, Fan et al. 1997, Fan & Steitz 1998, Peng et al. 1998), nELAVL2 (Antic et al. 1999, Jain et al. 1997), and nELAVL4 (Bolognani & Perrone-Bizzozero 2008)]. In vitro and tissue-culture experiments indicated that ELAVL could affect stability and/or translation of a large number of proteins, including tau, GAP-43, p21 (for review, see Hinman & Lou 2008) and even the neuronal splicing factor Nova1 (Ratti et al. 2008). Following the studies from White and colleagues in flies, a number of studies in mammalian cells also contributed to evidence that the ELAVL proteins could act as splicing factors, including in vitro RNA-protein binding assays, minigene overexpression studies, and analyses of conserved AU-rich elements adjacent to alternative exons (Wang et al. 2010a; Zhu et al. 2006, 2008; reviewed in Hinman & Lou 2008).

Together, these reports provided interesting suggestive data, but they were restricted by uncertainty regarding which targets were direct and physiologically relevant, given the concerns raised above in the section, Approaches to Studying Neuronal RNA Binding

DISCOVERY OF MAMMALIAN NEURON-SPECIFIC RNABPs THROUGH STUDY OF THE PARANEOPLASTIC NEUROLOGIC DISORDERS

In studies analogous to the discovery of snRNPs by Joan Steitz and colleagues made by studies of autoimmune lupus antisera (Lerner et al. 1981), two multigene families of neuron-specific RNA binding proteins were discovered through studies of autoimmune paraneoplastic neurologic disorder (PND) antisera. In the PNDs (as discussed in Darnell & Posner 2011), common types of cancers (small-cell lung, breast, ovarian, lymphoma, germ cell cancers, etc.) may induce expression of proteins that are normally restricted to the neurons. This expression triggers an antitumor immune response that can be clinically potent (Darnell & Posner 2003a,b; Darnell & DeAngelis 1993); in a subset of these patients, this immune response crosses into the nervous system, triggering neurologic symptoms and bringing patients to clinical attention. After the original demonstration that PND antisera could be used to screen expression cDNA libraries and identify the genes encoding the target antigens (Darnell et al. 1989, McKeever & Darnell 1992, Newman et al. 1995), this method was applied to identify the neuron-specific nElavl4 cDNA (Szabo et al. 1991), Nova1 (Buckanovich et al. 1993), and Nova2 (Yang et al. 1998) RNABPs.

ELAVL proteins (originally termed Hu proteins): show strong homology to the *Drosophila* neuron-specific RNABP ELAV (embryonic abnormal lethal) protein

Neuronal embryonic abnormal lethal-like (nELAVL): refers to the three neuron-specific paralogs of the mammalian ELAVL proteins

HITS-CLIP

HITS-CLIP describes the overlay of genome-wide data gathering and bioinformatic strategies with biochemical cross-linking strategies to analyze RNA-protein interactions in living tissues. The CLIP method was originally developed to analyze Nova-RNA interactions in the mouse brain (Ule et al. 2003). Fundamentally, the method uses ultraviolet-irradiation to penetrate live cells and cross-link, in situ, RNA-protein complexes that are in direct contact (~1 Å distances). Once cross-linked, RNA can be hydrolyzed to short fragments and protein complexes purified typically by immunoprecipitation. Epitope-tagged proteins may be used, but investigators must pay attention to expression levels to avoid overexpression artifacts.

Once purified, RNA-protein complexes are treated with proteinase K and linker ligation completed to allow RT-PCR amplification and sequencing. CLIP established that reverse transcriptase (RT) can read through sites of residual amino-acid-RNA cross-links; in fact, errors at these sites can be leveraged to map cross-link sites with single nucleotide resolution (Zhang & Darnell 2011). Alternatively, in iCLIP, RT arrest at these sites can be used to map cross-link sites (Wang et al. 2010c). Sequencing the RNA tags yields HITS-CLIP libraries representing genome-wide footprints of RNABP binding. Additional details on HITS-CLIP methods (Jensen & Darnell 2008, Konig et al. 2011, Ule et al. 2005a) and comparison with other related strategies such as PAR-CLIP (Hafner et al. 2010, Kishore et al. 2011) are described in several excellent papers and reviews (Blencowe et al. 2009, Calarco et al. 2011, Darnell 2010, Kishore et al. 2010, Konig et al. 2012, Sugimoto et al. 2012).

Protein Function. Until recently, a list of actual nELAVL target RNAs and the processes they regulated remained unclear; the most complete effort came from a list of RNAs coprecipitating with nELAVL4 in transgenic mice overexpressing nELAVL4 (Bolognani & Perrone-Bizzozero 2008). Nonetheless, this body of work stimulated great interest in the nELAVL proteins and was further fueled by functional studies, including the observations that the nELAVL3 and nELAVL4 proteins were necessary for neuronal maturation in mice (Akamatsu et al. 1999, 2005). This interest laid the groundwork for the generation of *nElavl3* (Ince-Dunn et al. 2012) and *nElavl4* (Akamatsu et al. 2005) null mice, which have proven critical to demarcating nELAVL function in vivo.

nElavl4 null mice were recently used to provide support for a noncanonical role for nELAVL4 in pancreatic beta cells (Lee et al. 2012). These interesting data need confirmation (RNA could not be detected in beta cells, although protein could be, raising concern about antibody specificity). Notably, the origin of pancreatic beta cells remains unclear; progenitor cells in the adult mouse pancreas can give rise to neurons and pancreatic cells, and beta cells express many neuronal proteins and pathways (Dor et al. 2004, Smukler et al. 2011), including GABA and synaptic-like microvesicles (Reetz et al. 1991). Examination of *nElavl4* null mice demonstrated abnormally high levels of insulin, and this result was correlated with the ability of nELAVL4 to bind to the transcript encoding insulin in vitro (Lee et al. 2012). Patients with paraneoplastic autoimmunity to the Hu antigen do not develop diabetes (Darnell & Posner 2011), even though beta cells are a common autoimmune target; thus, aspects of these observations remain puzzling.

A combination of experiments using *nElavl* null mice, HITS-CLIP, and bioinformatic analysis has recently provided a more precise in vivo picture of nELAVL-RNA interactions in neurons within the mouse brain (Ince-Dunn et al. 2012). In the mouse brain, nELAVL proteins were found to bind directly to 3′ UTR (untranslated region) elements and, to a lesser degree, intronic elements. Research revealed an interesting relationship with the original in vitro RNA selection studies, whereby the in vivo binding sites were U-rich elements harboring a purine, more commonly a G than an A residue (so rather than binding AU-rich elements, the nELAVL proteins bind purine/U-rich elements). Zhang & Darnell (2011) confirmed these binding sites by analyzing cross-link-induced mutations (Zhang & Darnell 2011) and found them to be functional. In particular, several intronic binding sites were able to mediate alternative splicing regulation, as evidenced by their ability to predict splicing changes in the nELAVL-null mouse brain. In addition, nELAVL 3′ UTR binding sites predicted changes in mRNA steady-state levels.

A major goal of identifying RNABP-RNA interaction sites is to gain insight into the relevant biology. Gene ontology (GO) analysis demonstrated prominent roles for nELAVL-splicing regulation of transcripts involved in synaptic cytoskeletal dynamics and nELAVL-steady-state regulation of transcripts involved in amino acid biosynthesis (Ince-Dunn et al. 2012). One outstanding target in both GO analyses was the pathway involved in glutamine synthesis, in particular the gene encoding glutaminase, which catalyzes formation of glutamate, the major excitatory neurotransmitter in the brain. Indeed, glutamate levels were reduced in the nELAV3 null mouse brain, and both these mice and haploinsufficient mice had spontaneous epilepsy. These studies illustrate how traditional and modern (HITS-CLIP) biochemistry, combined together with mouse genetics and bioinformatics, provides a potent means to discover the role of neuronal RNABP function in the brain.

Nova

The triad of biochemistry, mouse genetics, and bioinformatics to define neuronal RNABP function was first established in studies of the RNABP Nova (as reviewed in Licatalosi & Darnell 2010). As with the nElavl, the genes encoding the NOVA1 and NOVA2 proteins were identified using autoimmune sera from patients with paraneoplastic opsoclonus-myoclonus ataxia (Buckanovich et al. 1993, Yang et al. 1998). In this neurologic disorder, patients with lung or gynecologic tumors develop excessive motor movements, attributable neurologically (Darnell & Posner 2011) to a failure of motor inhibition, which sometimes progresses to encephalopathy (Hormigo et al. 1994, Luque et al. 1991). Characterization of *Nova1* and *Nova2* expression revealed expression that is extremely restricted to postmitotic neurons (Yano et al. 2010) in the central, but not the peripheral, nervous system; *Nova1* is expressed primarily in the hindbrain and ventral spinal cord, and *Nova2* is expressed primarily in the neocortex (Buckanovich et al. 1993, Yang et al. 1998, Racca et al. 2010).

Traditional biochemical studies established that Nova proteins bind to RNA in a sequence-specific manner and can regulate alternative splicing in vitro. Nova proteins function as RNABPs (Buckanovich & Darnell 1997, Buckanovich et al. 1996) and harbor three hnRNPK-homology (KH)-type RNA binding motifs. RNA selection defined their in vitro sequence targets as clusters of YCAY motifs (where Y is a pyrimidine) (Buckanovich & Darnell 1997, Jensen et al. 2000b, Yang et al. 1998). The mechanism of KH-domain sequence specificity was first revealed by X-ray crystallographic studies of NOVA-RNA complexes (Lewis et al. 1999, 2000; Jensen et al. 2000b; Teplova et al. 2011), which showed that the NOVA KH domains fold into a highly conserved structure in which side-chain amino acids are exactly positioned to provide Watson-Crick hydrogen bond donor and acceptor groups to specify precisely the CA dinucleotide and, to a lesser degree, restrict binding of the bounding nucleotides to pyrimidines.

These data were used to identify three Nova-regulated pre-mRNAs harboring YCAY clusters, targets that were confirmed with cellular and in vitro splicing assays. These targets included two transcripts encoding inhibitory neurotransmitter receptors encoding the $\alpha 2$ subunit of the glycine (GlyRα2) receptor (Buckanovich & Darnell 1997, Jensen et al. 2000b, Racca et al. 2010) and the $\gamma 2$ subunit of the $GABA_A$ ($GABA_A$ $\gamma 2$) receptor (Dredge & Darnell 2003, Jensen et al. 2000a). In addition, in vitro splicing assays demonstrated that NOVA1 inhibited splicing of its own pre-mRNA by binding YCAY elements to block U1A binding (Dredge et al. 2005, Ule et al. 2006).

However, critical analysis suggested two outstanding unresolved issues. First, biochemical assays lacked physiologic stoichiometry (overexpression assays in vitro or in vivo), leaving uncertainty about whether the splicing changes observed were relevant in vivo. This question led our lab to generate *Nova*-null mice (Jensen et al. 2000a, Yang et al. 1998) to reexplore splicing targets in vivo. Second, identification of the GlyRα2 and GABA$_A$ γ2 as Nova targets was intriguing in light of the fact that Nova was targeted in PND patients with a failure of motor inhibition, yet our approach to target identification had clearly not been systematic. This result prompted the use of genome-wide methods to assess Nova targets, culminating in the development of CLIP.

Initial genome-wide approaches to assess RNA splice variation in *Nova*-null mice used exon-junction splicing microarrays (Ule et al. 2005b). Analysis of these results using a bioinformatic algorithm (ASPIRE) designed to look for reciprocal splicing changes in wild-type and null mice helped to identify a robust set of ~50 target RNAs whose splicing was Nova-dependent. GO analysis of this list led to the first compelling evidence indicating that RNABPs could regulate a biologically coherent subset of transcripts in vivo—nearly all 50 transcripts encoded synaptic proteins (Calarco et al. 2011, Ule et al. 2005b, Ule & Darnell 2006). Moreover, identifying Nova target RNAs led to prediction and identification of functions for Nova in synaptic function, including roles in mediating inhibitory responses to LTP (Huang et al. 2005), in formation of the neuromuscular junction and motor neuron function (Ruggiu et al. 2009) and in Reelin signaling and neuronal migration (Park & Curran 2010, Yano et al. 2010) (see below).

An important adjunct to the biochemical and genetic analysis of Nova targets was a focused set of bioinformatic studies. Initially, these assessed whether the first 50 Nova targets harbored YCAY binding motifs and led to the unexpected discovery of a position-dependent map governing splicing regulation in neurons (Ule et al. 2006). Specifically, the position of binding within the primary transcript was a key determinant of the outcome of splicing regulation such that NOVA binding in upstream/within alternative exons mediated their exclusion, whereas downstream intronic binding enhanced alternative exon inclusion (Licatalosi et al. 2008, Licatalosi & Darnell 2010). Such positional effects are now recognized (Blencowe et al. 2009, Chen & Manley 2009, Corrionero & Valcarcel 2009, Llorian et al. 2010, Witten & Ule 2011) as a general feature of splicing control applicable to more than one dozen RNABPs (Charizanis et al. 2012, Chen & Manley 2009, Ince-Dunn et al. 2012, Kalsotra et al. 2008, Konig et al. 2010, Licatalosi et al. 2012, Licatalosi & Darnell 2006, Llorian et al. 2010, Tollervey et al. 2011, Wang et al. 2012, Xue et al. 2009, Yeo et al. 2009, Yuan et al. 2007, Zhang et al. 2008).

The effort to develop unbiased, genome-wide assessments that target transcripts directly regulated by NOVA led to the development of CLIP (Ule et al. 2003). These studies, subsequently combined with HITS-CLIP (Licatalosi et al. 2008), and more sophisticated computational methods (e.g., Bayesian network analysis of NOVA datasets; Zhang et al. 2010) led investigators to identify a robust set of ~700 alternative exons and a smaller number of alternative 3′ UTRs regulated by Nova in the mouse brain (as recently reviewed in Darnell 2006. 2010; Licatalosi & Darnell 2010). These data confirmed the biologic coherence of NOVA targets, indicating links between NOVA and a broader set of neurologic disorders, including autism, an observation strengthened by the experimentally demonstrated prediction that Nova regulates alternative splicing in a combinatorial manner with RBFOX1 (Zhang et al. 2010), itself a splicing factor implicated in autism (Voineagu et al. 2011).

Moreover, these studies laid the groundwork for applying genome-wide HITS-CLIP to help solve discrete biologic problems. The observation that *Nova*-null mice have neuronal migration defects similar to those seen with defects in the Reelin pathway led to a focused HITS-CLIP study on the developing neocortex (Park & Curran 2010, Yano et al. 2010). Focusing on transcripts encoding proteins in this pathway, HITS-CLIP identified one NOVA binding target—intronic YCAY elements within the transcript encoding Dab1, a key signaling protein in the Reelin pathway—that regulated an alternative exon encoding a new protein domain in Dab1. Functional studies, including in utero electroporation, demonstrated a specific role for NOVA regulation of this exon in neuronal migration.

The finding that NOVA regulated GlyRα2 splicing (Buckanovich & Darnell 1997, Jensen et al. 2000a) was linked to GlyRα2 mRNA localization in the dendrite (Racca et al. 1997) by analysis of nuclear and cytoplasmic HITS-CLIP. In this study, HITS-CLIP showed that intronic binding to regulate splicing could be coupled *in cis* with 3′ UTR binding to regulate mRNA localization and, presumably, dendritic protein translation (Racca et al. 2010). Most recently, NOVA HITS-CLIP was used to study transcripts whose steady-state mRNA levels were altered in *Nova*-null mice. This study led to the unexpected finding that large NOVA-dependent changes in protein production were regulated by intronic binding of NOVA to regulate the inclusion of cryptic exons that triggered nonsense-mediated mRNA decay (Eom et al. 2013). These included massive changes in the expression of a number of synaptic proteins, including several implicated in epilepsy, and indeed *Nova*-heterozygous mice were found to have spontaneous epilepsy. Taken together, these studies illustrate the potential of HITS-CLIP to provide sets of functionally relevant RNA-protein interactions that can predict the role of RNA regulation in neuronal biology, and they suggest links between neuronal RNA regulation of splicing and protein production to neuronal excitatory/inhibitory homeostasis.

NeuN/RbFox3

Investigators have only recently recognized that the neuronal marker NeuN is an RNABP in the RBFOX family. This finding was discovered by using the NeuN monoclonal antibody for immunoprecipitation-mass spectrometry analysis (Kim et al. 2009) and, independently, through expression of cDNA screening (Dredge & Jensen 2011). The RBFOX protein family has three paralogs, RBFOX1–3. Each protein has a single RRM-type RNA binding domain and has overlapping expression in neurons: RBFOX1 is expressed in neurons, heart, and muscle; RBFOX2 (also known as Rbm9) is expressed in neurons and other cell types, including hematopoietic and stem cells; and RBFOX3 is neuron-specific (Kuroyanagi 2009). To date, the best-studied paralog among these proteins is RBFOX1. RBFOX1 was originally identified in a yeast two-hybrid screen (Shibata et al. 2000) as a protein that interacts with the spinocerebellar ataxia type-2 gene product SCA2 (hence the original name for RBFOX1 was ataxin-2 binding protein 1, A2BP1); these interactions were subsequently confirmed for other family members (Lim et al. 2006). In vitro RNA selection with RBFOX1 identified a strong consensus binding sequence as (U)GCAUG (Jin et al. 2003).

HITS-CLIP analyses have been accomplished with RBFOX1 and RBFOX2. In embryonic stem (ES) cells, RBFOX2 bound to GCAUG clusters (Yeo et al. 2009), consistent with the traditional biochemical studies. The position of these clusters around alternative exons conformed to the position-dependent splicing map defined for Nova, consistent with earlier bioinformatics predictions made in analyzing UGCAUG binding sites (Castle et al. 2008) and supporting a role for RBFOX2 as a splicing factor in neurons that was predicted from biochemical analyses (Underwood et al. 2005). Studies also identified a number of binding clusters that did not harbor this element, and whether this relates to additional biologic roles for RBFOX2 or signal:noise problems in these experiments remains unclear. Many of the predicted RBFOX2 alternatively spliced target transcripts themselves encoded splicing factors (Yeo et al. 2009), suggesting the possibility of a higher-order regulatory network.

Biologic roles for RBFOX proteins in neurons have been studied in detail by Black and colleagues. Initial studies in tissue-culture cells identified position-dependent UGCAUG elements that regulated splicing of two Cav1.1 calcium channel exons, and regulation was abrogated by mutating the elements to

UGCGUG (Tang et al. 2009). Studies in P19 cells implicated a role for activity-dependent regulation of alternative splicing for several transcripts (Lee et al. 2009), findings supported by observing spontaneous seizures and widespread changes in alternative splicing in mice in which *RbFox1* had been specifically deleted in neurons with a nestin-Cre driver (Gehman et al. 2011). These splicing changes may be mediated, in part, by changes in *RbFox1* alternative splicing, which leads to a shift of the protein from the cytoplasm to the nucleus (Lee et al. 2009). In neurons, the net result of these splicing changes is to impact a biologically coherent set of transcripts encoding proteins mediating synaptic transmission and membrane excitability (Gehman et al. 2011). Because Geschwind and colleagues have implicated RBFOX1-mediated splicing defects as contributory to autism (Voineagu et al. 2011), it will be interesting to assess whether the activity-regulated splicing targets identified in mice overlay those associated with human disease.

Ptbp2

PTBP2 (also termed nPTB or brPTB) is a neuronal RNA binding protein harboring four RRM-type RNA binding motifs. PTBP2 was independently identified in a yeast two-hybrid screen for NOVA interacting proteins (Polydorides et al. 2000) and in a biochemical purification scheme to identify factors that contribute to alternative splicing of c-src in neurons (Markovtsov et al. 2000). The name derives from the high homology (73% identity) with the polypyrimidine tract-binding protein (PTB, now termed PTBP1), a well-studied protein that regulates alternative splicing among other aspects of RNA metabolism (Spellman et al. 2005). Although PTBP1 is expressed in most tissues, it is either not expressed or minimally expressed in the brain. Instead, the brain expresses PTBP2, the product of a distinct gene.

Much about PTBP2 function can be inferred from a wealth of studies on PTBP1. In vitro selection and biochemical studies identified UCUU elements in the context of a pyrimidine-rich region (Perez et al. 1997, Singh et al. 1995), as well as CUCUCU (CU-repeats) present in *c-src* pre-mRNA (Chan & Black 1997), as high-affinity binding sites. Whereas these studies suggested that PTBP1 functioned as a repressor of alternative splicing, splicing microarrays (Boutz et al. 2007, Llorian et al. 2010) and PTBP1 HITS-CLIP studies (Xue et al. 2009) revealed a smaller but significant role for PTBP1 as a splicing enhancer. Careful analysis of the latter studies (Llorian et al. 2010), along with biochemical validation, confirmed that PTBP1 generally acts according to the rules of the position-dependent splicing map identified for NOVA and other proteins.

Identification of these binding elements in a Nova-regulated transcript revealed that PTBP2 interacts with Nova functionally, antagonizing its ability to enhance GlyRα2 exon 3a inclusion, at least in cotransfection assays (Polydorides et al. 2000). However, these studies, and a series of careful in vitro studies assessing PTBP1/2 regulation of c-src splicing (Markovtsov et al. 2000), including the observation that PTBP1 can inhibit c-src splicing through binding to U1 snRNP (Sharma et al. 2011), do not address the role of PTBP2 in neurons.

Investigators first undertook such analyses by addressing the role of PTBP1/2 in neuronal differentiation. In different cell culture models, undifferentiated cells expressed PTBP1, which is proposed to bind PTBP2 pre-mRNA and trigger alternative splicing of an isoform harboring a nonsense-mediated decay isoform, thereby suppressing PTBP2 expression (Boutz et al. 2007). During neuronal differentiation, the switch to PTBP2 expression is proposed to be mediated by miR-124 induction in neural cells because miR-124 can suppress PTBP1 expression (Makeyev et al. 2007).

This elegant model for neuronal differentiation was updated following analysis of PTBP2 RNA targets and insight into PTBP2 function in the brain that came from HITS-CLIP studies (Licatalosi et al. 2012). Generation of a PTBP2 binding map confirmed many similarities with PTBP1, including binding to UCU-rich target sites, and use of the same position-dependent map governing the regulation of alternative splicing inhibition and enhancement. In some cases, in which transcripts such as *Actn1* were expressed in both brain and nonneuronal cells, PTBP2 cross-linked precisely to sites previously mapped as sites governing PTBP1 inhibition of alternative exon inclusion (Matlin et al. 2007). In addition, PTBP2 HITS-CLIP defined new sites in the c-src primary transcript associated with alternative splicing of the *c-src* N1 exon and a second developmentally regulated alternate "N2" exon (Licatalosi et al. 2012).

Licatalosi et al. (2012) found PTBP2 to be expressed early in development in both mitotic neuronal progenitors and in early postmitotic neurons. Moreover, they identified ectopic nests of neuronal progenitors in *Ptbp2*-null mice, suggesting that PTBP2 acts earlier than originally thought in the neuronal differentiation switch. Transcripts identified by HITS-CLIP as bound and regulated by PTBP2 in the E18.5 embryonic brain included RNAs encoding determinants of cell division, polarity, and cell fate, including *Prkci*, *Numb*, *Brat*, *Trim3*, *Prox1*, *Erbb*, and others (Licatalosi et al. 2012). Taken together, these suggest that PTBP2 may play a role in neuronal progenitors to inhibit cell division, maintain cell polarity, and mediate the temporal control of neurogenesis.

nSR100

Blencowe and colleagues set out to identify SR proteins expressed in a tissue-specific manner, using microarray profiling to assess expression of genes encoding RS domain proteins in 50 mouse cell lines and tissues (Calarco et al. 2009). One of these, termed nSR100, was found to have a highly neural-restricted pattern by analysis of RT-PCR and Western blot analysis of cell lines and by RT-PCR analysis of different mouse tissues, although such nonquantitative analysis leaves open the possibility of scattered and low-level expression in some other tissues, as evidenced by the authors' analysis of mRNA expression by microarray.

nSR100 has been previously reported to be a tumor antigen, reminiscent of the identification of the paraneoplastic antigens Nova and Hu/Elavl (Behrends et al. 2003). However, in this instance, nSR100 was identified in a screen with low titer (1:100 dilution) antisera taken from pediatric patients with medulloblastoma and was identified as an immunoreactive protein in 2/5 medulloblastoma patients and 2/40 healthy volunteers, such that the significance of this finding is uncertain. Nonetheless, this observation suggested a possible connection with neural differentiation, and indeed, nSR100 knockdown impaired neuritic extension in differentiating N2A cells, inhibited ESC differentiation into neurospheres, and disrupted neural differentiation in developing zebrafish (Calarco et al. 2009).

These biologic actions correlate with target transcripts whose splicing was regulated by nSR100. In particular, splicing microarrays were used to identify regulated transcripts and a pyrimidine-rich binding motif through which nSR100 can act, suggesting possible interactions with other splicing factors that bind such motifs. Indeed, nSR100 was found to promote the inclusion of the key *Ptbp2* exon 10, previously implicated in promoting neural differentiation (Calarco et al. 2009). These studies nicely illustrate the potential interconnections among neuronal-specific RNABPs to promote neuron-specific pathways of development.

RNA BINDING PROTEINS WITH SPECIAL ROLES IN NEURONS

This category of proteins is not entirely fair to demarcate because only a few from among a very large number of RNABPs are discussed, their relationship to neuron-specific biology is in some cases unclear, and space constraints severely limit their discussion. Nonetheless, several RNABPs are of particular interest here because their characterization in neurons supports the possibility that neurons may regulate RNA in unique ways. Several excellent reviews discuss RNABPs more generally (Calarco et al. 2011, Cooper et al. 2009, Poulos et al. 2011,

Richard 2010), including those focused on generally expressed RNABPs associated with motor neuron disorders that may have undefined neuronal actions (Battle et al. 2006, Da Cruz & Cleveland 2011). Additional aspects of RNA complexity that are of great interest in neurons, including alternative polyadenylation and RNA editing, but that go beyond the current discussion have been recently reviewed (Silberberg & Ohman 2011, Weill et al. 2012).

CELF

CELF proteins are a family of RRM-containing proteins with homology to ELAV. The proteins bind to pyrimidine-rich stretches and can activate splicing by competing with the PTB family of proteins (Chen & Manley 2009, Spellman et al. 2005). CELF1 (CUGBP1) is perhaps the best-studied member of this family because it was originally linked to binding of CUG triplet repeats in the pathogenesis of myotonic dystrophy (Cooper et al. 2009; although see MBNL below). Most CELF proteins are widely expressed and have been implicated in diverse functions, ranging from splicing regulation to mRNA localization, translation, stability, and processing (Barreau et al. 2006). As a group, CELF proteins may also have important actions in the brain; for example, CELF1 specifically has been reported to regulate alternative splicing of the tau exon 10 in the brain (Dhaenens et al. 2011). CELF4 is reported as being neuron-specific in the adult brain (Meins et al. 2002), although it is clearly expressed more widely early in development and demarcation of its expression has been complicated by cross-reactivity with other family members. Nonetheless, CELF4 haploinsufficient and null mice develop spontaneous seizures; this phenotype is evident only when CELF4 is deleted early in development, as deletion in adult mice using a floxed allele does not result in epilepsy (Wagnon et al. 2011).

CPEB

CPEB1 is the best characterized of a family of paralogous proteins that play important roles in translational regulation. CPEB proteins harbor 2 RRM-type RNA binding domains and act by binding defined 3′ UTR elements to regulate poly(A) tail length and thereby mRNA translation (Richter 2007). CPEB1 plays important roles in neuronal biology. The protein is present in neuronal dendrites, as well as the cell body, and it induces polyadenylation and translation of several brain transcripts (Richter 2010). Moreover, CPEB1-null mice have memory deficits and reduced LTP (Alarcon et al. 2004, Richter 2010). Burns et al. (2011) also recently linked CPEB1 to cell senescence, mediated in part through the fine titration of CPEB1 levels by miR-122 and consequent effects on p53 translation (Burns et al. 2011). Identification of the set of directly bound CPEB1 transcripts will be of great interest in further understanding the action of the RNABP in neuronal plasticity and, perhaps, to neuronal aging.

FMRP

Fragile X syndrome was the first human neurologic disease clearly linked to dysfunction of an RNABP (Bhakar et al. 2012, O'Donnell & Warren 2002), following the discoveries that triplet repeat expansions cause loss of expression of the *Fmr1* gene and that the protein product, FMRP, binds to ribohomopolymers (Siomi et al. 1993) and poly(A)+ mRNA (Ashley et al. 1993). Identification of FMRP RNA ligands using in vitro selection identified binding preferences [G-quadruplex RNA that binds the RGG domain (Darnell et al. 2001) and kissing complex RNAs that bind the KH domains (Darnell et al. 2005)], complex structures that have been difficult to use to predict target transcripts. FMRP HITS-CLIP recently identified a robust set of target RNAs bound by FMRP in mouse brain, where its expression is restricted largely to neurons (Christie et al. 2009). In contrast with other HITS-CLIP analyses of RNABPs, FMRP bound along the coding sequence of a discrete set of mRNAs (Darnell et al. 2011). This result in turn suggested a direct role for FMRP in translational regulation, consistent with its previously postulated function (Bear et al. 2004, Kelleher & Bear 2008). Development of an in vitro translation system, in which polyribosomes from wild-type or FMRP-mutant brain were used to assay FMRP-dependent translation on target transcripts, revealed that FMRP does indeed function to inhibit translation in the brain by causing stalling of ribosome elongation (Darnell et al. 2011). How precisely this translational inhibition is itself regulated in neurons remains uncertain but may involve FMRP dephosphorylation/phosphorylation (Lee et al. 2011, Narayanan et al. 2007) and/or interaction with miRNA-mediated controls (Edbauer et al. 2010, Jin et al. 2004, Muddashetty et al. 2011).

Analysis of the set of FMRP-regulated brain transcripts revealed a robust role in regulating expression of proteins in the pre- and post-synaptic proteome. Moreover, consistent with the observation that ~30% of children with fragile X syndrome also have autism spectrum disorders (Wang et al. 2010b), a high degree of overlap with autism candidate genes was found (Darnell et al. 2011). More recently, overlaying FMRP target transcripts with a data set of de novo mutations identified in autistic children revealed that 1 in 5 autism candidate mutations were also regulated by FMRP (Iossifov et al. 2012), a remarkable overlap that tightens our view of the pathways that may lead to autism. Together, these studies illustrate successful use of the paradigm leading from basic biochemistry to in vivo function and demonstrate a means by which careful analysis of RNABP-RNA interactions can lead to new disease insight.

Musashi

Musashi was discovered in a *Drosophila* screen for genes involved in sensory organ development—mutants failed to develop neurons from sensory organ precursor cells (Nakamura et al. 1994). The protein harbors an RRM domain, and subsequent careful analysis of its expression in *Drosophila* revealed it to be highly restricted to the nervous system. Within the mouse brain, expression is confined largely to periventricular neuronal progenitors, with very little immunohistochemical reactivity evident in postmitotic neurons (Sakakibara et al. 1996, Sakakibara & Okano 1997). These observations have suggested that Musashi may play a key role in mediating an RNA switch from neuronal progenitors to neurons (Okano et al. 2005), for example by regulating expression of key target RNAs such as those encoding members of the Notch/Numb signaling pathway (Imai et al. 2001, Okabe et al. 2001).

MBNL

The MBNL family of proteins has been widely studied in the context of the muscular disorder myotonic dystrophy (DM), where sequestration of these proteins by CUG repeats plays a key role in disease pathogenesis. MBNL proteins contain four CCCH zinc finger motifs and bind to YGCY motifs in pre-mRNA to regulate alternative splicing (Poulos et al. 2011). Researchers have made significant progress in understanding the muscle disease; most notably, the finding that mis-splicing of the chloride channel *Clcn1*, induced by sequestration of MBNL1, is a major contributor to the symptoms of myotonia in DM (Kanadia et al. 2003, Wheeler et al. 2007) and is a target for therapeutic intervention (Wheeler et al. 2009). However, only very recently have studies begun to consider the central nervous system features of DM. These include sleep disturbance, the most common CNS complaint, as well as sporadic instances of mental retardation, autism, and other cognitive difficulties (Charizanis et al. 2012).

To address the role of MBNL proteins in neurons, two groups recently completed analysis of MBNL-dependent splicing changes together with bioinformatic and HITS-CLIP studies to identify sets of directly bound MBNL-regulated transcripts in the brain (Charizanis et al. 2012, Wang et al. 2012). Swanson and colleagues studied MBNL2-null mice, revealing sleep abnormalities, memory loss, impaired NMDA receptor–dependent transmission, and hippocampal synaptic plasticity. These defects were correlated with aberrant splicing regulation of MBNL2 target transcripts, defined by exon junction microarray, HITS-CLIP, and validated as targets in both MBNL2-null mice and DM human tissue. These results led to a splicing map showing position-dependent splicing changes similar to that described for Nova and identified target RNAs that encode proteins involved in neuronal differentiation, development, axon guidance, and synaptic function (Charizanis et al. 2012). Burge and colleagues evaluated MBNL1 function in a similar set of studies using MBNL1 mutant mice, RNA-Seq, and HITS-CLIP (which they term CLIP-Seq) to identify transcripts whose splicing was directly regulated by MBNL1 and a splicing map, again consistent with prior position-dependent maps. The authors also performed crude subcellular purifications to identify transcripts bound by MBNL1 in 3′ UTRs (identified by CLIP) whose cellular localization was MBNL1-dependent.

ZBP

The ZBP proteins are a family of RNABPs harboring four KH-type RNA binding motifs that have been well studied for their ability to regulate dendritic localization of mRNAs within neurons. This ability was originally demonstrated in studies by Singer and colleagues, who demonstrated β-actin mRNA localization in neuronal dendrites (Gu et al. 2002) and subsequently linked it to β-actin translation (Huttelmaier et al. 2005). Characterization of a consensus binding motif for ZBP1, a bipartite "zip-code" with a defined spacer, has allowed investigators to predict a subset of candidate target transcripts (Patel et al. 2012). Extension of these studies more systematically into neurons is ongoing, as are physiologic correlates, such as observations of decreased axonal length and outgrowth in response to injury in ZBP1$^{+/-}$ mice (Donnelly et al. 2011). Analyzing the role of ZBP proteins in neurons provides one of the most robust systems developed for understanding links between mRNA localization in a translationally silent state and its switch to active translation following cell signaling, for example through protein phosphorylation (Huttelmaier et al. 2005).

Ago-miRNAs in Neurons

A burgeoning topic in neurobiology relates to the degree and mechanism by which miRNAs may play special roles in mediating neuronal biology. This topic is worthy of an entire review in and of itself, and fortunately several timely and interesting such pieces are available (Mendell & Olson 2012, O'Carroll & Schaefer 2012, Siegel et al. 2011). Several points are worth mentioning here regarding miRNA regulation specifically in neurons. First, from a methodologic point of view, technologies are now in place to analyze miRNA regulation using Ago HITS-CLIP in the brain (Chi et al. 2009), and it seems likely that Ago HITS-CLIP will be able to be applied to specific neuron populations (He et al. 2012).

Second, the biology of miRNA action in neurons is rapidly evolving, with several main themes emerging. Studies cataloging miRNA abundance has revealed that several are expressed specifically in the brain (Fiore et al. 2008), and some show enrichment in neuronal dendrites (Siegel et al. 2011). Analysis of one such localized miRNA, miR-134, initially raised the possibility that miRNAs may act within dendrites to modulate local mRNA translation (Schratt et al. 2006). This action on translation was evident after BDNF signaling, an interesting connection given that BDNF is well documented to trigger local protein synthesis in dendrites (Aakalu et al. 2001, Sutton et al. 2004). Another line of studies has indicated that miRNAs can rapidly turn over in neurons in response to stimuli, as first illustrated by Filipowicz and colleagues in the retina (Krol et al. 2010). These observations have been recently connected with the findings that BDNF rapidly induces expression of some miRNAs, as well as neuronal Dicer and Lin28a (an RNABP that can block miRNA production; Hagan et al. 2009, Heo et al. 2009), and that these coordinated effects may help mediate the specificity of translational response to BDNF (Huang et al. 2012).

Finally, an emerging story from *Drosophila* studies suggests that miRNAs may play unexpected roles in chronically suppressing expression of developmental genes in the adult brain and that such actions may be important to prevent brain aging and neurodegeneration (Liu et al. 2012). Such a role turns the traditional view of miRNAs in brain function—as mediators of developmental changes (miR-124, see discussion of PTBP2) or of rapid signaling responses—on its head, portraying them as long-term guardians against inappropriate gene expression.

FUTURE DIRECTIONS

Technologies are now emerging to analyze RNA regulation within specific neuronal populations. This is crucial because the brain is composed of many thousands of different cell types, and combining data from each obscures signal:noise. Indeed this point has been underscored by the wide appreciation of projects, such as the Gensat and Allen Brain Atlas projects, that enumerated neuron-specific transcripts and their subsequent use to drive Cre recombinase in specific neuronal cell types. Such technologies have been combined with epitope tagging of RNABPs, a method that works to purify cell-specific Ago-miRNA complexes (He et al. 2012) and, with epitope-tagged ribosomes, to assess cell-specific translation. For example, in bacTRAP translational profiling, cell-specific Cre drivers derived from GenSat to purify ribosomes from individual neuronal cell types (Doyle et al. 2008). Combined with drivers that discriminate D1 and D2 dopaminergic neurons, Heintz and colleagues identified transcripts specifically impacted by cocaine administration within D1 neurons that are likely to lead to changes in D1-specific synaptic responses (Heiman et al. 2008). In a second example, profiling of cortical layer Va, known to be responsive to antidepressant treatment, led to the identification of a neuron-specific protein, p11, necessary for physiologic actions on a specific serotonin receptor (HTR4) and behavioral responses to chronic administration of serotonin-specific reuptake inhibitors (Schmidt et al. 2012).

These observations suggest additional approaches that may be applied to HITS-CLIP profiling (Darnell 2010, Konig et al. 2012) or ribosomal footprinting (Ingolia et al. 2009) to develop detailed maps of functional RNA-protein interactions within individual neuronal cell types. Given the ability to map Ago-miRNA binding sites with HITS-CLIP, they also suggest possible means to map neuron cell-specific miRNA controls. Finally, as high-throughput sequencing becomes more potent, perhaps through the use of new direct RNA sequencing methods, these profiles may become applicable to subcellular compartments within neurons.

DISCLOSURE STATEMENT

The author is not aware of any affiliations, memberships, funding, or financial holdings that might be perceived as affecting the objectivity of this review.

ACKNOWLEDGMENTS

The work presented here incorporates the efforts of many tireless colleagues and hours of labor, for which the author is deeply indebted. Apologies to the many RNABPs that function in neurons but were not discussed here given space issues. This work was supported by NIH R01 NS34389, an NIH Transformative Research Award (NS081706), the Rockefeller University Hospital CTSA (UL1 RR024143), the Starr Cancer Consortium, the ALS Therapy Alliance, a Simons Foundation Research Award, and an Emerald Foundation Distinguished Investigator Award. RBD is an Investigator of the Howard Hughes Medical Institute.

LITERATURE CITED

Aakalu G, Smith WB, Nguyen N, Jiang C, Schuman EM. 2001. Dynamic visualization of local protein synthesis in hippocampal neurons. *Neuron* 30:489–502

Akamatsu W, Fujihara H, Mitsuhashi T, Yano M, Shibata S, et al. 2005. The RNA-binding protein HuD regulates neuronal cell identity and maturation. *Proc. Natl. Acad. Sci. USA* 102:4625–30

Akamatsu W, Okano HJ, Osumi N, Inoue T, Nakamura S, et al. 1999. Mammalian ELAV-like neuronal RNA-binding proteins HuB and HuC promote neuronal development in both the central and the peripheral nervous systems. *Proc. Natl. Acad. Sci. USA* 96:9885–90

Alarcon JM, Hodgman R, Theis M, Huang YS, Kandel ER, Richter JD. 2004. Selective modulation of some forms of schaffer collateral-CA1 synaptic plasticity in mice with a disruption of the CPEB-1 gene. *Learn. Mem.* 11:318–27

Amara SG, Jonas V, Rosenfeld MG, Ong ES, Evans R. 1982. Alternative RNA processing in calcitonin gene expression generates mRNAs encoding different polypeptide products. *Nature* 298:240–44

Antic D, Lu N, Keene JD. 1999. ELAV tumor antigen, Hel-N1, increases translation of neurofilament M mRNA and induces formation of neurites in human teratocarcinoma cells. *Genes Dev.* 13:449–61

Ashley CTJ, Wilkinson KD, Reines D, Warren ST. 1993. FMR1 protein: conserved RNP family domains and selective RNA binding. *Science* 262:563–66

Bakker CE, Verheij C, Willemsen R, van der Helm R, Oerlemans F, et al. 1994. Fmr1 knockout mice: a model to study fragile X mental retardation. The Dutch-Belgian Fragile X Consortium. *Cell* 78:23–33

Barreau C, Paillard L, Méreau A, Osborne HB. 2006. Mammalian CELF/Bruno-like RNA-binding proteins: molecular characteristics and biological functions. *Biochimie* 88:515–25

Battle DJ, Kasim M, Yong J, Lotti F, Lau CK, et al. 2006. The SMN complex: an assembly machine for RNPs. *Cold Spring Harb. Symp. Quant. Biol.* 71:313–20

Bear MF, Huber KM, Warren ST. 2004. The mGluR theory of fragile X mental retardation. *Trends Neurosci.* 27:370–77

Behrends U, Schneider I, Rossler S, Frauenknecht H, Golbeck A, et al. 2003. Novel tumor antigens identified by autologous antibody screening of childhood medulloblastoma cDNA libraries. *Int. J. Cancer* 106:244–51

Bhakar AL, Dolen G, Bear MF. 2012. The pathophysiology of fragile X (and what it teaches us about synapses). *Annu. Rev. Neurosci.* 35:417–43

Bier E, Ackerman L, Barbel S, Jan L, Jan YN. 1988. Identification and characterization of a neuron-specific nuclear antigen in *Drosophila*. *Science* 240:913–16

Blanchette M, Green RE, Brenner SE, Rio DC. 2005. Global analysis of positive and negative pre-mRNA splicing regulators in *Drosophila*. *Genes Dev.* 19:1306–14

Blencowe BJ, Ahmad S, Lee LJ. 2009. Current-generation high-throughput sequencing: deepening insights into mammalian transcriptomes. *Genes Dev.* 23:1379–86

Bolognani F, Perrone-Bizzozero NI. 2008. RNA-protein interactions and control of mRNA stability in neurons. *J. Neurosci. Res.* 86:481–89

Boutz PL, Stoilov P, Li Q, Lin CH, Chawla G, et al. 2007. A post-transcriptional regulatory switch in polypyrimidine tract-binding proteins reprograms alternative splicing in developing neurons. *Genes Dev.* 21:1636–52

Buckanovich RJ, Darnell RB. 1997. The neuronal RNA binding protein Nova-1 recognizes specific RNA targets in vitro and in vivo. *Mol. Cell. Biol.* 17:3194–201

Buckanovich RJ, Posner JB, Darnell RB. 1993. Nova, the paraneoplastic Ri antigen, is homologous to an RNA-binding protein and is specifically expressed in the developing motor system. *Neuron* 11:657–72

Buckanovich RJ, Yang YY, Darnell RB. 1996. The onconeural antigen Nova-1 is a neuron-specific RNA-binding protein, the activity of which is inhibited by paraneoplastic antibodies. *J. Neurosci.* 16:1114–22

Burd CG, Dreyfuss G. 1994. Conserved structures and diversity of functions of RNA-binding proteins. *Science* 265:615–21

Burns DM, D'Ambrogio A, Nottrott S, Richter JD. 2011. CPEB and two poly(A) polymerases control miR-122 stability and p53 mRNA translation. *Nature* 473:105–8

Calarco JA, Superina S, O'Hanlon D, Gabut M, Raj B, et al. 2009. Regulation of vertebrate nervous system alternative splicing and development by an SR-related protein. *Cell* 138:898–910

Calarco JA, Zhen M, Blencowe BJ. 2011. Networking in a global world: establishing functional connections between neural splicing regulators and their target transcripts. *RNA* 17:775–91

Campos AR, Rosen DR, Robinow SN, White K. 1987. Molecular analysis of the locus elav in *Drosophila melanogaster*: a gene whose embryonic expression is neural specific. *EMBO J.* 6:425–31

Castle JC, Zhang C, Shah JK, Kulkarni AV, Kalsotra A, et al. 2008. Expression of 24,426 human alternative splicing events and predicted cis regulation in 48 tissues and cell lines. *Nat. Genet.* 40:1416–25

Chan RC, Black DL. 1997. The polypyrimidine tract binding protein binds upstream of neural cell-specific c-src exon N1 to repress the splicing of the intron downstream. *Mol. Cell. Biol.* 17:4667–76

Charizanis K, Lee KY, Batra R, Goodwin M, Zhang C, et al. 2012. Muscleblind-like 2-mediated alternative splicing in the developing brain and dysregulation in myotonic dystrophy. *Neuron* 75:437–50

Chen CY, Shyu AB. 2011. Mechanisms of deadenylation-dependent decay. *Wiley Interdiscip. Rev. RNA* 2:167–83

Chen M, Manley JL. 2009. Mechanisms of alternative splicing regulation: insights from molecular and genomics approaches. *Nat. Rev. Mol. Cell Biol.* 10:741–54

Chi SW, Zang JB, Mele A, Darnell RB. 2009. Argonaute HITS-CLIP decodes microRNA-mRNA interaction maps. *Nature* 460:479–86

Christie SB, Akins MR, Schwob JE, Fallon JR. 2009. The FXG: a presynaptic fragile X granule expressed in a subset of developing brain circuits. *J. Neurosci.* 29:1514–24

Cooper TA, Wan L, Dreyfuss G. 2009. RNA and disease. *Cell* 136:777–93

Corrionero A, Valcarcel J. 2009. RNA processing: redrawing the map of charted territory. *Mol. Cell* 36:918–19

Da Cruz S, Cleveland DW. 2011. Understanding the role of TDP-43 and FUS/TLS in ALS and beyond. *Curr. Opin. Neurobiol.* 21:904–19

Darnell JC, Fraser CE, Mostovetsky O, Stefani G, Jones TA, et al. 2005. Kissing complex RNAs mediate interaction between the Fragile-X mental retardation protein KH2 domain and brain polyribosomes. *Genes Dev.* 19:903–18

Darnell JC, Jensen KB, Jin P, Brown V, Warren ST, Darnell RB. 2001. Fragile X mental retardation protein targets G Quartet mRNAs important for neuronal function. *Cell* 107:489–99

Darnell JC, Van Driesche SJ, Zhang C, Hung KY, Mele A, et al. 2011. FMRP stalls ribosomal translocation on mRNAs linked to synaptic function and autism. *Cell* 146:247–61

Darnell RB. 1996. Onconeural antigens and the paraneoplastic neurologic disorders: at the intersection of cancer, immunity, and the brain. *Proc. Natl. Acad. Sci. USA* 93:4529–36

Darnell RB. 2006. Developing global insight into RNA regulation. *Cold Spring Harb. Symp. Quant. Biol.* 71:321–27

Darnell RB. 2010. HITS-CLIP: panoramic views of protein-RNA regulation in living cells. *Wiley Interdiscip. Rev. RNA* 1:266–86

Darnell RB, DeAngelis LM. 1993. Regression of small-cell lung carcinoma in patients with paraneoplastic neuronal antibodies. *Lancet* 341:21–22

Darnell RB, Furneaux H, Posner JB. 1989. Characterization of antigens bound by CSF and serum of a patient with cerebellar degeneration: co-expression in Purkinje cells and tumor lines of neuroectodermal origin. *Neurology* 39:385

Darnell RB, Posner JB. 2003a. Observing the invisible: successful tumor immunity in humans. *Nat. Immunol.* 4:201

Darnell RB, Posner JB. 2003b. Paraneoplastic syndromes involving the nervous system. *N. Engl. J. Med.* 349:1543–54

Darnell RB, Posner JB. 2011. *Paraneoplastic Syndromes (Contemporary Neurology Series)*. New York: Oxford Univ. Press

Dean JL, Wait R, Mahtani KR, Sully G, Clark AR, Saklatvala J. 2001. The 3′ untranslated region of tumor necrosis factor alpha mRNA is a target of the mRNA-stabilizing factor HuR. *Mol. Cell. Biol.* 21:721–30

Dhaenens CM, Tran H, Frandemiche ML, Carpentier C, Schraen-Maschke S, et al. 2011. Mis-splicing of Tau exon 10 in myotonic dystrophy type 1 is reproduced by overexpression of CELF2 but not by MBNL1 silencing. *Biochim. Biophys. Acta* 1812:732–42

Donnelly CJ, Willis DE, Xu M, Tep C, Jiang C, et al. 2011. Limited availability of ZBP1 restricts axonal mRNA localization and nerve regeneration capacity. *EMBO J.* 30:4665–77

Dor Y, Brown J, Martinez OI, Melton DA. 2004. Adult pancreatic beta-cells are formed by self-duplication rather than stem-cell differentiation. *Nature* 429:41–46

Doyle JP, Dougherty JD, Heiman M, Schmidt EF, Stevens TR, et al. 2008. Application of a translational profiling approach for the comparative analysis of CNS cell types. *Cell* 135:749–62

Dredge BK, Darnell RB. 2003. Nova regulates GABA(A) receptor gamma2 alternative splicing via a distal downstream UCAU-rich intronic splicing enhancer. *Mol. Cell. Biol.* 23:4687–700

Dredge BK, Jensen KB. 2011. NeuN/Rbfox3 nuclear and cytoplasmic isoforms differentially regulate alternative splicing and nonsense-mediated decay of Rbfox2. *PLoS ONE* 6:e21585

Dredge BK, Stefani G, Engelhard CC, Darnell RB. 2005. Nova autoregulation reveals dual functions in neuronal splicing. *EMBO J.* 24:1608–20

Du H, Cline MS, Osborne RJ, Tuttle DL, Clark TA, et al. 2010. Aberrant alternative splicing and extracellular matrix gene expression in mouse models of myotonic dystrophy. *Nat. Struct. Mol. Biol.* 17:187–93

Edbauer D, Neilson JR, Foster KA, Wang CF, Seeburg DP, et al. 2010. Regulation of synaptic structure and function by FMRP-associated microRNAs miR-125b and miR-132. *Neuron* 65:373–84

Ellington AD, Szostak JW. 1990. In vitro selection of RNA molecules that bind specific ligands. *Nature* 346:818–22

Eom T, Zhang C, Wang H, Lay K, Fak J, Noebels JL, Darnell RB. 2013. NOVA-dependent regulation of cryptic NMD exons controls synaptic protein levels after seizure. *eLIFE* 2:e00178

Fan XC, Myer VE, Steitz JA. 1997. AU-rich elements target small nuclear RNAs as well as mRNAs for rapid degradation. *Genes Dev.* 11:2557–68

Fan XC, Steitz JA. 1998. Overexpression of HuR, a nuclear-cytoplasmic shuttling protein, increases the in vivo stability of ARE-containing mRNAs. *EMBO J.* 17:3448–60

Fiore R, Siegel G, Schratt G. 2008. MicroRNA function in neuronal development, plasticity and disease. *Biochim. Biophys. Acta* 1779:471–78

Gao FB, Carson CC, Levine T, Keene JD. 1994. Selection of a subset of mRNAs from combinatorial 3′ untranslated region libraries using neuronal RNA-binding protein Hel-N1. *Proc. Natl. Acad. Sci. USA* 91:11207–11

Gehman LT, Stoilov P, Maguire J, Damianov A, Lin CH, et al. 2011. The splicing regulator Rbfox1 (A2BP1) controls neuronal excitation in the mammalian brain. *Nat. Genet.* 43:706–11

Green R, Ellington AD, Bartel DP, Szostak JW. 1991. In vitro genetic analysis: selection and amplification of rare functional nucleic acids. *Methods Compan. Methods Enzymol.* 2:75–86

Gu W, Pan F, Zhang H, Bassell GJ, Singer RH. 2002. A predominantly nuclear protein affecting cytoplasmic localization of beta-actin mRNA in fibroblasts and neurons. *J. Cell Biol.* 156:41–51

Hafner M, Landthaler M, Burger L, Khorshid M, Hausser J, et al. 2010. Transcriptome-wide identification of RNA-binding protein and microRNA target sites by PAR-CLIP. *Cell* 141:129–41

Hagan JP, Piskounova E, Gregory RI. 2009. Lin28 recruits the TUTase Zcchc11 to inhibit let-7 maturation in mouse embryonic stem cells. *Nat. Struct. Mol. Biol.* 16:1021–25

Hansen TV, Hammer NA, Nielsen J, Madsen M, Dalbaeck C, et al. 2004. Dwarfism and impaired gut development in insulin-like growth factor II mRNA-binding protein 1-deficient mice. *Mol. Cell. Biol.* 24:4448–64

He M, Liu Y, Wang X, Zhang MQ, Hannon GJ, Huang ZJ. 2012. Cell-type-based analysis of microRNA profiles in the mouse brain. *Neuron* 73:35–48

Heiman M, Schaefer A, Gong S, Peterson JD, Day M, et al. 2008. A translational profiling approach for the molecular characterization of CNS cell types. *Cell* 135:738–48

Heo I, Joo C, Kim YK, Ha M, Yoon MJ, et al. 2009. TUT4 in concert with Lin28 suppresses microRNA biogenesis through pre-microRNA uridylation. *Cell* 138:696–708

Hinman MN, Lou H. 2008. Diverse molecular functions of Hu proteins. *Cell Mol. Life Sci.* 65:3168–81

Holt CE, Bullock SL. 2009. Subcellular mRNA localization in animal cells and why it matters. *Science* 326:1212–16

Homyk TJ, Szidonya J, Suzuki DT. 1980. Behavioral mutants of *Drosophila melanogaster*. III. Isolation and mapping of mutations by direct visual observations of behavioral phenotypes. *Mol. Gen. Genet.* 177:553–65

Hormigo A, Dalmau J, Rosenblum MK, River ME, Posner JB. 1994. Immunological and pathological study of anti-Ri-associated encephalopathy. *Ann. Neurol.* 36:896–902
Huang CS, Shi SH, Ule J, Ruggiu M, Barker LA, et al. 2005. Common molecular pathways mediate long-term potentiation of synaptic excitation and slow synaptic inhibition. *Cell* 123:105–18
Huang YW, Ruiz CR, Eyler EC, Lin K, Meffert MK. 2012. Dual regulation of miRNA biogenesis generates target specificity in neurotrophin-induced protein synthesis. *Cell* 148:933–46
Huttelmaier S, Zenklusen D, Lederer M, Dictenberg J, Lorenz M, et al. 2005. Spatial regulation of beta-actin translation by Src-dependent phosphorylation of ZBP1. *Nature* 438:512–15
Imai T, Tokunaga A, Yoshida T, Hashimoto M, Mikoshiba K, et al. 2001. The neural RNA-binding protein Musashi1 translationally regulates mammalian numb gene expression by interacting with its mRNA. *Mol. Cell. Biol.* 21:3888–900
Ince-Dunn G, Okano HJ, Jensen KB, Park WY, Zhong R, et al. 2012. Neuronal Elav-like (Hu) proteins regulate RNA splicing and abundance to control glutamate levels and neuronal excitability. *Neuron* 75:1067–80
Ingolia NT, Ghaemmaghami S, Newman JR, Weissman JS. 2009. Genome-wide analysis in vivo of translation with nucleotide resolution using ribosome profiling. *Science* 324:218–23
Iossifov I, Ronemus M, Levy D, Wang Z, Hakker I, et al. 2012. De novo gene disruptions in children on the autistic spectrum. *Neuron* 74:285–99
Jain RG, Andrews LG, McGowan KM, Pekala PH, Keene JD. 1997. Ectopic expression of Hel-N1, an RNA-binding protein, increases glucose transporter (GluT1) expression in 3T3-L1 adipocytes. *Mol. Cell. Biol.* 17:954–62
Jensen KB, Darnell RB. 2008. CLIP: crosslinking and immunoprecipitation of in vivo RNA targets of RNA-binding proteins. *Methods Mol. Biol.* 488:85–98
Jensen KB, Dredge BK, Stefani G, Zhong R, Buckanovich RJ, et al. 2000a. Nova-1 regulates neuron-specific alternative splicing and is essential for neuronal viability. *Neuron* 25:359–71
Jensen KB, Musunuru K, Lewis HA, Burley SK, Darnell RB. 2000b. The tetranucleotide UCAY directs the specific recognition of RNA by the Nova K-homology 3 domain. *Proc. Natl. Acad. Sci. USA* 97:5740–45
Jin P, Alisch RS, Warren ST. 2004. RNA and microRNAs in fragile X mental retardation. *Nat. Cell Biol.* 6:1048–53
Jin Y, Suzuki H, Maegawa S, Endo H, Sugano S, et al. 2003. A vertebrate RNA-binding protein Fox-1 regulates tissue-specific splicing via the pentanucleotide GCAUG. *EMBO J.* 22:905–12
Johnson JM, Castle J, Garrett-Engele P, Kan Z, Loerch PM, et al. 2003. Genome-wide survey of human alternative pre-mRNA splicing with exon junction microarrays. *Science* 302:2141–44
Kalsotra A, Xiao X, Ward AJ, Castle JC, Johnson JM, et al. 2008. A postnatal switch of CELF and MBNL proteins reprograms alternative splicing in the developing heart. *Proc. Natl. Acad. Sci. USA* 105:20333–38
Kanadia RN, Urbinati CR, Crusselle VJ, Luo D, Lee YJ, et al. 2003. Developmental expression of mouse muscleblind genes Mbnl1, Mbnl2 and Mbnl3. *Gene Expr. Patterns* 3:459–62
Kelleher R Jr, Bear MF. 2008. The autistic neuron: troubled translation? *Cell* 135:401–6
Kim KK, Adelstein RS, Kawamoto S. 2009. Identification of neuronal nuclei (NeuN) as Fox-3, a new member of the Fox-1 gene family of splicing factors. *J. Biol. Chem.* 284:31052–61
Kishore S, Jaskiewicz L, Burger L, Hausser J, Khorshid M, Zavolan M. 2011. A quantitative analysis of CLIP methods for identifying binding sites of RNA-binding proteins. *Nat. Methods* 8:559–64
Kishore S, Luber S, Zavolan M. 2010. Deciphering the role of RNA-binding proteins in the post-transcriptional control of gene expression. *Brief Funct. Genomics* 9:391–404
Konig J, Zarnack K, Luscombe NM, Ule J. 2012. Protein-RNA interactions: new genomic technologies and perspectives. *Nat. Rev. Genet.* 13:77–83
Konig J, Zarnack K, Rot G, Curk T, Kayikci M, et al. 2010. iCLIP reveals the function of hnRNP particles in splicing at individual nucleotide resolution. *Nat. Struct. Mol. Biol.* 17:909–15
Konig J, Zarnack K, Rot G, Curk T, Kayikci M, et al. 2011. iCLIP—transcriptome-wide mapping of protein-RNA interactions with individual nucleotide resolution. *J. Vis. Exp.* 30:pii:2638
Koushika SP, Lisbin MJ, White K. 1996. Elav, a *Drosophila* neuron-specific protein, mediates the generation of an alternatively spliced neural protein isoform. *Curr. Biol.* 6:1634–41
Koushika SP, Soller M, White K. 2000. The neuron-enriched splicing pattern of *Drosophila* erect wing is dependent on the presence of ELAV protein. *Mol. Cell. Biol.* 20:1836–45

Krol J, Busskamp V, Markiewicz I, Stadler MB, Ribi S, et al. 2010. Characterizing light-regulated retinal microRNAs reveals rapid turnover as a common property of neuronal microRNAs. *Cell* 141:618–31

Kuroyanagi H. 2009. Fox-1 family of RNA-binding proteins. *Cell Mol. Life Sci.* 66:3895–907

Lee EK, Kim W, Tominaga K, Martindale JL, Yang X, et al. 2012. RNA-binding protein HuD controls insulin translation. *Mol. Cell* 45:826–35

Lee HY, Ge WP, Huang W, He Y, Wang GX, et al. 2011. Bidirectional regulation of dendritic voltage-gated potassium channels by the fragile X mental retardation protein. *Neuron* 72:630–42

Lee JA, Tang ZZ, Black DL. 2009. An inducible change in Fox-1/A2BP1 splicing modulates the alternative splicing of downstream neuronal target exons. *Genes Dev.* 23:2284–93

Lerner MR, Boyle JA, Hardin JA, Steitz JA. 1981. Two novel classes of small ribonucleoproteins detected by antibodies associated with lupus erythematosus. *Science* 211:400–2

Levine TD, Gao F, King PH, Andrews LG, Keene JD. 1993. Hel-N1: an autoimmune RNA-binding protein with specificity for 3′ uridylate-rich untranslated regions of growth factor mRNA's. *Mol. Cell. Biol.* 13:3494–504

Lewis HA, Chen H, Edo C, Buckanovich RJ, Yang YY, et al. 1999. Crystal structures of Nova-1 and Nova-2 K-homology RNA-binding domains. *Structure* 7:191–203

Lewis HA, Musunuru K, Jensen KB, Edo C, Chen H, et al. 2000. Sequence-specific RNA binding by a Nova KH domain: implications for paraneoplastic disease and the fragile X syndrome. *Cell* 100:323–32

Licatalosi DD, Darnell RB. 2006. Splicing regulation in neurologic disease. *Neuron* 52:93–101

Licatalosi DD, Darnell RB. 2010. RNA processing and its regulation: global insights into biological networks. *Nat. Rev. Genet.* 11:75–87

Licatalosi DD, Mele A, Fak JJ, Ule J, Kayikci M, et al. 2008. HITS-CLIP yields genome-wide insights into brain alternative RNA processing. *Nature* 456:464–69

Licatalosi DD, Yano M, Fak JJ, Mele A, Grabinski SE, et al. 2012. Ptbp2 represses adult-specific splicing to regulate the generation of neuronal precursors in the embryonic brain. *Genes Dev.* 26:1626–42

Lim J, Hao T, Shaw C, Patel AJ, Szabo G, et al. 2006. A protein-protein interaction network for human inherited ataxias and disorders of Purkinje cell degeneration. *Cell* 125:801–14

Liu N, Landreh M, Cao K, Abe M, Hendriks GJ, et al. 2012. The microRNA miR-34 modulates ageing and neurodegeneration in *Drosophila*. *Nature* 482:519–23

Llorian M, Schwartz S, Clark TA, Hollander D, Tan LY, et al. 2010. Position-dependent alternative splicing activity revealed by global profiling of alternative splicing events regulated by PTB. *Nat. Struct. Mol. Biol.* 17:1114–23

Luque FA, Furneaux HM, Ferziger R, Rosenblum MK, Wray SH, et al. 1991. Anti-Ri: an antibody associated with paraneoplastic opsoclonus and breast cancer. *Ann. Neurol.* 29:241–51

Lynch MA. 2004. Long-term potentiation and memory. *Physiol. Rev.* 84:87–136

Magavi SS, Macklis JD. 2008. Immunocytochemical analysis of neuronal differentiation. *Methods Mol. Biol.* 438:345–52

Makeyev EV, Zhang J, Carrasco MA, Maniatis T. 2007. The microRNA miR-124 promotes neuronal differentiation by triggering brain-specific alternative pre-mRNA splicing. *Mol. Cell* 27:435–48

Markovtsov V, Nikolic JM, Goldman JA, Turck CW, Chou MY, Black DL. 2000. Cooperative assembly of an hnRNP complex induced by a tissue-specific homolog of polypyrimidine tract binding protein. *Mol. Cell. Biol.* 20:7463–79

Matlin AJ, Southby J, Gooding C, Smith CW. 2007. Repression of alpha-actinin SM exon splicing by assisted binding of PTB to the polypyrimidine tract. *RNA* 13:1214–23

McKeever MO, Darnell RB. 1992. NAP, a human cerebellar degeneration antigen, is a novel, neuron specific, adaptin family member. *Soc. Neurosci. Abstr.* 18:1092

Meins M, Schlickum S, Wilhelm C, Missbach J, Yadav S, et al. 2002. Identification and characterization of murine Brunol4, a new member of the elav/bruno family. *Cytogenet. Genome Res.* 97:254–60

Meisner NC, Filipowicz W. 2011. Properties of the regulatory RNA-binding protein HuR and its role in controlling miRNA repression. *Adv. Exp. Med. Biol.* 700:106–23

Mendell JT, Olson EN. 2012. MicroRNAs in stress signaling and human disease. *Cell* 148:1172–87

Ming GL, Wong ST, Henley J, Yuan XB, Song HJ, et al. 2002. Adaptation in the chemotactic guidance of nerve growth cones. *Nature* 417:411–18

Mortazavi A, Williams BA, McCue K, Schaeffer L, Wold B. 2008. Mapping and quantifying mammalian transcriptomes by RNA-Seq. *Nat. Methods* 5:621–28

Muddashetty RS, Nalavadi VC, Gross C, Yao X, Xing L, et al. 2011. Reversible inhibition of PSD-95 mRNA translation by miR-125a, FMRP phosphorylation, and mGluR signaling. *Mol. Cell* 42:673–88

Musunuru K, Darnell RB. 2001. Paraneoplastic neurologic disease antigens: RNA-binding proteins and signaling proteins in neuronal degeneration. *Annu. Rev. Neurosci.* 24:239–62

Myer VE, Fan XC, Steitz JA. 1997. Identification of HuR as a protein implicated in AUUUA-mediated mRNA decay. *EMBO J.* 16:2130–39

Nagalakshmi U, Wang Z, Waern K, Shou C, Raha D, et al. 2008. The transcriptional landscape of the yeast genome defined by RNA sequencing. *Science* 320:1344–49

Nakamura M, Okano H, Blendy JA, Montell C. 1994. Musashi, a neural RNA-binding protein required for *Drosophila* adult external sensory organ development. *Neuron* 13:67–81

Narayanan U, Nalavadi V, Nakamoto M, Pallas DC, Ceman S, et al. 2007. FMRP phosphorylation reveals an immediate-early signaling pathway triggered by group I mGluR and mediated by PP2A. *J. Neurosci.* 27:14349–57

Newman LS, McKeever MO, Okano HJ, Darnell RB. 1995. ®-NAP, a cerebellar degeneration antigen, is a neuron-specific vesicle coat protein. *Cell* 82:773–83

O'Carroll D, Schaefer A. 2012. General principals of miRNA biogenesis and regulation in the brain. *Neuropsychopharmacology* 38:39–54

O'Donnell WT, Warren ST. 2002. A decade of molecular studies of fragile X syndrome. *Annu. Rev. Neurosci.* 25:315–38

Okabe M, Imai T, Kurusu M, Hiromi Y, Okano H. 2001. Translational repression determines a neuronal potential in *Drosophila* asymmetric cell division. *Nature* 411:94–98

Okano H, Kawahara H, Toriya M, Nakao K, Shibata S, Imai T. 2005. Function of RNA-binding protein Musashi-1 in stem cells. *Exp. Cell Res.* 306:349–56

Okano HJ, Darnell RB. 1997. A hierarchy of Hu RNA binding proteins in developing and adult neurons. *J. Neurosci.* 17:3024–37

Pan Q, Shai O, Misquitta C, Zhang W, Saltzman AL, et al. 2004. Revealing global regulatory features of mammalian alternative splicing using a quantitative microarray platform. *Mol. Cell* 16:929–41

Park TJ, Curran T. 2010. Alternative splicing disabled by Nova2. *Neuron* 66:811–13

Patel VL, Mitra S, Harris R, Buxbaum AR, Lionnet T, et al. 2012. Spatial arrangement of an RNA zipcode identifies mRNAs under post-transcriptional control. *Genes Dev.* 26:43–53

Peng SS, Chen CY, Xu N, Shyu AB. 1998. RNA stabilization by the AU-rich element binding protein, HuR, an ELAV protein. *EMBO J.* 17:3461–70

Perez I, Lin CH, McAfee JG, Patton JG. 1997. Mutation of PTB binding sites causes misregulation of alternative 3′ splice site selection in vivo. *RNA* 3:764–78

Polydorides AD, Okano HJ, Yang YY, Stefani G, Darnell RB. 2000. A brain-enriched polypyrimidine tract-binding protein antagonizes the ability of Nova to regulate neuron-specific alternative splicing. *Proc. Natl. Acad. Sci. USA* 97:6350–55

Poulos MG, Batra R, Charizanis K, Swanson MS. 2011. Developments in RNA splicing and disease. *Cold Spring Harb. Perspect. Biol.* 3(1):a000778

Racca C, Gardiol A, Eom T, Ule J, Triller A, Darnell RB. 2010. The neuronal splicing factor Nova co-localizes with target RNAs in the dendrite. *Front. Neural Circuits* 4:5

Racca C, Gardiol A, Triller A. 1997. Dendritic and postsynaptic localizations of glycine receptor alpha subunit mRNAs. *J. Neurosci.* 17:1691–700

Ratti A, Fallini C, Colombrita C, Pascale A, Laforenza U, et al. 2008. Post-transcriptional regulation of neuro-oncological ventral antigen 1 by the neuronal RNA-binding proteins ELAV. *J. Biol. Chem.* 283:7531–41

Reetz A, Solimena M, Matteoli M, Folli F, Takei K, De Camilli P. 1991. GABA and pancreatic beta-cells: colocalization of glutamic acid decarboxylase (GAD) and GABA with synaptic-like microvesicles suggests their role in GABA storage and secretion. *EMBO J.* 10:1275–84

Richard S. 2010. Reaching for the stars: linking RNA binding proteins to diseases. *Adv. Exp. Med. Biol.* 693:142–57

Richter JD. 2007. CPEB: a life in translation. *Trends Biochem. Sci.* 32:279–85

Richter JD. 2010. Translational control of synaptic plasticity. *Biochem. Soc. Trans.* 38:1527–30

Robinow S, Campos A, Yao K, White K. 1988. The elav gene product of *Drosophila*, required in neurons, has three RNP consensus motifs. *Science* 242:1570–72

Ruggiu M, Herbst R, Kim N, Jevsek M, Fak JJ, et al. 2009. Rescuing Z+ agrin splicing in Nova null mice restores synapse formation and unmasks a physiologic defect in motor neuron firing. *Proc. Natl. Acad. Sci. USA* 106:3513–18

Sakakibara S, Imai T, Hamaguchi K, Okabe M, Aruga J, et al. 1996. Mouse-musashi-1, a neural RNA binding protein highly enriched in the mammalian CNS stem cell. *Dev. Biol.* 176:230–42

Sakakibara S, Nakamura Y, Yoshida T, Shibata S, Koike M, et al. 2002. RNA-binding protein Musashi family: roles for CNS stem cells and a subpopulation of ependymal cells revealed by targeted disruption and antisense ablation. *Proc. Natl. Acad. Sci. USA* 99:15194–99

Sakakibara S, Okano H. 1997. Expression of neural RNA-binding proteins in the postnatal CNS: implications of their roles in neuronal and glial cell development. *J. Neurosci.* 17:8300–12

Schmidt EF, Warner-Schmidt JL, Otopalik BG, Pickett SB, Greengard P, Heintz N. 2012. Identification of the cortical neurons that mediate antidepressant responses. *Cell* 149:1152–63

Schratt GM, Tuebing F, Nigh EA, Kane CG, Sabatini ME, et al. 2006. A brain-specific microRNA regulates dendritic spine development. *Nature* 439:283–89

Schuman EM, Dynes JL, Steward O. 2006. Synaptic regulation of translation of dendritic mRNAs. *J. Neurosci.* 26:7143–46

Sharma S, Maris C, Allain FH, Black DL. 2011. U1 snRNA directly interacts with polypyrimidine tract-binding protein during splicing repression. *Mol. Cell* 41:579–88

Shibata H, Huynh DP, Pulst SM. 2000. A novel protein with RNA-binding motifs interacts with ataxin-2. *Hum. Mol. Genet.* 9:1303–13

Siegel G, Saba R, Schratt G. 2011. MicroRNAs in neurons: manifold regulatory roles at the synapse. *Curr. Opin. Genet. Dev.* 21:491–97

Silberberg G, Ohman M. 2011. The edited transcriptome: novel high throughput approaches to detect nucleotide deamination. *Curr. Opin. Genet. Dev.* 21:401–6

Singh R, Valcarcel J, Green MR. 1995. Distinct binding specificities and functions of higher eukaryotic polypyrimidine tract-binding proteins. *Science* 268:1173–76

Siomi H, Siomi MC, Nussbaum RL, Dreyfuss G. 1993. The protein product of the fragile X gene, FMR1, has characteristics of an RNA-binding protein. *Cell* 74:291–98

Smukler SR, Arntfield ME, Razavi R, Bikopoulos G, Karpowicz P, et al. 2011. The adult mouse and human pancreas contain rare multipotent stem cells that express insulin. *Cell Stem Cell* 8:281–93

Soller M, White K. 2003. ELAV inhibits 3′-end processing to promote neural splicing of ewg pre-mRNA. *Genes Dev.* 17:2526–38

Soller M, White K. 2005. ELAV multimerizes on conserved AU4-6 motifs important for ewg splicing regulation. *Mol. Cell. Biol.* 25:7580–91

Spellman R, Rideau A, Matlin A, Gooding C, Robinson F, et al. 2005. Regulation of alternative splicing by PTB and associated factors. *Biochem. Soc. Trans.* 33:457–60

Steward O, Schuman EM. 2001. Protein synthesis at synaptic sites on dendrites. *Annu. Rev. Neurosci.* 24:299–325

Sugimoto Y, Konig J, Hussain S, Zupan B, Curk T, et al. 2012. Analysis of CLIP and iCLIP methods for nucleotide-resolution studies of protein-RNA interactions. *Genome Biol.* 13:R67

Sugnet CW, Srinivasan K, Clark TA, O'Brien G, Cline MS, et al. 2006. Unusual intron conservation near tissue-regulated exons found by splicing microarrays. *PLoS Comput. Biol.* 2:e4

Sutton MA, Wall NR, Aakalu GN, Schuman EM. 2004. Regulation of dendritic protein synthesis by miniature synaptic events. *Science* 304:1979–83

Szabo A, Dalmau J, Manley G, Rosenfeld M, Wong E, et al. 1991. HuD, a paraneoplastic encephalomyelitis antigen contains RNA-binding domains and is homologous to Elav and sex lethal. *Cell* 67:325–33

Tang ZZ, Zheng S, Nikolic J, Black DL. 2009. Developmental control of CaV1.2 L-type calcium channel splicing by Fox proteins. *Mol. Cell. Biol.* 29:4757–65

Teplova M, Malinina L, Darnell JC, Song J, Lu M, et al. 2011. Protein-RNA and protein-protein recognition by dual KH1/2 domains of the neuronal splicing factor Nova-1. *Structure* 19:930–44

Tollervey JR, Curk T, Rogelj B, Briese M, Cereda M, et al. 2011. Characterizing the RNA targets and position-dependent splicing regulation by TDP-43. *Nat. Neurosci.* 14:452–58

Tsai DE, Harper DS, Keene JD. 1991. U1-snRNP-A protein selects a ten nucleotide consensus sequence from a degenerate RNA pool presented in various structural contexts. *Nucleic Acids Res.* 19:4931–36

Tuerk C, Gold L. 1990. Systematic evolution of ligands by exponential enrichment: RNA ligands to bacteriophage T4 DNA polymerase. *Science* 249:505–10

Ule J, Darnell RB. 2006. RNA binding proteins and the regulation of neuronal synaptic plasticity. *Curr. Opin. Neurobiol.* 16:102–10

Ule J, Jensen K, Mele A, Darnell RB. 2005a. CLIP: a method for identifying protein-RNA interaction sites in living cells. *Methods* 37:376–86

Ule J, Jensen KB, Ruggiu M, Mele A, Ule A, Darnell RB. 2003. CLIP identifies Nova-regulated RNA networks in the brain. *Science* 302:1212–15

Ule J, Stefani G, Mele A, Ruggiu M, Wang X, et al. 2006. An RNA map predicting Nova-dependent splicing regulation. *Nature* 444:580–86

Ule J, Ule A, Spencer J, Williams A, Hu JS, et al. 2005b. Nova regulates brain-specific splicing to shape the synapse. *Nat. Genet.* 37:844–52

Underwood JG, Boutz PL, Dougherty JD, Stoilov P, Black DL. 2005. Homologues of the *Caenorhabditis elegans* Fox-1 protein are neuronal splicing regulators in mammals. *Mol. Cell. Biol.* 25:10005–16

Voineagu I, Wang X, Johnston P, Lowe JK, Tian Y, et al. 2011. Transcriptomic analysis of autistic brain reveals convergent molecular pathology. *Nature* 474:380–84

Wagnon JL, Mahaffey CL, Sun W, Yang Y, Chao HT, Frankel WN. 2011. Etiology of a genetically complex seizure disorder in Celf4 mutant mice. *Genes Brain Behav.* 10:765–77

Wang ET, Cody NA, Jog S, Biancolella M, Wang TT, et al. 2012. Transcriptome-wide regulation of pre-mRNA splicing and mRNA localization by muscleblind proteins. *Cell* 150:710–24

Wang ET, Sandberg R, Luo S, Khrebtukova I, Zhang L, et al. 2008. Alternative isoform regulation in human tissue transcriptomes. *Nature* 456:470–76

Wang H, Molfenter J, Zhu H, Lou H. 2010a. Promotion of exon 6 inclusion in HuD pre-mRNA by Hu protein family members. *Nucleic Acids Res.* 38:3760–70

Wang LW, Berry-Kravis E, Hagerman RJ. 2010b. Fragile X: leading the way for targeted treatments in autism. *Neurotherapeutics* 7:264–74

Wang Z, Kayikci M, Briese M, Zarnack K, Luscombe NM, et al. 2010c. iCLIP predicts the dual splicing effects of TIA-RNA interactions. *PLoS Biol.* 8:1–16

Weill L, Belloc E, Bava FA, Mendez R. 2012. Translational control by changes in poly(A) tail length: recycling mRNAs. *Nat. Struct. Mol. Biol.* 19:577–85

Wheeler TM, Lueck JD, Swanson MS, Dirksen RT, Thornton CA. 2007. Correction of ClC-1 splicing eliminates chloride channelopathy and myotonia in mouse models of myotonic dystrophy. *J. Clin. Invest.* 117:3952–57

Wheeler TM, Sobczak K, Lueck JD, Osborne RJ, Lin X, et al. 2009. Reversal of RNA dominance by displacement of protein sequestered on triplet repeat RNA. *Science* 325:336–39

Witten JT, Ule J. 2011. Understanding splicing regulation through RNA splicing maps. *Trends Genet.* 27:89–97

Wolf HK, Buslei R, Schmidt-Kastner R, Schmidt-Kastner PK, Pietsch T, et al. 1996. NeuN: a useful neuronal marker for diagnostic histopathology. *J. Histochem. Cytochem.* 44:1167–71

Xue Y, Zhou Y, Wu T, Zhu T, Ji X, et al. 2009. Genome-wide analysis of PTB-RNA interactions reveals a strategy used by the general splicing repressor to modulate exon inclusion or skipping. *Mol. Cell* 36:996–1006

Yang YY, Yin GL, Darnell RB. 1998. The neuronal RNA-binding protein Nova-2 is implicated as the autoantigen targeted in POMA patients with dementia. *Proc. Natl. Acad. Sci. USA* 95:13254–59

Yano M, Hayakawa-Yano Y, Mele A, Darnell RB. 2010. Nova2 regulates neuronal migration through an RNA switch in disabled-1 signaling. *Neuron* 66:848–58

Yao KM, Samson ML, Reeves R, White K. 1993. Gene elav of *Drosophila melanogaster*: a prototype for neuronal-specific RNA binding protein gene family that is conserved in flies and humans. *J. Neurobiol.* 24:723–39

Yeo GW, Coufal NG, Liang TY, Peng GE, Fu XD, Gage FH. 2009. An RNA code for the FOX2 splicing regulator revealed by mapping RNA-protein interactions in stem cells. *Nat. Struct. Mol. Biol.* 16:130–37

Yuan Y, Compton SA, Sobczak K, Stenberg MG, Thornton CA, et al. 2007. Muscleblind-like 1 interacts with RNA hairpins in splicing target and pathogenic RNAs. *Nucleic Acids Res.* 35:5474–86

Zang JB, Nosyreva ED, Spencer CM, Volk LJ, Musunuru K, et al. 2009. A mouse model of the human Fragile X syndrome I304N mutation. *PLoS Genet.* 5:e1000758

Zhang C, Darnell RB. 2011. Mapping in vivo protein-RNA interactions at single-nucleotide resolution from HITS-CLIP data. *Nat. Biotechnol.* 29:607–14

Zhang C, Frias MA, Mele A, Ruggiu M, Eom T, et al. 2010. Integrative modeling defines the Nova splicing-regulatory network and its combinatorial controls. *Science* 329:439–43

Zhang C, Zhang Z, Castle J, Sun S, Johnson J, et al. 2008. Defining the regulatory network of the tissue-specific splicing factors Fox-1 and Fox-2. *Genes Dev.* 22:2550–63

Zhang X, Poo MM. 2002. Localized synaptic potentiation by BDNF requires local protein synthesis in the developing axon. *Neuron* 36:675–88

Zhu H, Hasman RA, Barron VA, Luo G, Lou H. 2006. A nuclear function of Hu proteins as neuron-specific alternative RNA processing regulators. *Mol. Biol. Cell* 17:5105–14

Zhu H, Hinman MN, Hasman RA, Mehta P, Lou H. 2008. Regulation of neuron-specific alternative splicing of neurofibromatosis type 1 pre-mRNA. *Mol. Cell. Biol.* 28:1240–51

Muscarinic Signaling in the Brain

Alexander Thiele

Institute of Neuroscience, Henry Wellcome Building,
Newcastle University, Newcastle upon Tyne, NE2 4HH, United Kingdom; email:
alex.thiele@ncl.ac.uk

Annu. Rev. Neurosci. 2013. 36:271–94

The *Annual Review of Neuroscience* is online at
neuro.annualreviews.org

This article's doi:
10.1146/annurev-neuro-062012-170433

Keywords

attention, interneurons, pyramidal cells, oscillation, synchrony

Abstract

Muscarinic signaling affects attention, action selection, learning, and memory through multiple signaling cascades, which act at different timescales and which alter ion channels in cell type–specific manners. The effects of muscarinic signaling differ between cortical layers and between brain areas. Muscarinic signaling adds flexibility to the processing mode of neuronal networks, thereby supporting processing according to task demands. This review outlines possible scenarios to describe how it contributes to cellular mechanisms of attention and how it affects channeling of information in different neuronal circuits.

Contents

CHOLINERGIC SIGNALING IN THE CENTRAL NERVOUS SYSTEM

Cholinergic projection neurons are confined to only a few, well-defined clusters (**Figure 1**). The main clusters of cholinergic projection neurons have been labeled Ch1–Ch6 (Mesulam et al. 1983b), or Ch1–Ch8 (Yeomans 2012). Cholinergic neurons within each of these clusters project to distinct parts of the brain (Mesulam et al. 1983a,b, 1984). Their contribution to cognition is thus partially linked to the function of the target locations. Ch1–Ch4 are located in the basal forebrain (BF), and neurons within these clusters support attention, learning, and memory. Ch1 (medial septum) and Ch2 (within the vertical limb of the diagonal band of Broca) project mainly to the hippocampus and the medial parts of the cortex (Paré et al. 1988). Cholinergic neurons in Ch1 and Ch2 additionally affect the reticular thalamic, the mediodorsal, and anteroventral/anteromedial thalamic nuclei (Steriade et al. 1988). Ch3 (within the horizontal limb of the diagonal band of Broca) projects to the olfactory bulb. Ch4 corresponds largely to the nucleus basalis magnocellularis (termed the nucleus basalis Meynert in primates) and is the principal source of cholinergic innervation of the cortex and amygdala. Cholinergic projection neurons in Ch5 (pedunculopontine tegmental nucleus) and Ch6 (laterodorsal tegmental nucleus) reside in the caudal midbrain. They innervate the thalamus (Steriade et al. 1988), the pontine reticular formation, the ventral midbrain, the ventral tegmental area, and the substantia nigra (Yeomans 2012). Cholinergic neurons in Ch5 and Ch6 contribute to arousal, to sleep, and to the control of dopaminergic cell groups (Yeomans 2012). Ch7 neurons overlap with the habenula and project to the interpeduncular brain stem nuclei. Ch8 is located largely in the parabigeminal nucleus and innervates the superior colliculus. In addition to containing cholinergic neurons, Ch1–Ch8 contain GABAergic and glutamatergic projecting neurons, which target the areas that also receive input from the cholinergic neurons (Freund & Gulyas 1991, Henny & Jones 2008, Paré et al. 1988, Rye et al. 1984, Steriade et al. 1988). The exact function of the noncholinergic projection neurons remains speculative (Freund & Gulyas 1991, Hassani et al. 2009), and their existence complicates the interpretation of nonspecific lesions to Ch1–Ch8 (Everitt & Robbins 1997).

Ch1–Ch8: neuronal clusters that release acetylcholine to modulate information processing in target areas by acting on nicotinic and muscarinic receptors

Basal forebrain (BF): contains a large number of cholinergic neurons, which are the main source of acetylcholine input to the cortex and the hippocampus

Cholinergic signaling is not restricted to the cell groups described above. Local cholinergic neurons can be found in many brain areas. These are particularly numerous in the basal ganglia, which do not receive external cholinergic inputs (Calabresi et al. 2000). Cholinergic interneurons are found in the rodent cortex (Mesulam et al. 1983a,b), where they often coexpress vasointestinal polypeptide (VIP) (Eckenstein & Baughman 1984), and presumably corelease γ-aminobutyric acid (GABA), but von Engelhardt et al. (2007) have reported even pure cholinergic interneurons. Cholinergic interneurons are absent in cortical areas of

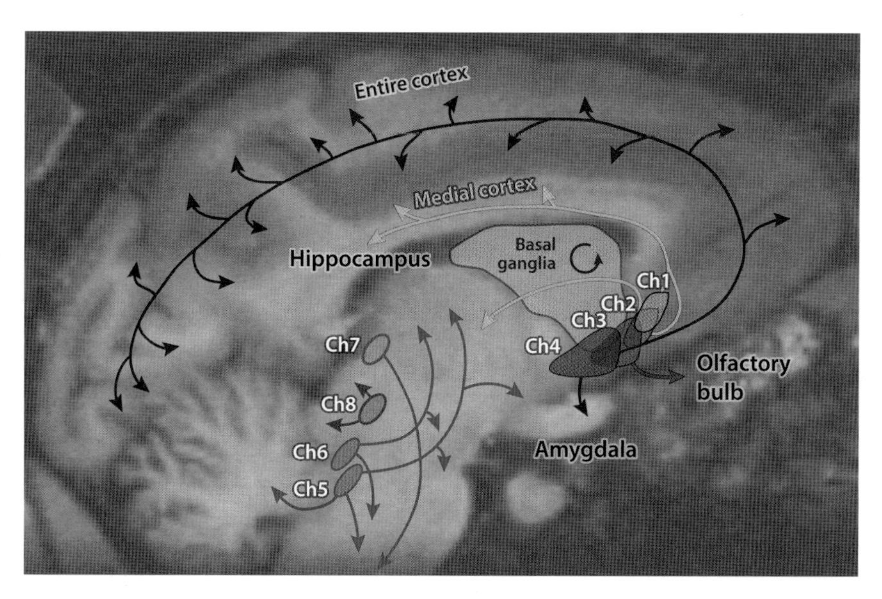

Figure 1

Main sources of acetylcholine in the brain. The figure shows a sagittal slice of a macaque MRI (5 mm from the midline). Superimposed are approximate locations of the main cholinergic cell clusters (Ch1–Ch8) and their predominant projection sites (*color coded and indicated by the respective arrows*). Recipient structures outside the image plane are indicated by black fonts with white outlines. The circular arrow inside the basal ganglia indicates that most cholinergic drive to the basal ganglia derives from local cholinergic interneurons.

macaque monkeys (Mesulam et al. 1984), but cholinergic interneurons are found again in the human and chimpanzee prefrontal cortex, whereby the human cortex shows the highest density (Raghanti et al. 2008).

The cholinergic projection and interneurons affect nicotinic and muscarinic receptors upon acetylcholine (ACh) release. This review focuses on contributions of muscarinic signaling to attention at the behavioral and cellular levels. The focus on muscarinic signaling by no means implies that nicotinic receptors are less involved, but even a complete review of muscarinic functions is beyond the scope of this article.

MUSCARINIC RECEPTORS

Muscarinic receptors are members of the class of heptahelical G protein–coupled receptors (GPCRs). Five main subtypes of muscarinic receptors (M1–M5) have been identified. Although these are found abundantly throughout the central and peripheral nervous system, different subtypes have unique, if overlapping, distributions in different areas (**Figure 2**). For example, M1, M2, and M4 are abundant in the cortex and striatum, but M2 is expressed more prominently in the cortex and M4 is expressed more prominently in the striatum (Bubser et al. 2012, Levey et al. 1991). M1–M4 are prominently expressed in the hippocampus. The BF mostly expresses M2 receptors. M5 receptors are relatively rare in the cortex but are numerous on dopaminergic neurons in the substantia nigra and the ventral tegmental area (Bubser et al. 2012, Levey et al. 1991).

The muscarinic receptor subtypes can be subdivided into two main families, which exert their functions by activating heterotrimeric G proteins containing α, β, and γ subunits (**Figure 3**). Each subunit occurs in different isoforms. The α-subunit isoform determines to a large extent the signaling specificity and the cellular effects that follow receptor activation (Tedford & Zamponi 2006). Consequently, G proteins are named after the α-subunit. The

G protein–coupled receptors (GPCRs): receptors that sense specific molecules outside the cell. Activation starts intracellular signaling cascades, which alter cellular integration properties

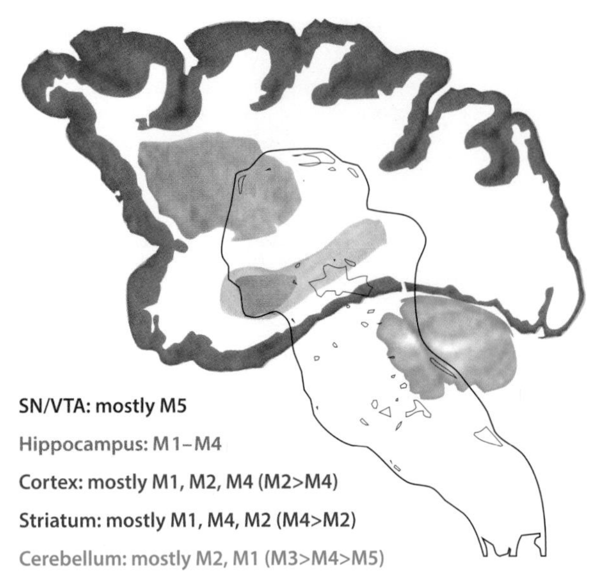

Figure 2

Schematic representation of the distribution of the main muscarinic receptor types in the cortex, basal ganglia (striatum), hippocampus, cerebellum, and substantia nigra/ventral tegmental area (SN/VTA). M1–M5 indicate the receptor types that are found most prominently in the respective areas. This dominance is indicative and does not imply that other receptors are absent. In general, the nondominant receptors are simply expressed at lower levels. Text color corresponds to the respectively colored areas.

M1-type receptors: receptors of the M1-, M3-, M5-subfamily, which interact with Gq/11 G proteins (and mostly act postsynaptically)

M2-type receptors: receptors of the M2-, M4-subfamily, which interact with $G_{i/o}$ G proteins and act pre- and postsynaptically

GPCR signaling: G protein activation initiated second messenger signaling cascade. These alter the neuronal signaling properties

human genome encodes 16 different α-subunits (Simon et al. 1991), grouped into 5 (Tedford & Zamponi 2006), or sometimes 4, different families (Wickman & Clapham 1995). Muscarinic receptors couple to members of the $G_{q/11}$ and the $G_{i/o}$ family, whereby odd-numbered muscarinic receptors (M1, M3, M5, referred to as M1-type) couple preferentially to the $G_{q/11}$ family, and even-numbered receptors (M2, M4, referred to as M2-type) couple preferentially to the $G_{i/o}$ family (Brown 2010, Caulfield & Birdsall 1998, Tedford & Zamponi 2006).

Upon binding ACh, G proteins associate with the muscarinic receptor, and the α-subunit exchanges GDP (guanosine diphosphate) for GTP (guanosine-5′-triphosphate) and dissociates from the β/γ-subunits (Tedford & Zamponi 2006) (**Figure 3**). Both the α-GTP subunit and the β/γ-subunit can interact with effector proteins within the signaling cascade, referred to as GPCR signaling. The α-GTP subunit affects primary effector proteins (see below) in the cytosol, whereas the β/γ-subunit remains anchored to the membrane (Simon et al. 1991, Tedford & Zamponi 2006). G proteins interact with their effector and/or target proteins through a second messenger and through direct (membrane-delimited) pathways (**Figure 3**). The main primary effector proteins influenced by muscarinic receptors through GPCR are adenylyl cyclase (AC), phospholipase C (PLCβ), phospholipase A2, and phospholipase D (Felder 1995). Generally, the $G_{q/11}$ α-subunit activates PLCβ, whereas the $G_{i/o}$ α-subunit inactivates AC (Caulfield & Birdsall 1998, Peralta et al. 1988). Consequently, the odd-numbered and even-numbered M-receptors affect different primary effector proteins and different second-messenger cascades, and they produce different effects on neuronal response properties and signaling. However, the division of labor is not perfect. M1-type receptors can cause limited AC inhibition (Caulfield 1993, Felder 1995). Furthermore, AC activity can be bimodally regulated by muscarinic receptors depending on the G protein subtype activated (Onali & Olianas 1995).

One of the pathways most commonly described in relation to neuronal $G_{q/11}$ signaling involves activation of the PLCβ pathway. PLCβ activation causes hydrolysis of phosphoinositol-1,4,5-biphosphate (PIP2) into two second messengers, namely inositol-1,4,5-triphosphate (IP3) and diacylglycerol (DAG). IP3 causes intracellular stores to release Ca^{2+} (Felder 1995) and then to activate Ca^{2+}-calmodulin, whereas DAG activates protein kinase C (PKC).

M2-type receptor activation inhibits AC by means of $G_{i/o}$ α-subunit signaling, thereby reducing the availability of the second-messenger cyclic AMP (adenosine monophosphate). This affects a variety of neuronal processes, an important example being neuronal plasticity (Kandel 2012). The more immediate effects of M2-type receptor activation on neuronal activity are mediated by $G_{\beta\gamma}$ subunit signaling (fast membrane-delimited pathway), which directly

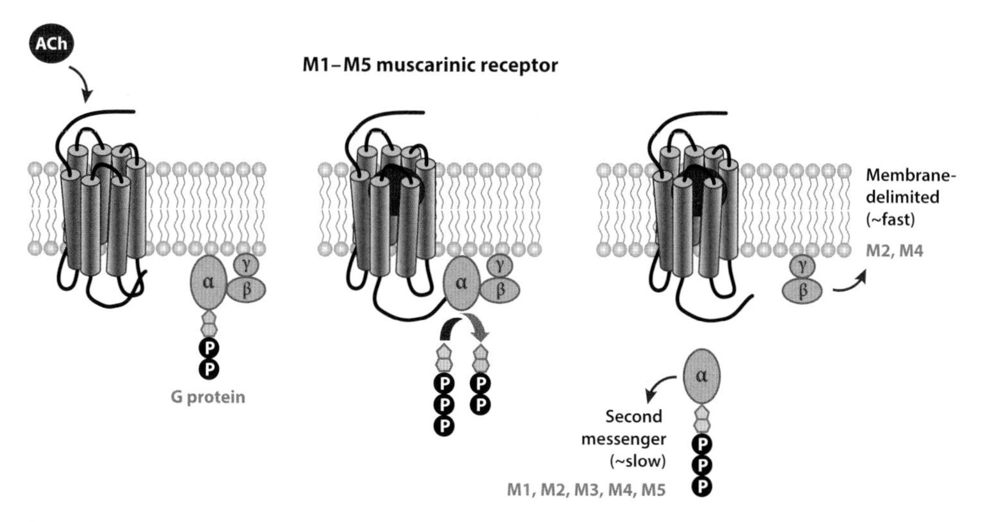

Figure 3

Muscarinic receptors and the first steps of their signaling cascade. Upon binding of acetylcholine (ACh), the receptor can interact with a G protein, consisting of 3 subunits (α, β, and γ, respectively). In the inactive form, the α-subunit binds GDP. Upon interaction with the M-receptor, GDP can be exchanged for GTP, which allows the α-subunit to dissociate from the β/γ-subunit. The α-subunit can then interact with other molecules in the cytosol and trigger various cascades. In the case of G proteins that interact with M2, M4 receptors (Gi/o), the β/γ-subunit can interact with membrane-bound proteins and activate the so-called membrane-delimited pathway (see text for additional details).

affects transmembrane ion channels (Wickman & Clapham 1995), e.g., opening of the inward rectifying potassium channel (GIRK). Fast muscarinic signaling through the $G_{\beta\gamma}$ subunit also affects N- and P/Q-type Ca^{2+} channels, causing, e.g., presynaptic inhibition (Brown 2010). The complexity and intricacies of the different signaling pathways and their potential interactions are highlighted by the fact that M2-type-mediated activation of GIRKs is opposed by M1-type activation. GIRKs are usually kept open when bound to PIP2. M1-type-induced hydrolysis of PIP2 thus results in GIRK closure (Kobrinsky et al. 2000). Also, M2-type receptors can activate PLC (normally the domain of M1-type receptor activation) through $G_{\beta\gamma}$ subunits. Activating PLC thus opens the IP3/DAG pathway to M2-type receptor activation (Felder 1995). GPCRs can affect cell responses and cell-cell interactions via a multitude of additional pathways and regulatory systems (see e.g., Perez & Karnik 2005).

Muscarinic Receptor Regulation of Ion Channels

Muscarinic receptor subtypes have been identified heterosynaptically as pre- and postsynaptic receptors (Rouse et al. 1997), whereas presynaptic autoreceptor function is restricted to M2-type receptors (Zhang et al. 2002). Thus, all receptor subtypes can alter the excitability of neurons and the release probability of various transmitters, but only M2-type receptors exhibit cholinergic feedback control over ACh release. Within the central nervous system, M-receptor activation is best known for its influence on K^+, Ca^{2+}, and unselective cation channels (**Figure 4**, **Table 1**). However, Na^+ and Cl^- channels can also be affected. The consequences on neuronal activity are summarized in **Table 1** and in **Figure 4**.

Potassium channel modulation. The M1-type signaling cascade affects neuronal activity by closing voltage-gated K^+ channels (K_M)

Membrane-delimited pathway: signaling pathway where the signaling agents are restricted to be within or near the cell membrane

GIRK: G protein–coupled inward rectifying K^+ channel

Heterosynaptic: events that occur at synapses between different neurons, i.e., transmitter from one neuron affects a different neuron

Presynaptic receptors: receptors located presynaptically, which influence the amount of transmitter released upon arrival of an action potential

Table 1 Muscarinic effects on different ion channels and the consequences for neuronal activity

Channel[a]	Receptor-channel effect[b]	Cellular effects[c]	References
K^+ channels			
K_M	M1-type → closure	Depolarization, reduction of spike frequency adaptation, increased excitability, improved integration properties of EPSP and IPSP, increased spontaneous activity	Brown 2010, McCormick & Prince 1986, McCormick & Williamson 1989, Womble & Moises 1992
K_{leak}	M1-type → closure	Increased input resistance of cells; inputs to distal parts of the dendritic tree are more likely to impact on integration at the axon hillock	Womble & Moises 1992
K_{sAHP}	M1-type → closure	Reduced afterhyperpolarization, reduced spike frequency adaptation	Ghamari-Langroudi & Bourque 2004, McCormick et al. 1993
KIR2	M1-type → closure	Depolarization (reduced hyperpolarization)	Carr & Surmeier 2007
SK_{Ca}	M1-type → opening	Transient hyperpolarization, possibly increased response reliability	Gulledge et al. 2007, Gulledge & Stuart 2005
SK_{Ca}	M1-type → reduced sensitivity of SK_{Ca} to Ca^{2+}, reduced likelihood of opening	Reduced transient hyperpolarization, increased Ca^{2+} transients through NMDA receptors and increased synaptic potentials and facilitation of long-term potentiation	Buchanan et al. 2010, Giessel & Sabatini 2010
GIRK/KIR	M2-type → opening	Neurons remain at relatively hyperpolarized level	Brown 2010
Ca^{2+} channels			
L-type	M2-type → inhibition/closure	Reduced ability of Ca^{2+} to trigger intracellular processes	Biscoe & Straughan 1966
P/Q, and N- type	M2-type → inhibition/closure M1-type → inhibition/closure	Reduced transmitter release, reduced action potential prolongation, and reduced afterhyperpolarization	Allen 1999, Allen & Brown 1993, Biscoe & Straughan 1966, Hasselmo & Bower 1992, Tedford & Zamponi 2006
T-type	M1-type → closure M3/M5 → activation	Altered rhythmic rebound burst firing and spindle waves associated with slow-wave sleep in thalamic reticular and relay neurons Altered dendritic integration and Ca^{2+} spiking in hippocampal pyramidal cells	Christie et al. 1995, Navaroli et al. 2012
Nonspecific cation channels			
TRPC channel (Ca^{2+} dependent)	M1-type → opening	Slow afterdepolarization	Haj-Dahmane & Andrade 1998, Yan et al. 2009
Ca^{2+}-independent nonspecific cation	M1-type → opening	Slow afterdepolarization	Egorov et al. 2003

(Continued)

Table 1 (*Continued*)

Channel[a]	Receptor-channel effect[b]	Cellular effects[c]	References
Na^+ channels			
Na^+ channel	M2-type → reduced and slowed inactivation PKC pathways (possibly M1-type) → reduced opening	Persistent Na^+ currents. Important role in synchronized activity in the striatum (Carrillo-Reid et al. 2009). Reduced Na^+ currents by PKC pathway	Chen et al. 2005, Ma et al. 1997

Abbreviations: EPSP, excitatory postsynaptic potential; GIRK: G protein–coupled inward rectifying K^+ channel; IPSP, inhibitory postsynaptic potential; KIR: inward rectifying K^+ channel; K_{leak}: leaky potassium channel; K_M: M-type (voltage-gated) potassium channel; K_{sAHP}: slow (non-SK) afterhyperpolarization K^+ channel; NMDA, *N*-methyl-D-aspartate; PKC, protein kinase C; SK_{Ca}: SK-type calcium-activated K^+ channel; TRPC: Ca^{2+}-activated nonspecific cation channel.

[a]Column 1 lists ion channels affected by muscarinic signaling.

[b]Column 2 lists the muscarinic receptor subtypes involved and the main effect the signaling cascade has on the channel of interest.

[c]Column 3 lists some of the ensuing effects on neuronal activity.

(Brown 2010, McCormick & Prince 1986, McCormick & Williamson 1989, Womble & Moises 1992). K_M is usually held open by binding to PIP2, and M1-$G_{q/11}$-PLC-induced hydrolysis of PIP2 will thus lead to channel closure (Gamper & Shapiro 2007, Suh & Hille 2005). M1-type signaling also leads to closure of leaky K^+ channels (K_{leak}) (Womble & Moises 1992), closure of slow afterhyperpolarization K^+ channels (K_{sAHP}) (Ghamari-Langroudi & Bourque 2004, McCormick et al. 1993), closure of inward rectifying K^+ type 2 (KIR2) channels (Carr & Surmeier 2007), and finally to opening (Gulledge et al. 2007, Gulledge & Stuart 2005) but also closing (Giessel & Sabatini 2010) of SK-type calcium-activated K^+ channels (SK_{Ca}). The M1-type-induced opening of SK_{Ca} is likely due to the release of Ca^{2+} from intracellular stores (Gulledge et al. 2007). However, M1-type activation simultaneously reduces the sensitivity of SK_{Ca} to Ca^{2+} through a casein kinase-2 pathway (Giessel & Sabatini 2010) or PKC pathway (Buchanan et al. 2010), and M1-type activation can thus also result in reduced SK_{Ca} opening.

The M2-type $G_{i/o}$ signaling cascade activates inward rectifying channels K^+ channels (GIRK/KIR) postsynaptically (Brown 2010).

Calcium channel modulation. Most voltage-gated Ca^{2+} channels are regulated by G proteins via fast (membrane-delimited) and slow (soluble second-messenger) mechanisms (Hille 1994, Wickman & Clapham 1995). P/Q-, N-, and L-type Ca^{2+} channels are part of the high voltage-activated (HVA) class of Ca^{2+} channels. HVA Ca^{2+} channels are inhibited mostly by M2-type $G_{i/o}$ fast ($G_{\beta\gamma}$) membrane-delimited mechanisms (Allen 1999, Allen & Brown 1993, Tedford & Zamponi 2006), but M1 activation can also contribute to presynaptic Ca^{2+} channel inhibition by M_1/PLC/Ca^{2+}-dependent PKC pathways (Salgado et al. 2007).

Low voltage-activated T-type Ca^{2+} channels can be blocked by M1-type $G_{q/11}$ mechanisms (Hildebrand et al. 2007), but they can also be activated by M3, M5 mechanisms (Pemberton et al. 2000).

Nonspecific cation channel modulation. M1-type activation can cause opening of Ca^{2+}-activated nonspecific (TRPC) cation channels (Haj-Dahmane & Andrade 1998, Yan et al. 2009) and of Ca^{2+}-independent nonspecific cation channels (Egorov et al. 2003).

Sodium channel modulation. Ma et al. (1997) have identified a $G_{i/o}$-$G_{\beta\gamma}$-dependent modulation of Na^+ channels that results in persistent Na^+ currents by reducing and slowing Na^+ channel inactivation (Ma et al. 1997). However, Na^+ current reduction can also occur through muscarinic receptor

Postsynaptic receptors: located postsynaptically. Their activation influences the postsynaptic neuronal membrane potential

Autoreceptor: presynaptic receptor that gets activated by a transmitter the host neuron releases. Serves an important feedback function

K_M: M-type (voltage-gated) potassium channel

Afterhyperpolarization (AHP): a hyperpolarization that follows the action potential

Soluble/diffusion pathway: G protein–coupled receptor signaling pathway that relies on signaling molecule interactions within a cell's cytosol

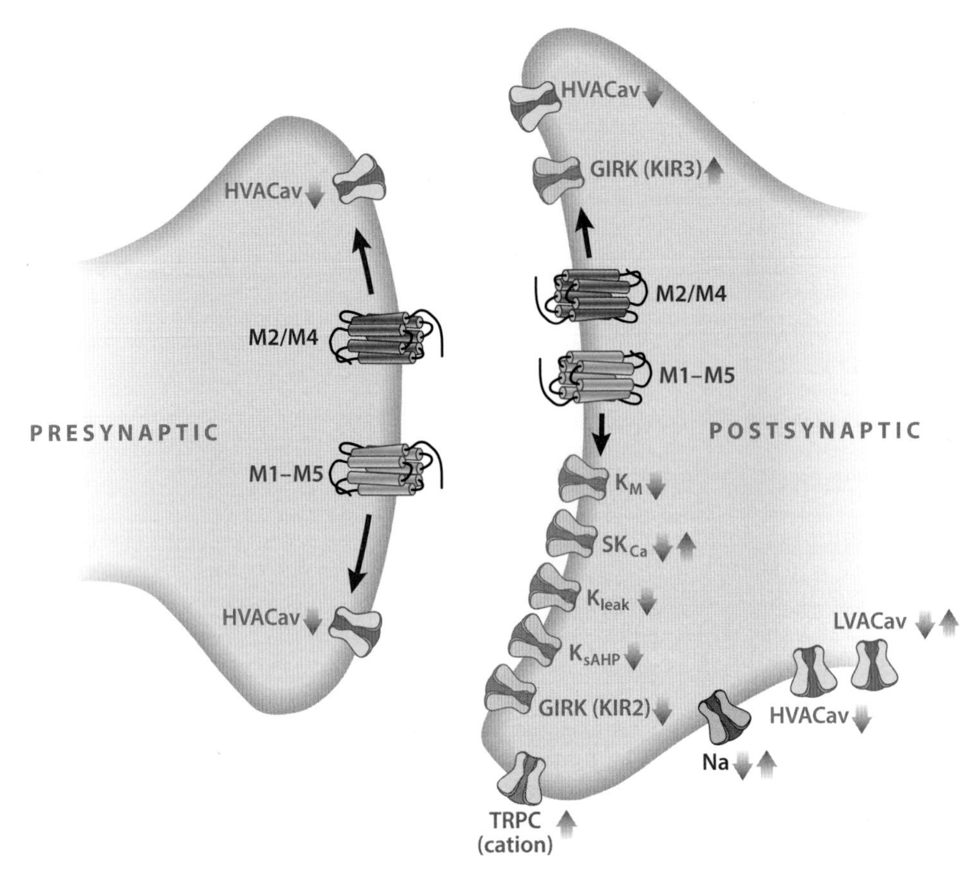

Figure 4

Overview of channels affected by the two main types of muscarinic signaling [*M1-type* (*M1, M3, M5*) *receptors are shown in green, M2-type* (*M2, M4*) *receptors are shown in blue*]. Channels most prevalently affected presynaptically are shown on the left, and channels most prevalently affected postsynaptically are shown on the right. Red arrows indicate which channels are affected by the specific type of muscarinic signaling. Gray gradient arrows indicate whether channels are opened (*upward-pointing arrow*) or closed (*downward-pointing arrow*) by muscarinic signaling. Note that other channels can be affected as well. The figure shows effects that are dominant in the cortex. Abbreviations: K_M, M-type (voltage-gated) potassium channel; GIRK, G protein–coupled inward rectifying K^+ channel; KIR, inward rectifying K^+ channel; SK_{Ca}, SK-type calcium-activated K^+ channel; K_{leak}, leaky potassium channel; K_{sAHP}, slow (non-SK) afterhyperpolarization K^+ channel; HVACav, high voltage-activated Ca^{2+} channels (L, P, N/Q-type); LVACav, low voltage-activated Ca^{2+} channels (T-type); TRPC, Ca^{2+}-activated nonspecific cation channel; Na, Na^+ channel.

activation via the PKC pathway (Chen et al. 2005).

It is worth highlighting that the actions of, e.g., M2 and M4 receptor activation on various ion channels and on cellular activity should in principle be similar, as they use largely similar signaling cascades. However, at least in certain neuronal types, M-receptor subtypes are further subdivided into specific micro domains, such that M2 receptors, for example, are involved in postsynaptic inhibition, whereas M4 receptors are involved only in presynaptic inhibition (Brown 2010).

INFLUENCE OF MUSCARINIC RECEPTORS ON COGNITION-BEHAVIORAL STUDIES

Muscarinic receptors influence multiple cognitive domains. Linking the contribution of individual muscarinic receptor subtypes to performance in specific domains is difficult because individual receptor subtypes do not perform a specific cognitive function. They shape neuronal and local network processing abilities. Depending on the network involved, a cognitive function will be supported. However, muscarinic receptor subtypes are not homogenously distributed within the brain, and therefore Eglen (2012) has reported some specificity of muscarinic receptor subtype contribution to individual cognitive domains.

Muscarinic Signaling and Attention

The involvement of cholinergic mechanisms in attentional processes is sometimes greeted with skepticism. This uncertainty may have historical reasons because ACh was thought to be involved in unspecific arousal rather than in the much more selective process of attention. This skepticism is also rooted in the comparatively diffuse innervation pattern the cortex receives from the BF. Investigators argue that attention has very good resolution, which cannot be provided by the cholinergic system. A few points are worth mentioning in relation to this argument. First, the cholinergic input to the cortex consists of modules that target selected cortical areas in task- and context-dependent manners (Zaborszky 2002). Second, the role an area plays in attention-demanding tasks can be differentiated at the muscarinic signaling level (Robbins et al. 1998); i.e., the function of the area in a cognitive task determines the contribution of muscarinic signaling not the fact that ACh is released within it. Third, muscarinic receptor densities and types differ among areas, layers, and cell types within (and between) areas (Disney & Aoki 2008, Disney et al. 2006, Gebhard et al. 1993, Mesulam & Geula 1992); therefore, a great level of signaling specificity can be achieved in each area, even though the release of the neurotransmitter involved may be relatively global. Fourth, the idea of global Ach release during attention-demanding tasks has been challenged because ACh release in the rat frontal cortex is locally controlled by glutamatergic synapses (Sarter et al. 2009). It remains to be determined whether similar local mechanisms exist in sensory areas, but Fournier et al. (2004) have established that cortical levels of ACh are adjusted in sensory areas according to the specificity of sensory inputs. All these arguments should temper the notion of very unspecific nature of cholinergic signaling.

Some studies have also argued that muscarinic signaling is slow, owing to the second-messenger systems involved. Signaling speed depends very much on the GPCR cascades involved, as evident in cone- and rod-mediated light signaling or in olfactory signal transduction (even though neither is muscarinically mediated). Moreover, it also depends on the signaling steps involved. Membrane-delimited processes are much faster than cytosolic pathways.

Finally, researchers often argue that ACh release occurs in a tonic manner that is not modulated in accordance with fast fluctuating behavioral demands. Recent measurements using cyclic voltammetry show that the ACh release closely follows the time course of attention-demanding events in the frontal cortex (Parikh et al. 2007), and increased levels of ACh are related to attentional effort rather than to attentional performance (Kozak et al. 2006). Even though we do not yet know whether this is mediated by muscarinic or nicotinic receptors (or both), these data demonstrate that cholinergic events follow behavioral demands.

Despite these arguments, future studies must delineate the spatial precision of cortical cholinergic innervation, delineate the speed of different GPCR signaling pathways under physiological conditions [in vitro experiments often use nonphysiological temperatures, thereby affecting enzymatic and channel kinetics (El-Fakahany & Richelson 1980, Thompson et al. 1985)], and determine the temporal precision of cholinergic release under different behavioral conditions.

Ample evidence demonstrates that cholinergic mechanisms contribute to attention. Unspecific BF lesions or more selective cholinergic BF lesions result in attentional deficits (reviewed, e.g., in Levin et al. 2011). Lesion studies cannot differentiate between the involvements of nicotinic versus muscarinic receptors. Muscarinic signaling in relation to attention often receives less coverage than nicotinic signaling does, but muscarinic contribution to attention is evident from studies where scopolamine was either systemically applied or locally infused into parietal areas of macaque monkeys performing covert top-down attention tasks. It resulted in increased reaction times for validly cued targets (Davidson et al. 1999, Davidson & Marrocco 2000). Scopolamine infused into the medial prefrontal cortex of rats performing the five-choice serial reaction time task (a test of visuospatial attention) affected choice accuracy and reaction times related to attentional performance but did not alter the number of omissions (Robbins et al. 1998). The latter were affected after scopolamine injections into the anterodorsal prefrontal cortex without producing any deficits in accuracy (Robbins et al. 1998), demonstrating site-specific contribution of muscarinic receptors to specific subtasks, provided that attention is taxed.

Scopolamine: unspecific blocker of muscarinic receptors

Covert top-down attention task: attention is deployed willfully to a location that does not coincide with gaze direction

Sustained attention: top-down attention to specific environmental features is maintained for the duration of the trial

Striate cortex: primary visual area of the cortex

Middle temporal area: a cortical area in primates; specialized in signaling the direction of visual motion

Systemic scopolamine application in rats performing a two-choice, discrete-trial signal-detection task decreased signal detection and increased the false-alarm rate, both signatures of reduced sustained attention (Bushnell et al. 1997). Similar results were found when rats performed a two-choice visual discrimination task where attentional demand was manipulated by altering the signal duration or intensity (Dillon et al. 2009, Hoff et al. 2007) and when mice performed a sustained visual attention task in a five-arm maze (Leblond et al. 2002).

Human studies suggest that muscarinic blockade reduces the ability to exploit knowledge about spatial probabilities of target occurrence, which has been interpreted as a failure to utilize attentional resources optimally (Dunne & Hartley 1986). Using two different forms of sustained attention tasks (digit vigilance and rapid visual information processing), Ellis et al. (2006) showed that muscarinic but not nicotinic blockade affected performance in both tasks Finally, muscarinic signaling in humans is responsible for maintaining attention but not for attention switching (Furey et al. 2008).

The analyses above indicate that muscarinic signaling affects cognitive abilities. How does that manifest at the level of neuronal activity? How does it affect local networks and large scale activity as evident by neuronal synchrony and oscillations?

MUSCARINIC CONTROL OF NEURONAL ACTIVITY AND LOCAL CIRCUITRY

Muscarinic Control of Cellular Activity

ACh can improve the orientation, direction, or spatial tuning in the striate cortex (V1) of cats and monkeys (Murphy & Sillito 1991; Roberts et al. 2005; Sato et al. 1987a,b; Sillito & Kemp 1983), but these findings are not universally supported (Soma et al. 2012, Thiele et al. 2012, Zinke et al. 2006). Recent data demonstrate that neurons in the middle temporal area (MT) are better able to discriminate between different directions of stimulus motion when ACh is applied (Thiele et al. 2012). Improved reliability of neuronal responses in rat V1 (Goard & Dan 2009) and decreased redundancy of information processing in rat V1 and in macaque MT (Goard & Dan 2009, Thiele et al. 2012) could be interpreted as an improved signal-to-noise ratio. Many of the above studies did not differentiate between muscarinic and nicotinic effects of ACh, but some did. The reduced redundancy in rat V1 reported on BF stimulation was mediated by muscarinic mechanisms within V1, whereas the increased response reliability in V1 was not (Goard & Dan 2009). Likewise, ACh application in macaque MT did not alter response reliability, but it reduced redundancy, suggesting that similar

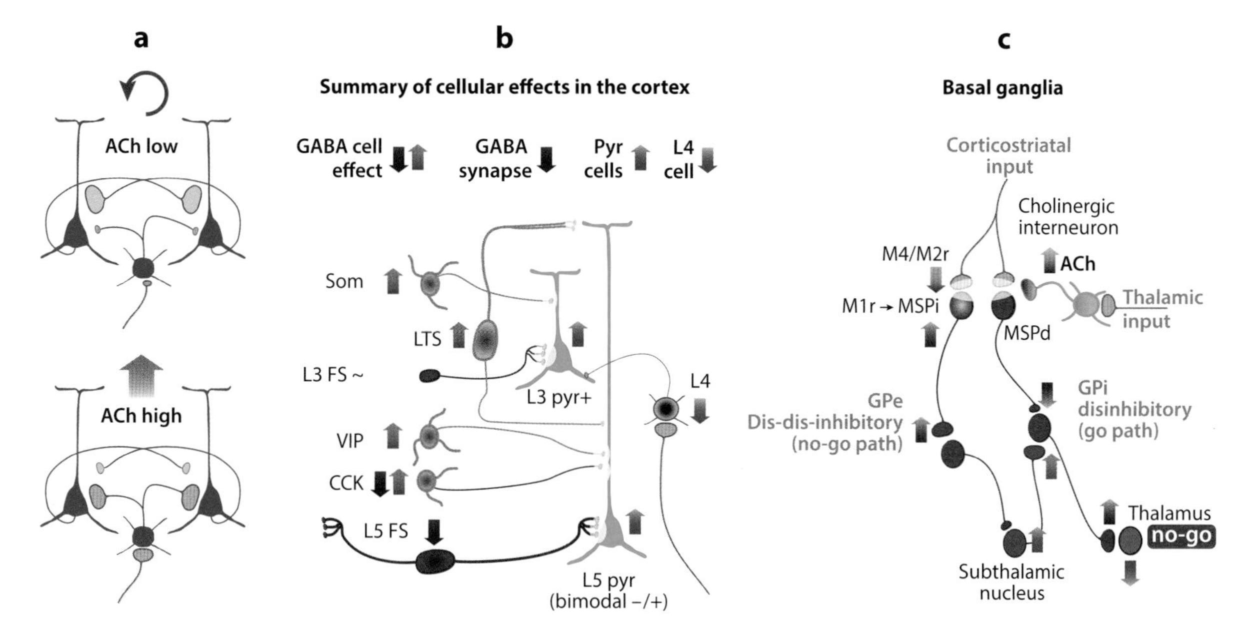

Figure 5

Cholinergic effects on cortical and striatal circuits. (*a*) Cholinergic effects on the processing of feedforward versus internal information. When acetylcholine (ACh) is low, synaptic efficacy of intrinsic connections is high, while feedforward synaptic efficacy is average. Therefore, information flow is dominated by recurrent processing. When ACh is high, feedforward synaptic efficacy is high (through presynaptic nicotinic mechanisms), while synaptic efficacy of intrinsic connections is reduced (through M2-type receptor mediated mechanisms). In this mode, processing is dominated by feedforward information. (*b*) Overview of muscarinic effects on cortical circuitry, which may enable efficient filtering of information. Arrows indicate whether neuronal activity is increased or decreased by muscarinic signaling. Reduced intensity of colors at synaptic sites indicates that the synaptic efficacy is reduced by muscarinic signaling. The diagram does not provide immediate insight into the likely consequences of the effects, but rather intends to give a graphic overview of the main effects reported in the literature. Additional detail can be found in **Table 2**. (*c*) Muscarinic effects on information processing in the striatum. If thalamic intralaminar neurons activate striatal cholinergic interneurons, the latter will switch from tonic to burst-pause firing mode. The brief increase in ACh causes fast reduction of corticostriatal synaptic efficacy (which informs about intended motor plans). As a consequence, disinhibition mediated by the direct pathway [medium spiny projection neurons (MSPd) projecting to the globus pallidus interna (GPi)] is less effective, and motor plans are stopped. This is followed by increased activity in medium spiny projection neurons of the indirect pathway (MSPi), mediated by M1-type receptors, resulting in further interruption of ongoing motor plans. Abbreviations: Som, somatostatin containing GABAergic interneuron; VIP, vasopressin containing GABAergic interneuron; LTS, low-threshold firing GABAergic interneurons; CCK, cholecystokinin containing GABAergic interneuron; FS, fast-spiking GABAergic interneuron; Pyr, pyramidal cell; L3, L4, L5, cortical layers 3,4,5; M1r, M1-type receptor mediated; M4/M2r, M2-type receptor mediated.

processes in macaque extrastriate and rodent V1 are at work.

A prominent model (**Figure 5*a***) of cholinergic signaling argues that ACh alters cortical network interaction by increasing the efficacy of feedforward/thalamocortical input connections onto excitatory neurons in layer IV (Barkai & Hasselmo 1994, Disney et al. 2007, Gil et al. 1997, Hasselmo & Bower 1992) through presynaptic nicotinic receptors, while suppressing the efficacy of lateral/intracortical connections through muscarinic receptors (Barkai & Hasselmo 1994, Gil et al. 1997, Hasselmo & Bower 1992, Hsieh et al. 2000, Kimura et al. 1999). These alterations could explain changes in spatial integration properties in primate V1 (Roberts et al. 2005, Silver et al. 2008), but multiple in vitro and some in vivo studies suggest that this model is overly simplistic when applied to neocortical processing, even though it may be adequate to describe alterations in paleo- and archicortical processing (Barkai &

Lateral/intracortical connections: connections between neurons located within the same area

Hasselmo 1994, Hasselmo 1999, Hasselmo & Bower 1992).

The idea that presynaptic (M2-type) muscarinic signaling suppresses the information from lateral intracortical connections and enhances local (possibly intracolumnar) processing is challenged by a study from Xiang et al. (Xiang et al. 1998). They demonstrate that low-threshold spiking (LTS) interneurons are activated through nicotinic mechanisms, whereas layer 5 fast-spiking (FS) interneurons are inhibited through muscarinic receptors. Based on the knowledge that LTS cells mostly convey inhibition vertically within a column and layer 5 FS cells convey inhibition between columns, ACh reduces intracolumnar communication through increased inhibition, and it reduces surround suppression in layer 5, as FS cells are suppressed. At face value, this directly contrasts what the previously stated model suggests. Investigators have not always found a muscarinic suppression of FS interneurons (Gulledge et al. 2007), which further complicates the issue.

Feedforward connections: connections from a lower to a higher level area (lower and higher relative to the processing hierarchy)

M-current: a voltage-gated potassium channel that raises the firing threshold of neurons; inactivates upon muscarinic signaling (M1-type) and upon hyperpolarization

Afterdepolarization (ADP): a depolarization that follows the action potential

The strengthening of feedforward connections through presynaptic nicotinic receptors that target spiny (excitatory) layer 4 neurons is undisputed (Disney et al. 2007, Gil et al. 1997, Hsieh et al. 2000, Kimura 2000). However, layer 4 spiny neurons in the rat barrel cortex are persistently hyperpolarized by M4 receptor activation, causing overall reduced efficacy of onward communication (Eggermann & Feldmeyer 2009). This finding is contrasted by depolarization of layer 2/3 and layer 5 pyramidal cells through M1 receptors (Eggermann & Feldmeyer 2009; McCormick & Prince 1985, 1986) (for more detail see **Table 2** and **Figure 5*b***). The combination of effects would ensure that weak inputs impinging on layer 4 cells are filtered out while stronger inputs are processed further. Thus, the role of muscarinic (and ACh) signaling in the cortex may be related more to efficient filtering of information, such that relevant information is processed while weak inputs are filtered out. This process would take place regardless of input source (i.e., whether arriving through feedforward or intracortical connections).

In relation to attention and other cognitive tasks, such a form of filtering would benefit from a reset mechanism that interrupts current output, and it would benefit from an increased ability to boost specific processing streams thereafter. Because layer 5 pyramidal cells provide the main outputs to subcortical structures involved in attention (pulvinar), action selection (basal ganglia), and motor circuits (superior colliculus and other brain stem structures), these cells should, and indeed do, show a form of reset behavior (Gulledge et al. 2007, 2009; McCormick & Prince 1985, 1986). Layer 5 pyramidal cells are transiently hyperpolarized, followed by lasting muscarinic depolarization through M-current and AHP current reduction, as well as afterdepolarization (ADP) current activation (Gulledge et al. 2007, 2009; McCormick & Prince 1985, 1986). Depolarization is accompanied by increased neuronal activity (Gulledge et al. 2007, 2009). The short latency inhibition is often assigned to nicotinic activation of inhibitory neurons (McCormick & Prince 1985, 1986). However, it can occur through M1 receptor–induced activation of SK_{Ca} channels (Gulledge et al. 2007, 2009). Layer 2/3 pyramidal cells do not show transient hyperpolarization, although they exhibit delayed depolarization (Gulledge et al. 2007, 2009; McCormick & Prince 1985, 1986). This ACh-induced phasic inhibition of layer 5 pyramidal cells could yield the postulated reset mechanisms, which may help an organism detect behaviorally relevant cues and integrate them appropriately into ongoing behavior (Hasselmo & Sarter 2011, Parikh et al. 2007, Parikh & Sarter 2008, Sarter et al. 2009). The depolarization that follows transient inhibition could trigger activity in layer 5 cells that reside in a column where the incoming information is sufficiently strong to overcome the depression of layer 4 neurons (Amar et al. 2010, Eggermann & Feldmeyer 2009). This action would provide the output to orchestrate behavior according to task demands. Muscarinic-induced depolarization in burst-spiking layer 5 cells (which project to the midbrain and pontine nuclei) also alters their

Table 2 Muscarinic effects on cellular activity and mechanisms involved

Cell type[a]	Layer[b]	Receptor[c]	Channel[d]	Effects	References
P	V, II, III	M1	K_M, K_{leak}, TRPC	Persistent depolarization	Eggermann & Feldmeyer 2009
P	V, deep III	M1	SK_{Ca}	Hyperpolarization (followed by depolarization if ACh applied persistently)	Gulledge et al. 2009, Gulledge & Stuart 2005
P	-	-	-	General depolarization	Levy et al. 2008
P	V	M1	K_M, K_{sAHP}, cation current	Depolarization; afterhyperpolarization replaced by afterdepolarization	Desai & Walcott 2006
P	-	-	-	Slow depolarization; increased input resistance; fast hyperpolarization (due to inhibitory interneuron activation)	McCormick & Prince 1985, 1986
P	-	-	K_M, K_{sAHP}	Reduced spike-frequency adaptation; slow depolarization	McCormick & Williamson 1989
P (burst spiking)	V	-	K current	Reduced K^+ conductance; switch from repetitive burst firing to single-spike activity	McCormick et al. 1993
P (regular spiking)	II/III, V	-	K_M; reduced Ca^{2+}-dependent K current	Decreased spike-frequency adaptation; increased responsiveness to depolarizing inputs	McCormick et al. 1993
P	V	M1	Reduced GIRK	Depolarization with tonic firing; increased temporal summation properties	Carr & Surmeier 2007
P	V	-	TRPC	Depolarization	Haj-Dahmane & Andrade 1998
P	-	M1?	K_M	Depolarization	Antal et al. 2010
P	V	M1?	TRPC	Persistent spiking activity	Rahman & Berger 2011
P	-	-	-	Increased activity in model of synaptic bombardment with preserved spike-timing precision	Tang et al. 1997
P	II/III	-	-	Short reduction in NMDA glutamatergic neurotransmission followed by lasting enhancement	Aramakis et al. 1997
P	V	-	-	Increased excitability of dendrite and increased conductance to soma	Mednikova et al. 1998
P	V	-	Slow Na^+-persistent current reduced	Reduced excitability when cells are close to threshold	Mittmann & Alzheimer 1998
P	V	M1, M2	K_{leak}, K_M, P/Q-type Ca^{2+}	Induction of dendritic Ca^{2+} spikes; increased EPSP; reduced IPSPs; changed inhibitory/excitatory balance with switch to strong output mode	Nuñez et al. 2012

(*Continued*)

Table 2 (*Continued*)

Cell type[a]	Layer[b]	Receptor[c]	Channel[d]	Effects	References
P	II/III	-	L-type Ca^{2+}, nonselective cation	Persistent firing	Zhang & Séguéla 2010
FS	II–V	-	-	No effect found	Gulledge et al. 2007
FS	Mostly V	-		Hyperpolarization	Xiang et al. 1998
FS	-	-	-	Depolarization	Levy et al. 2008
Non-FS	II–V	M1	KIR?, K_{sAHP}?	In ~45% of cells hyperpolarization	Gulledge et al. 2007
Non-FS	-	-	-	Depolarization	Levy et al. 2008
Som- or VIP-cells	-	-	-	Depolarization	Kawaguchi 1997
CCK	-	-	-	Small CCK cells depolarized; larger CCK cells hyperpolarized followed by depolarization	Kawaguchi 1997
Interneuron	-	-	-	Fast depolarization (decreased input resistance)	McCormick & Prince 1985, 1986
Spiny stellate	IV	M4	KIR	Persistent hyperpolarization	Eggermann & Feldmeyer 2009

Abbreviations: ACh, acetylcholine; CCK, cholecystokinine positive interneurons; EPSP, excitatory postsynaptic potential; FS, fast-spiking interneurons; GIRK: G protein–coupled inward rectifying K^+ channel; IPSP, inhibitory postsynaptic potential; KIR: inward rectifying K^+ channel; K_{leak}: leaky potassium channel; K_M: M-type (voltage-gated) potassium channel; K_{sAHP}: slow (non-SK) afterhyperpolarization K^+ channel; NMDA, *N*-methyl-D-aspartate; Non-FS, non-fast-spiking interneurons; P, pyramidal cells; SK_{Ca}: SK-type calcium-activated K^+ channel; Som, somatostatin positive interneurons; TRPC: Ca^{2+}-activated nonspecific cation channel; VIP, vasointestinal polypeptide positive interneurons. A dash indicates that data were not specified.

[a]Column 1 lists cortical cell types affected by muscarinic signaling.

[b]Column 2 lists the layers from which a given cell was recorded.

[c]Column 3 lists the channels that were identified (if any) to mediate the effects.

[d]Column 4 lists the main effects that were found at the cellular level.

firing pattern from burst to regular spiking. Such a transition may enable a cell to switch from event detection (signaled by all or none burst output) to a more linear transmission of information (McCormick et al. 1993), which is better suited to reflect potentially small fluctuations in the input; it could also be a sign of fine-grained cortical state changes associated with attention (Harris & Thiele 2011).

Layer 5 pyramidal cells are transiently hyperpolarized by phasic ACh pulses, and resetting through ACh could further be supported by fast activation of inhibitory interneurons. Such activation is absent in FS cells (Gulledge et al. 2007, 2009; Gulledge & Stuart 2005), but studies found short latency excitatory (nicotinic-mediated) responses in other classes of inhibitory interneurons (Christophe et al. 2002, Gulledge et al. 2007, Porter et al. 1999, Xiang et al. 1998). In addition to fast nicotinic activation, muscarinic-induced activation of GABAergic somatostatin- or VIP-immunoreactive cells and activation and deactivation of CCK cells (Kawaguchi 1997) occur. The overall increased inhibitory drive should cause reduced cellular activity in most cells, which could aid the filtering process. In contrast, most in vivo studies report increased neuronal activity upon ACh application in anesthetized (Murphy & Sillito 1991; Roberts et al. 2005; Sato et al. 1987a,b; Sillito & Kemp 1983; Thiele et al. 2012; but see Disney et al. 2012) or awake animals (Herrero et al. 2008), and they report reduced activity upon muscarinic blockade (Herrero et al. 2008). The finding that ACh overall increases firing rates in vivo, while not supporting the filtering hypothesis, supports the idea that ACh is involved in attentional signaling because attention usually increases neuronal activity (Deco & Thiele 2009, Reynolds & Chelazzi 2004).

Why does neuronal activity usually increase, rather than decrease, upon ACh application? In addition to nicotinic and M1-type-mediated activation of interneurons, muscarinic receptor signaling reduces the efficacy of GABAergic synapses in layer 3, 4 and 5 FS → pyramidal connections through a fast (membrane-delimited) M2-type pathway (Kruglikov & Rudy 2008), and it reduces the efficacy of GABAergic synapses in non-FS → pyramidal cell connections (Yamamoto et al. 2010); by comparison, the efficacy of GABAergic synapses of interneuron → interneuron connections was upregulated for relatively weak connections and downregulated for relatively strong connections (Yamamoto et al. 2010). Therefore, inhibition within the cortex could effectively be reduced by muscarinic signaling despite increased activity within certain inhibitory cells (**Figure 5*b***). Reduced overall inhibition does not easily fit into the proposed filtering framework, but the reduced GABAergic signaling that is prominent in FS cells could have an important role in resetting networks linked by gamma oscillatory activity, as outlined later.

Some of the above-described effects of cholinergic drive on cellular activity and cellular networks have been successfully used to model the contribution ACh makes to attentional rate enhancement (Deco & Thiele 2011). The model argues that ACh, through muscarinic signaling, allows feedback connections from higher areas to exert their influence: ACh sets the network balance, which is then exploited by selective feedback.

Cholinergic/muscarinic signaling is comparatively well understood in the basal ganglia, where it can be conceptualized within the active filtering theory, even if in a modified version. Cholinergic striatal interneurons normally fire tonically but revert to a burst-pause mode if activated by intralaminar thalamic nuclei neurons (Goldberg et al. 2012) (**Figure 5*c***). The increased activity in cholinergic interneurons results in a brief suppression of corticostriatal synaptic efficacy. Corticostriatal connections affect two types of striatal projection neurons: those that form part of the direct pathway and those that form part of the indirect pathway (**Figure 5*c***). The direct pathway enables currently desired motor behavior to be executed, whereas the indirect pathway inhibits motor behavior (Ding et al. 2010, Goldberg et al. 2012) (**Figure 5*c***). The short-term reduction of corticostriatal synaptic efficacy is mediated by M4/M2 receptors. Since M4/M2 receptors engage the membrane-delimited pathway the reduction of corticostriatal synaptic efficacy is fast and fairly short lasting (Goldberg et al. 2012). Following this reduction in synaptic efficacy, projection neurons of the indirect pathway show prolonged facilitation through M1-mediated actions. The combined effects cause rapid inhibition of ongoing behavior (corticostriatal inputs are made less efficient), followed by a lasting suppression of unwanted patterns of action through increased activity in projection neurons of the indirect pathway (**Figure 5*c***), which, as stated, is a form of efficient filtering.

Oscillatory activity: rhythmic activity in neuronal ensembles that is often linked to cognitive operations such as memory or attention

Feedback connections: Connections from a higher-level area to a lower-level area (higher and lower relative to the processing hierarchy)

Muscarinic Control of Neuronal Synchrony and Oscillatory Activity

Synchrony plays an important role in cognitive functions such as attention (Chalk et al. 2010, Fries et al. 2001, Gregoriou et al. 2009). Synchrony can occur in the absence (Abeles et al. 1993, de Oliveira et al. 1997) or presence of oscillatory activity (Engel et al. 2001). Disruption of oscillatory activity is associated with cognitive deficits, as seen in schizophrenia, autism, Alzheimer disease, and Parkinson disease (Uhlhaas & Singer 2006). Oscillatory activity involves many different cell types, transmitters, and receptors. Muscarinic receptors play an important role therein. Activation of muscarinic receptors supports the transitions from deactivated to activated brain states. The transition occurs when slow synchronization of large-scale population activity is replaced by smaller (and often faster) synchronous rhythms when subjects are in an alert state. The latter is dominated by oscillations in the theta-, beta-, and gamma-frequency range (Buzsáki et al. 1988, Metherate et al. 1992). The transition takes a rather strong form when sleep or drowsy states are followed by alert states, but we assume that the size of state changes forms a continuum to adapt neuronal processing to behavioral demands (Harris & Thiele 2011).

Oscillatory activity in different frequency ranges occurs throughout the brain, whereby different frequencies are proposed to serve different functions depending on the brain areas involved. For example, hippocampal theta oscillatory (4–10 Hz) activity is a hallmark of an active brain state, which may temporally organize interactions between different brain areas in line with behavioral demands (Sirota et al. 2008). Conversely, Mazaheri & Jensen (2011) have proposed that oscillatory activity in the cortex within lower frequencies [theta and alpha frequency band (4–13 Hz)] reflects an idle brain state. This view is not uncontested: Increased spiking and LFP oscillatory activity within the theta band occur in cortical areas during task performance (Bollimunta et al. 2011, Lakatos et al. 2005, Lee et al. 2005), and spike times relative to the phase of slow LFP fluctuations (~ low theta range) in V1 convey substantial information about visual animated scenes (Montemurro et al. 2008).

Oscillatory network activity depends on the activity in inhibitory interneurons (Gieselmann & Thiele 2008, Whittington et al. 1995), and these are influenced by muscarinic signaling (Disney & Aoki 2008, Disney et al. 2006, Kawaguchi 1997). Consequently, activation of muscarinic receptors elicits theta- (Lukatch & MacIver 1997) and gamma-frequency oscillations (Buhl et al. 1998). Theta oscillations in the hippocampus are subdivided into an atropine-insensitive type and an atropine-sensitive urethane anesthesia induced type (Buzsáki 2002, Vanderwolf 1988); i.e., although muscarinic signaling can induce hippocampal theta, it is not required. Similarly, cortical state changes and the associated decrease in low frequency and increase in gamma-frequency power (i.e., the brain activation) are sometimes, but not always, dependent on muscarinic signaling (Dringenberg & Vanderwolf 1998).

In anesthetized cats, activation of muscarinic receptors in the visual cortex by mesencephalic reticular formation stimulation enhances stimulus-driven neuronal synchronization in the gamma-frequency range (Rodriguez et al. 2004). Investigators have interpreted the increased oscillatory synchronization as a signature of attention. This interpretation is debatable because attention does not uniformly increase gamma-oscillatory activity across cortical areas (Chalk et al. 2010, Fries et al. 2001). Moreover, increasing the cholinergic drive does not alter attention-induced changes in gamma-frequency oscillations in the human visual cortex, but rather reduces low-frequency oscillations (Bauer et al. 2012). It remains to be determined in awake task performing preparations to what extent (local) muscarinic signaling is involved in attention-induced alterations of cortical oscillations.

The cellular effects described in the previous subsection suggest that muscarinic signaling should reduce gamma-frequency oscillations because it mostly reduces FS → pyramidal cell synaptic efficacy, and the strength of FS → pyramidal cell connectivity is regarded as critical in gamma oscillations (Borgers et al. 2008, Traub et al. 1996). If gamma oscillations were to be reduced, then this could be further interpreted as an involvement of muscarinic signaling in phase resetting and filtering, as proposed above. It would be an effective mechanism to temporarily interrupt established network communications and enable the formation of new functional networks, effectively reorganizing different ensembles to communicate within the framework of the gamma cycle (Fries et al. 2007). Muscarinic signaling, however, increases gamma-frequency oscillations (Buhl et al. 1998), although this notion is established only for sustained muscarinic activation.

It is unclear whether phasic application of ACh would transiently reduce and then increase gamma oscillations. Such a pattern would be a strong argument for phase resetting/filtering and support the role of gamma oscillations in the dynamic organization of neuronal ensemble interactions.

How could gamma activity increase (Buhl et al. 1998) despite reduced FS synaptic activity? When ambient ACh levels are low, pyramidal → FS interneuron synapses show strong adaptation during high-input regimes, whereas pyramidal → non-FS interneuron synapses show strong facilitation. When ambient ACh levels are high, both adaptation and facilitation are reduced (Levy et al. 2008). Thus, if gamma-oscillatory activity is generated in a reciprocally connected pyramidal-interneuron network, then the reduced adaptation in the pyramidal → FS connection would drive FS cells more strongly; this result could overcompensate for the reduced FS → pyramidal synaptic efficacy and increase gamma strength. This mechanism may also help delineate contributing ensembles more efficiently because cells that are only weakly participating may be dropped from the cyclic interactions.

CONCLUSIONS

Current data do not clearly delineate a unifying framework wherein cellular effects explain cortical influences on cognition. Muscarinic receptor activation can increase and decrease cellular activity, and it can increase and decrease synaptic efficacy in a cell-type-specific manner. The influential theory arguing that muscarinic receptors help reduce intracortical (inferential/feedback) processing, thereby allowing sensory information to dominate the processing machinery, seems overly simplistic when applied to neocortical circuits, but it may be a valid description of paleo- and allocortical circuits. An alternative theory, that muscarinic signaling enables efficient filtering of information in light of current task demands, has some appeal, and it may be adequate to describe cholinergic signaling within the basal ganglia. Within the context of cortical processing, the theory requires elaboration. Large amounts of fairly fragmented details make it difficult to propose a consistent, overarching framework within which to explain the contribution of muscarinic signaling to attention. Given the renewed interest in muscarinic signaling at the systems level, we hope this deficit will be overcome soon.

SUMMARY POINTS

1. Muscarinic signaling influences cellular activity by various, possibly interacting pathways affecting many different ion channels.
2. These effects differ in a cell-type-specific manner, which can vary within cell classes in different cortical layers (and possibly different areas).
3. It endows the muscarinic system to alter network activity in multiple, poorly understood, ways.
4. The current understanding of muscarinic-induced alteration of cortical processing gives rise to two main models: (*a*) Muscarinic signaling enables the cortex to rely mostly on information arriving from the sensory organs, while relying less on internal (inferential) processing; or (*b*) muscarinic signaling enables efficient filtering of information, irrespective of the source (whether arriving from the sensory organs or being generated internally), and enables a dynamic reorganization of functional connectivity.
5. Both models receive some support from existing data, but both models equally suffer from explanatory gaps.

FUTURE DIRECTIONS

1. What is the temporal and spatial precision of cholinergic/muscarinic signaling in different cortical areas under physiological conditions?
2. How does muscarinic signaling affect cellular interactions in different layers, and what are the consequences for different types of incoming information?
3. How does phasic cholinergic signaling affect oscillatory activity?
4. Are effects of muscarinic signaling on oscillatory activity layer dependent?
5. How does muscarinic signaling affect the communication between cortical areas?

DISCLOSURE STATEMENT

The author is not aware of any affiliations, memberships, funding, or financial holdings that might be perceived as affecting the objectivity of this review.

ACKNOWLEDGMENTS

This work is supported by the Wellcome Trust, the Biotechnology and Biological Sciences Research Council (BBSRC), and the Medical Research Council (MRC).

LITERATURE CITED

Abeles M, Bergman H, Margalit E, Vaadia E. 1993. Spatiotemporal firing patterns in the frontal cortex of behaving monkeys. *J. Neurophysiol.* 70:1629–38

Allen TG. 1999. The role of N-, Q- and R-type Ca^{2+} channels in feedback inhibition of ACh release from rat basal forebrain neurones. *J. Physiol.* 515(Pt. 1):93–107

Allen TG, Brown DA. 1993. M2 muscarinic receptor-mediated inhibition of the Ca^{2+} current in rat magnocellular cholinergic basal forebrain neurones. *J. Physiol.* 466:173–89

Amar M, Lucas-Meunier E, Baux G, Fossier P. 2010. Blockade of different muscarinic receptor subtypes changes the equilibrium between excitation and inhibition in rat visual cortex. *Neuroscience* 169:1610–20

Antal M, Acuna-Goycolea C, Pressler RT, Blitz DM, Regehr WG. 2010. Cholinergic activation of M2 receptors leads to context-dependent modulation of feedforward inhibition in the visual thalamus. *PLoS Biol.* 8:e1000348

Aramakis VB, Bandrowski AE, Ashe JH. 1997. Activation of muscarinic receptors modulates NMDA receptor-mediated responses in auditory cortex. *Exp. Brain Res.* 113:484–96

Barkai E, Hasselmo ME. 1994. Modulation of the input/output function of rat piriform cortex pyramidal cells. *J. Neurophysiol.* 72:644–58

Bauer M, Kluge C, Bach D, Bradbury D, Heinze HJ, et al. 2012. Cholinergic enhancement of visual attention and neural oscillations in the human brain. *Curr. Biol.* 22:397–402

Biscoe TJ, Straughan DW. 1966. Micro-electrophoretic studies of neurones in the cat hippocampus. *J. Physiol.* 183:341–59

Bollimunta A, Mo J, Schroeder CE, Ding M. 2011. Neuronal mechanisms and attentional modulation of corticothalamic alpha oscillations. *J. Neurosci.* 31:4935–43

Borgers C, Epstein S, Kopell NJ. 2008. Gamma oscillations mediate stimulus competition and attentional selection in a cortical network model. *Proc. Natl. Acad. Sci. USA* 105:18023–28

Brown DA. 2010. Muscarinic acetylcholine receptors (mAChRs) in the nervous system: some functions and mechanisms. *J. Mol. Neurosci.* 41:340–46

Bubser M, Byun N, Wood MR, Jones CK. 2012. Muscarinic receptor pharmacology and circuitry for the modulation of cognition. *Handb. Exp. Pharmacol.* (208):121–66

Buchanan KA, Petrovic MM, Chamberlain SE, Marrion NV, Mellor JR. 2010. Facilitation of long-term potentiation by muscarinic M(1) receptors is mediated by inhibition of SK channels. *Neuron* 68:948–63

Buhl EH, Tamas G, Fisahn A. 1998. Cholinergic activation and tonic excitation induce persistent gamma oscillations in mouse somatosensory cortex in vitro. *J. Physiol.* 513(Pt. 1):117–26

Bushnell PJ, Oshiro WM, Padnos BK. 1997. Detection of visual signals by rats: effects of chlordiazepoxide and cholinergic and adrenergic drugs on sustained attention. *Psychopharmacology* 134:230–41

Buzsáki G. 2002. Theta oscillations in the hippocampus. *Neuron* 33:325–40

Buzsáki G, Bickford RG, Ponomareff G, Thal LJ, Mandel R, Gage FH. 1988. Nucleus basalis and thalamic control of neocortical activity in the freely moving rat. *J. Neurosci.* 8:4007–26

Calabresi P, Centonze D, Gubellini P, Pisani A, Bernardi G. 2000. Acetylcholine-mediated modulation of striatal function. *Trends Neurosci.* 23:120–26

Carr DB, Surmeier DJ. 2007. M1 muscarinic receptor modulation of Kir2 channels enhances temporal summation of excitatory synaptic potentials in prefrontal cortex pyramidal neurons. *J. Neurophysiol.* 97:3432–38

Carrillo-Reid L, Tecuapetla F, Vautrelle N, Hernández A, Vergara R, et al. 2009. Muscarinic enhancement of persistent sodium current synchronizes striatal medium spiny neurons. *J. Neurophysiol.* 102:682–90

Caulfield MP. 1993. Muscarinic receptors—characterization, coupling and function. *Pharmacol. Ther.* 58:319–79

Caulfield MP, Birdsall NJ. 1998. International Union of Pharmacology. XVII. Classification of muscarinic acetylcholine receptors. *Pharmacol. Rev.* 50:279–90

Chalk M, Herrero JL, Gieselmann MA, Delicato LS, Gotthardt S, Thiele A. 2010. Attention reduces stimulus-driven gamma frequency oscillations and spike field coherence in V1. *Neuron* 66:114–25

Chen Y, Cantrell AR, Messing RO, Scheuer T, Catterall WA. 2005. Specific modulation of Na^{+} channels in hippocampal neurons by protein kinase C epsilon. *J. Neurosci.* 25:507–13

Christie BR, Eliot LS, Ito K, Miyakawa H, Johnston D. 1995. Different Ca^{2+} channels in soma and dendrites of hippocampal pyramidal neurons mediate spike-induced Ca^{2+} influx. *J. Neurophysiol.* 73:2553–57

Christophe E, Roebuck A, Staiger JF, Lavery DJ, Charpak S, Audinat E. 2002. Two types of nicotinic receptors mediate an excitation of neocortical layer I interneurons. *J. Neurophysiol.* 88:1318–27

Davidson MC, Cutrell EB, Marrocco RT. 1999. Scopolamine slows the orienting of attention in primates to cued visual targets. *Psychopharmacology* 142:1–8

Davidson MC, Marrocco RT. 2000. Local infusion of scopolamine into intraparietal cortex slows covert orienting in rhesus monkeys. *J. Neurophysiol.* 83:1536–49

Deco G, Thiele A. 2009. Attention: oscillations and neuropharmacology. *Eur. J. Neurosci.* 30:347–54

Deco G, Thiele A. 2011. Cholinergic control of cortical network interactions enables feedback-mediated attentional modulation. *Eur. J. Neurosci.* 34:146–57

de Oliveira SC, Thiele A, Hoffmann KP. 1997. Synchronization of neuronal activity during stimulus expectation in a direction discrimination task. *J. Neurosci.* 17:9248–60

Desai NS, Walcott EC. 2006. Synaptic bombardment modulates muscarinic effects in forelimb motor cortex. *J. Neurosci.* 26:2215–26

Dillon GM, Shelton D, McKinney AP, Caniga M, Marcus JN, et al. 2009. Prefrontal cortex lesions and scopolamine impair attention performance of C57BL/6 mice in a novel 2-choice visual discrimination task. *Behav. Brain Res.* 204:67–76

Ding JB, Guzman JN, Peterson JD, Goldberg JA, Surmeier DJ. 2010. Thalamic gating of corticostriatal signalling by cholinergic interneurons. *Neuron* 67:294–307

Disney AA, Aoki C. 2008. Muscarinic acetylcholine receptors in macaque V1 are most frequently expressed by parvalbumin-immunoreactive neurons. *J. Comp. Neurol.* 507:1748–62

Disney AA, Aoki C, Hawken MJ. 2007. Gain modulation by nicotine in macaque v1. *Neuron* 56:701–13

Disney AA, Aoki C, Hawken MJ. 2012. Cholinergic suppression of visual responses in primate V1 is mediated by GABAergic inhibition. *J. Neurophysiol.* 108(7):1907–23

Disney AA, Domakonda KV, Aoki C. 2006. Differential expression of muscarinic acetylcholine receptors across excitatory and inhibitory cells in visual cortical areas V1 and V2 of the macaque monkey. *J. Comp. Neurol.* 499:49–63

Dringenberg HC, Vanderwolf CH. 1998. Involvement of direct and indirect pathways in electrocorticographic activation. *Neurosci. Biobehav. Rev.* 22:243–57

Dunne MP, Hartley LR. 1986. Scopolamine and the control of attention in humans. *Psychopharmacology* 89: 94–97

Eckenstein F, Baughman RW. 1984. Two types of cholinergic innervation in cortex, one co-localized with vasoactive intestinal polypeptide. *Nature* 309:153–55

Eggermann E, Feldmeyer D. 2009. Cholinergic filtering in the recurrent excitatory microcircuit of cortical layer 4. *Proc. Natl. Acad. Sci. USA* 106:11753–58

Eglen RM. 2012. Overview of muscarinic receptor subtypes. *Handb. Exp. Pharmacol.* (208):3–28

Egorov AV, Angelova PR, Heinemann U, Müller W. 2003. Ca^{2+}-independent muscarinic excitation of rat medial entorhinal cortex layer V neurons. *Eur. J. Neurosci.* 18:3343–51

El-Fakahany E, Richelson E. 1980. Temperature dependence of muscarinic acetylcholine receptor activation, desensitization, and resensitization. *J. Neurochem.* 34:1288–95

Ellis JR, Ellis KA, Bartholomeusz CF, Harrison BJ, Wesnes KA, et al. 2006. Muscarinic and nicotinic receptors synergistically modulate working memory and attention in humans. *Int. J. Neuropsychopharmacol.* 9:175–89

Engel AK, Fries P, Singer W. 2001. Dynamic predictions: oscillations and synchrony in top-down processing. *Nat. Rev. Neurosci.* 2:704–16

Everitt BJ, Robbins TW. 1997. Central cholinergic systems and cognition. *Annu. Rev. Psychol.* 48:649–84

Felder CC. 1995. Muscarinic acetylcholine receptors: signal transduction through multiple effectors. *FASEB J.* 9:619–25

Fournier GN, Semba K, Rasmusson DD. 2004. Modality- and region-specific acetylcholine release in the rat neocortex. *Neuroscience* 126:257–62

Freund TF, Gulyas AI. 1991. GABAergic interneurons containing calbindin D28K or somatostatin are major targets of GABAergic basal forebrain afferents in the rat neocortex. *J. Comp. Neurol.* 314:187–99

Fries P, Nikolic D, Singer W. 2007. The gamma cycle. *Trends Neurosci.* 30:309–16

Fries P, Reynolds JH, Rorie AE, Desimone R. 2001. Modulation of oscillatory neuronal synchronization by selective visual attention. *Science* 291:1560–63

Furey ML, Pietrini P, Haxby JV, Drevets WC. 2008. Selective effects of cholinergic modulation on task performance during selective attention. *Neuropsychopharmacology* 33:913–23

Gamper N, Shapiro MS. 2007. Regulation of ion transport proteins by membrane phosphoinositides. *Nat. Rev. Neurosci.* 8:921–34

Gebhard R, Zilles K, Schleicher A, Everitt BJ, Robbins TW, Divac I. 1993. Distribution of seven major neurotransmitter receptors in the striate cortex of the New World monkey *Callithrix jacchus*. *Neuroscience* 56:877–85

Ghamari-Langroudi M, Bourque CW. 2004. Muscarinic receptor modulation of slow afterhyperpolarization and phasic firing in rat supraoptic nucleus neurons. *J. Neurosci.* 24:7718–26

Gieselmann MA, Thiele A. 2008. Comparison of spatial integration and surround suppression characteristics in spiking activity and the local field potential in macaque V1. *Eur. J. Neurosci.* 28:447–59

Giessel AJ, Sabatini BL. 2010. M1 muscarinic receptors boost synaptic potentials and calcium influx in dendritic spines by inhibiting postsynaptic SK channels. *Neuron* 68:936–47

Gil Z, Connors BW, Amitai Y. 1997. Differential regulation of neocortical synapses by neuromodulators and activity. *Neuron* 19:679–86

Goard M, Dan Y. 2009. Basal forebrain activation enhances cortical coding of natural scenes. *Nat. Neurosci.* 12:1444–49

Goldberg JA, Ding JB, Surmeier DJ. 2012. Muscarinic modulation of striatal function and circuitry. *Handb. Exp. Pharmacol.* (208):223–41

Gregoriou GG, Gotts SJ, Zhou H, Desimone R. 2009. High-frequency, long-range coupling between prefrontal and visual cortex during attention. *Science* 324:1207–10

Gulledge AT, Bucci DJ, Zhang SS, Matsui M, Yeh HH. 2009. M1 receptors mediate cholinergic modulation of excitability in neocortical pyramidal neurons. *J. Neurosci.* 29:9888–902

Gulledge AT, Park SB, Kawaguchi Y, Stuart GJ. 2007. Heterogeneity of phasic cholinergic signaling in neocortical neurons. *J. Neurophysiol.* 97:2215–29

Gulledge AT, Stuart GJ. 2005. Cholinergic inhibition of neocortical pyramidal neurons. *J. Neurosci.* 25:10308–20

Haj-Dahmane S, Andrade R. 1998. Ionic mechanism of the slow afterdepolarization induced by muscarinic receptor activation in rat prefrontal cortex. *J. Neurophysiol.* 80:1197–210

Harris KD, Thiele A. 2011. Cortical state and attention. *Nat. Rev. Neurosci.* 12:509–23

Hassani OK, Lee MG, Henny P, Jones BE. 2009. Discharge profiles of identified GABAergic in comparison to cholinergic and putative glutamatergic basal forebrain neurons across the sleep-wake cycle. *J. Neurosci.* 29:11828–40

Hasselmo ME 1999. Neuromodulation and the hippocampus: memory function and dysfunction in a network simulation. *Prog. Brain Res.* 121:3–18

Hasselmo ME, Bower JM. 1992. Cholinergic suppression specific to intrinsic not afferent fiber synapses in rat piriform (olfactory) cortex. *J. Neurophysiol.* 67:1222–29

Hasselmo ME, Sarter M. 2011. Modes and models of forebrain cholinergic neuromodulation of cognition. *Neuropsychopharmacology* 36:52–73

Henny P, Jones BE. 2008. Projections from basal forebrain to prefrontal cortex comprise cholinergic, GABAergic and glutamatergic inputs to pyramidal cells or interneurons. *Eur. J. Neurosci.* 27:654–70

Herrero JL, Roberts MJ, Delicato LS, Gieselmann MA, Dayan P, Thiele A. 2008. Acetylcholine contributes through muscarinic receptors to attentional modulation in V1. *Nature* 454:1110–14

Hildebrand ME, David LS, Hamid J, Mulatz K, Garcia E, et al. 2007. Selective inhibition of Cav3.3 T-type calcium channels by Galphaq/11-coupled muscarinic acetylcholine receptors. *J. Biol. Chem.* 282:21043–55

Hille B. 1994. Modulation of ion-channel function by G-protein-coupled receptors. *Trends Neurosci.* 17:531–36

Hoff EI, van Oostenbrugge RJ, Liedenbaum M, Steinbusch HW, Blokland A. 2007. Effects of right-hemisphere cortical infarction and muscarinic acetylcholine receptor blockade on spatial visual attention performance in rats. *Behav. Brain Res.* 178:62–69

Hsieh CY, Cruikshank SJ, Metherate R. 2000. Differential modulation of auditory thalamocortical and intracortical synaptic transmission by cholinergic agonist. *Brain Res.* 880:51–64

Kandel ER. 2012. The molecular biology of memory: cAMP, PKA, CRE, CREB-1, CREB-2, and CPEB. *Mol. Brain* 5:14

Kawaguchi Y. 1997. Selective cholinergic modulation of cortical GABAergic cell subtypes. *J. Neurophysiol.* 78:1743–47

Kimura F. 2000. Cholinergic modulation of cortical function: a hypothetical role in shifting the dynamics in cortical network. *Neurosci. Res.* 38:19–26

Kimura F, Fukuda M, Tsumoto T. 1999. Acetylcholine suppresses the spread of excitation in the visual cortex revealed by optical recording: possible differential effect depending on the source of input. *Eur. J. Neurosci.* 11:3597–609

Kobrinsky E, Mirshahi T, Zhang H, Jin T, Logothetis DE. 2000. Receptor-mediated hydrolysis of plasma membrane messenger PIP2 leads to K^+-current desensitization. *Nat. Cell Biol.* 2:507–14

Kozak R, Bruno JP, Sarter M. 2006. Augmented prefrontal acetylcholine release during challenged attentional performance. *Cereb. Cortex* 16:9–17

Kruglikov I, Rudy B. 2008. Perisomatic GABA release and thalamocortical integration onto neocortical excitatory cells are regulated by neuromodulators. *Neuron* 58:911–24

Lakatos P, Shah AS, Knuth KH, Ulbert I, Karmos G, Schroeder CE. 2005. An oscillatory hierarchy controlling neuronal excitability and stimulus processing in the auditory cortex. *J. Neurophysiol.* 94:1904–11

Leblond L, Beaufort C, Delerue F, Durkin TP. 2002. Differential roles for nicotinic and muscarinic cholinergic receptors in sustained visuo-spatial attention? A study using a 5-arm maze protocol in mice. *Behav. Brain Res.* 128:91–102

Lee H, Simpson GV, Logothetis NK, Rainer G. 2005. Phase locking of single neuron activity to theta oscillations during working memory in monkey extrastriate visual cortex. *Neuron* 45:147–56

Levey AI, Kitt CA, Simonds WF, Price DL, Brann MR. 1991. Identification and localization of muscarinic acetylcholine receptor proteins in brain with subtype-specific antibodies. *J. Neurosci.* 11:3218–26

Levin ED, Bushnell PJ, Rezvani AH. 2011. Attention-modulating effects of cognitive enhancers. *Pharmacol. Biochem. Behav.* 99:146–54

Levy RB, Reyes AD, Aoki C. 2008. Cholinergic modulation of local pyramid-interneuron synapses exhibiting divergent short-term dynamics in rat sensory cortex. *Brain Res.* 1215:97–104

Lukatch HS, MacIver MB. 1997. Physiology, pharmacology, and topography of cholinergic neocortical oscillations in vitro. *J. Neurophysiol.* 77:2427–45

Ma JY, Catterall WA, Scheuer T. 1997. Persistent sodium currents through brain sodium channels induced by G protein betagamma subunits. *Neuron* 19:443–52

Mazaheri A, Jensen O. 2011. Rhythmic pulsing: linking ongoing brain activity with evoked responses. *Front. Hum. Neurosci.* 4:177

McCormick DA, Prince DA. 1985. Two types of muscarinic response to acetylcholine in mammalian cortical neurons. *Proc. Natl. Acad. Sci. USA* 82:6344–48

McCormick DA, Prince DA. 1986. Mechanisms of action of acetylcholine in the guinea-pig cerebral cortex in vitro. *J. Physiol.* 375:169–94

McCormick DA, Wang Z, Huguenard J. 1993. Neurotransmitter control of neocortical neuronal activity and excitability. *Cereb. Cortex* 3:387–98

McCormick DA, Williamson A. 1989. Convergence and divergence of neurotransmitter action in human cerebral cortex. *Proc. Natl. Acad. Sci. USA* 86:8098–102

Mednikova YS, Karnup SV, Loseva EV. 1998. Cholinergic excitation of dendrites in neocortical neurons. *Neuroscience* 87:783–96

Mesulam MM, Geula C. 1992. Overlap between acetylcholinesterase-rich and choline acetyltransferase-positive (cholinergic) axons in human cerebral cortex. *Brain Res.* 577:112–20

Mesulam MM, Mufson EJ, Levey AI, Wainer BH. 1983a. Cholinergic innervation of cortex by the basal forebrain: cytochemistry and cortical connections of the septal area, diagonal band nuclei, nucleus basalis (substantia innominata), and hypothalamus in the rhesus monkey. *J. Comp. Neurol.* 214:170–97

Mesulam MM, Mufson EJ, Levey AI, Wainer BH. 1984. Atlas of cholinergic neurons in the forebrain and upper brainstem of the macaque based on monoclonal choline acetyltransferase immunohistochemistry and acetylcholinesterase histochemistry. *Neuroscience* 12:669–86

Mesulam MM, Mufson EJ, Wainer BH, Levey AI. 1983b. Central cholinergic pathways in the rat: an overview based on an alternative nomenclature (Ch1-Ch6). *Neuroscience* 10:1185–201

Metherate R, Cox CL, Ashe JH. 1992. Cellular bases of neocortical activation: modulation of neural oscillations by the nucleus basalis and endogenous acetylcholine. *J. Neurosci.* 12:4701–11

Mittmann T, Alzheimer C. 1998. Muscarinic inhibition of persistent Na^+ current in rat neocortical pyramidal neurons. *J. Neurophysiol.* 79:1579–82

Montemurro MA, Rasch MJ, Murayama Y, Logothetis NK, Panzeri S. 2008. Phase-of-firing coding of natural visual stimuli in primary visual cortex. *Curr. Biol.* 18:375–80

Murphy PC, Sillito AM. 1991. Cholinergic enhancement of direction selectivity in the visual cortex of the cat. *Neuroscience* 40:13–20

Navaroli VL, Zhao Y, Boguszewski P, Brown TH. 2012. Muscarinic receptor activation enables persistent firing in pyramidal neurons from superficial layers of dorsal perirhinal cortex. *Hippocampus* 22:1392–404

Nuñez A, Domínguez S, Buño W, Fernández de Sevilla D. 2012. Cholinergic-mediated response enhancement in barrel cortex layer V pyramidal neurons. *J. Neurophysiol.* 108:1656–68

Onali P, Olianas MC. 1995. Bimodal regulation of cyclic AMP by muscarinic receptors. Involvement of multiple G proteins and different forms of adenylyl cyclase. *Life Sci.* 56:973–80

Paré D, Smith Y, Parent A, Steriade M. 1988. Projections of brainstem core cholinergic and non-cholinergic neurons of cat to intralaminar and reticular thalamic nuclei. *Neuroscience* 25:69–86

Parikh V, Kozak R, Martinez V, Sarter M. 2007. Prefrontal acetylcholine release controls cue detection on multiple timescales. *Neuron* 56:141–54

Parikh V, Sarter M. 2008. Cholinergic mediation of attention: contributions of phasic and tonic increases in prefrontal cholinergic activity. *Ann. N. Y. Acad. Sci.* 1129:225–35

Pemberton KE, Hill-Eubanks LJ, Jones SVP. 2000. Modulation of low-threshold T-type calcium channels by the five muscarinic receptor subtypes in NIH 3T3 cells. *Pflugers Arch.* 440:452–61

Peralta EG, Ashkenazi A, Winslow JW, Ramachandran J, Capon DJ. 1988. Differential regulation of PI hydrolysis and adenylyl cyclase by muscarinic receptor subtypes. *Nature* 334:434–37

Perez DM, Karnik SS. 2005. Multiple signaling states of G-protein-coupled receptors. *Pharmacol. Rev.* 57:147–61

Porter JT, Cauli B, Tsuzuki K, Lambolez B, Rossier J, Audinat E. 1999. Selective excitation of subtypes of neocortical interneurons by nicotinic receptors. *J. Neurosci.* 19:5228–35

Raghanti MA, Stimpson CD, Marcinkiewicz JL, Erwin JM, Hof PR, Sherwood CC. 2008. Cholinergic innervation of the frontal cortex: differences among humans, chimpanzees, and macaque monkeys. *J. Comp. Neurol.* 506:409–24

Rahman J, Berger T. 2011. Persistent activity in layer 5 pyramidal neurons following cholinergic activation of mouse primary cortices. *Eur. J. Neurosci.* 34:22–30

Reynolds JH, Chelazzi L. 2004. Attentional modulation of visual processing. *Annu. Rev. Neurosci.* 27:611–47

Robbins TW, Granon S, Muir JL, Durantou F, Harrison A, Everitt BJ. 1998. Neural systems underlying arousal and attention. Implications for drug abuse. *Ann. N. Y. Acad. Sci.* 846:222–37

Roberts MJ, Zinke W, Guo K, Robertson R, McDonald JS, Thiele A. 2005. Acetylcholine dynamically controls spatial integration in marmoset primary visual cortex. *J. Neurophysiol.* 93:2062–72

Rodriguez R, Kallenbach U, Singer W, Munk MH. 2004. Short- and long-term effects of cholinergic modulation on gamma oscillations and response synchronization in the visual cortex. *J. Neurosci.* 24:10369–78

Rouse ST, Thomas TM, Levey AI. 1997. Muscarinic acetylcholine receptor subtype, m2: diverse functional implications of differential synaptic localization. *Life Sci.* 60:1031–38

Rye DB, Wainer BH, Mesulam MM, Mufson EJ, Saper CB. 1984. Cortical projections arising from the basal forebrain: a study of cholinergic and noncholinergic components employing combined retrograde tracing and immunohistochemical localization of choline acetyltransferase. *Neuroscience* 13:627–43

Salgado H, Bellay T, Nichols JA, Bose M, Martinolich L, et al. 2007. Muscarinic M2 and M1 receptors reduce GABA release by Ca^{2+} channel modulation through activation of PI3K/Ca^{2+}-independent and PLC/Ca^{2+}-dependent PKC. *J. Neurophysiol.* 98:952–65

Sarter M, Parikh V, Howe WM. 2009. Phasic acetylcholine release and the volume transmission hypothesis: time to move on. *Nat. Rev. Neurosci.* 10:383–90

Sato H, Hata Y, Hagihara K, Tsumoto T. 1987a. Effects of cholinergic depletion on neuron activities in the cat visual cortex. *J. Neurophysiol.* 58:781–94

Sato H, Hata Y, Masui H, Tsumoto T. 1987b. A functional role of cholinergic innervation to neurons in the cat visual cortex. *J. Neurophysiol.* 58:765–80

Sillito AM, Kemp JA. 1983. Cholinergic modulation of the functional organization of the cat visual cortex. *Brain Res.* 289:143–55

Silver MA, Shenhav A, D'Esposito M. 2008. Cholinergic enhancement reduces spatial spread of visual responses in human early visual cortex. *Neuron* 60:904–14

Simon MI, Strathmann MP, Gautam N. 1991. Diversity of G proteins in signal transduction. *Science* 252:802–8

Sirota A, Montgomery S, Fujisawa S, Isomura Y, Zugaro M, Buzsáki G. 2008. Entrainment of neocortical neurons and gamma oscillations by the hippocampal theta rhythm. *Neuron* 60:683–97

Soma S, Shimegi S, Osaki H, Sato H. 2012. Cholinergic modulation of response gain in the primary visual cortex of the macaque. *J. Neurophysiol.* 107:283–91

Steriade M, Pare D, Parent A, Smith Y. 1988. Projections of cholinergic and non-cholinergic neurons of the brainstem core to relay and associational thalamic nuclei in the cat and macaque monkey. *Neuroscience* 25:47–67

Suh BC, Hille B. 2005. Regulation of ion channels by phosphatidylinositol 4,5-bisphosphate. *Curr. Opin. Neurobiol.* 15:370–78

Tang AC, Bartels AM, Sejnowski TJ. 1997. Effects of cholinergic modulation on responses of neocortical neurons to fluctuating input. *Cereb. Cortex* 7:502–9

Tedford HW, Zamponi GW. 2006. Direct G protein modulation of Cav2 calcium channels. *Pharmacol. Rev.* 58:837–62

Thiele A, Herrero J, Distler C, Hoffmann KP. 2012. Contribution of cholinergic and GABAergic mechanisms to direction tuning, discriminability, response reliability and neuronal rate correlations in macaque middle temporal area. *J. Neurosci.* 32(47):16602–15

Thompson SM, Masukawa LM, Prince DA. 1985. Temperature dependence of intrinsic membrane properties and synaptic potentials in hippocampal CA1 neurons in vitro. *J. Neurosci.* 5:817–24

Traub RD, Whittington MA, Stanford IM, Jefferys JGR. 1996. A mechanism for generation of long-range synchronous fast oscillations in the cortex. *Nature* 383:621–24

Uhlhaas PJ, Singer W. 2006. Neural synchrony in brain disorders: relevance for cognitive dysfunctions and pathophysiology. *Neuron* 52:155–68

Vanderwolf CH. 1988. Cerebral activity and behavior: control by central cholinergic and serotonergic systems. *Int. Rev. Neurobiol.* 30:225–340

von Engelhardt J, Eliava M, Meyer AH, Rozov A, Monyer H. 2007. Functional characterization of intrinsic cholinergic interneurons in the cortex. *J. Neurosci.* 27:5633–42

Whittington MA, Traub RD, Jefferys JG. 1995. Synchronized oscillations in interneuron networks driven by metabotropic glutamate receptor activation. *Nature* 373:612–15

Wickman K, Clapham DE. 1995. Ion channel regulation by G proteins. *Physiol. Rev.* 75:865–85

Womble MD, Moises HC. 1992. Muscarinic inhibition of M-current and a potassium leak conductance in neurones of the rat basolateral amygdala. *J. Physiol.* 457:93–114

Xiang Z, Huguenard JR, Prince DA. 1998. Cholinergic switching within neocortical inhibitory networks. *Science* 281:985–88

Yamamoto K, Koyanagi Y, Koshikawa N, Kobayashi M. 2010. Postsynaptic cell type-dependent cholinergic regulation of GABAergic synaptic transmission in rat insular cortex. *J. Neurophysiol.* 104:1933–45

Yan HD, Villalobos C, Andrade R. 2009. TRPC channels mediate a muscarinic receptor-induced afterdepolarization in cerebral cortex. *J. Neurosci.* 29:10038–46

Yeomans JS. 2012. Muscarinic receptors in brain stem and mesopontine cholinergic arousal functions. *Handb. Exp. Pharmacol.* (208):243–59

Zaborszky L. 2002. The modular organization of brain systems. Basal forebrain: the last frontier. *Prog. Brain Res.* 136:359–72

Zhang W, Basile AS, Gomeza J, Volpicelli LA, Levey AI, Wess J. 2002. Characterization of central inhibitory muscarinic autoreceptors by the use of muscarinic acetylcholine receptor knock-out mice. *J. Neurosci.* 22:1709–17

Zhang Z, Séguéla P. 2010. Metabotropic induction of persistent activity in layers II/III of anterior cingulate cortex. *Cereb. Cortex* 20:2948–57

Zinke W, Roberts MJ, Guo K, McDonald JS, Robertson R, Thiele A. 2006. Cholinergic modulation of response properties and orientation tuning of neurons in primary visual cortex of anaesthetized Marmoset monkeys. *Eur. J. Neurosci.* 24:314–28

Mechanisms and Functions of Theta Rhythms

Laura Lee Colgin

Center for Learning and Memory, The University of Texas, Austin, Texas 78712-0805; email: colgin@mail.clm.utexas.edu

Annu. Rev. Neurosci. 2013. 36:295–312

First published online as a Review in Advance on May 29, 2013

The *Annual Review of Neuroscience* is online at neuro.annualreviews.org

This article's doi:
10.1146/annurev-neuro-062012-170330

Keywords

oscillations, memory, hippocampus, place cells, phase precession, CA1

Abstract

The theta rhythm is one of the largest and most sinusoidal activity patterns in the brain. Here I survey progress in the field of theta rhythms research. I present arguments supporting the hypothesis that theta rhythms emerge owing to intrinsic cellular properties yet can be entrained by several theta oscillators throughout the brain. I review behavioral correlates of theta rhythms and consider how these correlates inform our understanding of theta rhythms' functions. I discuss recent work suggesting that one function of theta is to package related information within individual theta cycles for more efficient spatial memory processing. Studies examining the role of theta phase precession in spatial memory, particularly sequence retrieval, are also summarized. Additionally, I discuss how interregional coupling of theta rhythms facilitates communication across brain regions. Finally, I conclude by summarizing how theta rhythms may support cognitive operations in the brain, including learning.

Contents

INTRODUCTION

Theta rhythms: ~4–12 Hz, nearly sinusoidal patterns of electrical activity that are associated with active behaviors and REM sleep

Execution of complex cognitive functions by the brain requires coordination across many neurons in multiple brain areas. Brain rhythms (or oscillations) provide a mechanism for such coordination by linking the activity of related ensembles of neurons. One of the most intriguing brain rhythms is the theta rhythm. The ~4–12-Hz theta rhythms were first discovered in the rabbit by Jung & Kornmuller (1938). Although researchers did not understand what theta rhythms signified, they did notice the large amplitude and nearly sinusoidal regularity of these rhythms (**Figure 1**). Scientific interest in theta rhythms continued to grow during the following decades, and investigators found that theta rhythms occurred in other species as well, including cats, rats, and monkeys (Green & Arduini 1954, Grastyan et al. 1959, Vanderwolf 1969).

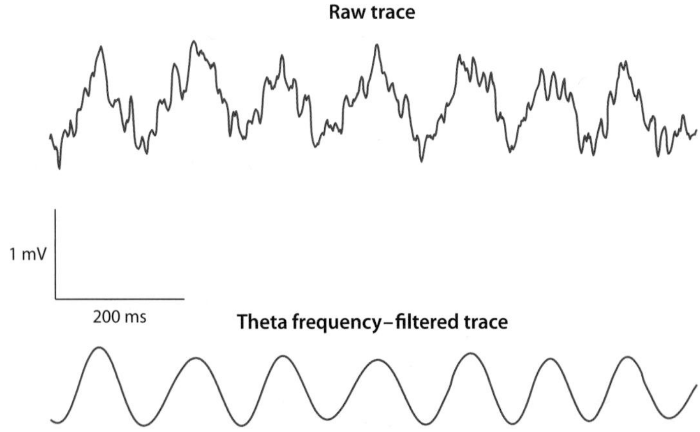

In 1972, Landfield and colleagues reported that the degree to which rats remembered an aversive foot shock was correlated with the amount of theta recorded from screws implanted in the rats' skulls above the cortex (Landfield et al. 1972). The link between theta rhythms and memory was controversial at the time, but many studies have since supported the conclusion that theta rhythms are important for different types of learning and memory (Berry & Thompson 1978, Winson 1978, Macrides et al. 1982, Mitchell et al. 1982, Mizumori et al. 1990, M'Harzi & Jarrard 1992, Klimesch et al. 1996, Osipova et al. 2006, Robbe & Buzsáki 2009, Rutishauser et al. 2010, Liebe et al. 2012), as well as for synaptic plasticity (Larson et al. 1986, Staubli & Lynch 1987, Greenstein et al. 1988, Pavlides et al. 1988, Orr et al. 2001, Hyman et al. 2003). However, it remains unclear why theta rhythms influence memory processing. Do theta rhythms promote memory, as they promote other cognitive states such as anxiety (Adhikari et al. 2010), by facilitating interregional interactions (Seidenbecher et al. 2003, Jones & Wilson 2005, Kay 2005, Benchenane et al. 2010, Hyman et al. 2010, Kim et al. 2011, Liebe et al. 2012)? Do theta rhythms affect memory directly by providing the correct timing required to induce changes in synaptic strength (Larson et al. 1986, Greenstein et al. 1988)? Do theta rhythms affect cognitive operations in general by providing a way to chunk information (Buzsáki 2006, Kepecs et al. 2006, Gupta et al. 2012)?

Here, I review the current status of research on theta rhythms. I begin by discussing mechanisms that generate theta rhythms.

Figure 1

Theta recorded from hippocampal subfield CA1 of a freely exploring rat (L.L. Colgin, unpublished data). A raw trace (*top*) and a theta frequency (4–12 Hz) band-pass filtered trace (*bottom*) are shown.

Understanding how theta rhythms are generated will provide clues regarding their function. I then review several proposed functions of theta rhythms. I discuss the effects of theta rhythms on the firing properties of hippocampal place cells, neurons that are activated in particular spatial locations (O'Keefe & Dostrovsky 1971). Functions of theta rhythms are also discussed with regard to cognition and behavior.

The question of whether theta rhythms exist in healthy humans was a subject of intense debate until fairly recently (Klimesch et al. 1994, Tesche & Karhu 2000). Consequently, research using animal models has established a longer-standing body of work on theta. This review focuses on studies of theta rhythms in lower mammals, primarily rats. I also place particular emphasis on theta rhythms in the hippocampus, a region that is critically involved in memory processing (Squire et al. 2004). Theta rhythms have been studied extensively in the hippocampus, and the hippocampal electroencephalogram displays prominent theta rhythms during active behaviors (**Figure 1**).

MECHANISMS OF THETA RHYTHMS

The Role of the Medial Septum in Theta Generation

Until recently, researchers generally agreed that the medial septum (MS) generates theta rhythms (see Vertes & Kocsis 1997 for a review) because lesioning or inactivating the MS disrupts theta (Green & Arduini 1954, Petsche et al. 1962, Mitchell et al. 1982, Mizumori et al. 1990). MS pacemaker cells are believed to be GABAergic inhibitory interneurons (Toth et al. 1997), which express hyperpolarization-activated and cyclic nucleotide-gated nonselective cation channels (HCN channels) (Varga et al. 2008). These HCN-expressing interneurons fire rhythmically at theta frequencies and are phase-locked to theta rhythms in the hippocampus (Hangya et al. 2009). Cholinergic neurons of the MS, on the other hand, do not fire rhythmically at theta frequencies (Simon et al. 2006) and are thus unlikely to act as theta pacemakers. Cholinergic neurons may instead modulate the excitability of other neurons in a way that promotes their theta rhythmic firing. Backprojections from the hippocampus to the MS may be important to keep the two regions coupled (Toth et al. 1993). The importance of other subcortical regions for theta generation has been addressed in another review (Vertes et al. 2004) and is not discussed here.

Recent work has brought into question the belief that the MS is responsible for theta generation. Goutagny and associates (2009) found that theta rhythms emerge in vitro in an intact hippocampus preparation lacking any connections with the MS (**Figure 2**). The in vitro theta rhythms appeared spontaneously (i.e., without application of any drugs). Also, theta activity persisted in hippocampal subfield CA1 after neighboring subfield CA3 was removed, indicating that an excitatory recurrent collateral network was not a required component of the theta-generating machinery. Local inactivation of an intermediate portion of the longitudinal axis did not eliminate theta from either the septal or the temporal pole. Instead, theta rhythms persisted in both septal and temporal regions but were no longer coherent. Theta in the septal hippocampus was faster than theta in the temporal hippocampus after the inactivation procedure. Whole-cell recordings of CA1 pyramidal cells revealed rhythmic inhibitory postsynaptic potentials, whereas rhythmic excitatory postsynaptic potentials were recorded in interneurons. These results suggest that theta rhythms are produced by local interactions between hippocampal interneurons and pyramidal cells.

Although the hippocampus may possess the machinery necessary to produce theta intrinsically in vitro, much evidence indicates that the MS is involved in theta generation in behaving animals. Lesioning or inactivating the MS disrupts theta in structures that receive MS projections, including the entorhinal cortex (EC) and the hippocampus (Green & Arduini 1954, Petsche et al. 1962, Mitchell et al. 1982, Mizumori et al. 1990, Brandon et al. 2011, Koenig et al. 2011). The MS transitions

Place cells: principal neurons of the hippocampus that fire selectively in specific spatial locations

Hippocampus: a region of the medial temporal lobe that is essential for spatial and episodic memory

Medial septum (MS): a subcortical region providing cholinergic and GABAergic inputs to cortical structures including the hippocampus and entorhinal cortex

HCN channels: hyperpolarization-activated and cyclic nucleotide-gated nonselective cation channels

EC: entorhinal cortex

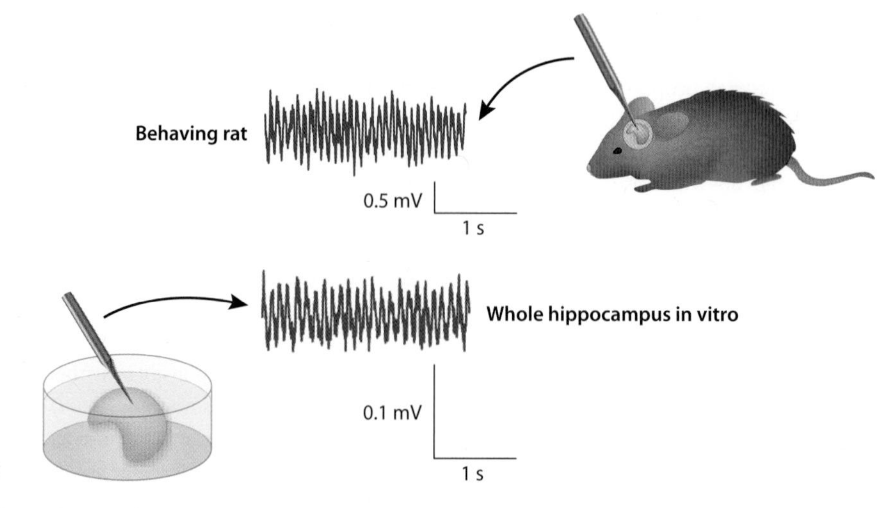

Figure 2

Theta rhythms recorded by Goutagny et al. (2009) in CA1 of a whole hippocampus in vitro preparation are highly similar to theta rhythms recorded from CA1 in vivo. Modified with permission from Colgin & Moser (2009).

to the theta state ~500 ms before theta appears in the hippocampus (Bland et al. 1999), supporting the idea that septohippocampal projections initiate theta rhythms. Spikes of MS pacemaker interneurons are maximally phase-locked to hippocampal theta occurring ~80 ms later (Hangya et al. 2009), supporting the idea that MS interneurons drive theta in the hippocampus (Toth et al. 1997), albeit with some delay. The substantial delay may reflect the time required to recruit a significant proportion of cells into a synchronized theta network (Hangya et al. 2009).

A Multitude of Theta Oscillators

Other mechanisms in addition to MS inputs appear to be involved in hippocampal theta rhythm generation in vivo. Current source density analyses of hippocampal local field potentials (LFPs) from freely exploring rats indicate that multiple current dipoles coexist during theta (Kamondi et al. 1998; see figure 4 in Buzsáki 2002). Presumably active current sinks are seen in stratum lacunosum-moleculare and stratum radiatum and are thought to reflect excitatory inputs from the EC and CA3, respectively. At the same time, a putative current source reflecting inhibitory inputs can be seen in stratum pyramidale. These inhibitory inputs are likely hippocampal interneurons firing rhythmically at theta frequencies during periods when they are released from theta inhibition imposed by MS interneurons (Toth et al. 1997). Following surgical removal of the EC, the stratum lacunosum-moleculare dipole disappears. Theta rhythms that remain after EC lesioning are suppressed by the cholinergic antagonist atropine, in contrast with atropine-resistant theta rhythms that normally occur during exploration (Kramis et al. 1975). Atropine-sensitive theta rhythms are normally observed during behavioral immobility and during anesthesia induced by urethane (Kramis et al. 1975), a drug that suppresses input from the EC (Ylinen et al. 1995). These findings suggest that the EC provides sufficient excitatory drive to entrain theta rhythms during active behaviors but that another source of excitatory drive is required in the absence of entorhinal inputs. This excitatory drive may come from MS cholinergic inputs, which produce slow depolarizations in hippocampal neurons (Madison et al. 1987).

LFP: local field potential

A problem with this interpretation is that theta rhythms in the whole hippocampus in vitro are not blocked by atropine, yet the isolated hippocampus receives no input from the EC (Goutagny et al. 2009). Thus, it is unclear from where the excitatory drive originates in the isolated hippocampal preparation. One possibility is that neurons are more excitable in this preparation than in other in vitro preparations. The potassium concentration used in the artificial cerebrospinal fluid (aCSF) in the Goutagny et al. (2009) study (3.5–5 mM) was higher than the potassium concentration (2.5 mM) used in typical hippocampal slice studies (e.g., Frerking et al. 2001, Brager & Johnston 2007). Studies of spontaneous sharp-wave ripples in hippocampal slices used similarly high potassium concentrations in the aCSF (4.25–4.75 mM; Kubota et al. 2003, Colgin et al. 2004, Ellender et al. 2010). It is not obvious, however, why a relatively high concentration of potassium in the aCSF would lead to theta rhythms in an isolated whole hippocampal preparation and to sharp-wave ripples in hippocampal slices. One difference between preparations is a more intact CA3 recurrent collateral system in the isolated whole hippocampus. However, CA3 was essential for sharp waves in slices (Colgin et al. 2004) but not for theta rhythms in the isolated whole hippocampus (Goutagny et al. 2009), indicating that differences in spontaneous in vitro activity are not due to preservation of CA3 recurrent collaterals. One likely explanation is that a higher number of connections between interneurons and pyramidal neurons are preserved in the whole hippocampus preparation. A high degree of pyramidal cell-interneuron interconnectivity may be essential for theta generation but not for sharp-wave ripples. Of particular interest in this regard are horizontal interneurons (Maccaferri 2005, Goutagny et al. 2009). Horizontal interneurons are activated almost exclusively by CA1 axon collaterals (Blasco-Ibáñez & Freund 1995). Horizontal interneurons fire spontaneously, without requiring fast excitatory input (Maccaferri & McBain 1996), and thus would be likely to fire in the intact hippocampal preparation. Moreover, horizontal interneurons show resonance at theta frequencies (Pike et al. 2000). Thus, horizontal interneurons may be essential for generating theta rhythms intrinsically in the hippocampus.

MEC: medial entorhinal cortex

A common feature of horizontal interneurons in the hippocampus and pacemaker interneurons in the MS is the expression of HCN channels (Maccaferri & McBain 1996, Varga et al. 2008). HCN channels are nonselective cation channels that are activated by hyperpolarization (Robinson & Siegelbaum 2003). Activation of HCN channels leads to slow depolarizing currents (I_h currents) that can drive the membrane back to threshold and trigger action potentials. In neurons that express HCN channels, a repeating sequence of events can emerge that consists of an action potential, afterhyperpolarization, depolarization via HCN channels, and another action potential that starts the cycle again. In this way, HCN channels may facilitate rhythmic firing in neurons. Many neurons that exhibit theta rhythmic firing express HCN channels (Maccaferri & McBain 1996, Dickson et al. 2000, Hu et al. 2002, Varga et al. 2008). Theta frequency membrane-potential oscillations are disrupted by pharmacological blockade (Dickson et al. 2000) or genetic deletion (Giocomo & Hasselmo 2009) of HCN channels. Moreover, the time course of activation and deactivation of I_h currents may determine the particular theta frequency of membrane-potential oscillations in medial entorhinal cortex (MEC) neurons; that is, faster time constants correspond to higher-frequency theta rhythms (Giocomo et al. 2007, Giocomo & Hasselmo 2008). Together, these findings suggest that HCN channels may be a key cellular mechanism that contributes to theta rhythm generation.

Effects of HCN channels provide one potential explanation of the seemingly contradictory findings that theta rhythms depend on MS inputs yet can be generated intrinsically in the hippocampus. Perhaps any network of neurons that expresses HCN channels is primed to participate in theta rhythms, provided that the neurons are sufficiently depolarized to initiate the cycle described above. Depolarization could come from a variety of sources, depending on experimental conditions. Rhythmically firing cells would recruit other cells, and theta rhythms would spread across the network. In this scenario, local neuronal ensembles would actually be separate oscillators; the range of synchronization would be determined by the extent of pacemaker projections. Although the simplicity of this explanation is attractive, theta rhythm generation appears to be more complicated, considering that theta rhythms in vivo are not decreased in HCN1 knockout mice (Nolan et al. 2004, Giocomo et al. 2011). Nevertheless, it seems likely that some combination of intrinsic conductances and network mechanisms allows theta oscillatory activity to originate from a variety of different sources.

Support for the idea of multiple theta oscillators comes from recent studies of theta oscillations in the hippocampus. In freely behaving rats, theta rhythms are not synchronized across the hippocampus. Instead, systematic phase shifts are observed across the septotemporal axis (Lubenov & Siapas 2009, Patel et al. 2012). That is, hippocampal theta rhythms are traveling waves that consistently propagate toward the temporal pole of the hippocampus. The mechanisms of traveling wave generation remain unknown, but investigators have proposed several possibilities (Lubenov & Siapas 2009, Patel et al. 2012). One possibility is that theta waves emerge first in the septal hippocampus because MS inputs reach the septal hippocampus first. This proposed mechanism is problematic for several reasons (Patel et al. 2012). A more likely mechanism is that traveling theta waves reflect interactions within a network of weakly coupled oscillators (Lubenov & Siapas 2009, Patel et al. 2012). Separate oscillators may relate to differences in intrinsic conductances between septal (also called dorsal) and temporal (also called ventral) hippocampi. Maurer et al. (2005) found that intrinsic theta oscillations have a lower frequency in the intermediate hippocampus than they do in the septal hippocampus in freely behaving rats. This finding is consistent with septal-to-temporal propagation of traveling waves (but see Marcelin et al. 2012 for contradictory in vitro results). Frequencies of intrinsic theta oscillations are lower in ventral MEC than in dorsal MEC (Giocomo et al. 2007, Giocomo & Hasselmo 2008), leading Lubenov & Siapas (2009) to predict that MEC theta is a traveling wave also.

Recordings of theta rhythms in the isolated whole hippocampus also provide support for the weakly coupled oscillators mechanism (Goutagny et al. 2009). Coherence between septal and temporal theta rhythms in CA1 decreased significantly when an intermediate CA1 location was inactivated. After the inactivation, septal and temporal theta rhythms became uncoupled, and the frequency of temporal theta rhythms decreased significantly. Prior to the inactivation, septal theta rhythms led temporal theta rhythms by ~50 ms. This finding suggests that the septal hippocampus entrains the temporal hippocampus during theta rhythms, an idea that is consistent with septotemporal propagation of traveling theta waves. These results suggest that theta oscillators are coupled by local interactions between pyramidal cells and interneurons, not by MS projections. However, different coupling mechanisms likely exist in behaving animals, including entrainment of hippocampal theta by MS and EC oscillators. Coupling of EC and hippocampal theta oscillators may involve excitatory inputs from the EC to the hippocampus (Buzsáki 2002) or long-range inhibitory connections between the EC and the hippocampus (Melzer et al. 2012).

Despite decades of research dedicated to uncovering the mechanisms of theta rhythm generation, new and surprising findings regarding theta mechanisms continue to be discovered. Achieving a deeper understanding of the mechanisms of theta rhythms is an essential step toward fully understanding the functions of this complex rhythm. Additionally, a thorough understanding of theta mechanisms will help investigators develop methods to block theta rhythm generation selectively, with minimal side effects (e.g., loss of neuromodulatory inputs, changes in mean firing rates). Such a selective blockade would allow researchers to assess how functions are altered when theta rhythms are not present and to reveal causal links between theta and its functions.

FUNCTIONS OF THETA RHYTHMS

Behavioral Correlates of Theta Rhythms

Since the early days of theta rhythms research, scientists have looked for clues regarding theta functions by determining the behavioral correlates of theta rhythms (Buzsáki 2005). Theta rhythms arise during movement (Vanderwolf 1969) and exhibit higher amplitudes during active movement than during passive movement (Terrazas et al. 2005). Theta rhythms are also present during behaviors associated with intake of stimuli. The frequency of theta matches the frequency of whisking (Berg & Kleinfeld 2003) and sniffing (Macrides et al. 1982) in rats, as well as saccadic eye movements in humans (Otero-Millan et al. 2008). Theta rhythms selectively correlate with active, and not passive, sampling of stimuli. In rats sampling behaviorally relevant stimuli or stimuli associated with reward, theta rhythms were temporally correlated with rhythmic sniffing (Macrides et al. 1982) and whisking (Ganguly & Kleinfeld 2004). However, no relationship was observed between theta and whisking when rats were aimlessly whisking in air (Berg et al. 2006).

Curiously, theta rhythms also occur during rapid eye movement (REM) sleep (Vanderwolf 1969, Winson 1974), but the role of theta in REM sleep is less clear. One possibility is that theta rhythms play a role in memory consolidation during REM sleep (Louie & Wilson 2001, Montgomery et al. 2008), but this idea is controversial (Vertes 2004). This review does not cover functions of theta during REM sleep and instead focuses on functions of theta during waking states.

Information Packaging by Theta Rhythms

The hippocampus is at the end of an information-processing stream that begins when information is taken in from the environment. The hippocampus receives convergent information from multiple areas and thus requires a mechanism to coordinate related activity. The relationship between theta rhythms and rhythmic behaviors involved in stimuli intake suggests that each theta cycle contains a discrete sample of related sensory information (Kepecs et al. 2006). Theta rhythms may also link this information from different sensory modalities with information related to motivational or emotional states. Each theta cycle may thus represent a discrete processing entity that contains information about the current conditions at any given moment. This idea has recently received empirical support from two studies.

In the first study, Jezek and associates (2011) used an innovative behavioral task to investigate how CA3 neuronal ensembles respond to abrupt changes in environmental stimuli. In this task, rats were trained in two boxes that were distinguished by different sets of light cues. Initially, the two boxes were placed in different locations, connected by a corridor through which the rats could freely pass. This setup ensured that distinct ensembles of place cells were active in the two boxes, indicating that separate representations for the two environments had formed (Colgin et al. 2010). At a later stage of training, investigators presented an identical box that contained both sets of light cues in a common central location, midway between the two original locations. The light cues could be switched on and off to present one or the other environment, and separate representations for each environment were maintained. By switching the light cues, the authors could instantaneously "teleport" the rat from one environment to the other. After the switch, the network usually transitioned to the spatial representation that matched the new set of cues. In some cases, though, the network would flicker back and forth for a few seconds between the two representations, the one that matched the current set of cues and the one that matched the previous set of cues. During these episodes

Theta phase precession: a phenomenon in which spikes from an individual cell occur on progressively earlier theta phases across successive theta cycles

of instability, the different representations were separated by the period of a theta cycle, indicating that theta rhythms modulated the switching between distinct representations. In ~99% of cases, each theta cycle contained information related to only one of the memory representations, which suggests that theta rhythms link representations of sets of stimuli related to one environment and segregate them from representations of the other environment. Thus, these results support the hypothesis that theta cycles package related information.

Another study provided support for the hypothesis that theta cycles chunk information during spatial memory processing. In this study, Gupta and colleagues (2012) recorded CA1 place cells in rats running on a T-maze with two choice points. The maze contained several landmarks including two reward locations on the return arm. The authors analyzed theta cycles during which three or more place cells were active to find evidence of place cell sequences (i.e., sequences of spikes with a significant spatial and temporal structure related to the structure of behaviors on the maze). Place cell sequences within a theta cycle were found to represent particular paths on the maze. Longer theta cycles were associated with longer path lengths, suggesting that the theta cycle was the processing unit used to organize representations of discrete paths. Each theta cycle represented a path that began just behind the animal's current location and ended just in front of the animal. Theta sequences represented more space ahead of an animal leaving a landmark and more space behind an animal approaching a landmark, meaning that segments of the maze between landmarks were overrepresented. Thus, individual theta cycles contained representations of task-relevant maze segments, which suggests that a path containing different yet related spatial locations may be more efficiently encoded within a theta cycle as a coherent spatial concept (e.g., path between two feeders).

Gupta and colleagues (2012) also showed that some theta cycles heavily represented upcoming locations, whereas other theta cycles were biased toward locations in the recent past. The authors hypothesized that the theta sequences coding upcoming locations reflect a look-ahead mode, which signifies anticipation of an upcoming location. Sequences coding locations in the recent past may reflect a look-behind mode, which encodes recent experiences. The sequences studied by Gupta et al. (2012) may provide insights about place cell sequences during a phenomenon known as theta phase precession, as explained in the next section.

Theta Phase Precession

In theta phase precession, spikes from a place cell initially occur at late theta phases as the rat enters the cell's place field and then occur at progressively earlier phases on subsequent theta cycles (O'Keefe & Recce 1993, Skaggs et al. 1996) (**Figure 3*a***). Because each theta cycle is associated with spikes from multiple place cells with sequentially occurring place fields, individual theta cycles contain compressed representations of space (Skaggs et al. 1996; "theta sequences" in Foster & Wilson 2007). The compressed representation consists of a code for current location (represented by the most active place cells) preceded by representations of recently visited locations on earlier theta phases and succeeded by representations of upcoming locations on later theta phases (Dragoi & Buzsáki 2006). An alternative viewpoint is that spikes on early theta phases represent the current location, whereas spikes on later theta phases represent upcoming locations (Lisman & Redish 2009). The findings by Gupta et al. (2012), showing that some theta cycles contain sequences that largely represent space behind the rat, whereas sequences on other theta cycles represent more space ahead of the rat, seem to support the former interpretation of theta phase precession (for additional discussion of the predictive component of phase precession, see Lisman & Redish 2009). Here, the discussion of theta phase precession focuses on CA1 because the majority of phase precession studies have been conducted in

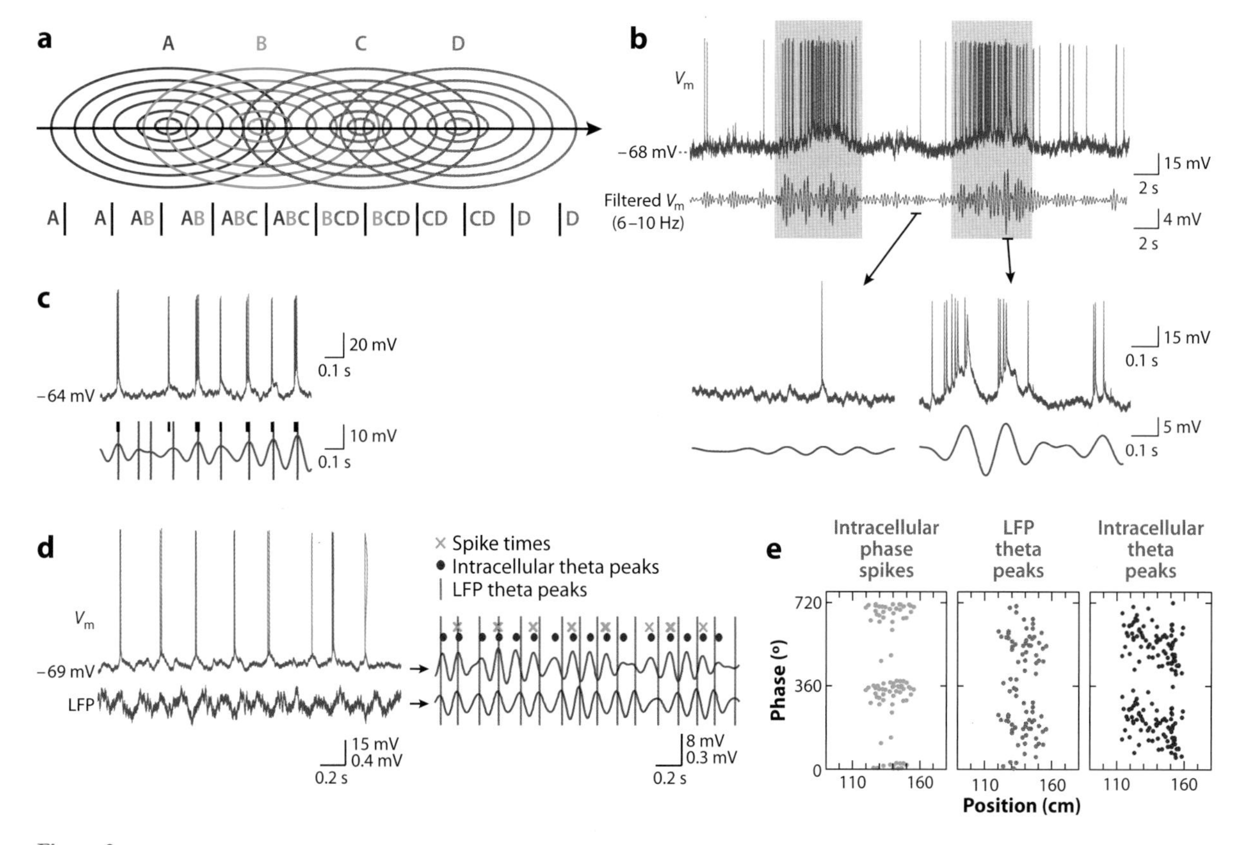

Figure 3

(*a*) Schematic of theta phase precession. Four partially overlapping place fields are shown for an animal running to the right. Repeated activation of place cell sequences on theta cycles (*vertical lines*) is depicted below. Reproduced with permission from Skaggs et al. (1996). (*b*) Example of unfiltered and theta-filtered (6–10 Hz) whole cell recordings from a mouse exploring a virtual environment during runs through the cell's place field (*gray*). Magnified recordings for the periods indicated with horizontal bars are shown below. (*c*) Example of raw (*top*) and theta-filtered membrane-potential recordings (*bottom*) illustrate phase-locking of spikes to intracellular theta. Black vertical lines indicate spike times. Peaks of filtered theta are indicated by purple vertical lines. (*d*) Raw (*left*) and theta-filtered (*right*) traces are shown for simultaneous whole cell (*top*) and local field potential (LFP) recordings (*bottom*). Note how spikes are locked to the peak of intracellular theta oscillations and show phase precession with respect to LFP theta. The top and bottom calibration bar labels indicate the scale for top and bottom traces, respectively. (*e*) Spikes' intracellular theta phases at positions along the virtual environment are plotted (*left*); note how theta phase-locking, but not theta phase precession, is seen. Spikes' phases with respect to LFP theta at positions in the virtual environment are plotted (*middle*); theta phase precession is apparent. The LFP theta phases associated with intracellularly detected theta peaks are plotted on the *y* axis, and corresponding positions are plotted on the *x* axis (*right*); note that intracellular theta shows phase precession with respect to LFP theta. Panels *b*, *c*, *d*, and *e* are modified with permission from Harvey et al. (2009).

CA1. However, it is important to note that theta phase precession has been reported in other regions as well, including the dentate gyrus (Yamaguchi et al. 2002, Mizuseki et al. 2009), CA3 (Mizuseki et al. 2009, 2012), the subiculum (Kim et al. 2012), the EC (Hafting et al. 2008, Mizuseki et al. 2009), and the ventral striatum (van der Meer & Redish 2011).

Many different models of theta phase precession propose various mechanisms for the phenomenon (O'Keefe & Recce 1993, Jensen & Lisman 1996, Tsodyks et al. 1996, Wallenstein & Hasselmo 1997, Kamondi et al. 1998, Magee 2001, Harris et al. 2002, Mehta et al. 2002, Maurer & McNaughton 2007, Navratilova et al. 2012). Thorough comparison and contrast of the different models are beyond the scope of this review. I refer the reader to several other articles that discuss strengths and weaknesses of the various models (Maurer & McNaughton 2007, Lisman & Redish 2009, Burgess & O'Keefe 2011, Navratilova et al. 2012). Here, I instead discuss how recent whole-cell recordings of place cells in mice running in a virtual reality environment revealed mechanisms of phase precession that must be accounted for in future models (Harvey et al. 2009). First, depolarization of the place cell membrane potential during the first half of the place field ramped up asymmetrically (**Figure 3*b***), similar to asymmetric excitatory input proposed in the phase precession model of Mehta and colleagues (2002). Next, the power of membrane-potential theta oscillations was approximately two times higher within the cell's place field than that outside it. Last, spikes were phase-locked to intracellularly recorded theta and showed phase precession relative to LFP theta (**Figure 3*c–e***), consistent with predictions of "dual oscillator" models (Burgess & O'Keefe 2011).

Although this intracellular report provides extremely useful information regarding the mechanisms of theta phase precession, some questions remain. The intrinsic membrane-potential theta oscillations are thought to reflect rhythmic basket cell inhibition at the soma (Kamondi et al. 1998). Yet, Harvey and associates (2009) report that the frequency and amplitude of intracellular theta oscillations increase in the cell's place field, and it is not obvious how such increases would involve basket cell inhibition. Membrane-potential theta oscillations likely reflect basket cell inhibition in anesthetized animals (Kamondi et al. 1998) or at times when the animal is not in a cell's place field. When an animal enters a cell's place field, however, strong depolarization in the apical dendrites likely induces higher-frequency theta oscillations that result from intrinsic dendritic properties (Kamondi et al. 1998, Burgess & O'Keefe 2011). These faster dendritic oscillations may then entrain theta recorded at the soma, which would explain why spikes within the place field were so consistently phase-locked to the somatic oscillation (Harvey et al. 2009), even though inputs carrying spatial information arrive at the dendrites. Further support for this idea comes from recordings of theta under anesthesia (Kamondi et al. 1998), a condition in which excitatory dendritic inputs are suppressed (Ylinen et al. 1995). Under these conditions, intracellular and LFP theta oscillations share the same frequency (see figure *2b* in Kamondi et al. 1998).

The question remains about how these proposed mechanisms of phase precession relate to the proposed sequence retrieval function of phase precession. One possibility is that gamma oscillations are involved in both theta phase precession and sequence retrieval. Gamma oscillations co-occur with theta oscillations (Buzsáki et al. 1983, Bragin et al. 1995) and may play a role in theta phase precession (Jensen & Lisman 1996, Dragoi & Buzsáki 2006). Spiking of some CA1 place cells is modulated by gamma oscillations during theta phase precession (Senior et al. 2008). Additionally, recent work indicates that theta-modulated gamma is the optimal pattern for dendritic signal propagation (Vaidya & Johnston 2012), raising the possibility that theta-modulated gamma inputs arriving in the dendrites (Bragin et al. 1995, Colgin et al. 2009) enhance membrane-potential oscillations during theta phase precession. Gamma oscillations have also been linked to sequence retrieval. Strong gamma rhythms occur when place cells represent upcoming sequences of locations as rats pause at decision points on a maze (Johnson & Redish 2007), and slow gamma rhythms promote sequence reactivation during nontheta states (Carr et al. 2012).

If gamma-mediated sequence retrieval occurs during theta phase precession, phase-phase coupling (Jensen & Colgin 2007) is expected between theta and gamma oscillations because spikes would be phase-locked to gamma while still maintaining phase precession relationships with theta. In support of this idea, Belluscio and colleagues (2012) recently reported phase-phase coupling between theta and gamma in the hippocampus. When they subdivided gamma into different frequency bands, the slow gamma band displayed the strongest phase-phase coupling (see figure 7*a* in Belluscio et al. 2012). Sequence reactivation in the absence of theta involved slow gamma oscillations but no other gamma frequencies (Carr et al. 2012). Slow gamma rhythms couple CA3 and CA1 (Colgin et al. 2009), and CA3 is thought to be critical for retrieving representations of upcoming locations (Jensen & Lisman 1996, Dragoi & Buzsáki 2006). Thus, slow gamma rhythms may transmit retrieved sequences of upcoming locations from CA3 to CA1 during theta phase precession; in this scheme, representations of consecutive locations would be activated on successive gamma cycles (Jensen & Lisman 1996, Dragoi & Buzsáki 2006). Inputs from the EC provide information about current location (Hafting et al. 2005) and likely serve as a cue for CA3 recall of upcoming locations (Jensen & Lisman 1996, Dragoi & Buzsáki 2006). Layer III EC inputs to CA1 may contribute to phase precession in the second half of the place field because CA3 cells do not receive layer III input and do not show phase precession across the full extent of the theta cycle (Mizuseki et al. 2012).

Work described above demonstrates how intrinsic cellular properties may combine with synaptic interactions to facilitate hippocampal functions such as sequence retrieval. Thus far, most results discussed here involve effects within an individual region, mainly CA1. The next section discusses how theta rhythms affect interactions across different brain regions.

Theta Coupling between Regions

Theta oscillations are seen in structures involved in initial stages of sensory processing as well as in regions further down the processing stream (Jung & Kornmuller 1938, Vanderwolf 1969, Jones & Wilson 2005, Kay 2005). Much evidence suggests that theta rhythms are involved in facilitating the transfer of information from one brain region to another during sensory information processing. Theta-enhanced transmission across brain regions may be important for several different functions. Here, I focus on the role of interregional theta coupling in memory operations involving the hippocampus.

mPFC: medial prefrontal cortex

Consistent with the hippocampus's key role in spatial memory processing, several studies have demonstrated links between interregional theta coupling and performance on a variety of spatial memory tasks. In a spatial working-memory task, medial prefrontal cortex (mPFC) neurons were more strongly phase-locked to CA1 theta rhythms during correct-choice trials than during error trials (Jones & Wilson 2005). Additionally, Jones & Wilson (2005) observed theta coherence between CA1 and mPFC during choice trials but not during forced-turn trials. Similarly, in a delayed nonmatch to position task, Hyman and colleagues (2010) found that 94% of theta-modulated mPFC neurons were significantly more phase-locked to hippocampal theta during correct trials than during error trials. Moreover, theta coherence between CA1 and mPFC increased significantly after rule acquisition at the choice point in a Y-maze with periodically switching reward contingency rules (Benchenane et al. 2010). In another study involving the hippocampus and the mPFC, the proportion of mPFC neurons that were phase-locked to hippocampal theta oscillations increased after successful learning of an object-place association (Kim et al. 2011).

The mPFC is not the only region that exhibits theta coupling with the hippocampus during spatial memory processing. Theta coherence between the striatum and the hippocampus increased during the period between the tone and the selected turn in a tone-cued T-maze task, but only in rats that successfully learned the task (DeCoteau et al. 2007). In this same task, coupling between striatal theta phase and hippocampal gamma amplitude was observed during the tone onset period (Tort et al. 2008). The authors suggested that striatal theta phase-hippocampal gamma amplitude coupling may signify times during decision making when the striatum accesses spatial information from the hippocampus.

The studies described thus far in this section involve spatial memory, but interregional theta coupling appears to be involved in other types of memory processing as well. The results of one study indicate that theta coordination of the hippocampus and the amygdala is important for fear memory retrieval. Theta coupling between the lateral amygdala and CA1 was significantly increased when conditioned fear stimuli were presented, and significant correlations between CA1 and lateral amygdala theta were seen in animals that displayed behavioral signs of fear (i.e., freezing; Seidenbecher et al. 2003). Interregional theta coupling also affects memory processing of nonaversive stimuli. Kay (2005) found positive correlations between performance on a two-odor discrimination task and theta coherence of the olfactory bulb and the hippocampus, suggesting that theta coupling enhances communication between the olfactory bulb and the hippocampus during olfactory memory processing. In a trace conditioning task using a visual conditioned stimulus, theta synchronization across different mPFC sites increased after investigators presented the conditioned stimulus only after learning had occurred (Paz et al. 2008). Because CA1 and the subiculum project to the mPFC (Swanson 1981, Jay & Witter 1991), the authors hypothesized that enhanced theta coupling with hippocampal inputs was responsible for the increased synchronization across mPFC.

The above-described studies show theta coordination during memory processing involving the hippocampus. However, theta coupling across regions appears to be a more general mechanism that is useful for other cognitive operations and physiological states, including visual short-term memory (Liebe et al. 2012) and anxiety (Adhikari et al. 2010). Reductions in theta coupling across regions may also relate to behavioral deficits in diseases such as schizophrenia (Dickerson et al. 2010, Sigurdsson et al. 2010). Remarkably, a new study using a rat model of schizophrenia found that interventions that restore normal performance on a cognitive control task also reestablish healthy theta synchrony between left and right hippocampi (Lee et al. 2012).

How does theta coupling enhance communication across brain areas? Theta synchronization of neurons in a given brain area likely leads to a more effective activation of downstream targets. Interregional theta coupling likely also ensures that downstream neurons will be excitable when inputs arrive. Theta's relatively slow time scale permits long synaptic delays and thus can feasibly sustain coupling across distributed brain regions. These points, taken together with the results reviewed above, support the conclusion that theta rhythms promote coordination across distributed brain areas during different types of information processing.

CONCLUDING REMARKS

Studies in recent years have provided breakthroughs in our understanding of the mechanisms and functions of theta rhythms. Theta-modulated neurons likely express intrinsic properties that prime them to produce theta oscillations in response to a variety of extrinsic inputs. Projections from interneurons in the MS are thought to pace theta in most neurons in the hippocampus at a given time. However, when an animal is in a cell's place field, excitatory CA3 and entorhinal inputs in the dendrites may produce stronger and faster theta oscillations that entrain the cell's firing. This flexibility in theta entrainment likely allows inputs to select the appropriate cell ensembles during particular cognitive tasks. Theta synchronization of related cell ensembles may then promote effective activation of target structures and thereby facilitate cognitive operations such as learning.

Although much progress has been made, many questions remain regarding the mechanisms and functions of theta rhythms. How important are intrinsic neuronal mechanisms for theta generation? It would be interesting to determine how theta rhythms in vivo are affected by manipulations that selectively and reversibly inhibit conductances that are tuned to theta. Regarding theta functions, major questions persist about how results from rodents pertain to other species. Theta rhythms are continuous in rodents during active behaviors (Buzsáki et al. 1983, Buzsáki 2002) but occur only in short bouts in humans (Kahana et al. 1999), bats (Yartsev et al. 2011), and monkeys (Killian et al. 2012). It seems unlikely that complex functions such as learning could be achieved by short theta bouts in bats and primates and yet require continuous theta in rats. One possibility is that the conditions used in typical primate experiments are not optimal for engaging theta machinery. Rodents in experimental settings often engage in behaviors that are similar to their natural behaviors (e.g., foraging for food). Human studies of theta often involve virtual-reality environments. These uncommon behaviors may not readily activate circuits involved in theta rhythm generation. Moreover, foraging tasks require rodents to use a combination of sensory cues (e.g., olfactory, tactile, visual) and thus would be expected to engage a mechanism that packages related information from different sensory modalities. In contrast, investigations of theta in primates typically employ tasks requiring only one sensory modality, vision. Perhaps theta-generating machinery is less easily triggered in such tasks. Wireless monitoring devices implanted in patients undergoing deep brain stimulation may provide a way for researchers to measure theta rhythms during ordinary human behaviors. However, a recent study has shown that the activity of hippocampal place cells in bats is not modulated by theta rhythms during flying, bats' natural exploratory behavior (Yartsev & Ulanovsky 2013). Moreover, bat MEC neurons, unlike rat MEC neurons, do not exhibit membrane-potential resonance at theta frequencies (Heys et al. 2013). The puzzles that remain regarding theta functions across species underscore the importance of ongoing theta research in humans and in animal models.

DISCLOSURE STATEMENT

The author is not aware of any affiliations, memberships, funding, or financial holdings that might be perceived as affecting the objectivity of this review.

ACKNOWLEDGMENTS

I thank the Klingenstein Fund and the Sloan Foundation for their support.

LITERATURE CITED

Adhikari A, Topiwala MA, Gordon JA. 2010. Synchronized activity between the ventral hippocampus and the medial prefrontal cortex during anxiety. *Neuron* 65:257–69

Belluscio MA, Mizuseki K, Schmidt R, Kempter R, Buzsáki G. 2012. Cross-frequency phase-phase coupling between theta and gamma oscillations in the hippocampus. *J. Neurosci.* 32:423–35

Benchenane K, Peyrache A, Khamassi M, Tierney PL, Gioanni Y, et al. 2010. Coherent theta oscillations and reorganization of spike timing in the hippocampal-prefrontal network upon learning. *Neuron* 66:921–36

Berg RW, Kleinfeld D. 2003. Rhythmic whisking by rat: retraction as well as protraction of the vibrissae is under active muscular control. *J. Neurophysiol.* 89:104–17

Berg RW, Whitmer D, Kleinfeld D. 2006. Exploratory whisking by rat is not phase locked to the hippocampal theta rhythm. *J. Neurosci.* 26:6518–22

Berry SD, Thompson RF. 1978. Prediction of learning rate from the hippocampal electroencephalogram. *Science* 200:1298–300

Bland BH, Oddie SD, Colom LV. 1999. Mechanisms of neural synchrony in the septohippocampal pathways underlying hippocampal theta generation. *J. Neurosci.* 19:3223–37

Blasco-Ibáñez JM, Freund TF. 1995. Synaptic input of horizontal interneurons in stratum oriens of the hippocampal CA1 subfield: structural basis of feed-back activation. *Eur. J. Neurosci.* 7:2170–80

Brager DH, Johnston D. 2007. Plasticity of intrinsic excitability during long-term depression is mediated through mGluR-dependent changes in I(h) in hippocampal CA1 pyramidal neurons. *J. Neurosci.* 27:13926–37

Bragin A, Jando G, Nadasdy Z, Hetke J, Wise K, Buzsáki G. 1995. Gamma (40–100 Hz) oscillation in the hippocampus of the behaving rat. *J. Neurosci.* 15:47–60

Brandon MP, Bogaard AR, Libby CP, Connerney MA, Gupta K, Hasselmo ME. 2011. Reduction of theta rhythm dissociates grid cell spatial periodicity from directional tuning. *Science* 332:595–99

Burgess N, O'Keefe J. 2011. Models of place and grid cell firing and theta rhythmicity. *Curr. Opin. Neurobiol.* 21:734–44

Buzsáki G. 2002. Theta oscillations in the hippocampus. *Neuron* 33:325–40

Buzsáki G. 2005. Theta rhythm of navigation: link between path integration and landmark navigation, episodic and semantic memory. *Hippocampus* 15:827–40

Buzsáki G. 2006. *Rhythms of the Brain*. New York: Oxford Univ. Press. 448 pp.

Buzsáki G, Leung LW, Vanderwolf CH. 1983. Cellular bases of hippocampal EEG in the behaving rat. *Brain Res.* 287:139–71

Carr MF, Karlsson MP, Frank LM. 2012. Transient slow gamma synchrony underlies hippocampal memory replay. *Neuron* 75:700–13

Colgin LL, Denninger T, Fyhn M, Hafting T, Bonnevie T, et al. 2009. Frequency of gamma oscillations routes flow of information in the hippocampus. *Nature* 462:353–57

Colgin LL, Kubota D, Jia Y, Rex CS, Lynch G. 2004. Long-term potentiation is impaired in rat hippocampal slices that produce spontaneous sharp waves. *J. Physiol.* 558:953–61

Colgin LL, Leutgeb S, Jezek K, Leutgeb JK, Moser EI, et al. 2010. Attractor-map versus autoassociation based attractor dynamics in the hippocampal network. *J. Neurophysiol.* 104:35–50

Colgin LL, Moser EI. 2009. Hippocampal theta rhythms follow the beat of their own drum. *Nat. Neurosci.* 12:1483–84

DeCoteau WE, Thorn C, Gibson DJ, Courtemanche R, Mitra P, et al. 2007. Learning-related coordination of striatal and hippocampal theta rhythms during acquisition of a procedural maze task. *Proc. Natl. Acad. Sci. USA* 104:5644–49

Dickerson DD, Wolff AR, Bilkey DK. 2010. Abnormal long-range neural synchrony in a maternal immune activation animal model of schizophrenia. *J. Neurosci.* 30:12424–31

Dickson CT, Magistretti J, Shalinsky MH, Fransén E, Hasselmo ME, Alonso A. 2000. Properties and role of I(h) in the pacing of subthreshold oscillations in entorhinal cortex layer II neurons. *J. Neurophysiol.* 83:2562–79

Dragoi G, Buzsáki G. 2006. Temporal encoding of place sequences by hippocampal cell assemblies. *Neuron* 50:145–57

Ellender TJ, Nissen W, Colgin LL, Mann EO, Paulsen O. 2010. Priming of hippocampal population bursts by individual perisomatic-targeting interneurons. *J. Neurosci.* 30:5979–91

Foster DJ, Wilson MA. 2007. Hippocampal theta sequences. *Hippocampus* 17:1093–99

Frerking M, Schmitz D, Zhou Q, Johansen J, Nicoll RA. 2001. Kainate receptors depress excitatory synaptic transmission at CA3->CA1 synapses in the hippocampus via a direct presynaptic action. *J. Neurosci.* 21:2958–66

Ganguly K, Kleinfeld D. 2004. Goal-directed whisking increases phase-locking between vibrissa movement and electrical activity in primary sensory cortex in rat. *Proc. Natl. Acad. Sci. USA* 101:12348–53

Giocomo LM, Hasselmo ME. 2008. Time constants of h current in layer II stellate cells differ along the dorsal to ventral axis of medial entorhinal cortex. *J. Neurosci.* 28:9414–25

Giocomo LM, Hasselmo ME. 2009. Knock-out of HCN1 subunit flattens dorsal-ventral frequency gradient of medial entorhinal neurons in adult mice. *J. Neurosci.* 29:7625–30

Giocomo LM, Hussaini SA, Zheng F, Kandel ER, Moser MB, Moser EI. 2011. Grid cells use HCN1 channels for spatial scaling. *Cell* 147:1159–70

Giocomo LM, Zilli EA, Fransén E, Hasselmo ME. 2007. Temporal frequency of subthreshold oscillations scales with entorhinal grid cell field spacing. *Science* 315:1719–22

Goutagny R, Jackson J, Williams S. 2009. Self-generated theta oscillations in the hippocampus. *Nat. Neurosci.* 12:1491–93

Grastyan E, Lissak K, Madarasz I, Donhoffer H. 1959. Hippocampal electrical activity during the development of conditioned reflexes. *Electroencephalogr. Clin. Neurophysiol.* 11:409–30

Green JD, Arduini AA. 1954. Hippocampal electrical activity in arousal. *J. Neurophysiol.* 17:533–57

Greenstein YJ, Pavlides C, Winson J. 1988. Long-term potentiation in the dentate gyrus is preferentially induced at theta rhythm periodicity. *Brain Res.* 438:331–34

Gupta AS, van der Meer MA, Touretzky DS, Redish AD. 2012. Segmentation of spatial experience by hippocampal theta sequences. *Nat. Neurosci.* 15:1032–39

Hafting T, Fyhn M, Bonnevie T, Moser MB, Moser EI. 2008. Hippocampus-independent phase precession in entorhinal grid cells. *Nature* 453:1248–52

Hafting T, Fyhn M, Molden S, Moser MB, Moser EI. 2005. Microstructure of a spatial map in the entorhinal cortex. *Nature* 436:801–6

Hangya B, Borhegyi Z, Szilagyi N, Freund TF, Varga V. 2009. GABAergic neurons of the medial septum lead the hippocampal network during theta activity. *J. Neurosci.* 29:8094–102

Harris KD, Henze DA, Hirase H, Leinekugel X, Dragoi G, et al. 2002. Spike train dynamics predicts theta-related phase precession in hippocampal pyramidal cells. *Nature* 417:738–41

Harvey CD, Collman F, Dombeck DA, Tank DW. 2009. Intracellular dynamics of hippocampal place cells during virtual navigation. *Nature* 461:941–46

Heys JG, MacLeod KM, Moss CF, Hasselmo ME. 2013. Bat and rat neurons differ in theta-frequency resonance despite similar coding of space. *Science* 340:363–67

Hu H, Vervaeke K, Storm JF. 2002. Two forms of electrical resonance at theta frequencies, generated by M-current, h-current and persistent Na+ current in rat hippocampal pyramidal cells. *J. Physiol.* 545:783–805

Hyman JM, Wyble BP, Goyal V, Rossi CA, Hasselmo ME. 2003. Stimulation in hippocampal region CA1 in behaving rats yields long-term potentiation when delivered to the peak of theta and long-term depression when delivered to the trough. *J. Neurosci.* 23:11725–31

Hyman JM, Zilli EA, Paley AM, Hasselmo ME. 2010. Working memory performance correlates with prefrontal-hippocampal theta interactions but not with prefrontal neuron firing rates. *Front. Integr. Neurosci.* 4:2

Jay TM, Witter MP. 1991. Distribution of hippocampal CA1 and subicular efferents in the prefrontal cortex of the rat studied by means of anterograde transport of Phaseolus vulgaris-leucoagglutinin. *J. Comp. Neurol.* 313:574–86

Jensen O, Colgin LL. 2007. Cross-frequency coupling between neuronal oscillations. *Trends Cogn. Sci.* 11:267–69

Jensen O, Lisman JE. 1996. Hippocampal CA3 region predicts memory sequences: accounting for the phase precession of place cells. *Learn. Mem.* 3:279–87

Jezek K, Henriksen EJ, Treves A, Moser EI, Moser MB. 2011. Theta-paced flickering between place-cell maps in the hippocampus. *Nature* 478:246–49

Johnson A, Redish AD. 2007. Neural ensembles in CA3 transiently encode paths forward of the animal at a decision point. *J. Neurosci.* 27:12176–89

Jones MW, Wilson MA. 2005. Theta rhythms coordinate hippocampal-prefrontal interactions in a spatial memory task. *PLoS Biol.* 3(12):e402

Jung R, Kornmuller AE. 1938. Eine Methodik der ableitung lokalisierter Potentialschwankungen aus subcorticalen Hirngebieten. *Arch. Psychiat. Nervenkrankh.* 109:1–30

Kahana MJ, Sekuler R, Caplan JB, Kirschen M, Madsen JR. 1999. Human theta oscillations exhibit task dependence during virtual maze navigation. *Nature* 399:781–84

Kamondi A, Acsády L, Wang XJ, Buzsáki G. 1998. Theta oscillations in somata and dendrites of hippocampal pyramidal cells in vivo: activity-dependent phase-precession of action potentials. *Hippocampus* 8:244–61

Kay LM. 2005. Theta oscillations and sensorimotor performance. *Proc. Natl. Acad. Sci. USA* 102:3863–68

Kepecs A, Uchida N, Mainen ZF. 2006. The sniff as a unit of olfactory processing. *Chem. Senses* 31:167–79

Killian NJ, Jutras MJ, Buffalo EA. 2012. A map of visual space in the primate entorhinal cortex. *Nature* 491:761–64

Kim J, Delcasso S, Lee I. 2011. Neural correlates of object-in-place learning in hippocampus and prefrontal cortex. *J. Neurosci.* 31:16991–7006

Kim SM, Ganguli S, Frank LM. 2012. Spatial information outflow from the hippocampal circuit: distributed spatial coding and phase precession in the subiculum. *J. Neurosci.* 32:11539–58

Klimesch W, Doppelmayr M, Russegger H, Pachinger T. 1996. Theta band power in the human scalp EEG and the encoding of new information. *NeuroReport* 7:1235–40

Klimesch W, Schimke H, Schwaiger J. 1994. Episodic and semantic memory: an analysis in the EEG theta and alpha band. *Electroencephalogr. Clin. Neurophysiol.* 91:428–41

Koenig J, Linder AN, Leutgeb JK, Leutgeb S. 2011. The spatial periodicity of grid cells is not sustained during reduced theta oscillations. *Science* 332:592–95

Kramis R, Vanderwolf CH, Bland BH. 1975. Two types of hippocampal rhythmical slow activity in both the rabbit and the rat: relations to behavior and effects of atropine, diethyl ether, urethane, and pentobarbital. *Exp. Neurol.* 49:58–85

Kubota D, Colgin LL, Casale M, Brucher FA, Lynch G. 2003. Endogenous waves in hippocampal slices. *J. Neurophysiol.* 89:81–89

Landfield PW, McGaugh JL, Tusa RJ. 1972. Theta rhythm: a temporal correlate of memory storage processes in the rat. *Science* 175:87–89

Larson J, Wong D, Lynch G. 1986. Patterned stimulation at the theta frequency is optimal for the induction of hippocampal long-term potentiation. *Brain Res.* 368:347–50

Lee H, Dvorak D, Kao HY, Duffy AM, Scharfman HE, Fenton AA. 2012. Early cognitive experience prevents adult deficits in a neurodevelopmental schizophrenia model. *Neuron* 75:714–24

Liebe S, Hoerzer GM, Logothetis NK, Rainer G. 2012. Theta coupling between V4 and prefrontal cortex predicts visual short-term memory performance. *Nat. Neurosci.* 15:456–62

Lisman J, Redish AD. 2009. Prediction, sequences and the hippocampus. *Philos. Trans. R. Soc. Lond. B* 364: 1193–201

Louie K, Wilson MA. 2001. Temporally structured replay of awake hippocampal ensemble activity during rapid eye movement sleep. *Neuron* 29:145–56

Lubenov EV, Siapas AG. 2009. Hippocampal theta oscillations are travelling waves. *Nature* 459:534–39

Maccaferri G. 2005. Stratum oriens horizontal interneurone diversity and hippocampal network dynamics. *J. Physiol.* 562:73–80

Maccaferri G, McBain CJ. 1996. The hyperpolarization-activated current (Ih) and its contribution to pacemaker activity in rat CA1 hippocampal stratum oriens-alveus interneurones. *J. Physiol.* 497(Pt. 1):119–30

Macrides F, Eichenbaum HB, Forbes WB. 1982. Temporal relationship between sniffing and the limbic theta rhythm during odor discrimination reversal learning. *J. Neurosci.* 2:1705–17

Madison DV, Lancaster B, Nicoll RA. 1987. Voltage clamp analysis of cholinergic action in the hippocampus. *J. Neurosci.* 7:733–41

Magee JC. 2001. Dendritic mechanisms of phase precession in hippocampal CA1 pyramidal neurons. *J. Neurophysiol.* 86:528–32

Marcelin B, Lugo JN, Brewster AL, Liu Z, Lewis AS, et al. 2012. Differential dorso-ventral distributions of Kv4.2 and HCN proteins confer distinct integrative properties to hippocampal CA1 pyramidal cell distal dendrites. *J. Biol. Chem.* 287:17656–61

Maurer AP, McNaughton BL. 2007. Network and intrinsic cellular mechanisms underlying theta phase precession of hippocampal neurons. *Trends Neurosci.* 30:325–33

Maurer AP, Vanrhoads SR, Sutherland GR, Lipa P, McNaughton BL. 2005. Self-motion and the origin of differential spatial scaling along the septo-temporal axis of the hippocampus. *Hippocampus* 15:841–52

Mehta MR, Lee AK, Wilson MA. 2002. Role of experience and oscillations in transforming a rate code into a temporal code. *Nature* 417:741–46

Melzer S, Michael M, Caputi A, Eliava M, Fuchs EC, et al. 2012. Long-range-projecting GABAergic neurons modulate inhibition in hippocampus and entorhinal cortex. *Science* 335:1506–10

M'Harzi M, Jarrard LE. 1992. Effects of medial and lateral septal lesions on acquisition of a place and cue radial maze task. *Behav. Brain Res.* 49:159–65

Mitchell SJ, Rawlins JN, Steward O, Olton DS. 1982. Medial septal area lesions disrupt theta rhythm and cholinergic staining in medial entorhinal cortex and produce impaired radial arm maze behavior in rats. *J. Neurosci.* 2:292–302

Mizumori SJ, Perez GM, Alvarado MC, Barnes CA, McNaughton BL. 1990. Reversible inactivation of the medial septum differentially affects two forms of learning in rats. *Brain Res.* 528:12–20

Mizuseki K, Royer S, Diba K, Buzsáki G. 2012. Activity dynamics and behavioral correlates of CA3 and CA1 hippocampal pyramidal neurons. *Hippocampus* 22:1659–80

Mizuseki K, Sirota A, Pastalkova E, Buzsáki G. 2009. Theta oscillations provide temporal windows for local circuit computation in the entorhinal-hippocampal loop. *Neuron* 64:267–80

Montgomery SM, Sirota A, Buzsáki G. 2008. Theta and gamma coordination of hippocampal networks during waking and rapid eye movement sleep. *J. Neurosci.* 28:6731–41

Navratilova Z, Giocomo LM, Fellous JM, Hasselmo ME, McNaughton BL. 2012. Phase precession and variable spatial scaling in a periodic attractor map model of medial entorhinal grid cells with realistic after-spike dynamics. *Hippocampus* 22:772–89

Nolan MF, Malleret G, Dudman JT, Buhl DL, Santoro B, et al. 2004. A behavioral role for dendritic integration: HCN1 channels constrain spatial memory and plasticity at inputs to distal dendrites of CA1 pyramidal neurons. *Cell* 119:719–32

O'Keefe J, Dostrovsky J. 1971. The hippocampus as a spatial map. Preliminary evidence from unit activity in the freely-moving rat. *Brain Res.* 34:171–75

O'Keefe J, Recce ML. 1993. Phase relationship between hippocampal place units and the EEG theta rhythm. *Hippocampus* 3:317–30

Orr G, Rao G, Houston FP, McNaughton BL, Barnes CA. 2001. Hippocampal synaptic plasticity is modulated by theta rhythm in the fascia dentata of adult and aged freely behaving rats. *Hippocampus* 11:647–54

Osipova D, Takashima A, Oostenveld R, Fernández G, Maris E, Jensen O. 2006. Theta and gamma oscillations predict encoding and retrieval of declarative memory. *J. Neurosci.* 26:7523–31

Otero-Millan J, Troncoso XG, Macknik SL, Serrano-Pedraza I, Martinez-Conde S. 2008. Saccades and microsaccades during visual fixation, exploration, and search: foundations for a common saccadic generator. *J. Vis.* 8:1–18

Patel J, Fujisawa S, Berenyi A, Royer S, Buzsáki G. 2012. Traveling theta waves along the entire septotemporal axis of the hippocampus. *Neuron* 75:410–17

Pavlides C, Greenstein YJ, Grudman M, Winson J. 1988. Long-term potentiation in the dentate gyrus is induced preferentially on the positive phase of theta-rhythm. *Brain Res.* 439:383–87

Paz R, Bauer EP, Pare D. 2008. Theta synchronizes the activity of medial prefrontal neurons during learning. *Learn. Mem.* 15:524–31

Petsche H, Stumpf C, Gogolak G. 1962. The significance of the rabbit's septum as a relay station between the midbrain and the hippocampus. I. The control of hippocampus arousal activity by the septum cells. *Electroencephalogr. Clin. Neurophysiol.* 14:202–11

Pike FG, Goddard RS, Suckling JM, Ganter P, Kasthuri N, Paulsen O. 2000. Distinct frequency preferences of different types of rat hippocampal neurones in response to oscillatory input currents. *J. Physiol.* 529 (Pt. 1):205–13

Robbe D, Buzsáki G. 2009. Alteration of theta timescale dynamics of hippocampal place cells by a cannabinoid is associated with memory impairment. *J. Neurosci.* 29:12597–605

Robinson RB, Siegelbaum SA. 2003. Hyperpolarization-activated cation currents: from molecules to physiological function. *Annu. Rev. Physiol.* 65:453–80

Rutishauser U, Ross IB, Mamelak AN, Schuman EM. 2010. Human memory strength is predicted by theta-frequency phase-locking of single neurons. *Nature* 464:903–7

Seidenbecher T, Laxmi TR, Stork O, Pape HC. 2003. Amygdalar and hippocampal theta rhythm synchronization during fear memory retrieval. *Science* 301:846–50

Senior TJ, Huxter JR, Allen K, O'Neill J, Csicsvari J. 2008. Gamma oscillatory firing reveals distinct populations of pyramidal cells in the CA1 region of the hippocampus. *J. Neurosci.* 28:2274–86

Sigurdsson T, Stark KL, Karayiorgou M, Gogos JA, Gordon JA. 2010. Impaired hippocampal-prefrontal synchrony in a genetic mouse model of schizophrenia. *Nature* 464:763–67

Simon AP, Poindessous-Jazat F, Dutar P, Epelbaum J, Bassant MH. 2006. Firing properties of anatomically identified neurons in the medial septum of anesthetized and unanesthetized restrained rats. *J. Neurosci.* 26:9038–46

Skaggs WE, McNaughton BL, Wilson MA, Barnes CA. 1996. Theta phase precession in hippocampal neuronal populations and the compression of temporal sequences. *Hippocampus* 6:149–72

Squire LR, Stark CE, Clark RE. 2004. The medial temporal lobe. *Annu. Rev. Neurosci.* 27:279–306

Staubli U, Lynch G. 1987. Stable hippocampal long-term potentiation elicited by 'theta' pattern stimulation. *Brain Res.* 435:227–34

Swanson LW. 1981. A direct projection from Ammon's horn to prefrontal cortex in the rat. *Brain Res.* 217:150–54

Terrazas A, Krause M, Lipa P, Gothard KM, Barnes CA, McNaughton BL. 2005. Self-motion and the hippocampal spatial metric. *J. Neurosci.* 25:8085–96

Tesche CD, Karhu J. 2000. Theta oscillations index human hippocampal activation during a working memory task. *Proc. Natl. Acad. Sci. USA* 97:919–24

Tort AB, Kramer MA, Thorn C, Gibson DJ, Kubota Y, et al. 2008. Dynamic cross-frequency couplings of local field potential oscillations in rat striatum and hippocampus during performance of a T-maze task. *Proc. Natl. Acad. Sci. USA* 105:20517–22

Toth K, Borhegyi Z, Freund TF. 1993. Postsynaptic targets of GABAergic hippocampal neurons in the medial septum-diagonal band of Broca complex. *J. Neurosci.* 13:3712–24

Toth K, Freund TF, Miles R. 1997. Disinhibition of rat hippocampal pyramidal cells by GABAergic afferents from the septum. *J. Physiol.* 500(Pt. 2):463–74

Tsodyks MV, Skaggs WE, Sejnowski TJ, McNaughton BL. 1996. Population dynamics and theta rhythm phase precession of hippocampal place cell firing: a spiking neuron model. *Hippocampus* 6:271–80

Vaidya SP, Johnston D. 2012. HCN channels contribute to the spatial synchrony of theta frequency synaptic inputs in CA1 pyramidal neurons. *Soc. Neurosci. Annu. Meet. Abstr.* 435.15 (Abstr.)

van der Meer MA, Redish AD. 2011. Theta phase precession in rat ventral striatum links place and reward information. *J. Neurosci.* 31:2843–54

Vanderwolf CH. 1969. Hippocampal electrical activity and voluntary movement in the rat. *Electroencephalogr. Clin. Neurophysiol.* 26:407–18

Varga V, Hangya B, Kranitz K, Ludanyi A, Zemankovics R, et al. 2008. The presence of pacemaker HCN channels identifies theta rhythmic GABAergic neurons in the medial septum. *J. Physiol.* 586:3893–915

Vertes RP. 2004. Memory consolidation in sleep: dream or reality. *Neuron* 44:135–48

Vertes RP, Hoover WB, Viana Di Prisco G. 2004. Theta rhythm of the hippocampus: subcortical control and functional significance. *Behav. Cogn. Neurosci. Rev.* 3:173–200

Vertes RP, Kocsis B. 1997. Brainstem-diencephalo-septohippocampal systems controlling the theta rhythm of the hippocampus. *Neuroscience* 81:893–926

Wallenstein GV, Hasselmo ME. 1997. GABAergic modulation of hippocampal population activity: sequence learning, place field development, and the phase precession effect. *J. Neurophysiol.* 78:393–408

Winson J. 1974. Patterns of hippocampal theta rhythm in the freely moving rat. *Electroencephalogr. Clin. Neurophysiol.* 36:291–301

Winson J. 1978. Loss of hippocampal theta rhythm results in spatial memory deficit in the rat. *Science* 201:160–63

Yamaguchi Y, Aota Y, McNaughton BL, Lipa P. 2002. Bimodality of theta phase precession in hippocampal place cells in freely running rats. *J. Neurophysiol.* 87:2629–42

Yartsev MM, Ulanovsky N. 2013. Representation of three-dimensional space in the hippocampus of flying bats. *Science* 340:367–72

Yartsev MM, Witter MP, Ulanovsky N. 2011. Grid cells without theta oscillations in the entorhinal cortex of bats. *Nature* 479:103–7

Ylinen A, Soltesz I, Bragin A, Penttonen M, Sik A, Buzsáki G. 1995. Intracellular correlates of hippocampal theta rhythm in identified pyramidal cells, granule cells, and basket cells. *Hippocampus* 5:78–90

Neural Basis of the Perception and Estimation of Time

Hugo Merchant,[1] Deborah L. Harrington,[2,3] and Warren H. Meck[4]

[1]Instituto de Neurobiología, UNAM, Campus Juriquilla, México; email: hugomerchant@unam.mx

[2]VA San Diego Healthcare System, San Diego, California 92161; email: dharrington@ucsd.edu

[3]Department of Radiology, University of California, San Diego, La Jolla, California 92093

[4]Department of Psychology and Neuroscience, Duke University, Durham, North Carolina 27701; email: meck@psych.duke.edu

Annu. Rev. Neurosci. 2013. 36:313–36

First published online as a Review in Advance on May 29, 2013

The *Annual Review of Neuroscience* is online at neuro.annualreviews.org

This article's doi: 10.1146/annurev-neuro-062012-170349

Keywords

temporal processing, interval tuning, medial premotor cortex, cortico-thalamic-basal ganglia circuit, striatal beat-frequency model

Abstract

Understanding how sensory and motor processes are temporally integrated to control behavior in the hundredths of milliseconds-to-minutes range is a fascinating problem given that the basic electrophysiological properties of neurons operate on a millisecond timescale. Single-unit recording studies in monkeys have identified localized timing circuits, whereas neuropsychological studies of humans who have damage to the basal ganglia have indicated that core structures, such as the cortico-thalamic-basal ganglia circuit, play an important role in timing and time perception. Taken together, these data suggest that a core timing mechanism interacts with context-dependent areas. This idea of a temporal hub with a distributed network is used to investigate the abstract properties of interval tuning as well as temporal illusions and intersensory timing. We conclude by proposing that the interconnections built into this core timing mechanism are designed to provide a form of degeneracy as protection against injury, disease, or age-related decline.

Contents

INTRODUCTION

Timing is everything. The flow of information through time structures how information is perceived, experienced, and remembered. Throughout normal development we gradually acquire a sense of duration and rhythm that is basic to many facets of behavior such as speaking, driving a car, dancing to or playing music, and performing physical activities (Allman et al. 2012, Meck 2003). Yet there is no specific biological system that senses time as there are for sight, hearing, and taste. As such, there has been an explosion of research into the neural underpinnings of timing. Initially, a driving force behind many studies was scalar timing theory, which defined sources and forms of timing variability that were derived from clock, memory, and decision processes (Gibbon et al. 1984). However, as an understanding of the neurobiological bases of timing developed, so did a neurophysiological model of timing (Matell & Meck 2004), which captured the intrinsic interactive nature of interval-timing circuits as well as Weber's law and the scalar property of interval timing (Brannon et al. 2008, Cheng & Meck 2007, Gu et al. 2013, Meck & Malapani 2004).

This article reviews recent progress toward elucidating the neural mechanisms of interval timing, wherein durations in the range of hundreds of milliseconds to multi-seconds are perceived, estimated, or reproduced. We begin by discussing advances in psychophysics and in cell-ensemble recording that address the fundamental debate about whether explicit timing is governed by a common system or by distributed context-dependent networks. Next, we review current research into the functional significance of neuroanatomical systems that govern interval timing. We then consider emerging investigations into brain mechanisms underlying distortions in temporal resolution, which emanate from intersensory timing or factors that produce illusions of time. This body of research indicates that interval timing emerges from interactions of a core timing center with distributed brain regions. To this end, the striatal-beat frequency model of interval timing (Matell & Meck 2004) is discussed, which captures the psychophysiological and neuroanatomical properties of timing networks.

EVIDENCE FAVORING A COMMON TIMING MECHANISM

The psychophysics of the perception and estimation of time started in the late nineteenth century (Fraisse 1984) and evolved as many timing tasks and species were studied to test the boundaries of categorical scaling (Penney et al. 2008) and the existence of one or multiple clocks (Buhusi & Meck 2009b). One central finding was that the variability of interval timing is proportional to the duration of the interval used. This scalar property implies that the standard deviation of the quantified intervals increases proportionally with the average of the intervals and follows Weber's law. Weber's law is given as $SD(T) = kT$, where k is a constant corresponding to the Weber fraction. The coefficients of variation (σ/μ) or the Weber fractions show similar values in the range of hundreds of milliseconds in a variety of tasks, sensory modalities, and species, suggesting a common timing mechanism for this, and perhaps other, time scales (Gibbon et al. 1997). The conceptual framework behind this hypothesis proposes that the temporal resources of a common internal clock are shared in a variety of timing tasks, thereby producing similar temporal variance (**Figure 1*a***).

Another major finding was that the overall variability in a timing task can be dissociated into time-dependent (e.g., clock) and time-independent (e.g., motor) sources (Repp 2005). Different quantitative and paradigmatic strategies were used to distinguish components of performance variability. The slope method, for instance, uses a generalized form of Weber's law, wherein investigators compute a linear regression between the variability and the squared interval duration. The resulting slope corresponds to the time-dependent component, whereas the intercept corresponds to the time-independent processes. Slopes of interval discrimination and synchronization-continuation tapping tasks are similar for a range of intervals from 325 to 550 ms (Ivry & Hazeltine 1995), supporting the view of a common clock in a variety of contexts (**Figure 2*a***). Moreover, temporal variability among an individual's performance correlates between different explicit timing tasks, including self-paced tapping tasks using the finger, foot, and heel (Keele et al. 1985), tapping and phasic figure drawing (Zelaznik et al. 2002), and duration-discrimination and tapping tasks (Keele et al. 1985). This correlation implies that participants who are good timers in one behavioral context are also good timers in another, again in support of a common timing mechanism.

The study of perceptual learning and generalization to other behaviors and modalities has provided important insights into the neural underpinnings of temporal processing (**Figure 1*b***). For example, intensive duration-discrimination learning can generalize across untrained auditory frequencies (Karmarkar & Buonomano 2003), sensory modalities, and stimulus locations (Nagarajan et al. 1998), and even from sensory to motor-timing tasks (Meegan et al. 2000). However, these studies found no reliable generalization toward untrained durations, which concurs with a study of duration-production learning that reported a smooth decrease in generalization as the untrained interval deviated from the trained duration and suggested the existence of neural circuits that are tuned to specific durations (Bartolo & Merchant 2009). Overall, these findings support the notion of a common or a partially overlapping distributed timing mechanism, but they also introduce the concept of duration-specific circuits. Of course, these two features are not mutually exclusive when a large neural network is considered (Karmarkar & Buonomano 2007, Matell & Meck 2004).

EVIDENCE FAVORING A UBIQUITOUS TIMING ABILITY OF CORTICAL NETWORKS

Other studies support the hypothesis that timing is an inherent computational ability of every cortical circuit and that it can be performed locally (**Figure 2*b***). This notion, based on network simulations, implies that cortical networks can tell time during perception tasks as a result of time-dependent changes in synaptic and cellular properties, which influence the population response to sensory events in a history-dependent manner (Karmakar & Buonomano

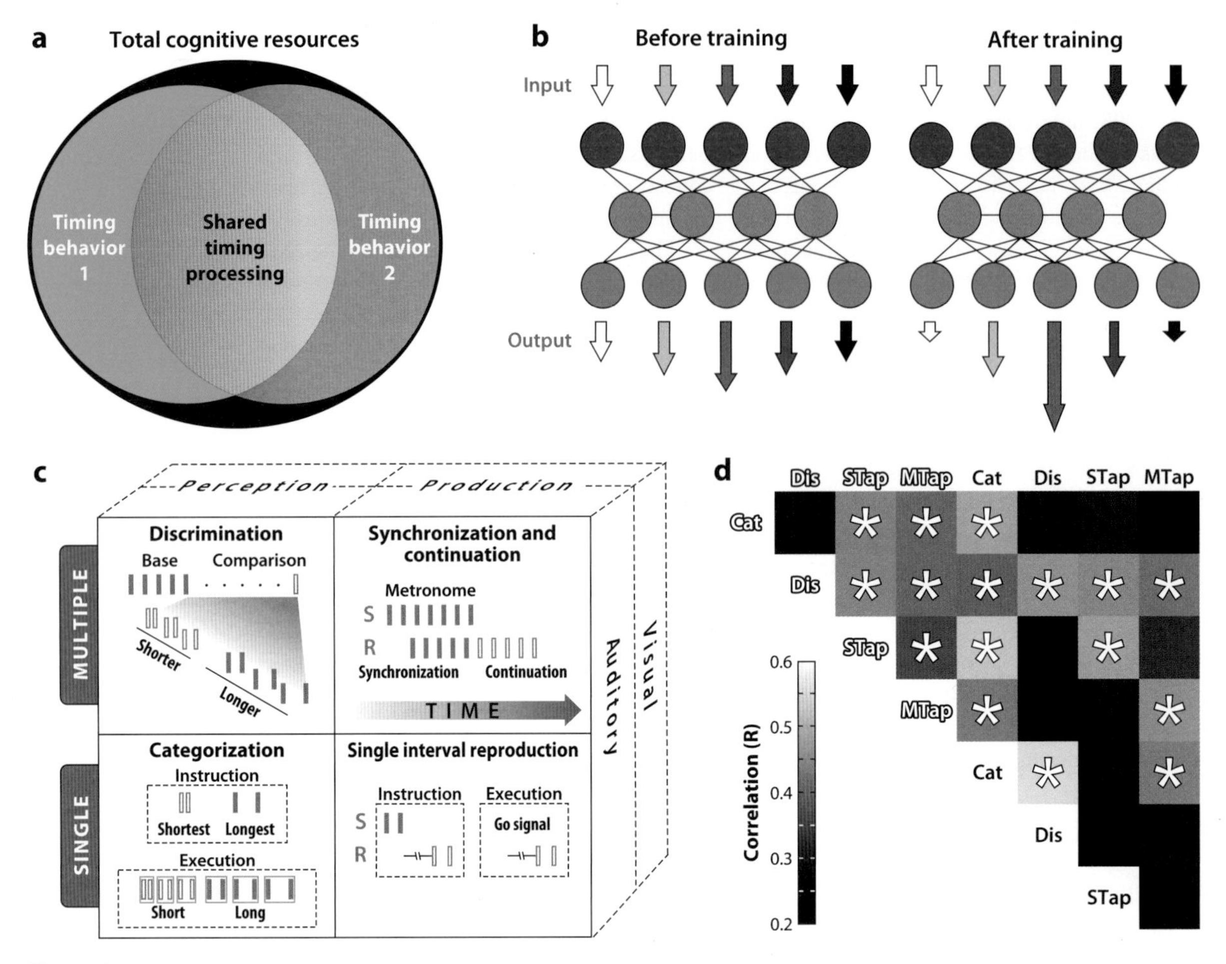

Figure 1

Psychophysical tools to study the neural underpinnings of interval timing using a black-box approach. (*a*) The hypothesis in some experimental psychology studies is that because some cognitive resources are shared across timing tasks there should be a common timing mechanism. (*b*) The characterization of the generalization properties of specific overlearned timing tasks across modalities, stimulation sites and properties, interval durations, and timing contexts; also has been used to test the existence of a common or multiple interval clocks. (*c*) Timing tasks. Four timing tasks, performed with auditory or visual interval markers, were used to evaluate the influence of three factors on timing performance: visual versus auditory modality, single versus multiple intervals, and perception versus production of the intervals. (*d*) Correlation matrix showing the Pearson R value in a grayscale matrix (*inset, bottom left*) for all possible pairwise task comparisons. Asterisks indicate significant correlations ($P < 0.05$) between specific pairs of tasks. Open and closed fonts correspond to tasks with auditory and visual markers, respectively. Abbreviations: Dis, discrimination; Cat, categorization; STap, single interval reproduction; MTap, synchronization and continuation task. Adapted from Merchant et al. 2008c.

2007). In contrast, motor timing is thought to depend on the activity of cortical recurrent networks with strong internal connections capable of self-sustained activity (Buonomano & Laje 2010). Chronic stimulation in cortical slices produces changes in the temporal structure of the cell activity that reflect the durations used during training (Johnson et al. 2010). Hence, this in vitro experimental evidence suggests that recurrent cortical circuits have inherent timing ability in the hundreds of milliseconds.

Psychophysical experiments also suggest that sensory timing is local. For example, the apparent duration of a visual stimulus can

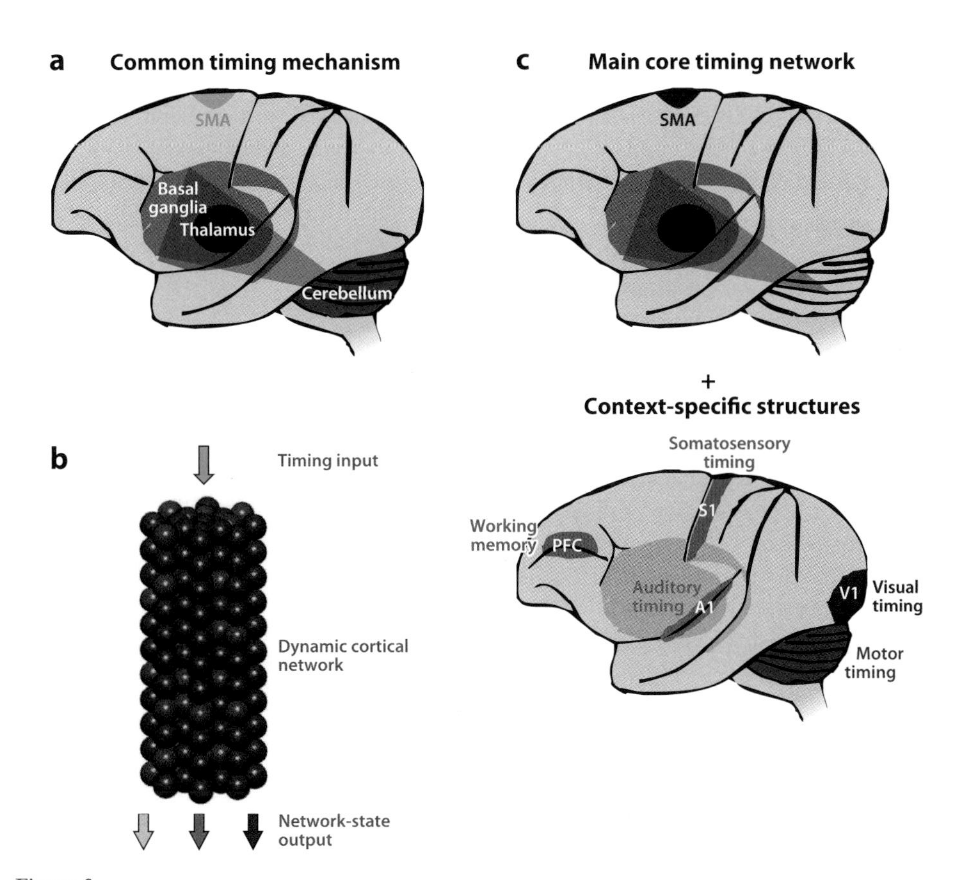

Figure 2

Three possible timing mechanisms. (*a*) Common timing mechanism. Psychophysical and lesion studies have suggested the existence of a dedicated timing mechanism that depends on one neural structure such as the cerebellum or the basal ganglia and that is engaged in temporal processing in a wide range of timing behaviors. However, fMRI studies have suggested that this general timing mechanism is distributed and depends on the activation of a large network including the supplementary motor area (SMA), the parietal and prefrontal cortices, as well as the basal ganglia and the cerebellum. (*b*) Ubiquitous timing. Modeling and cell-culture recordings have suggested that timing is an intrinsic property of cortical network dynamics and therefore that there is no dedicated neural circuit for temporal integration. (*c*) Partially shared timing mechanism. This model proposes that temporal estimation depends on the interaction of multiple areas, including regions that are consistently involved in temporal processing across timing context and that conform the main core timing network and areas that are activated in a context-dependent fashion. The main core timing network consists of the SMA and the basal ganglia. The interaction between the two sets of structures gives the specific temporal performance in a task. The red triangles correspond to the dopaminergic system innervating the basal ganglia.

be modified in a local region of the visual field by adapting to oscillatory motion or flicker, which suggests a spatially localized temporal mechanism for time perception of visual events (Burr et al. 2007, Johnston et al. 2006). Furthermore, learning to discriminate temporal modulation rates is accompanied not only by a specific learning transfer to duration discrimination, but also by an increase in the amplitude of the early auditory evoked responses to trained stimuli (van Wassenhove & Nagarajan 2007). These studies emphasize the concept of timing as a local, context-dependent process.

CTBG: cortico-thalamic-basal ganglia

MPC: medial premotor areas

SMA: supplementary motor area

AN INTERMEDIATE HYPOTHESIS: A MAIN CORE TIMING MECHANISM INTERACTS WITH CONTEXT-DEPENDENT AREAS

Other research suggests that a hybrid model may better account for temporal performance variability in different contexts. Merchant and colleagues (2008c) conducted a multidimensional analysis of performance variability on four timing tasks that differed in sensorimotor processing, the number of durations, and the modality of the stimuli that defined the intervals (**Figure 1*c***). Though variability increased linearly as a function of duration in all tasks, compliance with the scalar property was accompanied by a strong effect of the nontemporal variables on temporal accuracy (Merchant et al. 2008b,c). Intersubject analyses comparing performance variability between pairs of tasks revealed a complex set of correlations between many, but not all, tasks, irrespective of stimulus modality (**Figure 1*d***). These results can be interpreted neither as evidence for multiple context-dependent timing mechanisms, nor as evidence for a common timing mechanism. Rather, in accordance with neuroimaging research described below, the findings suggest a partially distributed timing mechanism, integrated by core structures such as the cortico-thalamic-basal ganglia (CTBG) circuit and areas that are selectively engaged by different behavioral contexts (Buhusi & Meck 2005, Coull et al. 2011) (**Figure 2*c***). Task-dependent areas may interact with the core timing system to produce the characteristic pattern of performance variability in a specific timing paradigm and the pattern of intertask correlations depicted in **Figure 1*d***.

NEUROPHYSIOLOGICAL BASIS OF TIME ESTIMATION: RAMPING ACTIVITY IN THE CORE TIMING CIRCUIT

Cell activity changes associated with temporal processing in behaving monkeys are found in the cerebellum (Perrett 1998), basal ganglia (Jin et al. 2009), thalamus (Tanaka 2007), posterior parietal cortex (Leon & Shadlen 2003), prefrontal cortex (Brody et al. 2003; Genovesio et al. 2006, 2009; Oshio et al. 2008), dorsal premotor cortex (Lucchetti & Bon 2001), motor cortex (Lebedev et al. 2008), and medial premotor areas (MPC), namely the supplementary (SMA) and presupplementary motor areas (preSMA) (Mita et al. 2009). These areas form different circuits that are linked to sensorimotor processing via the skeletomotor or oculomotor effector systems. Most studies report climbing activity during a variety of timing tasks. Therefore, the increase or decrease in instantaneous activity with the passage of time is a property present in many cortical and subcortical areas that may be involved in different aspects of temporal processing in the hundreds of milliseconds range.

Recently, the activity of MPC cells was recorded during a synchronization-continuation tapping task (SCT) that includes basic sensorimotor components of rhythmic behaviors (Zarco et al. 2009) (**Figure 1*c***). Different types of neurons exhibit ramping activity before or after the button press in the SCT (Merchant et al. 2011). A large group of cells shows ramps before movement onset that are similar across produced durations and the sequential structure of the task and are considered motor ramps (Perez et al. 2013) (**Figure 3*a***). Another cell population exhibits an increase in ramp duration but a decrease in slope as a function of the monkey's produced duration, reaching a particular discharge magnitude at a specific time before the button press. These cells are called relative-timing cells because their ramping profile appears to signal how much time is left to trigger the button press (**Figure 3*b***). Another group of cells shows a consistent increase followed by a decrease in their instantaneous discharge rate when the neural activity was aligned to the previous button press. In these absolute-timing cells, the duration of the up-down profile of activation increases as a function of the produced interval (**Figure 3*c***), whereas in the time-accumulator cells we see

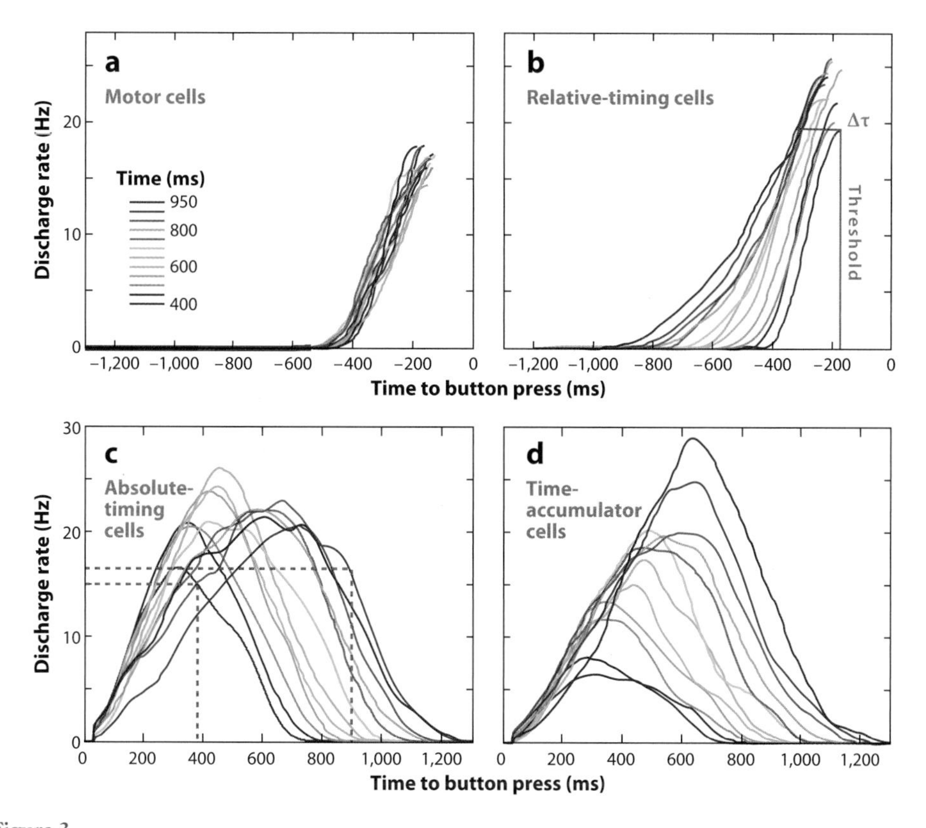

Figure 3

Ramp population functions for motor (*a*) and relative-timing (*b*) cells aligned to the next button press. Ramp population functions for absolute-timing (*c*) and time-accumulator (*d*) cells aligned to the previous button press. The color code in the inset of panel *a* corresponds to the duration of the produced intervals during the synchronization-continuation tapping task (SCT). The ramp population functions are equal to the total magnitude of individual ramps over time. Adapted from Merchant et al. 2011.

an additional increase in the magnitude of the ramps' peak (**Figure 3*d***). Therefore, these cells could be representing the passage of time since the previous movement, using two different encoding strategies: one functioning as an accumulator of elapsed time where the peak magnitude is directly associated with the time passed, and another where the duration of the activation period is encoding the length of the time passed since the previous movement. The noisy character of ramping activity implies that whatever is reading this temporal information downstream cannot be relying on single cells to quantify the passage of time or to produce accurately timed movements. Therefore, a population code is suggested for encoding elapsed time, by which the reading network adds the magnitudes of a population of individual ramps over time, resulting in a ramp population function (Merchant et al. 2011) (**Figure 3**).

The rhythmic structure of the SCT may impose the need to predict when to trigger the next tap to generate an interval, but also to quantify the time passed from the previous movement to have a cohesive mechanism to generate repetitive tapping behavior. Indeed, the cells encoding elapsed (absolute-timing) and remaining time (relative-timing) interacted during the repetitive phases of the SCT, supporting the proposed rhythmical timing mechanism (Merchant et al. 2011).

The ubiquitous increases or decreases in cell discharge rate as a function of time across different timing tasks and areas of the core CTBG timing circuit suggest that ramping activity is a fundamental element of the timing mechanism. Key characteristics of ramping activity are its instantaneous nature and the fact that it normally peaks at the time of an anticipated motor response. Although the absolute-timing and the time-accumulator cells (**Figure 3*c,d***) are encoding elapsed time since the previous motor event, ramping cells are engrained in the temporal construction of motor intentions and actions (Merchant et al. 2004). This integration is crucial because interval-timing tasks require a response to express a perceptual decision or produce a timed response. Therefore, the ramping activity may be part of the temporal apparatus that gaits the motor responses. However, more abstract tasks such as interval tuning may represent the more cognitive aspects of temporal processing that can be dissociated from the motor response and the corresponding ramping activity.

INTERVAL TUNING: AN ABSTRACT SIGNAL OF TEMPORAL COGNITION

Investigators have observed a graded modulation in the cell discharge rate as a function of event duration during the SCT in MPC. **Figure 4*a,c*** shows the activation profile of a cell in the preSMA of a monkey performing this task. The neuron shows a larger discharge rate for the longest durations, with a preferred interval around 900 ms (**Figure 4*e***). In fact, a large population of MPC cells is tuned to different signal durations during the SCT, with a distribution of preferred durations that covers all intervals in the hundreds of milliseconds. These observations suggest that the MPC contains a representation of event duration, where different populations of duration-tuned cells are activated depending on the duration of the produced interval (Merchant et al. 2012b). Most of these cells also showed selectivity to the sequential organization of the task, a property that has been described in sequential motor tasks in MPC (Tanji 2001). The cell in **Figure 4*a,c*** also shows an increase in activity during the last produced interval of the task's continuation phase. Again, at the cell-population level, all the possible preferred ordinal sequences were covered. These findings support the proposal that MPC can multiplex event duration with the number of elements in a sequence during rhythmic tapping (Merchant et al. 2013).

Researchers have evaluated the existence of a common timing mechanism or a set of context-dependent neural clocks on interval-selective cells, comparing their tuning properties during the execution of two tasks: the SCT and a single duration reproduction task (SIRT) (**Figure 1*c***) (Merchant et al. 2013). A large group of neurons showed similar preferred durations in both timing contexts. **Figure 4*a–d*** shows that the cell that is tuned to long durations during the SCT is similarly tuned during the SIRT. Furthermore, the tuning curves for the cells in the SCT and the SIRT are similar for auditory and visual markers (**Figure 4*e***). These findings confirm that MPC is part of a core timing circuit and suggest that interval tuning can be used to represent the duration of intervals in different timing tasks. Additional experiments are needed to determine whether tuning to event duration is an emergent property of MPC cells that depends on the local integration of graded inputs or occurs throughout the CTBG circuit (Matell et al. 2003, 2011). In summary, cell tuning is an encoding mechanism used by the cerebral cortex to represent different sensory, motor, and cognitive features, including event duration (Merchant et al. 2012). This signal must be integrated as a population code, in which the cells can vote in favor of their preferred duration to establish the interval for rhythmic tapping (Merchant et al. 2013).

NEUROANATOMICAL AND NEUROCHEMICAL SYSTEMS FOR INTERVAL TIMING

Turning to neuroanatomical and neurochemical levels of analysis, investigations in humans

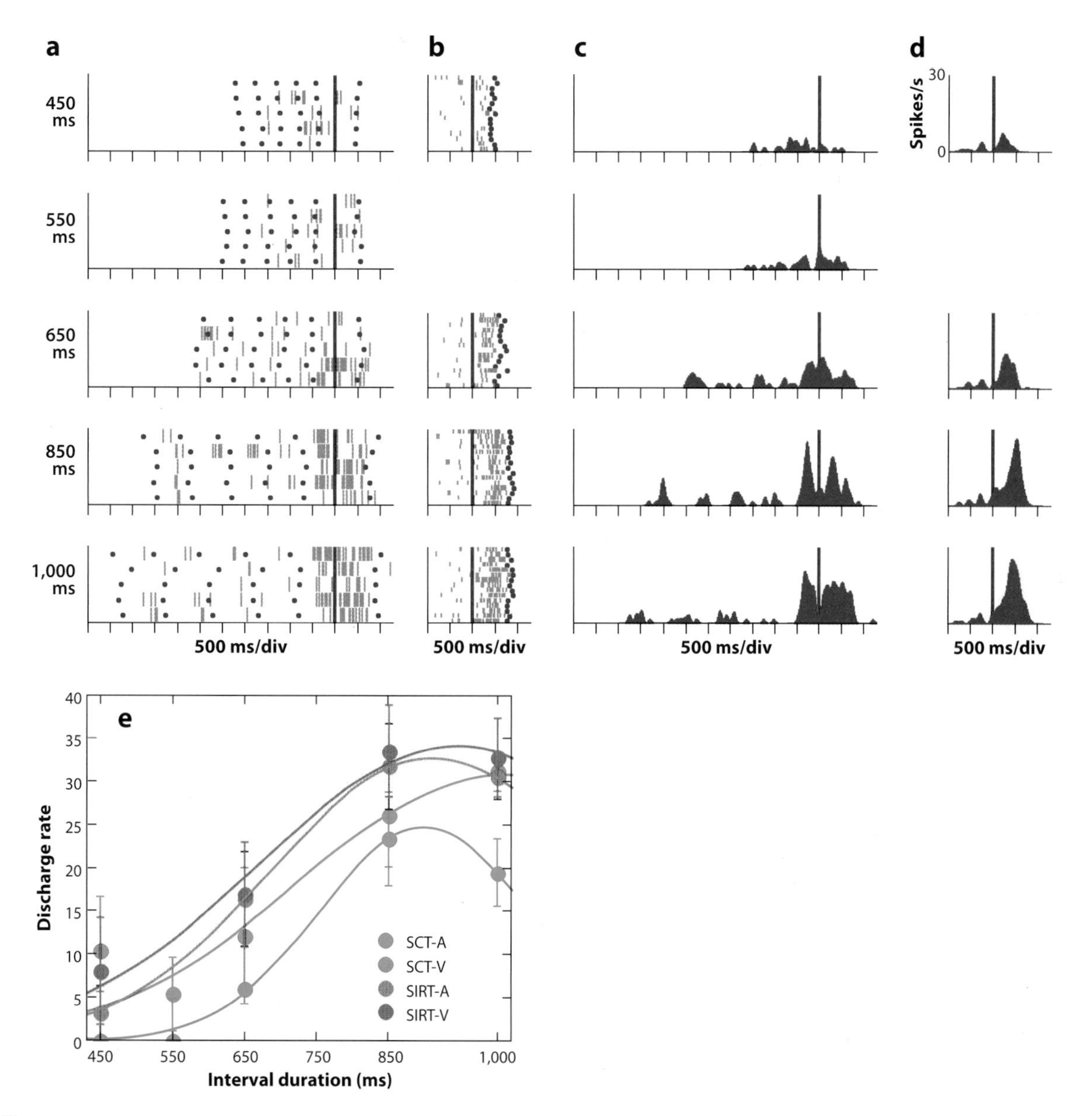

Figure 4

Interval tuning across tasks and sensory modalities. Responses of an interval-tuned cell with a long preferred interval across different temporal contexts. (*a*) Raster histograms (*blue*) aligned (*red line*) to the third tap of the continuation phase during the synchronization-continuation tapping task (SCT) in the visual condition. (*b*) Raster histograms (*blue*) aligned (*red line*) to the first tapping movement during the single duration reproduction task (SIRT) in the auditory condition. (*c, d*) Average spike-density functions of the responses shown in panels *a* and *b*, respectively. (*e*) Tuning functions for the same cell, where the mean (± SEM) of the discharge rate is plotted as a function of the target interval duration. The continuous lines correspond to the significant Gaussian fittings: SCT, visual SCT, auditory SIRT, visual SIRT.

DEGENERACY AND THE NEURAL REPRESENTATION OF TIME

Lewis & Meck (2012) have recently proposed a form of degeneracy in timing systems that parallels the degeneracy observed in many other aspects of brain function, as well as in the genetic code and immune systems (Mason 2010, Price & Friston 2002). Degeneracy is a type of functional compensation that is distinct from redundancy in that structurally different mechanisms are involved in degeneracy, whereas multiple copies of identical mechanisms are the basis for redundancy. The fundamental importance of timing to all perception and action necessitates degeneracy and may explain the ease with which timing can be performed by a range of different neural architectures (Wiener et al. 2011b). Degeneracy also allows for predictions about timing hierarchies and the sources and forms of variability in timing as a function of manipulations such as rTMS and reversible pharmacological treatments (Jahanshahi et al. 2010a,b; Teki et al. 2012). The applicability of this line of thought to time perception should be obvious: It is very difficult to abolish time perception completely, especially as a result of focal, unilateral brain lesions where, in addition to degeneracy, redundancy from the opposite hemisphere likely contributes to recovery (Aparicio et al. 2005; Coslett et al. 2010; Meck et al. 1984, 2013; Yin & Troger 2011).

and other animals emphasize the centrality of the striatum and dopamine (DA) neurotransmission in explicit timing (Agostino et al. 2011; Balci et al. 2013; Cheng et al. 2007; Coull et al. 2011, 2012; Gu et al. 2011; Hinton & Meck 1997, 2004; Höhn et al. 2011; Jones & Jahanshahi 2011; Lake & Meck 2012; Meck 1996, 2006a,b; Pleil et al. 2011). These findings are compatible with reports of timing dysfunction in disorders of the basal ganglia, including Parkinson's disease (PD) (Gu et al. 2013; Harrington et al. 1998b, 2011b; Jones et al. 2008; Koch et al. 2004, 2008; Meck & Benson 2002; Smith et al. 2007) and prodromal Huntington's disease (HD) (Paulsen et al. 2004, Rowe et al. 2010, Zimbelman et al. 2007). However, investigators have observed exceptions notably in PD, which has been studied the most extensively. For example, normal performance was reported on a test of motor timing (Spencer & Ivry 2005) and on several different tests of time perception (Wearden et al. 2008). The reasons for this variation are unclear but may be due to the insensitivity of tasks when temporal discriminations are easy and/or when feedback is frequently provided (Wearden et al. 2008). Another important consideration is that timing deficits in PD correlate with disease severity (Artieda et al. 1992). Numerous studies have tested early-stage PD patients, who, despite considerable DA cell loss, may have the capacity to compensate for cognitive difficulties. The cerebellum may be one compensatory route (Kotz & Schwartze 2011, Yu et al. 2007), possibly because it predicts and finely tunes behavioral states on the basis of an efferent copy of sensory and motor information. Cortical systems may support compensatory processing in PD, as well. These issues of degeneracy aside (see sidebar, Degeneracy and the Neural Representation of Time), the finding that subgroups of PD patients do and do not exhibit timing disturbances (Merchant et al. 2008a) resonates with the considerable heterogeneity in the disease's clinical symptoms, their day-to-day fluctuations, and individual differences in response to DA therapy.

DA: dopamine

PD: Parkinson's disease

Neurodegenerative disorders eventually alter cortical functioning, which may be another source of timing disturbances in PD and prodromal HD. Indeed, damage to the right hemisphere of the prefrontal (dorsolateral prefrontal and premotor) and the inferior parietal cortices disrupts time perception (Harrington et al. 1998b). Moreover, in patients with right but not left hemisphere damage, elevated temporal discrimination thresholds correlated with a weakened ability to reorient attention but did not correlate with deficits in pitch discrimination. This finding suggests that frontoparietal systems may govern attention and working memory, which interact with timekeeping processes (Lustig et al. 2005). Unfortunately, systematic investigations into cortical regions that are essential for timing in humans have been hampered by difficulties in obtaining sufficient samples of patients with focal damage, particularly to some brain regions now thought to be critical.

The advent of functional magnetic resonance imaging (fMRI) has been a welcome development to the study of timing. To date, most of this research has been conducted in healthy adults and has focused on regional activation patterns that are associated with various timing tasks. Research shows fairly good consensus that, apart from the basal ganglia, the SMA, an element of the CTBG (**Figure 2*c***), is routinely engaged during interval timing (Coull et al. 2011; Harrington et al. 2004, 2010; Meck et al. 2008; Wiener et al. 2010c). Carefully conducted meta-analyses further suggest that the rostral SMA, namely preSMA, may be more engaged by perceptually based timing within the suprasecond range, whereas caudal SMA may be more engaged by sensorimotor-based timing within the subsecond range (Schwartze et al. 2012). However, mounting evidence indicates that timing is governed by more distributed neural networks that are recruited depending on the behavioral context and stage of learning (Allman et al. 2012, MacDonald et al. 2012). Consequently, the major challenge has been to unravel their functional significance.

FUNCTIONAL SIGNIFICANCE OF BRAIN CIRCUITS THAT GOVERN INTERVAL TIMING

Temporal processing unavoidably engages a host of cognitive and sensorimotor processes that activate multiple brain regions. For this reason, it has been difficult in functional imaging to distinguish core-timing systems from those that support other interacting processes. One approach to the dilemma is to exploit the temporal resolution of event-related fMRI by linking the acquisition of brain images across time to different components of a task that are assumed to engage certain processes. The ordinal-comparison procedure commonly used to study time perception involves two steps: encoding an anchor or standard duration (encoding phase), followed by encoding a comparison interval and judging whether it is longer or shorter than the standard (decision phase) as illustrated in **Figure 5*a*** (Gu & Meck 2011, Harrington et al. 2004). A hypothetical hemodynamic response function associated with each of the phases is illustrated in **Figure 5*b***. One assumption is that activation in core-timing systems should be seen while encoding the standard and the comparison intervals. Activation in systems that support executive or decisional processes may be more apparent when comparing the two intervals during the decision phase when a judgment is made. To control for the engagement of cognitive and low-level sensorimotor processes, the time course of activation is compared with control tasks that contain the same stimuli but require different decisions. If auditory signals are timed, tasks that control for processing involved in perceptual decision making (pitch perception) or low-level sensorimotor processes (sensorimotor control tasks) may be used (**Figure 5*a***). An early fMRI study sought to distinguish timing-related brain activation using these methods (Rao et al. 2001). In the time-perception task, a 1,200-ms standard interval was compared with four shorter and longer intervals that were ±5% increments of the standard. For pitch perception, a 700-Hz standard tone was compared with 4 higher or lower comparison pitches. The results showed that caudate and putamen activation evolved early in association with standard-interval encoding and was sustained during the trial's decision phase. Activation of the right parietal cortex also began at the onset of the standard interval, possibly owing to its role in attention. In contrast, cerebellar activation unfolded later, just before and during movement execution, suggesting that it did not govern interval encoding. Subsequent studies using this general approach also linked striatal activation to interval timing (Coull et al. 2008, 2011; Harrington et al. 2004).

At the same time, independent empirical work (Lewis et al. 2004) and theoretical developments (Hazy et al. 2006) implicate the striatum in working memory. Owing to the short delays between the encoding and the decision phases of the time-perception tasks in earlier studies (Coull et al. 2008, Harrington et al. 2004, Rao et al. 2001), brain systems

fMRI: functional magnetic resonance imaging

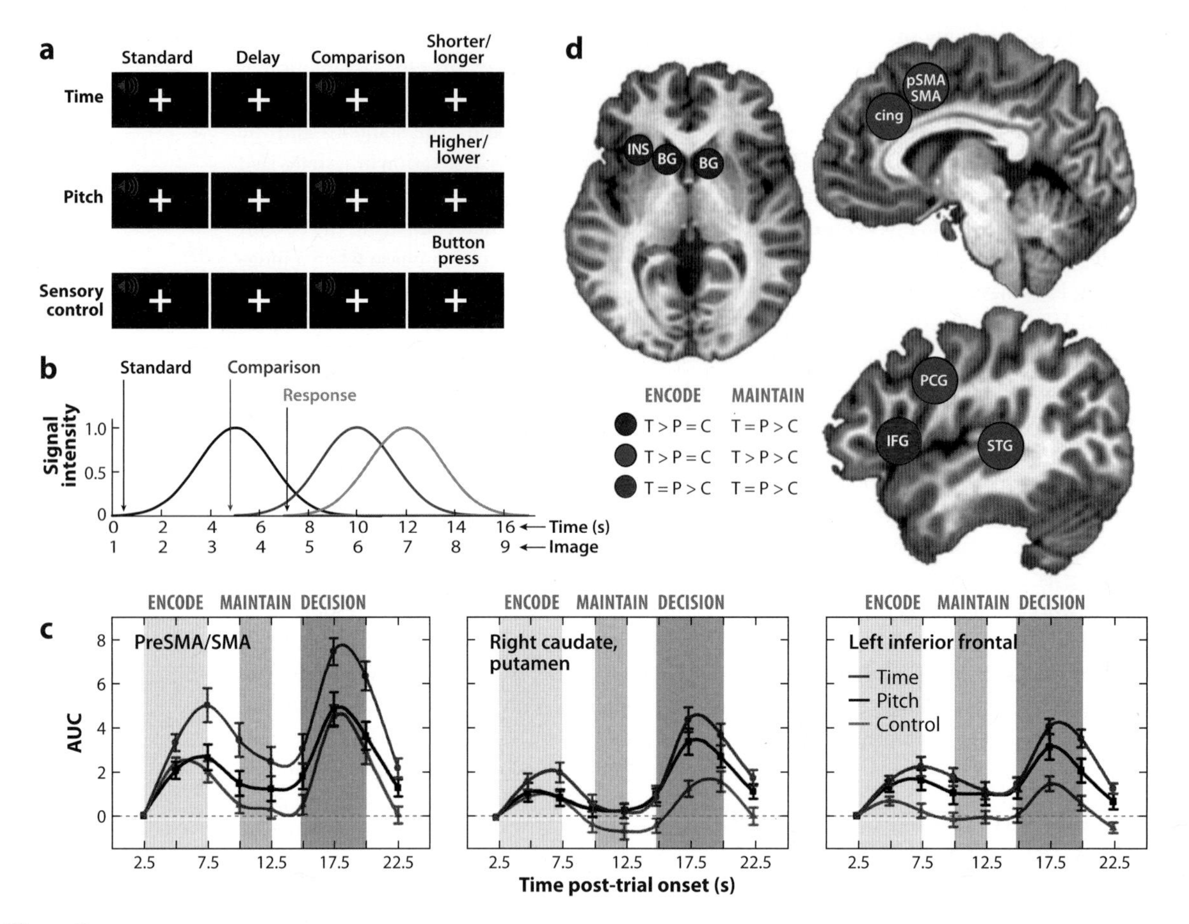

Figure 5

Functional significance of brain circuits that govern interval timing. (*a*) Illustration of the trial events in time-perception and control tasks. In the time-perception task, a standard and a comparison interval are successively presented and separated by a delay, which varies across studies. In the example, intervals are designated by tones. To control for brain activation related to cognitive and low-level sensorimotor processing, activation is compared with tasks composed of the same trial events but with different processing requirements. In the pitch perception task, the frequency of the two tones is discriminated. In the sensory control task, a button press is made following the presentation of the tones. A fixation cross remains on the screen during imaging. (*b*) The three hypothetical time-course functions illustrate the expected hemodynamic response associated with encoding the standard interval (*black*), encoding the comparison interval (*red*), and making a response (*gray*). Arrows leading from each event designate their onset. The hemodynamic response peaks 4 to 6 s after event onset. An image of the entire brain is acquired every 2 s. (*c*) Time course of activation (area under the curve; AUC) for the time-perception (*red*), pitch perception (*black*), and control (*gray*) tasks in representative regions (adapted from Harrington et al. 2010). Gray boxes on the *x* axis denote the epochs used to calculate AUC for a trial's encode, maintain, and decision phases. (*d*) Illustration of regional activation patterns for a trial's encode and maintain phases in the time (T), pitch (P), and control (C) tasks (Harrington et al. 2010). Activation in purple regions was related to interval encoding ($T > P = C$ for encode; $T = P > C$ for maintain). Activation in blue regions was related to accumulation and maintenance of sensory time codes ($T > P = C$ for encode and maintain). Activation in the red region was related to inhibitory control ($T = P > C$ for encode and maintain). For all regions except the IFG, activation was greater for the time than for the pitch and control tasks in the decision phase. Abbreviations: BG, basal ganglia; cing, anterior cingulate; IFG, inferior frontal gyrus; INS, anterior insula; PCG, precentral gyrus; preSMA, presupplementary motor area; SMA, supplementary motor area; STG, superior temporal gyrus.

involved in the maintenance of temporal information could not be ascertained. To address this issue, Harrington et al. (2010) inserted 10-s and 12.5-s delays between the standard and comparison interval, thereby permitting a better separation of activation associated with a trial's encoding, maintenance, and decision periods, as illustrated in **Figure 5*c***. Pairwise subtractions of brain activation during time discrimination (T), pitch discrimination (P), and the sensorimotor control (C) tasks were conducted for each period. Task difficulty did not differ between the time and the pitch tasks. The fMRI results showed that the striatum's pattern of timing-related activation during the different epochs could be distinguished from that found for most other brain regions. Striatal (bilateral caudate and putamen) activation was greater for the timing task than for the two control tasks (T > P = C) during encoding but did not differ between the time and pitch tasks (T = P > C) during maintenance (**Figure 5*c*,*d***). Thus, timing-related activity was specific to interval encoding, consistent with the positive correlation of striatal activation and timing proficiency noted during this same period (Coull et al. 2008, Harrington et al. 2004). This result also concurs with the view that passive maintenance of working memory is controlled via recurrent excitation of the prefrontal cortex and sensory neurons, and active maintenance over longer periods involves recurrent thalamocortical activity (Hazy et al. 2006). The only other region exhibiting the same activation pattern as did the striatum was the anterior insula. This region is situated to integrate processing from disparate domains (e.g., interoception, emotion, and cognition), including time (Kosillo & Smith 2010; Wittmann et al. 2010a,b), via its dense interconnections with most association centers and the basal ganglia. Moreover, anterior insula connectivity with frontal cognitive-control centers suggests that it also assists in the perceptual analysis of sensory information (Eckert et al. 2009). In contrast, preSMA/SMA, precentral, and superior temporal activation was greatest for the timing task (T > P = C) during both encoding and maintenance periods (**Figure 5*c*,*d***). This finding suggests that the preSMA and SMA modulate nontemporal aspects of processing, consistent with reports that timing-related activation is lost in the face of more difficult control tasks (Livesey et al. 2007). One prospect is that the SMA and association areas support accumulation and maintenance of sensory information, which is more demanding for interval timing. Another important finding was that activation of the inferior frontal gyrus did not differ between the time and pitch tasks during all three periods of the trial, although it was greater than in the control condition (T = P > C) (**Figure 5*c*,*d***). The inferior frontal gyrus is an inhibitory control center with fiber tracts to preSMA. This pathway may be a route to control the accumulation of sensory information into preSMA. Last, during the trial's decision phase, activation in almost all regions including the striatum was greater during the timing task (T > P > C), perhaps signifying the more significant integration demands of duration than pitch processing. In addition, timing-related activation of the cerebellum and a classic frontoparietal executive network did not emerge until the decision phase (T > P > C). The latter finding supports the positive correlation of frontoparietal region activity with time-discrimination difficulty (Harrington et al. 2004).

Taken together, interval timing was governed by distributed brain networks that flexibly altered activation, depending on task demands. Although some have argued that much brain activation during temporal processing relates to cognitive or sensory processes, timing emerges from the communication among brain regions rather than from processing in a single region. The use of control tasks can identify regional activity that is more dominant during timing; however, sensory and cognitive centers that are vital for interval timing can be overlooked. For example, using the subtractive method has led some to conclude that the inferior parietal cortex is not a key element of interval encoding (Coull et al. 2008, Harrington et al. 2010, Livesey et al. 2007). Yet applying repetitive transcranial magnetic stimulation (rTMS) to the supramarginal gyrus of the right hemisphere dilated perceived duration owing to its effect on interval encoding (Wiener et al. 2010a, 2012). This result indicates that the right supramarginal gyrus is a critical element of the neural circuitry that encodes time, which corroborates the detrimental effect of right parietal damage on time perception (Harrington et al. 1998b). Furthermore, as we begin to explore how the brain construes time under circumstances that alter the resolution of perceived duration, interactions among larger-scale brain networks will likely be considerably important.

ILLUSIONS OF TIME

An area of research that has potential to enrich our understanding of timing networks concerns illusions of time, which are important because they reveal how the brain normally organizes and interprets information depending on internal states, past experiences, or properties of stimuli. Psychophysical studies have long reported that the experience of time is not isomorphic to physical time, but rather depends on many factors. For example, emotionally aversive events are perceived as lasting longer than their physical duration (Cheng et al. 2008a, Droit-Volet & Meck 2007). Larger magnitude, more complex, or intense stimuli also expand perceived duration, whereas repeated, high-probability, and nonsalient stimuli compress time (Eagleman 2008, van Wassenhove et al. 2008). The mechanisms of temporal illusions continue to be debated, but the consensus indicates that attention and arousal are key factors. In pacemaker-counter models, attention and arousal are thought to speed up or slow down timing by closing or opening a switch that allows pulses generated from a "clock" to be accumulated and counted (Buhusi & Meck 2009a, Ulrich et al. 2006). For instance, heightened levels of physiological arousal induced by psychologically negative sounds expand subjective duration (Mella et al. 2011). Reducing the level of attention devoted to timing compresses perceived duration and attenuates activation in the CTBG timing-related circuit, but also in the frontal, temporal, and parietal cortices (Coull et al. 2004, 2011).

Despite this impressive body of work, little is understood about brain mechanisms that bring about temporal illusions. Emerging neuroimaging research suggests that the mechanisms may be partially context dependent. One fMRI study investigated the neural mechanisms of time dilation produced by a looming visual stimulus (Wittmann et al. 2010b), which captures attention possibly because it signals a potential intrinsic threat to organisms. Participants viewed a series of five discs; all but the fourth disc, which was a looming or receding disc, were static, and participants judged whether the fourth stimulus was longer or shorter than the others. Relative to the receding control condition, activation was greater in rostral-medial frontal areas and medial-posterior cortices, including the posterior cingulate. These areas are tightly interconnected with the limbic system, which governs a variety of functions including emotion and motivation processing. The results were attributed to the arousing effect of looming signals, which have an inherently emotional component.

Another study investigated mechanisms of time dilation produced by emotionally aversive stimuli (Dirnberger et al. 2012). Participants judged which of two pictures was displayed for a longer or shorter amount of time. Perceived duration was dilated when one of the pictures was aversive (aversive-neutral) relative to a control condition, in which both pictures were neutral. On a subsequent recognition test, overestimation of time was associated with better memory of a picture, indicating that time dilation enhanced memory encoding. A region-of-interest analysis revealed that activation of rostral-medial frontal areas (superior frontal, preSMA/SMA) and lateral inferior frontal cortex was greater for aversive-neutral than control pairs. Brain activation was further distinguished by the accuracy of time discriminations, whereby right amygdala, anterior insula, and putamen activation was greater on trials in which time was overestimated than on correct trials. Thus, when time was dilated, the limbic system (amygdala) and tightly interconnected regions, notably the anterior insula and medial frontal cortex, were more engaged. Taken together, both studies suggest that time dilation via stimuli that have an emotionally threatening connotation is partly brought about by heightened activity in elements of the limbic system and interconnecting medial cortical areas.

Temporal distortions also emerge in contexts that have no emotional overtone, such as the illusion that auditory signals are perceived as lasting longer than visual signals of the same physical duration when they are compared. A recent fMRI study sought to investigate the neural basis of the illusion (Harrington et al. 2011a), which is of interest because it may elucidate how synchrony is maintained across the senses. The audiovisual effect on perceived duration has been attributed to a pacemaker-accumulator system that runs faster for auditory than for visual signals, possibly owing to an attentional switch that permits a faster accumulation of pulses (Lustig & Meck 2011). In this study, participants judged whether an auditory (A) or visual (V) comparison interval was longer or shorter than a standard interval, which was either of the same or a different modality (**Figure 6*a***). Time was dilated relative to all other conditions when the duration of an A comparison interval was judged relative to a V standard (V-A), and time was compressed when the duration of a V comparison interval was judged relative to an A standard (A-V) (**Figure 6*b***). Regional analyses of brain activation showed that audiovisual distortions were governed by frontal cognitive-control centers (preSMA, middle/inferior frontal), where activation was greater when time was compressed, and higher association centers (superior temporal cortex, posterior insula, middle occipital cortex), where the level of activation was driven by the modality of the comparison interval (**Figure 6*c***). Although this study identified regional activation differences between time dilation and compression, timing emerges from communication among brain networks, to which conventional regional analyses of activation are insensitive. As such, the effective connectivity of these regions was examined to determine if the strength of their interactions with other brain regions differed between the time dilation and compression conditions. Effective connectivity was not found for frontal cortical areas, possibly because, as supramodal control centers, they flexibly direct attention and executive resources, though more so when time discriminations are demanding (A-V). Rather, connectivity of bilateral association areas with frontoparietal and temporal cortices and the striatum was typically stronger when perceived duration was dilated than when compressed (**Figure 6*d***). This result may be due to the salience of auditory signals in the context of timing (Repp & Penel 2002). This attention-based explanation would cause "clock pulses" to accumulate faster (pacemaker/accumulator models) or perhaps increase cortico-cortical oscillatory frequencies (Allman & Meck 2012), thereby producing an overestimation of time for auditory signals. Altogether, the finding reveals a basic principal of functional organization that produces distortions in the experience of time.

Notably, regional activation of the striatum did not differ between the time dilation and compression conditions (Harrington et al. 2011a). Although this result suggests that the integration of cortical oscillatory states by the striatum may not be faster for auditory than for visual signals, it leaves open the question of whether the strength of striatal connectivity is modulated by time dilation and compression (Matell & Meck 2004, N'Diaye et al. 2004). This area of inquiry is important for future research because measures of brain connectivity can be more sensitive to effects of psychological variables than are conventional regional analyses of activation. Indeed, it appears that the striatum may influence time dilation in emotionally aversive contexts (Dirnberger et al. 2012).

INTERSENSORY TIMING

The synthesis of temporal information across the senses is essential for perception and

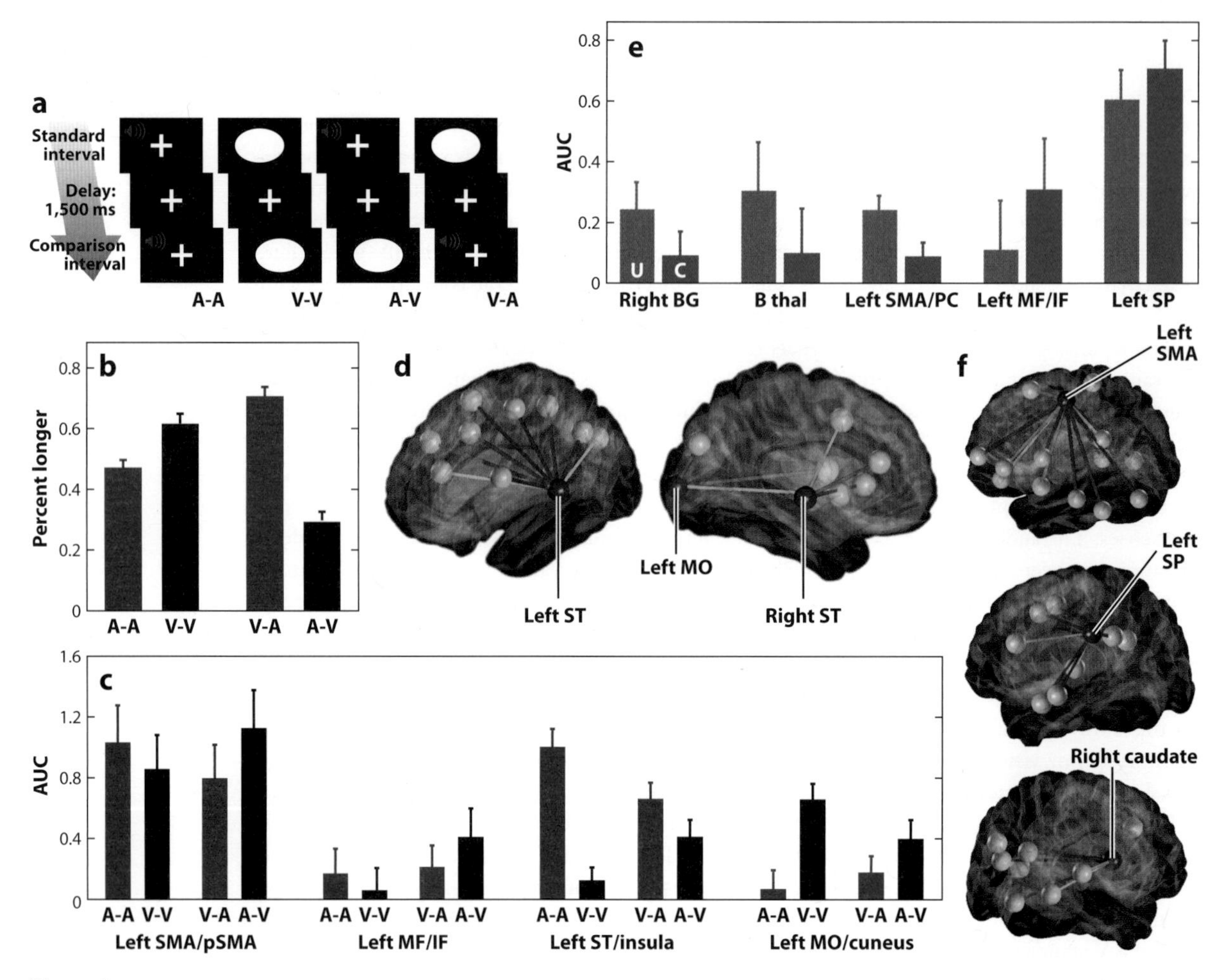

Figure 6

Brain networks that govern distortions in time and intersensory timing. The content contained in this figure is adapted from Harrington et al. (2011a). (*a*) Trial events for four conditions of a time-perception task. Pairs of auditory (A) and visual (V) stimuli were successively presented. The standard interval and comparison interval were of the same modality in the unimodal conditions (A-A, V-V) and were different in the cross-modal conditions (A-V, V-A). The four standard durations (1467, 1540, 1620, and 1710 ms) were each paired with three shorter and three longer comparison durations that were $\pm7\%$ increments of each standard interval. A fixation cross remained on the screen during imaging. (*b*) Mean percent longer responses. Relative to all other conditions, time was significantly dilated for the V-A condition and compressed for the A-V condition. (*c*, *d*) Results from analysis of time dilation and compression effects on brain activity. (*c*) Mean area under the curve (AUC) in representative regions showing an interaction of comparison interval modality $\times$ timing condition (unimodal versus cross-modal). Horizontal bars denote significant differences between conditions. (*d*) Connectivity map of the left and right superior temporal cortex and left middle occipital cortex with representative regions. Effective connectivity of the bilateral superior temporal cortex and insula and the left middle occipital cortex was typically stronger for the time dilation than for the compression condition. (*e*, *f*) Results from analyses of brain systems that govern intersensory timing. (*e*) Mean AUC in representative regions showing differences in activation between unimodal and cross-modal timing. (*f*) Connectivity maps of left SMA, left superior parietal cortex, and right caudate with representative regions. Effective connectivity for all regions was stronger for cross-modal than for unimodal timing. Abbreviations: BG, basal ganglia (caudate, putamen); IF, inferior frontal; MF, middle frontal; MO, middle occipital cortex; PC, precentral cortex; preSMA, presupplementary motor area; SMA, supplementary motor area; SP, superior parietal; ST, superior temporal; Thal, thalamus.

cognition, yet little is understood about how the brain maintains temporal synchrony among modalities. The striatum may be central to governing intersensory timing because it is involved in multisensory integration (Nagy et al. 2006) and is thought to integrate cortical oscillatory activity that comprises the clock signal (Coull et al. 2011). These proposals agree with a report that the bilateral striatum and also the thalamus and SMA exhibit greater activation during unimodal as compared with cross-modal timing (Harrington et al. 2011a) (**Figure 6*e***). The result was not compatible with an attention-switching model of striatal function (van Schouwenburg et al. 2010), which would predict greater activation in the cross-modal than in the unimodal condition. The activation pattern may develop if intersensory integration of time codes is less stable or noisier relative to intrasensory timing. By comparison, activation of a classic frontoparietal working-memory network was greater for cross-modal than for unimodal timing, likely because of greater attention and executive demands of intersensory timing. The effective connectivity of these regions was then examined to determine if the strength of their interactions with other brain regions differed between the timing conditions. All regions showed stronger connectivity for cross-modal than for unimodal timing, perhaps owing to the more effortful demands of multimodal temporal integration. **Figure 6*f*** illustrates that caudate, SMA, and superior parietal connectivity was stronger with frontal cognitive-control centers, association centers, visual areas, and a memory system (precuneus, posterior cingulate, parahippocampus). Thus, the synthesis of audiovisual time codes in core timing and attention networks involves interactions with extensive brain networks.

STRIATAL BEAT-FREQUENCY MODEL OF INTERVAL TIMING

The inherent interactive nature of timing networks revealed by the above studies is embodied by the striatal beat-frequency (SBF) model of interval timing (Allman & Meck 2012, Matell & Meck 2004, Oprisan & Buhusi 2011, van Rijn et al. 2011). The SBF model uses medium spiny neurons located in the dorsal striatum, which is typically thought to be involved in decision making and executive function. Each spiny neuron receives ~30,000 inputs from cortical neurons, and it is this level of convergence (many to one) that allows the medium spiny neurons to serve as coincidence detectors of activity patterns engaged in by the cortical neurons. One easily detectable activity pattern is the oscillatory firing pattern of cortical neurons that is typically synchronized to the onset of relevant stimuli by DA release from the substantia nigra pars compacta (Jahanshahi et al. 2006). Given this initial synchronization (i.e., temporary alignment of the downbeat of neural firing), the subsequent evolution of neural firing will reflect the inherent rhythmical structure of each neuron's tendency to fire as well as random drift or desynchronization of firing among individual neurons. Despite the variability in this pattern of neural firing, which grows as a function of the time since the initial synchronization, the medium spiny neurons can still detect different patterns of neural firing on the basis of the high degree of redundancy in the system due to the convergence of 30,000 inputs. This coincident detection involves the ability of medium spiny neurons to sense temporal patterns of simultaneous activity across their spatially arranged receptive fields. Individual synapses within these receptive fields are trained to detect and respond to specific patterns of oscillatory input on the basis of previous experience and the influence of long-term potentiation and depression—two well-known neurobiological mechanisms for the encoding of event durations (Matell & Meck 2000, 2004).

SBF: striatal beat-frequency (model)

Each time period from milliseconds to seconds or minutes to hours will be reflected by different patterns of neural activity that can be repeatedly reproduced as long as the initial synchronization retains its efficacy and the pattern is not reset or interrupted by subsequent stimulus onsets affecting that particular set of detectors. Multiple durations can be timed simultaneously by assuming multiple detectors or timers (i.e., spiny neurons) within the striatum that display chronotopy, a preference for particular ranges of durations. The readout of such a timing system is provided by frontal cortex monitoring of the firing activity of this chronotopically arranged time line—thereby completing the CTBG circuit. In this respect, the SBF model is an important advancement because of its veracity at both behavioral and physiological levels. Previous timing models either provided a good description of timing behaviors, but contained components that were inconsistent with the properties of the brain structures involved, or were neurobiologically feasible, but made inaccurate behavioral predictions. The computational version of the SBF model as described by Matell & Meck (2004) is constructed such that its mechanisms are consistent with neural regions thought to be involved in timing (e.g., frontal cortex and striatum), and its output is consistent with physiological recordings and behavioral results from interval-timing experiments (Matell et al. 2003, 2011; Meck et al. 2012). Indeed, striatal neurons fire in a peak-shaped manner centered on the target duration, following the predictions of the SBF model. Critically, the SBF model reproduces the scalar property, the hallmark of interval timing (Gibbon et al. 1984, 1997; Van Rijn et al. 2013). Preservation of this scalar property is critical for experimental manipulations thought to influence the timing system per se (i.e., deviations are considered diagnostic of effects on other systems that influence behavior). In the SBF model, the scalar property occurs because of variability in the firing patterns of striatal neurons and because cortical activity is assumed to be oscillatory, such that firing patterns at the harmonics (i.e., 1/2, 2/3, etc.) are similar but not identical to those occurring at the target duration. Teki et al. (2012) recently proposed a unified model of interval timing based on coordinated activity in the core striatal and ancillary olivocerebellar networks that takes advantage of the interconnections between these networks and the cerebral cortex. Timing in this model posits a type of degeneracy (Lewis & Meck 2012) that involves the initiation and maintenance of timing by beat-based striatal activation that is adjusted by olivocerebellar mechanisms that can substitute for the striatal timing system as a function of neural deactivation if needed (Allman & Meck 2012, Jahanshahi et al. 2010a) and/or by genetic modifications of neurotransmitter/receptor function (Liao & Cheng 2005, Sysoeva et al. 2010, Wiener et al. 2011a) and aging (Balci et al. 2009; Cheng et al. 2008b, 2011a).

CONCLUSIONS

Recent research concurs with modern neurophysiological models whereby the capacity to perceive and estimate time is thought to emerge from interactions of a core CTBG timing circuit with brain regions that provide signals needed to time events (Allman & Meck 2012, Lustig & Meck 2005). Important advances from cell recordings further indicate that elements of the CTBG not only display chronotopy, but also represent organizational features of the context that permits more abstract timing behaviors (Merchant et al. 2011). At the neuroanatomical level, basal ganglia and SMA functioning were dissociated by differential activity that was respectively linked to fluctuations in the task's interval timing and working-memory demands (Harrington et al. 2010). Thus, elements of the CTBG timing circuit display different context-dependent activation dynamics that warrant further inquiry. The functional significance of networks engaged by timing will also be advanced by studies of how the brain construes time in situations that influence the resolution of perceived duration. Emerging research indicates that temporal distortions produced by emotionally charged events or stimuli that capture attention are partly brought about by activity in emotion or association networks (Dirnberger et al. 2012, Cheng et al. 2011b, Harrington et al. 2011a, Wittmann et al. 2010b). Although the role of the striatum remains debated in these studies, temporal distortions that emanate from intersensory timing are driven by the CTBG circuit. Likewise, degeneracy in timing revealed by disease (Jahanshahi et al. 2010a) and individual differences in the expression of neurotransmitter function (Sysoeva et al. 2010, Wiener et al. 2011a) also hold promise for uncovering neurophysiological mechanisms of timing networks. More generally, it is important for future research to study brain connectivity, which more fully characterizes the communication of timing circuits with other brain networks (Cheng et al. 2011b; Harrington et al. 2011a,b; MacDonald & Meck 2004).

DISCLOSURE STATEMENT

The authors are not aware of any affiliations, memberships, funding, or financial holdings that might be perceived as affecting the objectivity of this review.

LITERATURE CITED

Agostino PV, Golombek DA, Meck WH. 2011. Unwinding the molecular basis of interval and circadian timing. *Front. Integr. Neurosci.* 5:64

Allman MJ, Meck WH. 2012. Pathophysiological distortions in time perception and timed performance. *Brain* 135:656–77

Allman MJ, Pelphrey KA, Meck WH. 2012. Developmental neuroscience of time and number: implications for autism and other neurodevelopmental disabilities. *Front. Integr. Neurosci.* 6:7

Aparicio P, Diedrichsen J, Ivry RB. 2005. Effects of focal basal ganglia lesions on timing and force control. *Brain Cogn.* 58:62–74

Artieda J, Pastor MA, Lacruz F, Obeso JA. 1992. Temporal discrimination is abnormal in Parkinson's disease. *Brain* 115:199–210

Balci F, Meck WH, Moore H, Brunner D. 2009. Timing deficits in aging and neuropathology. In *Animal Models of Human Cognitive Aging*, ed. JL Bizon, A Woods, pp. 161–201. Totowa, NJ: Humana

Balci F, Wiener M, Çavdaroğlu B, Coslett HB. 2013. Epistasis effects of dopamine genes on interval timing and reward magnitude in humans. *Neuropsychologia* 51:293–308

Bartolo R, Merchant H. 2009. Learning and generalization of time production in humans: rules of transfer across modalities and interval durations. *Exp. Brain Res.* 197:91–100

Brannon EM, Libertus ME, Meck WH, Woldorff MG. 2008. Electrophysiological measures of time processing in infant and adult brains: Weber's law holds. *J. Cogn. Neurosci.* 20:193–203

Brody CD, Hernández A, Zainos A, Romo R. 2003. Timing and neural encoding of somatosensory parametric working memory in macaque prefrontal cortex. *Cereb. Cortex* 13:1196–207

Buhusi CV, Meck WH. 2005. What makes us tick? Functional and neural mechanisms of interval timing. *Nat. Rev. Neurosci.* 6:755–65

Buhusi CV, Meck WH. 2009a. Relative time sharing: new findings and an extension of the resource allocation model of temporal processing. *Philos. Trans. R. Soc. Lond. B* 364:1875–85

Buhusi CV, Meck WH. 2009b. Relativity theory and time perception: single or multiple clocks? *PLoS ONE* 4(7):e6268

Buonomano DV, Laje R. 2010. Population clocks: motor timing with neural dynamics. *Trends Cogn. Sci.* 14:520–27

Burr D, Tozzi A, Morrone M. 2007. Neural mechanisms for timing visual events are spatially selective in real-world coordinates. *Nat. Neurosci.* 10:423–25

Cheng RK, Ali YM, Meck WH. 2007. Ketamine "unlocks" the reduced clock-speed effect of cocaine following extended training: evidence for dopamine-glutamate interactions in timing and time perception. *Neurobiol. Learn. Mem.* 88:149–59

Cheng RK, Dyke AG, McConnell MW, Meck WH. 2011a. Categorical scaling of duration as a function of temporal context in aged rats. *Brain Res.* 1381:175–86

Cheng RK, Jesuthasan S, Penney TB. 2011b. Time for zebrafish. *Front. Integr. Neurosci.* 5:40

Cheng RK, MacDonald CJ, Williams CL, Meck WH. 2008a. Prenatal choline supplementation alters the timing, emotion, and memory performance (TEMP) of adult male and female rats as indexed by differential reinforcement of low-rate schedule behavior. *Learn. Mem.* 15:153–62

Cheng RK, Meck WH. 2007. Prenatal choline supplementation increases sensitivity to time by reducing non-scalar sources of variance in adult temporal processing. *Brain Res.* 1186:242–54

Cheng RK, Scott AC, Penney TB, Williams CL, Meck WH. 2008b. Prenatal choline availability differentially modulates timing of auditory and visual stimuli in aged rats. *Brain Res.* 1237:167–75

Coslett HB, Wiener M, Chatterjee A. 2010. Dissociable neural systems for timing: evidence from subjects with basal ganglia lesions. *PLoS ONE* 5(4):e10324

Coull JT, Cheng RK, Meck WH. 2011. Neuroanatomical and neurochemical substrates of timing. *Neuropsychopharmacology* 36:3–25

Coull JT, Hwang HJ, Leyton M, Dagher A. 2012. Dopamine precursor depletion impairs timing in healthy volunteers by attenuating activity in putamen and SMA. *J. Neurosci.* 32:16704–15

Coull JT, Nazarian B, Vidal F. 2008. Timing, storage, and comparison of stimulus duration engage discrete anatomical components of a perceptual timing network. *J. Cogn. Neurosci.* 20:2185–97

Coull JT, Vidal F, Nazarian B, Macar F. 2004. Functional anatomy of the attentional modulation of time estimation. *Science* 303:1506–8

Dirnberger G, Hesselmann G, Roiser JP, Preminger S, Jahanshahi M, Paz R. 2012. Give it time: neural evidence for distorted time perception and enhanced memory encoding in emotional situations. *NeuroImage* 63:591–99

Droit-Volet S, Meck WH. 2007. How emotions colour our perception of time. *Trends Cogn. Sci.* 11:504–13

Eagleman DM. 2008. Human time perception and its illusions. *Curr. Opin. Neurobiol.* 18:131–36

Eckert MA, Menon V, Walczak A, Ahlstrom J, Denslow S, et al. 2009. At the heart of the ventral attention system: the right anterior insula. *Hum. Brain Mapp.* 30:2530–41

Fraisse P. 1984. Perception and estimation of time. *Annu. Rev. Psychol.* 35:1–36

Genovesio A, Tsujimoto S, Wise SP. 2006. Neuronal activity related to elapsed time in prefrontal cortex. *J. Neurophysiol.* 95:3281–85

Genovesio A, Tsujimoto S, Wise SP. 2009. Feature- and order-based timing representations in the frontal cortex. *Neuron* 63:254–66

Gibbon J, Church RM, Meck WH. 1984. Scalar timing in memory. *Ann. NY Acad. Sci.* 423:52–77

Gibbon J, Malapani C, Dale CL, Gallistel CR. 1997. Toward a neurobiology of temporal cognition: advances and challenges. *Curr. Opin. Neurobiol.* 7:170–84

Gu BM, Cheng RK, Yin B, Meck WH. 2011. Quinpirole-induced sensitization to noisy/sparse periodic input: temporal synchronization as a component of obsessive-compulsive disorder. *Neuroscience* 179:143–50

Gu BM, Jurkowski AJ, Lake JI, Malapani C, Meck WH. 2013. Bayesian models of interval timing and distortions in temporal memory as a function of Parkinson's disease and dopamine-related error processing. In *Time Distortions in Mind: Temporal Processing in Clinical Populations*, ed. A Vatakis, MJ Allman. Boston, MA: Brill. In press

Gu BM, Meck WH. 2011. New perspectives on Vierordt's law: memory-mixing in ordinal temporal comparison tasks. *Lect. Notes Comp. Sci.* 6789:67–78

Harrington DL, Boyd LA, Mayer AR, Sheltraw DM, Lee RR, et al. 2004. Neural representation of interval encoding and decision making. *Cogn. Brain Res.* 21:193–205

Harrington DL, Castillo GN, Fong CH, Reed JD. 2011a. Neural underpinnings of distortions in the experience of time across senses. *Front. Integr. Neurosci.* 5:32

Harrington DL, Castillo GN, Greenberg PA, Song DD, Lessig S, et al. 2011b. Neurobehavioral mechanisms of temporal processing deficits in Parkinson's disease. *PLoS ONE* 6:e17461

Harrington DL, Haaland KY, Hermanowicz N. 1998a. Temporal processing in the basal ganglia. *Neuropsychology* 12:3–12

Harrington DL, Haaland KY, Knight RT. 1998b. Cortical networks underlying mechanisms of time perception. *J. Neurosci.* 18:1085–95

Harrington DL, Zimbelman JL, Hinton SC, Rao SM. 2010. Neural modulation of temporal encoding, maintenance, and decision processes. *Cereb. Cortex* 20:1274–85

Hazy TE, Frank MJ, O'Reilly RC. 2006. Banishing the homunculus: making working memory work. *Neuroscience* 139:105–18

Hinton SC, Meck WH. 1997. How time flies: function and neural mechanisms of interval timing. *Adv. Psychol.* 120:409–57

Hinton SC, Meck WH. 2004. Frontal-striatal circuitry activated by human peak-interval timing in the supra-seconds range. *Cogn. Brain Res.* 21:171–82

Höhn S, Dallérac G, Faure A, Urbach Y, Nguyen HP, et al. 2011. Behavioral and in vivo electrophysiological evidence for presymptomatic alteration of prefronto-striatal processing in the transgenic rat model for Huntington disease. *J. Neurosci.* 31:8986–97

Ivry RB, Hazeltine RE. 1995. Perception and production of temporal intervals across a range of durations: evidence of a common timing mechanism. *J. Exp. Psychol. Hum. Percept. Perform.* 21:3–18

Jahanshahi M, Jones CR, Dirnberger G, Frith CD. 2006. The substantia nigra pars compacta and temporal processing. *J. Neurosci.* 26:12266–73

Jahanshahi M, Jones CR, Zijlmans J, Katsenschlager R, Lee L, et al. 2010a. Dopaminergic modulation of striato-frontal connectivity during motor timing in Parkinson's disease. *Brain* 133:727–45

Jahanshahi M, Wilkinson L, Gahir H, Dharminda A, Lagnado DA. 2010b. Medication impairs probabilistic classification learning in Parkinson's disease. *Neuropsychologia* 48:1096–103

Jin DZ, Fujii N, Graybiel AM. 2009. Neural representation of time in cortico-basal ganglia circuits. *Proc. Natl. Acad. Sci. USA* 106:19156–61

Johnson HA, Goel A, Buonomano DV. 2010. Neural dynamics of in vitro cortical networks reflects experienced temporal patterns. *Nat. Neurosci.* 13:917–19

Johnston A, Arnold DH, Nishida S. 2006. Spatially localized distortions of event time. *Curr. Biol.* 16:472–79

Jones CR, Malone TJ, Dirnberger G, Edwards M, Jahanshahi M. 2008. Basal ganglia, dopamine and temporal processing: performance on three timing tasks on and off medication in Parkinson's disease. *Brain Cogn.* 68:30–41

Jones CRG, Jahanshahi M. 2011. Dopamine modulates striato-frontal functioning during temporal processing. *Front. Integr. Neurosci.* 5:70

Karmarkar UR, Buonomano DV. 2003. Temporal specificity of perceptual learning in an auditory discrimination task. *Learn. Mem.* 10:141–47

Karmarkar UR, Buonomano DV. 2007. Timing in the absence of clocks: encoding time in neural network states. *Neuron* 53:427–38

Keele S, Nicoletti R, Ivry R, Pokorny R. 1985. Do perception and motor production share common timing mechanisms? A correlational analysis. *Acta Psychol.* 60:173–91

Koch G, Brusa L, Caltagirone C, Oliveri M, Peppe A, et al. 2004. Subthalamic deep brain stimulation improves time perception in Parkinson's disease. *NeuroReport* 15:1071–73

Koch G, Costa A, Brusa L, Peppe A, Gatto I, et al. 2008. Impaired reproduction of second but not millisecond time intervals in Parkinson's disease. *Neuropsychologia* 46:1305–13

Kosillo P, Smith AT. 2010. The role of the human anterior insular cortex in time processing. *Brain Struct. Funct.* 214:623–28

Kotz SA, Schwartze M. 2011. Differential input of the supplementary motor area to a dedicated temporal processing network: functional and clinical implications. *Front. Integr. Neurosci.* 5:86

Lake JI, Meck WH. 2012. Differential effects of amphetamine and haloperidol on temporal reproduction: dopaminergic regulation of attention and clock speed. *Neuropsychologia* 51:284–92

Lebedev MA, O'Doherty JE, Nicolelis MA. 2008. Decoding of temporal intervals from cortical ensemble activity. *J. Neurophysiol.* 99:166–86

Leon MI, Shadlen MN. 2003. Representation of time by neurons in the posterior parietal cortex of the macaque. *Neuron* 38:317–27

Lewis PA, Meck WH. 2012. Time and the sleeping brain. *Psychologist* 25:594–97

Lewis SJ, Dove A, Robbins TW, Barker RA, Owen AM. 2004. Striatal contributions to working memory: a functional magnetic resonance imaging study in humans. *Eur. J. Neurosci.* 19:755–60

Liao RM, Cheng RK. 2005. Acute effects of d-amphetamine on the differential reinforcement of low-rate (DRL) schedule behavior in the rat: comparison with selective dopamine receptor antagonists. *Chin. J. Physiol.* 48:41–50

Livesey AC, Wall MB, Smith AT. 2007. Time perception: manipulation of task difficulty dissociates clock functions from other cognitive demands. *Neuropsychologia* 45:321–31

Lucchetti C, Bon L. 2001. Time-modulated neuronal activity in the premotor cortex of macaque monkeys. *Exp. Brain Res.* 141:254–60

Lustig C, Matell MS, Meck WH. 2005. Not "just" a coincidence: frontal-striatal synchronization in working memory and interval timing. *Memory* 13:441–48

Lustig C, Meck WH. 2005. Chronic treatment with haloperidol induces working memory deficits in feedback effects of interval timing. *Brain Cogn.* 58:9–16

Lustig C, Meck WH. 2011. Modality differences in timing and temporal memory throughout the lifespan. *Brain Cogn.* 77:298–303

MacDonald CJ, Cheng RK, Meck WH. 2012. Acquisition of "Start" and "Stop" response thresholds in peak-interval timing is differentially sensitive to protein synthesis inhibition in the dorsal and ventral striatum. *Front. Integr. Neurosci.* 6:10

MacDonald CJ, Meck WH. 2004. Systems-level integration of interval timing and reaction time. *Neurosci. Biobehav. Rev.* 28:747–69

Mason PH. 2010. Degeneracy at multiple levels of complexity. *Biol. Theory* 5:277–88

Matell MS, Meck WH. 2000. Neuropsychological mechanisms of interval timing behaviour. *BioEssays* 22:94–103

Matell MS, Meck WH. 2004. Cortico-striatal circuits and interval timing: coincidence detection of oscillatory processes. *Cogn. Brain Res.* 21:139–70

Matell MS, Meck WH, Nicolelis MAL. 2003. Interval timing and the encoding of signal duration by ensembles of cortical and striatal neurons. *Behav. Neurosci.* 117:760–73

Matell MS, Shea-Brown E, Gooch C, Wilson AG, Rinzel J. 2011. A heterogeneous population code for elapsed time in rat medial agranular cortex. *Behav. Neurosci.* 125:54–73

Meck WH. 1996. Neuropharmacology of timing and time perception. *Cogn. Brain Res.* 3:227–42

Meck WH. 2003. *Functional and Neural Mechanisms of Interval Timing*. Boca Raton, FL: CRC Press

Meck WH. 2006a. Frontal cortex lesions eliminate the clock speed effect of dopaminergic drugs on interval timing. *Brain Res.* 1108:157–67

Meck WH. 2006b. Neuroanatomical localization of an internal clock: a functional link between mesolimbic, nigrostriatal, and mesocortical dopaminergic systems. *Brain Res.* 1109:93–107

Meck WH, Benson AM. 2002. Dissecting the brain's internal clock: how frontal-striatal circuitry keeps time and shifts attention. *Brain Cogn.* 48:195–211

Meck WH, Cheng RK, MacDonald CJ, Gainetdinov RR, Caron MG, Çevik MÖ. 2012. Gene-dose dependent effects of methamphetamine on interval timing in dopamine-transporter knockout mice. *Neuropharmacology* 62:1221–29

Meck WH, Church RM, Matell MS. 2013. Hippocampus, time, and memory—a retrospective analysis. *Behav. Neurosci.* In press

Meck WH, Church RM, Olton DS. 1984. Hippocampus, time, and memory. *Behav. Neurosci.* 98:3–22

Meck WH, Malapani C. 2004. Neuroimaging of interval timing. *Cogn. Brain Res.* 21:133–37

Meck WH, Penney TB, Pouthas V. 2008. Cortico-striatal representation of time in animals and humans. *Curr. Opin. Neurobiol.* 18:145–52

Meegan DV, Aslin RN, Jacobs RA. 2000. Motor timing learned without motor training. *Nat. Neurosci.* 3:860–62

Mella N, Conty L, Pouthas V. 2011. The role of physiological arousal in time perception: psychophysiological evidence from an emotion regulation paradigm. *Brain Cogn.* 75:182–87

Merchant H, Battaglia-Mayer A, Georgopoulos AP. 2004. Neural responses during interception of real and apparent circularly moving stimuli in motor cortex and area 7a. *Cereb. Cortex* 14:314–31

Merchant H, de Lafuente V, Peña-Ortega F, Larriva-Sahd J. 2012. Functional impact of interneuronal inhibition in the cerebral cortex of behaving animals. *Prog. Neurobiol.* 99:163–78

Merchant H, Luciana M, Hooper C, Majestic S, Tuite P. 2008a. Interval timing and Parkinson's disease: heterogeneity in temporal performance. *Exp. Brain Res.* 184:233–48

Merchant H, Pérez O, Zarco W, Gámez J. 2013. Interval tuning in the primate medial premotor cortex as a general timing mechanism. *J. Neurosci.* 33:9082–96

Merchant H, Zarco W, Bartolo R, Prado L. 2008b. The context of temporal processing is represented in the multidemensional relationships between timing tasks. *PLoS One* 3(9):e3169

Merchant H, Zarco W, Perez O, Prado L, Bartolo R. 2011. Measuring time with multiple neural chronometers during a synchronization-continuation task. *Proc. Natl. Acad. Sci. USA* 108:19784–89

Merchant H, Zarco W, Prado L. 2008c. Do we have a common mechanism for measuring time in the hundreds of millisecond range? Evidence from multiple timing tasks. *J. Neurophysiol.* 99:939–49

Mita A, Mushiake H, Shima K, Matsuzaka Y, Tanji J. 2009. Interval time coding by neurons in the presupplementary and supplementary motor areas. *Nat. Neurosci.* 12:502–7

Nagarajan SS, Blake DT, Wright BA, Byl N, Merzenich MM. 1998. Practice related improvements in somatosensory interval discrimination are temporally specific but generalize across skin location, hemisphere, and modality. *J. Neurosci.* 18:1559–70

Nagy A, Eordegh G, Paroczy Z, Markus Z, Benedek G. 2006. Multisensory integration in the basal ganglia. *Eur. J. Neurosci.* 24:917–24

N'Diaye K, Ragot R, Garnero L, Pouthas V. 2004. What is common to brain activity evokes by the perception of visual and auditory filled durations? A study with MEG and EEG co-recordings. *Cogn. Brain Res.* 21:250–68

Oprisan SA, Buhusi CV. 2011. Modeling pharmacological clock and memory patterns of interval timing in a striatal beat-frequency model with realistic, noisy neurons. *Front. Integr. Neurosci.* 5:52

Oshio K, Chiba A, Inase M. 2008. Temporal filtering by prefrontal neurons in duration discrimination. *Eur. J. Neurosci.* 28:2333–43

Paulsen JS, Zimbelman JL, Hinton SC, Langbehn DR, Leveroni CL, et al. 2004. fMRI biomarker of early neuronal dysfunction in presymptomatic Huntington's disease. *Am. J. Neuroradiol.* 25:1715–21

Penney TB, Gibbon J, Meck WH. 2008. Categorical scaling of duration bisection in pigeons (*Columba livia*), mice (*Mus musculus*), and humans (*Homo sapiens*). *Psychol. Sci.* 19:1103–9

Perez O, Kass R, Merchant H. 2013. Trial time warping to discriminate stimulus-related from movement-related neural activity. *J. Neurosci. Methods* 212:203–10

Perrett SP. 1998. Temporal discrimination in the cerebellar cortex during conditioned eyelid responses. *Exp. Brain Res.* 121:115–24

Pleil KE, Cordes S, Meck WH, Williams CL. 2011. Rapid and acute effects of estrogen on time perception in male and female rats. *Front. Integ. Neurosci.* 5:63

Price CJ, Friston KJ. 2002. Degeneracy and cognitive anatomy. *Trends Cogn. Sci.* 6:416–21

Rao SM, Mayer AR, Harrington DL. 2001. The evolution of brain activation during temporal processing. *Nat. Neurosci.* 4:317–23

Repp BH. 2005. Sensorimotor synchronization: a review of the tapping literature. *Psychon. Bull. Rev.* 12:969–92

Repp BH, Penel A. 2002. Auditory dominance in temporal processing: new evidence from synchronization with simultaneous visual and auditory sequences. *J. Exp. Psychol. Hum. Percept. Perform.* 28:1085–99

Rowe KC, Paulsen JS, Langbehn DR, Duff K, Beglinger LJ, et al. 2010. Self-paced timing detects and tracks change in prodromal Huntington disease. *Neuropsychology* 24:435–42

Schwartze M, Rothermich K, Kotz SA. 2012. Functional dissociation of pre-SMA and SMA-proper in temporal processing. *NeuroImage* 60:290–98

Smith JG, Harper DN, Gittings D, Abernethy D. 2007. The effect of Parkinson's disease on time estimation as a function of stimulus duration range and modality. *Brain Cogn.* 64:130–43

Spencer RM, Ivry RB. 2005. Comparison of patients with Parkinson's disease or cerebellar lesions in the production of periodic movements involving event-based or emergent timing. *Brain Cogn.* 58:84–93

Sysoeva OV, Tonevitsky AG, Wackermann J. 2010. Genetic determinants of time perception mediated by the serotonergic system. *PLoS ONE* 5(9):e12650

Tanaka M. 2007. Cognitive signals in the primate motor thalamus predict saccade timing. *J. Neurosci.* 27:12109–18

Tanji J. 2001. Sequential organization of multiple movements: involvement of cortical motor areas. *Annu. Rev. Neurosci.* 24:631–51

Teki S, Grube M, Griffiths TD. 2012. A unified model of time perception accounts for duration-based and beat-based timing mechanisms. *Front. Integr. Neurosci.* 5:90

Ulrich R, Nitschke J, Rammsayer T. 2006. Crossmodal temporal discrimination: assessing the predictions of a general pacemaker-counter model. *Percept. Psychophys.* 68:1140–52

van Rijn H, Gu BM, Meck WH. 2013. Dedicated clock/timing-circuit theories of interval timing. In *Neurobiology of Interval Timing*, ed. H Merchant, V de Lafuente. New York: Springer-Verlag. In press van Rijn H, Kononowicz TW, Meck WH, Ng KK, Penney TB. 2011. Contingent negative variation and its relation to time estimation: a theoretical evaluation. *Front. Integr. Neurosci.* 5:91

van Schouwenburg MR, den Ouden HE, Cools R. 2010. The human basal ganglia modulate frontal-posterior connectivity during attention shifting. *J. Neurosci.* 30:9910–18

van Wassenhove V, Buonomano DV, Shimojo S, Shams L. 2008. Distortions of subjective time perception within and across senses. *PLoS ONE* 3:e1437

van Wassenhove V, Nagarajan SS. 2007. Auditory cortical plasticity in learning to discriminate modulation rate. *J. Neurosci.* 27:2663–72

Wearden JH, Smith-Spark JH, Cousins R, Edelstyn NM, Cody FW, O'Boyle DJ. 2008. Stimulus timing by people with Parkinson's disease. *Brain Cogn.* 67:264–79

Wiener M, Hamilton R, Turkeltaub P, Matell MS, Coslett HB. 2010a. Fast forward: Supramarginal gyrus stimulation alters time measurement. *J. Cogn. Neurosci.* 22:23–31

Wiener M, Kliot D, Turkeltaub PE, Hamilton RH, Wolk DA, Coslett HB. 2012. Parietal influence on temporal encoding indexed by simultaneous transcranial magnetic stimulation and electroencephalography. *J. Neurosci.* 32:12258–67

Wiener M, Lohoff FW, Coslett HB. 2011a. Double dissociation of dopamine genes and timing in humans. *J. Cogn. Neurosci.* 23:2811–21

Wiener M, Matell MS, Coslett HB. 2011b. Multiple mechanisms for temporal processing. *Front. Integr. Neurosci.* 5:31

Wiener M, Turkeltaub P, Coslett HB. 2010c. The image of time: a voxel-wise meta-analysis. *NeuroImage* 49:1728–40

Wittmann M, Simmons AN, Aron JL, Paulus MP. 2010a. Accumulation of neural activity in the posterior insula encodes the passage of time. *Neuropsychologia* 48:3110–20

Wittmann M, van Wassenhove V, Craig AD, Paulus MP. 2010b. The neural substrates of subjective time dilation. *Front. Hum. Neurosci.* 4:2

Yin B, Troger AB. 2011. Exploring the 4th dimension: hippocampus, time, and memory revisited. *Front. Integr. Neurosci.* 5:36

Yu H, Sternad D, Corcos DM, Vaillancourt DE. 2007. Role of hyperactive cerebellum and motor cortex in Parkinson's disease. *NeuroImage* 35:222–33

Zarco W, Merchant H, Prado L, Mendez JC. 2009. Subsecond timing in primates: comparison of interval production between human subjects and Rhesus monkeys. *J. Neurophysiol.* 102:3191–202

Zelaznik HN, Spencer RMC, Ivry RB. 2002. Dissociation of explicit and implicit timing in repetitive tapping and drawing movements. *J. Exp. Psychol: Hum. Percept. Perform.* 28:575–88

Zimbelman JL, Paulsen JS, Mikos A, Reynolds NC, Hoffmann RG, Rao SM. 2007. fMRI detection of early neural dysfunction in preclinical Huntington's disease. *J. Int. Neuropsychol. Soc.* 13:758–69

RELATED RESOURCES

Buonomano DV. 2007. The biology of time across different scales. *Nat. Chem. Biol.* 3:594–97

Gorea A. 2011. Ticks per thought or thoughts per tick? A selective review of time perception with hints on future research. *J. Physiol.* 105:153–63

Grondin S. 2010. Timing and time perception: a review of recent behavioral and neuroscience findings and theoretical directions. *Atten. Percept. Psychophys.* 72:561–82

Macar F, Vidal F. 2009. Timing processes: an outline of behavioural and neural indices not systematically considered in timing models. *Can. J. Exp. Psychol.* 63: 227–39

Mauk MD, Buonomano DV. 2004. The neural basis of temporal processing. *Annu. Rev. Neurosci.* 27:304–40

Wittmann M. 2013. The inner sense of time: how the brain creates a representation of duration. *Nat. Rev. Neurosci.* 14:217–23

Cortical Control of Arm Movements: A Dynamical Systems Perspective

Krishna V. Shenoy,[1,2] Maneesh Sahani,[1,3] and Mark M. Churchland[4]

[1]Departments of Electrical Engineering, [2]Bioengineering, and Neurobiology, Bio-X and Neurosciences Programs, Stanford Institute for Neuro-Innovation and Translational Neuroscience, Stanford University, Stanford, California 94305; email: shenoy@stanford.edu

[3]Gatsby Computational Neuroscience Unit, University College London, London WC1N 3AR, United Kingdom; email: maneesh@gatsby.ucl.ac.uk

[4]Department of Neuroscience, Grossman Center for the Statistics of Mind, David Mahoney Center for Brain and Behavior Research, Kavli Institute for Brain Science, Columbia University Medical Center, New York, NY 10032; email: mc3502@columbia.edu

Annu. Rev. Neurosci. 2013. 36:337–59

First published online as a Review in Advance on May 29, 2013

The *Annual Review of Neuroscience* is online at neuro.annualreviews.org

This article's doi:
10.1146/annurev-neuro-062111-150509

Keywords

premotor cortex, primary motor cortex, neural control of movement, dimensionality reduction

Abstract

Our ability to move is central to everyday life. Investigating the neural control of movement in general, and the cortical control of volitional arm movements in particular, has been a major research focus in recent decades. Studies have involved primarily either attempts to account for single-neuron responses in terms of tuning for movement parameters or attempts to decode movement parameters from populations of tuned neurons. Even though this focus on encoding and decoding has led to many seminal advances, it has not produced an agreed-upon conceptual framework. Interest in understanding the underlying neural dynamics has recently increased, leading to questions such as how does the current population response determine the future population response, and to what purpose? We review how a dynamical systems perspective may help us understand why neural activity evolves the way it does, how neural activity relates to movement parameters, and how a unified conceptual framework may result.

Contents

INTRODUCTION

It is difficult to appreciate just how central movement is to everyday life until the ability to move is lost owing to neurological injury or disease. Moving is how we interact and communicate with the world. We move our legs and feet to walk, we move our arms and hands to manipulate the objects around us, and we move our tongues and vocal cords to speak. Movement is therefore also central to self-image and psychological well-being. Decades of research have explored the neural basis of movement preparation, generation, and control. In particular, a substantial body of knowledge about the cortical control of arm movements in rhesus macaques has grown from Evarts' pioneering research (e.g., Evarts 1964, 1968; Georgopoulos et al. 1982, 1986; Kalaska 2009; Tanji & Evarts 1976; Weinrich & Wise 1982; Wise 1985). This knowledge recently helped investigators to design cortically controlled neural prosthetic systems aimed at restoring motor function to paralyzed patients (for recent reviews, see, e.g., Green & Kalaska 2011, Hatsopoulos & Donoghue 2009).

M1: primary motor cortex

PMd: premotor cortex, dorsal aspect

Extensive as these discoveries have been, and encouraging as these medical applications are, our understanding of the neural control of movement remains incomplete. Indeed, there is remarkably little agreement regarding even the basic response properties of the motor cortex, including PMd and M1 (e.g., Churchland et al. 2010a; Churchland & Shenoy 2007b; Fetz 1992; Graziano 2009, 2011a; Hatsopoulos 2005; Mussa-Ivaldi 1988; Reimer & Hatsopoulos 2009; Scott 2000, 2008; Scott & Kalaska 1995; Todorov 2000 and associated articles). This lack of agreement contrasts starkly with, say, the primary visual cortex, where basic response properties have been largely agreed upon for decades. To understand the motor cortex is thus a major challenge, as well as an essential step toward designing more capable, accurate, and robust neural prostheses (e.g., Gilja et al. 2011, 2012; Shenoy et al. 2011).

Much of the controversy over motor cortex responses has hinged on the question of whether the cortical activity codes (or represents) muscle action on the one hand or higher-level movement parameters such as effector velocity on the other. **Figure 1** illustrates the dichotomy. Cortical activity passes, via the spinal cord, to the muscles, which contract to move the arm; but the temporal patterns of muscle activity and hand movement differ. Which signal is found in the cortex? Does the firing of cortical cells drive muscle contraction with little intervening translation, so that cortical activity resembles muscle activity; or does it encode the intended movement end point or path, to be transformed by the spinal cord into commands that contract the muscles? Studies correlating neural activity with electromyographic (EMG) muscle activity or with movement kinematics (factors such as velocity and position) have proven equivocal; investigators have seen both patterns (for a recent review, see Kalaska 2009). Just as critically, the activity of most neurons is poorly explained by either pattern (e.g., Churchland & Shenoy 2007b, Graziano

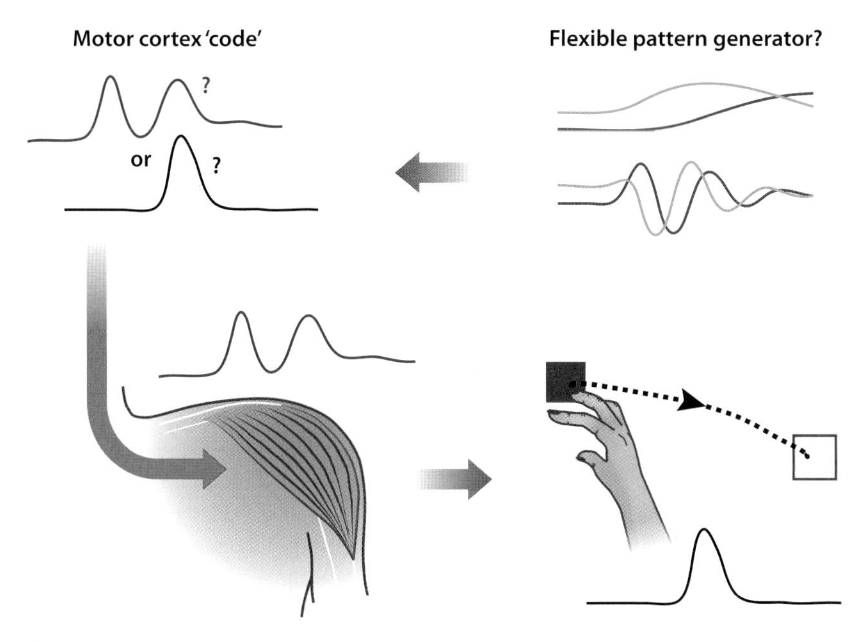

Figure 1

Schematic illustrating the focus of the representational perspective and of the dynamical systems perspective. The traditional perspective has concentrated on the representation or code employed by the motor cortex. For example, does the motor cortex (*upper left panel*) code muscle activity (*red trace*) or reach velocity (*black trace*)? Thus, the traditional perspective attempts to determine the output or controlled parameters of the motor cortex. The dynamical systems perspective focuses less on the output itself and more on how that output is created (*upper right panel*). It attempts to isolate the basic patterns (*blue*) from which the final output might be built. It further attempts to understand the dynamics that produced that set of patterns and the role of preparatory activity in creating the right set of patterns for a particular movement. The red trace indicates the activity of the deltoid versus time during a rightward reach (e.g., Churchland et al. 2012). The black trace is the hand velocity for that same reach; the black trace between the beginning and ending reach targets is the hand path. The light and dark blue traces (*upper right*) illustrate a potential dynamical basis set from which the red trace might be built.

2011b, Scott 2008). Thus, the controversy has continued.

In fact, determining the 'code' or 'representation' in motor cortex is but half the challenge. Whatever the cortical output, its temporal pattern must be generated by the circuitry of the cortex and reciprocally connected subcortical structures. Where is this flexible pattern generator—which can produce the wide variety of motor commands necessary to drive our large repertoire of movements—to be found? Is it upstream of M1, handing down a 'motor program' to be executed there (Miles & Evarts 1979), or is the pattern at least partly generated in M1 itself? Questions such as these suggest a different way to study the motor cortex, shifting the focus from the meaning of the output to the nature of the dynamical system that creates the required, precisely patterned, command (e.g., Graziano 2011b). A core prediction of this perspective is that activity in the motor system reflects a mix of signals: Some will be outputs to drive the spinal cord and muscles, but many will be internal processes that help to compose the outputs but are themselves only poorly described in terms of the movement. They may, for instance, reflect a much larger basis set of patterns from which the eventual commands are built (see **Figure 1**). Some of these internal signals may well correlate coarsely with movement parameters: For example, one of the blue traces in **Figure 1** resembles hand position, whereas

Dynamical system: a physical system whose future state is a function of its current state, its input, and possibly some noise

another resembles a filtered version of hand velocity. But such coincidental correlations may not generalize across different tasks and need not constitute a representation of the movement parameters that is actively used by the brain (e.g., Churchland & Shenoy 2007b, Fetz 1992, Todorov 2000). Indeed, the dynamical systems perspective predicts that the evolution of neural activity should be best captured not in terms of movement parameter evolution, but in terms of the dynamical rules by which the current state causes the next state.

The dynamical systems perspective is not new to motor neuroscience. Brown, a student of Sherrington, argued in 1914 that internal pattern generators are at least as important as feedforward reflex arcs (Brown 1914, Yuste et al. 2005). Since then, the approach has shaped our understanding of central pattern generators (e.g., Grillner 2006, Kopell & Ermentrout 2002) and the brain stem circuitry that guides eye movements (e.g., Lisberger & Sejnowski 1992, Skavenski & Robinson 1973). In studying the motor cortex, Fetz (1992) argued for the dynamical systems perspective 20 years ago in an article entitled, "Are Movement Parameters Recognizably Coded in the Activity of Single Neurons?" He noted that

> over the last three decades this formula [recording single neurons in behaving animals] has generated numerous papers illustrating neurons whose activity appears to code (i.e., to covary with) various movement parameters or representations of higher-order sensorimotor functions.... the search for neural correlates of motor parameters may actually distract us from recognizing the operation of radically different neural mechanisms of sensorimotor control. (p. 77)

The same point has been reiterated recently by Cisek 2006b, who summarizes that "the role of the motor system is to produce movement, not to describe it" (p. 2843). The dynamical systems perspective is also reflected in recent attempts to understand motor cortex as it relates to optimal feedback control (e.g., Scott 2004, Todorov & Jordan 2002). Indeed, the dynamical systems perspective may be experiencing a renaissance in neuroscience as a whole (e.g., Briggman et al. 2005, 2006; Broome et al. 2006; Mazor & Laurent 2005; Rabinovich et al. 2008; Stopfer et al. 2003; Yu et al. 2006), largely as the result of the widening adoption of multichannel recording techniques (e.g., Churchland et al. 2007, Harvey et al. 2012, Maynard et al. 1999), machine-learning based algorithms for estimating the population state from those recordings (e.g., Yu et al. 2009), and the computational resources necessary for data analysis and the exploration of plausible models. Just as importantly, there are growing bodies of neural data that are difficult to interpret from a purely representational framework but may be more approachable when dynamical systems concepts are brought to bear (e.g., Ganguli et al. 2008, Machens et al. 2010).

In this review we focus on one such body of literature, that from the field of motor control. We focus less on the role of dynamics in the context of sensory feedback (e.g., Scott 2004) and more on the internal neural dynamics that occur during movement preparation and the subsequent dynamics that translate preparatory activity into movement activity.

A DYNAMICAL SYSTEMS PERSPECTIVE OF MOTOR CONTROL

An Alternative to the Representational View

In principle, the representational and dynamical perspectives are compatible: The first seeks to determine the parameters controlled by cortical output, whereas the latter seeks to determine how that output is generated. However, in practice, adoption of the representational perspective has led to attempts to explain most neural activity in terms of tuning for movement parameters. That is, studies have sought to describe the firing (r) of each neuron (n) in the motor cortex as a function of various parameters ($param_i$) of an upcoming or concurrent movement:

$$r_n(t) = f_n(param_1(t), param_2(t), \ldots). \quad 1.$$

Admittedly, the available range of parameters is extensive, so such models may be adjusted to exhibit considerable richness (e.g., Fu et al. 1995, Hatsopoulos & Amit 2012, Pearce & Moran 2012, Reimer & Hatsopoulos 2009, Wang et al. 2010). Possible covariates include the intended target location, the kinematics of the hand or of the joints, the activity of individual muscles or synergistic groups, the activity of proprioceptors, the predicted end point error, and many others. These parameters may also be filtered, allowing for varying time lags, differentiation, or integration of the corresponding time-dependent signals. The common theme, however, is that neuronal activity should be understood in terms of such representational functions.

By contrast, the dynamical systems perspective stresses the view that the nervous system is a machine that must generate a pattern of activity appropriate to drive the desired movement. That is, the cortical activity [a time-varying vector $\mathbf{r}(t)$], when mapped to muscle activity [a time-varying vector $\mathbf{m}(t)$] by downstream circuitry,

$$\mathbf{m}(t) = G[\mathbf{r}(t)], \quad 2.$$

must produce forces that move the body in a way that achieves the organism's goals. The mapping $G[\,]$ captures the action of all the circuits that lie between the cortex and the muscles, which may themselves implement sophisticated controllers. The dimension of $\mathbf{m}(t)$, set approximately by the number of independent muscle groups or synergies, is much lower than that of $\mathbf{r}(t)$, the number of different neurons in the motor cortex. Thus it is unlikely that $G[\,]$ will be invertible. That is, knowledge of the final output alone (e.g., desired muscle activity or kinematics) may be insufficient to determine fully the pattern of neural activity that generated the output. This view thus accords with the observation that the apparent tuning of many neurons changes idiosyncratically with time (Churchland & Shenoy 2007b), with arm starting location (Caminiti et al. 1991), with posture (Kakei et al. 1999, Scott & Kalaska 1995), and with movement speed (Churchland & Shenoy 2007b). More broadly, it may help to understand why, despite many well-designed experiments, the issue of representation in the motor cortex has remained unresolved (e.g., Reimer & Hatsopoulos 2009, Scott 2008). In this view, a confusion of representation is not unexpected: the functions (f_n) of Equation 1 may not exist for any proposed set of movement parameters (Churchland et al. 2010a).

By moving the activity $\mathbf{r}(t)$ to the right-hand side of the equation, the dynamical systems perspective brings into focus the system that must generate that firing pattern (Graziano 2011b). Mathematically, population activity evolves with a derivative $\dot{\mathbf{r}}$, scaled by time constant τ, that is determined by the local circuitry of the motor cortex acting on its current activity through a function $h()$ and by inputs that arrive from other areas, $\mathbf{u}(t)$:

$$\tau\dot{\mathbf{r}}(t) = h(\mathbf{r}(t)) + \mathbf{u}(t). \quad 3.$$

With the appropriate input, this dynamical system causes the population activity to trace a path in time that maps through G to generate the correct movement. As this occurs, the neurons in the population may exhibit a variety of response patterns. Some patterns will directly influence the output of G, but others will reflect the act of pattern generation itself. A central aim of the dynamical systems approach is to understand these responses and thus to understand how the dynamics of a neural population produce the temporal patterns needed to drive movement.

Thus, the representational and dynamical perspectives often suggest very different forms of experiment and analysis. If seeking a representation, one asks which parameters are represented by neural activity, in which reference frames, and how these representations are transformed from one reference frame to another (e.g., Andersen & Buneo 2002, Batista et al. 2007, McGuire & Sabes 2011, Mullette-Gillman et al. 2009, Pesaran et al. 2006). Equation 1 suggests that the pattern of neural

State-space trajectory: evolution of network activity in a space where each axis captures one neuron's response, or a factor shared among neurons

activity in time, and across different movements, should resemble that of the encoded parameters. Thus by designing experiments in which movement variables (e.g., muscle activity) vary systematically, one searches for neural firing patterns that vary in the same way. Conversely, a failure to find neural activity that covaries systematically with muscle activity would be taken to falsify the hypothesis that the cortex is concerned with control of muscles (e.g., Hatsopoulos 2005) or at the very least to suggest a 'messiness' of representation.

From the dynamical standpoint, the possibility that some cortical activity patterns are only indirectly related to the movement and reflect instead internal states of the dynamical process dictates a different approach. At least three practical possibilities present themselves. First, one might seek evidence of this internal state-space via direct visualization (Yu et al. 2009) or by testing the prediction that the population response is more complex than expected given the final output (Churchland & Shenoy 2007b). Second, one might attempt to trace the causality of the dynamical system. One might ask how the population's premovement state is determined (e.g., Churchland et al. 2006c), how this state influences the subsequent neural activity (e.g., Churchland et al. 2010a, 2012), and how variability in this state influences both neural activity and the movement (e.g., Afshar et al. 2011, Churchland et al. 2006a). Finally, and perhaps most challenging, one might seek to characterize the function $h()$ of Equation 3 by mapping out attractor states (whether fixed-points, limit cycles, or more complex), probing the system's transient behavior (e.g., Buesing et al. 2012, Macke et al. 2011, Sussillo et al. 2013, Sussillo & Abbott 2009, Yu et al. 2006) and studying the effects of perturbations applied to the neural activity (e.g., Churchland & Shenoy 2007a, Diester et al. 2011, Gerits et al. 2012, O'Shea et al. 2013). These different investigations are reviewed in greater detail below.

The Population Dynamical State

A dynamical description of cortical function is inherently a description of activity at the population level. This notion is evident in Equations 2 and 3 above, neither of which can easily be separated into single-neuron components. Unfortunately, obtaining direct empirical access to the relevant scale of population activity is challenging. The full dynamical system is an extensively connected recurrent network of millions of neurons, coupled through input and feedback signals with much of the rest of the brain. The best current measurement technology can record either individual activity of no more than hundreds of neurons (using silicon or microwire arrays or calcium imaging) or aggregate signals that pool over thousands or more neurons at a time (local field potentials, fluorescence changes in voltage-sensitive dyes, or hemodynamic responses). Neither recording scale would seem suited to describing in detail the activity of the whole population. The unreliability of neuronal spiking introduces further challenges (e.g., Faisal et al. 2008, Manwani et al. 2002). Activity cannot be time-averaged on a scale longer than the dynamical time constant of Equation 3 without distorting the resulting dynamics. Similarly, one cannot average over repeated movements to construct a peri-stimulus-time histogram (PSTH) without suppressing intrinsic trial-to-trial variability, which is often of interest (e.g., Afshar et al. 2011, Yu et al. 2009).

These challenges can be addressed using at least two approaches. The first approach avoids the attempt to visualize or describe the dynamical process directly. Instead, hypotheses derived from the dynamical systems viewpoint are tested by assessing related predictions. One example is the prediction that trial-to-trial variance should fall as movement preparation brings the activity of the cortex to a suitable initial point from which appropriate movement activity can be generated (Churchland et al. 2006c, 2010b).

The second approach uses statistical methods to infer the population state from the available data and to examine how that state changes with time. Neurons within a single cortical population do not act alone; instead the coordinated firing of all the neurons presumably guides the evolution of activity within the population and the evolution of its outputs.

This coordination may be intuitively most clear in the context of representation. Activity in a sensory population that encodes the features of a stimulus will covary as those features change. If the number of features is fewer than the number of neurons in the population, then population-level activity must be confined to a space the dimension of which is lower than the number of neurons. Even if the stimulus were unknown, the relevant aspects of the population activity may still be read out by looking for this lower-dimensional coordination. The same idea applies when the low-dimensional structure derives from the population dynamics rather than from a stimulus representation.

There are at least two reasons to think that the essential dimensions of the dynamical state will be few and will be distributed across many, if not all, of the neurons within a local area. First, the tight recurrent connectivity of the network will naturally tend to spread activity between cells. Second, and more subtly, the need for the network to be robust against the very unreliability that hampers experimental observation favors redundant activation patterns. The vector $\mathbf{r}$ in Equations 2 and 3 spans many neurons, and we assume that independent noise in the activity of those neurons (or, indeed, injury to some of them) has only minimal impact on the output of the map G. Thus both G and the function h that determines the dynamics are likely to pool responses from many neurons, compressing the high-dimensional activity into a smaller set of meaningful degrees of freedom and thereby rejecting noise. These meaningful degrees of freedom may be viewed as defining a restricted space of lower dimension that is embedded within the space of all possible activity patterns. Because only the projection of the activity into this lower-dimensional space matters both to the dynamics of the area and to the influence it exerts on the muscles, the meaningful outputs of h must also be confined to this space. Thus, the projection of $\mathbf{r}$ into this lower-dimensional space defines a population dynamical state (see Definition of Terms).

The dimensions explored by the population dynamical state may depend on the type of movements being performed. Over the full repertoire of movements (e.g., Foster et al. 2012, Gilja et al. 2010, Szuts et al. 2011), the range of dimensions might number in the thousands or more. However, in limited experimental settings with well-controlled motor outputs, the state may be confined to many fewer degrees of freedom. If so, then it should be possible to access the population dynamical state by means of dimensionality reduction techniques applied to the recorded data (e.g., Yu et al. 2009). These methods trace out the trajectory followed by the dynamical system, often on a single trial. Such trajectories make it possible to observe the dynamics more directly—indeed, in some cases the dimensionality reduction itself depends on forming a simultaneous estimate of the dynamical equations—and also to ask qualitative questions about the nature of the dynamics. For example, does the population state observed before the arm moves relate sensibly to the state trajectory traced out during the subsequent movement?

DEFINITION OF TERMS

Population dynamical state: a set of coordinates, often represented as a vector, describing the instantaneous configuration of a dynamical system and that is sufficient to determine the future evolution of that system and its response to inputs. The population dynamical state of a neuronal network might be the vector of instantaneous firing of all its cells or may incorporate aspects of the neurons' biophysical states. It may also be a lower-dimensional projection of this network-wide description. See Dimensionality Reduction.

Dimensionality reduction: in this context, a technique for mapping the responses of many neurons onto a small number of variables that capture the basic patterns present in those responses. For example, the first variable/dimension might capture the response of a large proportion of neurons that all have very similar responses.

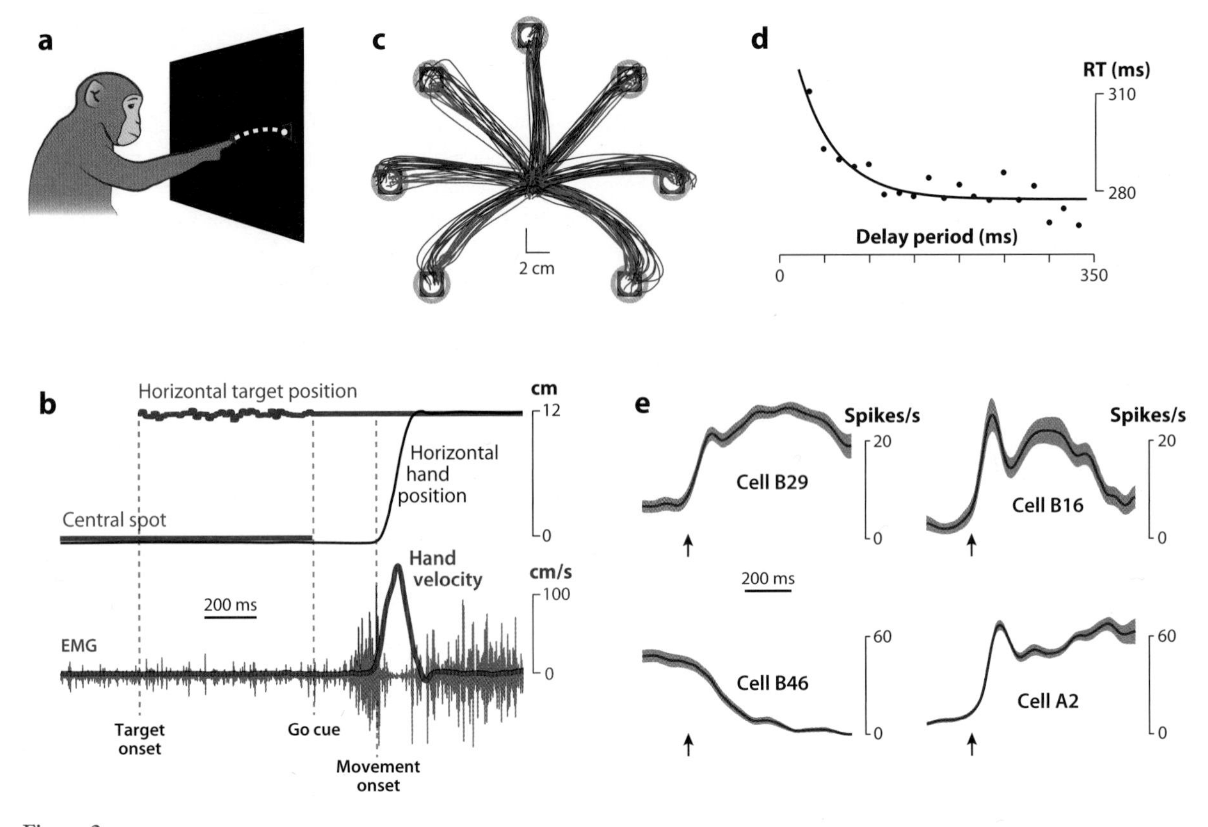

Figure 2

Overview of experimental paradigm, behavioral measurements, muscle measurements, and neural measurements. (*a*) Illustration of the instructed-delay task. Monkeys sit in a primate chair ~25 cm from a fronto-parallel display. A trial begins by fixating (eye) and touching (hand) a central target (*red filled square*) and holding for a few hundred milliseconds. A peripheral target (*red open square*) then appears, cuing the animal about where a movement must ultimately be made. After a randomized delay period (e.g., 0–1 s) a go cue is given (e.g., extinction of central fixation and touch targets) signaling that an arm movement to the peripheral target may begin. (*b*) Sample hand measurements and electromyographic (EMG) recordings for the same trial as in panel *a*. *Top*: Horizontal hand (*black*) and target (*red*) positions are plotted. For this experiment, the target jittered on first appearing and stabilized at the go cue. *Bottom*: Hand velocity superimposed on the voltage recorded from the medial deltoid. (*c*) Sample reach trajectories and end points in a center-out two-instructed-speed version of the instructed-delay task. Red and green traces/symbols correspond to instructed-fast and instructed-slow conditions. (*d*) Mean reaction time (RT) plotted versus delay-period duration. The line shows an exponential fit. (*e*) Examples of typical delay-period firing-rate responses in PMd. Mean ± Standard Error firing rates for four sample neurons are shown. Figure adapted from Churchland et al. (2006c).

CORTICAL ACTIVITY DURING MOVEMENT PREPARATION

Studies indicate that voluntary movements are prepared before they are executed (Day et al. 1989, Ghez et al. 1997, Keele 1968, Kutas & Donchin 1974, Riehle & Requin 1989, Rosenbaum 1980, Wise 1985). To build intuition, consider the sudden, rapid, and accurate movement needed to swat a fly. An immediate, unpremeditated attack could miss, allowing the fly to escape. Conversely, a short preparatory delay may permit the accuracy and velocity of movement to be improved, increasing the chances of success. In the laboratory, movement preparation has been studied by instructing a similar, but experimentally controlled, delay prior to a rapid, accurate movement (e.g., Mountcastle et al. 1975). **Figure 2*a–c*** illustrates the experimental design and task timing, along with sample hand position and EMG measurements.

Evidence that subjects use this instructed delay period to prepare a movement comes in part from the observation that reaction time (RT) is shorter on trials with a delay (e.g., Churchland et al. 2006c, Ghez et al. 1989, Riehle et al. 1997, Riehle & Requin 1989, Rosenbaum 1980). **Figure 2*d*** illustrates how RT first decreases and then plateaus with delay duration, suggesting that a time-consuming preparatory process has been given a head start during the initial ~200 ms of delay (e.g., Crammond & Kalaska 2000, Riehle & Requin 1989, Rosenbaum 1980).

Further evidence for movement preparation comes from neural recordings. Neurons in many cortical areas, including the parietal reach region (e.g., Snyder et al. 1997), PMd (e.g., Weinrich & Wise 1982), and M1 (e.g., Tanji & Evarts 1976), systematically modulate their activity during the delay. Thus, these motor-related areas appear to be engaged in computation prior to the movement (Crammond & Kalaska 2000). **Figure 2*e*** shows four PMd neurons that exemplify the range of delay-period firing patterns: Some neurons' firing rates increase, some decrease, and some stabilize after an initial transient, whereas others vary throughout. This variety of neural responses contrasts with the simple monotonic decline of behavioral RT (**Figure 2*d***) and complicates efforts to understand the role of this activity.

Preparatory Activity as a Subthreshold Representation

Early proposals extended the representational view with the suggestion that preparatory neural activity represents the desired movement at a subthreshold level with the same tuning as that used during movement but with lower overall firing rates (e.g., Tanji & Evarts 1976). This lower-intensity activity is thought not to evoke movement by itself, but instead to reduce the time taken to achieve the correct suprathreshold firing pattern, thus shortening RT. This hypothesis has been assumed by many models of reach generation (e.g., Bastian et al. 1998, Cisek 2006a, Erlhagen & Schöner 2002, Schöner 2004) and agrees with our understanding of the saccadic system (e.g., Hanes & Schall 1996).

Reaction time (RT): the time from the go cue until the start of the movement

Many studies, particularly those exploring summary measures such as the population vector, have indeed reported consistently tuned neural activity before and during movement (e.g., Bastian et al. 1998, 2003; Cisek 2006a; Erlhagen et al. 1999; Georgopoulos et al. 1989; Requin et al. 1988; Riehle & Requin 1989). However, some studies at the single-neuron level have come to the opposite conclusion: that preparatory and movement tuning are often dissimilar (e.g., Wise et al. 1986) and nearly uncorrelated on average (e.g., Churchland et al. 2010a, Crammond & Kalaska 2000, Kaufman et al. 2010, 2013). Attempts to verify a threshold mechanism have also proven inconclusive. Higher premovement firing rates are not consistently associated with shorter RTs (e.g., Bastian et al. 2003, Churchland et al. 2006c). Furthermore, responses of cortical inhibitory interneurons seem inconsistent (Kaufman et al. 2010) with the common hypothesis that subthreshold preparatory activity is released from inhibition to initiate the movement (Bullock & Grossberg 1988, Sawaguchi et al. 1996).

Preparatory Activity as the Initial Dynamical State

The dynamical systems view suggests a different purpose for preparatory activity. Equations 2 and 3 describe the evolution of neural activity and its translation to muscle activity and thus to movement. The population state trajectory, and thus the movement produced, will clearly depend on the dynamics by which the population state evolves, captured by the function $h()$. It may also be affected by descending input or feedback [$\mathbf{u}(t)$] and by any sources of noise (e.g., van Beers et al. 2004). Finally, and crucially for our current purposes, the trajectory will depend on the population state $\mathbf{r}(t_0)$ at the time (t_0) that movement-related activity begins to be generated. Thus, all else remaining equal, different initial states will lead to different movements. This suggests that one role of preparation is to bring the population dynamical state to an initial value from which accurate movement-related activity will follow efficiently.

In general, more than one initial population dynamical state may lead to a movement that is sufficiently accurate: for example, a reach adequate to earn a reward. Assuming smoothness in the dynamics, and in the mapping to muscle activity and thus to kinematics, we might expect preparation for each movement to be associated with a compact subregion of the space of all possible population states (**Figure 3*a***). State-space trajectories [$\mathbf{r}(t)$ for $t > t_0$] originating from different points in this subregion may vary; however, for the reach to be successful, such variation must (*a*) be confined to dimensions that are discounted by the mapping to muscles, (*b*) perturb the movement by too little to affect the desired outcome, or (*c*) be contained by compensatory changes in the external input provided by other areas, including corrective feedback signals.

Thus, the representational and dynamical perspectives both suggest that different movements should require different preparatory activity. Indeed, premovement firing is found to vary with every movement parameter studied so far (Cisek 2006b), including direction (e.g., Kurata 1989, Wise 1985), distance (e.g., Crammond & Kalaska 2000), speed (Churchland et al. 2006b), and curvature (Hocherman & Wise 1991); even apparently random variability in the preparatory state correlates with variability in the subsequent movement (Churchland et al. 2006a). However, short of a rapid de- and re-coding of activity between preparation and movement, the representational view predicts that preparatory and movement tuning should be congruent, which contrasts with the single-unit data as reviewed above. If the link between pre- and peri-movement activity were simply dynamical, on the other hand, then there would be no reason to necessarily expect such congruence.

Two recent studies have extended further support for the dynamical view. First, Churchland et al. (2010a) showed that although preparatory activity does indeed covary with movement parameters such as direction, distance, speed, and curvature, it is more closely related to the pattern of cortical neural activity during the movement—as would be expected if the premovement population state led directly to the subsequent trajectory of movement-period neural activity, and only indirectly to the movement. Second, Kaufman et al. (2011) observed that preparatory states associated with different reaches were arranged along dimensions orthogonal to the dimensions of activity that correlate with changes in muscle force during movement. This result is consistent with a view in which preparatory activity does not itself engage changes in muscle output through the mapping $G[\ ]$ but nonetheless leads to movement control signals that do. In a representational picture, where prepatory and movement activity are similarly tuned, such orthogonality would be unexpected.

The Dynamics of Preparation

The end point of motor preparation is hypothesized to be an initial population dynamical state, from which the movement-period neural activity evolves to generate the desired movement. How is the correct initial state achieved between the times when the subject first sees the target and subsequently is told to move? Clearly the dynamics of movement preparation cannot be the same as the dynamics of movement activity. During movement preparation, the dynamical system must bring the population state toward the optimal preparatory region (as in **Figure 3*a***) not away from it. Is it possible to detect signatures of this convergent dynamical process?

Activity in the experimental premovement period starts from a baseline condition, in which the only behavioral constraints are that the eyes remain fixated and the hand remains still (**Figure 2*a***). There is little to prevent motor cortical activity in this state varying substantially across trials. During preparation, the activity then approaches the preparatory state, while avoiding the premature generation of movement. Again, because the intervening

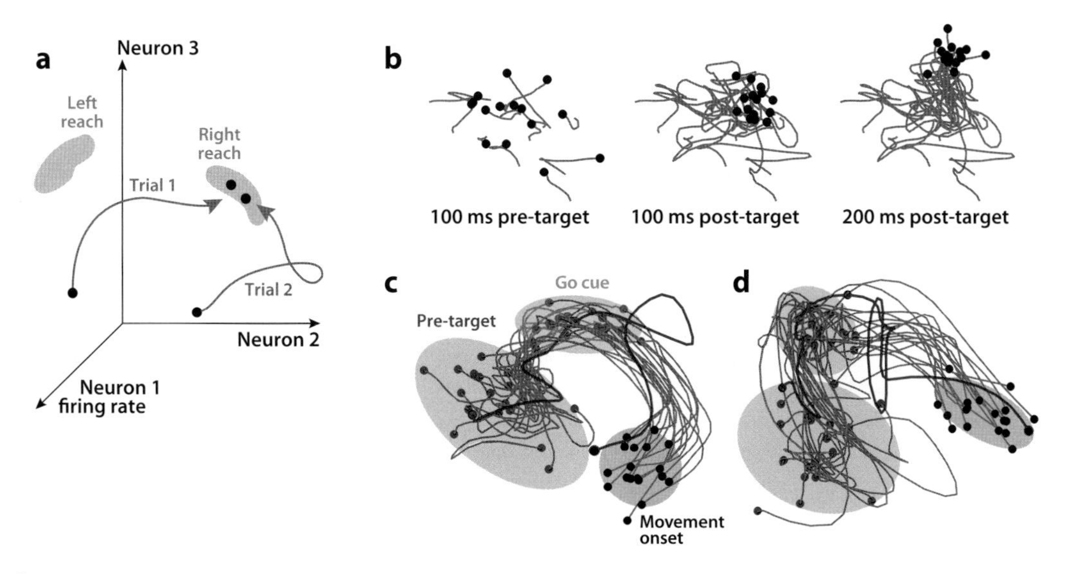

Figure 3

Schematic illustration of the optimal subspace hypothesis and single-trial neural trajectories computed using Gaussian process factor analysis (GPFA). (*a*) The configuration of firing rates is represented in a state-space, with the firing rate of each neuron contributing an axis, only three of which are drawn. For each possible movement, we hypothesize that there exists a subspace/subregion of states that are optimal in the sense that they will produce the desired result, with a minimal reaction time, when the movement is triggered. Different movements will have different optimal subspaces (*shaded areas*). The goal of motor preparation would be to optimize the configuration of firing rates so that it lies within the optimal subspace for the desired movement. For different trials (*arrows*), this process may take place at different rates, along different paths, and from different starting points. Figure from Churchland et al. (2006c). (*b*) Projections of PMd activity into a two-dimensional state-space. Each black point represents the location of neural activity on one trial. Gray traces show trajectories from 200 ms before target onset until the indicated time. The stimulus was a reach target (135°, 60 mm distant), with no reach allowed until a subsequent go cue; 15 (of 47) randomly selected trials are shown. (*c*) Trajectories were plotted until movement onset. Blue dots indicate 100 ms before stimulus (reach target) onset. No reach was allowed until after the go cue (*green dots*), 400–900 ms later. Activity between the blue and green dots thus relates to movement planning. Movement onset (*black dots*) was ~300 ms after the go cue. For display, 18 randomly selected trials are plotted, plus one hand-selected outlier trial (*red*, trial ID 211). Covariance ellipses were computed across all 47 trials. This is a two-dimensional projection of a ten-dimensional latent space. In the full space, the black ellipse is far from the edge of the blue ellipse. This projection was chosen to preserve accurate relative sizes (on a per-dimension basis) of the true ten-dimensional volumes of the ellipsoids. Data are from the G20040123 dataset. (*d*) Data are presented as in panel *c*, with the same target location, but for data from another day's data set (G20040122; *red* outlier trial: ID 793). Figure panels *b–d* adapted from Churchland et al. (2010b).

states do not themselves engage muscles, they may well be less constrained than those traversed during the movement's active phase. (See schematic trials 1 and 2 in **Figure 3*a*.**) The final preparatory state, however, is constrained by the need to generate the correct movement. Thus, we might expect that as preparation progresses, the relative variability across different trials should fall. Such a decrease has indeed been identified in the Fano factor of individual neurons in both the premotor cortex and motor cortex (Churchland et al. 2006c, Mandelblat-Cerf et al. 2009, Rickert et al. 2009). The decline in variability is also apparent in the population dynamical state directly visualized via dimensionality reduction (Churchland et al. 2010b), as shown in **Figure 3*b***. As predicted, the reduction in variance comes primarily from convergence in the low-dimensional population dynamical state rather than in the spiking noise of each cell (Churchland et al. 2010b). Finally, variability is only partially reduced when incomplete information about the target is provided (Rickert et al. 2009). These findings support the hypothesis that motor preparation requires network activity to converge to a relatively tight set of population dynamical states. As an aside, a similar decline in neural variability is present in a variety of different cortical areas whenever a relevant stimulus is presented (Churchland et al. 2010b). These findings suggest that many different cortical computations may involve attractor-like dynamics. Nonetheless, the significance of such computation must depend on the function of the area. In visual cortex, the decline in variability may reflect the formation of a more consistent representation of the visual stimulus. In the motor areas considered here, the evidence (discussed below) indicates that the decline in variability reflects convergence to a preparatory state that has motor consequences.

What are the consequences if the convergence of the preparatory state is not complete at the time of the go cue? Instructed-delay experiments in which accuracy was emphasized have gathered some data to address this question. First, Churchland et al. (2006c) found that neural variability was lower among trials with short RT, in which motor preparation was likely to have been complete at the time of the go cue, than among trials with longer RT, in which motor preparation may have been incomplete or not quite accurate. This result is consistent with lower variability indicating greater preparatory accuracy (i.e., closer to the optimal preparatory state). Second, when subthreshold electrical microstimulation disrupted the preparatory state, RT was increased (Churchland & Shenoy 2007a). This effect could reflect additional time taken to recover the appropriate preparatory state (see also Ames et al. 2012 for another possibility). The effect was specific (Churchland & Shenoy 2007a). First, RT was more strongly affected when the microstimulation targeted the premotor cortex (where preparatory activity is more common) rather than the motor cortex (where preparatory activity is less common). Second, effects were seen only when the preparatory state was disrupted around the time of the go cue; disruption of the preparatory state before it was needed had little impact on RT. Third, the impact on reach RT was much greater than the impact on saccadic RT, consistent with the role of the premotor cortex in preparing reaches rather than saccades and inconsistent with the possibility that microstimulation simply distracted the animal. Finally, O'Shea et al. (2013) recently found that optical stimulation of optogenetically transfected PMd neurons during the preparatory period similarly increases RT.

THE TRANSITION TO MOVEMENT

By itself, the idea that motor cortical activity represents or codes movement parameters (Equation 1) does not constrain the relationship between preparatory and movement activity. Nonetheless, this transition has frequently been thought to depend on the strengthening of a representation until it crosses a firing-rate threshold, by analogy to the oculomotor system. However, direct evidence for such a threshold has been lacking in the case of reaches (Bastian et al. 2003, Churchland et al. 2006c). By contrast, under the dynamical systems perspective (Equations 2 and 3), the transition to movement is a transition between two different types of network dynamics, most likely to be mediated by a change in the external input term $\mathbf{u}(t)$ of Equation 3. Preparatory dynamics, which brings the population to a suitable state of readiness, then gives way to the dynamics that generate movement. As the movement is triggered, the population dynamical state departs from the prepared initial state and follows a trajectory through state-space. It is that state-space trajectory—determined by the initial state, the neural dynamics, and any feedback—that drives the movement.

The transition from preparatory to movement dynamics may be directly observable. Petreska et al. (2011) used an unsupervised machine-learning technique to study changes in dynamics within multielectrode neuronal data gathered while animals performed instructed-delay reaches. They observed stereotyped dynamical transitions occurring at times shortly after target presentation as well as between the go cue and the beginning of movement. The timing of the dynamical transition that followed the go cue was correlated with the timing of subsequent movement onset. Indeed, this transition—the identity and timing of which were determined by the neural data alone—predicted trial-to-trial variation in RT much better than did alternatives based on a threshold applied to overall firing or to the length of the population vector, even when that threshold was chosen with direct reference to the behavior.

Afshar et al. (2011) addressed a related issue. They reasoned that natural variability might occasionally displace the population dynamical state from the average point of preparation toward the direction in which that state will need to evolve when the movement is to be initiated. Any such variability could actually be beneficial to the initiation of movement and might reduce RT. Indeed, Afshar et al. found that the displacement of the dynamical state at the time of the go cue in the direction defined by the movement-period activity was negatively correlated with RT. Furthermore, RT was even lower if the preparatory state happened to be moving in that direction at the time of the go cue. Thus, although previous studies (Churchland et al. 2006c, Rickert et al. 2009) stressed the importance of an accurate and consistent preparatory state (RTs being lower on average when neural activity is near that state), displacement from the preparatory state can, in fact, result in a lower RT when the displacement is in the direction that is to be traversed during movement.

CORTICAL ACTIVITY DURING MOVEMENT

The dynamical systems perspective focuses on the population dynamical state and its evolution. Testing specific hypotheses therefore often requires direct visualization of that state. The traces in **Figure 3*c*,*d*** illustrate the trajectories of the population dynamical state on 19 trials from just before target onset to the moment when movement begins. After target onset, the dynamical state approaches a preparatory region and its variability falls. Then, following the go cue, the neural state moves rapidly away from the preparatory state in a curved trajectory. Some trial-to-trial variability is evident even after the go cue. In particular, for two outlier trials, the neural state wanders before falling back on track. On these trials, the monkey hesitated for an abnormally long few hundred milliseconds before beginning to reach. These observations underscore the ability of dimensionality reduction methods, when applied to data collected from multielectrode arrays, to reveal single-trial (and potentially rare) phenomena that would normally have been lost to averaging or discarded owing to abnormal behavior. That said, for this highly practiced task, such trials were rare (about 0.1%). On the vast majority of trials, the population state evolved along a stereotyped curved trajectory. How can we characterize that trajectory: its shape, its time evolution, and the principles that give rise to it?

This relates to Fetz's original question, "Are movement parameters recognizably coded in the activity of single neurons?" If they are, then the neural state-space trajectory should reflect the trajectory of the represented parameters. For example, consider the model in which neural activity is cosine-tuned for reach velocity (Moran & Schwartz 1999). This relationship can be written in matrix form as $\mathbf{r}(t-\tau) = \mathbf{M}\mathbf{v}(t)$, where $\mathbf{r}(t)$ is an $n \times 1$ vector describing the firing rate of each neuron, $\mathbf{v}(t)$ is the three-dimensional reach velocity vector at time t, τ is the lag by which neural activity leads movements, and $\mathbf{M}$ is an $n \times 3$ matrix describing (in each row) each neuron's preferred direction. Under this model, the population state would be three dimensional, with those dimensions capturing the neural representation of velocity. The population vector is a dimensionality reduction method made specifically for just such a situation.

However, a simple velocity-tuned model is inadequate to fully capture the richness of the neural responses. **Figure 4** illustrates the PSTH responses of two typical neurons recorded from the motor cortex

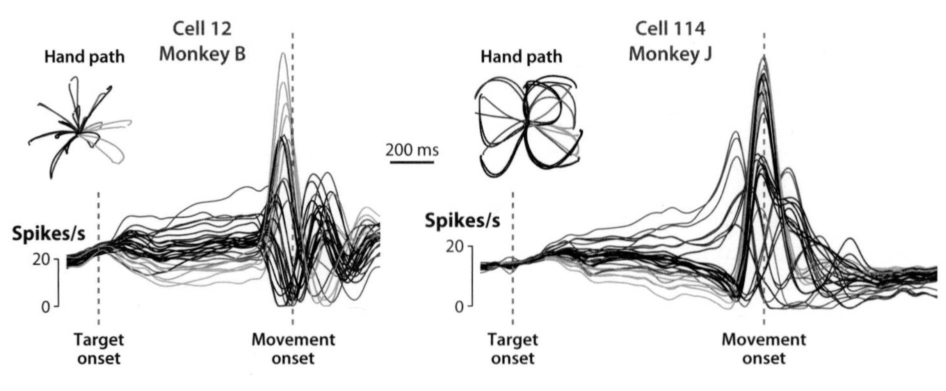

Figure 4

Peri-stimulus-time histogram (PSTHs) and arm-movement kinematics. PSTH from two sample neurons (*left and right panels*), including hand paths for sample reaches (*insets*). Monkey B performed a standard center-out reaching task, with two distances and two instructed speeds. Monkey J performed a more complex version of this task where some reaches were required to curve around a barrier. Traces are colored red to green on the basis of the relative level of preparatory activity for each condition. Insets show hand trajectories and are color-coded to reveal the directional nature of preparatory tuning. Note that what was preferred during the preparatory period was typically not preferred during the movement period; e.g., the first neuron shows a preference for left and up during the preparatory period but a preference for right and down by movement onset. Figure adapted from Churchland et al. (2010a).

of two monkeys. Monkey B performed a standard center-out reaching task, with two distances and two instructed speeds. Monkey J performed a similar task, but in it some reaches were required to curve around a barrier.

Four features of the neural responses are relevant to the controversy over what is being coded. First, the same neurons exhibit both preparatory- and movement-period activity, yet tuning during the preparatory period often differs from that during the movement period (e.g., cell 12 prefers up-left during the preparation, but down-right by movement onset) (Churchland et al. 2010a). Second, the movement-period responses are complex and multiphasic (Churchland et al. 2010a, 2012; Churchland & Shenoy 2007b). Third, the responses of different neurons are heterogeneous, even in the same animal and the same local region of the cortex (Churchland et al. 2010a, Churchland & Shenoy 2007b, Fetz 1992). In dynamical systems terms, the neural responses occupy a relatively high-dimensional space, on the order of 15–30 dimensions (Churchland & Shenoy 2007b). Thus, if neurons are to represent movement parameters, there must be many such parameters (e.g., Pearce & Moran 2012). Finally, neural firing fluctuates over 400–800 ms, even when the reaches themselves are quite brief (e.g., the reaches for Monkey B lasted ~150–300 ms). Thus if movement parameters are represented directly, there must be some unexpected temporal multiplexing (e.g., Fu et al. 1995).

In large part because the neural responses are complex, our field does not yet agree on the relationship between movement-period neural responses and movement itself; the nature of G in Equation 2 remains mostly unresolved (although, see, e.g., Fetz et al. 2000). Yet some recent progress has been made in characterizing the nature of h in Equation 3 and in describing the dynamics that generate movement-period neural responses. First, the collective activity of motor cortical neurons is better described by a model in which activity is driven by a low-dimensional dynamical model, relative to a model in which firing coordination emerges from direct connections between recorded cells (Macke et al. 2011).

Second, the dynamics at play during movement appear to have some simple aspects. In

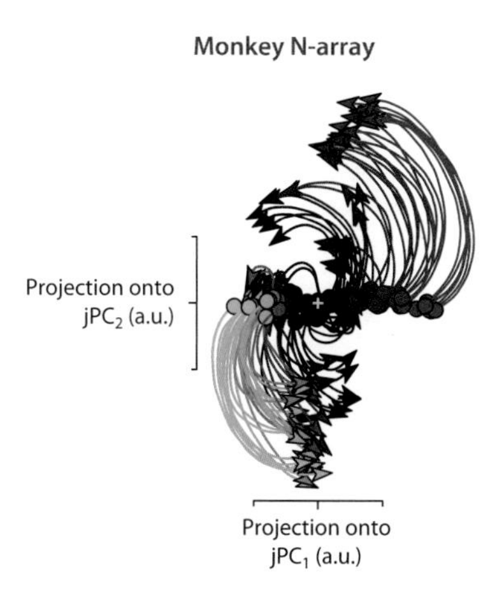

Figure 5

Projections of the neural population response, produced by applying jPCA to the first six principal components of the data. Two-dimensional projection using 218 single- and multiunit isolations, 108 conditions employing straight and curved reaches, from the N-array data set. Each trace is the average trajectory for one of 108 conditions. Trial-averaged neural trajectories are colored red to green on the basis of the level of preparatory activity for that projection. Each trace (one condition) plots the first 200 ms of movement-related activity away from the preparatory state (*circles*). Figure adapted from Churchland et al. (2012).

many lower-dimensionality projections, one of which is shown in **Figure 5**, the neural trajectory simply rotates with a phase and amplitude set by the preparatory state (Churchland et al. 2012). The rotational trajectories of the neural state resemble what is seen during rhythmic movement, even though the reaches were not overtly rhythmic. These trajectories suggest that the role of motor cortex may be most naturally thought of in the context of pattern generation. Consistent with this idea, EMG activity was well fit by the sum of two rhythmic components that were fixed in frequency but varied in phase and amplitude across the different movements (Churchland et al. 2012).

It should be stressed that the neural dynamics have aspects that are not well fit by a simple linear model (Churchland et al. 2012, Sussillo et al. 2013). However, a simple linear time-invariant system accounted for a large proportion of the variance (48.5% of variance explained over the nine data sets tested) (Elsayed et al. 2013). Furthermore, the linear component was almost entirely normal and rotational (linear systems constrained to be normal and rotational performed 93.2% and 91.3% as well as did an unconstrained linear system) (Elsayed et al. 2013).

Although purely rotational dynamics are only an approximation to the true nonlinear dynamics, the observation that neural pattern generation involves rotations of the neural state illustrates two key points of the dynamical systems perspective. First, the goal of the preparatory state is not to act as an overt representation, but rather to set the amplitude and phase of the subsequent rotations. Second, those state-space rotations produce, in the temporal domain, brief sinusoidal oscillations that form a natural basis set for building more complex patterns (e.g., the blue traces in **Figure 1** can be linearly combined to fit the red EMG trace quite well). Thus, the neural dynamics and the resulting patterns can be understood in simple terms, even though they do not constitute an overt representation of movement parameters.

The observed similarity between the average trajectories during rhythmic and nonrhythmic movement might come about because the nervous system has redeployed old principles for a new purpose. Alternatively, rotations are a common dynamical motif and produce a natural basis set (Rokni & Sompolinsky 2012). Thus we see more than one potential explanation for the key features of the observed dynamics. These key features are that dynamics are similar across different reach types, have a strong rotational component, and have their phase and amplitude determined by the neural state achieved during movement preparation. A number of dynamical models, possibly including control-theory style models, may be able to account for these features (e.g., Scott 2004, Todorov & Jordan 2002). That said, the data are inconsistent with many classes of dynamics (Churchland et al. 2012). Because the observed dynamics are similar across reach types, they are not consistent with a system that is dominated by reach-specific inputs (e.g., a dynamical system that converts velocity commands to muscle activity). The rotational patterns are also not consistent with rise-to-threshold or burst-generator models. This discrepancy is important because many previous intuitions regarding single-neuron activity derived from such models. Most centrally, these intuitions included the expectation that the preparatory- and movement-period preferred direction should be closely related. In contrast, the empirical preferred direction is in a constant state of flux, a natural consequence of the underlying rotations (Churchland et al. 2010a, 2012). This observation illustrates how the complex responses of individual neurons can often hide simple underlying structure—structure that is readily interpretable from a state-space, dynamical systems perspective.

DISCUSSION

The past 50 years have witnessed remarkable progress in our understanding of the cortical control of arm movements. Many fundamental principles have been discovered, and this basic scientific knowledge has led to rapid advances, including early clinical trials of cortically controlled neural prostheses for paralyzed participants (Collinger et al. 2013; Hochberg et al. 2006, 2012). Despite this considerable progress, it remains arguable whether an adequate conceptual framework has yet been identified, around which experimental, computational, and theoretical research can be oriented. Indeed without an adequate conceptual framework, it is unclear how a unified and comprehensive understanding for cortical motor control could be assembled or even recognized. This sentiment, initially expressed by Fetz in his 1992 article, appears to be of at least as great a concern today as it was 20 years ago. The following excerpts from recent articles serve as examples.

> A shift in how to examine the motor system occurred in the 1980s from a problem of control back to a problem of what variables were coded in the activity of neurons.... [P]erhaps it is time to re-evaluate what we are learning about M1 function from continuing to ask what coordinate frames or neural representations can be found in M1. Perhaps it is time to stop pursuing the penultimate goal of identifying the coordinate frame(s) represented in the discharge patterns of M1 and again move back to the question of control. (Scott 2008, p. 1220)

> Neurophysiological experiments have revealed neural correlates of many arm movement parameters, ranging from the spatial kinematics of hand path trajectories to muscle activation patterns. However, there is still no broad consensus on the role of the motor cortex in the control of voluntary movement. The answer to that question will depend as much on further theoretical insights into the computational architecture of the motor system as on the design of the definitive neurophysiological experiment. (Kalaska 2009, p. 172)

> An epic, twenty-year battle was fought over the cortical representation of movement. Do motor cortex neurons represent the direction of the hand during reaching, or do they represent other features of movement such as joint rotation or muscle output? As vigorous as this debate may have been, it still did not address the nature of the network within the motor cortex. Indeed, it tended to emphasize the properties of individual neurons rather than network properties.... The battles over the cortical representation of movement never satisfactorily addressed those questions. (Graziano 2011b, p. 388)

It appears that the field has reached a point where a new way of conceptualizing cortical motor control is needed. We have reviewed here a relatively new dynamical systems framework that appears to have several of the desired attributes. As such, the dynamical systems framework may help deepen our understanding of the neural control of movement. It may help do so by (*a*) making relatively few assumptions (e.g., being agnostic to tuning curves, specifically their lack of invariance and generalization); (*b*) observing and documenting the dominant, and perhaps essential, features of neural state-space trajectories; (*c*) providing single-trial neural correlates of behavior that offer insights beyond those available using average relationships alone; (*d*) generating hypotheses that can be answered in the quantitative terms of population dynamical states and evolution rules (equations of motion), without the need to ascribe representational meaning to the detailed response of single neurons; and (*e*) being entirely open to the nature of the dynamics uncovered (e.g., ranging from pattern generators through sophisticated feedback controllers).

The dynamical systems framework is not single-neuron nihilistic: It does not ignore or attempt to average away the complex features of single-neuron responses. Indeed, we hope that by capturing the underlying dynamics it will become possible to explain the many seemingly odd aspects of individual-neuron responses. The dynamical systems perspective also provides a clear road map and goal, which is to quantify the dynamical systems instantiated by neural circuits. This mathematical quantification comes in two inter-related parts: the state-space neural trajectories (a focus of this review article) and the form and meaning of the evolution rule or equations of motion (a focus of ongoing research; e.g., Abeles et al. 1995, Churchland et al. 2012, Petreska et al. 2011, Rabinovich et al. 2008, Seidemann et al. 1996, Smith et al. 2004, Vaadia et al. 1995, Yu et al. 2006). It appears possible that three primary dynamical systems underlie reaching arm movements: one to prepare the neural state in an appropriate manner; a second system to use this computationally optimized starting point to generate movement activity, muscle contraction, and thus movement itself; and a third system that uses feedback for control. Much future research is certainly needed to explore this possibility, to extend and relate it to numerous behaviors beyond the instructed-delay point-to-point reaching task, and to see whether the dynamical systems perspective can ultimately help provide a more comprehensive understanding of cortical motor control.

FUTURE ISSUES

The predictions, experiments, and analyses described above stem from a dynamical systems perspective, and to some degree their confirmation argues for that perspective. Yet many central questions remain largely unaddressed. What is the nature of the relevant dynamics (h in Equation 3), and why are they what they are? Do they relate to the dynamics of movement-generating circuits in simpler organisms? Do they reflect sophisticated mechanisms of online control and feedback (e.g., Scott 2004, Todorov & Jordan 2002)? How and why do dynamics change as a function of input from other brain areas (e.g., resting versus planning versus moving)? What is the nature of the circuitry, both local and feedback, that produces these dynamics? What is the mapping between the population dynamical state and muscular activity (G in Equation 2)? Answering such questions will likely depend on progress in three domains of research: first, the increased ability to resolve dynamical structure in neural data; second, the increased ability to perturb the population dynamical state while observing dynamical structure; and third, the increased ability to relate state-space trajectories to externally measurable parameters.

For example, it is becoming possible to employ optogenetic techniques to briefly increase (or decrease) the firing rate of excitatory or inhibitory neurons in the cortex of the rhesus monkey (e.g., Diester et al. 2011, O'Shea et al. 2013). This can be accomplished at various times relative to withholding, preparing, or generating arm movements while simultaneously observing the resulting perturbation and recovery of the population dynamical state. This class of pump-probe experiment should enable more quantitative measurement of the neural dynamics in operation during various phases of the behavioral task and should help illuminate the nature and operation of the neural circuitry underlying these neural dynamics.

SUMMARY POINTS

1. Movement preparation has long been thought to be a critical step in generating movement. Recent work supports this idea and argues that achieving the correct preparatory state is important for producing the desired movement.
2. Measurements of the preparatory state predict both reaction time and trial-to-trial movement variability. Disruption of the preparatory state increases reaction time.
3. Preparatory activity is not a subthreshold version of movement activity but instead appears to serve as an initial state for dynamics that engage shortly before movement onset.
4. The onset of these dynamics is tied to movement onset rather than to the go cue and is predictive of trial-by-trial reaction time.
5. Neural responses during the movement appear complex but have at least some simple aspects: Dynamics can be approximated by a linear differential equation in which the same dynamics govern many reaching movements.
6. Because dynamics are similar across conditions, the pattern of movement-related activity is determined largely by the preparatory state.
7. The best linear approximation (to the true nonlinear dynamics) is dominated by rotational dynamics. Preparatory activity sets the amplitude and phase of the movement-period rotations.
8. The resulting firing rate patterns form a natural basis set for building more complex patterns such as muscle activity.

DISCLOSURE STATEMENT

The authors are not aware of any affiliations, memberships, funding, or financial holdings that might be perceived as affecting the objectivity of this review.

ACKNOWLEDGMENTS

We apologize in advance to all the investigators whose research could not be appropriately cited owing to space limitations. This work was supported by Burroughs Wellcome Fund Career Awards in the Biomedical Sciences (to K.V.S. and M.M.C.), DARPA REPAIR N66001-10-C-2010 (to K.V.S. and M.S.), NSF-NIH CRCNS NINDS R01NS054283 (to K.V.S. and M.S.), an NIH Director's Pioneer Award 1DP1OD006409 (to K.V.S.), the Gatsby Charitable Foundation (to M.S.), Searle Scholars Program (to M.M.C.), and NIH Director's New Innovator Award DP2NS083037 (to M.M.C.).

LITERATURE CITED

Abeles M, Bergman H, Gat I, Meilijson I, Seidemann E, et al. 1995. Cortical activity flips among quasi-stationary states. *Proc. Natl. Acad. Sci. USA* 92:8616–20

Afshar A, Santhanam G, Yu BM, Ryu SI, Sahani M, Shenoy KV. 2011. Single-trial neural correlates of arm movement preparation. *Neuron* 71:555–64

Ames KC, Ryu SI, Shenoy KV. 2012. Neural dynamics of reaching following incomplete or incorrect planning. *Front. Neurosci. Conf.: Comput. Syst. Neurosci.* (*COSYNE*). Abstr. T–5

Andersen RA, Buneo CA. 2002. Intentional maps in posterior parietal cortex. *Annu. Rev. Neurosci.* 25:189–220

Bastian A, Riehle A, Erlhagen W, Schöner G. 1998. Prior information preshapes the population representation of movement direction in motor cortex. *Neuroreport* 9:315–19

Bastian A, Schöner G, Riehle A. 2003. Preshaping and continuous evolution of motor cortical representations during movement preparation. *Eur. J. Neurosci.* 18:2047–58

Batista AP, Santhanam G, Yu BM, Ryu SI, Afshar A, Shenoy KV. 2007. Reference frames for reach planning in macaque dorsal premotor cortex. *J. Neurophysiol.* 98:966–83

Briggman KL, Abarbanel HD, Kristan WB Jr. 2005. Optical imaging of neuronal populations during decision-making. *Science* 307:896–901

Briggman KL, Abarbanel HD, Kristan WB Jr. 2006. From crawling to cognition: analyzing the dynamical interactions among populations of neurons. *Curr. Opin. Neurobiol.* 16:135–44

Broome BM, Jayaraman V, Laurent G. 2006. Encoding and decoding of overlapping odor sequences. *Neuron* 51:467–82

Brown TG. 1914. On the nature of the fundamental activity of the nervous centres; together with an analysis of the conditioning of rhythmic activity in progression, and a theory of the evolution of function in the nervous system. *J. Physiol.* 48:18–46

Buesing L, Macke JH, Sahani M. 2012. Learning stable, regularised latent models of neural population dynamics. *Network* 23:24–47

Bullock D, Grossberg S. 1988. Neural dynamics of planned arm movements: emergent invariants and speed-accuracy properties during trajectory formation. *Psychol. Rev.* 95:49–90

Caminiti R, Johnson PB, Galli C, Ferraina S, Burnod Y. 1991. Making arm movements within different parts of space: the premotor and motor cortical representation of a coordinate system for reaching to visual targets. *J. Neurosci.* 11:1182–97

Churchland MM, Afshar A, Shenoy KV. 2006a. A central source of movement variability. *Neuron* 52:1085–96

Churchland MM, Cunningham JP, Kaufman MT, Foster JD, Nuyujukian P, et al. 2012. Neural population dynamics during reaching. *Nature* 487:51–56

Churchland MM, Cunningham JP, Kaufman MT, Ryu SI, Shenoy KV. 2010a. Cortical preparatory activity: representation of movement or first cog in a dynamical machine? *Neuron* 68:387–400

Churchland MM, Santhanam G, Shenoy KV. 2006b. Preparatory activity in premotor and motor cortex reflects the speed of the upcoming reach. *J. Neurophysiol.* 96:3130–46

Churchland MM, Shenoy KV. 2007a. Delay of movement caused by disruption of cortical preparatory activity. *J. Neurophysiol.* 97:348–59

Churchland MM, Shenoy KV. 2007b. Temporal complexity and heterogeneity of single-neuron activity in premotor and motor cortex. *J. Neurophysiol.* 97:4235–57

Churchland MM, Yu BM, Cunningham JP, Sugrue LP, Cohen MR, et al. 2010b. Stimulus onset quenches neural variability: a widespread cortical phenomenon. *Nat. Neurosci.* 13:369–78

Churchland MM, Yu BM, Ryu SI, Santhanam G, Shenoy KV. 2006c. Neural variability in premotor cortex provides a signature of motor preparation. *J. Neurosci.* 26:3697–712

Churchland MM, Yu BM, Sahani M, Shenoy KV. 2007. Techniques for extracting single-trial activity patterns from large-scale neural recordings. *Curr. Opin. Neurobiol.* 17:609–18

Cisek P. 2006a. Integrated neural processes for defining potential actions and deciding between them: a computational model. *J. Neurosci.* 26:9761–70

Cisek P. 2006b. Preparing for speed. Focus on "Preparatory activity in premotor and motor cortex reflects the speed of the upcoming reach." *J. Neurophysiol.* 96:2842–43

Collinger JL, Wodlinger B, Downey JE, Wang W, Tyler-Kabara EC, et al. 2013. High-performance neuroprosthetic control by an individual with tetraplegia. *Lancet* 381:557–64

Crammond DJ, Kalaska JF. 2000. Prior information in motor and premotor cortex: activity during the delay period and effect on pre-movement activity. *J. Neurophysiol.* 84:986–1005

Day BL, Rothwell JC, Thompson PD, Maertens de Noordhout A, Nakashima K, et al. 1989. Delay in the execution of voluntary movement by electrical or magnetic brain stimulation in intact man. Evidence for the storage of motor programs in the brain. *Brain* 112(Pt. 3):649–63

Diester I, Kaufman MT, Mogri M, Pashaie R, Goo W, et al. 2011. An optogenetic toolbox designed for primates. *Nat. Neurosci.* 14:387–97

Elsayed G, Kaufman MT, Ryu SI, Shenoy KV, Churchland MM, Cunningham JP. 2013. Characterization of dynamical activity in motor cortex. *Front. Neurosci. Conf.: Comput. Syst. Neurosci. (COSYNE)*. Abstr. III-61

Erlhagen W, Bastian A, Jancke D, Riehle A, Schöner G. 1999. The distribution of neuronal population activation (DPA) as a tool to study interaction and integration in cortical representations. *J. Neurosci. Methods* 94:53–66

Erlhagen W, Schöner G. 2002. Dynamic field theory of movement preparation. *Psychol. Rev.* 109:545–72

Evarts EV. 1964. Temporal patterns of discharge of pyramidal tract neurons during sleep and waking in the monkey. *J. Neurophysiol.* 27:152–71

Evarts EV. 1968. Relation of pyramidal tract activity to force exerted during voluntary movement. *J. Neurophysiol.* 31:14–27

Faisal AA, Selen LPJ, Wolpert DM. 2008. Noise in the nervous system. *Nat. Rev. Neurosci.* 9:292–303

Fetz EE. 1992. Are movement parameters recognizably coded in the activity of single neurons? *Behav. Brain Sci.* 15:679–90

Fetz EE, Perlmutter SI, Prut Y. 2000. Functions of mammalian spinal interneurons during movement. *Curr. Opin. Neurobiol.* 10:699–707

Foster JD, Nuyujukian P, Freifeld O, Ryu SI, Black MJ, Shenoy KV. 2012. A framework for relating neural activity to freely moving behavior. *Conf. Proc. IEEE Eng. Med. Biol. Soc.* 2012:2736–39

Fu QG, Flament D, Coltz JD, Ebner T. 1995. Temporal encoding of movement kinematics in the discharge of primate primary motor and premotor neurons. *J. Neurophysiol.* 73:836–54

Ganguli S, Bisley JW, Roitman JD, Shadlen MN, Goldberg ME, Miller KD. 2008. One-dimensional dynamics of attention and decision making in LIP. *Neuron* 58:15–25

Georgopoulos AP, Crutcher MD, Schwartz AB. 1989. Cognitive spatial-motor processes 3. Motor cortical prediction of movement direction during an instructed delay period. *Exp. Brain Res.* 75:183–94

Georgopoulos AP, Kalaska JF, Caminiti R, Massey JT. 1982. On the relations between the direction of two-dimensional arm movements and cell discharge in primate motor cortex. *J. Neurosci.* 2:1527–37

Georgopoulos AP, Schwartz AB, Kettner RE. 1986. Neuronal population coding of movement direction. *Science* 233:1416–19

Gerits A, Farivar R, Rosen BR, Wald LL, Boyden ES, Vanduffel W. 2012. Optogenetically induced behavioral and functional network changes in primates. *Curr. Biol.* 22:1722–26

Ghez C, Favilla M, Ghilardi MF, Gordon J, Bermejo R, Pullman S. 1997. Discrete and continuous planning of hand movements and isometric force trajectories. *Exp. Brain Res.* 115:217–33

Ghez C, Hening W, Favilla M. 1989. Gradual specification of response amplitude in human tracking performance. *Brain Behav. Evol.* 33:69–74

Gilja V, Chestek CA, Diester I, Henderson JM, Deisseroth K, Shenoy KV. 2011. Challenges and opportunities for next-generation intracortically based neural prostheses. *IEEE Trans. Biomed. Eng.* 58:1891–99

Gilja V, Chestek CA, Nuyujukian P, Foster J, Shenoy KV. 2010. Autonomous head-mounted electrophysiology systems for freely behaving primates. *Curr. Opin. Neurobiol.* 20:676–86

Gilja V, Nuyujukian P, Chestek C, Cunningham J, Yu B, et al. 2012. A high-performance neural prosthesis enabled by control algorithm design. *Nat. Neurosci.* 15:1752–57

Graziano MSA. 2008. *The Intelligent Movement Machine: An Ethological Perspective on the Primate Motor System*, Vol. 224. Oxford/New York: Oxford Univ. Press

Graziano MSA. 2011a. Cables vs. networks: old and new views on the function of motor cortex. *J. Physiol.* 589:2439

Graziano MSA. 2011b. New insights into motor cortex. *Neuron* 71:387–88

Green AM, Kalaska JF. 2011. Learning to move machines with the mind. *Trends Neurosci.* 34:61–75

Grillner S. 2006. Biological pattern generation: the cellular and computational logic of networks in motion. *Neuron* 52:751–66

Hanes DP, Schall JD. 1996. Neural control of voluntary movement initiation. *Science* 274:427–30

Harvey CD, Coen P, Tank DW. 2012. Choice-specific sequences in parietal cortex during a virtual-navigation decision task. *Nature* 484:62–68

Hatsopoulos NG. 2005. Encoding in the motor cortex: Was Evarts right after all? Focus on "motor cortex neural correlates of output kinematics and kinetics during isometric-force and arm-reaching tasks." *J. Neurophysiol.* 94:2261–62

Hatsopoulos NG, Amit Y. 2012. Synthesizing complex movement fragment representations from motor cortical ensembles. *J. Physiol. Paris* 106:112–19

Hatsopoulos NG, Donoghue JP. 2009. The science of neural interface systems. *Annu. Rev. Neurosci.* 32:249–66

Hochberg LR, Bacher D, Jarosiewicz B, Masse NY, Simeral JD, et al. 2012. Reach and grasp by people with tetraplegia using a neurally controlled robotic arm. *Nature* 485:372–75

Hochberg LR, Serruya MD, Friehs GM, Mukand JA, Saleh M, et al. 2006. Neuronal ensemble control of prosthetic devices by a human with tetraplegia. *Nature* 442:164–71

Hocherman S, Wise SP. 1991. Effects of hand movement path on motor cortical activity in awake, behaving rhesus monkeys. *Exp. Brain Res.* 83:285–302

Kakei S, Hoffman DS, Strick PL. 1999. Muscle and movement representations in the primary motor cortex. *Science* 285:2136–39

Kalaska JF. 2009. From intention to action: motor cortex and the control of reaching movements. *Adv. Exp. Med. Biol.* 629:139–78

Kaufman MT, Churchland MM, Santhanam G, Yu BM, Afshar A, et al. 2010. Roles of monkey premotor neuron classes in movement preparation and execution. *J. Neurophysiol.* 104:799–810

Kaufman MT, Churchland MM, Shenoy KV. 2011. Cortical preparatory activity avoids causing movement by remaining in a muscle-neutral space. *Front. Neurosci. Conf.: Comput. Syst. Neurosci. (COSYNE)*. Abstr. II-61

Keele SW. 1968. Movement control in skilled motor performance. *Psychol. Bull.* 70:387–403

Kopell N, Ermentrout GB. 2002. Mechanisms of phase-locking and frequency control in pairs of coupled neural oscillators. In *Handbook on Dynamical Systems*, Vol. 2: *Toward Applications*, ed. B Fiedler, pp. 3–54. Philadelphia: Elsevier

Kurata K. 1989. Distribution of neurons with set- and movement-related activity before hand and foot movements in the premotor cortex of rhesus monkeys. *Exp. Brain Res.* 77:245–56

Kutas M, Donchin E. 1974. Studies of squeezing: handedness, responding hand, response force, and asymmetry of readiness potential. *Science* 186:545–48

Lisberger SG, Sejnowski TJ. 1992. Motor learning in a recurrent network model based on the vestibulo-ocular reflex. *Nature* 360:159–61

Machens CK, Romo R, Brody CD. 2010. Functional, but not anatomical, separation of "what" and "when" in prefrontal cortex. *J. Neurosci.* 30:350–60

Macke JH, Büsing L, Cunningham JP, Yu BM, Shenoy KV, and Sahani M. 2011. Empirical models of spiking in neural populations. See Shawe-Taylor et al. 2011, pp. 1350–58

Mandelblat-Cerf Y, Paz R, Vaadia E. 2009. Trial-to-trial variability of single cells in motor cortices is dynamically modified during visuomotor adaptation. *J. Neurosci.* 29:15053–62

Manwani A, Steinmetz PN, Koch C. 2002. The impact of spike timing variability on the signal-encoding performance of neural spiking models. *Neural Comput.* 14:347–67

Maynard EM, Hatsopoulos NG, Ojakangas CL, Acuna BD, Sanes JN, et al. 1999. Neuronal interactions improve cortical population coding of movement direction. *J. Neurosci.* 19:8083–93

Mazor O, Laurent G. 2005. Transient dynamics versus fixed points in odor representations by locust antennal lobe projection neurons. *Neuron* 48:661–73

McGuire LM, Sabes PN. 2011. Heterogeneous representations in the superior parietal lobule are common across reaches to visual and proprioceptive targets. *J. Neurosci.* 31:6661–73

Miles FA, Evarts EV. 1979. Concepts of motor organization. *Annu. Rev. Psychol.* 30:327–62

Moran DW, Schwartz AB. 1999. Motor cortical representation of speed and direction during reaching. *J. Neurophysiol.* 82:2676–92

Mountcastle VB, Lynch JC, Georgopoulos A, Sakata H, Acuna C. 1975. Posterior parietal association cortex of the monkey: command functions for operations within extrapersonal space. *J. Neurophysiol.* 38:871–908

Mullette-Gillman OA, Cohen YE, Groh JM. 2009. Motor-related signals in the intraparietal cortex encode locations in a hybrid, rather than eye-centered reference frame. *Cereb. Cortex* 19:1761–75

Mussa-Ivaldi FA. 1988. Do neurons in the motor cortex encode movement direction? An alternative hypothesis. *Neurosci. Lett.* 91:106–11

O'Shea D, Goo W, Kalanithi P, Diester I, Ramakrishnan C, et al. 2013. Neural dynamics following optogenetic disruption of motor preparation. *Front. Neurosci. Conf.: Comput. Syst. Neurosci. (COSYNE)*. Abstr. III-60

Pearce TM, Moran DW. 2012. Strategy-dependent encoding of planned arm movements in the dorsal premotor cortex. *Science* 337:984–88

Pesaran B, Nelson MJ, Andersen RA. 2006. Dorsal premotor neurons encode the relative position of the hand, eye, and goal during reach planning. *Neuron* 51:125–34

Petreska B, Yu BM, Cunningham JP, Santhanam G, Ryu SI, et al. 2011. Dynamical segmentation of single trials from population neural data. See Shawe-Taylor et al. 2011, pp. 756–64

Rabinovich M, Huerta R, Laurent G. 2008. Transient dynamics for neural processing. *Science* 321:48–50

Reimer J, Hatsopoulos NG. 2009. The problem of parametric neural coding in the motor system. *Adv. Exp. Med. Biol.* 629:243–59

Requin J, Riehle A, Seal J. 1988. Neuronal activity and information processing in motor control: from stages to continuous flow. *Biol. Psychol.* 26:179–98

Rickert J, Riehle A, Aertsen A, Rotter S, Nawrot MP. 2009. Dynamic encoding of movement direction in motor cortical neurons. *J. Neurosci.* 29:13870–82

Riehle A, Grün S, Diesmann M, Aertsen A. 1997. Spike synchronization and rate modulation differentially involved in motor cortical function. *Science* 278:1950–53

Riehle A, Requin J. 1989. Monkey primary motor and premotor cortex: single-cell activity related to prior information about direction and extent of an intended movement. *J. Neurophysiol.* 61:534–49

Rokni U, Sompolinsky H. 2012. How the brain generates movement. *Neural Comput.* 24:289–331

Rosenbaum DA. 1980. Human movement initiation: specification of arm, direction, and extent. *J. Exp. Psychol. Gen.* 109:444–74

Sawaguchi T, Yamane I, Kubota K. 1996. Application of the GABA antagonist bicuculline to the premotor cortex reduces the ability to withhold reaching movements by well-trained monkeys in visually guided reaching task. *J. Neurophysiol.* 75:2150–56

Schöner G. 2004. Dynamical systems approaches to understanding the generation of movement by the nervous system. In *Progress in Motor Control*, Vol. 3, ed. ML Latash, MF Levin. Champaign, IL: Hum. Kinet.

Scott SH. 2000. Population vectors and motor cortex: neural coding or epiphenomenon? *Nat. Neurosci.* 3:307–8

Scott SH. 2004. Optimal feedback control and the neural basis of volitional motor control. *Nat. Rev. Neurosci.* 5:532–46

Scott SH. 2008. Inconvenient truths about neural processing in primary motor cortex. *J. Physiol.* 586:1217–24

Scott SH, Kalaska JF. 1995. Changes in motor cortex activity during reaching movements with similar hand paths but different arm postures. *J. Neurophysiol.* 73:2563–67

Seidemann E, Meilijson I, Abeles M, Bergman H, Vaadia E. 1996. Simultaneously recorded single units in the frontal cortex go through sequences of discrete and stable states in monkeys performing a delayed localization task. *J. Neurosci.* 16:752–68

Shawe-Taylor J, Zemel RS, Bartlett P, Pereira F, Weinberger KQ, eds. 2011. *Advances in Neural Information Processing Systems* (*NIPS*), Vol. 24. Red Hook, NY: Curran

Shenoy KV, Kaufman MT, Sahani M, Churchland MM. 2011. A dynamical systems view of motor preparation: implications for neural prosthetic system design. *Prog. Brain Res.* 192:33–58

Skavenski AA, Robinson DA. 1973. Role of abducens neurons in vestibuloocular reflex. *J. Neurophysiol.* 36:724–38

Smith AC, Frank LM, Wirth S, Yanike M, Hu D, et al. 2004. Dynamic analysis of learning in behavioral experiments. *J. Neurosci.* 24:447–61

Snyder LH, Batista AP, Andersen RA. 1997. Coding of intention in the posterior parietal cortex. *Nature* 386:167–70

Stopfer M, Jayaraman V, Laurent G. 2003. Intensity versus identity coding in an olfactory system. *Neuron* 39:991–1004

Sussillo D, Abbott LF. 2009. Generating coherent patterns of activity from chaotic neural networks. *Neuron* 63:544–57

Sussillo D, Churchland MM, Kaufman MT, Shenoy KV. 2013. A recurrent neural network that produces EMG from rhythmic dynamics. *Front. Neurosci. Conf.: Comput. Syst. Neurosci.* (*COSYNE*). Abstr. III-67

Szuts TA, Fadeyev V, Kachiguine S, Sher A, Grivich MV, et al. 2011. A wireless multi-channel neural amplifier for freely moving animals. *Nat. Neurosci.* 14:263–69

Tanji J, Evarts EV. 1976. Anticipatory activity of motor cortex neurons in relation to direction of an intended movement. *J. Neurophysiol.* 39:1062–68

Todorov E. 2000. Direct cortical control of muscle activation in voluntary arm movements: a model. *Nat. Neurosci.* 3:391–98

Todorov E, Jordan MI. 2002. Optimal feedback control as a theory of motor coordination. *Nat. Neurosci.* 5:1226–35

Vaadia E, Haalman I, Abeles M, Bergman H, Prut Y, et al. 1995. Dynamics of neuronal interactions in monkey cortex in relation to behavioural events. *Nature* 373:515–18

van Beers RJ, Haggard P, Wolpert DM. 2004. The role of execution noise in movement variability. *J. Neurophysiol.* 91:1050–63

Wang W, Chan SS, Heldman DA, Moran DW. 2010. Motor cortical representation of hand translation and rotation during reaching. *J. Neurosci.* 30:958–62

Weinrich M, Wise SP. 1982. The premotor cortex of the monkey. *J. Neurosci.* 2:1329–45

Wise SP. 1985. The primate premotor cortex: past, present, and preparatory. *Annu. Rev. Neurosci.* 8:1–19

Wise SP, Weinrich M, Mauritz KH. 1986. Movement-related activity in the premotor cortex of rhesus macaques. *Prog. Brain Res.* 64:117–31

Yu BM, Afshar A, Santhanam G, Ryu SI, Shenoy KV, Sahani M. 2006. Extracting dynamical structure embedded in neural activity. In *Advances in Neural Information Processing Systems (NIPS), Vol. 18*, ed. Y Weiss, B Scholkopf, J Platt, pp. 1545–52. Cambridge, MA: MIT Press

Yu BM, Cunningham JP, Santhanam G, Ryu SI, Shenoy KV, Sahani M. 2009. Gaussian-process factor analysis for low-dimensional single-trial analysis of neural population activity. *J. Neurophysiol.* 102:614–35

Yuste R, MacLean JN, Smith J, Lansner A. 2005. The cortex as a central pattern generator. *Nat. Rev. Neurosci.* 6:477–83

The Genetics of Hair Cell Development and Regeneration

Andrew K. Groves,[1,2] Kaidi D. Zhang,[3] and Donna M. Fekete[3,4]

[1]Departments of Neuroscience and Molecular and Human Genetics and [2]Program in Developmental Biology, Baylor College of Medicine, Houston, Texas 77030; email: akgroves@bcm.edu

[3]Department of Biological Sciences and [4]Purdue University Center for Cancer Research, Purdue University, West Lafayette, Indiana 47907; email: dfekete@purdue.edu

Annu. Rev. Neurosci. 2013. 36:361–81

First published online as a Review in Advance on May 29, 2013

The *Annual Review of Neuroscience* is online at neuro.annualreviews.org

This article's doi:
10.1146/annurev-neuro-062012-170309

Keywords

cochlea, Atoh1, miRNA, Notch, cell cycle

Abstract

Sensory hair cells are exquisitely sensitive vertebrate mechanoreceptors that mediate the senses of hearing and balance. Understanding the factors that regulate the development of these cells is important, not only to increase our understanding of ear development and its functional physiology but also to shed light on how these cells may be replaced therapeutically. In this review, we describe the signals and molecular mechanisms that initiate hair cell development in vertebrates, with particular emphasis on the transcription factor Atoh1, which is both necessary and sufficient for hair cell development. We then discuss recent findings on how microRNAs may modulate the formation and maturation of hair cells. Last, we review recent work on how hair cells are regenerated in many vertebrate groups and the factors that conspire to prevent this regeneration in mammals.

Contents

INTRINSIC AND EXTRINSIC SIGNALS THAT DRIVE HAIR CELL INDUCTION

The inner ear develops from a thickened patch of cranial ectoderm, the otic placode, and then invaginates or cavitates to form a spherical otocyst. At this point, the ear primordium has already received signals that confer regional identity and polarity to the otocyst (Groves & Fekete 2012). These signals lead to the formation of multiple prosensory patches, each of which will give rise to a particular inner-ear sensory organ: the three cristae at the base of each semicircular canal, the two maculae, and the auditory epithelium of the cochlea (Alsina et al. 2009, Bok et al. 2007). These patches are characterized by the expression of the transcription factor Sox2, and mice carrying hypomorphic mutations for *Sox2* lack all sensory cells in the inner ear (Kiernan et al. 2005). These sensory patches then differentiate to produce the hair cells and supporting cells of each inner-ear sensory organ.

The Temporal and Spatial Regulation of Hair Cell Differentiation

After *Sox2*$^{+}$ prosensory tissue has been induced in each sensory organ, the prosensory domain begins to exit the cell cycle and terminally differentiate into hair cells. In most vertebrates, including the vestibular system of mammals, exit from the cell cycle and the appearance of the first markers of hair cells are tightly coupled. In the vestibular system, differentiation typically begins near the center of each prosensory patch and expands out over an extended period of time. For example, the first hair cells appear in the future striolar region of the mouse utricle at embryonic day (E) 11 (Raft et al. 2007); however, more than half of the total hair cells are generated after birth and small numbers of hair cells are generated from mitotic progenitors between postnatal days 12 and 14 (Burns et al. 2012b, Kirkegaard & Nyengaard 2005). In the case of the chicken hearing organ, the basilar papilla, the first hair cells (and supporting cells) are born on the superior side of the basal cochlea and down the middle of the apical cochlea beginning at E6. Over the next three days, this cell cycle withdrawal spreads inferiorly and proximally in the base and inferiorly, superiorly, and distally in the apex (Katayama & Corwin 1989).

The mammalian organ of Corti has a strikingly different arrangement of hair cells and supporting cells compared with all other vertebrate sensory patches. Instead of a quasi-hexagonal arrangement, where each hair cell is surrounded by between 4 and 8 supporting cells depending on its position in the sensory epithelium (Goodyear & Richardson 1997), hair cells and supporting cells are arranged in uniform rows and invariant proportions along the length of the cochlear duct (Kelley 2006). This serially repeating pattern is generated by a highly unusual pattern of cell cycle exit and differentiation. In the mouse, the prosensory domain of the future organ of Corti begins to exit the cell cycle in the apical tip of the cochlea at E12 (Lee et al. 2006, Matei et al. 2005, Ruben 1967), and a wave of cell cycle exit then proceeds along the prosensory domain from apex to base over the next 48–60 h; some cells in the most basal region still incorporate mitotic labels at E14.5–E15.0 (Lee et al. 2006). Starting at about E13.5, cells in the midbasal region of the cochlea begin to differentiate into hair cells by expressing the transcription factor Atoh1 (Chen et al. 2002), and this region of differentiating cells spreads down to the apex over the next three to four days. Thus, the first cells to exit the cell cycle in the apex of the cochlear duct are the last ones to differentiate terminally into hair cells five days later, whereas the last cells to exit the cell cycle in the midbasal region are some of the first to differentiate into hair cells (**Figure 1**). This dramatic temporal and spatial uncoupling of cell cycle exit and differentiation has no parallel in any other vertebrate tissue. When maturation is complete, numerous morphological, physiological, and molecular properties of the cochlear duct and its resident cells vary systematically along this longitudinal axis and are responsible for the gradient of selectivity to sounds of different frequencies (**Figure 1**). In addition, the differentiating hair cells adopt a polarized orientation of their

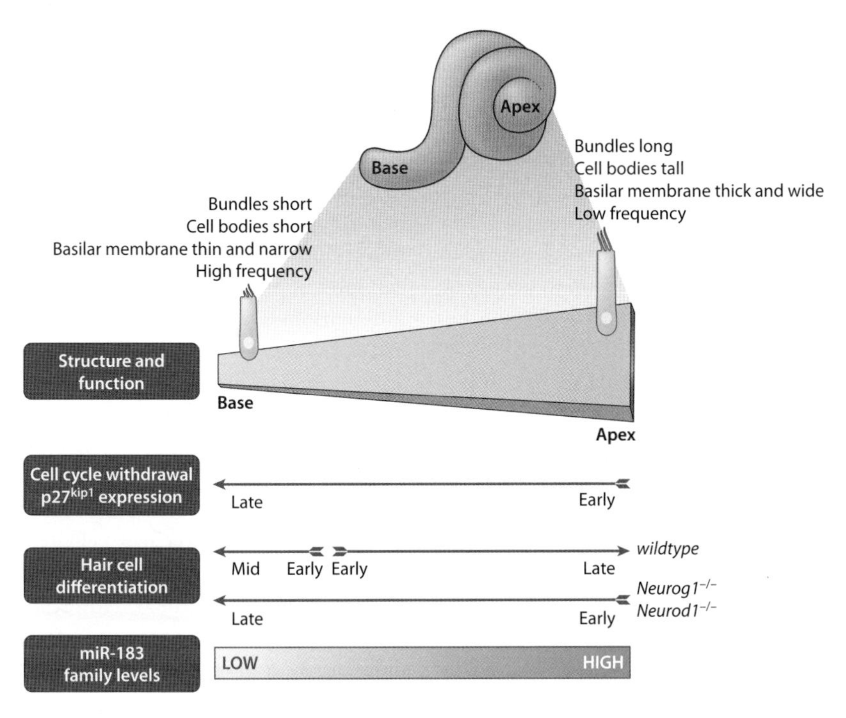

Figure 1

Longitudinal gradients of the mammalian organ of Corti in normal and mutant mice. The cochlea coils from base to apex and exhibits systematic gradients in the dimensions of its fluid-filled chambers; the width and thickness of the basilar membrane (shown uncoiled) are also shown. Sitting on the basilar membrane is the delicate organ of Corti. The physical dimensions of hair cells (shown schematically for one outer hair cell at each extreme) and supporting cells (not shown) also change systematically. Arrows depict temporal gradients in cell cycle exit, $p27^{kip1}$ expression, and hair cell differentiation in normal and mutant mice. In the mature organ of Corti, the miR-183 family is expressed in a gradient from base (lowest levels) to apex (highest levels).

stereociliary bundles that requires signaling from the planar cell polarity pathway (Jones & Chen 2007, May-Simera & Kelley 2012, Montcouquiol et al. 2006, Rida & Chen 2009).

The mechanisms that propagate the apical-basal gradient of cell cycle exit and the midbasal-apical gradient of differentiation in the mammalian cochlea are poorly understood. The cyclin-dependent kinase inhibitor $p27^{kip1}$ is upregulated in the cochlea in an apical-basal gradient concomitant with the gradient of cell cycle exit (**Figure 1**) (Chen & Segil 1999, Lee et al. 2006) and is necessary for the correct sequence of cell cycle exit and generation of appropriate numbers of hair cells and supporting cells (Chen & Segil 1999, Kanzaki et al. 2006, Lowenheim et al. 1999). Its expression is regulated at both the transcriptional and protein levels (Lee et al. 2006), but no signals have been identified that directly regulate its expression in the cochlea. The wave of hair cell differentiation does not appear to require contact with the underlying mesenchyme because it can proceed normally when the embryonic cochlear epithelium is maintained alone in organ culture (Montcouquiol & Kelley 2003). Moreover, there is no absolute requirement for a fully intact cochlear epithelium to propagate differentiation signals in a planar fashion because the correct basal-apical sequence of differentiation can be recapitulated even when the cochlear epithelium is cut into a series of pieces along its length and cultured separately (Montcouquiol & Kelley 2003). Genetic experiments suggest a role for the transcription factors Neurog1 and Neurod1 in these gradients. The cochlear prosensory region of *Neurog1* mutant mice shows a premature exit from the cell cycle of ~24–36 h earlier than normal, and hair cell differentiation, assayed by expression of Atoh1, commences in the apical, not the basal, region of the *Neurog1* mutant cochlea (Matei et al. 2005). This disordered pattern of hair cell differentiation is also seen in *Neurod1* mutant mice (**Figure 1**) (Jahan et al. 2010). It is important to note that neither Neurod1 nor Neurog1 are expressed in the cochlea at detectable levels during this period of differentiation, which suggests that they play an indirect role in triggering the timing of prosensory domain cell cycle exit and differentiation. Both transcription factors are necessary for the differentiation of the spiral ganglion because both *Neurod1* and *Neurog1* mutant mice have spiral ganglia that are either greatly reduced or completely absent. This finding suggests the developing spiral ganglion may regulate the tempo of cochlear hair cell differentiation.

In addition to the basal-apical gradient of differentiation in the cochlea, hair cell differentiation also proceeds in a neural-abneural direction at any given position along the length of the cochlear duct: Inner hair cells differentiate before outer hair cells, and pillar cells separate the two differentiating populations. Although birds do not display the same dramatic discontinuity and separation between inner and outer hair cells, they do show a neural-abneural gradient in hair cell length between tall and short hair cells (Gleich et al. 2004). At present it is not clear whether the temporal separation between inner and outer hair cell differentiation in the organ of Corti reflects distinct sets of inductive signals or a common inducing signal that is interpreted at different times or by different cellular contexts in the inner and outer hair cell regions. For example, recent evidence suggests that a gradient of bone morphogenic protein (BMP) signaling normally spreads from the abneural to the neural side of the cochlear duct and that perturbation of this gradient can change the numbers of both inner and outer hair cells (Groves & Fekete 2012, Hwang et al. 2010, Ohyama et al. 2010, Puligilla et al. 2007). However, several mouse mutants display specific defects in the differentiation of outer versus inner hair cells. For example, *Emx2* mutant mice develop inner hair cells but not outer hair cells or Deiters' cells (Holley et al. 2010), and *Jag1* mutant mice develop multiple rows of inner hair cells but also lack outer hair cells and Deiters' cells along parts of the cochlea (Brooker et al. 2006, Kiernan et al. 2006). *FGF20* mutant mice also lack outer hair cells and Deiters' cells, a phenotype that seems to be caused by a failure of the prosensory domain to differentiate correctly (Huh et al. 2012). These data suggest that regions containing inner versus outer hair cells may differentiate as distinct developmental units.

Atoh1: The Key Initiator of Hair Cell Development

The first transcription factor to be expressed in differentiating hair cell progenitors is the basic helix-loop-helix (bHLH) transcription factor Atoh1. *Atoh1* is expressed prior to all other known hair cell genes and is later downregulated in maturing hair cells before they begin to participate in hearing and balance (Mulvaney & Dabdoub 2012). *Atoh1* mutant mice lack hair cells in all sensory regions of the ear (Bermingham et al. 1999); this deficiency is due to hair cell progenitors that fail to differentiate and then die (Chen et al. 2002, Pan et al. 2011). Together, these expression and loss-of-function data suggest that Atoh1 is required to initiate the hair cell differentiation program but is not required for the function of mature hair cells. Atoh1 is also sufficient to generate hair cells when ectopically expressed in some locations in the developing inner ear (Izumikawa et al. 2005, Kawamoto et al. 2003, Kelly et al. 2012, Liu et al. 2012a, Zheng & Gao 2000).

Despite the central role played by Atoh1 in the first steps of hair cell differentiation, surprisingly little is known about the factors that regulate its expression in hair cell progenitors. Manipulation of the BMP, fibroblast growth factor (FGF), sonic hedgehog (Shh), and Wnt signaling pathways can alter the numbers of hair cells in inner-ear sensory organs (Driver et al. 2008, Hayashi et al. 2008, Huh et al. 2012, Kamaid et al. 2010, Li et al. 2005, Pirvola et al. 2002, Pujades et al. 2006, Stevens et al. 2003), although it is far less clear to what extent these effects are due to the direct regulation of *Atoh1* transcription, to indirect transcriptional regulation by inducing other factors that can themselves regulate *Atoh1*, or to posttranslational modification of the Atoh1 protein (Mulvaney & Dabdoub 2012). Far more attention has focused on regions of the *Atoh1* locus that regulate its transcription, particularly on two enhancer elements located 3.4 kb downstream of the *Atoh1* coding region (**Figure 2**) (Helms et al. 2000). These enhancers can drive reporter gene expression in the nervous system and in hair cells of transgenic mice (Lumpkin et al. 2003) and are strongly conserved in mammals (Helms et al. 2000). Atoh1 positively autoregulates its own expression by binding to an E-box motif in this enhancer complex (Helms et al. 2000). This autoregulation is necessary for the continued expression of *Atoh1* in hair cells because *Atoh1* mutant mice cannot maintain expression in hair cells of a green fluorescent protein (GFP) reporter transgene containing the autoregulatory enhancer (Raft et al. 2007). However, we do not know whether other regulatory sequences in this enhancer initiate expression of *Atoh1* before Atoh1 itself binds to its own enhancer or whether other enhancers located elsewhere in the *Atoh1* locus are responsible for the first burst of *Atoh1* transcription.

The autoregulatory enhancer region of *Atoh1* contains candidate binding sites for a menagerie of transcriptional activators and repressors, some of which bind directly to the enhancer by biochemical assays (**Figure 2**) (Akazawa et al. 1995, Briggs et al. 2008, D'Angelo et al. 2010, Ebert et al. 2003, Mutoh et al. 2006, Shi et al. 2010). Some of these transcription factors can also activate or repress *Atoh1* expression in vivo or in vitro, although it is not always clear whether these factors work by directly binding the *Atoh1* enhancer or are acting indirectly or through other, as yet unidentified, elements in the *Atoh1* locus. Recent studies have demonstrated that Six1 and its transcriptional coactivator Eya1, which are expressed in the prosensory domain of the cochlea, can bind directly to the *Atoh1* autoregulatory enhancer (Ahmed et al. 2012). These two factors are sufficient to induce *Atoh1* expression when electroporated into the greater epithelial ridge adjacent to the organ of Corti (Ahmed et al. 2012). This activation of *Atoh1* can be potentiated by the transcription factor Sox2, which is strongly expressed in all

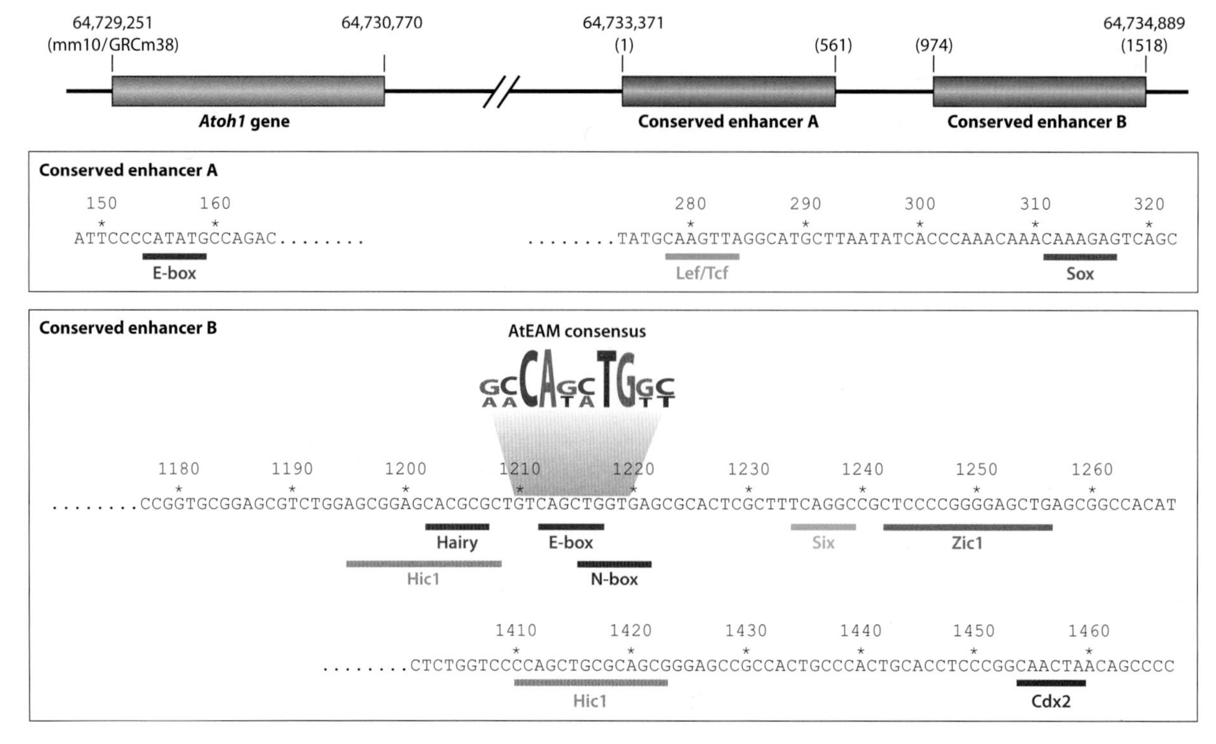

Figure 2

A diagram of the mouse *Atoh1* locus showing the position of its 3′ autoregulatory enhancer. Numbers refer to the position on mouse chromosome 6, according to the current build of the mouse genome (mm10/GRCm38). The *Atoh1* autoregulatory enhancer consists of two conserved elements, A and B. Transcription factor binding sites that have been experimentally verified are shown in color on relevant regions of the enhancer sequence. Two candidate binding sites, a hairy-preferred motif and an N-box, are shown in black; although these have not been experimentally verified, they have been featured in many previous reviews of this regulatory region. The *Atoh1* E-box associated motif consensus sequence for Atoh1 binding identified by Klisch et al. (2011) is shown aligned to the likely E-box site in enhancer B.

prosensory domains of the inner ear and is necessary for each prosensory domain to differentiate (Kiernan et al. 2005). Sox2 is likely also sufficient to activate *Atoh1* and induce ectopic hair formation in the chick otocyst (Neves et al. 2012). *Sox2* is rapidly downregulated in hair cells as they differentiate; this downregulation appears to be required for further maturation of hair cells because sustained expression of *Sox2* in Atoh1-expressing cells blocks the induction of later hair cell markers such as Myosin 7a (Ahmed et al. 2012, Puligilla et al. 2010).

The Transcriptional Control of Hair Cell Differentiation by Atoh1

Despite the generally accepted importance of Atoh1 in hair cell development, we know very little about the molecular basis by which Atoh1 regulates hair cell differentiation. It is conceivable that Atoh1 acts as a delegator, sitting atop a large hierarchy of regulatory transcription factors, each devoted to controlling a different aspect of hair cell differentiation. At the other extreme, Atoh1 might act as a micromanager, by directly regulating many different aspects of hair cell development and maturation. Moreover, it is not clear whether Atoh1 simply regulates generic aspects of cell identity shared by hair cells in all inner-ear sensory organs, or whether it can regulate genes that distinguish between different classes of hair cells, such as inner hair cells versus outer hair cells in the cochlea or type I versus type II vestibular hair cells.

Current efforts to identify direct targets of transcription factors typically involve chromatin immunoprecipitation of the transcription factor bound to DNA in the cells of interest, followed by high-throughput sequencing of the DNA (ChIP-Seq). The application of these approaches to finding Atoh1 targets in hair cells has been confounded by the relatively small number of hair cells in mammalian sensory organs and by the large number of other cell types in the cochlea. However, Klisch et al. (2011) recently used ChIP-sequencing to identify Atoh1 targets in cerebellar granule cells, which can be obtained in large numbers and which represent ~50% of the total cells in a neonatal mouse cerebellum. This work supports the idea that Atoh1 is more of a micromanager than a delegator: It regulates genes associated with not only transcription, but also cell division, chromosomal organization, metabolism, cell migration, and cell adhesion (Klisch et al. 2011). Clearly many of these genes are unlikely to be expressed in postmitotic and nonmigratory hair cells, and indeed, a recent comparison of the ChIP-sequencing data from the cerebellum with gene expression data sets from cerebellar granule neurons, dorsal spinal interneurons, and hair cells have identified only three genes—*Rab15*, *Selm* and *Atoh1* itself—that are candidates to be direct targets of Atoh1 in all three cell types (Lai et al. 2011). However, the ChIP-sequencing results from the cerebellum have also led to the identification of a consensus Atoh1 binding motif (G/A,C/A,CA,G/T,C/A,TG,G/T,C/T) that is longer and more specific than the generic E-box sequence (CANNTG) bound by many bHLH transcription factors (**Figure 2**) (Klisch et al. 2011). This *Atoh1* E-box associated motif (AtEAM) is present in the majority of sequences bound by Atoh1 in the cerebellum and thus may be useful to predict candidate Atoh1 targets in hair cells. The significance of the specificity of the Atoh1-binding AtEAM site was recently demonstrated by replacing the *Atoh1* coding sequence with that of the closely related bHLH factor *Neurog1*, which is expressed in precursors of inner ear neurons but not hair cells. Mice homozygous for this replacement allele have a severely disrupted organ of Corti containing very few hair cells (Jahan et al. 2012). The surviving hair cells were extremely immature and lacked significant stereocilia, although this phenotype was not as severe as that seen in *Atoh1* null mice. This finding suggests that bHLH transcription factors that are closely related to Atoh1 are unable to activate sufficient Atoh1 target genes correctly to promote correct hair cell differentiation and survival, even when expressed in the correct cells at the correct time.

ChIP-Seq: chromatin immunoprecipitation and deep sequencing

miRNA: microRNA

MODULATION OF HAIR CELL GENE EXPRESSION THROUGH microRNAs

MicroRNAs (miRNAs) are widely appreciated for their role in posttranscriptional repression of gene expression (Hobert 2007, 2008; Kloosterman & Plasterk 2006; Takacs & Giraldez 2010). miRNAs are single-stranded RNAs, usually 21–23 nucleotides in length, that are generated from long RNA polymerase II transcripts and then processed through a series of double-stranded-specific ribonuclease III enzymes such as Drosha and Dicer (Esquela-Kerscher & Slack 2006, Kosik 2006, Winter et al. 2009). Mature miRNAs ultimately bind to target messenger RNA (mRNA) transcripts, which either destabilizes the targets or lowers their translation efficiency (Filipowicz et al. 2008, Guo et al. 2010, Lewis & Steel 2010, Liu et al. 2008). The resulting repression in protein levels rarely exceeds 50%, at least for the small number of miRNAs that have been tested to date (Baek et al. 2008, Bartel 2009, Selbach et al. 2008). Thus, miRNAs are modulators of protein expression levels rather than "off" switches. Several comprehensive reviews have recently highlighted the literature related to inner-ear miRNAs (Friedman et al. 2009, Rudnicki & Avraham 2012, Soukup 2009, Weston & Soukup 2009). In this section, we narrow the focus to the expression and manipulation of miRNAs and their targets in hair cells and hearing.

Manipulating miRNA Levels in Developing Hair Cells

To evaluate the role of miRNAs in hair cell development and maintenance, the miRNA processing enzyme, Dicer, has been conditionally deleted in the mouse inner ear. Several different Cre driver lines have been used to eliminate a floxed Dicer allele at different developmental time points. Foxg1-Cre and Pax2-Cre lines excise Dicer at the otocyst stages (Soukup et al. 2009), Atoh1-Cre recombines the locus about the time the hair cells are born (Weston et al. 2011), and Pou4f3-Cre recombines shortly thereafter (Friedman et al. 2009). These conditional mutants each show defects in hair cell development; the severity depends on the stage of Dicer excision and probably also on the amount and duration of residual miRNA expression that persists after Cre onset (see discussion by Soukup 2009). In general, however, deleting Dicer in hair cells leads to disrupted hair bundle morphology and subsequent hair cell death.

A major caveat of this approach is that phenotypes resulting from Dicer knockouts may not necessarily reflect a worst-case scenario that reveals the full role played by miRNAs. This is because some miRNAs, as well as the transcripts they repress, may act in direct opposition to others that are coexpressed at the same time and place. As a result, the loss of a single species of miRNA (or miRNA family) may throw the system further out of equilibrium than would the loss of two opposing miRNAs or the blockade of all miRNA processing. Such a scenario is seen in the wing imaginal disc of the fly larvae, where results from the absence of miR-9 alone are more severe than those from the deletion of Dicer (Bejarano et al. 2010).

Let us consider now the necessity of miRNA processing for proper hair cell development and maturation. The miR-183 family includes three members (miR-96, miR-182, and miR-183) that are expressed at high levels in young hair cells, are undetectable in supporting cells, and are also present in the inner-ear ganglion neurons (Li et al. 2010a, Weston et al. 2006). The literature is divided about whether these miRNAs remain confined to hair cells in the postnatal organ of Corti (Sacheli et al. 2009, Weston et al. 2011). Applying whole-mount in situ hybridization methods to the cochlear epithelium, Weston and colleagues reported that all 3 miR-183 family members behave similarly and remain restricted to hair cells. Expression is stronger in the base than the apex at P0 but then disappears from the base to establish a gradient from apex (high) to base (low) that persists into adulthood (**Figure 1**) (Sacheli et al. 2009, Weston et al. 2011).

One is thus tempted to speculate that the molecular basis of anatomical variation and frequency selectivity that is systematically arrayed along the length of the cochlea (shown schematically in **Figure 1**) is modulated, and perhaps even enforced, by systematic differences in the levels of the miR-183 family and its targets. If so, then loss of this gene family should lead to weaker longitudinal gene expression gradients. A counterintuitive result is obtained when comparing the expression patterns of the miR-183 family with the phenotype observed when Dicer is deleted selectively in hair cells using the Atoh1-Cre driver line. Here, Dicer deletion using Atoh1-Cre enhances transcript gradients (Sacheli et al. 2009, Weston et al. 2011). This observation leads to the conclusion that, in toto, hair-cell-expressed miRNAs serve to repress longitudinal transcript gradients rather than to promote them. Resolution of this paradox must await experiments that selectively delete only the miR-183 family to discover whether they share this *en masse* effect of miRNA loss. So far, studies in zebrafish have revealed that the number of hair cells in developing macular sensory organs is affected by levels of miR-183 family members: Overexpression of miR-96 or -182 initially yields more hair cells, whereas knockdown of miR-96, -182, and/or -183 generates fewer hair cells (Li et al. 2010a). Although these studies emphasize the impact of the miR-183 family on hair cell fate specification, they cannot address the question of longitudinal maturation gradients because zebrafish do not have a cochlea.

miR-96 Mutations Underlie Inherited Hearing Loss in Mice and Humans

Progressive, nonsyndromic autosomal semidominant or dominant hearing loss has been associated with miR-96 mutations in *Diminuendo* mice and in DFNA50 human families, respectively (Lewis et al. 2009, Mencia et al. 2009). *Diminuendo* hair cells are born and initiate differentiation but become stalled at an early postnatal stage and fail to differentiate fully into inner and outer hair cells (Kuhn et al. 2011). It is uncertain to what degree this phenotype reflects the loss-of-function of the wild type allele versus a gain-of-function effect of the novel allele in repressing a new set of target genes. One fact arguing in favor of the former mechanism is that two human families have unique point mutations in the seed region of miR-96 [and thus could interact with unique sets of potential new targets (Mencia et al. 2009)], whereas a third family has a mutation in the precursor miRNA transcript that results in poor expression levels of the mature miR-96 (Solda et al. 2012). Despite these allele differences among the human DFNA50 families and the *Diminuendo* mice, progressive hearing loss is a common outcome. To address the underlying cause of the hair cell phenotype in *Diminuendo* homozygous mice, cochlear tissue was subjected to gene expression profiling in an effort to identify target transcripts (Lewis et al. 2009). As might be expected, genes with 3′UTRs carrying heptamer sequences complementary to the seed region of miR-96 were enriched in the pool of transcripts that were upregulated in the organ of Corti in *Diminuendo* homozygotes. Likewise, the binding site for the *Diminuendo* allele of miR-96 was enriched in the pool of downregulated transcripts. However, binding sites from this latter group of mouse genes were not conserved in human and rat orthologs. Ultimately, distinguishing gain-of-function from loss-of-function mechanisms may be revealed by genetic rescue of *Diminuendo* hair cells.

Evaluation of Putative Targets for Hair-Cell-Enriched miRNAs

Although the miR-183 family holds particular interest for understanding hair cell biology, the members of this family are also expressed in developing retina and olfactory epithelium (Wienholds et al. 2005, Xu et al. 2007). Furthermore, they are misregulated in certain tumor tissues and cancer cell lines (Guttilla & White 2009, Li et al. 2010b, Lin et al. 2010, Liu et al. 2012b, Lowery et al. 2010, Mihelich et al. 2011, Sarver et al. 2010, Segura et al. 2009, Wang et al. 2012a, Yan et al. 2012, Yu et al. 2010). These discoveries have sparked efforts to identify target transcripts and explore their potential functions in other systems, thereby generating lists of genes that investigators eventually could evaluate as targets in hair cells. Bioinformatic analysis is useful as a preliminary target screening tool. Target-search algorithms, such as PicTar, TargetScan, and Microcosm, favor transcripts containing sequences in their 3′ untranslated regions (3′UTRs) that are complementary to the seed regions of miRNAs, particularly nucleotides 2 through 7 or 8 (Lewis et al. 2003, 2005). Other criteria can include the number of putative miRNA binding sites as well as their evolutionary conservation. Each of the common target-prediction programs can generate lists of genes numbering in the hundreds for a single miRNA. Ultimately, though, the translational repression of potential targets requires validation.

Researchers employ in vitro assays to evaluate protein levels directly (such as with immunostains or western blots) or to monitor translational activity indirectly using luciferase reporters. As one example of the latter method, **Figure 3** shows results from luciferase reporter assays performed in human embryonic kidney cells (HEK293T cells) for 20 human 3′UTRs, the score values of which varied widely as predicted targets for miR-182. Normalized luciferase activity was reduced by at least 30% in the presence of double-stranded (ds)-miR-182 mimic for 14 tested targets. The strength of the scores generated from both TargetScan and PicTar algorithms showed moderately good correlations with the magnitude of the knockdowns in the luciferase assays. One exception was seen for the *MYOSIN1C* (*MYO1C*) 3′UTR reporter, which showed relatively greater repression by miR-182 in comparison with the other potential targets (**Figure 3*c,d***).

Combining these results with other assays reported in the literature, we compiled a list of validated targets of miR-96, miR-182, and miR-183 (**Table 1**). A number of these transcripts are expressed in hair cells, as determined by deep sequencing of purified hair cells [Shared Harvard Inner-Ear Laboratory Database (SHIELD); **https://shield.hms.harvard.edu/gene_search.html**]. For these genes, the miR-183 family likely acts to modulate target protein levels, possibly to initiate or maintain a mature hair cell phenotype. Others have speculated that whereas miR-183 may promote hair cell differentiation, miR-96 and miR-182 may instead promote a proliferative state (Hei et al. 2011). This idea is based on their association with oncogenesis in other cell types, as well as temporal changes in their expression levels as sphere-forming cochlear progenitors are

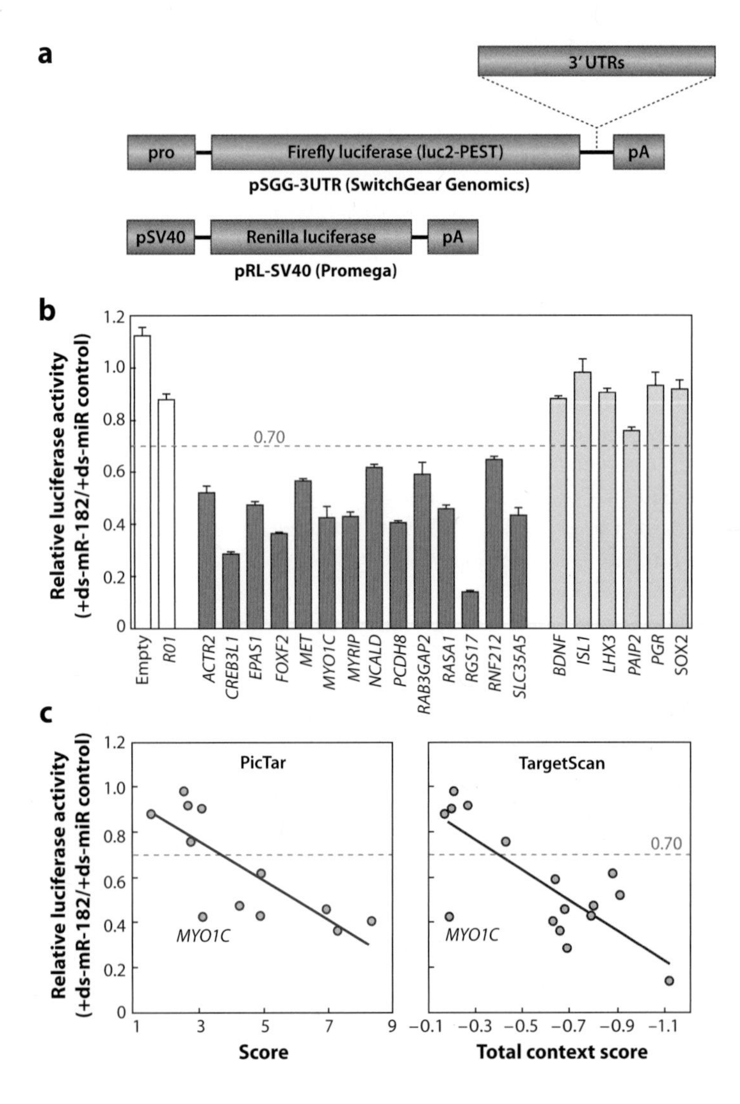

Figure 3

Luciferase assays for validating potential targets of miR-182. (*a*) Plasmids for in vitro luciferase assays. Human 3′UTRs were placed downstream of the coding region for destabilized Firefly luciferase (purchased from SwitchGear Genomics), and these test plasmids were then cotransfected with Renilla luciferase plasmids into HEK293T cells, along with double-stranded (ds)-miRNA mimics (Thermo Scientific Dharmacon). (*b*) Relative luciferase activity. At 24 h post transfection, luminescence originating from Renilla protein was used to normalize the transfection efficiency in each well of a 96-well plate using the Dual-Glo Luciferase Assay System (Promega). Then, the luminescence from Firefly luciferase in the presence of miRIDIAN ds-miRNA-182 mimic was compared with that obtained in the presence of a miRIDIAN ds-miRNA mimic negative control #1 (5′-UCACAACCUCCUAGAAAGAGUAGA-3′) provided by the manufacturer. Values are shown as means plus standard errors from at least six replicates performed over at least two independent experiments. White bars are negative controls. Dark green bars are constructs with luminescence ratios below 0.70 (*gray dashed line*); in this group the knockdowns by miR-182 are statistically significant in comparison with a plasmid carrying the R01 3′UTR that has no predicted binding sites for miR-182 (t-test, $p < 0.05$). Light green bars are constructs that were not significantly different from R01. (*c*) Relative luciferase activity shown as a function of target prediction scores for PicTar and TargetScan. Note that each program outputs a score for only a subset of the 20 tested 3′UTRs. Linear regression analysis of the data gives modest correlation coefficients for TargetScan ($R^2 = 0.5961$) and PicTar ($R^2 = 0.5851$).

Table 1 Validated targets of hair-cell-enriched miRNAs based on bioactivity assays[a,b]

miR-96 target (model organism)	Reference	miR-182 target (model organism)	Reference
ACVR2B (h)	Solda et al. 2012	**ACTR2** (h)	**Figure 3**
ADCY6 (h)	Xu et al. 2007, Jalvy-Delvaille et al. 2012[f]	ADCY6 (h)	Xu et al. 2007, Jalvy-Delvaille et al. 2012[f]
AQP5 (m, h)	Lewis et al. 2009, Mencia et al. 2009	***ARRDC3*** (m)	Zhu et al. 2011
ARRDC3 (m)	Zhu et al. 2011	***BCL2*** (h)	Yan et al. 2012
AVIL (m)[c]	Lewis et al. 2009	BRCA1 (h)	Moskwa et al. 2011
CACNB4 (h)	Solda et al. 2012	***CASP2*** (m)	Zhu et al. 2011
CASP2 (m)	Zhu et al. 2011	***CCND2*** (h)	Yan et al. 2012
CELSR2 (m, h)	Lewis et al. 2009, Mencia et al. 2009	**CLOCK** (h)	Saus et al. 2010
COL2A1 (h)[d]	Mencia et al. 2009	***CREB3L1*** (h)	**Figure 3**
FMNL2 (h)[e]	Mencia et al. 2009	***EPAS1*** (h)	**Figure 3**
FN1 (h)[f]	Jalvy-Delvaille et al. 2012	FOXF2 (h)	**Figure 3**
FOXO1 (h)	Guttilla & White 2009, Jalvy-Delvaille et al. 2012[f]	***FOXO1*** (h)	Guttilla & White 2009; not a target in Jalvy-Delvaille et al. 2012[f]
FOXO3 (h)	Lin et al. 2010	FOXO3 (h)	Segura et al. 2009
GPC3 (h)[f]	Jalvy-Delvaille et al. 2012	***MET*** (h)	**Figure 3**
HTR1B (h)	Jensen & Covault 2009, Jensen et al. 2010	***MITF*** (h)	Xu et al. 2007, Segura et al. 2009, Yan et al. 2012
KRAS (h)	Yu et al. 2010	**MTSS1** (h)	Wang et al. 2012a
LMX1A (h)[e]	Mencia et al. 2009	***MYO1C*** (h)	**Figure 3**
MITF (h)	Xu et al. 2007	MYRIP (h)	**Figure 3**
MYLK (h)[e]	Mencia et al. 2009	**NCALD** (h)	**Figure 3**
MYO1B (h)[d]	Mencia et al. 2009	**PCDH8** (h)	**Figure 3**
MYRIP (m, h)	Lewis et al. 2009, Mencia et al. 2009, Solda et al. 2012	**RAB3GAP2** (h)	**Figure 3**
NEUROD4 (m)	Zhu et al. 2011	***RASA1 (h)***	**Figure 3**
NR3C1 (m)	Riester et al. 2012	RGS17 (h)	Sun et al. 2010; **Figure 3**
ODF2 (m, h)	Lewis et al. 2009, Mencia et al. 2009	RNF212 (h)	**Figure 3**
PGR (h, rhesus, not m)	Liu et al. 2012a	**SLC30A1** (h)[f]	Mihelich et al. 2011
RAD51 (h)[g]	Wang et al. 2012c	***SLC30A7*** (h)[f]	Mihelich et al. 2011
REV1 (h)	Wang et al. 2012c	**SLC35A5** (h)	**Figure 3**
RYK (m, h)	Lewis et al. 2009, Mencia et al. 2009	***SLC39A1*** (h)[f]	Mihelich et al. 2011
SEMA6D (h)[c,d]	Mencia et al. 2009	***SLC39A7*** (h)[f]	Mihelich et al. 2011
SLC19A2 (h)[e]	Mencia et al. 2009	**SOX2** (h)[c]	Weston et al. 2011
SLC39A1 (h)[f]	Mihelich et al. 2011	**SPAST** (h)[f]	Henson et al. 2012
SLC39A3 (h)[f]	Mihelich et al. 2011	TBX1 (m)	Wang et al. 2012b
SLC39A7 (h)[f]	Mihelich et al. 2011		

(Continued)

Table 1 (*Continued*)

miR-96 target (model organism)	Reference	miR-183 target (model organism)	Reference
SPAST (h)[f]	Henson et al. 2012	***EGR1*** (h)	Sarver et al. 2010
ZIC1 (h)[d]	Mencia et al. 2009	***ITGB1*** (h)	Li et al. 2010a
		KIF2A (h)	Li et al. 2010a
		PDCD4 (h)	Li et al. 2010c

[a]Abbreviations: GFP, green fluorescent protein; h, human; m, mouse; miRNA, microRNA.
[b]Bold font indicates genes expressed in cochlear hair cells (GFP+) at greater than 50 reads at P0, P4, and/or P7 in Pou4f3-GFP mice (SHIELD database); italic font indicates genes repressed at least 30% in P0 cochlear hair cells (GFP+) versus other cochlear cells (GFP-) in Pou4f3-GFP mice (these values were only checked for genes in bold font; SHIELD database).
[c]These may not be valid targets because they were repressed less than 30% in the presence of the miRNA.
[d]Tested as a potential target for hsa-mR-96(13G > A) but also repressed by wild type hsa-miR-96.
[e]Tested as a potential target for hsa-mR-96(14C > A) but also repressed by wild type hsa-miR-96.
[f]Not validated by luciferase assay.
[g]miR-96 binding site is present in the coding region, not in the 3′UTR.

grown under differentiation conditions for two weeks: miR-183 peaks at six days as the other two fall precipitously. However, the percentage of cells differentiating into hair-cell-like cells was not evaluated in these cultures and is likely to be below 0.3% (Diensthuber et al. 2009). Resolving the potentially conflicting activities of miR-183 family members will require a more comprehensive accounting and verification of their direct targets, as well as in vivo validations such as miR-specific knockouts.

THE GENETIC MECHANISMS AND CONSTRAINTS OF HAIR CELL REGENERATION

All nonmammalian vertebrates examined to date can regenerate their hair cells. This process can occur either as part of normal turnover [for example, in the constant replacement of lateral line hair cells in bony fish or in the vestibular organs of birds (Ma & Raible 2009, Roberson et al. 1992)] or following loss of hair cells after injury in fish, amphibians, lizards, and birds (Avallone et al. 2008, Corwin & Cotanche 1988, Ma & Raible 2009, Ryals & Rubel 1988, Straube & Tanaka 2006). Supporting cells can trans-differentiate directly into hair cells following injury; this has been best characterized in birds and begins in the first 24 h after hair cell loss (Cafaro et al. 2007). The potential depletion of supporting cells by this regenerative strategy is offset by a robust proliferation of supporting cells that begins several days after damage, and at least some of these cell divisions are likely asymmetric, generating both hair cells and supporting cells (Stone & Cotanche 2007). Mammals, in contrast, have a very limited capacity for hair cell regeneration (Groves 2010). Although a very small amount of proliferation can be seen in the mature vestibular system of mammals following damage, together with a small number of morphologically immature hair cells (Forge et al. 1993, 1998; Kawamoto et al. 2009), the degree of vestibular hair cell regeneration seen in mammals is a fraction of that seen in other vertebrates. Moreover, there is almost no evidence of proliferation or hair cell recovery following damage to the mature organ of Corti (Roberson & Rubel 1994, Yamasoba & Kondo 2006). Nevertheless, the widespread occurrence of hair cell loss in aging humans has prompted many comparisons between mammals and nonmammalian vertebrates to identify genetic or biochemical pathways that underlie regeneration but may lie dormant in mammals.

To What Extent Is Hair Cell Regeneration a Recapitulation of Hair Cell Development?

As discussed above (see Intrinsic and Extrinsic Signals That Drive Hair Cell Induction), *Atoh1* is both necessary and in some circumstances sufficient for hair cell differentiation. Its role appears to be confined to establishing rather than maintaining hair cell fate, which is rapidly downregulated in hair cells as they mature. A small number of *Atoh1*-expressing cells can be observed in sensory organs of nonmammalian vertebrates that exhibit normal hair cell turnover, such as fish lateral line organs (Ma et al. 2008) or bird vestibular organs (Cafaro et al. 2007). This observation suggests that *Atoh1* is reactivated in supporting cells during the process of regeneration, an idea borne out by the observation that *Atoh1* is rapidly and broadly upregulated in fish- and bird-supporting cells following damage (Cafaro et al. 2007, Lewis et al. 2012, Ma et al. 2008, Ma & Raible 2009). Recent experiments using adenovirally transduced reporters of *Atoh1* enhancer activity in cultures of the chicken basilar papilla treated with ototoxic antibiotics suggest that only ~50% of supporting cells that reactivate the *Atoh1* enhancer go on to differentiate into hair cells as defined by the expression of Myosin6 (Lewis et al. 2012). It is possible that some cells that begin to express Atoh1 are eventually diverted back to a supporting cell fate by processes such as Notch-mediated lateral inhibition. Supporting this, the Notch ligand *Delta1* is one of the first genes to be reactivated during hair cell regeneration in birds and fish (Ma et al. 2008, Stone & Rubel 1999). Moreover, pharmacological inhibition of Notch signaling in drug-damaged chick basilar papilla cultures increased the proportion of hair cells that derived from cells activating the *Atoh1* enhancer from ~50% to 75% (Lewis et al. 2012).

Although only very limited hair cell regeneration is seen in the vestibular system of mammals, recent work suggests that the potential of mammalian sensory tissue to respond to damage may be greater than previously realized. Transduction of *Atoh1* reporter constructs into cultures of the drug-damaged adult mouse utricle showed that an average of almost 200 cells per utricle activated the *Atoh1* enhancer after 18 days of culture (Lin et al. 2011). However, very few (less than 5%) of these cells expressed Atoh1 protein or hair cell markers such as Myosin7a. Similar results have also been observed with an in vivo ototoxic lesion to the mouse vestibular system (Kawamoto et al. 2009). As in birds, the efficiency of hair cell generation in mouse utricular cultures could be significantly increased—by almost 30-fold—by blocking Notch signaling (Lin et al. 2011), although the number of hair cells generated by the damaged mouse utricle is still far less than that seen in birds even after Notch inhibition.

These results suggest that hair cell regeneration has at least some superficial similarities with hair cell development: *Atoh1* is activated rapidly and in a manner that is at least partly subject to regulation by the Notch pathway (Puligilla & Kelley 2009). Moreover, mammalian vestibular tissue can initiate at least some of these regenerative steps, albeit at a much lower efficiency than in other vertebrates. One significant difference is that regenerating hair cells are derived from supporting cells rather than from prosensory progenitors. At present, the lack of unambiguous markers to distinguish these two cell populations confounds the question of whether supporting cells undergo partial dedifferentiation into a prosensory cell state during the course of regeneration or whether they directly activate *Atoh1* and become hair cells from their mature state.

Age-Dependent Obstacles to Hair Cell Regeneration in Mammals

Several studies suggest that young mammals possess at least some of the molecular components to respond to damage by replacing hair cells or triggering proliferation of supporting cells but that this potential is rapidly lost with age. For example, cochlear hair cells can be replaced in the embryonic organ of Corti after laser ablation but not in neonatal mice

(Kelley et al. 1995). Genetic ablation of neonatal mouse utricular hair cells with diphtheria toxin leads to significant mitotic replacement of hair cells, most likely by supporting cells, but this phenomenon is no longer observed in five-day-old mice (Burns et al. 2012a). Supporting cells purified from the neonatal mouse cochlea are capable of reentering the cell cycle and generating Atoh1$^+$ hair cells in culture, but this capacity is lost by two weeks after birth (White et al. 2006). Similarly, the ability of cochlear or vestibular sensory tissue to generate proliferative nonadherent sphere cultures also declines with age (Oshima et al. 2007).

What are the age-dependent constraints on supporting cell proliferation and hair cell replacement in mammals? Recent work from Corwin and colleagues has established a strong correlation between (*a*) the decline in the ability of supporting cells to spread and to proliferate in normal or damaged utricular explants and (*b*) the establishment of thick bands of F-actin and E-cadherin in the apical junctions between supporting cells (Burns et al. 2008). In contrast, the cortical actin bands in chicken utricular supporting cells remain thin throughout the animal's life. In the absence of perturbation studies, it is not clear at present whether the establishment of thick F-actin/E-cadherin-containing bands in mammals directly regulates the capacity of supporting cells to divide, or whether it is simply an indicator of more global changes occurring in supporting cells with age that might also include their inability to generate hair cells. In this regard, it is interesting to note that supporting cells in the striolar region of the utricle—which has the capacity to activate Atoh1 expression in the adult mouse after damage and inhibition of Notch signaling—contain significantly less E-cadherin and have thinner actin bands than do their extrastriolar counterparts (Collado et al. 2011).

In addition to undergoing changes in cell structure, the mature sensory epithelium of mammals likely also exhibits changes in responsiveness to cell-cell signaling. In particular, several recent studies suggest that the Notch signaling pathway may no longer be instrumental in maintaining the balance between hair cells and supporting cells in older animals. For example, blocking Notch signaling in the undamaged chicken basilar papilla or the undamaged zebrafish larval lateral line has no effect on supporting cell proliferation or hair cell generation (Daudet et al. 2009, Ma et al. 2008), which suggests either that Notch signaling is not necessary to maintain the mature array of hair cells and supporting cells or that other factors can compensate for the loss of Notch activity. Moreover, although pharmacological or genetic inhibition of Notch signaling can rapidly upregulate *Atoh1* and promote the expression of hair cell markers such as Myosin6 and Myosin7a in perinatal mouse cochlear or utricular cultures, these effects decline rapidly with age, both in vitro and in vivo (Doetzlhofer et al. 2009, Hori et al. 2007, Lin et al. 2011).

As described above, some of the age-dependent changes in the ability of mammalian sensory epithelium to regenerate may be due to a failure to reactivate expression of Atoh1 after it is downregulated in early postnatal life. To bypass these age-dependent limitations, several studies have used adenoviral or transgenic expression of *Atoh1* to generate new hair cells (e.g., Izumikawa et al. 2005). Two recent studies used transgenic techniques to activate *Atoh1* expression either throughout the inner ear (Kelly et al. 2012) or specifically in subpopulations of supporting cells (Liu et al. 2012a). In either case, expression of *Atoh1* in supporting cells or in nonsensory cochlear epithelium induced new hair cells, some of which assembled stereociliary bundles and displayed voltage-dependent currents (Kelly et al. 2012, Liu et al. 2012a). However, in both studies, the ability of *Atoh1* to promote ectopic hair cell formation in all regions of the cochlea declined rapidly and was severely compromised by two weeks of age, the time at which hearing begins in mice (Kelly et al. 2012, Liu et al. 2012a). This failure was also seen when investigators activated *Atoh1* in supporting cells of adult mice in which hair cells had been ablated with ototoxic drugs (Liu et al. 2012a).

There are many possible reasons why the mature mammalian ear becomes refractory to overexpression of Atoh1. In a recent review of Atoh1 function, Mulvaney & Dabdoub (2012) discuss a number of ways in which Atoh1 activity and binding to its E-box DNA targets may be regulated at the posttranslational level, and each of these may contribute to the failure of Atoh1 to activate hair cell–specific gene expression in the adult mammal. For example, the C-terminus of Atoh1 has a conserved serine-rich domain that contains potential phosphorylation sites that could regulate its activity (Mulvaney & Dabdoub 2012). Basic helix-loop-helix proteins such as Atoh1 require bHLH binding partners such as E47 to bind correctly to their E-box targets. Competition for the partners of Atoh1 by inhibitory Id HLH family members, or by other bHLH proteins, may reduce or abolish the transcriptional activity of Atoh1 in adult tissue. Atoh1 activity can also be modulated in different cell types by forming complexes with different bHLH factors that promote binding to different sets of E-boxes. The expression of such factors in adult supporting cells may direct Atoh1 away from the promoters of hair cell genes and hence may block its hair cell–promoting activity. Finally, other classes of transcription factors may cooperate with Atoh1 in young sensory tissue but are no longer expressed in older tissue, thus depriving Atoh1 of transcriptional partners.

A final possibility for why Atoh1 is unable to promote hair cell formation in the mature mammalian inner ear is that direct transcriptional targets of Atoh1 become epigenetically modified in supporting cells with age and are therefore unavailable for transcription, even in the presence of exogenous Atoh1. In such a model, some form of epigenetic reprogramming would be required to render mammalian supporting cells competent to respond to Atoh1 again. Whether this could be achieved by altering the activity of histone-modifying complexes or would require defined reprogramming factors [for example, similar to those required to reprogram fibroblasts into neurons (Lujan et al. 2012, Vierbuchen et al. 2010, Yang et al. 2011)] is not clear. A better understanding of the epigenetic regulation of Atoh1 targets in the inner ear is hampered by a lack of good candidates for genes directly regulated by Atoh1. As discussed above, two recent studies used ChIP-Seq, RNA-seq, and mining of microarray databases to identify and validate a set of potential direct targets of Atoh1 in cerebellar granule cells and dorsal spinal cord interneurons (Klisch et al. 2011, Lai et al. 2011). However, of these, only three—*Rab15*, *Selm*, and *Atoh1* itself—have been expressed in hair cells.

DISCLOSURE STATEMENT

The authors are not aware of any affiliations, memberships, funding, or financial holdings that might be perceived as affecting the objectivity of this review.

LITERATURE CITED

Ahmed M, Wong EY, Sun J, Xu J, Wang F, Xu PX. 2012. Eya1-Six1 interaction is sufficient to induce hair cell fate in the cochlea by activating Atoh1 expression in cooperation with Sox2. *Dev. Cell* 22:377–90

Akazawa C, Ishibashi M, Shimizu C, Nakanishi S, Kageyama R. 1995. A mammalian helix-loop-helix factor structurally related to the product of Drosophila proneural gene atonal is a positive transcriptional regulator expressed in the developing nervous system. *J. Biol. Chem.* 270:8730–38

Alsina B, Giraldez F, Pujades C. 2009. Patterning and cell fate in ear development. *Int. J. Dev. Biol.* 53:1503–13

Avallone B, Fascio U, Balsamo G, Marmo F. 2008. Gentamicin ototoxicity in the saccule of the lizard *Podarcis Sicula* induces hair cell recovery and regeneration. *Hear. Res.* 235:15–22

Baek D, Villen J, Shin C, Camargo FD, Gygi SP, Bartel DP. 2008. The impact of microRNAs on protein output. *Nature* 455:64–71

Bartel DP. 2009. MicroRNAs: target recognition and regulatory functions. *Cell* 136:215–33

Bejarano F, Smibert P, Lai EC. 2010. miR-9a prevents apoptosis during wing development by repressing Drosophila LIM-only. *Dev. Biol.* 338:63–73

Bermingham NA, Hassan BA, Price SD, Vollrath MA, Ben-Arie N, et al. 1999. *Math1*: an essential gene for the generation of inner ear hair cells. *Science* 284:1837–41

Bok J, Chang W, Wu DK. 2007. Patterning and morphogenesis of the vertebrate inner ear. *Int. J. Dev. Biol.* 51:521–33

Briggs KJ, Eberhart CG, Watkins DN. 2008. Just say no to ATOH: how HIC1 methylation might predispose medulloblastoma to lineage addiction. *Cancer Res.* 68:8654–56

Brooker R, Hozumi K, Lewis J. 2006. Notch ligands with contrasting functions: Jagged1 and Delta1 in the mouse inner ear. *Development* 133:1277–86

Burns JC, Christophel JJ, Collado MS, Magnus C, Carfrae M, Corwin JT. 2008. Reinforcement of cell junctions correlates with the absence of hair cell regeneration in mammals and its occurrence in birds. *J. Comp. Neurol.* 511:396–414

Burns JC, Cox BC, Thiede BR, Zuo J, Corwin JT. 2012a. In vivo proliferative regeneration of balance hair cells in newborn mice. *J. Neurosci.* 32:6570–77

Burns JC, On D, Baker W, Collado MS, Corwin JT. 2012b. Over half the hair cells in the mouse utricle first appear after birth, with significant numbers originating from early postnatal mitotic production in peripheral and striolar growth zones. *J. Assoc. Res. Otolaryngol.* 13:609–27

Cafaro J, Lee GS, Stone JS. 2007. Atoh1 expression defines activated progenitors and differentiating hair cells during avian hair cell regeneration. *Dev. Dyn.* 236:156–70

Chen P, Johnson JE, Zoghbi HY, Segil N. 2002. The role of Math1 in inner ear development: uncoupling the establishment of the sensory primordium from hair cell fate determination. *Development* 129:2495–505

Chen P, Segil N. 1999. p27(Kip1) links cell proliferation to morphogenesis in the developing organ of Corti. *Development* 126:1581–90

Collado MS, Thiede BR, Baker W, Askew C, Igbani LM, Corwin JT. 2011. The postnatal accumulation of junctional E-cadherin is inversely correlated with the capacity for supporting cells to convert directly into sensory hair cells in mammalian balance organs. *J. Neurosci.* 31:11855–66

Corwin JT, Cotanche DA. 1988. Regeneration of sensory hair cells after acoustic trauma. *Science* 240:1772–74

D'Angelo A, Bluteau O, Garcia-Gonzalez MA, Gresh L, Doyen A, et al. 2010. Hepatocyte nuclear factor 1alpha and beta control terminal differentiation and cell fate commitment in the gut epithelium. *Development* 137:1573–82

Daudet N, Gibson R, Shang J, Bernard A, Lewis J, Stone J. 2009. Notch regulation of progenitor cell behavior in quiescent and regenerating auditory epithelium of mature birds. *Dev. Biol.* 326:86–100

Diensthuber M, Oshima K, Heller S. 2009. Stem/progenitor cells derived from the cochlear sensory epithelium give rise to spheres with distinct morphologies and features. *J. Assoc. Res. Otolaryngol.* 10:173–90

Doetzlhofer A, Basch ML, Ohyama T, Gessler M, Groves AK, Segil N. 2009. Hey2 regulation by FGF provides a Notch-independent mechanism for maintaining pillar cell fate in the organ of Corti. *Dev. Cell* 16:58–69

Driver EC, Pryor SP, Hill P, Turner J, Ruther U, et al. 2008. Hedgehog signaling regulates sensory cell formation and auditory function in mice and humans. *J. Neurosci.* 28:7350–58

Ebert PJ, Timmer JR, Nakada Y, Helms AW, Parab PB, et al. 2003. Zic1 represses Math1 expression via interactions with the Math1 enhancer and modulation of Math1 autoregulation. *Development* 130:1949–59

Esquela-Kerscher A, Slack FJ. 2006. Oncomirs—microRNAs with a role in cancer. *Nat. Rev. Cancer* 6:259–69

Filipowicz W, Bhattacharyya SN, Sonenberg N. 2008. Mechanisms of post-transcriptional regulation by microRNAs: Are the answers in sight? *Nat. Rev. Genet.* 9:102–14

Forge A, Li L, Corwin JT, Nevill G. 1993. Ultrastructural evidence for hair cell regeneration in the mammalian inner ear. *Science* 259:1616–19

Forge A, Li L, Nevill G. 1998. Hair cell recovery in the vestibular sensory epithelia of mature guinea pigs. *J. Comp. Neurol.* 397:69–88

Friedman LM, Dror AA, Mor E, Tenne T, Toren G, et al. 2009. MicroRNAs are essential for development and function of inner ear hair cells in vertebrates. *Proc. Natl. Acad. Sci. USA* 106:7915–20

Gleich O, Fischer FP, Köppl C, Manley GA. 2004. Hearing organ evolution and specialization: Archosaurs. In *Evolution of the Vertebrate Auditory System*, ed. GA Manley, AN Popper, RR Fay, pp. 224–55. New York: Springer-Verlag

Goodyear R, Richardson G. 1997. Pattern formation in the basilar papilla: evidence for cell rearrangement. *J. Neurosci.* 17:6289–301

Groves AK. 2010. The challenge of hair cell regeneration. *Exp. Biol. Med.* 235:434–46

Groves AK, Fekete DM. 2012. Shaping sound in space: the regulation of inner ear patterning. *Development* 139:245–57

Guo H, Ingolia NT, Weissman JS, Bartel DP. 2010. Mammalian microRNAs predominantly act to decrease target mRNA levels. *Nature* 466:835–40

Guttilla IK, White BA. 2009. Coordinate regulation of FOXO1 by miR-27a, miR-96, and miR-182 in breast cancer cells. *J. Biol. Chem.* 284:23204–16

Hayashi T, Ray CA, Bermingham-McDonogh O. 2008. Fgf20 is required for sensory epithelial specification in the developing cochlea. *J. Neurosci.* 28:5991–99

Hei R, Chen J, Qiao L, Li X, Mao X, et al. 2011. Dynamic changes in microRNA expression during differentiation of rat cochlear progenitor cells in vitro. *Int. J. Pediatr. Otorhinolaryngol.* 75:1010–14

Helms AW, Abney AL, Ben-Arie N, Zoghbi HY, Johnson JE. 2000. Autoregulation and multiple enhancers control Math1 expression in the developing nervous system. *Development* 127:1185–96

Henson BJ, Zhu W, Hardaway K, Wetzel JL, Stefan M, et al. 2012. Transcriptional and post-transcriptional regulation of SPAST, the gene most frequently mutated in hereditary spastic paraplegia. *PLoS One* 7:e36505

Hobert O. 2007. miRNAs play a tune. *Cell* 131:22–24

Hobert O. 2008. Gene regulation by transcription factors and microRNAs. *Science* 319:1785–86

Holley M, Rhodes C, Kneebone A, Herde MK, Fleming M, Steel KP. 2010. Emx2 and early hair cell development in the mouse inner ear. *Dev. Biol.* 340:547–56

Hori R, Nakagawa T, Sakamoto T, Matsuoka Y, Takebayashi S, Ito J. 2007. Pharmacological inhibition of Notch signaling in the mature guinea pig cochlea. *NeuroReport* 18:1911–14

Huh SH, Jones J, Warchol ME, Ornitz DM. 2012. Differentiation of the lateral compartment of the cochlea requires a temporally restricted FGF20 signal. *PLoS Biol.* 10:e1001231

Hwang CH, Guo D, Harris MA, Howard O, Mishina Y, et al. 2010. Role of bone morphogenetic proteins on cochlear hair cell formation: analyses of Noggin and Bmp2 mutant mice. *Dev. Dyn.* 239:505–13

Izumikawa M, Minoda R, Kawamoto K, Abrashkin KA, Swiderski DL, et al. 2005. Auditory hair cell replacement and hearing improvement by Atoh1 gene therapy in deaf mammals. *Nat. Med.* 11:271–76

Jahan I, Pan N, Kersigo J, Calisto LE, Morris KA, et al. 2012. Expression of Neurog1 instead of Atoh1 can partially rescue organ of Corti cell survival. *PLoS One* 7:e30853

Jahan I, Pan N, Kersigo J, Fritzsch B. 2010. Neurod1 suppresses hair cell differentiation in ear ganglia and regulates hair cell subtype development in the cochlea. *PLoS One* 5:e11661

Jalvy-Delvaille S, Maurel M, Majo V, Pierre N, Chabas S, et al. 2012. Molecular basis of differential target regulation by miR-96 and miR-182: the Glypican-3 as a model. *Nucleic Acids Res.* 40:1356–65

Jensen KP, Covault J. 2010. Human miR-1271 is a miR-96 paralog with distinct non-conserved brain expression pattern. *Nucleic Acids Res.* 39:701–11

Jensen KP, Covault J, Conner TS, Tennen H, Kranzler HR, Furneaux HM. 2009. A common polymorphism in serotonin receptor 1B mRNA moderates regulation by miR-96 and associates with aggressive human behaviors. *Mol. Psychiatry* 14:381–89

Jones C, Chen P. 2007. Planar cell polarity signaling in vertebrates. *BioEssays* 29:120–32

Kamaid A, Neves J, Giraldez F. 2010. Id gene regulation and function in the prosensory domains of the chicken inner ear: a link between Bmp signaling and Atoh1. *J. Neurosci.* 30:11426–34

Kanzaki S, Beyer LA, Swiderski DL, Izumikawa M, Stover T, et al. 2006. p27(Kip1) deficiency causes organ of Corti pathology and hearing loss. *Hear. Res.* 214:28–36

Katayama A, Corwin JT. 1989. Cell production in the chicken cochlea. *J. Comp. Neurol.* 281:129–35

Kawamoto K, Ishimoto S, Minoda R, Brough DE, Raphael Y. 2003. Math1 gene transfer generates new cochlear hair cells in mature guinea pigs in vivo. *J. Neurosci.* 23:4395–400

Kawamoto K, Izumikawa M, Beyer LA, Atkin GM, Raphael Y. 2009. Spontaneous hair cell regeneration in the mouse utricle following gentamicin ototoxicity. *Hear. Res.* 247:17–26

Kelley MW. 2006. Regulation of cell fate in the sensory epithelia of the inner ear. *Nat. Rev. Neurosci.* 7:837–49

Kelley MW, Talreja DR, Corwin JT. 1995. Replacement of hair cells after laser microbeam irradiation in cultured organs of corti from embryonic and neonatal mice. *J. Neurosci.* 15:3013–26

Kelly MC, Chang Q, Pan A, Lin X, Chen P. 2012. Atoh1 directs the formation of sensory mosaics and induces cell proliferation in the postnatal mammalian cochlea in vivo. *J. Neurosci.* 32:6699–710

Kiernan AE, Pelling AL, Leung KK, Tang AS, Bell DM, et al. 2005. Sox2 is required for sensory organ development in the mammalian inner ear. *Nature* 434:1031–35

Kiernan AE, Xu J, Gridley T. 2006. The Notch ligand JAG1 is required for sensory progenitor development in the mammalian inner ear. *PLoS Genet.* 2:e4

Kirkegaard M, Nyengaard JR. 2005. Stereological study of postnatal development in the mouse utricular macula. *J. Comp. Neurol.* 492:132–44

Klisch TJ, Xi Y, Flora A, Wang L, Li W, Zoghbi HY. 2011. In vivo Atoh1 targetome reveals how a proneural transcription factor regulates cerebellar development. *Proc. Natl. Acad. Sci. USA* 108:3288–93

Kloosterman WP, Plasterk RH. 2006. The diverse functions of microRNAs in animal development and disease. *Dev. Cell* 11:441–50

Kosik KS. 2006. The neuronal microRNA system. *Nat. Rev. Neurosci.* 7:911–20

Kuhn S, Johnson SL, Furness DN, Chen J, Ingham N, et al. 2011. miR-96 regulates the progression of differentiation in mammalian cochlear inner and outer hair cells. *Proc. Natl. Acad. Sci. USA* 108:2355–60

Lai HC, Klisch TJ, Roberts R, Zoghbi HY, Johnson JE. 2011. In vivo neuronal subtype-specific targets of Atoh1 (Math1) in dorsal spinal cord. *J. Neurosci.* 31:10859–71

Lee YS, Liu F, Segil N. 2006. A morphogenetic wave of p27Kip1 transcription directs cell cycle exit during organ of Corti development. *Development* 133:2817–26

Lewis BP, Burge CB, Bartel DP. 2005. Conserved seed pairing, often flanked by adenosines, indicates that thousands of human genes are microRNA targets. *Cell* 120:15–20

Lewis BP, Shih IH, Jones-Rhoades MW, Bartel DP, Burge CB. 2003. Prediction of mammalian microRNA targets. *Cell* 115:787–98

Lewis MA, Quint E, Glazier AM, Fuchs H, De Angelis MH, et al. 2009. An ENU-induced mutation of miR-96 associated with progressive hearing loss in mice. *Nat. Genet.* 41:614–18

Lewis MA, Steel KP. 2010. MicroRNAs in mouse development and disease. *Semin. Cell Dev. Biol.* 21:774–80

Lewis RM, Hume CR, Stone JS. 2012. Atoh1 expression and function during auditory hair cell regeneration in post-hatch chickens. *Hear. Res.* 289:74–85

Li H, Corrales CE, Wang Z, Zhao Y, Wang Y, et al. 2005. BMP4 signaling is involved in the generation of inner ear sensory epithelia. *BMC Dev. Biol.* 5:16

Li H, Kloosterman W, Fekete DM. 2010a. MicroRNA-183 family members regulate sensorineural fates in the inner ear. *J. Neurosci.* 30:3254–63

Li J, Fu H, Xu C, Tie Y, Xing R, et al. 2010b. miR-183 inhibits TGF-beta1-induced apoptosis by downregulation of PDCD4 expression in human hepatocellular carcinoma cells. *BMC Cancer* 10:354

Lin H, Dai T, Xiong H, Zhao X, Chen X, et al. 2010. Unregulated miR-96 induces cell proliferation in human breast cancer by downregulating transcriptional factor FOXO3a. *PLoS One* 5:e15797

Lin V, Golub JS, Nguyen TB, Hume CR, Oesterle EC, Stone JS. 2011. Inhibition of Notch activity promotes nonmitotic regeneration of hair cells in the adult mouse utricles. *J. Neurosci.* 31:15329–39

Liu X, Fortin K, Mourelatos Z. 2008. MicroRNAs: biogenesis and molecular functions. *Brain Pathol.* 18:113–21

Liu Z, Dearman JA, Cox BC, Walters BJ, Zhang L, et al. 2012a. Age-dependent in vivo conversion of mouse cochlear pillar and Deiters' cells to immature hair cells by Atoh1 ectopic expression. *J. Neurosci.* 32:6600–10

Liu Z, Liu J, Segura MF, Shao C, Lee P, et al. 2012b. MiR-182 overexpression in tumourigenesis of high-grade serous ovarian carcinoma. *J. Pathol.* 228:204–15

Lowenheim H, Furness DN, Kil J, Zinn C, Gultig K, et al. 1999. Gene disruption of p27(Kip1) allows cell proliferation in the postnatal and adult organ of Corti. *Proc. Natl. Acad. Sci. USA* 96:4084–88

Lowery AJ, Miller N, Dwyer RM, Kerin MJ. 2010. Dysregulated miR-183 inhibits migration in breast cancer cells. *BMC Cancer* 10:502

Lujan E, Chanda S, Ahlenius H, Sudhof TC, Wernig M. 2012. Direct conversion of mouse fibroblasts to self-renewing, tripotent neural precursor cells. *Proc. Natl. Acad. Sci. USA* 109:2527–32

Lumpkin EA, Collisson T, Parab P, Omer-Abdalla A, Haeberle H, et al. 2003. Math1-driven GFP expression in the developing nervous system of transgenic mice. *Gene Expr. Patterns* 3:389–95

Ma EY, Raible DW. 2009. Signaling pathways regulating zebrafish lateral line development. *Curr. Biol.* 19:R381–86

Ma EY, Rubel EW, Raible DW. 2008. Notch signaling regulates the extent of hair cell regeneration in the zebrafish lateral line. *J. Neurosci.* 28:2261–73

Matei V, Pauley S, Kaing S, Rowitch D, Beisel KW, et al. 2005. Smaller inner ear sensory epithelia in Neurog 1 null mice are related to earlier hair cell cycle exit. *Dev. Dyn.* 234:633–50

May-Simera H, Kelley MW. 2012. Planar cell polarity in the inner ear. *Curr. Top. Dev. Biol.* 101:111–40

Mencia A, Modamio-Høybjør S, Redshaw N, Morin M, Mayo-Merino F, et al. 2009. Mutations in the seed region of human miR-96 are responsible for nonsyndromic progressive hearing loss. *Nat. Genet.* 41:609–13

Mihelich BL, Khramtsova EA, Arva N, Vaishnav A, Johnson DN, et al. 2011. miR-183–96–182 cluster is overexpressed in prostate tissue and regulates zinc homeostasis in prostate cells. *J. Biol. Chem.* 286:44503–11

Montcouquiol M, Crenshaw EB 3rd, Kelley MW. 2006. Noncanonical Wnt signaling and neural polarity. *Annu. Rev. Neurosci.* 29:363–86

Montcouquiol M, Kelley MW. 2003. Planar and vertical signals control cellular differentiation and patterning in the mammalian cochlea. *J. Neurosci.* 23:9469–78

Moskwa P, Buffa FM, Pan Y, Panchakshari R, Gottipati P, et al. 2011. miR-182-mediated downregulation of BRCA1 impacts DNA repair and sensitivity to PARP inhibitors. *Mol. Cell* 41:210–20

Mulvaney J, Dabdoub A. 2012. Atoh1, an essential transcription factor in neurogenesis and intestinal and inner ear development: function, regulation, and context dependency. *J. Assoc. Res. Otolaryngol.* 13:281–93

Mutoh H, Sakamoto H, Hayakawa H, Arao Y, Satoh K, et al. 2006. The intestine-specific homeobox gene Cdx2 induces expression of the basic helix-loop-helix transcription factor Math1. *Differentiation* 74:313–21

Neves J, Uchikawa M, Bigas A, Giraldez F. 2012. The prosensory function of Sox2 in the chicken inner ear relies on the direct regulation of Atoh1. *PLoS One* 7:e30871

Ohyama T, Basch ML, Mishina Y, Lyons KM, Segil N, Groves AK. 2010. BMP signaling is necessary for patterning the sensory and nonsensory regions of the developing mammalian cochlea. *J. Neurosci.* 30:15044–51

Oshima K, Grimm CM, Corrales CE, Senn P, Martinez Monedero R, et al. 2007. Differential distribution of stem cells in the auditory and vestibular organs of the inner ear. *J. Assoc. Res. Otolaryngol.* 8:18–31

Pan N, Jahan I, Kersigo J, Kopecky B, Santi P, et al. 2011. Conditional deletion of Atoh1 using Pax2-Cre results in viable mice without differentiated cochlear hair cells that have lost most of the organ of Corti. *Hear. Res.* 275:66–80

Pirvola U, Ylikoski J, Trokovic R, Hebert JM, McConnell SK, Partanen J. 2002. FGFR1 is required for the development of the auditory sensory epithelium. *Neuron* 35:671–80

Pujades C, Kamaid A, Alsina B, Giraldez F. 2006. BMP-signaling regulates the generation of hair-cells. *Dev. Biol.* 292:55–67

Puligilla C, Dabdoub A, Brenowitz SD, Kelley MW. 2010. Sox2 induces neuronal formation in the developing mammalian cochlea. *J. Neurosci.* 30:714–22

Puligilla C, Feng F, Ishikawa K, Bertuzzi S, Dabdoub A, et al. 2007. Disruption of fibroblast growth factor receptor 3 signaling results in defects in cellular differentiation, neuronal patterning, and hearing impairment. *Dev. Dyn.* 236:1905–17

Puligilla C, Kelley MW. 2009. Building the world's best hearing aid; regulation of cell fate in the cochlea. *Curr. Opin. Genet. Dev.* 19:368–73

Raft S, Koundakjian EJ, Quinones H, Jayasena CS, Goodrich LV, et al. 2007. Cross-regulation of Ngn1 and Math1 coordinates the production of neurons and sensory hair cells during inner ear development. *Development* 134:4405–15

Rida PC, Chen P. 2009. Line up and listen: planar cell polarity regulation in the mammalian inner ear. *Semin. Cell Dev. Biol.* 20:978–85

Riester A, Issler O, Spyroglou A, Rodrig SH, Chen A, Beuschlein F. 2012. ACTH-dependent regulation of microRNA as endogenous modulators of glucocorticoid receptor expression in the adrenal gland. *Endocrinology* 153:212–22

Roberson DF, Weisleder P, Bohrer PS, Rubel EW. 1992. Ongoing production of sensory cells in the vestibular epithelium of the chick. *Hear. Res.* 57:166–74

Roberson DW, Rubel EW. 1994. Cell division in the gerbil cochlea after acoustic trauma. *Am. J. Otol.* 15:28–34

Ruben RJ. 1967. Development of the inner ear of the mouse: a radioautographic study of terminal mitoses. *Acta Otolaryngol.* 220(Suppl.):1–44

Rudnicki A, Avraham KB. 2012. microRNAs: the art of silencing in the ear. *EMBO Mol. Med.* 4:849–59

Ryals BM, Rubel EW. 1988. Hair cell regeneration after acoustic trauma in adult Coturnix quail. *Science* 240:1774–76

Sacheli R, Nguyen L, Borgs L, Vandenbosch R, Bodson M, et al. 2009. Expression patterns of miR-96, miR-182 and miR-183 in the development inner ear. *Gene Expr. Patterns* 9:364–70

Sarver AL, Li L, Subramanian S. 2010. MicroRNA miR-183 functions as an oncogene by targeting the transcription factor EGR1 and promoting tumor cell migration. *Cancer Res.* 70:9570–80

Saus E, Soria V, Escaramis G, Vivarelli F, Crespo JM, et al. 2010. Genetic variants and abnormal processing of pre-miR-182, a circadian clock modulator, in major depression patients with late insomnia. *Hum. Mol. Genet.* 19:4017–25

Segura MF, Hanniford D, Menendez S, Reavie L, Zou X, et al. 2009. Aberrant miR-182 expression promotes melanoma metastasis by repressing FOXO3 and microphthalmia-associated transcription factor. *Proc. Natl. Acad. Sci. USA* 106:1814–19

Selbach M, Schwanhausser B, Thierfelder N, Fang Z, Khanin R, Rajewsky N. 2008. Widespread changes in protein synthesis induced by microRNAs. *Nature* 455:58–63

Shi F, Cheng YF, Wang XL, Edge AS. 2010. Beta-catenin up-regulates Atoh1 expression in neural progenitor cells by interaction with an Atoh1 3′ enhancer. *J. Biol. Chem.* 285:392–400

Solda G, Robusto M, Primignani P, Castorina P, Benzoni E, et al. 2012. A novel mutation within the MIR96 gene causes non-syndromic inherited hearing loss in an Italian family by altering pre-miRNA processing. *Hum. Mol. Genet.* 21:577–85

Soukup GA. 2009. Little but loud: Small RNAs have a resounding affect on ear development. *Brain Res.* 1277:104–14

Soukup GA, Fritzsch B, Pierce ML, Weston MD, Jahan I, et al. 2009. Residual microRNA expression dictates the extent of inner ear development in conditional Dicer knockout mice. *Dev. Biol.* 328:328–41

Stevens CB, Davies AL, Battista S, Lewis JH, Fekete DM. 2003. Forced activation of Wnt signaling alters morphogenesis and sensory organ identity in the chicken inner ear. *Dev. Biol.* 261:149–64

Stone JS, Cotanche DA. 2007. Hair cell regeneration in the avian auditory epithelium. *Int. J. Dev. Biol.* 51:633–47

Stone JS, Rubel EW. 1999. Delta1 expression during avian hair cell regeneration. *Development* 126:961–73

Straube WL, Tanaka EM. 2006. Reversibility of the differentiated state: regeneration in amphibians. *Artif. Organs* 30:743–55

Sun Y, Fang R, Li C, Li L, Li F, et al. 2010. Hsa-mir-182 suppresses lung tumorigenesis through down regulation of RGS17 expression in vitro. *Biochem. Biophys. Res. Commun.* 396:501–7

Takacs CM, Giraldez AJ. 2010. MicroRNAs as genetic sculptors: fishing for clues. *Semin. Cell Dev. Biol.* 21:760–67

Vierbuchen T, Ostermeier A, Pang ZP, Kokubu Y, Sudhof TC, Wernig M. 2010. Direct conversion of fibroblasts to functional neurons by defined factors. *Nature* 463:1035–41

Wang J, Li J, Shen J, Wang C, Yang L, Zhang X. 2012a. MicroRNA-182 downregulates metastasis suppressor 1 and contributes to metastasis of hepatocellular carcinoma. *BMC Cancer* 12:227

Wang XR, Zhang XM, Du J, Jiang H. 2012b. MicroRNA-182 regulates otocyst-derived cell differentiation and targets T-box1 gene. *Hear. Res.* 286:55–63

Wang Y, Huang JW, Calses P, Kemp CJ, Taniguchi T. 2012c. MiR-96 downregulates REV1 and RAD51 to promote cellular sensitivity to cisplatin and PARP inhibition. *Cancer Res.* 72:4037–46

Weston MD, Pierce ML, Jensen-Smith HC, Fritzsch B, Rocha-Sanchez S, et al. 2011. MicroRNA-183 family expression in hair cell development and requirement of microRNAs for hair cell maintenance and survival. *Dev. Dyn.* 240:808–19

Weston MD, Pierce ML, Rocha-Sanchez S, Beisel KW, Soukup GA. 2006. MicroRNA gene expression in the mouse inner ear. *Brain Res.* 1111:95–104

Weston MD, Soukup GA. 2009. MicroRNAs sound off. *Genome Med.* 1:59

White PM, Doetzlhofer A, Lee YS, Groves AK, Segil N. 2006. Mammalian cochlear supporting cells can divide and trans-differentiate into hair cells. *Nature* 441:984–87

Wienholds E, Kloosterman WP, Miska E, Alvarez-Saavedra E, Berezikov E, et al. 2005. MicroRNA expression in zebrafish embryonic development. *Science* 309:310–11

Winter J, Jung S, Keller S, Gregory RI, Diederichs S. 2009. Many roads to maturity: microRNA biogenesis pathways and their regulation. *Nat. Cell Biol.* 11:228–34

Xu S, Witmer PD, Lumayag S, Kovacs B, Valle D. 2007. MicroRNA (miRNA) transcriptome of mouse retina and identification of a sensory organ-specific miRNA cluster. *J. Biol. Chem.* 282:25053–66

Yamasoba T, Kondo K. 2006. Supporting cell proliferation after hair cell injury in mature guinea pig cochlea in vivo. *Cell Tissue Res.* 325:23–31

Yan D, Dong XD, Chen X, Yao S, Wang L, et al. 2012. Role of microRNA-182 in posterior uveal melanoma: regulation of tumor development through MITF, BCL2 and cyclin D2. *PLoS One* 7:e40967

Yang N, Ng YH, Pang ZP, Sudhof TC, Wernig M. 2011. Induced neuronal cells: how to make and define a neuron. *Cell Stem Cell* 9:517–25

Yu S, Lu Z, Liu C, Meng Y, Ma Y, et al. 2010. miRNA-96 suppresses KRAS and functions as a tumor suppressor gene in pancreatic cancer. *Cancer Res.* 70:6015–25

Zheng JL, Gao WQ. 2000. Overexpression of Math1 induces robust production of extra hair cells in postnatal rat inner ears. *Nat. Neurosci.* 3:580–86

Zhu Q, Sun W, Okano K, Chen Y, Zhang N, et al. 2011. Sponge transgenic mouse model reveals important roles for the microRNA-183 (miR-183)/96/182 cluster in postmitotic photoreceptors of the retina. *J. Biol. Chem.* 286:31749–60

Neuronal Computations in the Olfactory System of Zebrafish

Rainer W. Friedrich

Friedrich Miescher Institute for Biomedical Research, 4058 Basel, Switzerland;
email: Rainer.Friedrich@fmi.ch

Annu. Rev. Neurosci. 2013. 36:383–402

First published online as a Review in Advance on May 29, 2013

The *Annual Review of Neuroscience* is online at neuro.annualreviews.org

This article's doi:
10.1146/annurev-neuro-062111-150504

Keywords

olfactory bulb, olfactory cortex, decorrelation, multiplexing, activity pattern, neurophysiology

Abstract

The main olfactory system encodes information about molecules in a combinatorial fashion by distributed spatiotemporal activity patterns. As activity propagates from sensory neurons to the olfactory bulb and to higher brain areas, odor information is processed by multiple transformations of these activity patterns. This review discusses neuronal computations associated with such transformations in the olfactory system of zebrafish, a small vertebrate that offers advantages for the quantitative analysis and manipulation of neuronal activity in the intact brain. The review focuses on pattern decorrelation in the olfactory bulb and on the readout of multiplexed sensory representations in the telencephalic area Dp, the homolog of the olfactory cortex. These computations are difficult to study in larger species and may provide insights into general information-processing strategies in the brain.

Contents

INTRODUCTION

The world of odorous molecules is high dimensional and highly discontinuous, although a parametric description is still lacking. This odor space is sampled by odorant receptors that define discrete input channels into the brain (Axel 1995, Buck 2000). The number of odorant receptors can be more than 1,000 in some species but is still small in comparison to the number of discernible odorants, implying that information about most odors is encoded in a combinatorial fashion. Unlike in other sensory systems, responses of individual input channels to a large set of stimuli cannot be fully described by low-dimensional receptive fields. Hence, the neural space to encode odor information also covers many dimensions. As information flows through the olfactory pathway, odor-encoding activity patterns undergo multiple transformations. Such transformations not only filter sensory input but also establish representations of higher-order stimulus features, make these representations accessible to other circuits, and store them in memory. Transformations of activity patterns therefore provide direct insight into the computations underlying neuronal information processing.

To understand how evolution shaped the function of a sensory system, it is useful to consider the system's performance as a whole (Barlow 1961). Subjects can discriminate among similar odorants but also assign them to a common category, suggesting that odor discrimination, precise identification, and generalization are important tasks in olfaction (Gottfried 2009). Another important aspect of olfactory processing appears to be concentration invariance because the perceived quality of an odor can remain stable over a substantial concentration range (Gross-Isseroff & Lancet 1988). Speed seems to be important in some animals but less so in others because behavioral discrimination times vary widely among species. Olfactory systems are usually, although not always, poor in identifying components within mixtures (Livermore & Laing 1996, Wiltrout et al. 2003), indicating that segmentation of mixtures into their components is not a priority. Rather, olfaction generates gestalt-like percepts whose characters do not change fundamentally with the number or diversity of compounds that constitute an odor. Odor perception is often said to be closely associated with emotions, although it is not clear whether this relationship is stronger than in other sensory systems or whether it reflects the fact that odors cannot easily be verbalized. Olfactory processing is also closely associated with memory and may be continuously adapted to changes in the environment by experience and plasticity including adult neurogenesis (Wilson & Stevenson 2003, Wilson & Sullivan 2011). Unfortunately, the statistical properties of natural odors such as the identity and co-occurrence of odorants in natural environments are still poorly understood. It therefore remains unclear to what extent odorant receptors and neuronal circuits are adapted to ecologically relevant subregions of odor space. Although such adaptations likely exist (Khan et al. 2007), natural and artificial odors appear to be detected and discriminated with similar sensitivity and precision. Generalist olfactory systems can therefore operate throughout broad domains of odor space.

One approach toward an integrated understanding of olfaction and other brain functions is to identify elementary computations involved in the transformations of sensory inputs into perceptions, memory, and behavioral reactions. Insights into elementary computations are important to develop theoretical concepts of brain function and provide a foundation to study the underlying mechanisms. This review focuses on computations that are performed as odor-encoding activity patterns are transformed at successive stages of the olfactory pathway, with an emphasis on computations that may be of broader relevance. The review focuses specifically on results from adult zebrafish, an animal model that offers not only advanced molecular and genetic tools but also advantages for quantitative analyses of neuronal circuits (Friedrich et al. 2010). Owing to space constraints, exciting results from other species cannot be discussed in detail, pheromone processing is not addressed, and descriptions of brain regions are limited to a minimum.

THE OLFACTORY SYSTEM: BASIC ORGANIZATION AND FUNCTIONAL PROPERTIES

General Organization of the Olfactory System

Odor information detected by olfactory sensory neurons (OSNs) in the nose reaches higher brain areas including the cortex via an obligatory processing center, the olfactory bulb (OB). OSNs terminate in the OB in an array of glomeruli, each of which receives convergent input from OSNs expressing the same odorant receptor (Axel 1995, Buck 2000). Neuronal circuits in the OB consist of principal neurons, the mitral/tufted cells, and a large number of interneurons that mediate feedforward and feedback inhibition over multiple spatial scales (Shepherd et al. 2004, Wilson & Mainen 2006). Output of the OB is conveyed in parallel to multiple higher brain areas, including the piriform cortex, the anterior olfactory nucleus, the olfactory tubercle, the cortical amygdala, and the entorhinal cortex in mammals. Most of these areas are interconnected and project back to the OB. The largest target of the OB is the piriform cortex, a paleocortical area that may be further subdivided into functionally different subregions (Neville & Haberly 2004). The first olfactory processing center of insects, the antennal lobe, shows anatomical, molecular, and functional similarities to the OB, possibly as a result of convergent evolution. Output of the antennal lobe is conveyed by projection neurons (PNs) to two targets: the mushroom bodies and the lateral protocerebrum (Laurent 2002, Laurent et al. 2001, Masse et al. 2009).

OSN: olfactory sensory neuron

OB: olfactory bulb

PN: projection neuron

Approaches to Study Olfactory Processing in Zebrafish

Zebrafish offer a variety of methods and advantages to study olfactory processing. First, natural odorants for teleosts include amino acids (Carr 1988), which comprise molecules with varying degrees of similarity and constitute a

MC: mitral cell

well-defined subregion of odor space. Second, the head of an adult zebrafish can be maintained ex vivo without anesthesia, which facilitates physiological measurements in the intact brain (Zhu et al. 2012). Third, the small size of the zebrafish brain allows investigators to measure odor responses from ~10% of all neurons in an OB, and from even larger fractions of neurons in selected subregions, by multiphoton calcium imaging. Patterns of firing rate changes underlying these calcium signals can then be reconstructed by temporal deconvolution (Yaksi et al. 2007, Yaksi & Friedrich 2006) and analyzed by multivariate methods. Fourth, sensory input across the array of glomeruli can be imaged selectively after loading OSN axons with voltage- or calcium-sensitive dyes (Friedrich & Korsching 1997, 1998; Fuss & Korsching 2001). Fifth, on the basis of imaging results, electrophysiological recordings can be targeted to regions of the OB that are activated by specific odors. Sixth, the small brain size and advanced molecular methods enable genetic tagging of neuron types and precise optogenetic manipulations of activity patterns in space and time (Zhu et al. 2009, 2012). Seventh, the availability of detailed data at the single-neuron and population level provides a basis for realistic simulations and mathematical analyses of circuit function (Wiechert et al. 2010).

The Olfactory System of Zebrafish

The zebrafish is an advanced teleost found in still or slow waters such as rice fields in Southeast Asia (Spence et al. 2008). As in most teleosts, water flow through the nose is laminar and unidirectional, entering and leaving through separate nares. When zebrafish are not moving, nonneuronal cells bearing motile cilia maintain a constant low flow over the sensory epithelium that is apparently not under neural control. Additional water flow is generated by swimming. Zebrafish therefore do not sniff and may actively control odor delivery to the nose only by moving through the environment.

Odorant receptors are encoded in zebrafish by up to 300 genes from multiple families (Shi & Zhang 2009). The functional architecture of the zebrafish OB is similar to that of other vertebrates except for some specializations common to teleosts (Byrd & Brunjes 1995, Fuller et al. 2006, Satou 1990, Yoshihara 2009). The adult OB contains at least 140 stereotyped glomeruli and a substantial number of loosely defined, small axonal termination fields in a lateral domain (Baier & Korsching 1994, Braubach et al. 2012). For simplicity, we refer to these terminations also as glomeruli, and we collectively refer to all output neurons as mitral cells (MCs). In total, the adult OB contains ~25,000–30,000 neurons, including ~1,500 MCs (Wiechert et al. 2010). At least 70% of these MCs appear to receive input from a single glomerulus (Fuller et al. 2006). Zebrafish MCs lack extensive basal dendrites but can nevertheless interact over many glomerular diameters via interneurons (Wiechert et al. 2010).

Odor responses of MCs in adult zebrafish have been studied primarily in the lateral OB using amino acid stimuli at an intermediate concentration (10 μM). Individual MCs were spontaneously active and responded to subsets of amino acids with excitation, inhibition, or sometimes successive epochs of both (Friedrich & Laurent 2001, 2004; Yaksi et al. 2007). The mean firing rate across MCs in the lateral OB increased only slightly during an odor response. The tuning width of MCs is difficult to compare between zebrafish and other vertebrates because different stimulus sets were used in different studies. However, major differences are not obvious. Many PNs in the antennal lobe of insects, in contrast, appear to be more broadly tuned (Perez-Orive et al. 2002, Turner et al. 2008).

During the first few hundred milliseconds of an odor response, researchers observed that activity was redistributed across the MC population. At the level of individual MCs, this reorganization of activity patterns is reflected by odor- and neuron-dependent firing-rate modulations (Friedrich & Laurent 2001, 2004). Within less than one second, MC activity then approached a more stable pattern (steady state). Odor stimuli in these studies took at least

400–600 ms to reach a plateau concentration, suggesting that more rapid stimuli would produce faster dynamics. Moreover, pattern dynamics was compressed in time when odor concentration was increased. During an odor response, subsets of MCs rhythmically synchronized their action potentials and phase-locked to an oscillation in the local field potential (LFP) with a frequency near 20 Hz. Oscillation power increased as the number and/or response intensity of activated glomeruli increased, but the preferred phase of spikes relative to the oscillation was independent of these parameters (Friedrich et al. 2004). Hence, odor-evoked activity in the zebrafish OB is temporally patterned on multiple time scales, as observed in a wide range of other species (Laurent 2002). Output of the OB is conveyed by MC axons to multiple pallial and subpallial target areas in the telencephalon. The zebrafish homolog of the olfactory cortex is the posterior zone of the dorsal telencephalon (Dp) (Mueller et al. 2011, Wullimann & Mueller 2004), which is discussed in more detail below.

CHEMOTOPIC MAPS

Topographic maps exist in many sensory and motor areas of the brain, but possible computational functions of such maps are often unclear. In the olfactory system, spatial order is established in the input layer of the OB by the convergence of OSNs onto stereotyped glomeruli (Axel 1995, Buck 2000). This organization by itself does not, however, imply a systematic "chemotopic" mapping of molecular features along spatial coordinates in the brain. Insight into the functional topography of the glomerular array has been obtained by mapping glomerular responses to odors. Most odors activate scattered subsets of glomeruli, but the spatial distribution of these glomeruli is not entirely random (Wilson & Mainen 2006). In zebrafish, different natural classes of odors (amino acids, bile acids, nucleotides) preferentially stimulated glomeruli in segregated domains of the OB (lateral, medial, posterior-lateral, respectively), and amino acids sharing secondary chemical features (e.g., aromatic or basic side chains) activated clusters of glomeruli within the lateral domain (Friedrich & Korsching 1997, 1998). In first approximation, the spatial organization of glomeruli may therefore be described as chemotopic and hierarchical: Broad odor classes are mapped onto domains (primary chemotopy) and secondary molecular features are mapped onto smaller subregions (secondary chemotopy), but the identity of individual odorants is encoded by complex, distributed patterns.

Other analyses, however, result in a more complex picture. Although subsets of glomeruli collectively responded to secondary features of amino acids, individual glomeruli were not specific for these features. Rather, glomerular response profiles were often only moderately similar when assessed using a broader set of stimuli. Secondary chemotopy was therefore recognized only by multivariate analyses of ensemble responses using factor analysis or other methods (Friedrich & Korsching 1997). The mean correlation between response profiles decreased, on average, with the distance between glomeruli. However, response profiles of nearby glomeruli varied greatly in similarity, and neighboring glomeruli with very different response profiles were frequently observed. Glomerular response patterns are therefore also fractured and discontinuous in space, consistent with recent observations in the mammalian OB (Soucy et al. 2009, Wilson & Mainen 2006). The possible function of fractured topographic maps in the OB is still poorly understood.

The primary chemotopic map in zebrafish is preserved largely at the level of MCs, but the secondary chemotopic map was observed only transiently after stimulus onset. Chemotopy is less prominent in the interneuron population and almost completely absent in two major target areas of the OB: the ventral nucleus of the ventral telencephalon (Vv) and Dp (Yaksi et al. 2007, 2009). The absence of topographic constraints in higher brain areas may allow for the convergence of processing channels to integrate information about different molecular features.

Olfactory cortex: pallial brain regions that receive direct input from the olfactory bulb; in mammals, includes the piriform cortex, anterior olfactory nucleus, cortical amygdala, and entorhinal cortex

Dp: posterior zone of the dorsal telencephalic hemisphere; presumably homologous to piriform cortex in mammals

PATTERN EQUALIZATION

Although the total input from OSNs can vary over many orders of magnitude, the total firing rate across MCs or PNs is confined to a relatively narrow range, indicating that the OB and antennal lobe equalize activity patterns. In zebrafish, an equalization of response patterns has been observed by comparing responses of OSNs and MCs to the same odors in the steady state (Friedrich & Laurent 2004). Response equalization is likely to involve gain-control mechanisms within glomeruli and interglomerular inhibition that increases with overall stimulus intensity (Cleland et al. 2007, McGann et al. 2005, Tabor et al. 2008, Tabor & Friedrich 2008, Tan et al. 2010). Hence, pattern equalization may be related to normalization, a mathematically defined computation that scales a neuron's (or a population's) output as a function of its driving input and the mean activity of other neurons (Carandini & Heeger 2011).

Mushroom body: in the insect brain, receives direct input from the antennal lobe, contains Kenyon cells, and is implicated in learning and memory

Classifier: maps data to distinct categories; in neuroscience, usually assigns activity patterns to distinct outputs such as responses of "readout neurons"

Decorrelation: reduces the correlation between two patterns; refers here to the decorrelation of activity patterns across a population of neurons

Normalization has been implicated in a variety of neuronal processing tasks, particularly in the visual system (Carandini & Heeger 2011). In the antennal lobe of *Drosophila*, PN output activity is normalized by a saturating nonlinearity at the OSN-to-PN synapse and by widespread inhibition (Bhandawat et al. 2007, Olsen et al. 2010, Olsen & Wilson 2008). Another operation related to normalization occurs in the mushroom body (Papadopoulou et al. 2011). Modeling demonstrated that the normalization of PN activity facilitates concentation-invariant odor classification by simple linear classifiers or by basic models of higher-order olfactory neurons, mostly by equalizing total population activity across stimuli (Luo et al. 2010, Olsen et al. 2010). Without equalization, a simple classifier may confuse a correct input pattern with incorrect input patterns that are more intense. In addition, normalization or equalization can have other benefits (Carandini & Heeger 2011). For example, normalization can decorrelate some patterns and facilitate their discrimination, but normalization can also increase pattern correlations. Indeed, normalization does not decorrelate activity patterns across neurons in *Drosophila* (Olsen et al. 2010). Effects of normalization therefore depend on the statistical properties of activity patterns.

PATTERN DECORRELATION

Pattern Decorrelation: Basic Considerations

Odorants that appear chemically similar to one another, as judged by a chemist, often activate overlapping sets of glomeruli. Hence, chemically related stimuli evoke correlated input patterns in the OB, which may be an inevitable consequence of the mechanisms governing ligand-receptor interactions. Glomerular activation patterns representing chemically related odors therefore form clusters in coding space that can be directly visualized using dimension-reduction methods such as principal component analysis (PCA) (Friedrich & Korsching 1997, Friedrich & Laurent 2001). Correlations among input patterns may convey useful information if chemically similar odorants share biological significance (e.g., fruity odors signal food). For many computations, however, input correlations are not desired, for example because overlap between patterns complicates the discrimination of similar stimuli (e.g., apples versus pears within the fruity odors). In general, correlated (clustered) representations make inefficient use of coding space and require precise classifiers for discrimination (**Figure 1*a***). A decorrelation of input patterns is therefore a useful early step in many pattern classification procedures.

Decorrelation per se does not increase the information content of activity patterns or a perfect classifier's ability to discriminate between patterns. Rather, decorrelation increases the tolerance region for suboptimal classifiers, as illustrated by a simple example in **Figure 1*b***. Decorrelation therefore facilitates the search for a classifier with acceptable performance and generates robustness. In the nervous system, this process may facilitate learning and subsequent identification of stimuli and allow for the use of simpler classification strategies.

Mechanisms of decorrelation can be adaptive or nonadaptive. Adaptive mechanisms are tuned to specific inputs on the basis of prior knowledge. When applied to random inputs, however, linear adaptive methods will, on average, fail to decrease correlations (Wiechert et al. 2010). Nonadaptive decorrelation mechanisms, in contrast, do not require prior knowledge and may decorrelate any inputs. One possibility to achieve a nonadaptive decorrelation is to transform activity patterns into sparse patterns of higher dimensionality, but doing so requires large numbers of neurons (Cortes & Vapnik 1995, Laurent 2002, Marr 1969).

It is also instructive to distinguish between channel decorrelation and pattern decorrelation. Channel decorrelation decreases correlations between processing channels, such as glomeruli or neurons, and thereby reduces redundancy. Hence, channel decorrelation increases coding capacity with respect to stimulus space and results in "efficient coding" (Barlow 1961). Pattern decorrelation, in contrast, reduces correlations between patterns across processing channels, i.e., activity patterns across neuronal populations. Assuming that information is encoded in such population activity patterns, pattern decorrelation makes representations more distinct. The effect of pattern decorrelation may therefore be described as "informative coding." Under special cirumstances, channel decorrelation can result in pattern decorrelation or vice versa (see Wick et al. 2010 and references therein), but in principle they can occur independently.

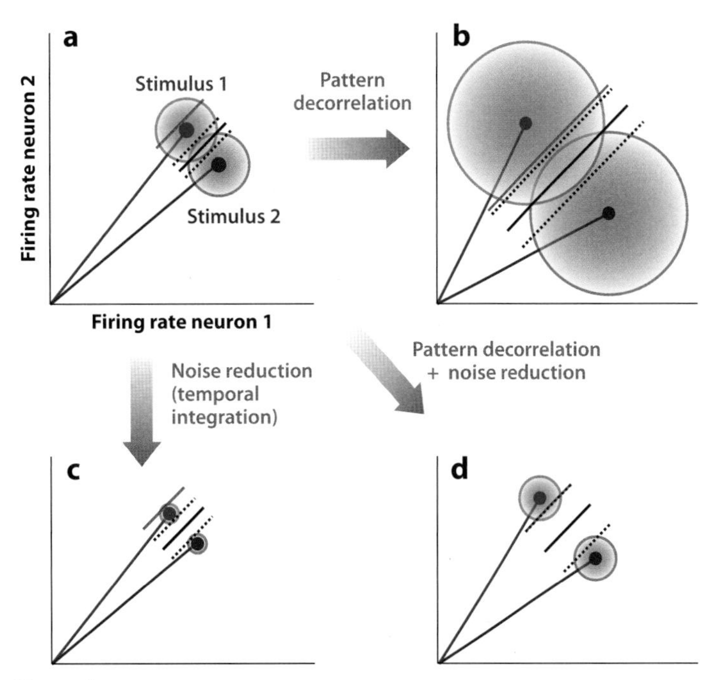

Figure 1

Pattern decorrelation and noise reduction. (*a*) Blue and purple points are activity patterns evoked by two similar stimuli in a two-dimensional coding space (two neurons). Shaded circular regions represent noise (trial-to-trial variability of response patterns); angular separation reflects pattern correlation. Black line shows an optimal classifier; dashed lines define a hypothetical limit of acceptable classification error (tolerance range). Gray line represents an imperfect classifier with a fixed offset from the optimal classifier. (*b*) Decorrelated patterns with the same relative overlap of noise areas. Note that classification error by the optimal classifier is the same as in panel *a*, but the tolerance range (distance between dashed lines) is increased and misclassification by the imperfect classifier (*gray line*) is reduced. (*c*) Noise reduction without decorrelation improves classification by an optimal classifier, slightly increases the tolerance range, and slightly improves the performance of the imperfect classifier. (*d*) Combination of noise reduction and decorrelation partially improves optimal pattern classification, tolerance range, and the performance of a suboptimal classifier. This scenario most closely reflects observations in the zebrafish olfactory bulb.

Pattern Decorrelation in the Zebrafish Olfactory Bulb

In zebrafish, chemically related amino acids evoked overlapping activity patterns across glomeruli and initially also across MCs. However, as MC activity was reorganized, most of these correlations decreased (**Figure 2*a***). This decorrelation can be visualized in a lower-dimensional coding space as a divergence of trajectories representing dynamic activity patterns (**Figure 2*b***) (Laurent et al. 2001, Niessing & Friedrich 2010). Concomitantly, correlations between initially dissimilar patterns increased slightly so that correlation coefficients converged toward low positive values around 0.3. Transformations of MC activity patterns during the initial phase of an odor response, therefore, decluster odor representations and distribute them more evenly in

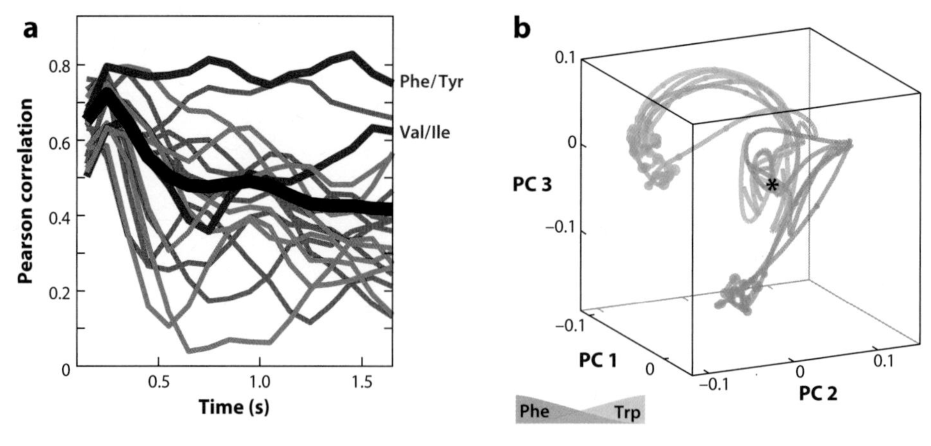

Figure 2

Discrete pattern decorrelation. (*a*) Pairwise correlation between mitral cell (MC) activity patterns evoked by amino acids as a function of time. Only pairs of patterns with high initial correlations (mean correlation 100–300 ms after response onset ≥0.6) were included. Red lines correspond to odor pairs used in a behavioral discrimination task (Miklavc & Valentinčič 2012). Thick red lines correspond to the two odor pairs, Phe/Tyr and Val/Ile, that were not discriminated significantly by the population of fish tested. Gray lines correspond to odor pairs that were not tested in the behavioral task. The black line shows the average. Activity was measured by loose-patch recordings of action potentials (raw data from Friedrich et al. 2004). (*b*) MC activity patterns evoked by a series of stimuli that morphed one amino acid (Phe) into a similar one (Trp). Each trajectory shows the evolution of an activity pattern from the resting state (*) to a steady state (*large dots*), projected onto the first three principal components. Time is indicated by increasing size of dots. The divergence of trajectories toward distinct steady states indicates pattern decorrelation. Note that decorrelation occurs abruptly within the morphing series and that trajectories diverge already close to the resting state. For details, see Niessing & Friedrich (2010); figure also modified from Niessing & Friedrich (2010). Abbreviations: Ile, isoleucine; Phe, phenylalanine; Tyr, tyrosine; Val, valine.

coding space (Friedrich et al. 2004, Friedrich & Laurent 2001, Niessing & Friedrich 2010).

The observed time course of decorrelation varied depending on the odor pair. In general, most of the decorrelation occurred within 600–800 ms after response onset (**Figure 2*a***) (Friedrich & Laurent 2001). Subsequent analyses showed that pattern decorrelation occurs within specific ensembles of MCs. When these ensembles were analyzed in isolation, decorrelation was already pronounced after 400 ms (Friedrich 2006, Friedrich & Laurent 2004). These times are probably not fixed but may vary with the time course of the stimulus and odor concentration.

Multiphoton calcium imaging revealed that MC activity patterns initially overlapped among strongly active and spatially clustered MCs. This overlap subsequently decreased because odor-dependent subsets of these MCs became less active or silent, resulting in a local, but not a global, sparsening of MC activity. Concomitantly, the activity of inhibitory interneurons increased and became more dense. These and other results indicate that decorrelation involves odor-specific patterns of inhibition that diversify MC activity (Tabor et al. 2008, Yaksi et al. 2007).

Whereas decorrelation facilitates pattern classification by simple or suboptimal classifiers, classification by an optimal classifier would be improved most effectively by noise reduction (**Figure 1*c***), for example through temporal integration. Consistent with such a mechanism, the trial-to-trial variability of MC firing rates decreased as the steady state was approached (Friedrich & Laurent 2001, 2004), particularly among MCs that contributed to decorrelation (Friedrich et al. 2004). Concomitantly, the reliability of odor identification by a template-matching procedure based on time-binned MC activity patterns improved. Information about an odor is therefore accumulated during a response by at least two processes: Responses of individual MCs become less variable, and the network evolves toward an increasingly distinct state (**Figure 1*d***).

Discrete Decorrelation and Concentration Invariance

Perception is sensitive to small differences in a stimulus but also exhibits stability, for example against changes in odor intensity. Sensory representations should therefore be robust against noisy fluctuations and possibly invariant with respect to some stimulus parameters. Pattern decorrelation may undermine such stability if any small difference between inputs is amplified. To examine stability of MC activity patterns in the steady state, odor stimuli were systematically changed in small steps (Niessing & Friedrich 2010). Varying the concentration of amino acid odors evoked heterogeneous changes in individual MCs' responses, but pronounced pattern decorrelation was rarely observed, with a few exceptions. MC activity patterns were therefore relatively stable against changes in odor concentration, consistent with concentration-invariant odor quality perception (Gross-Isseroff & Lancet 1988). This consistency is remarkable because odor-evoked glomerular activity patterns change substantially with odor concentration, although in a systematic fashion (Friedrich & Korsching 1997).

The molecular identity of an odor was varied by morphing one amino acid odor into another through a systematic series of binary mixtures. MC activity patterns remained similar within a subrange of a morphing series but then changed abruptly at defined transition points. Decorrelation is therefore not chaotic but occurs at defined transition zones in coding space that separate more stable states. Abrupt transitions between activity patterns were driven by coordinated response changes among small, odor-specific MC ensembles. As a consequence, the possible number of discrete stable states is very high (Niessing & Friedrich 2010). This finding also implies that transitions between MC activity patterns differ from abrupt changes in hippocampal place cell activity during global remapping (Wills et al. 2005). Moreover, this finding emphasizes the need for exhaustive sampling methods because small yet important neuronal ensembles may otherwise be missed. Although abrupt transitions between activity patterns are indicative of attractor networks, they can also be produced by other mechanisms such as an amplification of step-like transitions in input patterns (Wiechert et al. 2010).

Functional Relevance of Pattern Decorrelation

It has been proposed that insights into the relationship between pattern decorrelation and odor discrimination may be obtained by comparing the time course of decorrelation to reaction times in behavioral odor discrimination tasks. However, such comparisons are problematic because olfactory reaction times have not been measured in zebrafish and vary substantially among species (Abraham et al. 2004, Friedrich 2006, Khan et al. 2007, Wise & Cain 2000). In rodents and humans, discrimination time increased significantly with task difficulty, which may reflect a speed-accuracy trade-off (Abraham et al. 2004, Rinberg et al. 2006b, Uchida & Mainen 2003, Wise & Cain 2000). Although consistent with a role for pattern decorrelation in discrimination, these results could also reflect other processing strategies such as temporal integration. A general problem associated with comparing activity measurements in naïve animals to behavioral reaction times is that behavioral training changes odor processing and most likely enhances the discriminability of odor representations (Chapuis & Wilson 2011, Gottfried 2009, Li et al. 2008, Rabin 1988). Moreover, the reliability of early sensory representations may not be the limiting factor governing discrimination behavior, at least under certain conditions (Kepecs et al. 2008).

More direct insight into a behavioral function of pattern decorrelation may be obtained by comparing the ability to discriminate odor pairs with the associated activity patterns. Miklavc & Valentinčič (2012) trained adult zebrafish to associate one of five amino acid odorants with a food reward and subsequently tested behavioral discrimination of the rewarded odorant against 16–18 other amino acids, yielding a total of 62 pairwise discrimination tests (Miklavc & Valentinčič 2012). Fish failed to discriminate only 2 of these 62 amino acid pairs: phenylalanine (Phe)/tyrosine (Tyr) and valine (Val)/isoleucine (Ile). Initial correlations between the corresponding activity patterns were high but not unusual. However, unlike most other patterns, activity patterns evoked by Phe/Tyr and Val/Ile were not efficiently decorrelated (Friedrich et al. 2004, Friedrich & Korsching 1997, Friedrich & Laurent 2001) (**Figure 2*a***). Furthermore, a specific zebrafish strain that decorrelated activity patterns evoked by Phe/Tyr also discriminated these odors in the behavioral test (T. Valentinčič, unpublished observations). These results are consistent with the hypothesis that pattern decorrelation supports odor discrimination.

Because decorrelation facilitates the search for an appropriate classifier (**Figure 1*b***), it may also be expected to enhance the rate of learning rather than maximal acuity. Consistent with this hypothesis, more difficult odor discriminations require longer training, at least in some tasks (Abraham et al. 2004). Furthermore, decorrelation appears to be an important preprocessing step for pattern storage by autoassociative networks. Such networks are likely to mediate memory storage in the hippocampus (Marr 1971, McNaughton & Morris 1987, Rolls & Kesner 2006) and have been proposed to operate also in the olfactory cortex (Haberly 2001, Haberly & Bower 1989, Hasselmo et al. 1990, Wilson & Sullivan 2011). Autoassociative networks can perform two opposing computations, pattern separation and pattern completion, which would be greatly facilitated by partially decorrelated input. Moreover, decorrelated inputs reduce the danger of catastrophic interference, a process that results in the sudden erasure of stored patterns by newly learned patterns in artificial neuronal networks (French 1999). Decorrelation in the OB may therefore be an important preprocessing step for higher-order associative computations.

Mechanisms of Pattern Decorrelation

Decorrelation is sometimes mingled with contrast enhancement, although these are different computations. Contrast enhancement shifts response intensities (firing rates) of individual neurons toward extremes by amplifying large responses, reducing small responses, or both. It has been suggested that the contrast of MC activity patterns is enhanced by topographically organized input and lateral inhibition (Yokoi et al. 1995), but this hypothesis has not been confirmed experimentally (Fantana et al. 2008). More recently, investigators have proposed nontopographic models of global contrast enhancement (Arevian et al. 2008, Cleland & Sethupathy 2006) that are related to normalization (Carandini & Heeger 2011). Global contrast enhancement can decorrelate patterns that overlap in their weakly active elements but may increase correlations when patterns overlap in their strongly active elements (**Figure 3*a***). The effect of contrast enhancement on pattern correlations therefore depends on properties of the patterns.

In the zebrafish OB, the major contributions to high initial pattern correlations came from strongly active MCs (**Figure 3*b***). Moreover, bath application of dopamine, which enhanced pattern contrast by modulating input-output functions of MCs, failed to reduce correlations (Bundschuh et al. 2012). Pattern decorrelation is therefore not mediated by global contrast enhancement in zebrafish, although an involvement of local contrast enhancement cannot be ruled out. More generally, these and further results indicate that decorrelation cannot be explained by scaling of MC activity based on the mean population activity across glomeruli or MCs. Rather, decorrelation appears to rely on more complex features of activity patterns,

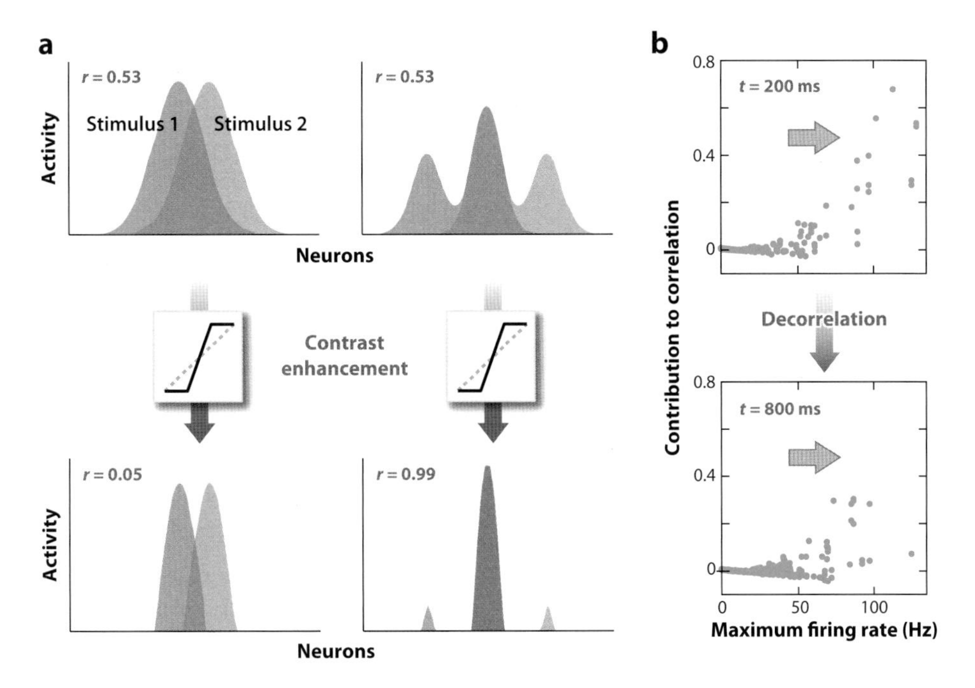

Figure 3

Contrast enhancement and pattern decorrelation. (*a*) Effect of contrast enhancement on pairs of activity patterns (*blue*, *light purple*) that overlap either in weakly active neurons (*left*) or in their strongly active neurons (*right*). *r*, Pearson correlation coefficient. (*b*) Contribution of individual MCs in the zebrafish OB to the correlation between activity patterns (100 ms time bins) as a function of firing rate. Each data point shows one neuron's contribution to the correlation between one pair of patterns (*y*-axis) and the maximal firing rate of this neuron evoked by the corresponding odors (*x*-axis). Only patterns that were initially highly correlated (mean correlation during the first 200 ms ≥0.65) were included. Activity was measured by loose-patch recordings of action potentials (raw data from Friedrich et al. 2004). (*Top*) Before decorrelation. (*Bottom*) After decorrelation. Note that the major contribution to the initial correlation came from highly active neurons (*top*, *dark yellow arrow*). After decorrelation, the contribution of these neurons was reduced (*bottom*) because the firing rates evoked by the corresponding odor pairs diverged (not shown).

which can convey more detailed information about an odor. It is therefore important to consider activity patterns and network interactions at the neuron-by-neuron level.

A theoretical study by Koulakov & Rinberg (2011) proposed that reciprocal connectivity in the OB creates sparse incomplete representations of MC firing patterns in granule cells that are removed from OB output by inhibitory feedback. The remaining MC output then transmits an error signal—the mismatch between MC activation and the inhibitory feedback pattern—that may be less correlated than the input. This study predicts that granule cell activity patterns are extremely sparse and specific, which is not consistent with experimental results from zebrafish (Yaksi et al. 2007) and mice (Tan et al. 2010). It is therefore unclear whether the underlying mathematical model is adequate. Moreover, it has not been demonstrated that this model results in pattern decorrelation. A related theoretical study building on concepts of predictive coding suggests that reciprocal connectivity successively removes from the output pattern components that correspond to factors, leaving behind the residuals of a factor analysis (S. Druckmann, T. Hu, & D. Chklovskii, manuscript in preparation). This general idea is reminiscent of experimental observations in zebrafish (Friedrich & Laurent 2001, Yaksi et al. 2007), but further analyses are required to compare the model to experimental data.

Multiplexing: simultaneous transmission of different signals; refers here to the simultaneous transmission of information about different stimulus features

Analytical approaches and simulations showed that generic networks of stochastically connected threshold-linear analog neurons (SNOREs) with uniform synaptic weights decorrelate, on average, any positively correlated, normally distributed input patterns by a mechanism referred to as recurrence-enhanced threshold-induced decorrelation (reTIDe) (Wiechert et al. 2010). The initial step in reTIDe is to threshold the input. The threshold is an elementary nonlinearity that represents action potential initiation and produces a decorrelation (see also de la Rocha et al. 2007). This decorrelation is then amplified by feeding the thresholded output patterns back into the network through recurrent connections until a steady state is reached. Hence, pattern decorrelation emerges naturally from generic properties of neurons and their connectivity. reTIDe is robust against small fluctuations in the input that may represent noise. Because reTIDe is nonadaptive and performs true pattern decorrelation, it is particularly useful when input patterns and their statistical properties are unpredictable.

reTIDe is pronounced when connectivity is sparse and baseline activity is high, two characteristics of the OB whose functions have remained elusive (Fantana et al. 2008, Rinberg et al. 2006a, Wiechert et al. 2010). In a simplified computational model of the zebrafish OB, decorrelation could be attributed predominantly to reTIDe, while topographic lateral inhibition played a minor role. Consistent with the prediction that decreasing baseline activity should increase pattern correlations, a small increase in pattern correlations was observed in the presence of dopamine, which hyperpolarized MCs (Bundschuh et al. 2012). Furthermore, reTIDe predicts that decorrelation should be enhanced when inhibitory synaptic weights are increased (Wiechert et al. 2010). This prediction is generally consistent with behavioral results from mice (Abraham et al. 2010), although it may not be diagnostic for reTIDe. reTIDe is therefore a plausible mechanism for pattern decorrelation in the OB.

MULTIPLEXED ODOR REPRESENTATIONS IN THE OLFACTORY BULB

Oscillatory synchronization can dynamically define ensembles of neurons that collectively transmit specific information (Laurent 2002, Laurent et al. 2001, Singer 1999). To explore this possibility in the zebrafish OB, Friedrich et al. (2004) recorded odor-evoked spikes of MCs together with the LFP and sorted them into synchronized and nonsynchronized subpopulations on the basis of their phase relationship to the LFP oscillation (Friedrich et al. 2004). Virtual activity patterns consisting only of synchronized spikes converged, rather than diverged, in response to chemically related odors. Synchronized spikes, therefore, did not contribute to pattern decorrelation but instead increased correlations. When synchronized spikes were removed from the ensemble, pattern decorrelation was enhanced and also faster (Friedrich et al. 2004, Friedrich & Laurent 2001). Because synchronized spikes were a minority, decorrelation, rather than pattern convergence, dominated when all spikes were combined irrespective of synchrony. Hence, odor-dependent subsets of MCs, defined by their transient oscillatory synchronization, simultaneously participate in different computations.

Whereas decorrelated activity patterns were particularly informative about the precise identity of an odor, correlated patterns of synchronized spikes conveyed information about chemical categories of odors. Information about similarity in odor space is therefore not lost during decorrelation but maintained by ensembles of synchronized MCs. As a consequence, the population of MCs encodes complementary information about odors in a multiplexed format (Friedrich et al. 2004). This may be an efficient strategy to transmit preprocessed information about different stimulus features to multiple higher brain areas.

NEURONAL COMPUTATIONS IN Dp, THE HOMOLOG OF OLFACTORY CORTEX

Functional Organization of Dp

Computational functions of higher olfactory brain areas such as olfactory cortex have not been studied as extensively as the OB but have recently become a focus of interest. The piriform cortex is a paleocortical area that receives divergent, nontopographic input from the OB and contains widespread intracortical connections among pyramidal neurons (Davison & Ehlers 2011, Franks et al. 2011, Johnson et al. 2000, Miyamichi et al. 2011, Neville & Haberly 2004). In addition, the piriform cortex contains different types of inhibitory interneurons that remain more local (Neville & Haberly 2004). The teleost homolog of the olfactory cortex, or more specifically the piriform cortex, is Dp (Mueller et al. 2011, Wullimann & Mueller 2004). Because pallial brain areas of teleosts are relocated by morphogenetic movements during ontogeny and do not follow the inside-out pattern of cortical development, their gross histological appearance differs from cortices in other vertebrate classes. Nevertheless, immunocytochemical, electrophysiological, and optical approaches indicate that Dp shares functional characteristics with the piriform cortex (see references below; Y.-P. Zhang Schärer, J. Shum, M. Zou, & R.W. Friedrich, unpublished data). These include (*a*) mono- and polysynaptic convergence of diverse input channels from the OB (Apicella et al. 2010, Miyamichi et al. 2011, Yaksi et al. 2009); (*b*) broadly tuned excitatory and inhibitory synaptic input from intrinsic neurons (Franks et al. 2011; Poo & Isaacson 2009, 2011; Schärer et al. 2012); (*c*) short-term plasticity of synaptic input (Stokes & Isaacson 2010); (*d*) different classes of interneurons mediating feedforward and feedback inhibition (Stokes & Isaacson 2010); (*e*) divergent projections from the OB without pronounced topography (Miyamichi et al. 2011, Miyasaka et al. 2009); and (*f*) scattered, but not extremely sparse, activity patterns and pronounced mixture suppression (Blumhagen et al. 2011, Schärer et al. 2012, Stettler & Axel 2009, Yaksi et al. 2009). Thus the piriform cortex and Dp likely perform similar functions, consistent with their close phylogenetic relationship.

Representations of Olfactory Objects in Dp

The main excitatory input to MCs comes from OSNs expressing a single odorant receptor, suggesting that individual MCs convey information about a limited set of molecular features. Excitatory input to neurons in piriform cortex, in contrast, is often dominated by intracortical connections, rather than by sensory afferents from the OB (Poo & Isaacson 2011). Cortical circuits are therefore thought to mediate associative interactions between processing channels and to establish synthetic representations of odor objects (Wilson & Stevenson 2003, Wilson & Sullivan 2011).

In the OB of zebrafish, responses of individual MCs to binary odor mixtures were often closely related to the response evoked by one of the components alone (Tabor et al. 2004). Activity patterns across MCs are therefore mixture specific, but subsets of MCs still convey substantial information about individual components. Dp neurons, in contrast, showed pronounced mixture interactions, particularly mixture suppression (Yaksi et al. 2009). Hence, transformations of activity patterns in Dp discard component information and establish specific representations of odor mixtures, consistent with synthetic object representations. Similar observations were made in the piriform cortex (Chapuis & Wilson 2011, Stettler & Axel 2009).

The prominent intracortical excitatory connectivity also gave rise to the hypothesis that the piriform cortex functions as an autoassociative network to store and recall odor-encoding activity patterns (Haberly 2001, Haberly & Bower 1989, Wilson & Sullivan 2011). This far-reaching notion is based on theoretical work (Kanerva 1988, Kohonen 1984, Marr 1971), modeling (Hasselmo et al. 1990), and some experimental results (Barnes et al. 2008, Chapuis & Wilson 2011, Choi et al. 2011, Wilson 2003). Because autoassociative pattern storage changes synaptic weights within the network, object representations may change continuously as a function of experience, even on short timescales. Olfactory object representations—and possibly olfactory perception—may thus be inseparable from olfactory memory (Gottfried 2009, Wilson & Stevenson 2003, Wilson & Sullivan 2011).

Retrieval of Information from a Multiplexed Code

The multiplexed transmission of information about different stimulus features by MCs raises the questions of whether and how this information is retrieved in higher brain areas (Friedrich et al. 2004). Information about precise odor identity is efficiently encoded by activity patterns across nonsynchronized MC ensembles in the steady state, but the retrieval of this information may be complicated by synchronized MCs because synchronization usually enhances the impact of synaptic input onto a target neuron (Bruno & Sakmann 2006, Perez-Orive et al. 2002). The impact of MC synchrony on Dp neurons was therefore examined using an optogenetic approach that allowed researchers to manipulate oscillatory synchrony among MCs without major effects on firing rate (Blumhagen et al. 2011, Zhu et al. 2012). Surprisingly, firing rate responses of Dp neurons were almost insensitive to oscillatory synchrony because they behaved as passive low-pass filters and because collective synaptic currents were slow.

Further results showed that firing rates of Dp neurons during an odor response were determined by the amplitude of a slow depolarization rather than by oscillatory membrane potential fluctuations (Blumhagen et al. 2011). The amplitude of this depolarization depended on the relative strength of excitatory and inhibitory synaptic inputs, which effectively clamped the membrane potential to an odor-dependent value that often remained subthreshold (Yaksi et al. 2009). Moreover, action potentials of Dp neurons were rarely observed during the dynamic phase of MC input but occurred primarily during the steady state. Even when fast input was evoked by optogenetic stimulation, responses of Dp neurons were substantially delayed. Hence, neuronal circuits in Dp perform two filtering operations that (*a*) attenuate the impact of synchronized input and (*b*) delay responses until MC input approaches the decorrelated steady state. Temporal filtering therefore tunes Dp neurons to those features of MC activity patterns that are particularly informative about precise odor identity. Consistent with these filtering operations, Blumhagen et al. (2011) found that correlations between odor-evoked activity patterns in Dp were low.

It is still unclear how the output of Dp or the piriform cortex is further processed and how it influences perception. One possibility is that perception is directly determined by output from Dp and, thus, depends on steady state odor representations in the OB. Another hypothesis is that Dp is involved primarily in the storage of odor information and has a more indirect influence on perception. In this scenario, steady state MC input could support specific synaptic modifications in auto-associative networks by entraining neuronal circuits to stable and informative activity patterns. Training of auto-associative networks on the steady state of an odor response may improve pattern recognition also during the transient phase because informative components of steady-state patterns emerge rapidly, well before the steady state is reached (Niessing & Friedrich 2010). Both these possibilities are currently speculative and not mutually exclusive.

Although oscillatory membrane potential fluctuations of Dp neurons during an odor response were small, they nevertheless influenced the timing of action potentials during suprathreshold responses. The oscillatory timing of MC spikes is, therefore, not lost in Dp. Indeed, oscillatory synchronization among a subset of inputs to Dp may even entrain the output of Dp neurons when other inputs are not synchronized to this oscillation. Moreover, odors evoked prominent oscillatory

LFP activity in another telencephalic brain area (Blumhagen et al. 2011). Information conveyed by synchronized MC ensembles may therefore be processed in other brain areas in parallel or through interactions with Dp.

OLFACTORY PROCESSING STRATEGIES: SOME OPEN QUESTIONS AND SPECULATIONS

Unlike Dp neurons, Kenyon cells in the mushroom body of locusts are exquisitely sensitive coincidence detectors and respond primarily to input from synchronized PN ensembles (Laurent 2002, Laurent et al. 2001, Laurent & Naraghi 1994, Perez-Orive et al. 2002). Spiking depends on the amplitude of oscillatory membrane-potential fluctuations that are driven by synchronized PN input; nonsynchronized fluctuations are averaged out by the high convergence of PNs (Jortner et al. 2007). Kenyon cells respond to odors with one or a few spikes at defined times during the odor-specific evolution of PN response patterns, particularly during the early phase (Perez-Orive et al. 2002). Dp neurons, in contrast, behave as low-pass filters, are driven to spike threshold by a large and slow depolarization, and fire predominantly during the steady state of MC input (Blumhagen et al. 2011, Yaksi et al. 2009). Hence, locusts and zebrafish both establish distributed and specific representations of odors in higher brain areas but use different mechanisms of pattern processing. Although the reasons for these differences are a matter of speculation, three possibilities are discussed below.

First, differences in readout strategies may reflect adaptations to odor representations at the previous processing stage. In the antennal lobe of locusts, PN responses are dense, and synchronization conveys significant information about odor identity (Perez-Orive et al. 2002, Stopfer et al. 1997, Wehr & Laurent 1996). The detection of synchronized PN ensembles is, therefore, an efficient strategy to create informative and sparse odor representations across Kenyon cells (Laurent 2002, Laurent et al. 2001). Moreover, odor representations across PNs appear to be most informative during the dynamic phase (Mazor & Laurent 2005), favoring readout mechanisms that operate on short timescales (individual oscillation cycles). In zebrafish, in contrast, odor identity is efficiently encoded by nonsynchronized MC activity in the steady state, favoring low-pass filtering and delayed readout to retrieve information about odor identity.

Second, differences between olfactory encoding and decoding strategies may reflect different temporal constraints on olfaction. Zebrafish may afford to use steady states for olfactory computations because fast odor fluctuations are rare and unlikely to convey significant information in their natural environment. In the presence of fast odor fluctuations, however, information may have to be extracted from dynamic representations because steady states are rarely reached. In addition, the sniffing rhythm may impose strict time windows for neuronal computations in some species (Wachowiak 2011). Another possible constraint could be the size of the system, particularly the number of input channels. Because zebrafish have relatively few odorant receptors and MCs, odors may initially be represented with high overlap so that subsequent pattern analysis requires extensive processing. More input channels would allow for more decorrelated and possibly sparser representations already at early stages and should facilitate higher-order processing. Such a strategy may be particularly useful for macrosmatic species, which depend on fast scent tracking for survival.

Third, different processing strategies in insects and vertebrates may reflect different functions of higher olfactory brain areas. Mushroom bodies mediate associative learning and adjust responses of output neurons to odors by adapting the synaptic weights of their Kenyon cell inputs (Heisenberg 2003). The high-dimensional and sparse format of odor representations supports this function because it facilitates readout by simple classifiers such as individual or small numbers of neurons (Laurent 2002, Perez-Orive et al. 2002). The functions of Dp and the olfactory cortex, in contrast, are less clear, partly because targeted inactivation or ablation studies are still lacking. Considering the obvious and possibly fundamental differences in the basic architecture of mushroom bodies and the olfactory cortex, it would not be surprising if their basic tasks turn out to be at least partially different.

Concluding Remarks

This review discusses selected computations that have been explored in the olfactory system of adult zebrafish. We have focused on computations that cannot be understood on the basis of coarse features of activity patterns such as the mean activity levels of neuronal populations. Rather, mechanistic insights into computations such as pattern decorrelation or multiplexing require detailed, quantitative, and dense information about the neuron-by-neuron structure and dynamics of activity patterns. Because this information is difficult to obtain in large brains, the zebrafish is an excellent vertebrate model in which to study such computations. Further progress in understanding complex neuronal computations will critically depend on quantitative and exhaustive reconstructions of neuronal wiring diagrams (Denk et al. 2012). Moreover, important progress will undoubtedly come from functional manipulations of specific types and ensembles of neurons, particularly from studies that use light. Because physically small brains and genetic approaches are advantageous to address these issues, the zebrafish is again a promising model system. Insight into the olfactory system will be instructive to understand the function of other brain areas, even if the underlying computations turn out to be different in general or in detail.

DISCLOSURE STATEMENT

The author is not aware of any affiliations, memberships, funding, or financial holdings that might be perceived as affecting the objectivity of this review.

ACKNOWLEDGMENTS

I thank Andreas Lüthi and members of the Friedrich lab, particularly Gilad Jacobson, for discussions and comments on the manuscript. I am grateful for financial support by the Novartis Research Foundation, the Max Planck Society, the Swiss Nationalfonds (SNF), the German Research Foundation (DFG), the European Union (EU), the Human Frontiers Science Program (HFSP), and other funding agencies.

LITERATURE CITED

Abraham NM, Egger V, Shimshek DR, Renden R, Fukunaga I, et al. 2010. Synaptic inhibition in the olfactory bulb accelerates odor discrimination in mice. *Neuron* 65:399–411

Abraham NM, Spors H, Carleton A, Margrie TW, Kuner T, Schaefer AT. 2004. Maintaining accuracy at the expense of speed: stimulus similarity defines odor discrimination time in mice. *Neuron* 44:865–76

Apicella A, Yuan Q, Scanziani M, Isaacson JS. 2010. Pyramidal cells in piriform cortex receive convergent input from distinct olfactory bulb glomeruli. *J. Neurosci.* 30:14255–60

Arevian AC, Kapoor V, Urban NN. 2008. Activity-dependent gating of lateral inhibition in the mouse olfactory bulb. *Nat. Neurosci.* 11:80–87

Axel R. 1995. The molecular logic of smell. *Sci. Am.* 273:130–37

Baier H, Korsching S. 1994. Olfactory glomeruli in the zebrafish olfactory system form an invariant pattern and are identifiable across animals. *J. Neurosci.* 14:219–30

Barlow HB. 1961. Possible principles underlying the transformations of sensory messages. In *Sensory Communication*, ed. WA Rosenblith, pp. 217–34. Cambridge, MA: MIT Press

Barnes DC, Hofacer RD, Zaman AR, Rennaker RL, Wilson DA. 2008. Olfactory perceptual stability and discrimination. *Nat. Neurosci.* 11:1378–80

Bhandawat V, Olsen SR, Gouwens NW, Schlief ML, Wilson RI. 2007. Sensory processing in the *Drosophila* antennal lobe increases reliability and separability of ensemble odor representations. *Nat. Neurosci.* 10:1474–82

Blumhagen F, Zhu P, Shum J, Scharer YP, Yaksi E, et al. 2011. Neuronal filtering of multiplexed odour representations. *Nature* 479:493–98

Braubach OR, Fine A, Croll RP. 2012. Distribution and functional organization of glomeruli in the olfactory bulbs of zebrafish (*Danio rerio*). *J. Comp. Neurol.* 520:2317–39

Bruno RM, Sakmann B. 2006. Cortex is driven by weak but synchronously active thalamocortical synapses. *Science* 312:1622–27

Buck LB. 2000. The molecular architecture of odor and pheromone sensing in mammals. *Cell* 100:611–18

Bundschuh ST, Zhu P, Scharer YP, Friedrich RW. 2012. Dopaminergic modulation of mitral cells and odor responses in the zebrafish olfactory bulb. *J. Neurosci.* 32:6830–40

Byrd CA, Brunjes PC. 1995. Organization of the olfactory system in the adult zebrafish: histological, immunohistochemical, and quantitative analysis. *J. Comp. Neurol.* 358:247–59

Carandini M, Heeger DJ. 2011. Normalization as a canonical neural computation. *Nat. Rev. Neurosci.* 13:51–62

Carr WES. 1988. The molecular nature of chemical stimuli in the aquatic environment. In *Sensory Biology of Aquatic Animals*, ed. J Atema, RR Fay, AN Popper, WN Tavolga, pp. 3–27. New York: Springer

Chapuis J, Wilson DA. 2011. Bidirectional plasticity of cortical pattern recognition and behavioral sensory acuity. *Nat. Neurosci.* 15:155–61

Choi GB, Stettler DD, Kallman BR, Bhaskar ST, Fleischmann A, Axel R. 2011. Driving opposing behaviors with ensembles of piriform neurons. *Cell* 146:1004–15

Cleland TA, Johnson BA, Leon M, Linster C. 2007. Relational representation in the olfactory system. *Proc. Natl. Acad. Sci. USA* 104:1953–58

Cleland TA, Sethupathy P. 2006. Non-topographical contrast enhancement in the olfactory bulb. *BMC Neurosci.* 7:7

Cortes C, Vapnik V. 1995. Support-vector networks. *Mach. Learn.* 20:273–97

Davison IG, Ehlers MD. 2011. Neural circuit mechanisms for pattern detection and feature combination in olfactory cortex. *Neuron* 70:82–94

de la Rocha J, Doiron B, Shea-Brown E, Josić K, Reyes A. 2007. Correlation between neural spike trains increases with firing rate. *Nature* 448:802–6

Denk W, Briggman KL, Helmstaedter M. 2012. Structural neurobiology: missing link to a mechanistic understanding of neural computation. *Nat. Rev. Neurosci.* 13:351–58

Fantana AL, Soucy ER, Meister M. 2008. Rat olfactory bulb mitral cells receive sparse glomerular inputs. *Neuron* 59:802–14

Franks KM, Russo MJ, Sosulski DL, Mulligan AA, Siegelbaum SA, Axel R. 2011. Recurrent circuitry dynamically shapes the activation of piriform cortex. *Neuron* 72:49–56

French RM. 1999. Catastrophic forgetting in connectionist networks. *Trends Cogn. Sci.* 3:128–35

Friedrich RW. 2006. Mechanisms of odor discrimination: neurophysiological and behavioral approaches. *Trends Neurosci.* 29:40–47

Friedrich RW, Habermann CJ, Laurent G. 2004. Multiplexing using synchrony in the zebrafish olfactory bulb. *Nat. Neurosci.* 7:862–71

Friedrich RW, Jacobson GA, Zhu P. 2010. Circuit neuroscience in zebrafish. *Curr. Biol.* 20:R371–81

Friedrich RW, Korsching SI. 1997. Combinatorial and chemotopic odorant coding in the zebrafish olfactory bulb visualized by optical imaging. *Neuron* 18:737–52

Friedrich RW, Korsching SI. 1998. Chemotopic, combinatorial, and noncombinatorial odorant representations in the olfactory bulb revealed using a voltage-sensitive axon tracer. *J. Neurosci.* 18:9977–88

Friedrich RW, Laurent G. 2001. Dynamic optimization of odor representations in the olfactory bulb by slow temporal patterning of mitral cell activity. *Science* 291:889–94

Friedrich RW, Laurent G. 2004. Dynamics of olfactory bulb input and output activity during odor stimulation in zebrafish. *J. Neurophysiol.* 91:2658–69

Fuller CL, Yettaw HK, Byrd CA. 2006. Mitral cells in the olfactory bulb of adult zebrafish (*Danio rerio*): morphology and distribution. *J. Comp. Neurol.* 499:218–30

Fuss SH, Korsching SI. 2001. Odorant feature detection: activity mapping of structure response relationships in the zebrafish olfactory bulb. *J. Neurosci.* 21:8396–407

Gottfried JA. 2009. Function follows form: ecological constraints on odor codes and olfactory percepts. *Curr. Opin. Neurobiol.* 19:422–29

Gross-Isseroff R, Lancet D. 1988. Concentration-dependent changes of perceived odor quality. *Chem. Senses* 13:191–204

Haberly LB. 2001. Parallel-distributed processing in olfactory cortex: new insights from morphological and physiological analysis of neuronal circuitry. *Chem. Senses* 26:551–76

Haberly LB, Bower JM. 1989. Olfactory cortex: model circuit for study of associative memory? *Trends Neurosci.* 7:258–64

Hasselmo ME, Wilson MA, Anderson BP, Bower JM. 1990. Associative memory function in piriform (olfactory) cortex: computational modeling and neuropharmacology. *Cold Spring Harb. Symp. Quant. Biol.* 55:599–610

Heisenberg M. 2003. Mushroom body memoir: from maps to models. *Nat. Rev. Neurosci.* 4:266–75

Johnson DM, Illig KR, Behan M, Haberly LB. 2000. New features of connectivity in piriform cortex visualized by intracellular injection of pyramidal cells suggest that "primary" olfactory cortex functions like "association" cortex in other sensory systems. *J. Neurosci.* 20:6974–82

Jortner RA, Farivar SS, Laurent G. 2007. A simple connectivity scheme for sparse coding in an olfactory system. *J. Neurosci.* 27:1659–69

Kanerva P. 1988. *Sparse Distributed Memory*. Cambridge, MA: MIT Press

Kepecs A, Uchida N, Zariwala HA, Mainen ZF. 2008. Neural correlates, computation and behavioural impact of decision confidence. *Nature* 455:227–31

Khan RM, Luk CH, Flinker A, Aggarwal A, Lapid H, et al. 2007. Predicting odor pleasantness from odorant structure: pleasantness as a reflection of the physical world. *J. Neurosci.* 27:10015–23

Kohonen T. 1984. *Self-Organization and Associative Memory*. Berlin: Springer-Verlag

Koulakov AA, Rinberg D. 2011. Sparse incomplete representations: a potential role of olfactory granule cells. *Neuron* 72:124–36

Laurent G. 2002. Olfactory network dynamics and the coding of multidimensional signals. *Nat. Rev. Neurosci.* 3:884–95

Laurent G, Naraghi M. 1994. Odorant-induced oscillations in the mushroom bodies of the locust. *J. Neurosci.* 14:2993–3004

Laurent G, Stopfer M, Friedrich RW, Rabinovich M, Volkovskii A, Abarbanel HDI. 2001. Odor coding as an active, dynamical process: experiments, computation and theory. *Annu. Rev. Neurosci.* 24:263–97

Li W, Howard JD, Parrish TB, Gottfried JA. 2008. Aversive learning enhances perceptual and cortical discrimination of indiscriminable odor cues. *Science* 319:1842–45

Livermore A, Laing DG. 1996. Influence of training and experience on the perception of multicomponent odor mixtures. *J. Exp. Psychol.* 22:267–77

Luo SX, Axel R, Abbott LF. 2010. Generating sparse and selective third-order responses in the olfactory system of the fly. *Proc. Natl. Acad. Sci. USA* 107:10713–18

Marr D. 1969. A theory of cerebellar cortex. *J. Physiol.* 202:437–70

Marr D. 1971. Simple memory: a theory for archicortex. *Philos. Trans. R. Soc. Lond. B* 262:23–81

Masse NY, Turner GC, Jefferis GS. 2009. Olfactory information processing in *Drosophila*. *Curr. Biol.* 19:R700–13

Mazor O, Laurent G. 2005. Transient dynamics versus fixed points in odor representations by locust antennal lobe projection neurons. *Neuron* 48:661–73

McGann JP, Pirez N, Gainey MA, Muratore C, Elias AS, Wachowiak M. 2005. Odorant representations are modulated by intra- but not interglomerular presynaptic inhibition of olfactory sensory neurons. *Neuron* 48:1039–53

McNaughton BL, Morris RGM. 1987. Hippocampal synaptic enhancement and information storage within a distributed memory system. *Trends Neurosci.* 10:408–15

Miklavc P, Valentinčič T. 2012. Chemotopy of amino acids on the olfactory bulb predicts olfactory discrimination capabilities of zebrafish *Danio rerio*. *Chem. Senses* 37:65–75

Miyamichi K, Amat F, Moussavi F, Wang C, Wickersham I, et al. 2011. Cortical representations of olfactory input by trans-synaptic tracing. *Nature* 472:191–96

Miyasaka N, Morimoto K, Tsubokawa T, Higashijima S, Okamoto H, Yoshihara Y. 2009. From the olfactory bulb to higher brain centers: genetic visualization of secondary olfactory pathways in zebrafish. *J. Neurosci.* 29:4756–67

Mueller T, Dong Z, Berberoglu MA, Guo S. 2011. The dorsal pallium in zebrafish, *Danio rerio* (Cyprinidae, Teleostei). *Brain Res.* 1381:95–105

Neville KR, Haberly LB. 2004. Olfactory cortex. See Shepherd 2004, pp. 415–54

Niessing J, Friedrich RW. 2010. Olfactory pattern classification by discrete neuronal network states. *Nature* 465:47–52

Olsen SR, Bhandawat V, Wilson RI. 2010. Divisive normalization in olfactory population codes. *Neuron* 66:287–99

Olsen SR, Wilson RI. 2008. Lateral presynaptic inhibition mediates gain control in an olfactory circuit. *Nature* 452:956–60

Papadopoulou M, Cassenaer S, Nowotny T, Laurent G. 2011. Normalization for sparse encoding of odors by a wide-field interneuron. *Science* 332:721–25

Perez-Orive J, Mazor O, Turner GC, Cassenaer S, Wilson RI, Laurent G. 2002. Oscillations and sparsening of odor representations in the mushroom body. *Science* 297:359–65

Poo C, Isaacson JS. 2009. Odor representations in olfactory cortex: "sparse" coding, global inhibition, and oscillations. *Neuron* 62:850–61

Poo C, Isaacson JS. 2011. A major role for intracortical circuits in the strength and tuning of odor-evoked excitation in olfactory cortex. *Neuron* 72:41–48

Rabin MD. 1988. Experience facilitates olfactory quality discrimination. *Percept. Psychophys.* 44:532–40

Rinberg D, Koulakov A, Gelperin A. 2006a. Sparse odor coding in awake behaving mice. *J. Neurosci.* 26:8857–65

Rinberg D, Koulakov A, Gelperin A. 2006b. Speed-accuracy tradeoff in olfaction. *Neuron* 51:351–58

Rolls ET, Kesner RP. 2006. A computational theory of hippocampal function, and empirical tests of the theory. *Prog. Neurobiol.* 79:1–48

Satou M. 1990. Synaptic organization, local neuronal circuitry, and functional segregation of the teleost olfactory bulb. *Prog. Neurobiol.* 34:115–42

Schärer YP, Shum J, Moressis A, Friedrich RW. 2012. Dopaminergic modulation of synaptic transmission and neuronal activity patterns in the zebrafish homolog of olfactory cortex. *Front. Neural Circuits* 6:76

Shepherd GM, ed. 2004. *The Synaptic Organization of the Brain.* New York: Oxford Univ. Press

Shepherd GM, Chen WR, Greer CA. 2004. Olfactory bulb. See Shepherd 2004, pp. 165–216

Shi P, Zhang J. 2009. Extraordinary diversity of chemosensory receptor gene repertoires among vertebrates. *Results Probl. Cell Differ.* 47:1–23

Singer W. 1999. Neuronal synchrony: a versatile code for the definition of relations? *Neuron* 24:49–65

Soucy ER, Albeanu DF, Fantana AL, Murthy VN, Meister M. 2009. Precision and diversity in an odor map on the olfactory bulb. *Nat. Neurosci.* 12:210–20

Spence R, Gerlach G, Lawrence C, Smith C. 2008. The behaviour and ecology of the zebrafish, *Danio rerio*. *Biol. Rev. Camb. Philos. Soc.* 83:13–34

Stettler DD, Axel R. 2009. Representations of odor in the piriform cortex. *Neuron* 63:854–64

Stokes CC, Isaacson JS. 2010. From dendrite to soma: dynamic routing of inhibition by complementary interneuron microcircuits in olfactory cortex. *Neuron* 67:452–65

Stopfer M, Bhagavan S, Smith BH, Laurent G. 1997. Impaired odour discrimination on desynchronization of odour-encoding neural assemblies. *Nature* 390:70–74

Tabor R, Friedrich RW. 2008. Pharmacological analysis of ionotropic glutamate receptor function in neuronal circuits of the zebrafish olfactory bulb. *PLoS ONE* 3:e1416

Tabor R, Yaksi E, Friedrich RW. 2008. Multiple functions of GABA(A) and GABA(B) receptors during pattern processing in the zebrafish olfactory bulb. *Eur. J. Neurosci.* 28:117–27

Tabor R, Yaksi E, Weislogel JM, Friedrich RW. 2004. Processing of odor mixtures in the zebrafish olfactory bulb. *J. Neurosci.* 24:6611–20

Tan J, Savigner A, Ma M, Luo M. 2010. Odor information processing by the olfactory bulb analyzed in gene-targeted mice. *Neuron* 65:912–26

Turner GC, Bazhenov M, Laurent G. 2008. Olfactory representations by *Drosophila* mushroom body neurons. *J. Neurophysiol.* 99:734–46

Uchida N, Mainen ZF. 2003. Speed and accuracy of olfactory discrimination in the rat. *Nat. Neurosci.* 6:1224–29

Wachowiak M. 2011. All in a sniff: olfaction as a model for active sensing. *Neuron* 71:962–73

Wehr M, Laurent G. 1996. Odor encoding by temporal sequences of firing in oscillating neural assemblies. *Nature* 384:162–66

Wick SD, Wiechert MT, Friedrich RW, Riecke H. 2010. Pattern orthogonalization via channel decorrelation by adaptive networks. *J. Comput. Neurosci.* 28:29–45

Wiechert MT, Judkewitz B, Riecke H, Friedrich RW. 2010. Mechanisms of pattern decorrelation by recurrent neuronal circuits. *Nat. Neurosci.* 13:1003–10

Wills TJ, Lever C, Cacucci F, Burgess N, O'Keefe J. 2005. Attractor dynamics in the hippocampal representation of the local environment. *Science* 308:873–76

Wilson DA. 2003. Rapid, experience-induced enhancement in odorant discrimination by anterior piriform cortex neurons. *J. Neurophysiol.* 90:65–72

Wilson DA, Stevenson RJ. 2003. The fundamental role of memory in olfactory perception. *Trends Neurosci.* 26:243–47

Wilson DA, Sullivan RM. 2011. Cortical processing of odor objects. *Neuron* 72:506–19

Wilson RI, Mainen ZF. 2006. Early events in olfactory processing. *Annu. Rev. Neurosci.* 29:163–201

Wiltrout C, Dogra S, Linster C. 2003. Configurational and nonconfigurational interactions between odorants in binary mixtures. *Behav. Neurosci.* 117:236–45

Wise PM, Cain WS. 2000. Latency and accuracy of discriminations of odor quality between binary mixtures and their components. *Chem. Senses* 25:247–65

Wullimann MF, Mueller T. 2004. Teleostean and mammalian forebrains contrasted: evidence from genes to behavior. *J. Comp. Neurol.* 475:143–62

Yaksi E, Friedrich RW. 2006. Reconstruction of firing rate changes across neuronal populations by temporally deconvolved Ca^{2+} imaging. *Nat. Methods* 3:377–83

Yaksi E, Judkewitz B, Friedrich RW. 2007. Topological reorganization of odor representations in the olfactory bulb. *PLoS Biol.* 5:e178

Yaksi E, von Saint Paul F, Niessing J, Bundschuh ST, Friedrich RW. 2009. Transformation of odor representations in target areas of the olfactory bulb. *Nat. Neurosci.* 12:474–82

Yokoi M, Mori K, Nakanishi S. 1995. Refinement of odor molecule tuning by dendrodendritic synaptic inhibition in the olfactory bulb. *Proc. Natl. Acad. Sci. USA* 92:3371–75

Yoshihara Y. 2009. Molecular genetic dissection of the zebrafish olfactory system. *Results Probl. Cell Differ.* 47:1–24

Zhu P, Fajardo O, Shum J, Zhang Schärer YP, Friedrich RW. 2012. High-resolution optical control of spatiotemporal neuronal activity patterns in zebrafish using a digital micromirror device. *Nat. Protoc.* 7:1410–25

Zhu P, Narita Y, Bundschuh ST, Fajardo O, Schärer YP, et al. 2009. Optogenetic dissection of neuronal circuits in zebrafish using viral gene transfer and the Tet system. *Front. Neural Circuits* 3:21

Transformation of Visual Signals by Inhibitory Interneurons in Retinal Circuits

Pablo D. Jadzinsky and Stephen A. Baccus

Department of Neurobiology, Stanford University School of Medicine, Stanford, California 94305; email: baccus@stanford.edu, jadz@stanford.edu

Annu. Rev. Neurosci. 2013. 36:403–28

First published online as a Review in Advance on May 29, 2013

The *Annual Review of Neuroscience* is online at neuro.annualreviews.org

This article's doi:
10.1146/annurev-neuro-062012-170315

Keywords

interneurons, computational models, neural circuits

Abstract

One of the largest mysteries of the brain lies in understanding how higher-level computations are implemented by lower-level operations in neurons and synapses. In particular, in many brain regions inhibitory interneurons represent a diverse class of cells, the individual functional roles of which are unknown. We discuss here how the operations of inhibitory interneurons influence the behavior of a circuit, focusing on recent results in the vertebrate retina. A key role in this understanding is played by a common representation of the visual stimulus that can be applied at different stages. By considering how this stimulus representation changes at each location in the circuit, we can understand how neuron-level operations such as thresholds and inhibition yield circuit-level computations such as how stimulus selectivity and gain are controlled by local and peripheral visual stimuli.

Contents

INTRODUCTION

Understanding how the parts of a complex system combine to perform a function is a difficult problem. Component parts often have very simple functions, and it is only when one considers how many of these pieces interact, as well as the overall function, that one can understand the role of a component of a system. Imagining the parts of an automobile, one could describe the function of a bolt as "to fasten something" in the same way as one can describe the function of an inhibitory neuron as "to inhibit something." But between a bolt and a moving car are many possible different higher functions (such as automatic windows and transmission), and a full understanding must connect the important properties of the part with the requirements of the higher function. Complicating things further, a part may have properties not relevant to a higher function within the system, such as the color of a bolt or its part number. In addition, a part may have many uses, but focusing on a particular function greatly targets a search for relevant parts and excludes others. Therefore, a key starting point is understanding the probable higher-level function.

The problem of understanding how a profusion of mechanisms acts together pervades the study of biological systems, in particular the nervous system. Over the past century, the retina has served as a proving ground for ideas of how neural circuits perform computations,

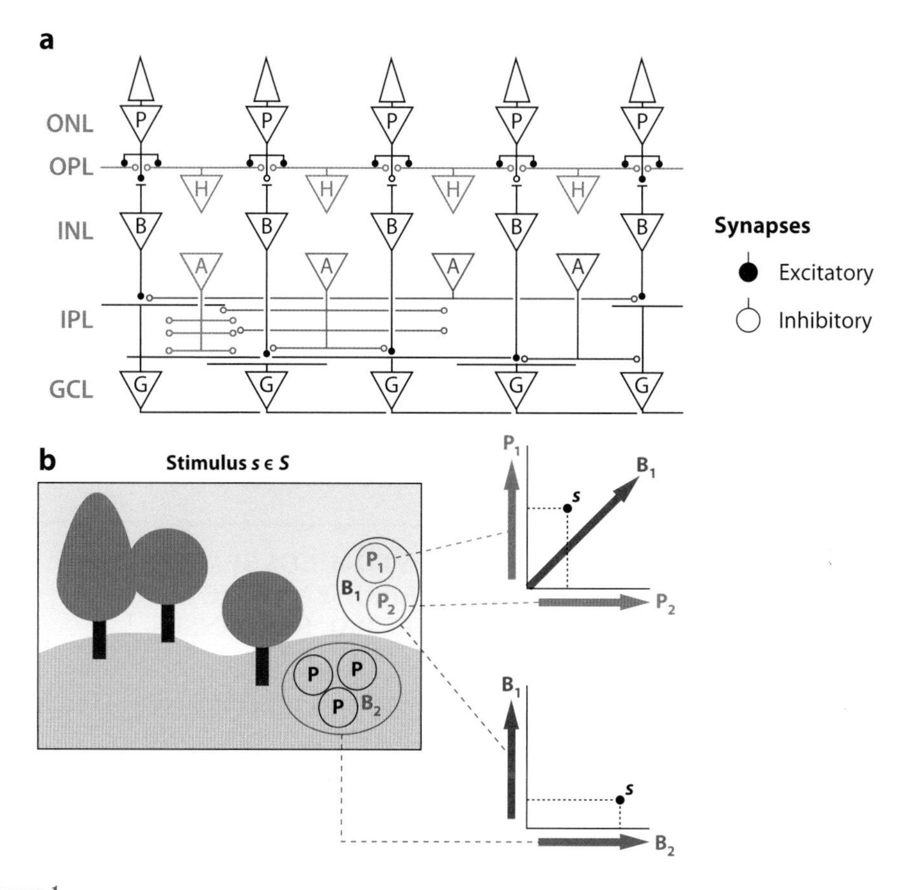

Figure 1

A stimulus space representation of retinal visual input. (*a*) Schematic diagram of the retina. Signals travel from photoreceptors (P) to bipolar cells (B) to ganglion cells (G) and are modified by inhibitory interneuron horizontal cells (H) and amacrine cells (A). Three layers of cell bodies are found in the outer nuclear layer (ONL), inner nuclear layer (INL), and ganglion cell layer (GCL). Synaptic connections are made in the outer and inner plexiform layers (OPL and IPL). A few different classes of amarine cells are schematically depicted, with dendrites stratified in one or multiple sublamina in the IPL. (*b*) The diagram shows one stimulus frame, s, depicted in terms of the familiar two-dimensional visual space. The stimulus frame s is just one possible point in a much higher-dimensional stimulus space S. Signals from two photoreceptors (P_1 and P_2), which depolarize in response to a dark stimulus, form a two-dimensional slice within stimulus space, S. In the plot at upper right, each photoreceptor's preferred stimulus (a dark spot in its receptive field) is represented as an orthogonal axis in stimulus space. The stimulus frame s is shown as a point whose coordinates in the $P_1 \times P_2$ space are determined by the intensity on each photoreceptor. An Off-type bipolar cell (B_1), which also depolarizes in response to a dark stimulus, sums over P_1 and P_2, yielding a preferred stimulus represented by a vector sum of the P_1 and P_2 dimensions. The plot at lower right shows a two-dimensional slice of stimulus space defined by the preferred stimuli of bipolar cells B_1 and B_2. Because the $B_1 \times B_2$ space is different from the $P_1 \times P_2$ space, the point s appears at a different location, by virtue of a brighter intensity over B_1 than B_2.

enabling in certain cases a connection of mechanism with function (Gollisch & Meister 2010, Masland 2012). Although there are only five basic cell types (**Figure 1*a***), the existence of more than 50 subtypes suggests a diversity of components capable of diverse functions.

The function of a sensory system is to provide information to drive behaviors. In some

Stimulus space: the multidimensional space defined by considering each receptor activated by the stimulus and each interval of time to be a separate dimension

cases, the behavioral relevance of retinal responses is clear, as in the visual detection of a moving object (Lettvin et al. 1959). In other cases, the relationship between the output of the retina and behavior is less apparent, as in the case of retinal ganglion cells that reverse polarity (Geffen et al. 2007). However, because of the experimenter's ability to gain exacting control over the visual stimulus in retinal experimental preparations, the mathematical function of a cell's response can be precise, even if the qualitative behavioral function is mysterious.

In the retina as in the rest of the nervous system, one set of components that has been difficult to connect to higher-level function is inhibitory interneurons. The anatomy and biochemistry of these cells are diverse (MacNeil & Masland 1998), but for most, their relationship with a higher-level function is unknown. Here we consider the progress toward understanding how inhibitory neurons in the retina are used in a computation.

HOW CAN THE FUNCTION OF AN INTERNEURON BE UNDERSTOOD?

Compared with a principal or projection neuron such as a retinal ganglion cell that delivers the output of a neural circuit, an interneuron by definition resides within the circuit. Of course, this distinction is somewhat artificial: If one defines the circuit to be the entire brain, then all projection neurons take the role of interneurons, and their role in brain function may become equally mysterious. But if one accepts that the brain can be divided into anatomical parts with distinct functions, then interneurons are cells whose input and output reside in a brain region of indivisible function.

Given that a projection neuron's function in a circuit is to deliver its output, to understand the functional role of a projection neuron, one need (only) create a mathematical description of the relationship between stimulus and response. This effort neglects the circuitry that gave rise to the cell's computation but at least allows the function to be understood precisely. In contrast, to understand the function of an interneuron, one needs to understand not only how the stimulus creates the cell's response, but also how the cell's output changes the overall behavior of the circuit (De Vries et al. 2011, Lehky & Sejnowski 1988, Manu & Baccus 2011). Thus we need to understand (*a*) how the stimulus produces the interneuron's response, (*b*) how the stimulus produces the circuit's output, and (*c*) how the interneuron's output contributes to the circuit's output. Each of these requirements has different experimental approaches, illustrating the difficulty of understanding an interneuron's functional role in a circuit.

A STIMULUS SPACE REPRESENTATION OF NEURAL RESPONSES

To understand how simple operations can be combined for an overall function, it is helpful to have a common frame of reference that can be compared at different levels in the circuit. Each photoreceptor represents a separate input to the visual system. In the laboratory, the input to each of these N receptors can be controlled independently, so we consider them to be separate dimensions in a high-dimensional stimulus space, which is distinct from visual space (**Figure 1*b***). In addition, stimuli change over time, and thus for each photoreceptor we view each of T time points as a separate T dimension. Thus each visual stimulus becomes a single point in an $N \times T$ dimensional space. Not all these points are equally likely, however, because natural visual scenes have strong spatial and temporal correlations, meaning that the different dimensions of stimulus space are not independent (Ruderman & Bialek 1994, Simoncelli & Olshausen 2001, Van Hateren 1993).

Classically, the receptive field has been characterized by probing each of the receptors independently and then computing a weight representing the contribution of each receptor (Barlow 1953, Kuffler 1953). An assumption that a cell simply sums the separate inputs according to these weights is known as a linear model of the receptive field. If one considers that each receptor (and each time point) represents an independent direction in stimulus space, then in cases where a large, unbiased set of stimuli are probed (e.g., white noise), the linear receptive field typically represents the direction in stimulus space to which the cell is, on average, most sensitive.

The linear receptive field, however, does not describe the entire relationship between stimulus and response (Schwartz & Rieke 2011). When visual stimuli have a broad range of temporal frequencies, as occurs in natural images, retinal ganglion cell responses are sparse in time, consisting of periods of silence lasting hundreds of milliseconds punctuated by brief signals consisting of one to several spikes (Berry II, Warland, & Meister 1997). As such, ganglion cells signal only in certain regions of the stimulus space and are silent at other times. Lacking a sharp threshold, a linear model would not predict that the cell would be silent most of the time.

A now-standard extension of a model containing a linear receptive field is a linear-nonlinear (LN) model, which passes the stimulus through a linear spatiotemporal filter followed by a threshold. This classic model (Chichilnisky 2001, Rodieck & Stone 1965) contains both the linear receptive field that defines the preferred stimulus feature (direction in stimulus space) and a threshold to generate sparse responses but has no further complexity.

But for many neurons, an LN model is insufficient to describe the response to a broad range of stimuli. For example, when multiple neural pathways combine, each with their own preferred stimulus feature and threshold nonlinearity, this arrangement causes the cell's response to depend on more than a single stimulus feature. As described below, this variation in the response as a function of multiple stimulus features is a key function of inhibitory interneurons and multiple neural pathways that converge on a ganglion cell. Here we consider functional transformations observed in the retina and relate them to the circuitry thought to implement those computations, focusing on the contributions of inhibitory interneurons.

Linear receptive field: the stimulus direction defined by the linear spatiotemporal filter in a linear model that best captures the cell's response

Linear-nonlinear (LN) model: a spatiotemporal filter followed by a nonlinear function such as a threshold that is independent of time

Linear spatiotemporal filter: a weighted sum of input, with a separate weight applied to each spatial location of the input and interval of time

SIGN INVERSION IN AII AMACRINE CELLS

Perhaps the simplest function of an inhibitory neuron is to produce a sign inversion, transmitting a negative version of the input signal with little distortion. The mammalian rod pathway contains a well-known example of this process. Signals travel from rods to On-type bipolar cells to AII amacrine cells. The amacrine cell then transmits a negative version of the signal through an inhibitory chemical synapse to Off-type cone bipolar and ganglion cells. Although AII amacrine cells produce action potentials, they do not strongly rectify their input because they also transmit signals at hyperpolarized potentials through their inhibitory synapses (Habermann et al. 2003). In addition to the sign-inverted signal, AII amacrine cells also transmit their signal in a sign-conserving manner through gap junctions to On cone bipolar cells. Thus the AII amacrine cell transmits both a positive and a negative version of the same signal, both of which are largely nonrectified, to On and Off cone bipolar cell terminals (Demb & Singer 2012). Thus the AII amacrine cell appears to act largely as a relay, rather than conveying a negative signal to compare with a separate positive signal, as in the case of the horizontal cell described below.

PREFERRED STIMULUS DIRECTION

As noted above, sensitivity to a stimulus is typically defined by a preferred direction in stimulus space. Adding two neural pathways can yield a new preferred direction, effectively rotating the preferred direction. To consider how this effect occurs, **Figure 2** shows how the selectivities of photoreceptor and horizontal cell pathways are combined together, yielding a final result in a bipolar cell. The excitatory stimulus direction is conveyed directly from photoreceptors. The inhibitory stimulus direction is conveyed through the horizontal cell network, which shares signals among horizontal cells through electrical synapses such that the preferred stimulus of the network equals a weighted sum of all photoreceptors that feed into the network (Thoreson & Mangel 2012). Because the pathways at the bipolar cell dendritic field are summed approximately linearly, the preferred direction of the final output is the vector sum of its two inputs. The simple diagram in **Figure 2** indicates that the horizontal cell direction is largely independent of that of any photoreceptor and, in particular, those in the bipolar cell's receptive field center. Although this independence is not complete because the horizontal cell itself receives input from photoreceptors in the center, a more accurate model would place the horizontal cell stimulus dimension axis at an angle other than 90°. More generally, one can consider that when

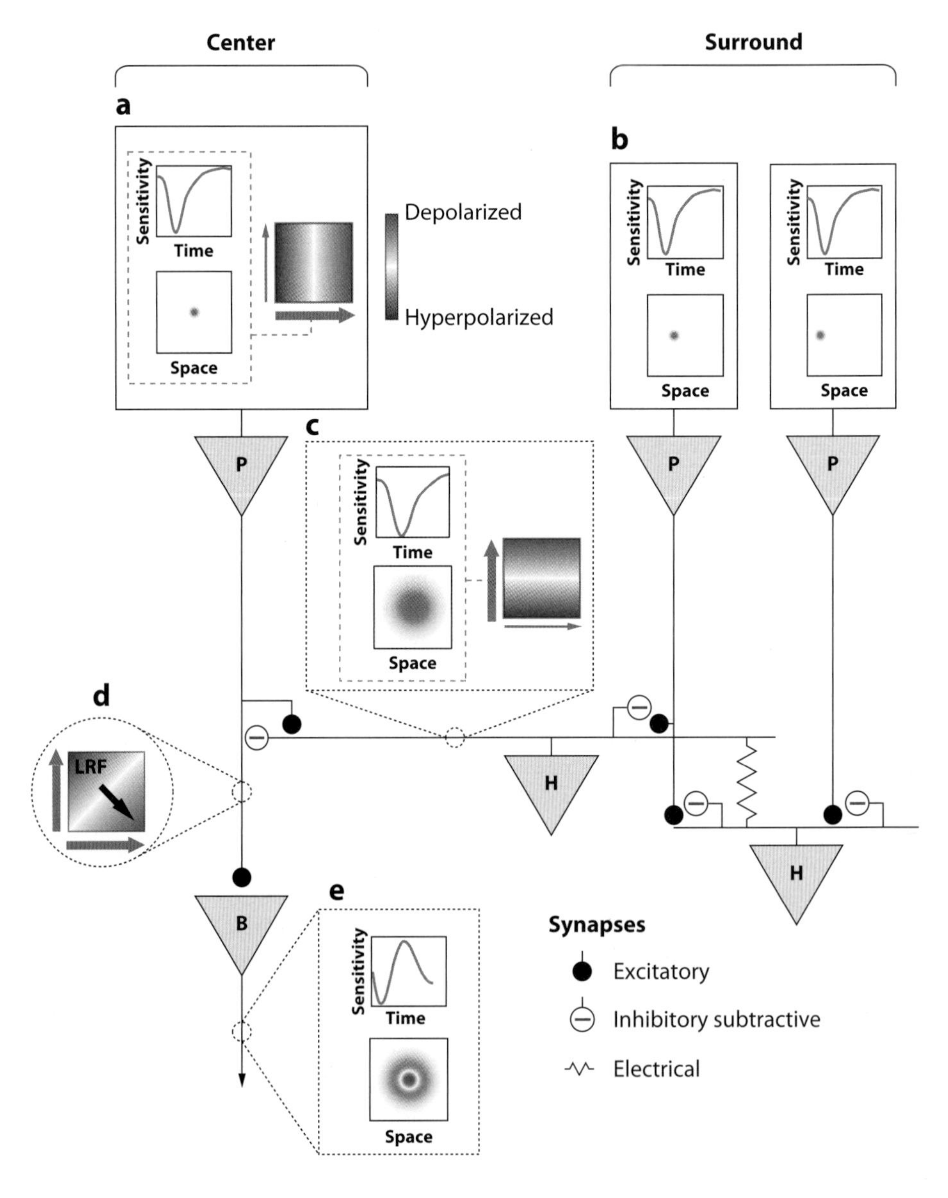

the outputs of multiple pathways with different preferred stimuli are summed or subtracted, the sensitivity to the stimulus will rotate in stimulus space according to this weighted sum.

Studies of the effects of horizontal cells indicate that they primarily contribute a linear subtractive feedback to the cone terminal (Kraaij et al. 2012, van Hateren 2005). Thus, their effects can be considered to implement primarily a rotation of the receptive field's direction, without other strong nonlinear effects such as a dynamic change in the sensitivity to the stimulus. The particular change in temporal sensitivity of the receptive field is to create a more biphasic response. This change in the receptive field reduces sensitivity to slower changes in light intensity, analogous to how the receptive field surround reduces sensitivity to uniform regions of intensity.

In natural visual scenes, light intensity is similar in nearby regions of space and successive points in time. Given this predictable structure, inhibition in the linear receptive field surround acts to subtract a prediction of the intensity at a point in space and time computed by measuring the surrounding area of space and the preceding interval of time (Hosoya et al. 2005, Srinivasan et al. 1982). In this way, the bipolar cell conveys not the original light intensity, but the deviation from the expected light intensity, yielding a more efficient representation of the stimulus.

AMACRINE CELL ROTATION OF CHROMATIC PREFERENCE IN THE S-CONE PATHWAY

Recently, two papers proposed a new role for an amcrine cell in chromatic processing (**Figure 3**). Although previously, amacrine cells were not known to contribute to color selectivity, an amacrine cell was described with selectivity for S-cone stimuli, receiving input from short-wavelength On bipolar cells (Chen & Li 2012). Other work showed that Off-type short-wavelength ganglion cells actually receive input from the On pathway that is delivered by inhibitory amacrine cells (Sher & Devries 2012). It seems likely that the two papers describe the same circuit.

Like the rod pathway mentioned above, Off-type S-cone ganglion cells receive input from the On pathway, which is then inverted through an amacrine cell. The circuitry of the S-cone amacrine cell is similar to that of the AII amacrine cell because it inverts a signal from the On pathway and delivers output to the Off pathway. However, because the S-cone amacrine cell delivers its output to a ganglion cell that receives M-cone input, in its computation the S-cone amacrine cell acts as a relay and also rotates the preferred stimulus direction of the target neuron, similar to the effect of the horizontal cell on the preferred stimulus. Thus, the S-cone amacrine cell has a dual role. Its role as a relay—inverting the sign of its input without

Figure 2

Rotation of the linear receptive field by the inhibitory surround. (*a*) The preferred stimulus feature of a photoreceptor (P) is a decrease in intensity over a brief interval of time and a small spot in space. This feature is represented by the horizontal axis (*green arrow*) of a two-dimensional slice of stimulus space. The color at each point in stimulus space indicates the depolarizing or hyperpolarizing response to each stimulus at this stage in the circuit, showing that the photoreceptor responds to changes in intensity along only one direction of stimulus space (*thick arrow*) but not another direction (*thin arrow*). (*b*) Photoreceptors in the surrounding region have a different preferred spatial stimulus. (*c*) Horizontal cells (H), by summing the inputs of many photoreceptors, create a vector sum of the stimulus directions encoded by photoreceptors. This combined input is largely orthogonal to any one photoreceptor's preferred stimulus direction. Orange and green arrows indicate stimulus directions conveyed by excitatory and inhibitory inputs, respectively. When viewed across the two-dimensional stimulus space represented by the central photoreceptor in panel *a* and the horizontal cell population, the horizontal cell responds only to one direction (*thick arrow*). (*d*) After sign inversion through inhibitory synaptic feedback onto the photoreceptor terminal, the combination in the bipolar cell (B) is a vector sum of the central photoreceptor and the horizontal cell output. The response across the two-dimensional stimulus space varies linearly along a single direction, the linear receptive field (LRF).

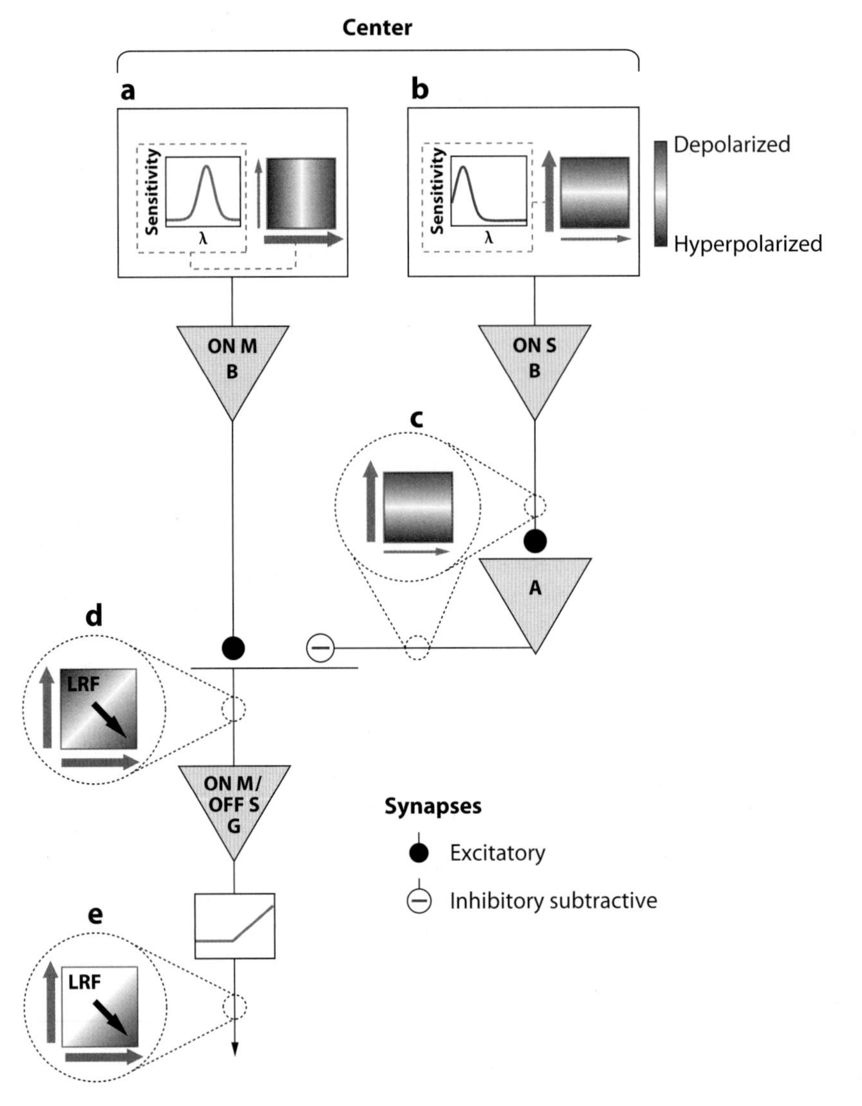

Figure 3

Sign inversion in the S-cone pathway. The circuit for an On M/Off S ganglion cell, which has On-type responses to medium-wavelength stimuli and an Off-type response to short (S) wavelength stimuli. (*a*) The preferred chromatic stimulus of an On M bipolar cell is shown as sensitivity to wavelength (λ). This preferred direction in stimulus space is indicated by the horizontal (*thick*) arrow. In the two-dimensional stimulus space shown, the vertical dimension indicates the spectral preference of an On S bipolar cell, shown as the thick arrow in panel *b*. Each bipolar cell also has its own spatiotemporal receptive field (*not shown*). The stimulus directions are shown here to be orthogonal in a two-dimensional space, although this is not strictly true because the wavelength ranges of M and S cones partially overlap. (*c*) Instead of using an Off bipolar cell, the Off S response of the On S bipolar cell is inverted through an intervening On S amacrine cell. (*d*, *e*) The sum of the two pathways yields a linear response whose direction of greatest sensitivity is a vector sum of the two input pathways, the linear receptive field (LRF) in inset *d* and *e*.

significant distortion—derives from lack of a strong nonlinearity in its input and output. Its role in defining the preferred stimulus derives from the different preferred directions between its output and that of the target ganglion cell.

CONTROL OF LINEAR TEMPORAL FILTERING WITHOUT SPATIAL POOLING

Although horizontal cell feedback acts to create a more biphasic response in bipolar cells, because the horizontal cell pools photoreceptor inputs, it also changes spatial filtering in bipolar cells. Must spatial and temporal filtering be coupled? If a target computation was temporal filtering without spatial pooling, one might think that a separate feedback cell for each input cell would be necessary to prevent mixing of signals.

However, the A17 amacrine cell, which makes feedback synapses to mammalian rod bipolar cells, creates temporal filtering without spatial pooling by having synaptic contacts that act independently. The A17 cell is unusual because it consists of many distinct varicosities separated by tens of microns (Grimes et al. 2010). Compartmental modeling indicates that these varicosities act separately and electrotonically distant from each other.

Feedback synapses have two potential roles: One is to change the pathway's linear filtering properties, and the other is to control the gain of the pathway, such that a strong input would lead to a future weak response. The A17 pathway appears to act primarily to change the linear filtering properties without a strong effect on gain. Because $GABA_C$ receptors are found only on bipolar cell synaptic terminals (Feigenspan et al. 1993), investigators can make a relatively specific pharmacological manipulation of inhibitory synapses onto bipolar cell terminals that include the inputs to A17 amacrine cells (Eggers & Lukasiewicz 2006). Blocking $GABA_C$ receptors changes the filtering properties, leading to a more monophasic response (Dong & Hare 2003). However, dynamic changes in gain of the rod bipolar cell pathway are largely unaffected (Dunn & Rieke 2008).

Thus, both horizontal cell and A17 feedback synapses appear mainly to control temporal filtering but do not act to change the gain of the signal dynamically on the basis of previous input. An open question, therefore, is whether reciprocal feedback in general acts to change the gain of retinal input or simply to affect linear filtering. Because a feedback dynamic control of gain creates a complex time-dependent nonlinearity, a general finding that feedback acts simply to control linear filtering would greatly simplify the understanding of retinal feedback synapses. However, at the present time sufficient experiments have not been performed to draw a general conclusion.

Push-pull amplification: a rectified signal x is combined with the negative of x according to $N(x)$ - $N(-x)$, where $N()$ is a threshold rectifier, yielding approximately the original signal x

Crossover inhibition: an inhibitory connection between the On and Off pathways (or vice versa), used to implement push-pull amplification

SIGN INVERSION LINEARIZES THE RESPONSE

Encoding a signal using a threshold nonlinearity necessarily involves a loss of information about the region of the signal that falls below the threshold. In some cases, that loss of information is an advantageous computation, for example when a cell acts as a feature detector, signaling only at the strong presence of a particular stimulus feature (Gollisch & Meister 2010, Zhang et al. 2012). However, in many cases it would be useful to transmit all regions of the signal equally for further processing or comparison with light intensity in other spatial regions. Neglecting differences in noise at different levels of the signal, a transformation that would convey equal information about all signal levels would be a linear representation of the stimulus. In some ganglion cells, even though incoming signals from both On and Off pathways are rectified, the signals are recombined to yield a more linear response. For example, certain On-type ganglion cells receive direct input from On bipolar cells, but Off bipolar cell input is first inverted through an inhibitory amacrine cell (**Figure 4**). This circuit organization, named push-pull amplification or crossover inhibition (Werblin 2010) (reflecting an inhibitory input

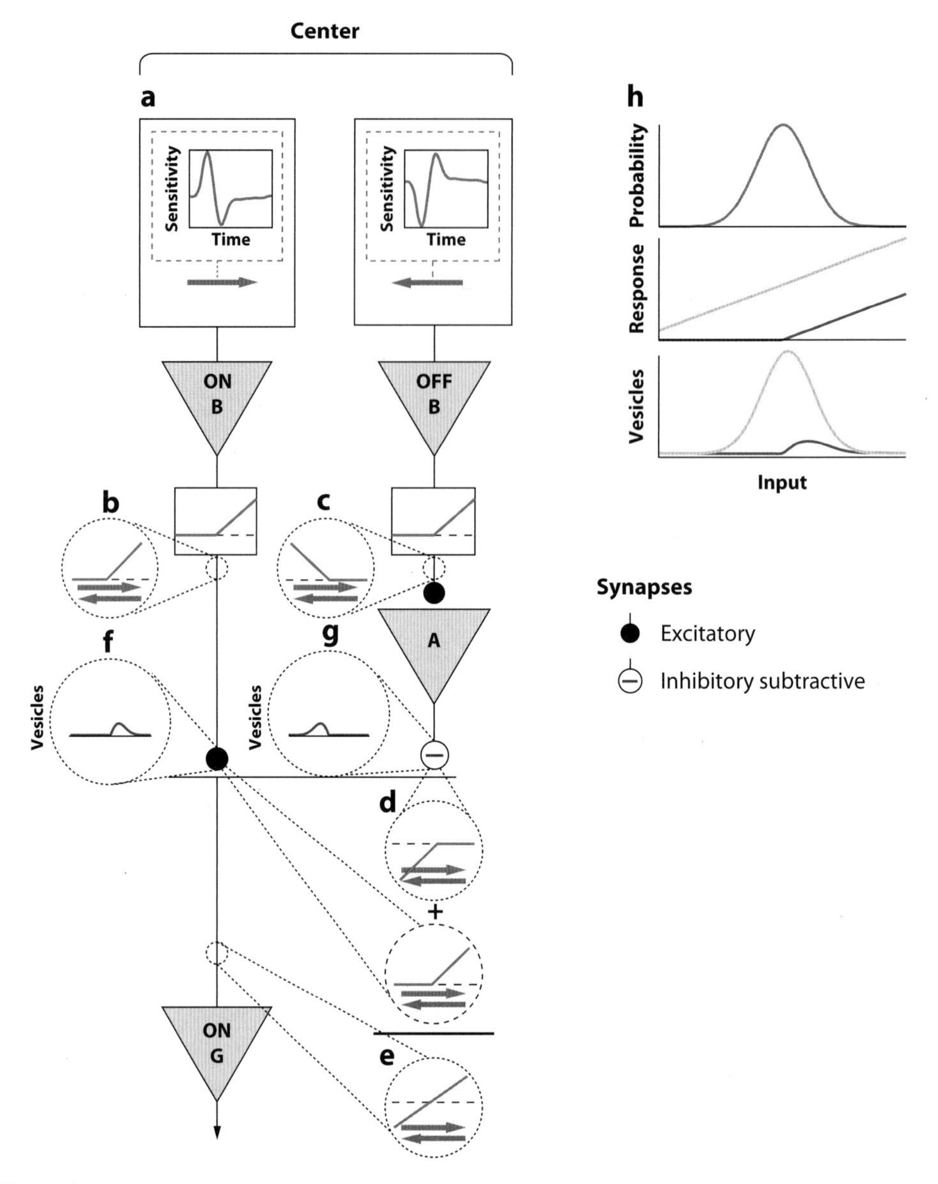

Figure 4

Linearization and efficiency using push-pull amplification. (*a*) Two pathways from On and Off bipolar cells are combined in a single On ganglion cell. The two bipolar cells have nearly the same temporal filter, but one has an opposite sign from the other. Contrary to previous examples, the two inputs have nearly opposite, not orthogonal, preferred directions in stimulus space represented by the arrows. (*b*, *c*) At the output of each bipolar cell, the signal is passed through a threshold, such that only half of the signal is represented in the one-dimensional stimulus space (*circles*). (*d*) Sign inversion through an amacrine cell inverts the Off pathway to create a response that has the same sign with respect to the light input as does the On pathway. (*e*) When the two pathways are combined, a linear representation of the signal is restored. Overall, the operation performed is $G(s) = N(s) - N(-s) \cong s$, where s is the stimulus, $G(s)$ is the ganglion cell response, and $N(s)$ is a rectified function that is linear for positive inputs and zero for negative inputs. (*f*, *g*) The rate of vesicles at each intensity level is shown for the On (*f*) and Off (*g*) pathways. (*h*, *top*) Probability distribution of a stimulus, showing most values close to the mean; see linear (*gray*) and rectified (*blue*) response functions. (*Bottom*) Estimated total vesicle release at each input value for linear and nonlinear response functions.

that crosses from the On pathway to the Off pathway or vice versa), yields an output that is much more linear than each of the two contributing pathways.

If one considers a pair of rectified On and Off bipolar cells, each bipolar cell transmits information about a different range of the signal. Thus, although a fixed nonlinearity will necessarily discard a signal, by preserving the signal in two pathways and then recombining them in their original relationship, the effect of the nonlinearity is reduced. This similar arrangement is seen in electronic push-pull amplifiers, which compensate for the inherent rectification in transistors (Horowitz & Hill 1989). Just as in the retina, one pathway conveys the positive rectified signal, and one conveys the negative rectified signal. Because this plan avoids the effect of the threshold in synaptic transmission, one can also view this circuit as extending the dynamic range of an output pathway using a combination of two input pathways, each with a more narrow dynamic range (Manookin et al. 2008).

EFFICIENCY OF PUSH-PULL AMPLIFICATION USING CROSSOVER INHIBITION

But why have a nonlinearity in the first place, only to correct its effects? One might consider that the nonlinearity is unavoidable, an inherent part of synaptic transmission. However, this is not always the case, as for example in the case of AII amacrine cells, whose resting potential lies in the linear range with respect to synaptic transmission (Habermann et al. 2003). An important functional implication is that the organization of crossover inhibition reduces the amount of vesicle release needed to convey a signal, thus making more efficient use of synaptic transmission. To support this conclusion, one can simply compute the average vesicle release needed to convey a signal through a synapse with a single pathway that rests in the linear range of the synapse or with two rectified pathways that are recombined using crossover inhibition (**Figure 4*a*,*b***). In natural visual scenes, because low spatial and temporal frequencies are prevalent, the most common inputs are near the mean light intensity (Laughlin 1981). Thus, with a single pathway, the most common output will be at half of the mean output value. However, with two rectified pathways, the mean output will be much lower, using fewer synaptic vesicles to transmit the signal (Harris et al. 2012).

DENDRITIC SPATIOTEMPORAL FILTERING UNDERLYING DIRECTION SELECTIVITY

One of the most well-studied examples of how a cell achieves selectivity for a particular subset of stimulus space occurs in direction-selective ganglion cells, which exhibit a strong selectivity for a specific direction of motion. Numerous studies have focused on the underlying mechanism of direction selectivity (Vaney et al. 2012), showing that the starburst amacrine cell (SAC) plays a clear role in the computation of direction selectivity. Calcium imaging studies have revealed that the dendrites of SACs act as independent processing elements with responses that differ from each other and from those of the soma. SAC dendrites are direction selective in the radial direction, such that outward image motion away from the soma, but not inward image motion toward the soma, causes depolarization (Euler et al. 2002). Because these dendrites process the signal locally and are in different locations with respect to the soma, different dendrites in the same cell prefer different directions of motion with respect to retinal coordinates. To generate a final representation selective for a particular direction, asymmetric connections from SAC dendrites to direction selective (DS) ganglion cells deliver inhibition tuned to the particular null direction of the ganglion cell.

SACs thus appear to be the first source of direction selectivity generated from asymmetric excitatory and inhibitory input on to SAC dendrites. Excitation arrives in a region near the soma, whereas delayed inhibition is generated

OMS: object motion sensitive

by more peripheral stimuli, in part from other SACs (Lee & Zhou 2006). Because SACs form inhibitory connections at their tips, inhibition onto each dendrite comes from a location that is more distant from the soma than is the source of excitation. This spatial arrangement of inputs creates in each dendrite a sensitivity for central stimuli that arrive prior to peripheral stimuli, yielding the spatiotemporal sensitivity for outward motion. At the stage of the SAC dendrites, the response seems consistent with a linear spatiotemporal filter passed through a threshold, although measurements have not been made of more complex nonlinearities such as adaptive changes in sensitivity over time.

In addition to GABAergic inhibition, cholinergic excitatory input from SACs (Fried et al. 2002) and glutamatergic input to DS ganglion cells (Vaney & Taylor 2002) both appear to be asymmetric. However, Vaney et al. (2012) recently proposed that inadequate voltage clamping in DS ganglion cells may have improperly attributed strong asymmetry in inhibition to other cholinergic and excitatory synaptic inputs. This issue awaits resolution. Other mechanisms that appear to play a role in direction selectivity include dendritic spiking in DS ganglion cells, which may have a role in transmitting local computations from dendrite to axon (Oesch et al. 2005). In addition, Gavrikov et al. (2006) proposed that a chloride gradient that exists along SACs contributes to the preference for centripetal motion.

Nearly all studies of direction selectivity focus on the responses of just two stimuli: motion in the most sensitive (preferred) direction, and motion in the opposite, least-sensitive (null) direction. However, DS cells exhibit broad tuning for motion, approximated by a cosine function (Nowak et al. 2011). This directional tuning is maintained across different speeds and luminance. One expects that in natural visual stimuli the abundance of mechanisms regulating direction selectivity exists not merely to discriminate between two stimuli, but also to maintain the directional stimulus preference across a range of environmental conditions.

FEATURE-SPECIFIC LATERAL INHIBITION CREATES OBJECT MOTION SENSITIVITY

Retinal ganglion cells signal a change in light intensity within their receptive fields. Moving objects are one source of a change in intensity, but eye movements are another. Signaling the presence of a moving object requires a neural circuit to discriminate between these two different sources of retinal motion. Object motion sensitive (OMS) ganglion cells perform this task; they signal the presence of local changes in intensity caused by a moving object but remain silent in response to global motion arising from small fixational eye movements that cause image motion approximated by a random walk (Baccus et al. 2008; Olveczky et al. 2003, 2007). Note that fixational eye movements are different from microsaccades, which have separate neural origin and statistics.

In the OMS circuit (**Figure 5**), the central input to OMS ganglion cells rectifies the output of each bipolar cell. Any bipolar cell that is depolarized creates an excitatory synaptic input, and rectification prevents bipolar cell hyperpolarization from canceling the effects of depolarization. Pooling over the outputs of many such bipolar cells renders the response largely insensitive to the specific spatial pattern and conveys primarily the speed of the input.

Inhibition from the background is delivered to OMS cells from polyaxonal amacrine cells, which have receptive field centers similar in size to OMS ganglion cells but convey their output over long distances across the retina. Polyaxonal amacrine cells also receive excitatory input from rectified, fast bipolar cells. Thus, polyaxonal amacrine cells locally compute the same feature sensed by OMS cells. The combined output of many polyaxonal amacrine cells pools this feature across the retina.

To distinguish differential motion from global motion, input from polyaxonal amacrine cells is delivered from distant areas of the retina. The inhibition from polyaxonal amacrine cells acts only for a brief (<100 ms) window of time, thus common timing in the excitatory and

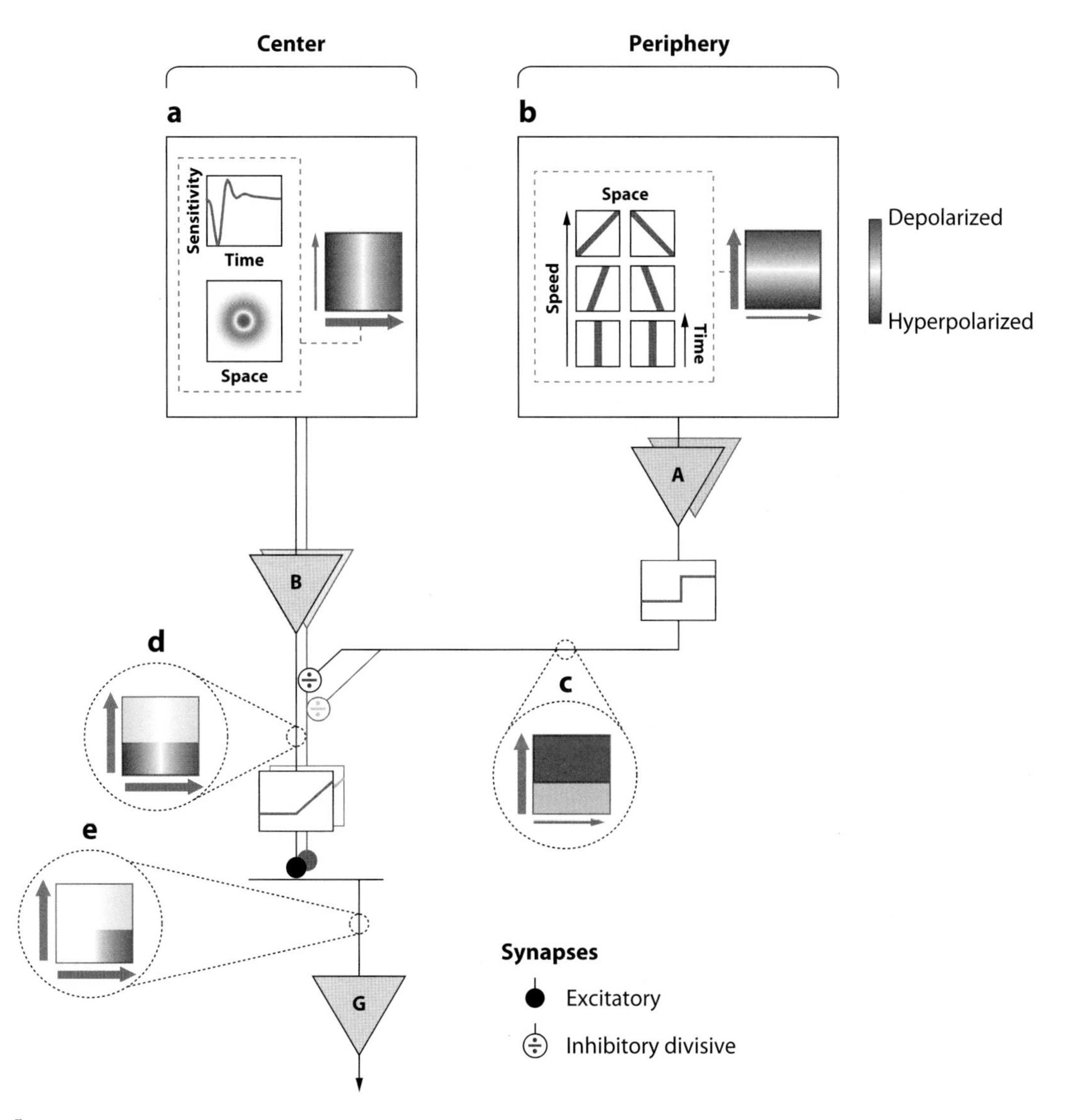

Figure 5

Object motion sensitivity. The OMS circuit signals when image motion is different between the center and the periphery. (*a*) In the center, bipolar cells filter the local image using a linear spatiotemporal receptive field that is strongly biphasic, conveying changes in intensity. This preferred direction in stimulus space is indicated by the horizontal arrow. Each bipolar cell output is rectified by passing through a threshold, such that the sum of bipolar cell output signals local changes in intensity but not the specific spatial pattern of active bipolar cells. Thus, excitation conveys a temporal sequence specific to the motion trajectory. (*b*) In the periphery, the population of polyaxonal amacrine cells computes a sum of rectified local intensity changes, using spatiotemporal properties (*not shown*) nearly identical to that of the central region. This nonlinear transformation approximates the local speed of the image, which is represented by a vertical arrow in the two-dimensional (2D) stimulus space. The amacrine cell spiking output delivers a thresholded version of the speed (*c, blue region of the 2D stimulus space*). The amacrine cell population then inhibits the central bipolar cell terminals with a divisive operation, reducing the gain of the central input. (*d*) In the 2D stimulus space defined by a single bipolar cell filter (*horizontal arrow*) and the speed of the background (*vertical arrow*), the bipolar cell input changes only with the local stimulus intensity. The bipolar cell terminal changes its membrane potential only when the background moves at a low speed. (*e*) The ganglion cell membrane potential then sums this local differential motion signal over multiple bipolar cells. Only the 2D stimulus space containing one of these bipolar outputs is shown.

IPL: inner plexiform layer

inhibitory input is a critical aspect of the computation (Baccus 2007). This inhibition is consistent with a model of a divisive effect, reducing the gain of the central input (Baccus et al. 2008). Because the inhibitory input is delivered presynaptically, differential motion sensitivity is first computed at the bipolar cell synaptic terminal.

The overall organization of the OMS circuit can be considered in two stages. The input stage computes localized rectified changes in light intensity. This rectification, combined with the transient responses of bipolar cells in the circuit, transforms a local measurement of image speed into a temporal sequence specific to the local motion trajectory. At the output stage, transient polyaxonal cells thus cancel out any input that is synchronized with background motion. This arrangement implements an operation similar to typical lateral inhibition but one that is performed not on the original light intensity, as in the case of a horizontal cell, but on a particular local feature.

Summation over smaller rectified subunits is the hallmark of the Y-type ganglion cell (Crook et al. 2008, Shapley & Victor 1980). In OMS cells, this property is key to generating sensitivity to motion but not the specific spatial pattern. Although cells sensitive to differential motion have been seen in salamander, rabbit and mouse (Olveczky et al. 2003, Zhang et al. 2012), it remains to be seen whether Y-cells in other species such as cat and primate also have properties of object motion sensitivity. Because the inhibitory circuitry responsible for object motion sensitivity spans a long distance, one key factor to revealing this property in the retina of a larger animal may be to experiment with an intact retina.

SUPPRESSION OF FAST GLOBAL IMAGE SHIFTS IN THE INNER PLEXIFORM LAYER

At the other end of the spectrum of eye movements, certain rabbit ganglion cells are suppressed by rapid peripheral stimuli simulating saccades (Roska & Werblin 2003). The inhibitory suppression is sensitive to tetrodotoxin (TTX) and blockers of GABAergic transmission, which in the mammalian retina are delivered by polyaxonal amacrine cells (Pourcho & Goebel 1983). These ganglion cells ramify in the central strata of the inner plexiform layer (IPL) bounded by direction-selective ganglion cells. This central region of the IPL is also occupied by polyaxonal amacrine cells (Völgyi et al. 2001), so it seems likely that these amacrine cells deliver inhibition generated by fast shifts in the image. This anatomical correspondence indicates the connection between suppression of a particular region of stimulus space and a layer of the retina. Polyaxonal amacrine cells are narrowly stratified, and different layers of the inner plexiform appear to carry signals with different timing (Baccus 2007). For both object motion sensitivity and suppression of fast image shifts, achieving a particular type of suppression may simply involve arborization of both inhibitory and target neurons in the appropriate sublamina of the inner plexiform layer.

Figure 6

Polarity reversal. This ganglion cell circuit rapidly switches between an On-center and Off-center cell, depending on the peripheral input. (*a*) The spatiotemporal filter of an Off bipolar cell, represented by the horizontal arrow in the two-dimensional (2D) stimulus space. (*b*) The filter of an On bipolar cell, whose preferred direction in stimulus space (*horizontal arrow*) is opposite to that of the Off bipolar cell. (*c*) The periphery measures the speed of the background in a manner similar to that of the OMS circuit (see **Figure 5**). The speed of the background is represented by the vertical arrow in the 2D stimulus space. (*d*) Motion in the background activates A_1, inhibiting the Off bipolar cell. As seen in the 2D stimulus space, when the input on the vertical (peripheral speed) axis is large, the response for a given input on the horizontal (bipolar cell) axis is less. (*e*) The signal from the background is inverted through a second amacrine cell A_2, yielding a disinhibitory (depolarizing) input to the On bipolar cell terminal. Thus, peripheral motion increases the On bipolar cell output. (*f*) The combination of opposite effects on the On and Off pathways yields a response feature whose sign can be reversed, controlled by the background speed. (*g*) Response curves showing the output firing rate as a function of the input in the center at low and high background speeds. Curves are one-dimensional slices through panel (*f*) at the speeds indicated by the gray colored bars.

FAST MODULATION OF STIMULUS PREFERENCE BY EYE MOVEMENTS

The preferred stimulus of a sensory neuron is often considered to be fixed, or at most modified gradually owing to adaptation arising from a strong presence of one type of stimulus (Barlow 1961, Hosoya et al. 2005). However, when the background suddenly changes, as would occur during a saccade, certain salamander ganglion

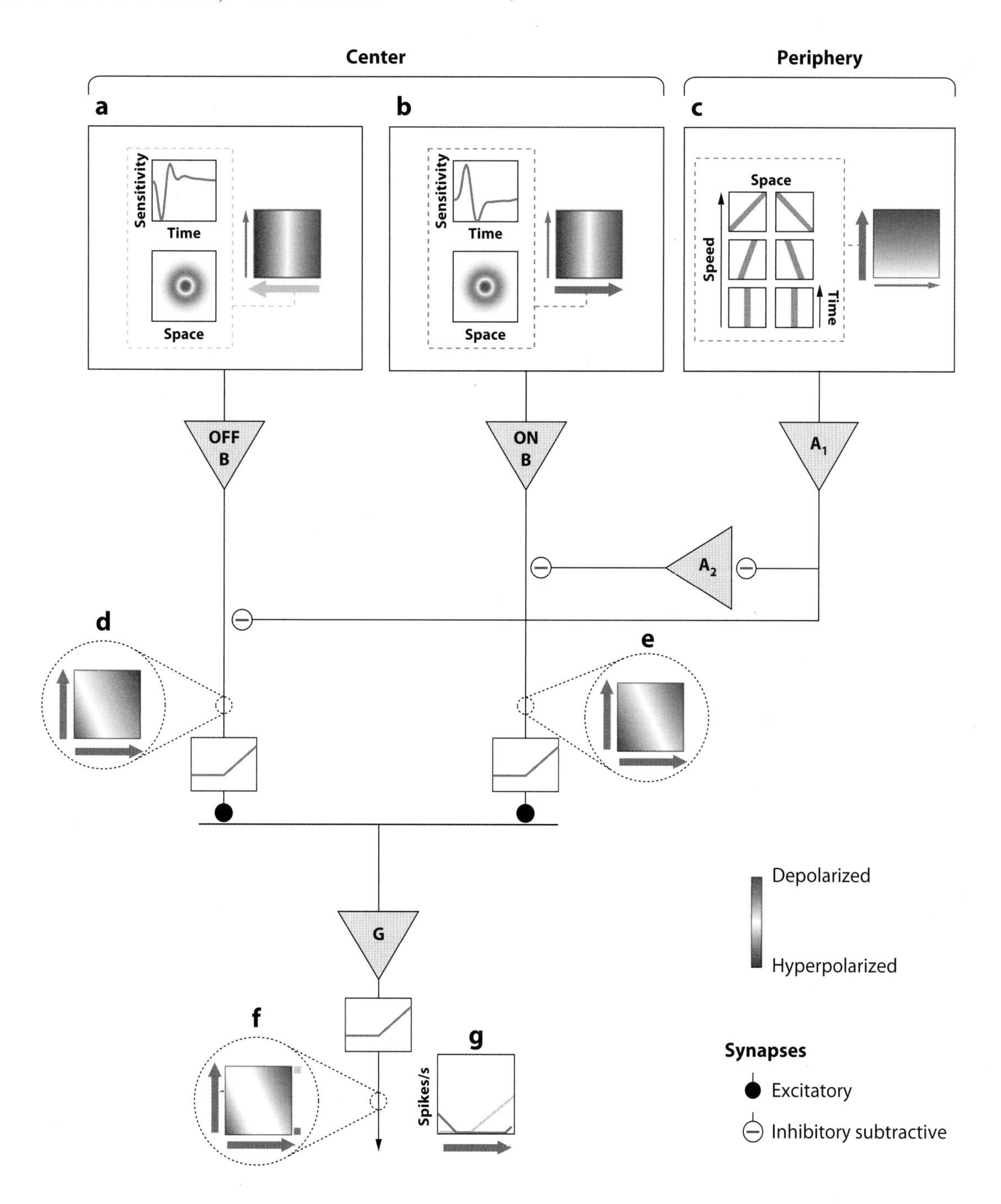

cells change their response properties from being primarily Off-type cells to being primarily On-type cells for a brief (~100 ms) interval of time (Geffen et al. 2007). This response property can be explained by ganglion cells with input from both On and Off pathways, each of which experiences different effects from background input. Amacrine cells whose receptive fields are in the periphery inhibit the Off pathway but disinhibit the On pathway (**Figure 6**). Injection of current into amacrine cells of various types identified a transient On-type amacrine cell that affected individual ganglion cells by enhancing sensitivity of the On pathway relative to the Off pathway. One would expect that this cell type would act through a second intervening amacrine cell to produce disinhibition of the On pathway. Although this effect was described as a reversal of polarity, because the On and Off pathways have slightly different temporal properties, a change in the weight of inputs will change the time course of the receptive field, rotating the stimulus direction of greatest sensitivity. More generally, this type of organization could change the shape of the spatiotemporal receptive field by selecting among different inputs.

INTERACTION OF LOCAL AND WIDE-FIELD INHIBITION IN THE LOCAL EDGE DETECTOR

Classic lateral inhibition involves the comparison of stimulus intensities in neighboring regions. Early studies of rabbit retinal ganglion cells revealed a type of neuron that responded to small but not large spatial stimuli. However, this inhibition was present regardless of whether patches of the stimulus were spatially uniform or textured, and thus the response differed from the standard linear center-surround receptive field (Levick 1967). This type of neuron was called the local edge detector (LED). Similar to the OMS cell, both center and peripheral regions have nonlinear responses, and lateral inhibition decreases the central input. Thus, in both cases lateral inhibition is applied after a more complex feature has been computed. This process causes small textured stimuli to generate responses, whereas large textured stimuli do not. Also similar to the OMS circuit, the circuitry for the LED has both excitatory central input and peripheral inhibition that are delivered onto the bipolar cell presynaptic terminal (**Figure 7**). In addition, responses to fine textures are generated by summing bipolar responses that are transmitted through a threshold.

Several differences can also be seen between the OMS and LED circuits. LED cells have a more sluggish response and appear to lack synchronization in precise timing between central and peripheral input. Consequently, LED cells experience only minor suppression when probed with global and differential motion (Olveczky et al. 2003). In addition, for extended textured stimuli the edges have a stronger effect, indicating that the spatial scale

Figure 7

Local edge detector. This circuit responds when a texture is presented in a central region, but not if a texture also stimulates the background. (*a*, *b*) Central input arises from On and Off bipolar cells, each with its own spatiotemporal filter represented by the horizontal axis (*thick arrow*) in the two-dimensional (2D) stimulus space. As an approximation, On and Off bipolar cells are shown as having preferred features that are opposite in sign. Each pathway is then rectified, conveying one half of the signal for this single direction in stimulus space (*d*). (*c*, *e*) The background takes a local measurement of the change in intensity by taking a rectified sum of multiple bipolar cells (*not shown*), conveyed through a population of amacrine cells that may comprise multiple different types (A_1). This creates a nonlinear transformation to a new direction in stimulus space (*vertical arrow*) that represents a changing background texture. (*f*, *g*) This inhibition is delivered presynaptically to the central bipolar cell terminals, such that input along the background stimulus direction reduces the gain of the response in the On (*f*) and Off (*g*) pathways. (*h*) The On and Off pathways are then combined so that either pathway creates a change in response as shown in the 2D stimulus space. A local population of amacrine cells (A_2) acts to reduce the effect of background input. This additional transformation of the stimulus provided by A_2 is not shown. (*i*) After rectification by the spiking output of the ganglion cell, either On or Off input increases the firing rate; but when background input is large along the vertical axis, the central input has little effect.

of inhibition for the LED circuit is narrower or stronger than that in the OMS circuit (Russell & Werblin 2010). In the LED circuit, central inhibition suppresses peripheral inhibition, although we do not know whether this property is also present in the OMS circuit.

The OMS and LED circuits perform computations that go well beyond a simple LN

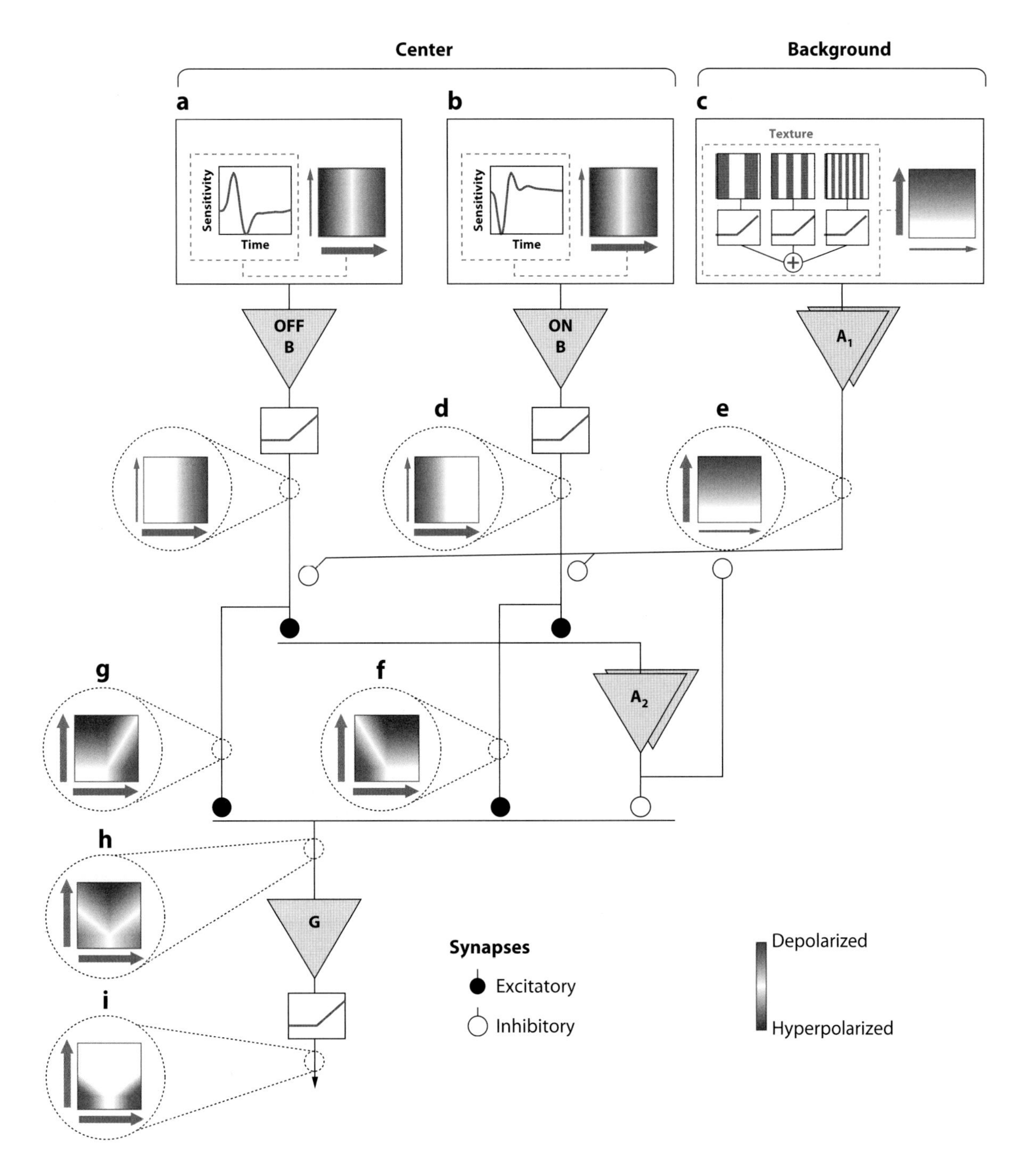

model. Using a common representation of the response with respect to the stimulus at each stage in the circuit allows us to understand how each processing element contributes to the overall transformation.

LOCAL MODULATION BY SUSTAINED AMACRINE CELLS

The sensitivity of a cell to the stimulus is not fixed but will change depending on the recent stimulus. For example, during adaptation, strong stimuli will reduce sensitivity, thus avoiding saturation and creating a more efficient use of a cell's dynamic range (Fairhall et al. 2001, Laughlin 1989, Van Hateren 1993). However, changes in sensitivity can also result not from direct input, but through a second modulatory pathway. The importance of this modulation may result from the statistics of natural images such that a strong stimulus of one type may predict the presence of a second type, allowing the visual system to suppress the effects of predictable input (Simoncelli & Olshausen 2001). Object motion sensitivity and fast image shift suppression are examples of an extreme modulatory effect, in which one pathway can completely silence another, but how can more gradual changes in gain be achieved?

Amacrine cells can be broadly arranged into two categories, sustained and transient, although the division between these groups is not strict, and each group is composed of a number of different cell types (Baccus 2007). Large, wide field cells are always transient, and narrow cells are typically sustained. Researchers have investigated sustained Off-type amacrine cells in the salamander using simultaneous intracellular and multielectrode recording (De Vries et al. 2011, Manu & Baccus 2011).

These amacrine cells release tonic inhibition; hyperpolarizing them with current injection activates nearby ganglion cells in the absence of a visual stimulus. When steady current pulses are combined with a visual stimulus, these amacrine cells change the sensitivity but not the temporal filtering of nearby ganglion cells (De Vries et al. 2011). This change in sensitivity can be considered primarily a modulatory effect because it does not change the direction in stimulus space (feature) encoded by the ganglion cell but merely the sensitivity to that direction.

This effect is not spatially homogeneous: The amacrine pathway produces a local reduction in the gain of the ganglion cell but causes a peripheral increase in sensitivity, which likely arises through disinhibition (**Figure 8**). Because the effect occurs at a scale smaller than the ganglion cell receptive field, transmission is, in part, likely to be presynaptic. Pharmacological experiments injecting current into this group of amacrine cells indicate that this modulatory effect is mediated most likely by metabotropic $GABA_B$ receptors, which may contribute to the delay in amacrine transmission.

Because these amacrine cells inhibit both On- and Off-type ganglion cells, the function appears to be a generalized local modulation in sensitivity. On the assumption that nearby stimuli are correlated, a localized stimulus in one region will likely predict that a nearby stimulus is imminent. In this regard, the peripheral increase in sensitivity may anticipate the appearance of a future stimulus.

Further studies of sustained amacrine cells considered more fully the timing of the response of the amacrine cell, the timing of transmission, and other input to the ganglion cell (Manu & Baccus 2011). By recording the voltage fluctuations of an amacrine cell and then playing back a rescaled version of these fluctuations, the output of the interneuron was either amplified or canceled. Although sustained amacrine cells are inhibitory, increasing the amplitude of the amacrine light response using current injection increased the activity of nearby ganglion cells. This apparent paradox was explained according to two combined effects. A nonlinear response function interposed between the two cells caused disinhibition produced by hyperpolarization of the amacrine cell to be larger than inhibition produced by amacrine depolarization. In addition,

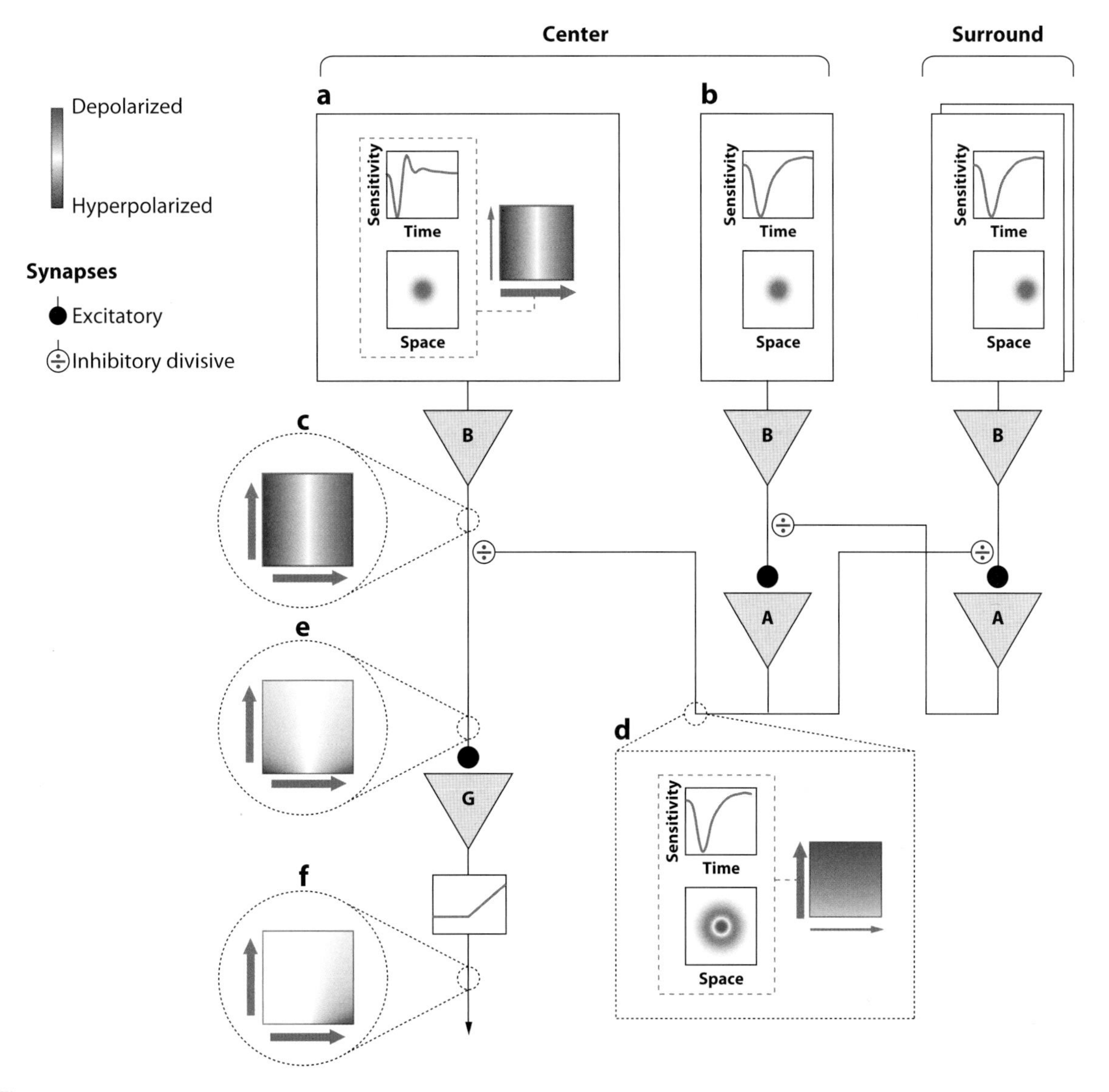

Figure 8

Sustained Off-type amacrine cell. (*a*, *c*) Central input is conveyed through a fast, biphasic bipolar cell, whose preferred feature is represented in the horizontal axis of the two-dimensional (2D) stimulus space shown. (*b*) A second pathway with a slower filter activates a linear sustained amacrine cell (A). Stimuli in a nearby surrounding region act through an intervening amacrine cell, creating a disinhibitory (net excitatory) effect. (*d*) Thus, the stimulus feature conveyed by the amacrine cell transmits local inhibition but peripheral disinhibition. Amacrine cells conveying the disinhibitory effect may be the same types as or different from those conveying the local inhibitory effect. Because the amacrine cell response is slower, the direction conveyed by the amacrine cell (*vertical arrow*) is largely orthogonal to that of the bipolar cell. (*e*) The output of the amacrine cell is delivered presynaptically, reducing the sensitivity to the ganglion cell's preferred feature through a divisive effect. (*f*) After rectification, the ganglion cell has a lower sensitivity to its preferred feature when the input along the amacrine feature (*vertical axis*) is larger.

because of the timing of the amacrine cell's visual response, on average the amacrine cell hyperpolarized just before other cells delivered excitatory input to the ganglion cell. Thus, at rest, the tonic inhibitory output of the amacrine cell was near its maximal effect, and reducing this inhibition caused a strong influence on ganglion cell firing. These results illustrate that understanding the role of an interneuron in a circuit requires not only a measurement from a single cell, but also information about an interneuron's input, output, and other circuit activity.

A POTENTIAL ROLE FOR INHIBITION IN SENSITIZATION

A fundamental problem faced by all adaptive systems—neural, biochemical, and electronic—is that the more a system optimizes for one condition, the more it is vulnerable to failure when conditions suddenly change. In the amphibian and mammalian retina, a class of ganglion cells experiences a new form of plasticity termed sensitization, whereby cells raise their sensitivity following a strong stimulus (Kastner & Baccus 2011). This behavior opposes the known dynamics of typical adapting ganglion cells, which lower their sensitivity following a strong stimulus. Sensitizing cells, in turn, are more prone to saturation at high contrast. The opposing dynamics in adapting and sensitizing cells preserve sensitivity to the stimulus when the environment changes; one population maintains the ability to respond when the other fails.

The circuit underlying sensitization is unknown but must have more complex dynamics than that of a single adapting pathway that reduces its sensitivity in response to a large input. A computational model indicates that sensitization could be produced by two adapting pathways: one inhibitory and one excitatory (**Figure 9**). High contrast causes adaptation in the inhibitory pathway; when the contrast decreases, a residual reduction of inhibition causes a temporary increase in sensitivity. We do not currently know the identity of such inhibitory neurons. But because amacrine cells are known to adapt and horizontal cells are not (Baccus & Meister 2002, Ozuysal & Baccus 2012, Rieke 2001), amacrine cells are suspected

Figure 9

Sensitization. The sensitizing circuit produces an elevated sensitivity after a high-contrast stimulus ends. This elevated sensitivity then decays to a lower value after several seconds. (*a*) The stimulus flickers randomly with a constant mean and a contrast (σ) that changes from high to low. Colored bars indicate time intervals when the low contrast (L_{early}) ends and several seconds later (L_{late}). (*b*) Two adapting pathways, one excitatory (*orange*) and one inhibitory (*green*), have a similar stimulus preference indicated by the temporal filter (*vertical axis*). In the two-dimensional (2D) stimulus space, the horizontal axis indicates the time of the stimulus spanning 20 s. High contrast (0–10 s) spans a wider range of inputs along the vertical axis than does low contrast (10–20 s). The color scale is shown in discrete levels to emphasize differences in response to different regions of the stimulus space. (*c*) In the inhibitory pathway, the stimulus is rectified, such that during both low and high contrast, only input above a threshold produces a response at this level. (*d*) The signal passes through an adaptive stage represented by a simplified feedforward divisive operation such that the stimulus amplitude integrated over a time interval of seconds (exponential filter) reduces the sensitivity to the preferred feature. This process can be seen by noting that during L_{early}, the response colors vary more slowly along the vertical axis than during L_{late}. (*e*) This yields a response at the output of the inhibitory pathway that decays at high contrast and recovers at low contrast. (*f*) One-dimensional response curve showing that with respect to the preferred stimulus feature, inhibition has a higher threshold during L_{early} than L_{late}. This inhibition is passed through a synaptic delay (*g*) and delivered prior to a threshold in the excitatory pathway. This delay causes the output of the inhibitory pathway to have a different preferred feature than the excitatory pathway, such that only the average output of the inhibitory pathway is conveyed. (*h*, *i*) As a result of adaptation in the inhibitory pathway, during L_{early} a residual lack of inhibition causes a higher sensitivity to the preferred feature (colors change more quickly along the vertical axis during L_{early}). (*j*, *k*) After the excitatory pathway itself adapts to its input from its own feedforward adaptive stage, the result is adaptation at high contrast and a residual sensitization of the firing rate at low contrast. The 2D stimulus space shows that along the vertical axis, the response has a higher sensitivity during L_{early} than during L_{late}. (*l*) One-dimensional response curves along the preferred stimulus feature during L_{early} and L_{late}.

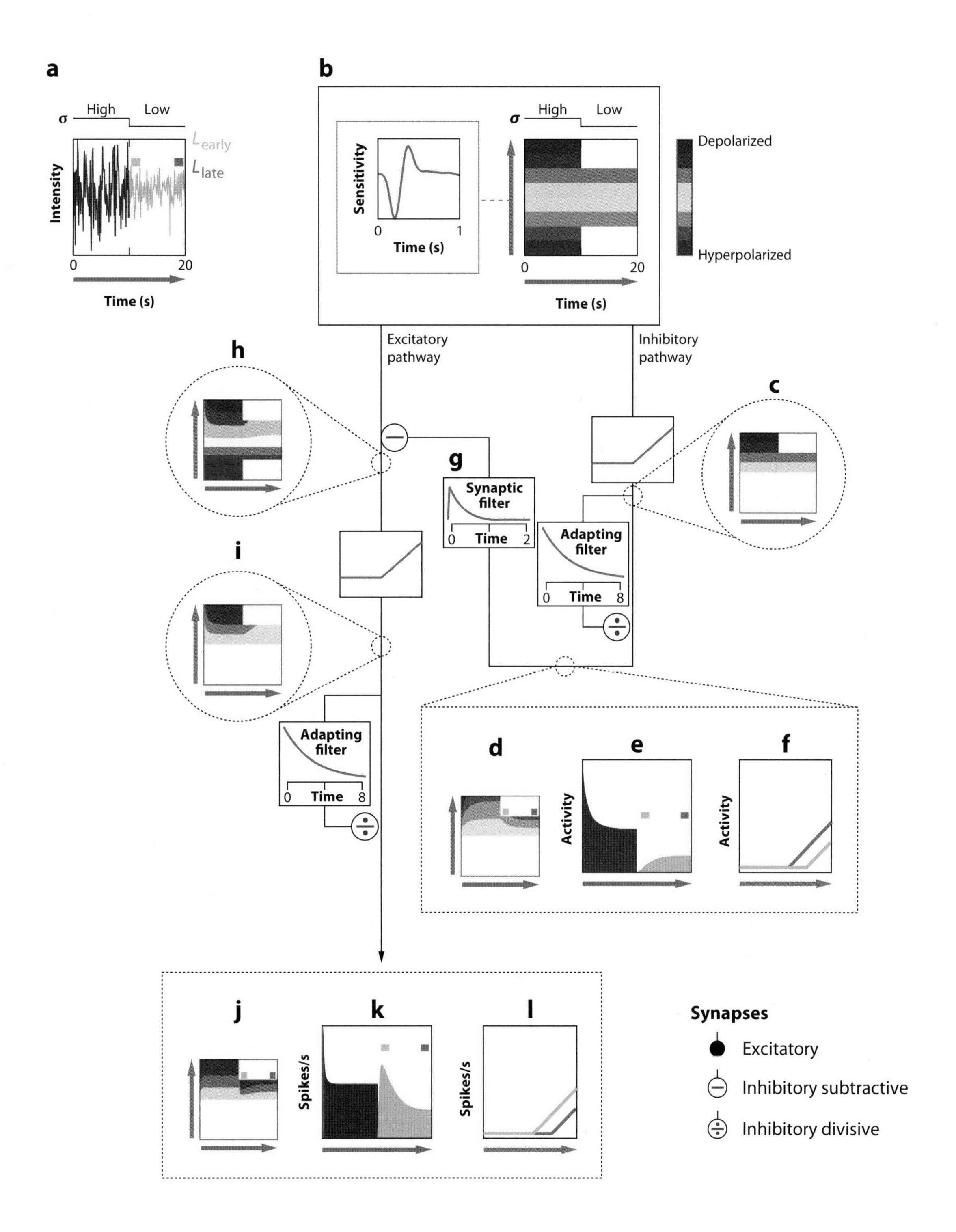
a
σ
High
Low
L_{early}
L_{late}
Intensity
0
20
Time (s)
b
Sensitivity
0
1
Time (s)
σ
High
Low
0
20
Time (s)
Depolarized
Hyperpolarized
Excitatory pathway
Inhibitory pathway
h
c
g
Synaptic filter
0
Time
2
Adapting filter
0
Time
8
i
Adapting filter
0
Time
8
d
e
f
Activity
Activity
j
k
l
Spikes/s
Spikes/s
Synapses
Excitatory
Inhibitory subtractive
Inhibitory divisive

to be responsible for sensitization. In general, little is known about the functional role of dynamics and plasticity in retinal inhibitory interneurons.

DO NEURAL CIRCUITS FOLLOW GENERAL RULES?

Although the study of neural circuits has lasted decades, it is unclear whether each circuit solves a problem in a different way, or if general rules apply to other parts of the nervous system. Specific examples in the retina suggest such general rules. Comparing sustained and transient amacrine cells, studies indicate that the two cells have different roles. Because transient amacrine cells are often strongly rectified and depolarize sparsely in time, they suppress a limited subset of stimulus space. Furthermore, because wide-field transient amacrine cells such as polyaxonal amacrine cells are narrowly stratified (Pang et al. 2002), they inhibit only a select set of bipolar or ganglion cells. In comparison, sustained cells, including amacrine and horizontal cells, release tonic inhibition, acting over more points in time than do transient cells. In addition, horizontal cells target photoreceptors, and many sustained amacrine cells have a diffuse stratification in the IPL that acts on many types of bipolar and ganglion cells. Combined with the release of tonic inhibition, the diffuse anatomical connections of sustained amacrine cells cause these cells to affect the processing of a greater number of stimuli. Thus sustained cells seem to play a greater role than transient cells in defining the primary direction of sensitivity and modulation under a wide range of stimuli. One circuit function hypothesis suggests that these two different functional roles generalize in sensory circuits, implemented by two general classes of inhibitory neurons (Manu & Baccus 2011).

Certain inhibitory cells such as horizontal cells and the A17 amacrine cell have a more linear subtractive effect, but others such as the salamander sustained Off amacrine cell or polyaxonal amacrine cells have an effect that is divisive, at least in part (Baccus 2007, de Vries et al. 2011). One important function of a subtractive effect is to rotate the preferred stimulus direction, whereas a divisive effect can modulate the gain of another stimulus direction with less of an effect on the direction of the preferred stimulus. Both subtractive and divisive effects create a more efficient representation of the stimulus (Simoncelli & Olshausen 2001, Srinivasan et al. 1982). In the examples discussed here, the two effects are implemented by different inhibitory interneurons, and we do not know whether this distinction generalizes.

COMPLEX COMPUTATIONS FROM SIMPLE OPERATIONS

The functions described above combine simple operations such as linear filtering, thresholds, sign inversion, division, and local gain control to create more complex computations. To relate the higher-level function to these lower-level properties, mathematical descriptions of the response play a key role. For research at a lower mechanistic level, if an explicit goal is to connect components to such a portable mathematical description, studies will be directed toward mechanisms that are most likely to have an understandable role in a higher function. Furthermore, such a mathematical model is more likely to generalize between neural circuits. Creating such a rich description will aid in the comparison between models at different levels of integration. A probe with a simple stimulus such as a flash yields only a single response point in a much larger stimulus space. By exploring a wider range of stimuli and responses, a more detailed description of both higher-level functions and cellular mechanisms will enable a better match between higher-level transformations and elemental biophysical operations.

SUMMARY POINTS

1. A common mathematical representation of the response in multiple stimulus dimensions is useful to understand how lower-level operations in neurons construct a circuit's overall output.
2. Local dendritic processing plays a key role in generating direction-selective responses in starburst amacrine cells and in changing temporal filtering without spatial pooling in the A17 amacrine cell.
3. Wide-field and polyaxonal amacrine cells generate global inhibition from eye movements to generate object motion sensitivity, to suppress responses during saccadic eye movements, and to change the weighting of synaptic inputs to specific ganglion cell types.
4. The inputs to ganglion and amacrine cells can transform the original stimulus space into a nonlinear representation of speed or texture. Simple lateral inhibition then generates object motion sensitivity or the detection of textured edges.
5. Modulatory effects cause one stimulus feature to change the gain of another. Two separate functional classes of amacrine cells appear to have different roles. Wide-field amacrine cells depolarize sparsely and are narrowly stratified, targeting only a subset of ganglion cells. These cells appear to suppress responses to a particular defined region of stimulus space. Local sustained amacrine cells release tonic inhibition and target many ganglion cell types and thus appear to act more generally by controlling the gain of the response to a wide range of stimuli.

FUTURE ISSUES

1. Feedback inhibition results when an inhibitory neuron inhibits one of its synaptic inputs. Although it would simplify understanding if feedback inhibition only produced a change in temporal filtering, we do not know whether feedback from amacrine cells also causes a more complex dynamic change in gain.
2. More than 20 amacrine cells currently have unknown functions. Many of these cells inhibit not only bipolar and ganglion cells, but also other amacrine cells. The effects of amacrine to amacrine inhibition on higher-level processing are largely unknown.
3. Little is known about the functional role of dynamics and plasticity in amacrine cells. Sensitization in retinal ganglion cells occurs when a strong stimulus causes a future increase in sensitivity to weaker stimuli. Although studies have proposed a model for a role of adapting inhibitory neurons, the identity of those neurons and their contribution to sensitization are unknown.

DISCLOSURE STATEMENT

The authors are not aware of any affiliations, memberships, funding, or financial holdings that might be perceived as affecting the objectivity of this review.

ACKNOWLEDGMENTS

We thank D.B. Kastner for helpful comments on the manuscript. S.A.B. was supported by grants from the National Institutes of Health (National Eye Institute), the McKnight Foundation, and E. Matilda Ziegler Foundation for the Blind. P.D.J. was supported by a fellowship from the Walter and Idun Berry Foundation.

LITERATURE CITED

Baccus SA. 2007. Timing and computation in inner retinal circuitry. *Annu. Rev. Physiol.* 69:271–90

Baccus SA, Meister M. 2002. Fast and slow contrast adaptation in retinal circuitry. *Neuron* 36(5):909–19

Baccus SA, Olveczky BP, Manu M, Meister M. 2008. A retinal circuit that computes object motion. *J. Neurosci.* 28(27):6807–17

Barlow H. 1953. Summation and inhibition in the frog's retina. *J. Physiol.* 119:69–88

Barlow H. 1961. Possible principles underlying the transformation of sensory messages. In *Sensory Communication*, ed. WA Rosenblith, pp. 217–34. Cambridge, MA: MIT Press

Berry MJ II, Warland DK, Meister M. 1997. The structure and precision of retinal spike trains. *Proc. Natl. Acad. Sci. USA* 94(10):5411–16

Chen S, Li W. 2012. A color-coding amacrine cell may provide a blue-Off signal in a mammalian retina. *Nat. Neurosci.* 15(7):954–56

Chichilnisky EJ. 2001. A simple white noise analysis of neuronal light responses. *Network* 12(2):199–213

Crook JD, Peterson BB, Packer OS, Robinson FR, Troy JB, Dacey DM. 2008. Y-cell receptive field and collicular projection of parasol ganglion cells in macaque monkey retina. *J. Neurosci.* 28(44):11277–91

Demb JB, Singer JH. 2012. Intrinsic properties and functional circuitry of the AII amacrine cell. *Vis. Neurosci.* 29(1):51–60

De Vries SEJ, Baccus SA, Meister M. 2011. The projective field of a retinal amacrine cell. *J. Neurosci.* 31(23):8595–604

Dong CJ, Hare WA. 2003. Temporal modulation of scotopic visual signals by A17 amacrine cells in mammalian retina in vivo. *J. Neurophysiol.* 89(4):2159–66

Dunn FA, Rieke F. 2008. Single-photon absorptions evoke synaptic depression in the retina to extend the operational range of rod vision. *Neuron* 57(6):894–904

Eggers ED, Lukasiewicz PD. 2006. GABA(A), GABA(C) and glycine receptor-mediated inhibition differentially affects light-evoked signalling from mouse retinal rod bipolar cells. *J. Physiol.* 572(Pt. 1):215–25

Euler T, Detwiler PB, Denk W. 2002. Directionally selective calcium signals in dendrites of starburst amacrine cells. *Nature* 418:845–52

Fairhall AL, Lewen GD, Bialek W, de Ruyter van Steveninck RR. 2001. Efficiency and ambiguity in an adaptive neural code. *Nature* 412:787–92

Feigenspan A, Wässle H, Bormann J. 1993. Pharmacology of GABA receptor Cl- channels in rat retinal bipolar cells. *Nature* 361:159–62

Fried SI, Münch TA, Werblin FS. 2002. Mechanisms and circuitry underlying directional selectivity in the retina. *Nature* 420:411–14

Gavrikov KE, Nilson JE, Dmitriev AV, Zucker CL, Mangel SC. 2006. Dendritic compartmentalization of chloride cotransporters underlies directional responses of starburst amacrine cells in retina. *Proc. Natl. Acad. Sci. USA* 103(49):18793–98

Geffen MN, De Vries SEJ, Meister M. 2007. Retinal ganglion cells can rapidly change polarity from Off to On. *PLoS Biol.* 5(3):e65

Gollisch T, Meister M. 2010. Eye smarter than scientists believed: neural computations in circuits of the retina. *Neuron* 65(2):150–64

Grimes WN, Zhang J, Graydon CW, Kachar B, Diamond JS. 2010. Retinal parallel processors: more than 100 independent microcircuits operate within a single interneuron. *Neuron* 65(6):873–85

Habermann CJ, O'Brien BJ, Wässle H, Protti DA. 2003. AII amacrine cells express L-type calcium channels at their output synapses. *J. Neurosci.* 23(17):6904–13

Harris JJ, Jolivet R, Attwell D. 2012. Synaptic energy use and supply. *Neuron* 75(5):762–77

Horowitz P, Hill W. 1989. *The Art of Electronics*. Cambridge, UK: Cambridge Univ. Press. 2nd ed.

Hosoya T, Baccus SA, Meister M. 2005. Dynamic predictive coding by the retina. *Nature* 436:71–77

Kastner DB, Baccus SA. 2011. Coordinated dynamic encoding in the retina using opposing forms of plasticity. *Nat. Neurosci.* 14(10):1317–22

Kraaij DA, Spekreijse H, Kamermans M. 2012. The open- and closed-loop gain-characteristics of the cone/horizontal cell synapse in goldfish retina. *J. Neurophysiol.* 84:1256–65

Kuffler SW. 1953. Discharge patterns and functional organization of mammalian retina. *J. Neurophysiol.* 16(1):37–68

Laughlin SB. 1981. A simple coding procedure enhances a neuron's information capacity. *Z. Naturforsch.* 36:910–12

Laughlin SB. 1989. The role of sensory adaptation in the retina. *J. Exp. Biol.* 146:39–62

Lee S, Zhou ZJ. 2006. The synaptic mechanism of direction selectivity in distal processes of starburst amacrine cells. *Neuron* 51(6):787–99

Lehky S, Sejnowski T. 1988. Network model of shape-from-shading: neural function arises from both receptive and projective fields. *Nature* 333:452–54

Lettvin JY, Maturana HR, McCulloch WS, Pitts WH. 1959. What the frog's eye tells the frog's brain. *Proc. IRE* 47:1940–51

Levick WR. 1967. Receptive fields and trigger features of ganglion cells in the visual streak of the rabbit's retina. *J. Physiol.* 188(3):285–307

MacNeil MA, Masland RH. 1998. Extreme diversity among amacrine cells: implications for function. *Neuron* 20(5):971–82

Manookin MB, Beaudoin DL, Ernst ZR, Flagel LJ, Demb JB. 2008. Disinhibition combines with excitation to extend the operating range of the OFF visual pathway in daylight. *J. Neurosci.* 28(16):4136–50

Manu M, Baccus SA. 2011. Disinhibitory gating of retinal output by transmission from an amacrine cell. *Proc. Natl. Acad. Sci. USA* 108(45):18447–52

Masland RH. 2012. The tasks of amacrine cells. *Vis. Neurosci.* 29(01):3–9

Nowak P, Dobbins AC, Gawne TJ, Grzywacz NM, Amthor FR. 2011. Separability of stimulus parameter encoding by on-off directionally selective rabbit retinal ganglion cells. *J. Neurophysiol.* 105(5):2083–99

Oesch N, Euler T, Taylor WR. 2005. Direction-selective dendritic action potentials in rabbit retina. *Neuron* 47(5):739–50

Ölveczky BP, Baccus SA, Meister M. 2003. Segregation of object and background motion in the retina. *Nature* 423(6938):401–8

Ölveczky BP, Baccus SA, Meister M. 2007. Retinal adaptation to object motion. *Neuron* 56(4):689–700

Ozuysal Y, Baccus SA. 2012. Linking the computational structure of variance adaptation to biophysical mechanisms. *Neuron* 73(5):1002–15

Pang JJ, Gao F, Wu SM. 2002. Segregation and integration of visual channels: layer-by-layer computation of ON-OFF signals by amacrine cell dendrites. *J. Neurosci.* 22(11):4693–701

Pourcho RG, Goebel DJ. 1983. Neuronal subpopulations in cat retina which accumulate the GABA agonist, (3H)muscimol: a combined Golgi and autoradiographic study. *J. Comp. Neurol.* 219(1):25–35

Rieke F. 2001. Temporal contrast adaptation in salamander bipolar cells. *J. Neurosci.* 21(23):9445–54

Rodieck RW, Stone J. 1965. Analysis of receptive fields of cat retinal ganglion cells. *J. Neurophysiol.* 28(5):832–49

Roska B, Werblin F. 2003. Rapid global shifts in natural scenes block spiking in specific ganglion cell types. *Nat. Neurosci.* 6(6):600–8

Ruderman D, Bialek W. 1994. Statistics of natural images: scaling in the woods. *Phys. Rev. Lett.* 73(6):814–17

Russell TL, Werblin FS. 2010. Retinal synaptic pathways underlying the response of the rabbit local edge detector. *J. Neurophysiol.* 103(5):2757–69

Schwartz G, Rieke F. 2011. Perspectives on: information and coding in mammalian sensory physiology. Nonlinear spatial encoding by retinal ganglion cells: when $1 + 1 \neq 2$. *J. Gen. Physiol.* 138(3):283–90

Shapley RM, Victor JD. 1980. The effect of contrast on the non-linear response of the Y cell. *J. Physiol.* 302:535–47

Sher A, DeVries SH. 2012. A non-canonical pathway for mammalian blue-green color vision. *Nat. Neurosci.* 15(7):952–53

Simoncelli EP, Olshausen BA. 2001. Natural image statistics and neural representation. *Annu. Rev. Neurosci.* 24:1193–216

Srinivasan MV, Laughlin SB, Dubs A. 1982. Predictive coding: a fresh view of inhibition in the retina. *Proc. R. Soc. Lond. Ser. B.* 216:427–59

Thoreson WB, Mangel SC. 2012. Lateral interactions in the outer retina. *Prog. Retin. Eye Res.* 31(5):407–41

Vaney DI, Sivyer B, Taylor WR. 2012. Direction selectivity in the retina: symmetry and asymmetry in structure and function. *Nat. Rev. Neurosci.* 13(3):194–208

Vaney DI, Taylor WR. 2002. Direction selectivity in the retina. *Curr. Opin. Neurobiol.* 12(4):405–10

van Hateren H. 2005. A cellular and molecular model of response kinetics and adaptation in primate cones and horizontal cells. *J. Vis.* 5:331–47

Van Hateren JH. 1993. Spatiotemporal contrast sensitivity of early vision. *Vis. Res.* 33(2):257–67

Völgyi B, Xin D, Amarillo Y, Bloomfield SA. 2001. Morphology and physiology of the polyaxonal amacrine cells in the rabbit retina. *J. Comp. Neurol.* 440:109–25

Werblin FS. 2010. Six different roles for crossover inhibition in the retina: correcting the nonlinearities of synaptic transmission. *Vis. Neurosci.* 27(1–2):1–8

Zhang Y, Kim I-J, Sanes JR, Meister M. 2012. The most numerous ganglion cell type of the mouse retina is a selective feature detector. *Proc. Natl. Acad. Sci. USA* 109(36):E2391–98

Electrical Compartmentalization in Dendritic Spines

Rafael Yuste

HHMI, Departments of Biological Sciences and Neuroscience, and Kavli Institute for Brain Science, Columbia University, New York, NY 10027; email: rafaelyuste@columbia.edu

Annu. Rev. Neurosci. 2013. 36:429–49

First published online as a Review in Advance on May 29, 2013

The *Annual Review of Neuroscience* is online at neuro.annualreviews.org

This article's doi:
10.1146/annurev-neuro-062111-150455

Keywords

NMDA, computation, cortex, imaging, uncaging, network, emergent

Abstract

Most excitatory inputs in the CNS contact dendritic spines, avoiding dendritic shafts, so spines must play a key role for neurons. Recent data suggest that, in addition to enhancing connectivity and isolating synaptic biochemistry, spines can behave as electrical compartments independent from their parent dendrites. It is becoming clear that, although spines experience voltages similar to those of dendrites during action potentials (APs), spines must sustain higher depolarizations than do dendritic shafts during excitatory postsynaptic potentials (EPSPs). Synaptic potentials are likely amplified at the spine head and then reduced as they invade the dendrite through the spine neck. These electrical changes, probably due to a combination of passive and active mechanisms, may prevent the saturation of dendrites by the joint activation of many inputs, influence dendritic integration, and contribute to rapid synaptic plasticity. The electrical properties of spines could enable neural circuits to harness a high connectivity, implementing a "synaptic democracy," where each input can be individually integrated, tallied, and modified in order to generate emergent functional states.

Contents

INTRODUCTION

Many neurons throughout different brain regions are covered with dendritic spines (Ramón y Cajal 1888), small dendritic appendages composed of a spine head (~<1 μm in diameter), which typically accommodates an excitatory synapse, and a thin spine neck (<0.1 μm thick and ~1 μm long) that connects the spine to the dendritic shaft (Gray 1959). Interestingly, most excitatory contacts choose to terminate on spines rather than on their adjacent dendritic shafts (Arellano et al. 2007, Harris & Kater 1994); spines must therefore play an important role in neuronal function because otherwise these inputs could directly contact the dendrites. Speculation about this special role that spines must play has centered on the potential functions of spines in enhancement of structural connectivity or in biochemical compartmentalization (Koch 1999, Peters & Kaiserman-Abramof 1969, Shepherd 1996, Swindale 1981, Yuste 2010). Indeed, spines are very small and extremely numerous and can be arranged in helicoidal patterns (O'Brien & Unwin 2006), as if they were systematically sampling the neighboring axons and helping the circuit become more distributed (Yuste 2011). Also, spines compartmentalize calcium and provide the biochemical isolation necessary for input-specific synaptic plasticity (Koch 1999, Yuste et al. 2000, Yuste & Denk 1995). Nevertheless, input-specific biochemical isolation can occur without spines (Goldberg et al. 2003, Soler-Llavina & Sabatini 2006), so spines are not strictly necessary to implement local biochemical domains, raising the issue that they could carry out a more specific function in the neuron.

As an alternative function, spines could be electrical compartments; i.e., their special morphology could enable synaptic inputs to generate and experience different membrane potential dynamics than if they were situated on the dendritic shaft. This idea was first proposed by Ramón y Cajal (1904), who suggested that spines could store electric energy, and has been endorsed by many investigators since then (Chang 1952, Jack et al. 1975, Llinás & Hillman 1969, Rall 1974, Rall & Rinzel 1971, Rall & Segev 1988, Segev & Rall 1988, Shepherd et al. 1985). According to this view, the main function of spines is electrical rather than structural or biochemical. This article reviews this hypothesis by first focusing on the computational models of the electrical properties of spines, then by reviewing recent data consistent with the idea that spines can indeed behave as electrical compartments, and finally by discussing potential biophysical mechanisms responsible for this process. I conclude by commenting on functional consequences of the electrical properties of spines, highlighting the role that spines could play in building widely distributed and plastic circuits as biological analogs of neural networks, computational systems where the functional states are represented by the emergent dynamics of activity of many or all the neurons.

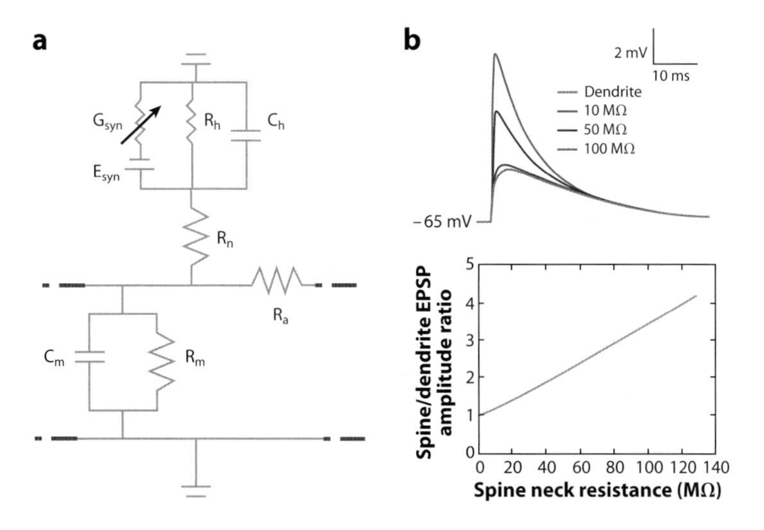

Figure 1

Passive electrical model of a dendritic spine. (*a*) Electrical circuit diagram of a passive spine with a synapse of conductance, G_{syn}, and reversal potential, E_{syn}. Cable parameters of the spine are represented by its input resistance (R_h), spine head capacitance (C_h), and neck resistance (R_n); adjacent dendrite is diagrammed by its axial resistance (R_a), membrane capacitance (C_m), and membrane resistance (R_m). (*b*) Numerical simulations demonstrate amplification of the excitatory postsynaptic potential (EPSP) at the spine head and subsequent reduction of EPSPs by spines, due to this cable structure. Note how increasing spine neck resistance results in larger EPSPs at the spine (*top*); the peak EPSP spine/dendrite voltage ratio is proportional to neck resistance (*bottom*). Adapted from Tsay & Yuste (2004).

ELECTRICAL MODELS OF SPINES

Due of the dearth of experimental data from living spines, until very recently the discussion of the electrical properties of spines was based solely on theoretical models using either analytical calculations or numerical simulations. This ample literature, extending over many decades, unfortunately has not always been in agreement, probably because even simple cable models need to assume values for experimental variables that are not yet measured. For example, we still do not know the values of the parameters that determine the electrical behavior of spines, or even dendrites, such as their input resistance (R_{sp} or R_h), membrane resistance (R_m), membrane capacitance (C_m, C_{sp}, or C_h), cytoplasmic or axial resistance (R_a), neck resistance (R_{neck}), spine synaptic conductance G_{syn}, or spine-reversal potential E_{syn} (**Figure 1**). Even the spine membrane's basic structure, lipid composition, and precise complement of conductances still remain relatively unknown. For this reason, computational models of spines have been ahead of the experimental measurements and should be interpreted as the exploration of a spectrum of potential scenarios, rather than as the ground truth. The next paragraphs briefly summarize the highlights of some of these models, which are discussed in more detail elsewhere (Shepherd 1996, Tsay & Yuste 2004, Yuste 2010).

Passive Models of Spines

Theoretical models of spines can be grouped into passive or active, depending on whether they incorporate voltage-dependent conductances. Passive models have used cable theory to explore the electrical consequences of the peculiar morphologies of spines, highlighting the electrical asymmetry created by a small spine connected to a large dendrite (Jack et al. 1975; Koch & Poggio 1983a,b; Llinás &

EPSP: excitatory postsynaptic potential

Hillman 1969; Rall 1970, 1978; Rall & Rinzel 1971) (**Figure 1**). The spine behaves as a sealed cable end with high input resistance (R_{sp}) and very small local capacitance (C_{sp}). Thus dendritic voltage pulses invade the spine without significant attenuation, whereas excitatory postsynaptic potentials (EPSPs) then attenuate as they propagate from the spine toward the dendrite. At the same time, the high spine input resistance can locally enhance the voltage of the EPSPs, when compared with EPSPs injected into the dendritic shaft. Finally, significant decreases in the driving force of the EPSP may occur, owing to the large effect that even small conductances can have on the ionic composition of the small volume of the spine. This driving force decrease could potentially even lead to the collapse of Na^+ gradients across the spine membrane.

In addition to an EPSP attenuation based on this cable impedance mismatch, the spine neck could also have a high electrical resistance (R_{neck}), which would further reduce the amplitude of synaptic inputs at the dendrite (Chang 1952, Jack et al. 1975, Llinás & Hillman 1969, Rall 1974, Rall & Rinzel 1971). Moreover, because the resistance of the spine neck should be proportional to its length, by altering their neck length, spines could potentially modulate the magnitude of this attenuation, thus controlling synaptic strength (Crick 1982, Rall 1978). It is still unknown what is the exact value of the spine neck resistance (R_n) or of its inverse, the neck conductance (G_n), whose ratio to the synaptic conductance determines the amount of electrical filtering. Initial estimates from passive models based on ultrastructural reconstructions and diffusional coupling suggested that G_n ranges from 7 to 138 nanosiemens (nS) (Harris & Stevens 1988, Svoboda et al. 1996). These values are much higher than those reported for synaptic conductances (0.05–0.1 nS; see Bekkers et al. 1990), although synaptic conductances in spines may be underestimated because they rely on somatic voltage clamp measurements (Williams & Mitchell 2008). This large difference between spine and synaptic conductances has been used to predict a relatively modest filtering of EPSP by spines and to argue that spines are not electrical compartments, but solely chemical ones (Koch & Zador 1993, Svoboda et al. 1996, Wickens 1988). At the same time, as we argue below, spine neck resistance could be significant enough to partly isolate spine voltages from dendritic ones. A final prediction from passive cable models is that spines, by their sheer numbers, may add a significant amount of membrane to the dendrite (Wilson 1986). This could lower the input impedance of the neuron and increase its overall capacitance, altering the temporal dynamics of input integration (Jaslove 1992).

INHIBITION IN SPINES

In addition to excitatory synapses, spines occasionally have symmetric (inhibitory) ones (Arellano et al. 2007, Jones & Powell 1969), which can originate from specific interneuron subtypes (DeFelipe et al. 1989). What is the function of this spine-specific inhibition? This inhibition probably curtails excitatory inputs, perhaps implementing logical gates (Shepherd & Brayton 1987). Moreover, extrasynaptic GABA receptors could be present on spines, even without an inhibitory synapse. In fact, a specific $GABA_A$ receptor ($\alpha 4\beta\delta$) is present on many dendritic spines of mouse CA1 hippocampal pyramidal cells. These receptors are sensitive to low levels of ambient GABA (<1 μM) and could shunt the current necessary to activate NMDA receptors (Shen et al. 2010).

Active Models of Spines

Passive cable models are only a first approximation to the physiological situation of the neuron because dendrites are active structures, endowed with voltage-dependent conductances (Llinás & Nicholson 1971, Stuart et al. 1999, Stuart & Sakmann 1994, Yuste & Tank 1996), and these conductances can override the effect of cable properties. For example, active conductances could allow spines to electrically isolate synaptic inputs (Diamond et al. 1970) and become electrogenic, particularly if the spine input resistance is high (Jack et al. 1975). If spines had sodium channels, they could act as synaptic amplifiers (Coss & Perkel 1985, Perkel 1982, Perkel & Perkel 1985), generating dendritic action potentials (APs), which could spread to neighboring spines (**Figure 2**) perhaps in a saltatory fashion (Baer & Rinzel 1991; Miller et al. 1985; Rall & Segev 1987, 1988; Segev & Rall 1988; Shepherd et al. 1985; Tsay & Yuste 2002). Moreover, electrogenic spines could also implement logical operations: for example, AND gates, when two EPSPs occur simultaneously on two active spines; OR gates, when a single active spine can saturate the dendrite; and AND-NOT gates, when an EPSP and an inhibitory postsynaptic potential (IPSP) coincide at a different dendritic position (Koch et al. 1983, Shepherd & Brayton 1987).

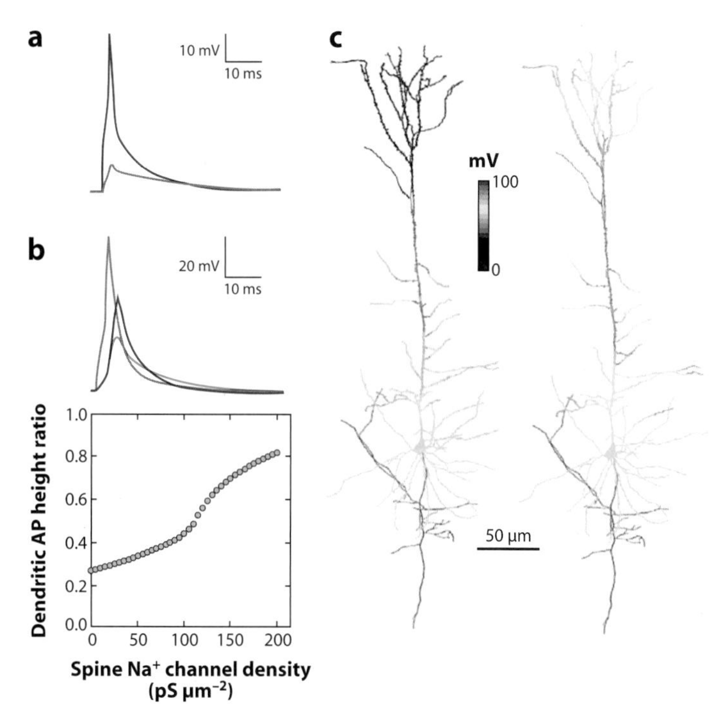

Figure 2

Active electrical models of spines. (*a*) Amplification of excitatory postsynaptic potentials (EPSPs) by spine sodium channels. Single-spine model with high sodium-channel densities in spines [~7,000 millisiemens per centimeter square (mS/cm^2)] elicits localized action-potential (AP) responses in the spine (*red*) but not in the dendrite (*gray*). For this to occur, spine neck resistances must provide some electrical decoupling (>100 MΩ). (*b*) Moderate sodium-channel densities in many spines increase the efficacy of backpropagation. Gray: AP at soma; blue: AP in apical dendrite (200 μm away), with weakly active spines (GNa = 40 mS/cm^2); red: AP in apical dendrite with more active spines (GNa = 200 mS/cm^2). (*c*) Peak voltage response to a backpropagating AP in a model with active spines. Backpropagation exhibits a decremental invasion of the apical tree when sodium-channel densities in spines are low (110 mS/cm^2) (*left*); but with higher densities in spines (200 mS/cm^2), a backpropagating AP fully invades the dendritic tree (*right*). Adapted from Tsay & Yuste 2004.

EVIDENCE OF ELECTRICAL COMPARTMENTALIZATION BY SPINES

Traditionally, the study of spines was purely anatomical, using fixed samples, because their small size makes them, even to this date, inaccessible to electrical recordings. However, the introduction of novel optical imaging techniques enabled for the first time functional measurements from living spines in vitro and in vivo. In particular, two-photon microscopy has allowed the imaging of calcium dynamics from spines under synaptic or AP stimulation, demonstrating that spines are invaded by backpropagating APs, are endowed with voltage-sensitive calcium channels (Yuste & Denk 1995) and, therefore, have active conductances (**Figure 3**).

Similar experiments revealed that spine *N*-methyl-D-aspartate receptors (NMDARs) flux significant amounts of calcium under minimal quantal synaptic stimulation (Koester & Sakmann 1998, Kovalchuk et al. 2000, Yuste et al. 1999), even when the somatic depolarization is very small [<1 millivolts (mV)] (**Figure 4**). These large calcium accumulations were unexpected because investigators assumed that NMDARs should be mostly blocked by Mg^{2+} under these small depolarizations. Although some residual unblocked NMDARs may exist at rest, these data first suggested that the voltages generated by EPSPs at the spine could be significantly larger than those measured in the soma or at the dendrite.

Two-photon glutamate uncaging produced further evidence that spines can behave as separate electrical compartments, by demonstrating that the somatic amplitude of the potentials generated by activating a spine was inversely proportional to the length of its spine neck (Araya et al. 2006b). Thus, whereas spines with short necks generated larger somatic potentials,

AP: action potential

NMDAR: *N*-methyl-D-aspartate receptor

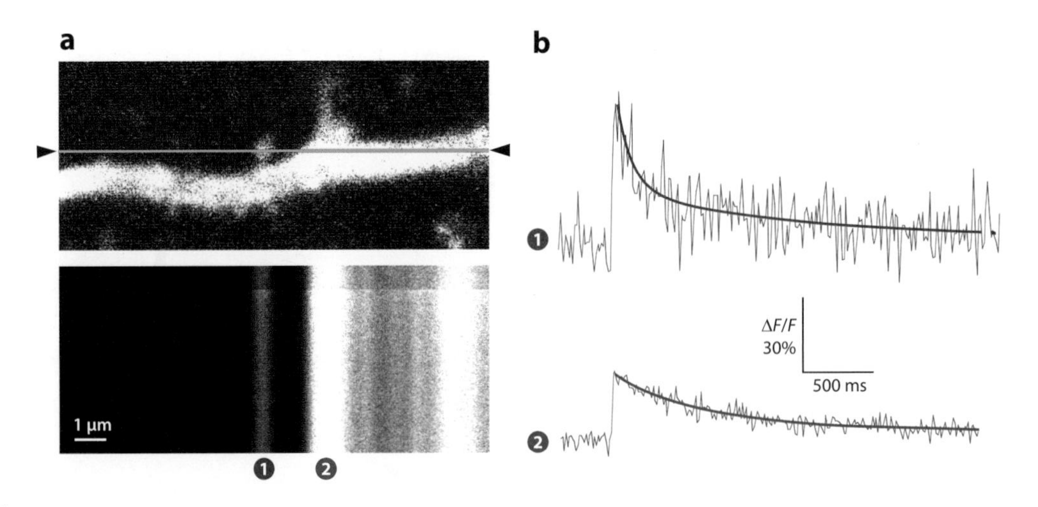

Figure 3

Imaging action potential (AP) invasion of spines. (*a*) Top: Two-photon image of a dendrite from a neocortical pyramidal neuron filled with a calcium indicator. Bottom: Line scan through a spine and nearby dendritic shaft (*between black arrowheads*) during invasion by a somatic AP. The backpropagating AP induces calcium accumulations in the spine (1) and dendritic shaft (2). (*b*) Kinetics of AP-triggered calcium accumulations in spine (1) and dendrite (2). Note the lack of significant delay between spine and dendrite accumulations, indicative of a local calcium influx at the spines triggered by the AP invasion. Adapted from Holthoff et al. 2002.

those with long necks generated smaller, or even undetectable, depolarizations (**Figure 5**). These data suggested the possibility that significant electrical reduction of synaptic potentials occurs at the spine neck. Although glutamate uncaging stimulates spines artificially, spines with longer necks also appear to have smaller postsynaptic potentials than do spines with shorter necks in physiological activation of synapses under minimal axonal stimulation (R. Araya, T. Vogels, and R. Yuste, submitted).

Additional support for electrical compartmentalization by spines comes from two-photon uncaging data that show that voltage-gated conductances can be differentially activated in a spine and its neighboring dendritic shaft, something that should not occur if both compartments were isopotential. Indeed, after glutamate uncaging on spines, calcium channels in the spine head activate differently than do those in the neighboring dendrite (Bloodgood et al. 2009). Moreover, the diffusional isolation generated by the spine neck can be regulated by neuronal activity and becomes significant enough to isolate the spine electrically (Bloodgood & Sabatini 2005). Also, bath application of the sodium-channel blocker TTX selectively blocks glutamate uncaging potentials generated on the spines but does not affect those in the neighboring dendritic shaft (Araya et al. 2007). These last experiments indicate that there must be sodium channels in spines and also that these channels can be activated independently from those in the neighboring dendrite. Moreover, the spine voltages reached after glutamate uncaging must be sufficiently high (>15 mV depolarizations) to activate sodium channels. Finally, a recent study, also using two-photon uncaging of glutamate, provides complementary evidence for electrical compartmentalization by spines (Harnett et al. 2012). Using calcium imaging, and assuming that the calcium influx is proportional to the local voltage, the authors estimate the ratio of spine to dendritic voltages during the uncaging potential by activating AMPA receptors (under blockade of NMDAR and sodium channels). They report up to a 45-fold amplification of the uncaging voltages in the spines, as compared with those in adjacent dendrites, as well as high spine neck resistances.

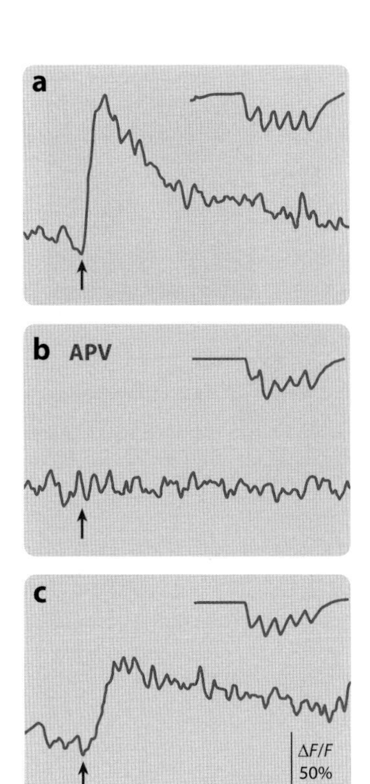

Figure 4

Activation of spine *N*-methyl-D-aspartate receptors (NMDARs) by subthreshold synaptic potentials. (*a*) Calcium accumulations in a spine in response to a train of EPSCs (*red arrow*) (*Inset*). (*b*) Identical stimulation in the presence of 100 μM APV, an NMDA receptor blocker. Calcium accumulations during EPSCs are completely blocked. (*c*) After washout of APV, the EPSC-induced calcium influx recovers. These results first suggested that the EPSP voltage is significantly higher in the spine than at the dendritic shaft or soma. Adapted from Yuste et al. 1999.

These results together indicate that spines may become electrical compartments when activated synaptically. However, this conclusion is still based on indirect evidence and should be supported by direct measurements of membrane potential in spines during EPSPs. Different types of voltage-imaging techniques (Peterka et al. 2011), either with fluorescence voltage-sensitive dyes (Cohen 1989) or with second-harmonic generation (SHG) chromophores (Lewis et al. 1999, Millard et al. 2005, Nemet et al. 2004), are enabling researchers to measure, for the first time, spine voltages. In a recent series of papers, investigators have directly demonstrated the invasion of backpropagated APs from dendrites into spine heads, revealing that the AP amplitude at the spine is similar to that of their parent dendritic shafts (Nuriya et al. 2006, Palmer & Stuart 2009, Holthoff et al. 2010, Acker et al. 2011) (**Figure 6**). Finally, voltage measurements under large-scale synaptic stimulation to dendrites indicate that some spines, but not all, can sustain substantially higher voltages than can their neighboring dendritic shafts (Palmer & Stuart 2009), although voltage measurements of "quantal" EPSP in individual spines still remain elusive.

POTENTIAL MECHANISMS OF ELECTRICAL COMPARTMENTALIZATION

That spines and dendrites can sustain different membrane potentials, at least in some circumstances such as glutamate uncaging or synaptic activation, indicates that they can be electrically isolated. What are the mechanisms responsible for this electrical compartmentalization? They may comprise passive electrical properties of the spine, active conductances, or, perhaps more likely, a combination of both.

As mentioned above, passive electrical alteration of spine potentials could be due to the impedance mismatch between the spine and the dendrite, created by the spine's high resistance and low capacitance. The spine neck may act effectively as a diode, propagating dendritic voltages into the spine without decrement yet diminishing spine voltages as they invade the dendrite. This mechanism likely depends on the dendritic diameter: Spines on thinner dendrites would become more isopotential than would those on thicker dendrites. In addition, a passive filtering mechanism could occur if the spine neck resistance is high and directly reduces spine potentials, explaining the dependence of voltage reduction on the spine neck length (Araya et al. 2006b; **Figure 5**). Although simulations based on ultrastructural reconstructions

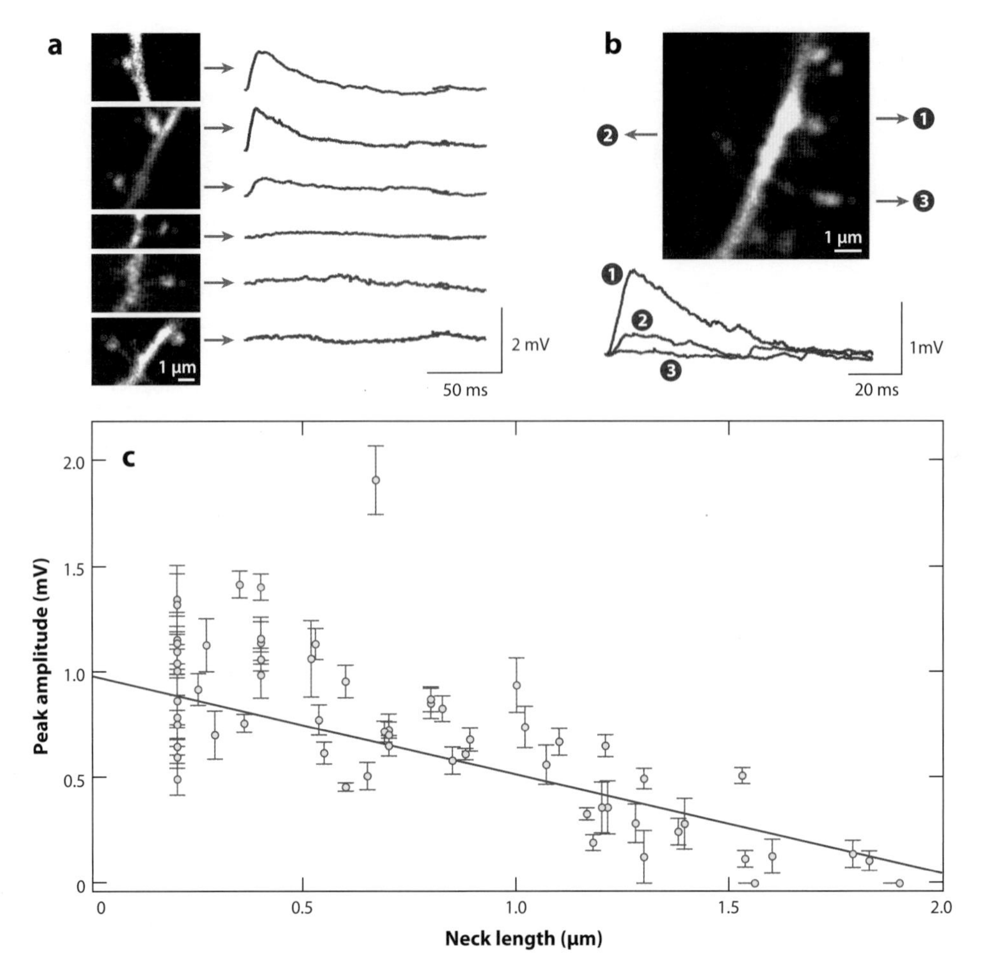

Figure 5

Effect of spine neck on spine potentials generated by two-photon glutamate uncaging. (*a*) Uncaging potentials from spines with short and long necks. Red dots indicate the site of uncaging, and red traces correspond to average uncaging potentials generated by different spines and measured at the soma. (*b*) Activation of three neighboring spines (1, 2, and 3) with different neck lengths. Note the large difference in their uncaging potentials at the soma. The shorter spine generates the stronger response, whereas activating the longer spine has no effect. Scale 1 µm. (*c*) Inverse correlation between uncaging potentials (peak amplitude) versus neck length. Line is linear regression of the data. Adapted from Araya et al. 2006a.

indicate that the spine neck resistance may not be significant to filter spine potentials (Harris & Stevens 1988, Koch & Zador 1993), calculations based on diffusional coupling vary widely (Bloodgood & Sabatini 2005, Svoboda et al. 1996); and some indicate Giga-Ohm neck resistances that could implement voltage filtering (Bloodgood & Sabatini 2005). Moreover, the spine neck, which is often depicted as a cylinder, is not a simple anatomical structure but often reveals constrictions where both membranes touch, as well as intracellular organelles that obstruct it (Arellano et al. 2007). The electrical consequences of these "pinches" and "plugs" are unknown but could significantly restrict ionic diffusion and transfer of electrical charges, particularly if the spine neck membrane lipids or proteins are charged and can screen ions.

Other mechanisms underlying the electrical compartmentalization in spines may derive from the presence of active conductances in the spine head or in the neck. In fact, dendrites are full of active conductances (Stuart et al. 1999), and it seems unlikely that spines would exclude dendritic channels, particularly because spines can emerge in a matter of seconds (Dunaevsky et al. 1999, Engert & Bonhoeffer 1999, Fischer et al. 1998, Kwon & Sabatini 2011). Indeed, evidence for the existence of active conductances in spines is accumulating from structural techniques, proteomics, and functional imaging assays. In addition to glutamatergic and γ-amino-butyric-acid (GABA)-ergic receptors on postsynaptic membranes (Nusser et al. 1997, 1998), ultrastructural techniques have also revealed sodium- and calcium-channel subunits in the spine cytoplasm (Caldwell et al. 2000, Mills et al. 1994; although see Lorincz & Nusser 2010). Researchers have also found potassium and nonselective channels in spine heads and necks, including SK2, G protein–coupled inwardly rectifying potassium channel (GIRK), Kv4.2, and hyperpolarization-activated cyclic nucleotide-gated (HCN) channels (Allen et al. 2011, Kim et al. 2007, Lin et al. 2008, Lujan et al. 2009, Ngo-Anh et al. 2005, Wang et al. 2007). Moreover, because postsynaptic densities (PSDs) are restricted to spines (Gray 1959), their biochemical analysis should reveal the spine protein complement. Indeed, proteomics analysis of PSD fractions demonstrates a large diversity of receptors and conductances, including subunits from essentially all dendritic channel families, including sodium channels (Cheng et al. 2006, Grant et al. 2004, Husi et al. 2000, Li et al. 2004, Walikonis et al. 2000, Yoshimura et al. 2004).

In addition to these structural methods, functional evidence from two-photon imaging also supports the existence of active conductances in spines. Aside from revealing glutamate receptors (Koester & Sakmann 1998, Kovalchuk et al. 2000, Matsuzaki et al. 2004, Yuste et al. 1999, Yuste & Denk 1995), calcium imaging has also revealed several types of voltage-sensitive calcium channels (VSCCs) (Yuste & Denk 1995), including T-type, L-type, N-type, R-type, P/Q-type, and low-voltage-activated (LVA) Ca^{2+} channels. (Bloodgood et al. 2009; Bloodgood & Sabatini 2007a,b; Sabatini et al. 2002; Sabatini & Svoboda 2000). Sodium imaging of spines under backpropagating APs has indicated that local sodium influx occurs in some spines (Rose et al. 1999), consistent with the specific effect of TTX on spine uncaging potentials (Araya et al. 2007) (**Figure 7**).

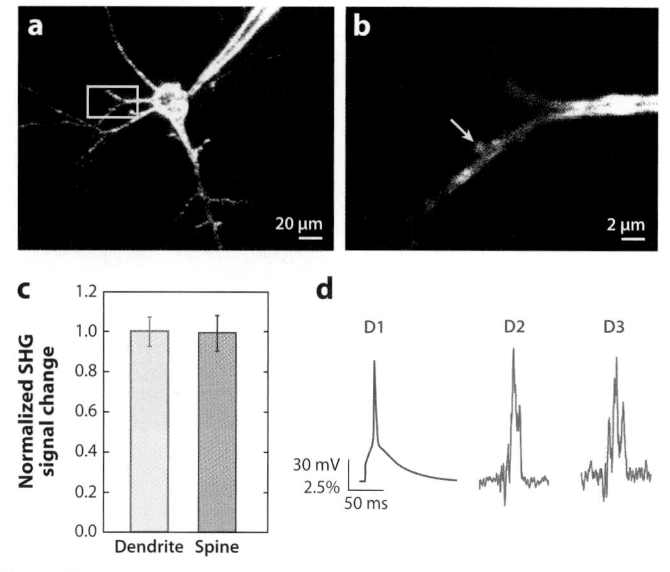

Figure 6

Imaging voltage in spines using second-harmonic generation (SHG). (*a*) SHG image of pyramidal neuron filled with the voltage-dependent SHG chromophore FM 4–64. (*b*) High-resolution image of a dendritic spine on the basal dendrite boxed in panel *a*. (*c*) Similar SHG voltage responses upon depolarizing voltage steps at spines and their parent dendrites. (*d*) SHG voltage measurements of spines during backpropagating APs. A single AP was initiated by current injection at the soma (D1), and SHG signals changes with similar amplitudes were measured at soma (D2) and dendritic spines (D3). Adapted from Nuriya et al. 2006.

FUNCTIONAL CONSEQUENCES OF ELECTRICAL COMPARTMENTALIZATION OF SPINES

As discussed, mounting evidence indicates that spines are not always isopotential with the dendritic shaft and that spines must be significantly more depolarized than dendritic shafts when activated by synaptic inputs. The exact spine

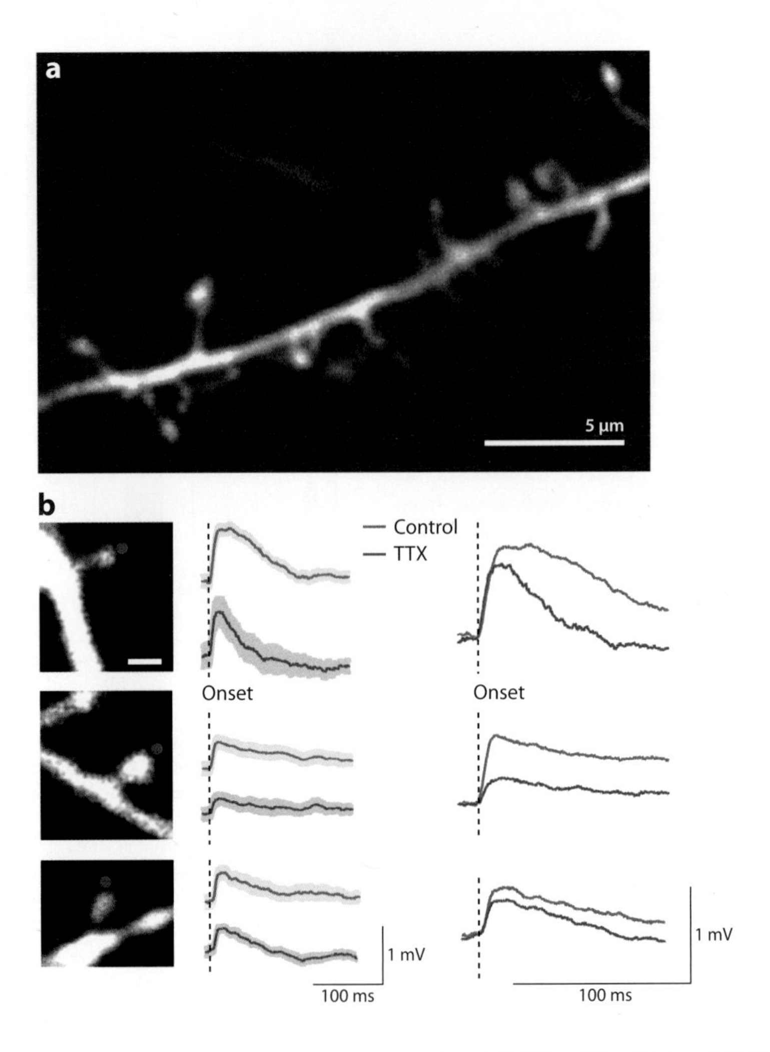

Figure 7

The sodium-channel blocker TTX reduces spine uncaging potentials. (*a*) Dendrite from a layer 5 pyramidal neuron, the spines of which were stimulated with two-photon glutamate uncaging. (*b*) Glutamate uncaging experiments. (*Left*) Red dots indicate the site of laser uncaging. (*Center*) Uncaging potentials under control conditions (*blue traces*) and in the presence of TTX (*red traces*). Dashed line indicates the time of uncaging onset. Thicker traces are an average of 10–15 depolarizations, and shaded areas illustrate standard error of the mean (SEM). (*Right*) Average uncaging potentials are superimposed. Note how TTX attenuates spine uncaging potentials. In similar experiments, TTX had no effect on uncaging potentials on dendritic shafts. Adapted from Araya et al. 2007.

voltage during uncaging activation or EPSPs could be as high as 20 mV, whereas at the dendritic shaft it is diminished to ~1 mV (Araya et al. 2006b, 2007; Harnett et al. 2012; Palmer & Stuart 2009). It seems paradoxical that once synaptic transmission is successful and activates postsynaptic glutamate receptors, its resulting depolarization becomes strongly curtailed. What could be the purpose of this reduction in synaptic potentials? Why would excitatory inputs need to be diminished as they traverse through the spine neck toward the dendritic shaft? In this section I briefly discuss the potential functional consequences of electrically isolating spines (see also Yuste 2010, 2011).

Enhancement of Input Integration

One functional advantage of electrically isolating spines, regardless of how it occurs, could be the local amplification of synaptic currents. Passive or active mechanisms could enhance the effect of EPSP at the spine, so a similar presynaptic dose of a neurotransmitter could generate a larger postsynaptic depolarization than if there was no local amplification of EPSPs. This could be advantageous for the neuron, by economizing postsynaptic receptors and ensuring a more reliable activation of receptors and conductances. For example, an enhanced spine depolarization could release the Mg^{2+} block of NMDA receptors (**Figure 4**), enable subthreshold calcium accumulations, and activate calcium, sodium, or potassium conductances (**Figures 3** and **7**). These processes could contribute to biochemical signaling and significantly shape postsynaptic depolarizations, thus providing potential mechanisms to modulate and modify it. For example, they could enable cooperative interaction between neighboring spines (Harnett et al. 2012).

But if the purpose of electrically isolating the spine is to amplify EPSPs, why then proceed to reduce the size of EPSPs as they invade the dendrite? One potential explanation of this paradox is that EPSP filtering prevents the electrical saturation of the neuron, a problem that must be particularly acute when neurons need to integrate thousands of inputs. Indeed, spines are normally found in neurons that receive large numbers of excitatory inputs, and electrically isolating spines could be a defense mechanism to prevent large conductance changes in the dendrite that could render it unexcitable, as in a shorted circuit (Llinás & Hillman 1969). Alternatively, spines could simply diminish the depolarization generated by each input so that more of them can be integrated before the neuron fires an AP. But if this is true, why not simply make synapses weaker directly, by reducing the number of glutamate receptors? Perhaps the initial amplification followed by filtering could permit the integration of more inputs without also increasing synaptic noise, because reducing the already very small number of receptors in each synapse could lead to excessive variability. Another advantage of using more receptors per synapse offers the possibility of encoding a large dynamic range of synaptic strengths. An alternative functional reason for EPSP amplitude reduction could be to devalue the additional electrical filtering caused by the dendritic tree, generating a dendrite where integration of the synapse would not matter; this process would help standardize the amplitude and kinetics of all EPSPs regardless of the synapse's position (Andersen et al. 1980, Gulledge et al. 2012, Konur et al. 2003, Magee & Cook 2000, Yuste 2011). Finally, an advantage of electrically isolating spines could be to enable the linear summation of EPSPs, as some models have predicted (Grunditz et al. 2008, Jack et al. 1975). In fact, excitatory inputs in spiny neurons are often integrated linearly (Cash & Yuste 1999); linear summation is found when glutamate uncaging stimulates neighboring spines but, interestingly, is not found when this stimulation is performed in the dendritic shafts (Araya et al. 2006a) (**Figure 8**). Altogether, a strategy with local amplification of EPSPs, followed by their attenuation and linear summation, would be of great functional advantage to the neuron, as a sort of biophysical homeostasis, if it needed to integrate a maximum number of inputs and add their values accurately without interference or cross talk between them. It makes perfect sense.

Regulation of Synaptic Plasticity

An alternative functional advantage for the electrical compartmentalization of spines is that its regulation enables precise control of synaptic strength. This process could occur by modifying the amplification of EPSPs at the spine head, by altering the activation of spine conductances, or by altering the spine neck/dendritic shaft electrical coupling by either active or passive mechanisms.

The idea that changes in the spine neck control synaptic plasticity was proposed already in the first discussions of spine electrical properties (Chang 1952, Rall 1978). In fact, ample

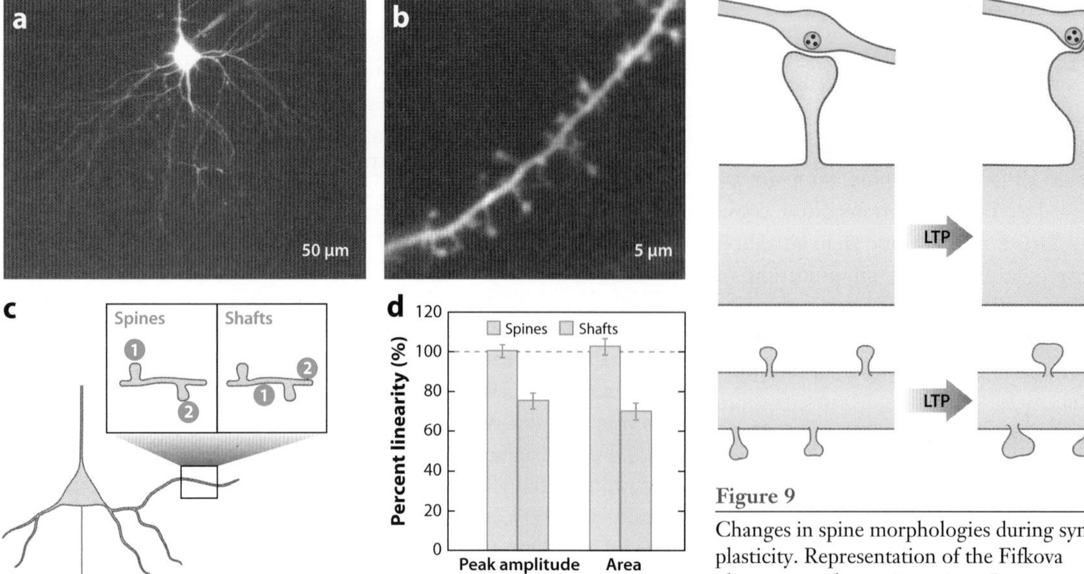

Figure 8

Linear summation of uncaging potentials on spines but not on dendritic shafts. (*a*) Layer 5 pyramidal cell filled with a fluorescence dye. (*b*) Basal dendrite selected for uncaging. (*c*) Protocol for testing input summation. Blue and red dots indicate the site of uncaging in spines or shaft locations, respectively. Uncaging was performed first at each spine or shaft location (1 or 2) and then simultaneously in both spines or both shaft locations (1 + 2). (*d*) Linearity of summation of uncaging potentials, measured as a ratio of the peak amplitude, or area, of the combined stimulation to the expected values calculated by summing the two independent stimulations. Note how summation of two spine potentials is linear, but summation of two shaft potentials is already reduced by 30%. A stronger shunting is expected if more shaft locations are stimulated, generating a short-circuited dendrite. Adapted from Araya et al. 2006b.

Figure 9

Changes in spine morphologies during synaptic plasticity. Representation of the Fifkova ultrastructural reconstructions. Long-term potentiation (LTP) results in larger spine heads and shorter and wider spine necks. Adapted from Yuste 2010.

experimental evidence shows that long-term synaptic plasticity is associated with morphological changes in spines, which could generate changes in electrical compartmentalization (Yuste & Bonhoeffer 2001). For example, LTP leads to increases in spine head size and to spine neck shortening and widening (Fifkova & Anderson 1981, Matsuzaki et al. 2004, Van Harreveld & Fifkova 1975) (**Figure 9**). This morphological plasticity could increase the number of glutamate receptors in the PSD and decrease the spine neck resistance, both mechanisms leading to larger EPSPs. Moreover, spines are constantly experiencing morphological plasticity in vitro and in vivo (Bonhoeffer & Yuste 2002). This "motility" is actin based (Fischer et al. 1998) and can lead to major changes in spine shapes, with elongation, shortening, or even complete disappearance of the spine neck in a matter of seconds (Dunaevsky et al. 1999) (**Figure 10**). This morphological plasticity likely has an effect on synaptic strength, perhaps due to changes in the electrical properties of spines. This scenario highlights the potential importance of the cell-biological mechanisms that control the tension, shape, and motility of cellular protrusions in the function of the neuron (Hall 1994).

CONCLUSION: A SYNAPTIC DEMOCRACY FOR EMERGENT COMPUTATIONS

Spines are clearly electrical compartments, although the exact mechanisms and functional purpose of this are still unclear. This electrical compartmentalization may be advantageous for the neuron to integrate and better control the plasticity of large numbers of synaptic inputs. This could enable the circuit to function as a distributed neural network with a

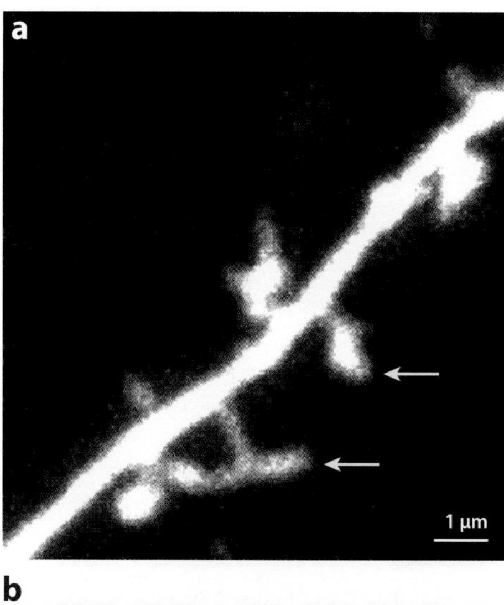

Figure 10

Spine motility. (*a*) Two-photon image of a pyramidal neuron from mouse visual cortex, labeled with green fluorescence protein (GFP). Diverse morphologies of dendritic protrusions can be observed. (*b*, *c*) Morphological changes in spines. Outlines of two spines (*yellow arrows in panel a*), shown at 2.5-min intervals. Note how the spines experience major changes in shape. Adapted from Konur & Yuste 2004a.

large synaptic matrix, where inputs would be integrated independently and linearly, as in a giant synaptic democracy (Yuste 2011). This view incorporates into a single functional framework traditional proposals for the function of spines, such as their role in the structural enhancement of connectivity or in the biochemical isolation necessary for input-specific synaptic plasticity (Chklovskii et al. 2002, Koch 1999, Peters & Kaiserman-Abramof 1969, Swindale 1981, Yuste 2010), because all these functions are actually necessary to create, and fully exploit, a distributed connectivity in the circuit (Yuste 2011). Spines being found in CNS circuits with large connectivity matrixes may not be coincidental because they may not be necessary for neurons that need to integrate few inputs, which can be accommodated in separate dendrites (Purves & Hume 1981).

Moreover, a distributed circuit could enable the brain to implement computational strategies where functional states are encoded at an emergent level of function, one based on the coordinated activity of many neurons, rather than at the single cell level. Indeed, the more distributed a circuit is, the less important the role that an individual neuron has. From this circuit-level viewpoint, spines would represent the anatomical signature of distributed neural networks, serving as their basic units of neuronal integration and plasticity and enabling them to function robustly. In fact, by ensuring the faithful integration and modification of all inputs, spines could greatly simplify and make more robust the function of these distributed circuits. Neural networks built with these simple elements, where each neuron adds input to lead to spiking, can implement Boolean logic (McCulloch & Pitts 1943), matrix multiplication, or vector remapping (Pellionisz & Llinas 1979) and perform relatively sophisticated computations such as associative memory (Hopfield 1982), optimization, or decision making (Hopfield & Tank 1986). It is fascinating to think that spines may illustrate an underlying simplicity hidden in the apparently complex morphological and functional design of neurons (Mead 1989).

NEW TECHNIQUES TO STUDY DENDRITIC SPINES

New methods are suggested to be more important for science than new discoveries or even new ideas (Brenner 2002). Indeed, research on spines illustrates very clearly the critical dependence of science on the introduction of novel methods, from their first description by the application of then-novel Golgi stain (Ramón y Cajal 1888), to the demonstration that they serve as synaptic units by the introduction of electron microscopy (Gray 1959), to the

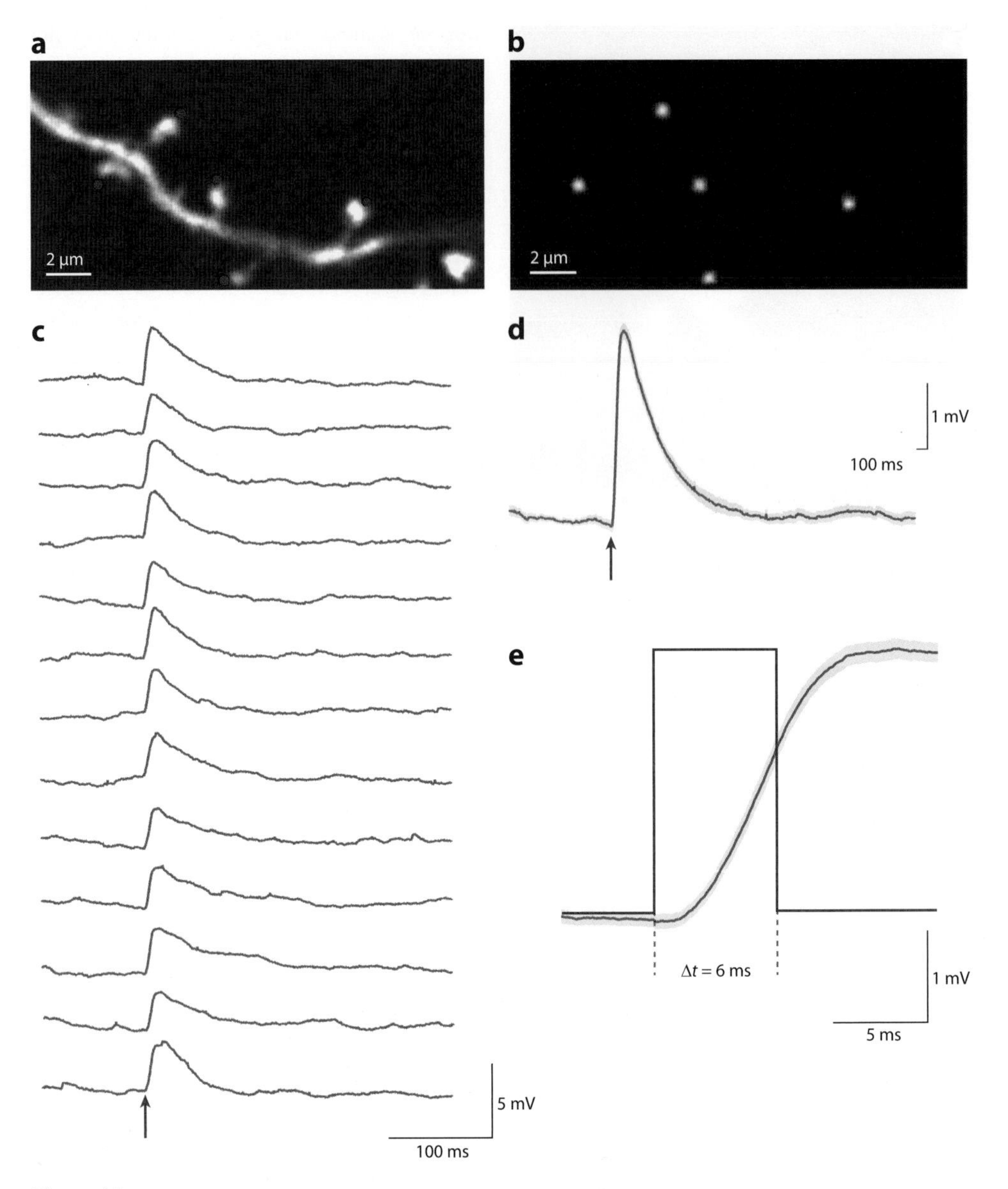

Figure 11

New techniques: holographic stimulation of groups of spines with a spatial light modulator (SLM). (*a*) Basal dendrite from a layer-5 pyramidal neuron in a mouse neocortical slice. Red dots indicate the sites chosen for simultaneous two-photon glutamate SLM uncaging. (*b*) Diffraction pattern of five uncaging spots next to the five spine heads, generated by sending to the SLM a Fourier transform of the image with the uncaging locations selected in panel *a*; (*c*) Whole-cell recording from the soma of the same cell during the experiment. Individual uncaging potentials generated after simultaneously uncaging glutamate (*red arrow*) next to the five spines shown in panels *a–d*. Average of the uncaging potentials shown in panel *c*. (*e*) Higher-resolution plot of response onset. Red trace represents the uncaging laser pulse. Blue trace is the average uncaging potential in panel *d*. Light blue areas in panels *d* and *e* are ± SEM. With SLM two-photon uncaging, one can "play the piano" with spines, activating them in an arbitrary spatiotemporal pattern. Adapted from Nikolenko et al. 2008.

THE MYSTERY OF THE LONG SPINES

The possibility that EPSPs are electrically reduced in the spine neck as they arrive into the dendritic shaft has interesting implications for spines that have longer necks because those spines appear to generate no significant depolarization at the soma (Araya et al. 2006b). Are long spines electrically silent, as reserved connections? Perhaps their necks become shorter and wider during synaptic plasticity protocols (**Figure 9**), "plugging in" the presynaptic neuron they represent into the circuit and enabling fast circuit switching. Interestingly, human neurons have not only higher spine densities but also spines with characteristically long necks (Benavides-Piccione et al. 2002, 2012) (**Figure 12**). If long-necked spines are reserve connections, they could reflect a higher degree of synaptic connectivity and plasticity by our brains. Indeed, Ramón y Cajal himself pointed out that human neurons are particularly spiny, and part of his interest in spines was motivated by his lifelong quest to understand the physical basis of human intelligence (Ramón y Cajal 1904, 1923). Moreover, longer "humanized" spines can be induced in mice by manipulating human-specific gene paralogs (Charrier et al. 2012). Finally, spines with abnormally long necks are prominent in mental retardation patients (Purpura 1974) and could be related to cognitive impairments.

more recent discoveries of their biochemical compartmentalization and motility using two-photon microscopy (Yuste & Denk 1995). Thus, our understanding of spine function may be further advanced by a series of upcoming novel technologies: novel voltage-imaging methods to image spine voltages (Nuriya et al. 2006, Peterka et al. 2011), two-photon optogenetics activation of individual spines (Packer et al. 2012), holographic stimulation of multiple spines using spatial light modulator (SLM) uncaging (Nikolenko et al. 2008) (**Figure 11**), or stimulated-emission-depletion (STED) microscopy and other super-resolution techniques to image the fine structure of living spines (Nagerl et al. 2008).

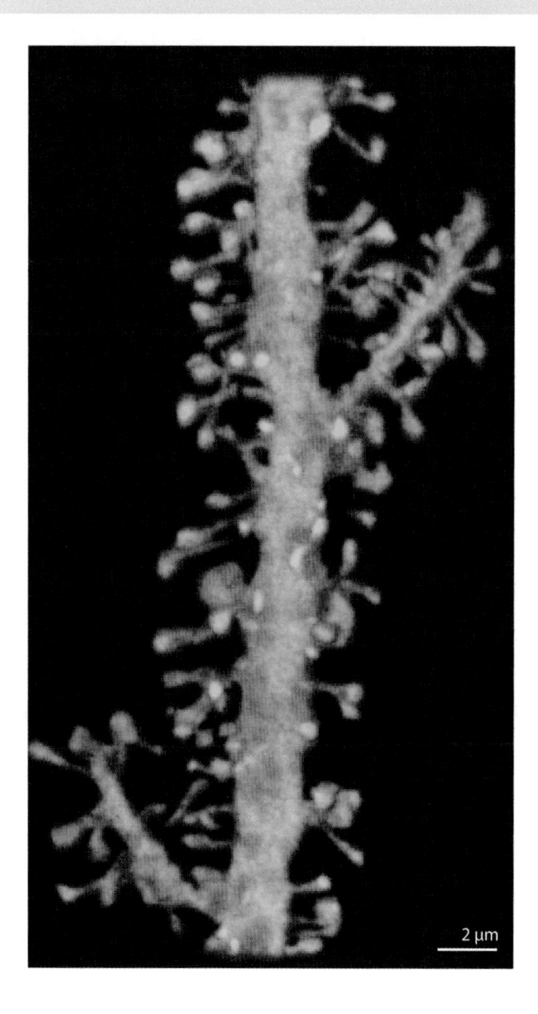

Figure 12

Long-necked spines in a human pyramidal neuron. Intracellularly injected layer-3 pyramidal neuron of a sample from a human cingulate cortex. Confocal image of an apical dendritic segment at ~100 μm from soma. Note the large density of spines and also the particularly long spine necks, both typical characteristics of human pyramidal neurons. Adapted from Yuste 2010. See also Benavides-Piccione et al. (2002, 2012).

SUMMARY POINTS

1. Dendritic spines can sustain membrane potential that differs from that of the dendrite.
2. This electrical compartmentalization could be due to passive and active biophysical mechanisms.
3. The electrical isolation of spines could help the neuron integrate and independently modulate the strength of large numbers of synaptic inputs and implement emergent-level computations.

DISCLOSURE STATEMENT

The author is not aware of any affiliations, memberships, funding, or financial holdings that might be perceived as affecting the objectivity of this review.

ACKNOWLEDGMENTS

The author thanks L. Abbott, P. Adams, S. Golob, and L. Luo for their comments and the Kavli Institute for Brain Science, NEI, NINDS, NIHM, NIDA, Keck Foundation, and NARSAD for their support. This material is based on work supported by, or in part by, the US Army Research Laboratory and the US Army Research Office under contract number W911NF-12-1-0594.

LITERATURE CITED

Acker CD, Yan P, Loew LM. 2011. Single-voxel recording of voltage transients in dendritic spines. *Biophys. J.* 101:L11–13

Allen D, Bond CT, Lujan R, Ballesteros-Merino C, Lin MT, et al. 2011. The SK2-long isoform directs synaptic localization and function of SK2-containing channels. *Nat. Neurosci.* 14:744–49

Andersen P, Jansen JKS, eds. 1970. *Excitatory Synaptic Mechanisms. Proc. 5th Int. Meet. Neurobiol.* Oslo: Universitetsforlaget

Andersen P, Silfvenius H, Sundberg SH, Sveen O. 1980. A comparison of distal and proximal dendritic synapses on CA1 pyramids in guinea-pig hippocampal slices in vitro. *J. Physiol.* 307:273–99

Araya R, Eisenthal KB, Yuste R. 2006a. Dendritic spines linearize the summation of excitatory potentials. *Proc. Natl. Acad. Sci. USA* 103:18779–804

Araya R, Jiang J, Eisenthal KB, Yuste R. 2006b. The spine neck filters membrane potentials. *Proc. Natl. Acad. Sci. USA* 103:17961–66

Araya R, Nikolenko V, Eisenthal KB, Yuste R. 2007. Sodium channels amplify spine potentials. *Proc. Natl. Acad. Sci. USA* 104:12347–52

Arellano JI, Benavides-Piccione R, DeFelipe J, Yuste R. 2007. Ultrastructure of dendritic spines: correlation between synaptic and spine morphologies. *Front. Neurosci.* 1:131–43

Baer SM, Rinzel J. 1991. Propagation of dendritic spikes mediated by excitable spines: a continuum theory. *J. Neurophysiol.* 65:874–90

Bekkers JM, Richerson GB, Stevens CF. 1990. Origin of variability in quantal size in cultured hippocampal neurons and hippocampal slices. *Proc. Natl. Acad. Sci. USA* 87:5359–62

Benavides-Piccione R, Ballesteros-Yañez I, DeFelipe J, Yuste R. 2002. Cortical area and species differences in dendritic spine morphology. *J. Neurocytol.* 31:337–46

Benavides-Piccione R, Fernaud-Espinosa I, Robles V, Yuste R, Defelipe J. 2012. Age-based comparison of human dendritic spine structure using complete three-dimensional reconstructions. *Cerebral Cortex.* In press

Bloodgood BL, Giessel AJ, Sabatini BL. 2009. Biphasic synaptic Ca influx arising from compartmentalized electrical signals in dendritic spines. *PLoS Biol.* 7:e1000190

Bloodgood BL, Sabatini BL. 2005. Neuronal activity regulates diffusion across the neck of dendritic spines. *Science* 310:866–69

Bloodgood BL, Sabatini BL. 2007a. Ca(2+) signaling in dendritic spines. *Curr. Opin. Neurobiol.* 17:345–51

Bloodgood BL, Sabatini BL. 2007b. Nonlinear regulation of unitary synaptic signals by CaV(2.3) voltage-sensitive calcium channels located in dendritic spines. *Neuron* 53:249–60

Bonhoeffer T, Yuste R. 2002. Spine motility. Phenomenology, mechanisms, and function. *Neuron* 35:1019–27

Brenner S. 2002. Life sentences: ontology recapitulates philology. *Genome Biol.* 3:COMMENT1006

Caldwell J, Schallwe K, Lasher R, Peles E, Levisnon S. 2000. Sodium channel Na(v)1.6 is localized at nodes of ranvier, dendrites, and synapses. *Proc. Natl. Acad. Sci. USA* 97:5616–20

Cash S, Yuste R. 1999. Linear summation of excitatory inputs by CA1 pyramidal neurons. *Neuron* 22:383–94

Chang HT. 1952. Cortical neurons with particular reference to the apical dendrite. *Cold Spring Harb. Symp. Quant. Biol.* 17:189–202

Charrier C, Joshi K, Coutinho-Budd J, Kim JE, Lambert N, et al. 2012. Inhibition of SRGAP2 function by its human-specific paralogs induces neoteny during spine maturation. *Cell* 149:923–35

Cheng D, Hoogenraad CC, Rush J, Ramm E, Schlager MA, et al. 2006. Relative and absolute quantification of postsynaptic density proteome isolated from rat forebrain and cerebellum. *Mol. Cell Proteomics* 5:1158–70

Chklovskii DB, Schikorski T, Stevens CF. 2002. Wiring optimization in cortical circuits. *Neuron* 34:341–47

Cohen L. 1989. Special topic: Optical approaches to neuron function. *Annu. Rev. Physiol.* 51:487–90

Coss RG, Perkel DH. 1985. The function of dendritic spines: a review of theoretical issues. *Behav. Neural Biol.* 44:151–85

Crick F. 1982. Do spines twitch? *Trends Neurosci.* 5:44–46

DeFelipe J, Hendry SH, Jones EG. 1989. Synapses of double bouquet cells in monkey cerebral cortex visualized by calbindin immunoreactivity. *Brain Res.* 503:49–54

Diamond J, Gray EG, Yasargil GM. 1970. The function of dendritic spines: a hypothesis. See Andersen & Jansen 1970, pp. 213–22

Dunaevsky A, Tashiro A, Majewska A, Mason C, Yuste R. 1999. Developmental regulation of spine motility in the mammalian central nervous system. *Proc. Natl. Acad. Sci. USA* 96:13438–43

Engert F, Bonhoeffer T. 1999. Dendritic spine changes associated with hippocampal long-term synaptic plasticity. *Nature* 399:66–70

Fifkova E, Anderson CL. 1981. Stimulation-induced changes in dimensions of stalks of dendritic spines in the dentate molecular layer. *Exp. Neurol.* 74:621–27

Fischer M, Kaech S, Knutti D, Matus A. 1998. Rapid actin-based plasticity in dendritic spine. *Neuron* 20:847–54

Goldberg JH, Tamas G, Aronov D, Yuste R. 2003. Calcium microdomains in aspiny dendrites. *Neuron* 40:807–21

Grant SGN, Husi H, Choudhary J, Cumiskey M, Blackstock W, Armstrong JD. 2004. The organization and integrative function of the post-synaptic proteome. In *Excitatory-Inhibitory Balance, Synapses, Circuits, Systems*, ed. TK Hensch, M Fagiolini, pp. 13–44. New York: Kluwer Acad./Plenum

Gray EG. 1959. Electron microscopy of synaptic contacts on dendritic spines of the cerebral cortex. *Nature* 183:1592–94

Grunditz A, Holbro N, Tian L, Zuo Y, Oertner TG. 2008. Spine neck plasticity controls postsynaptic calcium signals through electrical compartmentalization. *J. Neurosci.* 28:13457–66

Gulledge AT, Carnevale NT, Stuart GJ. 2012. Electrical advantages of dendritic spines. *PLoS One* 7:e36007

Hall A. 1994. Small GTP-binding proteins and the regulation of the actin cytoskeleton. *Annu. Rev. Cell Biol.* 10:31–54

Harnett MT, Makara JK, Spruston N, Kath WL, Magee JC. 2012. Synaptic amplification by dendritic spines enhances input cooperativity. *Nature* 491:599–602

Harris KM, Kater SB. 1994. Dendritic spines: cellular specializations imparting both stability and flexibility to synaptic function. *Annu. Rev. Neurosci.* 17:341–71

Harris KM, Stevens JK. 1988. Dendritic spines of rat cerebellar Purkinje cells: serial electron microscopy with reference to their biophysical characteristics. *J. Neurosci.* 8:4455–69

Holthoff K, Tsay D, Yuste R. 2002. Calcium dynamics in spines depend on their dendritic position. *Neuron* 33:425–37

Holthoff K, Zecevic D, Konnerth A. 2010. Rapid time course of action potentials in spines and remote dendrites of mouse visual cortex neurons. *J. Physiol.* 588:1085–96

Hopfield JJ. 1982. Neural networks and physical systems with emergent collective computational abilities. *Proc. Natl. Acad. Sci. USA* 79:2554–58

Hopfield JJ, Tank DW. 1986. Computing with neural circuits: a model. *Science* 233:625–33

Husi H, Ward MA, Choudhary JS, Blackstock WP, Grant SG. 2000. Proteomic analysis of NMDA receptor-adhesion protein signaling complexes. *Nat. Neurosci.* 3:661–69

Jack JJB, Noble D, Tsien RW. 1975. *Electric Current Flow in Excitable Cells*. London: Oxford Univ. Press

Jaslove SW. 1992. The integrative properties of spiny distal dendrites. *Neuroscience* 47:495–519

Jones EG, Powell TP. 1969. Morphological variations in the dendritic spines of the neocortex. *J. Cell Sci.* 5:509–29

Kim J, Jung S, Clemens A, Petralia R, Hoffman D. 2007. Regulation of dendritic excitability by activity-dependent trafficking of the A-type K+ channel subunit Kv4.2 in hippocampal neurons. *Neuron* 54: 933–47

Koch C. 1999. Dendritic spines. In *Biophysics of Computation*, ed. C Koch, pp. 280–308. New York: Oxford Univ. Press

Koch C, Poggio T. 1983a. Electrical properties of dendritic spines. *Trends Neurosci.* 6:80–83

Koch C, Poggio T. 1983b. A theoretical analysis of electrical properties of spines. *Proc. R. Soc. Lond. B* 213: 455–77

Koch C, Poggio T, Torre V. 1983. Nonlinear interactions in a dendritic tree: localization, timing and role in information processing. *Proc. Natl. Acad. Sci. USA* 80:2799–802

Koch C, Zador A. 1993. The function of dendritic spines: devices subserving biochemical rather than electrical compartmentalization. *J. Neurosci.* 13:413–22

Koester HJ, Sakmann B. 1998. Calcium dynamics in single spines during coincident pre- and postsynaptic activity depend on relative timing of back-propagating action potentials and subthreshold excitatory postsynaptic potentials. *Proc. Natl. Acad. Sci. USA* 95:9596–601

Konur S, Rabinowitz D, Fenstermaker VL, Yuste R. 2003. Systematic regulation of spine sizes and densities in pyramidal neurons. *J. Neurobiol.* 56:95–112

Konur S, Yuste R. 2004. Developmental regulation of spine and filopodial motility in primary visual cortex: reduced effects of activity and sensory deprivation. *J. Neurobiol.* 59:236–46

Kovalchuk Y, Eilers J, Lisman J, Konnerth A. 2000. NMDA receptor-mediated subthreshold Ca(2+) signals in spines of hippocampal neurons. *J. Neurosci.* 20:1791–99

Kwon HB, Sabatini BL. 2011. Glutamate induces de novo growth of functional spines in developing cortex. *Nature* 474:100–4

Lewis A, Khatchatouriants A, Treinin M, Chen Z, Peleg G, et al. 1999. Second-harmonic generation of biological interfaces: probing the membrane protein bacteriorhodopsin and imaging membrane potential around GFP molecules at specific sites in neuronal cells of *C. elegans*. *Chem. Phys.* 245:133–44

Li KW, Hornshaw MP, Van der Schors RC, Watson R, Tate S, et al. 2004. Proteomics analysis of rat brain postsynaptic density. *J. Biol. Chem.* 279:987–1002

Lin MT, Lujan R, Watanabe M, Adelman JP, Maylie J. 2008. SK2 channel plasticity contributes to LTP at Schaffer collateral-CA1 synapses. *Nat. Neurosci.* 11:170–77

Llinás R, Hillman DE. 1969. Physiological and morphological organization of the cerebellar circuits in various vertebrates. In *Neurobiology of Cerebellar Evolution and Development*, ed. R Llinas, pp. 43–73. Chicago: Am. Med. Assoc. Educ. Res. Found.

Llinás R, Nicholson C. 1971. Electroresponsive properties of dendrites and somata in alligator Purkinje cells. *J. Neurophysiol.* 34:532–51

Lorincz A, Nusser Z. 2010. Molecular identity of dendritic voltage-gated sodium channels. *Science* 328:906–9

Lujan R, Maylie J, Adelman JP. 2009. New sites of action for GIRK and SK channels. *Nat. Rev. Neurosci.* 10:475–80

Magee JC, Cook EP. 2000. Somatic EPSP amplitude is independent of synapse location in hippocampal pyramidal neurons. *Nat. Neurosci.* 3:895–903

Matsuzaki M, Ellis-Davies GC, Nemoto T, Miyashita Y, Iino M, Kasai H. 2001. Dendritic spine geometry is critical for AMPA receptor expression in hippocampal CA1 pyramidal neurons. *Nat. Neurosci.* 4:1086–92

Matsuzaki M, Honkura N, Ellis-Davies GC, Kasai H. 2004. Structural basis of long-term potentiation in single dendritic spines. *Nature* 429:761–66

McCulloch WS, Pitts W. 1943. A logical calculus of the ideas immanent in nervous activity. *Bull. Math. Biol.* 52:99–115; discussion 73–97

Mead C. 1989. *Analog VLSI and Neural Systems*. Reading, MA: Addison-Wesley

Millard AC, Lewis A, Loew L. 2005. Second harmonic imaging of membrane potential. In *Imaging in Neuroscience and Development*, ed. R Yuste, A Konnerth, pp. 463–74. Cold Spring Harbor, NY: Cold Spring Harbor Press

Miller JP, Rall W, Rinzel J. 1985. Synaptic amplification by active membrane in dendritic spines. *Brain Res.* 325:325–30

Mills LR, Niesen CE, So AP, Carlen PL, Spigelman I, Jones OT. 1994. N-type Ca2+ channels are located on somata, dendrites, and a subpopulation of dendritic spines on live hippocampal pyramidal neurons. *J. Neurosci.* 14:6815–24

Nagerl UV, Willig KI, Hein B, Hell SW, Bonhoeffer T. 2008. Live-cell imaging of dendritic spines by STED microscopy. *Proc. Natl. Acad. Sci. USA* 105:18982–87

Nemet BA, Nikolenko V, Yuste R. 2004. Second harmonic imaging of membrane potential of neurons with retinal. *J. Biomed. Opt.* 9:873–81

Ngo-Anh TA, Bloodgood BL, Lin M, Sabatini BL, Maylie J, Adelman JP. 2005. SK channels and NMDA receptors form a Ca(2+)-mediated feedback loop in dendritic spines. *Nat. Neurosci.* 8:642–49

Nikolenko V, Watson BO, Araya R, Woodruff A, Peterka DS, Yuste R. 2008. SLM microscopy: scanless two-photon imaging and photostimulation with spatial light modulators. *Front. Neural Circuits* 2:1–14

Nuriya M, Jiang J, Nemet B, Eisenthal KB, Yuste R. 2006. Imaging membrane potential in dendritic spines. *Proc. Natl. Acad. Sci. USA* 103:786–90

Nusser Z, Cull-Candy S, Farrant M. 1997. Differences in synaptic GABA(A) receptor number underlie variation in GABA mini amplitude. *Neuron* 19:697–709

Nusser Z, Lujan R, Laube G, Roberts J, Molnar E, Somogyi P. 1998. Cell type and pathway dependence of synaptic AMPA receptor number and variability in the hippocampus. *Neuron* 21:545–59

O'Brien J, Unwin N. 2006. Organization of spines on the dendrites of Purkinje cells. *Proc. Natl. Acad. Sci. USA* 103:1575–80

Packer AM, Peterka DS, Hirtz JJ, Prakash R, Deisseroth K, Yuste R. 2012. Two-photon optogenetics of dendritic spines and neural circuits. *Nat. Methods* 9:1202–5

Palmer LM, Stuart GJ. 2009. Membrane potential changes in dendritic spines during action potentials and synaptic input. *J. Neurosci.* 29:6897–903

Pellionisz A, Llinas R. 1979. Brain modeling by tensor network theory and computer simulation. The cerebellum: distributed processor for predictive coordination. *Neuroscience* 4:323–48

Perkel DH. 1982. Functional role of dendritic spines. *J. Physiol.* 78:695–99

Perkel DH, Perkel DJ. 1985. Dendritic spines: role of active membrane in modulating synaptic efficacy. *Brain Res.* 325:331–35

Peterka DS, Takahashi H, Yuste R. 2011. Imaging voltage in neurons. *Neuron* 69:9–21

Peters A, Kaiserman-Abramof IR. 1969. The small pyramidal neuron of the rat cerebral cortex. The synapses upon dendritic spines. *Z. Zellforsch. Mikrosk. Anat.* 100:487–506

Purpura D. 1974. Dendritic spine "dysgenesis" and mental retardation. *Science* 186:1126–28

Purves D, Hume RI. 1981. The relation of postsynaptic geometry to the number of presynaptic axons that innervate autonomic ganglion cells. *J. Neurosci.* 1:441–52

Rall W. 1970. Cable properties of dendrites and effects of synaptic location. See Andersen & Jansen 1970, pp. 175–87

Rall W. 1974. Dendritic spines, synaptic potency and neuronal plasticity. In *Cellular Mechanisms Subserving Changes in Neuronal Activity*, ed. CD Woody, KA Brown, TJ Crow, JD Knispel, pp. 13–21. Los Angeles: Brain Inf. Serv.

Rall W. 1978. Dendritic spines and synaptic potency. In *Studies in Neurophysiology*, ed. R Porter, pp. 203–9. New York: Cambridge Univ. Press

Rall W, Rinzel J. 1971. Dendritic spine function and synaptic attenuation calculations. *Soc. Neurosci. Abstr.* 1:64

Rall W, Segev I. 1987. Functional possibilities for synapses on dendrites and on dendritic spines. In *Synaptic Function*, ed. GE Edelman, WF Gall, WM Cowan, pp. 605–37. New York: Wiley

Rall W, Segev I. 1988. Excitable dendritic spine clusters: nonlinear synaptic processing. In *Computer Simulation in Brain Science*, ed. RMJ Cotterill, pp. 26–43. Cambridge, UK: Cambridge Univ. Press

Ramón y Cajal S. 1888. Estructura de los centros nerviosos de las aves. *Rev. Trim. Histol. Norm. Patol.* 1:1–10

Ramón y Cajal S. 1904. *Textura del Sistema Nerviosa del Hombre y los Vertebrados*, Vols. 2. Madrid: Moya

Ramón y Cajal S. 1923. *Recuerdos de mi Vida: Historia de mi Labor Científica.* Madrid: Alianza Ed.

Rose C, Kovalchuk Y, Eilers J, Konnerth A. 1999. Two-photon Na+ imaging in spines and fine dendrites of central neurons. *Pflugers Arch.* 439:201–7

Sabatini BL, Oertner TG, Svoboda K. 2002. The life cycle of Ca(2+) ions in dendritic spines. *Neuron* 33:439–52

Sabatini BL, Svoboda K. 2000. Analysis of calcium channels in single spines using optical fluctuation analysis. *Nature* 408:589–93

Segev I, Rall W. 1988. Computational study of an excitable dendritic spine. *J. Neurophysiol.* 60:499–523

Shen H, Sabaliauskas N, Sherpa A, Fenton AA, Stelzer A, et al. 2010. A critical role for alpha4betadelta GABAA receptors in shaping learning deficits at puberty in mice. *Science* 327:1515–18

Shepherd GM. 1996. The dendritic spine: a multifunctional integrative unit. *J. Neurophysiol.* 75:2197–210

Shepherd GM, Brayton RK. 1987. Logic operations are properties of computer-simulated interactions between excitable dendritic spines. *Neuroscience* 21:151–65

Shepherd GM, Brayton RK, Miller JP, Segev I, Rinzel J, Rall W. 1985. Signal enhancement in distal cortical dendrites by means of interactions between active dendritic spines. *Proc. Natl. Acad. Sci. USA* 82:2192–95

Soler-Llavina GJ, Sabatini BL. 2006. Synapse-specific plasticity and compartmentalized signaling in cerebellar stellate cells. *Nat. Neurosci.* 9:798–806

Stuart G, Spruston N, Hausser M, eds. 1999. *Dendrites*. New York: Oxford Univ. Press

Stuart GJ, Sakmann B. 1994. Active propagation of somatic action potentials into neocortical pyramidal cell dendrites. *Nature* 367:69–72

Svoboda K, Tank DW, Denk W. 1996. Direct measurement of coupling between dendritic spines and shafts. *Science* 272:716–19

Swindale NV. 1981. Dendritic spines only connect. *Trends Neurosci.* 4:240–41

Tsay D, Yuste R. 2002. Role of dendritic spines in action potential backpropagation: a numerical simulation study. *J. Neurophysiol.* 88:2834–45

Tsay D, Yuste R. 2004. On the electrical function of spines. *Trends Neurosci.* 27:77–83

Van Harreveld A, Fifkova E. 1975. Swelling of dendritic spines in the fascia dentata after stimulation of the perforant fibers as a mechanism of post-tetanic potentiation. *Exp. Neurol.* 49:736–49

Walikonis RS, Jensen ON, Mann M, Provance DW Jr, Mercer JA, Kennedy MB. 2000. Identification of proteins in the postsynaptic density fraction by mass spectrometry. *J. Neurosci.* 20:4069–80

Wang M, Ramos BP, Paspalas CD, Shu Y, Simen A, et al. 2007. Alpha2A-adrenoceptors strengthen working memory networks by inhibiting cAMP-HCN channel signaling in prefrontal cortex. *Cell* 129:397–410

Wickens J. 1988. Electrically coupled but chemically isolated synapses: dendritic spines and calcium in a rule for synaptic modification. *Prog. Neurobiol.* 31:507–28

Williams SR, Mitchell SJ. 2008. Direct measurement of somatic voltage clamp errors in central neurons. *Nat. Neurosci.* 11:790–98

Wilson CJ. 1986. Three dimensional analysis of dendritic spines by means of HVEM. *J. Electron. Microsc.* 35(Suppl.):1151–55

Yoshimura Y, Yamauchi Y, Shinkawa T, Taoka M, Donai H, et al. 2004. Molecular constituents of the post-synaptic density fraction revealed by proteomic analysis using multidimensional liquid chromatography-tandem mass spectrometry. *J. Neurochem.* 88:759–68

Yuste R. 2010. *Dendritic Spines*. Cambridge, MA: MIT Press

Yuste R. 2011. Dendritic spines and distributed circuits. *Neuron* 71:772–81

Yuste R, Bonhoeffer T. 2001. Morphological changes in dendritic spines associated with long-term synaptic plasticity. *Annu. Rev. Neurosci.* 24:1071–89

Yuste R, Denk W. 1995. Dendritic spines as basic units of synaptic integration. *Nature* 375:682–84

Yuste R, Majewska A, Cash SS, Denk W. 1999. Mechanisms of calcium influx into hippocampal spines: heterogeneity among spines, coincidence detection by NMDA receptors, and optical quantal analysis. *J. Neurosci.* 19:1976–87

Yuste R, Majewska A, Holthoff K. 2000. From form to function: calcium compartmentalization in dendritic spines. *Nat. Neurosci.* 3:653–59

Yuste R, Tank DW. 1996. Dendritic integration in mammalian neurons, a century after Cajal. *Neuron* 16:701–16

Prefrontal Contributions to Visual Selective Attention

Ryan F. Squire,[1] Behrad Noudoost,[1] Robert J. Schafer,[1] and Tirin Moore[1,2]

[1]Department of Neurobiology and [2]Howard Hughes Medical Institute, Stanford University School of Medicine, Stanford, California 94305; email: tirin@stanford.edu

Annu. Rev. Neurosci. 2013. 36:451–66

The *Annual Review of Neuroscience* is online at neuro.annualreviews.org

This article's doi:
10.1146/annurev-neuro-062111-150439

Keywords

cognition, neural circuitry, executive functions, visual perception

Abstract

The faculty of attention endows us with the capacity to process important sensory information selectively while disregarding information that is potentially distracting. Much of our understanding of the neural circuitry underlying this fundamental cognitive function comes from neurophysiological studies within the visual modality. Past evidence suggests that a principal function of the prefrontal cortex (PFC) is selective attention and that this function involves the modulation of sensory signals within posterior cortices. In this review, we discuss recent progress in identifying the specific prefrontal circuits controlling visual attention and its neural correlates within the primate visual system. In addition, we examine the persisting challenge of precisely defining how behavior should be affected when attentional function is lost.

Contents

NEURAL CORRELATES OF VISUAL ATTENTION

Attention describes the basic cognitive function in which behaviorally relevant information is selected in favor of irrelevant information for further sensory processing and for the guidance of behavioral responses. As William James (1890, pp. 403–4) classically described, attention involves the "...withdrawal from some things in order to deal effectively with others." Neurophysiological investigations have established that in the visual modality, attention involves the amplification of neuronal representations corresponding to selected targets at the expense of other representations (Noudoost et al. 2010). Investigators have observed attention-dependent modulation of visually driven neural activity across multiple stages of the primate visual system, primarily in the form of increases in firing rate (Reynolds & Chelazzi 2004). In more recent years, neurophysiological studies have found that, in addition to increasing firing rates, attention may enhance signaling efficacy for targets by changing other parameters of visually driven neural activity. These parameters include decreases in the trial-to-trial variability of spiking activity (Mitchell et al. 2007), decreases in low-frequency correlated variability across neuronal ensembles (Cohen & Maunsell 2009), and increases in the phase locking of spiking responses with particular frequency components of local field potentials (LFPs) (Fries et al. 2001), as well as shifts and reductions in the breadth of visual receptive fields (RFs) (Connor et al. 1997). Although we still need to determine which (if any) of these parameters, including firing rate, is most strongly correlated with the behavioral effects of attention, it is nonetheless clear that attentional deployment is associated with robust changes to the signaling properties of neurons within the visual system.

Quite separate from the question of how attention changes the fidelity of visual signals in the brain is the more basic question of which mechanism initiates those changes. Although it is overwhelmingly apparent that attention modulates the visual responses of neurons across many stages of the visual system, the neural circuits underlying that modulation remain unknown. The causal basis of visual selective attention has been an area of much research for well over a century, primarily in the context of lesion studies, yet its origin continues to be one of the more fundamental unresolved questions. As we discuss in the closing section of this review, part of the reason for this missing information is the difficulty in defining precisely what a loss of attention should look like. As a result, for example, competing hypotheses of a critical role of the parietal cortex (Bisley & Goldberg 2010), or of midbrain or thalamic structures (e.g., Shipp 2004), or of the prefrontal cortex (PFC) (Miller & Cohen 2001) in visual attention have persisted in parallel, and almost in spite of one another, for decades. Given that attention has many forms (e.g., spatial versus feature-based), these candidate structures may, in fact, contribute to these different forms in complementary (e.g., Buschman & Miller 2007) or redundant ways. Yet, whether any one of them, or perhaps even some other structure, is indeed a source of attentional modulation of visual signals remains to be determined. In this review, we focus on evidence that supports a causal role of PFC in visual selective attention. Although it seems unlikely that PFC is both a necessary and a sufficient source of attentional modulation in all its forms, recent work has nonetheless offered important new insights into what is likely a fundamental contribution of PFC to visual attention.

The PFC consists of a multitude of frontal cortical areas anterior to primary and association motor cortices, and it has long been implicated in high-level cognitive functions. In humans, the PFC occupies a much larger proportion of the cerebral cortex than in other species, which prompts the notion that it may contribute more to cognitive capacities unique to humans (Fuster 1995). Consistent with this notion is the evidence that subregions within the PFC play a crucial role in attention, including visual attention. In particular, two regions, the dorsolateral prefrontal cortex (dlPFC) and the frontal eye field (FEF), comprising Brodmann areas 8, 45 and 46, have been implicated (**Figure 1**). Beginning with the lesion studies in monkeys conducted by David Ferrier in the late nineteenth century, researchers have known that damage to this region of the PFC resulted in behavioral deficits believed to be consistent with a loss of attentional control (Ferrier 1876). More recently, studies in monkeys have yielded similar observations (Wardak et al. 2006), as have studies of human subjects with PFC damage (Knight et al. 1995, Rueckert & Grafman 1996). Moreover, similar to what is observed within the posterior visual cortex, neurophysiological studies of dlPFC and FEF have identified neural correlates of covert visual attention in both of these areas. As in posterior visual areas, visually driven responses in both areas are enhanced when attention is directed to stimuli within the neuronal RF (Lebedev et al. 2004, Thompson et al. 2005, Buschman & Miller 2007, Armstrong et al. 2009, Gregoriou et al. 2009). This enhancement is evident whether attention is directed voluntarily (top-down) (Buschman & Miller 2007, Armstrong et al. 2009, Gregoriou et al. 2009) or shifted to stimuli as the result of their greater salience

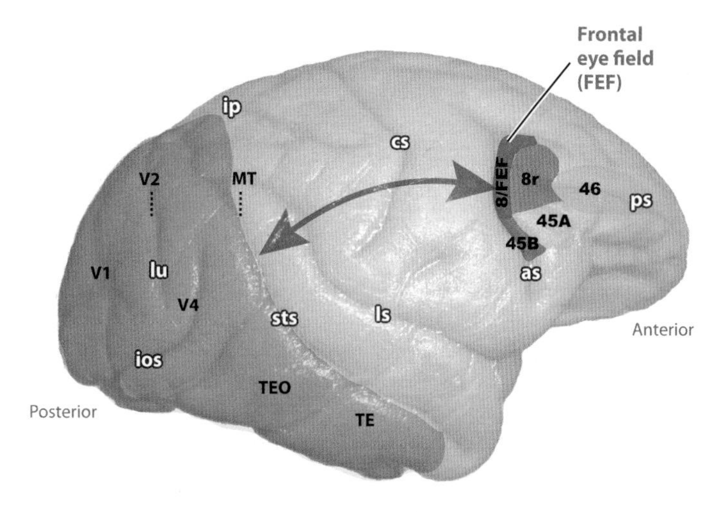

Figure 1

The prefrontal cortex (PFC) of the macaque monkey and its connections with the posterior visual cortex. Shown in the lateral view of the monkey cerebral cortex (right hemisphere) are areas of the PFC with known involvement in visual attention, particularly the frontal eye field (FEF, area 8) in the anterior bank of the arcuate sulcus and the dorsolateral prefrontal cortex (dlPFC) in the principal sulcus (area 46). The arrow between the PFC and the posterior visual cortex depicts the direct and indirect connections of the PFC to areas within the extrastriate cortex [V2, V4, MT (middle temporal), etc.]. Dotted lines denote cortical areas that are mostly buried within a sulcus. Abbreviations: as, arcuate sulcus; cs, central sulcus; ios, inferior occipital sulcus; ip, intraparietal sulcus; lu, lunate sulcus; ls, lateral sulcus; ps, principal sulcus; sts, superior temporal sulcus. PFC areas are shown as defined in Gerbella et al. 2010.

(bottom-up) (Buschman & Miller 2007). Combined with the lesion results, the evidence of neural correlates in these areas has prompted attempts to address the causal role of the PFC in visual attention using more circuit-specific approaches, particularly in the FEF.

GAZE CONTROL, VISUAL ATTENTION, AND THE FEF

Before studies well established that covert attention involves a widespread modulation of visual activity, researchers knew that saccadic eye movements (overt attention) involve their own form of visual response modulation. In their early studies of the visual and motor properties of neurons within the superficial layers of the superior colliculus (SC) in alert monkeys,

Wurtz & Goldberg (1972a,b) observed that the neuronal responses elicited by visual stimuli were enhanced when monkeys used those stimuli as the targets of saccades. Similar effects were observed subsequently within the FEF (Goldberg & Bushnell 1981) and within the posterior parietal cortex (Mountcastle et al. 1975). Later studies demonstrated that the presaccadic visual enhancement is also observed within the posterior visual cortex, specifically in area V4 (Fischer & Boch 1981), and the inferior temporal cortex (Chelazzi et al. 1993). These neurophysiological observations appear consistent with the long-appreciated relationship between visual spatial attention and gaze control. Gaze shifts, which are most often achieved by saccades, appear to occur in conjunction with shifts of visual attention, as shown by the decrement in target-detection thresholds observed near the end points of upcoming saccades (Hoffman & Subramaniam 1995, Peterson et al. 2004). In the 1980s, Rizzolatti and colleagues proposed a "premotor theory of attention," which hypothesized that the mechanisms responsible for spatial attention and the mechanisms involved in programming saccades are the same, but that in the covert case "the eyes are blocked at a certain peripheral stage" (Rizzolatti et al. 1987, p. 37). Subsequent studies demonstrated that visual detection and discrimination are in fact facilitated at the end points of saccades, even when subjects are instructed to attend elsewhere (Shepherd et al. 1986, Hoffman & Subramaniam 1995, Deubel & Schneider 1996). These results led to the hypothesis that the selection of objects for perceptual processing and the preparation of appropriate motor responses are controlled by a common mechanism. Later neurophysiological experiments provided crucial tests of that hypothesis.

The FEF provides an interface between the saccadic system, the representation of visual stimuli within posterior cortices, and the executive control functions of PFC. The FEF is the region from which contraversive saccadic eye movements can be elicited with electrical stimulation (Robinson & Fuchs 1969, Bruce et al. 1985). Subthreshold FEF stimulation, i.e., stimulation with currents below that required to evoke a saccade, nonetheless increases the likelihood that an animal will subsequently initiate the saccade represented by neurons at the stimulation site. That is, subthreshold FEF stimulation can bias saccade planning (Schiller & Tehovnik 2001). The dynamics of FEF spiking responses recorded during saccade tasks reveal a continuum of visual and movement functions among neurons within the FEF. Some neurons exhibit purely visual activity in response to the onset of a stimulus (visual neurons), and others respond exclusively before a saccade is initiated (movement neurons), although most FEF neurons (visuomovement neurons) exhibit a combination of visual and movement properties (Bruce & Goldberg 1985, Sommer & Wurtz 2000). The aforementioned finding of attentional modulation among FEF neurons includes the observation that only visual and visuomovement neurons, and not movement neurons, are modulated during covert attention (Thompson et al. 2005, Gregoriou et al. 2012).

Prompted by the psychophysical evidence of a link between saccades and attention, Moore and Fallah (Moore & Fallah 2001, Moore & Fallah 2004) examined whether manipulating neural activity within the FEF could affect the deployment of spatial attention. The authors stimulated FEF sites using subthreshold currents while monkeys monitored a target stimulus among distracters for a small change in luminance. On trials in which microstimulation occurred, monkeys were able to detect smaller luminance changes than they could on control trials. This effect was spatially and temporally specific: An increase in sensitivity was observed only if the target location matched the end point of saccades evoked from the microstimulation site, and the effect was strongest when onset of microstimulation immediately preceded, and temporally overlapped, the luminance change. Moreover, the magnitude of the change in sensitivity produced by microstimulation was comparable to removing the distracters altogether.

In addition to the behaviorally defined effects of FEF microstimulation on attention, a series of related studies found that FEF microstimulation alters the visual responses of neurons within the posterior visual cortex, specifically area V4. Moore & Armstrong (2003) found that subthreshold microstimulation of the FEF enhanced the visual responses in V4 neurons at retinotopically corresponding locations, whereas responses at other locations were suppressed. This modulation was stronger in the presence of distracters and was critically dependent on an overlap in the RF of the V4 neuron and end point of saccades evoked from the microstimulation site. The enhancement also depended on the placement of the visual stimulus precisely at the end point of evoked saccades and not merely anywhere within the larger V4 receptive field (Armstrong et al. 2006). In addition, the microstimulation-driven enhancement was larger for the V4 neuron's preferred stimulus than for a nonpreferred stimulus, resulting in an increase in the ability of a V4 cell to discriminate between preferred and nonpreferred stimuli (Armstrong & Moore 2007). Placing both a preferred and nonpreferred stimulus within a V4 neuron's RF produces a response that is intermediate in magnitude between its responses to either stimulus alone (Reynolds et al. 1999). The responses of V4 neurons to such competing RF stimuli could be biased toward one stimulus or the other with FEF microstimulation, depending on which stimulus was aligned with the stimulated FEF vector. This effect mirrors the known influence of voluntary attention on the responses of visual cortical neurons to competing RF stimuli (Moran & Desimone 1985, Reynolds et al. 1999). A subsequent study by Ekstrom et al. 2009 examined the influence of FEF microstimulation on visual cortical activity using functional magnetic resonance imaging (fMRI), thus allowing them to see effects in all visual areas. They observed modulations in visually driven BOLD responses throughout the visual cortex, including V1, and found that the impact of microstimulation depended on the presence of distracters.

CONTROL OF VISUAL CORTICAL MODULATION BY FEF NEURONS

The fact that FEF microstimulation produces attention-like modulation within area V4 suggests that FEF neurons themselves are the source of that modulation. But such a conclusion cannot be made. Electrical stimulation is known to antidromically activate neurons within areas projecting to the stimulated site, in addition to orthodromic activation of downstream neurons, and stimulation may even activate cells in remote regions whose axons pass in proximity to the electrode tip (see Clark et al. 2011 for review). Thus, the possibility remains that neurons antidromically activated by stimulation, for example those within parietal area LIP (lateral intraparietal) (Bisley & Goldberg 2010), are in fact those directly responsible for producing the observed modulation. In addition, the more recent observation of an increased coupling of spiking and gamma-band local field activity between the FEF and V4 during attention (Gregoriou et al. 2009) has been interpreted by some as evidence of a direct effect of FEF neurons on V4 (e.g., Anderson et al. 2011). Moreover, a Granger causality analysis of the direction of gamma-band local field potential increases suggests an early causal influence of the FEF on V4 and a later causal influence of the latter on the former (Gregoriou et al. 2009). Though elegant, these results, like the stimulation results, leave open the possibility that neurons within one or more other structures not being studied are in fact the ones driving modulation within both V4 and the FEF. Resolving the question of a direct influence of FEF neurons on visual cortical modulation instead requires testing whether changes in FEF neuronal activity are sufficient to bring about that modulation. Such a test was recently carried out by Noudoost & Moore (2011a).

Experimental and clinical evidence suggests that dopamine (DA) within the PFC plays an important role in cognitive functions, including attention (Ernst et al. 1998, Castellanos & Tannock 2002, Robbins & Arnsten 2009). Noudoost & Moore (2011a) exploited this evidence and hypothesized that perhaps DAergic activity within the FEF mediates the apparent influence that FEF neurons have on signals within the visual cortex. They reasoned that if DA plays a role in visual attention, then changes in DAergic activity within the FEF should in some way alter signals within the visual cortex. DAergic innervation of the PFC originates from neurons within the ventral midbrain, including those within the ventral tegmental area (VTA) (Björklund & Dunnett 2007). Compared with other subtypes, D1 receptors (D1Rs) are more abundant in the PFC and are believed to play a more prominent role in regulating cognitive functions (Lidow et al. 1991, Goldman-Rakic et al. 1992, Santana et al. 2009). Although the effects of DA on PFC neuron activity are rather complex, evidence from a variety of experimental approaches suggests that when acting via D1Rs, DA can alter the strength and reliability of converging excitatory (glutamatergic) synapses (Gao et al. 2001). This property suggests a means by which D1Rs could mediate the selection and maintenance of particular FEF signals and the influence of those signals on other areas.

To address the direct role of FEF neuronal activity in the modulation of visual cortical signals as well as the role of DA in mediating that modulation, Noudoost & Moore (2011a) studied the impact of manipulating D1R-mediated activity within the FEF on the visual responses of extrastriate area V4 neurons (**Figure 2*a***). Manipulation of D1R-mediated

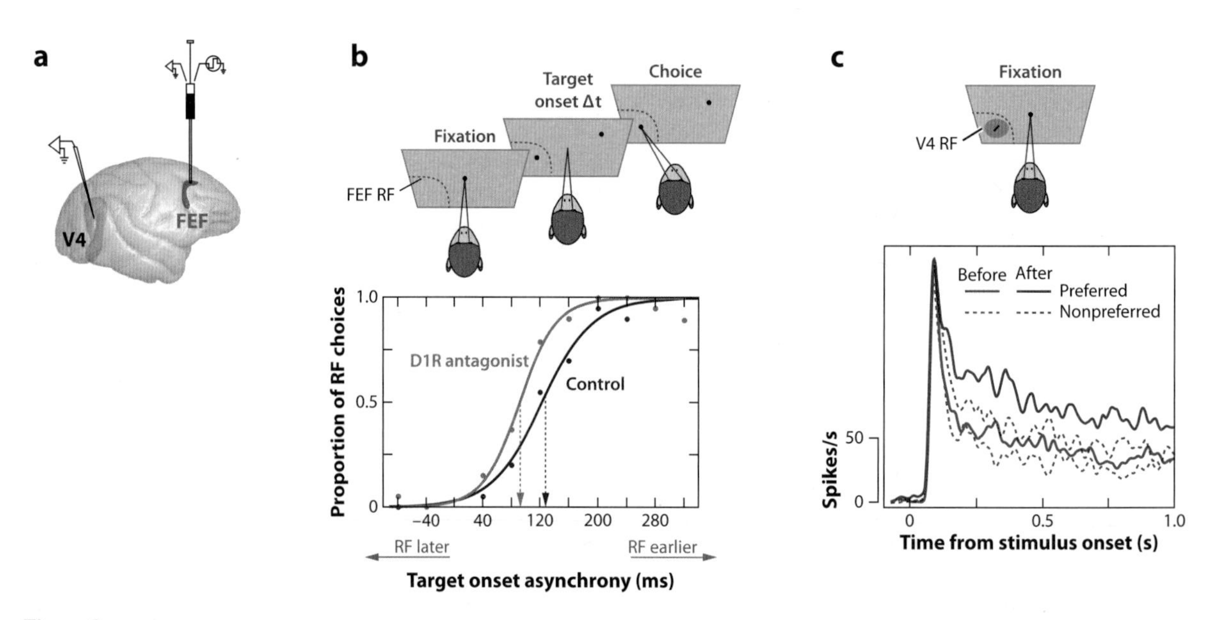

Figure 2

Dopamine-mediated frontal eye field (FEF) control of saccadic target selection and visual cortical responses. (*a*) Local manipulation of D1 receptor (D1R)-mediated activity within the FEF during single-neuron electrophysiology in area V4. Lateral view of the macaque brain depicts the location of a recording microinjectrode within the FEF and of recording sites within area V4. (*b*) A free-choice saccade task was used to measure the monkey's tendency to make saccades to targets within the part of visual space represented by neurons at the drug infusion site (FEF RF) versus targets at a location in the opposite hemifield. In the task, the two targets appeared at varying temporal onset asynchronies (Δt). The receptive field (RF) target appeared either earlier or later than the target outside the RF. The monkey's bias toward either target was measured as the asynchrony at which targets were chosen with equal probability (*dotted arrows in bottom plot*). Following a local infusion of a D1R antagonist into the FEF, there was a leftward shift in the psychometric curve (*gold*), indicating an increase in the tendency to make saccades to targets within the FEF RF. (*c*) Visual responses of a V4 neuron with an RF within the FEF RF; responses were measured during passive fixation. The plot shows mean visual responses over time to oriented bar stimuli presented at the preferred (*solid lines*) or nonpreferred (*dotted lines*) orientation both before (*blue*) and after (*red*) the FEF D1R manipulation. Adapted from Noudoost & Moore 2011a.

FEF activity was achieved via small (a microliter or less) injections of the selective D1 antagonist SCH23390 into sites within the FEF. The authors then measured the spiking responses of area V4 neurons that had RFs within the part of space affected by the D1R manipulation. Thus, measurements of visually driven V4 activity could be made before and after manipulating the D1R-mediated activity of FEF neurons projecting to the recorded V4 neurons. In addition, given the evidence mentioned above that attentional deployment tends to coincide with the preparation of saccades, the authors also measured the effects of the D1R manipulation on the selection of visual stimuli as targets for saccades (**Figure 2*b***). They observed that visual stimuli presented within the part of space affected by the D1R manipulation were consistently more likely to be selected as saccadic targets compared with control trials. Thus, the manipulation increased saccadic target selection. Most importantly, within area V4, the authors observed that responses to visual stimuli were altered in three important ways. First, the manipulation produced an enhancement in the magnitude of responses to visual stimulation (**Figure 2*c***). Second, visual responses became more stimulus selective. Third, visual responses became less variable across trials. Notably, all three of the observed changes in V4 visual activity are known effects of visual attention (Motter 1993, McAdams & Maunsell 1999, Mitchell et al. 2007). Moreover, the magnitude of the observed modulation was nearly equal to that seen in attention studies. Thus, manipulation of D1R-mediated FEF activity not only increased saccadic target selection within the corresponding part of space but also increased the magnitude, selectivity, and reliability of V4 visual responses. In essence, the manipulation effectively elicited correlates of covert attention within the extrastriate cortex in the absence of a behavioral task.

The effects of the FEF D1R manipulation on V4 neurons show that changes to FEF neuronal activity are sufficient to exert a long-range influence on signals within the visual cortex. In addition, the above effects show that DA, acting via D1Rs, is involved in the FEF's influence on visual cortical signals as well as its influence on saccadic preparation. Because a wealth of evidence implicates D1Rs in the neural mechanisms of spatial working memory, specifically in regulating the persistent activity of neurons within the dlPFC (Williams & Goldman-Rakic 1995), the above results suggest that D1Rs are part of a common mechanism underlying spatial attention and spatial working memory (Noudoost & Moore 2011b). Like dlPFC neurons, FEF neurons also exhibit persistent, delay-period activity, even in tasks not involving saccades (Armstrong et al. 2009) (**Figure 3**). Persistent activity within the PFC is thought to be generated by recurrent glutamatergic connections between prefrontal pyramidal neurons (Gao et al. 2001, Seamans & Yang 2004). DAergic modulation of persistent activity within the PFC appears to be achieved by the influence of D1Rs on these recurrent connections. The above results suggest a model in which D1Rs contribute to signatures of attention within the visual cortex by a mechanism similar to their influence on persistent activity, namely by modulating long-range, recurrent connections between the FEF and the visual cortex (**Figure 4**). Consistent with this idea is the finding that FEF neurons exhibiting persistent activity tend to exhibit greater attentional modulation than do those without (Armstrong et al. 2009). In the model, attention (and/or saccadic preparation) is directed toward particular locations according to the pattern of activity across the map of visual space within the FEF, similar to what Bisley & Goldberg (2010) proposed for parietal area LIP. Cortical columns with greater activity would then correspond to locations of greater attentional deployment (and/or saccadic preparation) and consequently higher gain of spatially overlapping visual cortical signals, compared with nonoverlapping signals. A possible role of DA would be to control the extent of the FEF gain modulation, effectively setting the breadth of the so-called attentional window. Thus, optimum DA levels would translate into larger differences between

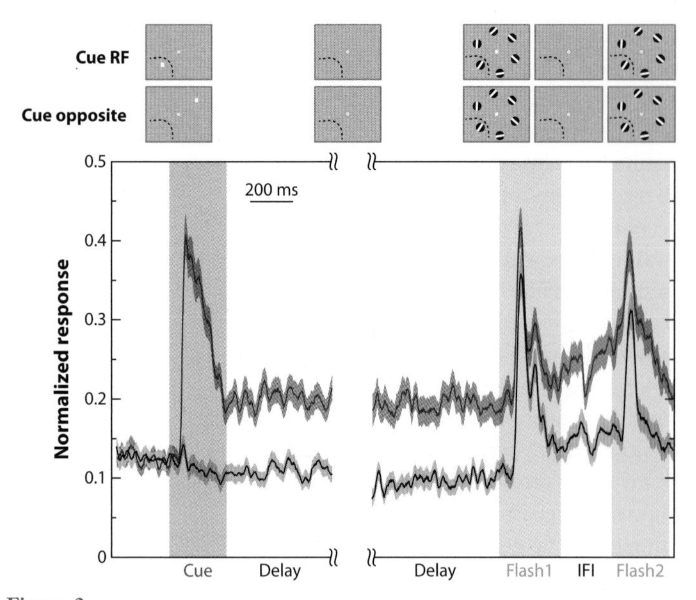

Figure 3

Modulation of frontal eye field (FEF) neuronal responses in monkeys performing an attention task. In the task, monkeys fixated on a central fixation spot (*yellow dot in gray panels*) and depressed a manual lever. The brief appearance of a peripheral cue (*white square, first panels*) instructed the monkey that, after a delay (~1 s), a change in the orientation of a flashed grating stimulus may occur at the cued location (50% of trials). During the grating flash epoch that followed the delay, an array of six oriented gratings was flashed twice, and the monkey was rewarded for releasing the lever if the grating at the cued location changed its orientation and for holding the lever if a change did not occur. The diagram depicts trials without a change. The five remaining gratings were distracters. The neuronal response histograms below show the average response of a population of 106 FEF neurons on correct trials in which monkeys were cued to attend either to the location coinciding with the receptive field (RF) of a FEF neuron (*red*) or to a location in the opposite hemifield (*gray*). The dotted half circle in each panel depicts the neuronal RF. Note that the average neuronal response not only signals the visual appearance of the brief cue but also continues to encode the attended location throughout the trial. Data are from only trials on which the grating orientation did not change. Adapted from Armstrong et al. 2009.

attended and unattended stimuli, whereas suboptimal DA levels would mean small differences and perhaps a less stable attentional focus. At least superficially, such a role of DA in attentional deployment would be consistent with the perceptual deficits characteristic of attention deficit hyperactivity disorder (ADHD) patients (Mason et al. 2003), who generally exhibit abnormal PFC DA levels (Ernst et al. 1998).

Noudoost & Moore (2011a) also observed that manipulation of D2 receptor (D2R)-mediated activity increased saccadic target selection in a manner equivalent to the D1R manipulation. However, only the D1R manipulation produced attention-like effects within area V4; the D2R manipulation exerted no measurable effects on the visual activity of V4 neurons. Thus, in addition to being dissociable at the level of functional subclasses of FEF neurons (Thompson et al. 2005, Gregoriou et al. 2009), the control of visual attention and saccadic target selection appear to be dissociable at the level of DA receptor subtypes. This dissociability appears to result, at least in part, from differing patterns of D1R and D2R expression across cortical laminae (**Figure 4**). Within the cortex, D1Rs exhibit a bilaminar pattern of expression, appearing in both supragranular and infragranular layers. In contrast, D2Rs are less abundant and tend to be expressed primarily within infragranular layers (Lidow et al. 1991, Santana et al. 2009). In the FEF, the principal source of output to the brain stem oculomotor nuclei and to the SC emanates from pyramidal neurons in layer V (Segraves & Goldberg 1987), where both DA receptor subtypes are expressed. This finding proposes how both the D1R and D2R manipulations could have altered saccadic target selection. On the other hand, FEF neurons projecting to the posterior visual cortex reside primarily within superficial layers II and III (Pouget et al. 2009), where D1Rs are the dominant receptor subtype and are thus more likely to mediate the FEF's influence on visual cortical activity.

OPERANT CONTROL OF FEF NEURONS AND VISUAL ATTENTION

The aforementioned attention studies demonstrate that when monkeys are engaged in a learned task in which they must deploy attention, the activity of visual and visuomovement FEF neurons (as in many other areas) reflects that deployment. They do not,

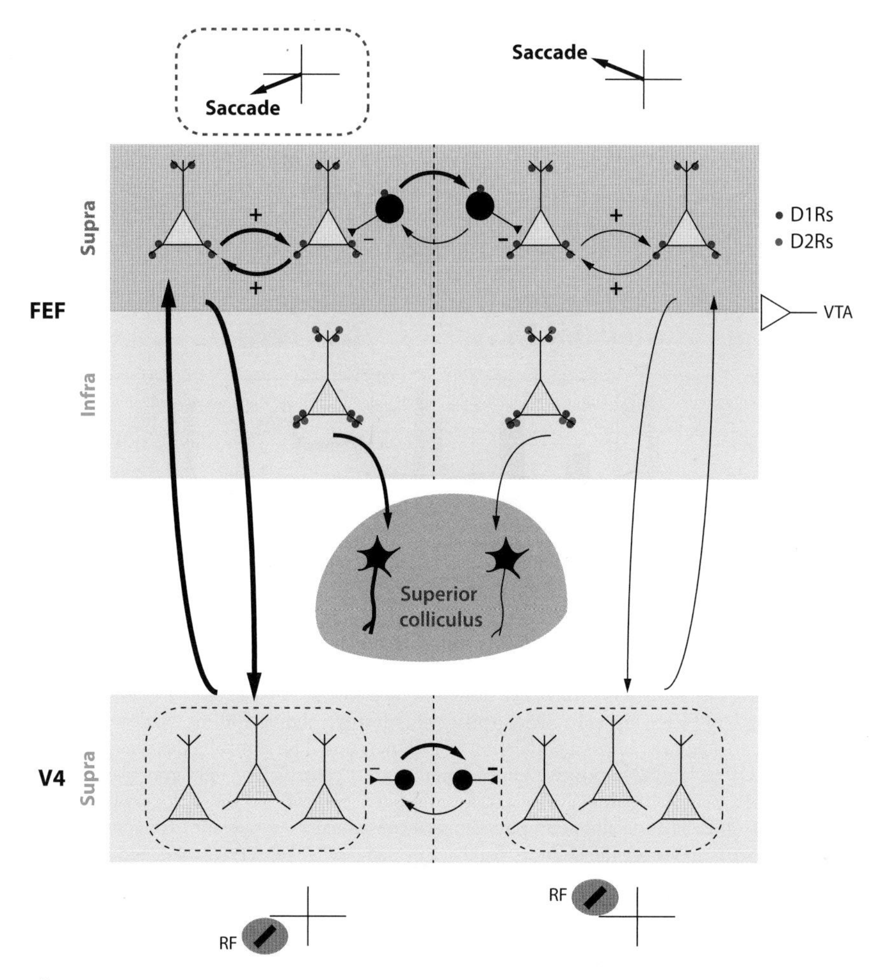

Figure 4

Possible influence of D1 receptors (D1Rs) on recurrent networks within the frontal eye field (FEF) and between the FEF and V4. The diagram depicts two adjacent FEF or V4 columns representing different, but adjacent, locations in saccadic or visual space, respectively. The columns are assumed to interact competitively (*black inhibitory neurons*). Positive arrows between FEF neurons within the same column depict the recurrent excitatory connections thought to underlie the persistence of spatial signals during remembered saccades or locations. Recurrent activity between the FEF and V4 is proposed to underlie the influence of FEF on the gain of visual inputs within V4. Diffuse dopaminergic input from the ventral tegmental area (VTA) (*input at right*) to the FEF (all columns) may modulate recurrence both within the FEF and between FEF and V4 through D1Rs and may influence competition between spatial representations. For example, increases in recurrence in a particular column while remembering or attending to a corresponding location (*dotted rectangle, thicker arrows at left*) may be modulated by dopamine levels. Biases in competitive interactions between columns within the visual cortex can also be achieved by experimental manipulation of D1R-mediated FEF activity. Also shown are the projections from infragranular FEF neurons to the superior colliculus (SC). Red circles represent D1Rs, and blue circles indicate D2Rs. Note the localization of D2Rs primarily in infragranular, SC-projecting, layers, which is consistent both with anatomical evidence and with the observation that changes in D2R-mediated FEF activity affect only saccadic target selection and not visual cortical activity. Adapted from Noudoost & Moore 2011b.

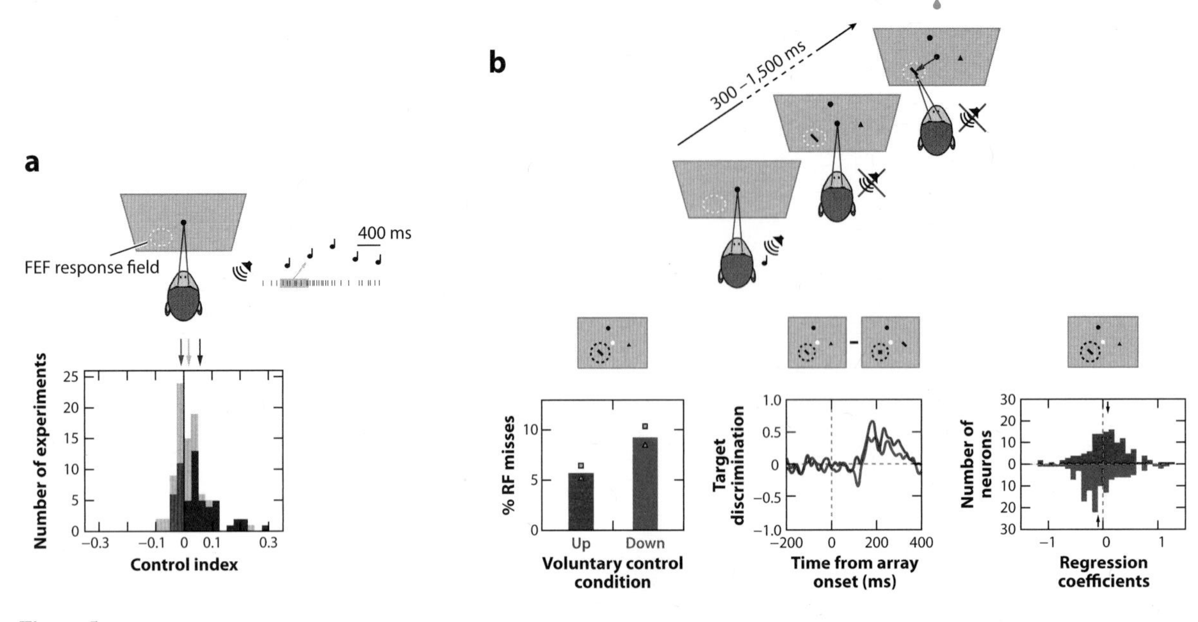

Figure 5

Operant control of frontal eye field (FEF) neurons and its effects on selective attention measured behaviorally and neurophysiologically. (*a*) In the operant control task, the monkey fixated a central spot on an otherwise blank video display and was rewarded for increasing or decreasing the firing rate of FEF neurons. The dotted circle shows the FEF receptive field (RF). Speaker icon and musical notes depict auditory feedback of FEF neuronal activity (spike train) during a sliding 500-ms window (*gray rectangle*). Bottom plot shows a histogram of operant control indices across a population of FEF neurons. The control index measures the change in FEF firing rate in the rewarded direction (UP or DOWN); positive values denote correct control. The light gray histogram shows all experiments, the purple histogram shows experiments with individually significant positive control, and the dark gray histogram shows experiments with significant negative control. (*b*) Behavioral and neurophysiological consequences of operant FEF control. (*Top*) Visual-search probe trials, in which a search array appeared, the auditory feedback ceased (*red "X" on speaker icon*), and the monkey was rewarded (*blue droplet*) for directing a saccade toward an oriented bar target. (*Bottom left*) Mean proportion of target misses opposite the RF was increased during DOWN operant control of FEF activity in both monkeys (*square and triangle symbols*). (*Bottom middle*) Target discrimination by FEF neurons was increased during upward (*red*) operant control relative to downward (*blue*). (*Bottom right*) Correlation of spontaneous activity with FEF responses to the target array. The direction of operant control determined the sign of the relationship between baseline and target-driven FEF activity. Adapted from Schafer & Moore (2011).

however, show that changes in endogenously generated neuronal activity in the absence of an explicit task are sufficient to bring about that deployment. Schafer & Moore (2011) tested the hypothesis that behaviorally conditioned, voluntary changes in FEF neuronal activity are sufficient to bring about the deployment of visual attention. The authors took advantage of the evidence from previous studies that demonstrated humans' and monkeys' ability to manipulate activity voluntarily within motor-related brain structures, even in the absence of movement (Fetz & Finocchio 1975). Schafer & Moore (2011) used similar operant training techniques to examine the impact of voluntary control of FEF activity on visually driven behavior. Monkeys were given real-time auditory feedback based on the firing rate of FEF activity and rewarded for either increasing or decreasing that activity (in alternating UP and DOWN blocks of trials), while maintaining central fixation on a blank visual display (**Figure 5*a***). In each behavioral trial that lasted several seconds, monkeys were rewarded every time neuronal activity measured within a moving time window exceeded (for UP trials) or dropped below (for DOWN trials) an arbitrary spike rate threshold. Reward therefore depended solely on the rate of neural activity during fixation and not on any explicit behavioral task.

The authors made several important observations. First, monkeys were indeed able to alter the average firing rate of FEF neurons significantly in UP versus DOWN trials at a majority of recorded sites. Second, they found that the magnitude of voluntary modulation was uncorrelated with the visual or motor properties of the individual FEF neurons being recorded. That is, neurons that responded to visual stimuli, but not in advance of saccades, were just as likely to be operantly controlled as neurons with little or no visual activity but with a great amount of saccade-related activity. Third, the authors observed significantly greater power in the gamma band of FEF local field potentials (LFPs). Fourth, and perhaps most importantly, the authors probed the behavioral and neurophysiological consequences of operant control of FEF activity. They introduced probe trials in which the monkey performed a visual search task while exerting operant control over FEF activity. Partway through randomly chosen trials, the auditory feedback would cease and a search array appeared on the screen consisting of different shapes of equal area. The monkey was then rewarded for making a saccade to the oriented bar (the target) or withholding a saccade if the search target was absent from the array (**Figure 5*b***). Saccades to other shapes were counted as incorrect and were not rewarded. The authors reasoned that if the monkeys' strategy for altering FEF firing rates was one of general vigilance or arousal, any effects of UP versus DOWN modulation on behavior should generalize across target locations. Instead, they found behavioral effects of operant control that were limited to trials in which the target appeared in the RF. Specifically, when the target appeared in the RF, monkeys were less likely to detect the target (i.e., they had more misses) on the DOWN trials than they were on the UP trials. Unlike the effects on search performance, however, neither saccade probability nor saccade metrics were affected by operant conditioning, demonstrating a dissociation between the attention-related and motor-preparatory effects of FEF activity. In addition, the authors found that FEF neurons could discriminate targets from distracters better during UP trials compared with DOWN trials. This change in target discriminability was dependent on the direction of operant control and not on spontaneous fluctuations in firing rate. Lastly, splitting UP and DOWN trials revealed a positive correlation between preprobe spontaneous firing rate and neuronal responses to the target in the RF during UP trials but revealed a negative correlation during DOWN trials. Thus, the direction of operant control seemed to determine the nature of the relationship between spontaneous and target-driven neural activity. Taken together, the above results show that endogenous, voluntary changes in FEF neural activity are sufficient to bring about both the behavioral and the neurophysiological effects of visual attention and that explicit learning of an attention task is not required.

USING LOSS OF FUNCTION TO IDENTIFY SOURCES OF ATTENTIONAL CONTROL

Understanding what controls a particular behavior ultimately requires demonstrating a specific loss in that behavior when the suspected underlying mechanism is damaged. But what constitutes a loss of attention? Researchers have proposed a number of recent models to account for the effects of attention on visual signals (Desimone & Duncan 1995, Reynolds et al. 2000, Reynolds & Heeger 2009, Lee & Maunsell 2010). These models provide a useful framework for understanding the interaction of attentional control with the encoding of visual information, thus potentially allowing one to distinguish between deficits in either process. Attention is generally thought to affect the competitive interactions inherent in visual processing. For example, although a representation with larger stimulus drive will tend to exert greater suppression on its competitors than one with lower stimulus drive, attention

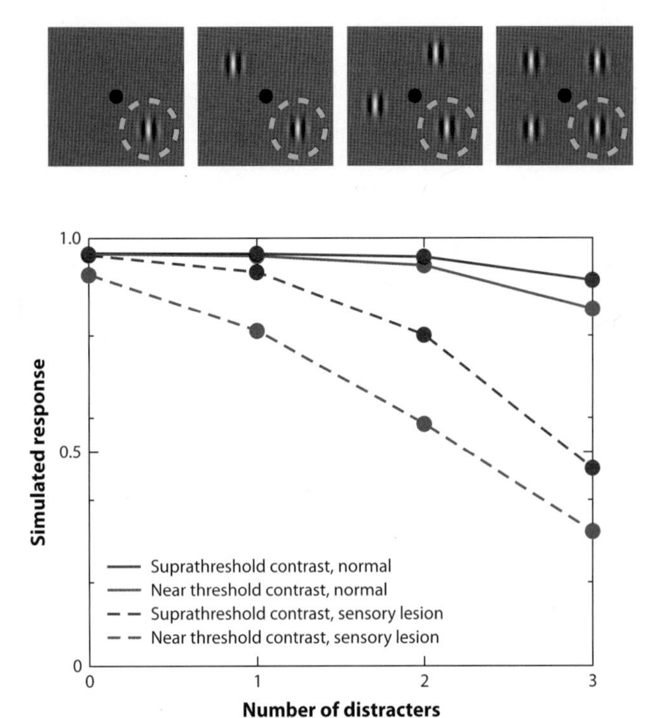

to one stimulus or the other is expected to mitigate (or exacerbate) that effect. However, when attention is held constant, decreases in the strength of a representation can be brought about not only by reduced competitiveness with that of other stimuli, but also by reduced stimulus drive, including reduced drive due to brain damage. For example, cortical visual representations are distributed across a large set of highly connected retinotopic maps, and damage to a portion of one of them (e.g., V4) should result in significant (yet incomplete) loss of stimulus drive at retinotopically corresponding portions in the others [e.g., V2, MT, MST (medial superior temporal area)]. This loss in stimulus drive should result in a competitive disadvantage at that spatial location within the intact maps. As a result, because stimulus drive and attention both interact competitively, the above models suggest that altering stimulus drive in such a way should affect the magnitude of neural responses in a manner consistent with a loss of attention, even when attention is functioning normally.

This idea can be illustrated using one of the above-mentioned models, for example, with the normalization model described by Reynolds & Heeger (2009). In their model, stimulus drive combines with an attentional gain signal (the "attention field"), and then competitive interactions between multiple stimuli reciprocally inhibit one another as a function of their activity (normalization). Using this model, we simulated a lesion in the sensory drive independent of attentional gain and examined its effects on downstream encoding (**Figure 6**). The simulation shows that weakening the stimulus drive can produce a deficit in the sensory encoding that is dependent on the presence of other competing stimuli. In the absence of competing stimuli (distracters), there may be no observable deficit in encoding, particularly when stimuli are sufficiently suprathreshold (e.g., at high contrast). But with multiple competing or less salient stimuli, a significant deficit emerges. Thus, even when attentional gain remains intact, a diminution in the strength of visual input could be sufficient to produce a distracter-dependent deficit. Unfortunately, distracter dependency is often what is used to define a loss in attentional function (e.g., De Weerd et al. 1999, Wardak et al. 2006, Lovejoy & Krauzlis 2010). Such a definition may thus be overly broad in many cases (Desimone & Duncan 1995) and may make it difficult to distinguish a role of brain structures in visual attention and a role in visual processing, as in the case of area V4 (e.g., Schiller & Lee 1991, De Weerd et al. 1999).

What then should an attention deficit look like? The answer to this question will only grow in importance as we contemplate future studies investigating the neural mechanisms that are causally related to the filtering and selection of behaviorally relevant sensory information (visual or otherwise). Carefully defined attention deficits are particularly important for studies that employ cell- and circuit-specific experimental tools (Fenno et al. 2011) and for those that employ animal models with more rudimentary forms of attention-related behavior than primates (Muir et al. 1992). Subsequent attempts to pinpoint the neurons, neural circuits,

Figure 6

Distracter-dependent deficits from a loss of visual inputs in a normalization model of attention. (*Top, left to right*) Schematic representations of a visual display with a target stimulus (Gabor grating) in isolation or with one, two, or three distracters. In each display, the gray dot depicts the point of fixation during a covert attention task. The dotted green line depicts the region of space affected by the lesion (the scotoma), the attended location, and the receptive field of the simulated neuron. Both the location of attention and the target stimulus are constant across the different distracter conditions, and the distracters have the same contrast as the target. (*Bottom*) Responses of a simulated neuron as a function of the number of distracters in the display. Without a lesion (*blue and red solid lines*), the responses remain nearly constant across the number of distracters. With a visual lesion (*dotted lines*), which weakens the stimulus drive while leaving the attention gain intact, responses become compromised as more distracters are added to the display. The distracter-dependent deficits are less severe for high-contrast stimuli (*red lines*) than for near threshold, low-contrast stimuli (*blue lines*). Model: The set of Matlab routines provided by the normalization model of attention (NMA) (Reynolds & Heeger 2009) was used to implement the present simulation. In the NMA, stimulus-driven inputs (stimulus drive) are multiplied by an attention gain signal (attention field), after which competitive interactions between all units reciprocally inhibit one another as a function of their activities (normalization) to yield the responses of the neurons in the model. Here, four significant additions were made. First, the script for figure3F.m (Reynolds & Heeger 2009) was augmented to simulate multiple distracters. Second, for more control over the spread and magnitude of the inhibition underlying the competitive interactions between neurons (i.e., normalization), an additional variable, *IxKernelHeight*, was added. *IxKernelHeight* allowed the user, along with the preexisting variable *IxWidth*, to independently control the height and width of the Gaussian inhibition kernel, respectively. Third, to simulate a lesion in the stimulus drive, independent of the attention gain, the stimulus drive was multiplied by a lesion matrix before any other computations were performed. This produced a lesioned stimulus drive matrix that the model used in place of the original stimulus drive. The lesion matrix consisted of values ranging between 0 and 1, and it can be visualized as an inverted Gaussian (specified by the variables *lesionWidth* and *lesionDepth*) centered on the spatial center of the lesion. Multiplying the lesion matrix with the stimulus drive to create the lesioned stimulus drive is equivalent to a gain down-modulation of the original stimulus drive. Fourth, to simulate model neurons with saturating responses, final responses in the present simulation were computed by passing the output of the NMA through a simple sigmoidal function: Final Response $= 1/(1 + e^{(\beta^*(H-x))})$, where e is the natural number, β modifies the slope of the sigmoid, H is equal to the half-max value of the sigmoid, and x is equal to the output of the NMA. Parameters: $Apeak = 2$; $AxWidth = 7$; $\beta = 0.75$; $H = 7$; $IxKernelHeight = 0.0175$; $IxWidth = 100$; $lesionWidth = 7$; $lesionDepth = 0.5$; $stimWidth = 7$.

and neural computations that confer a nervous system with the unique capacity to distinguish a target from distracters will need to determine more definitively when a behavioral phenotype reflects a loss in that capacity rather than a loss in some other function. In this review of recent evidence for the PFC's contribution to visual attention, we have highlighted studies that demonstrate experimentally produced benefits in visual processing, either by neurons (Noudoost & Moore 2011a, Schafer & Moore 2011) or in behavioral performance (Schafer & Moore 2011). Such benefits in processing, separate from processing itself, appear to be what most specifically defines attention. We therefore suggest that the loss of function one should expect when attentional mechanisms are absent is a loss of such benefits rather than a deficit in sensory processing per se. We assume that independent of the absolute level of perceptual performance, or even the degree of distracter dependency, a loss of attentional control should result in performance that cannot be improved by attentional cues. Remarkably, most, if not all, of the experimental literature is devoid of such results.

DISCLOSURE STATEMENT

The authors are not aware of any affiliations, memberships, funding, or financial holdings that might be perceived as affecting the objectivity of this review.

ACKNOWLEDGMENTS

This work was supported by NIH EY014924. The authors thank J.H. Reynolds, for discussion and comments on our use of the normalization model, and members of the Moore lab, for insightful discussions.

LITERATURE CITED

Anderson JC, Kennedy H, Martin KA. 2011. Pathways of attention: synaptic relationships of frontal eye field to V4, lateral intraparietal cortex, and area 46 in macaque monkey. *J. Neurosci.* 31:10872–81

Armstrong KM, Chang MH, Moore T. 2009. Selection and maintenance of spatial information by frontal eye field neurons. *J. Neurosci.* 29:15621–29

Armstrong KM, Fitzgerald JK, Moore T. 2006. Changes in visual receptive fields with microstimulation of frontal cortex. *Neuron* 50:791–98

Armstrong KM, Moore T. 2007. Rapid enhancement of visual cortical response discriminability by microstimulation of the frontal eye field. *Proc. Natl. Acad. Sci. USA* 104:9499–504

Bisley JW, Goldberg ME. 2010. Attention, intention, and priority in the parietal lobe. *Annu. Rev. Neurosci.* 33:1–21

Björklund A, Dunnett SB. 2007. Dopamine neuron systems in the brain: an update. *Trends Neurosci.* 30:194–202

Bruce CJ, Goldberg ME. 1985. Primate frontal eye fields. I. Single neurons discharging before saccades. *J. Neurophysiol.* 53:603–35

Bruce CJ, Goldberg ME, Bushnell MC, Stanton GB. 1985. Primate frontal eye fields. II. Physiological and anatomical correlates of electrically evoked eye movements. *J. Neurophysiol.* 54:714–34

Buschman TJ, Miller EK. 2007. Top-down versus bottom-up control of attention in the prefrontal and posterior parietal cortices. *Science* 315:1860–62

Castellanos FX, Tannock R. 2002. Neuroscience of attention-deficit/hyperactivity disorder: the search for endophenotypes. *Nat. Rev. Neurosci.* 3:617–28

Chelazzi L, Miller EK, Duncan J, Desimone R. 1993. A neural basis for visual search in inferior temporal cortex. *Nature* 363:345–47

Clark KL, Armstrong KM, Moore T. 2011. Probing neural circuitry and function with electrical microstimulation. *Proc. Biol. Sci.* 278:1121–30

Cohen MR, Maunsell JHR. 2009. Attention improves performance primarily by reducing interneuronal correlations. *Nat. Neurosci.* 12:1594–600

Connor CE, Preddie DC, Gallant JL, Van Essen DC. 1997. Spatial attention effects in macaque area V4. *J. Neurosci.* 17:3201–14

De Weerd P, Peralta MR 3rd, Desimone R, Ungerleider LG. 1999. Loss of attentional stimulus selection after extrastriate cortical lesions in macaques. *Nat. Neurosci.* 2:753–58

Desimone R, Duncan J. 1995. Neural mechanisms of selective visual attention. *Annu. Rev. Neurosci.* 18:193–222

Deubel H, Schneider WX. 1996. Saccade target selection and object recognition: evidence for a common attentional mechanism. *Vis. Res.* 36:1827–37

Ekstrom LB, Roelfsema PR, Arsenault JT, Kolster H, Vanduffel W. 2009. Modulation of the contrast response function by electrical microstimulation of the macaque frontal eye field. *J. Neurosci.* 29:10683–94

Ernst M, Zametkin AJ, Matochik JA, Jons PH, Cohen RM. 1998. DOPA decarboxylase activity in attention deficit hyperactivity disorder adults. A [fluorine-18]fluorodopa positron emission tomographic study. *J. Neurosci.* 18:5901–7

Fenno L, Yizhar O, Deisseroth K. 2011. The development and application of optogenetics. *Annu. Rev. Neurosci.* 34:389–412

Ferrier D. 1876. *Functions of the Brain*. London: Smith, Elder and Co.

Fetz EE, Finocchio DV. 1975. Correlations between activity of motor cortex cells and arm muscles during operantly conditioned response patterns. *Exp. Brain Res.* 23:217–40

Fischer B, Boch R. 1981. Enhanced activation of neurons in prelunate cortex before visually guided saccades of trained rhesus monkeys. *Exp. Brain Res.* 44:129–37

Fries P, Reynolds JH, Rorie AE, Desimone R. 2001. Modulation of oscillatory neuronal synchronization by selective visual attention. *Science* 291:1560–63

Fuster JM. 1995. *Memory in the Cerebral Cortex*. Cambridge, MA: MIT Press

Gao WJ, Krimer LS, Goldman-Rakic PS. 2001. Presynaptic regulation of recurrent excitation by D1 receptors in prefrontal circuits. *Proc. Natl. Acad. Sci. USA* 98:295–300

Gerbella M, Belmalih A, Borra E, Rozzi S, Luppino G. 2010. Cortical connections of the macaque caudal ventrolateral prefrontal areas 45A and 45B. *Cereb. Cortex* 20:141–68

Goldberg ME, Bushnell MC. 1981. Behavioral enhancement of visual responses in monkey cerebral cortex. II. Modulation in frontal eye fields specifically related to saccades. *J. Neurophysiol.* 46:773–87

Goldman-Rakic PS, Lidow MS, Smiley JF, Williams MS. 1992. The anatomy of dopamine in monkey and human prefrontal cortex. *J. Neural Transm. Suppl.* 36:163–77

Gregoriou GG, Gotts SJ, Desimone R. 2012. Cell-type-specific synchronization of neural activity in FEF with V4 during attention. *Neuron* 73:581–94

Gregoriou GG, Gotts SJ, Zhou HH, Desimone R. 2009. High-frequency, long-range coupling between prefrontal and visual cortex during attention. *Science* 324:1207–10

Hoffman JE, Subramaniam B. 1995. The role of visual attention in saccadic eye movements. *Percept. Psychophys.* 57:787–95

James W. 1890. *Principles of Psychology*, Vol. 1. New York: H. Holt

Knight RT, Grabowecky MF, Scabini D. 1995. Role of human prefrontal cortex in attention control. *Adv. Neurol.* 66:21–34; discussion 34–36

Lebedev MA, Messinger A, Kralik JD, Wise SP. 2004. Representation of attended versus remembered locations in prefrontal cortex. *PLoS Biol.* 2:e365

Lee J, Maunsell JH. 2010. The effect of attention on neuronal responses to high and low contrast stimuli. *J. Neurophysiol.* 104:960–71

Lidow MS, Goldman-Rakic PS, Gallager DW, Rakic P. 1991. Distribution of dopaminergic receptors in the primate cerebral cortex: quantitative autoradiographic analysis using [3H]raclopride, [3H]spiperone and [3H]Sch23390. *Neuroscience* 40:657–71

Lovejoy LP, Krauzlis RJ. 2010. Inactivation of primate superior colliculus impairs covert selection of signals for perceptual judgments. *Nat. Neurosci.* 13:261–66

Mason DJ, Humphreys GW, Kent LS. 2003. Exploring selective attention in ADHD: visual search through space and time. *J. Child Psychol. Psychiatry* 44:1158–76

McAdams CJ, Maunsell JH. 1999. Effects of attention on orientation-tuning functions of single neurons in macaque cortical area V4. *J. Neurosci.* 19:431–41

Miller EK, Cohen JD. 2001. An integrative theory of prefrontal cortex function. *Annu. Rev. Neurosci.* 24:167–202

Mitchell JF, Sundberg KA, Reynolds JH. 2007. Differential attention-dependent response modulation across cell classes in macaque visual area V4. *Neuron* 55:131–41

Moore T, Armstrong KM. 2003. Selective gating of visual signals by microstimulation of frontal cortex. *Nature* 421:370–73

Moore T, Fallah M. 2001. Control of eye movements and spatial attention. *Proc. Natl. Acad. Sci. USA* 98:1273–76

Moore T, Fallah M. 2004. Microstimulation of the frontal eye field and its effects on covert spatial attention. *J. Neurophysiol.* 91:152–62

Moran J, Desimone R. 1985. Selective attention gates visual processing in the extrastriate cortex. *Science* 229:782–84

Motter BC. 1993. Focal attention produces spatially selective processing in visual cortical areas V1, V2, and V4 in the presence of competing stimuli. *J. Neurophysiol.* 70:909–19

Mountcastle VB, Lynch JC, Georgopolous A, Sakata H, Acuna C. 1975. Posterior parietal association cortex of the monkey: command functions for operations within extrapersonal space. *J. Neurophysiol.* 38:871–908

Muir JL, Robbins TW, Everitt BJ. 1992. Disruptive effects of muscimol infused into the basal forebrain on conditional discrimination and visual attention: differential interactions with cholinergic mechanisms. *Psychopharmacology (Berl.)* 107:541–50

Noudoost B, Chang MH, Steinmetz NA, Moore T. 2010. Top-down control of visual attention. *Curr. Opin. Neurobiol.* 20:183–90

Noudoost B, Moore T. 2011a. Control of visual cortical signals by prefrontal dopamine. *Nature* 474:372–75

Noudoost B, Moore T. 2011b. The role of neuromodulators in selective attention. *Trends Cogn. Sci.* 15:585–91

Peterson MS, Kramer AF, Irwin DE. 2004. Covert shifts of attention precede involuntary eye movements. *Percept. Psychophys.* 66:398–405

Pouget P, Stepniewska I, Crowder EA, Leslie MW, Emeric EE, et al. 2009. Visual and motor connectivity and the distribution of calcium-binding proteins in macaque frontal eye field: implications for saccade target selection. *Front. Neuroanat.* 3:2

Reynolds JH, Chelazzi L. 2004. Attentional modulation of visual processing. *Annu. Rev. Neurosci.* 27:611–47

Reynolds JH, Chelazzi L, Desimone R. 1999. Competitive mechanisms subserve attention in macaque areas V2 and V4. *J. Neurosci.* 19:1736–53

Reynolds JH, Heeger DJ. 2009. The normalization model of attention. *Neuron* 61:168–85

Reynolds JH, Pasternak T, Desimone R. 2000. Attention increases sensitivity of V4 neurons. *Neuron* 26:703–14

Rizzolatti G, Riggio L, Dascola I, Umiltá C. 1987. Reorienting attention across the horizontal and vertical meridians: evidence in favor of a premotor theory of attention. *Neuropsychologia* 25:31–40

Robbins TW, Arnsten AF. 2009. The neuropsychopharmacology of fronto-executive function: monoaminergic modulation. *Annu. Rev. Neurosci.* 32:267–87

Robinson DA, Fuchs AF. 1969. Eye movements evoked by stimulation of frontal eye fields. *J. Neurophysiol.* 32:637–48

Rueckert L, Grafman J. 1996. Sustained attention deficits in patients with right frontal lesions. *Neuropsychologia* 34:953–63

Santana N, Mengod G, Artigas F. 2009. Quantitative analysis of the expression of dopamine D1 and D2 receptors in pyramidal and GABAergic neurons of the rat prefrontal cortex. *Cereb. Cortex* 19:849–60

Schafer RJ, Moore T. 2011. Selective attention from voluntary control of neurons in prefrontal cortex. *Science* 332:1568–71

Schiller PH, Lee K. 1991. The role of the primate extrastriate area V4 in vision. *Science* 251:1251–53

Schiller PH, Tehovnik EJ. 2001. Look and see: how the brain moves your eyes about. *Prog. Brain Res.* 134:127–42

Seamans JK, Yang CR. 2004. The principal features and mechanisms of dopamine modulation in the prefrontal cortex. *Prog. Neurobiol.* 74:1–58

Segraves MA, Goldberg ME. 1987. Functional properties of corticotectal neurons in the monkey's frontal eye field. *J. Neurophysiol.* 58:1387–419

Shepherd M, Findlay JM, Hockey RJ. 1986. The relationship between eye movements and spatial attention. *Q. J. Exp. Psychol. A* 38:475–91

Shipp S. 2004. The brain circuitry of attention. *Trends Cogn. Sci.* 8:223–30

Sommer MA, Wurtz RH. 2000. Composition and topographic organization of signals sent from the frontal eye field to the superior colliculus. *J. Neurophysiol.* 83:1979–2001

Thompson KG, Bichot NP, Sato TR. 2005. Frontal eye field activity before visual search errors reveals the integration of bottom-up and top-down salience. *J. Neurophysiol.* 93:337–51

Wardak C, Ibos G, Duhamel JR, Olivier E. 2006. Contribution of the monkey frontal eye field to covert visual attention. *J. Neurosci.* 26:4228–35

Williams GV, Goldman-Rakic PS. 1995. Modulation of memory fields by dopamine D1 receptors in prefrontal cortex. *Nature* 376:572–75

Wurtz RH, Goldberg ME. 1972a. Activity of superior colliculus in behaving monkey. IV. Effects of lesions on eye movements. *J. Neurophysiol.* 35:587–96

Wurtz RH, Goldberg ME. 1972b. The primate superior colliculus and the shift of visual attention. *Invest. Ophthalmol.* 11:441–50

Gene Therapy for Blindness

José-Alain Sahel[1–5] and Botond Roska[6]

[1]Inserm, UMR_S 968, [2]UPMC, University of Paris 06, Institut de la Vision, Paris, 75012, France; email: j.sahel@gmail.com

[3]CNRS, UMR_7210, Paris, F-75012, France

[4]Fondation Ophtalmologique Adolphe de Rothschild, Paris, 75019, France

[5]Centre d'Investigation Clinique 503, Inserm-Center Hospitalier National d'Ophtalmologie des Quinze-Vingts, Paris, 75012, France

[6]Neural Circuit Laboratories, Friedrich Miescher Institute for Biomedical Research, Basel, 4058, Switzerland; email: botond.roska@fmi.ch

Annu. Rev. Neurosci. 2013. 36:467–88

First published online as a Review in Advance on May 31, 2013

The *Annual Review of Neuroscience* is online at neuro.annualreviews.org

This article's doi:
10.1146/annurev-neuro-062012-170304

Keywords

eye, retina, optogenetics, virus, gene replacement, neuroprotection

Abstract

Sight-restoring therapy for the visually impaired and blind is a major unmet medical need. Ocular gene therapy is a rational choice for restoring vision or preventing the loss of vision because most blinding diseases originate in cellular components of the eye, a compartment that is optimally suited for the delivery of genes, and many of these diseases have a genetic origin or genetic component. In recent years we have witnessed major advances in the field of ocular gene therapy, and proof-of-concept studies are under way to evaluate the safety and efficacy of human gene therapies. Here we discuss the concepts and recent advances in gene therapy in the retina. Our review discusses traditional approaches such as gene replacement and neuroprotection and also new avenues such as optogenetic therapies. We conjecture that advances in gene therapy in the retina will pave the way for gene therapies in other parts of the brain.

Contents

AN EYE-OPENER: THE TRANSLATIONAL SUCCESS OF *RPE65* GENE REPLACEMENT

Over the past decade, considerable progress has been made in gene therapy for monogenic inherited blinding diseases, as epitomized by the advances achieved in the treatment of one form of Leber congenital amaurosis (LCA). LCA is the most severe retinal dystrophy and leads to a major visual impairment at birth or before the age of one year. LCA is caused by mutations in any one of at least 17 genes encoding proteins involved in a variety of retinal functions (den Hollander et al. 2008, Falk et al. 2012). To date, a form of LCA that is caused by loss-of-function mutations in the retinal pigment epithelium (RPE)-specific *RPE65* gene is the most extensively studied, both in animal models and in humans. In animal models, proof-of-principle for gene-replacement therapy was first demonstrated by a team led by Jean Bennett. The team restored visual function to Briard dogs affected by naturally occurring *RPE65* mutations using adeno-associated virus (AAV)-mediated delivery of *RPE65* to the eye (Acland et al. 2001). Three years after a single subretinal administration of AAV-*RPE65*, retinal responses remained stable, providing evidence of long-term restoration of photoreceptor function (Acland et al. 2005). Remarkably, the improvements have remained stable for more than ten years. In humans, phase I clinical trials simultaneously conducted by three different groups demonstrated that AAV-mediated *RPE65* gene therapy was safe and led to a slight improvement in vision, mostly under dark-adapted conditions (Bainbridge et al. 2008, Hauswirth et al. 2008, Maguire et al. 2008). Investigators have subsequently reported stable clinical benefits in more than 30 patients, with improvement in bright and dim light vision and no severe adverse effects (Cideciyan et al. 2009a,b; Simonelli et al. 2010). Moreover, readministration of AAV-*RPE65* to the contralateral eye in three LCA patients 1.7 to 3.3 years after the initial injection of *RPE65* gene–based treatment in one eye was shown to be safe and effective (Bennett et al. 2012). The pioneering *RPE65* gene therapy trials provided promise for gene therapy in different forms of LCA (Pawlyk et al. 2005, Sun et al. 2010) and other retinal diseases.

WHY THE EYE?

The number of visually impaired people throughout the world is estimated to be 285 million, of whom 39 million are legally blind (Mariotti 2012). Most of these patients suffer from diseases that affect cell types in the eye. The eye has unique characteristics, compared with other tissues and organs, that make it particularly suited for gene therapy (Bainbridge et al. 2006). First, it is a small, closed compartment. Because of this, a long-lasting, high viral concentration can be achieved by injecting only a small amount of a virus, and systemic dissemination and risk for adverse systemic effects are minimal. Owing to internal compartmentalization within the eye, a virus can be selectively delivered to different ocular structures, such as the anterior chamber, the vitreous cavity, or the subretinal space. Second, most cell types of the eye are stable, and many of them are evolutionarily highly conserved across mammals and even across vertebrates. Since most cell types do not divide, the risk for malignant transformation is reduced, and one can achieve sustained gene expression because transgenes are not diluted in cell division. Nevertheless, division of glial cells is a possibility (Bhatia et al. 2011); therefore, the use of vectors that do not integrate into the genome is desirable. Because gene expression in cell types is often conserved, one can screen cell-type targeting vectors in nonprimate mammals before testing them in primates. Third, the eye is partially shielded from the actions of the immune system by a blood-retinal barrier (Streilein et al. 1997). This feature, together with a gamut of others, including the local inhibition of immune responses and the systemic induction of immunosuppressive regulatory T-cells by eye-specific mechanisms, contributes to a phenomenon known as ocular immune privilege (Caspi 2010, Streilein 2003), which ensures partial protection against immune responses directed against gene products and vector antigens. Fourth, numerous animal models of inherited retinal diseases have already been developed in rodents, cats, and dogs, which facilitates preclinical assessment of therapeutic efficacy (Fletcher et al. 2011). Fifth, the optical transparency of the eye, together with recent advances in in vivo imaging techniques such as scanning fluorescent ophthalmoscopy, optical coherence tomography, autofluorescence, and adaptive optics, allows not only for direct noninvasive visualization of reporter gene expression from targeting viruses in animal models but also for accurate evaluation of the gene therapy outcomes in both animal models and human patients. Sixth, the untreated contralateral fellow eye is potentially useful as a control in clinical trials.

THE RETINA: AN IMAGE-PROCESSING MACHINE

The retina can be viewed as a parallel image processor that acquires images via a mosaic of photoreceptors and that extracts various visual features from the acquired images (Azeredo da Silveira & Roska 2011, Gollisch & Meister 2010, Masland 2001, Wässle 2004) (**Figure 1**). Rod photoreceptors respond directly to light at lower intensities and cone photoreceptors at higher intensities. The cellular infrastructure that underlies parallel processing consists of mosaics of local neuronal circuits. The retina has ~20 such circuit mosaics, built from more than 60 cell types, that independently extract different features from the visual world. Each mosaic has an associated mosaic of output cells, the ganglion cells, which relay the computed feature to higher brain centers. Here we briefly summarize current knowledge about the characteristics of the retinal circuit that lead to a better understanding of different approaches in retinal gene therapy. Each cone in the retina is connected to ~10 types of cone bipolar cells, and each of these bipolar cells is connected to

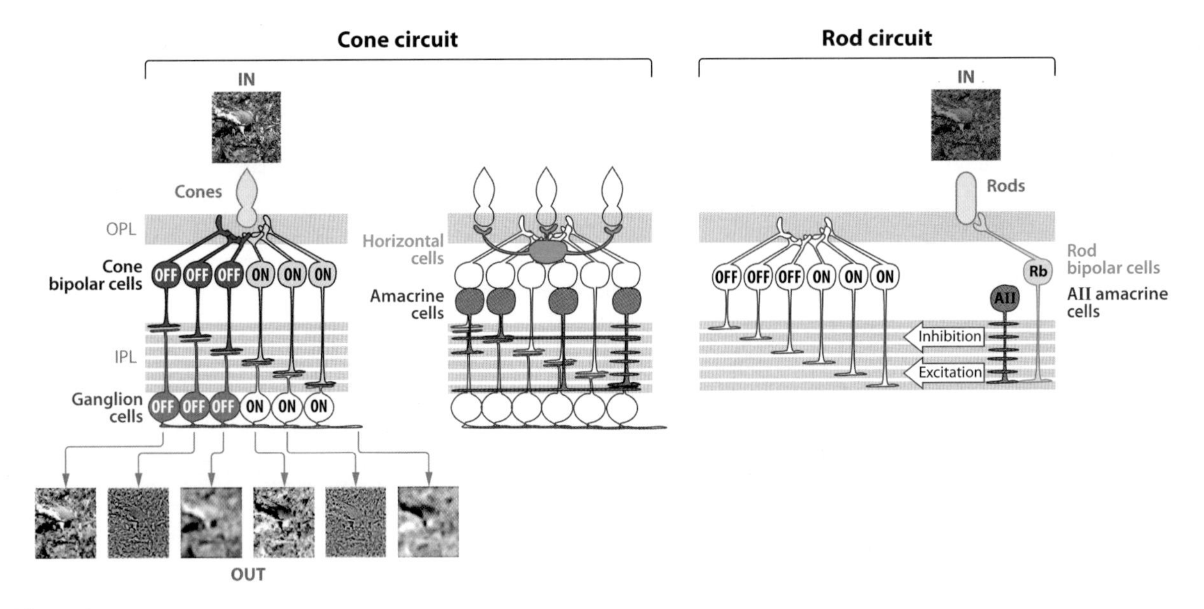

Figure 1

Basic elements of cone- and rod-driven circuits. (*Left*) Cone circuit, (*right*) rod circuit. ON and OFF refer to cells that are activated by light contrast increments and decrements, respectively. For rods, only the rod-rod bipolar (Rb) circuit is shown. This rod circuit joins the cone circuit at the level of cone bipolar cell axon terminals. The AII amacrine cell is excitatory toward ON cone bipolar cells and inhibitory toward OFF cone bipolar cells. IPL, inner plexiform layer; OPL, outer plexiform layer.

several types of ganglion cells. Cones, bipolar cells, and ganglion cells use the excitatory neurotransmitter glutamate to communicate. The axon terminals of bipolar cells and the dendrites of ganglion cells are restricted to narrow laminae in the inner retina, forming ~10 layers. Communication between cones and bipolar cells is modified by the inhibitory horizontal cells, and communication between bipolar cells and ganglion cells is modified by a large variety of inhibitory amacrine cells. Cones respond to light by lowering their membrane voltage; i.e., they hyperpolarize. Half of the cone bipolar cells also hyperpolarize (OFF cells), whereas the other half increase their membrane voltage, depolarizing when light intensity increases (ON cells). The polarity of the ganglion cell responses is determined by the polarity of the bipolar cells from which they receive input. Each rod is connected to a special bipolar cell type called the rod bipolar cell. Rod bipolar cells "talk" to the so-called AII amacrine cells, which then provide excitatory input to the axon terminals of ON cone bipolar cells and inhibitory input to OFF cone bipolar cell terminals. Rods are hyperpolarized by light, whereas rod bipolar cells and AII amacrine cells are depolarized: These are therefore ON cells. The key point here is that AII amacrine cells can modulate both ON and OFF cone bipolar cells, but with opposite effects. The retina also incorporates different glial cell types, the most well studied being the Muller cells, which have important roles in a variety of homeostatic processes as well as in responses to injury and disease (Bringmann & Wiedemann 2012). Retinal cells are arranged in mosaics, covering the entire retina. The only exception to the mosaic arrangement is a special area of the retina in some primates and in a few predatory birds and reptiles. This area is called the fovea (Hendrickson 1992) and is the place with the highest cone density. The human fovea, also called macula, has no rods within its center, and the only cellular compartment that is organized in a mosaic fashion is the cone outer segment. Foveal cone cell bodies are piled on top of each other, whereas cell bodies of all other cell types are shuffled to the side, forming a concentric ring of cell bodies (Hendrickson 1992).

TARGETING RETINAL CELL TYPES: THE PROBLEM OF DIVERSITY

Gene therapy appears to be conceptually simple: A gene is delivered to a tissue where the lack of function of the mutated gene leads to loss of tissue function. However, when the target tissue has tens of different cell types, and only a few of them express the disease-associated gene, cell-type targeting is needed. One cannot treat a part of the brain the same way as one would tissues with simple architectures, such as the liver, which is made up of only a few cell types. Achieving cell-type-restricted gene expression is a largely unsolved problem (Busskamp et al. 2012, Busskamp & Roska 2011). The key for targeting lies in two components: one that permits the efficient entry and intracellular transport of the vector to a specific cell type and a second that restricts expression to a given cell type. This second factor often takes the form of a cell-type-specific promoter. In addition to cell targeting, two other very important variables to control for are the desired gene-expression level and its cell-to-cell variation. First we discuss permissive factors for gene delivery, followed by strategies to restrict expression, and then we comment on controlling the mean and variance of gene-expression levels.

Permissive Factors

Gene delivery can be based on viral or nonviral vectors. Viral vectors are engineered from adeno-associated viruses (AAVs), lenti viruses (LVs), and herpes viruses (HVs) (Colella & Auricchio 2012). Nonviral methods include gene delivery based on naked DNA, oligonucleotides, DNA enclosed in cationic liposomes, and DNA associated with polymers (Li & Huang 2000). The most widely used vectors for retinal gene delivery are AAVs (Colella & Auricchio 2010). AAVs can be modified to fine-tune a number of permissive components. These components do not give specificity to the AAV, in terms of targeting specific cell types, but do permit high expression levels or expression in a particular time frame. Permissive components include capsid types responsible for viral entry, capsid residues responsible for intracellular AAV processing, and single- or double-stranded AAV forms, as well as the sequence of the gene of interest. An additional permissive property is the site of injection in the eye: whether it be the anterior chamber, vitreous cavity, or subretinal space. AAVs are available with more than 100 different natural capsids, and thousands of genetically modified capsids have been created through rational mutagenesis of the capsid (Petrs-Silva et al. 2009, 2011; Zhong et al. 2008) or selected through combinatorial screens (Bartel et al. 2012, Dalkara et al. 2011, Klimczak et al. 2009). AAVs with different capsids are called serotypes. Each serotype has a different efficiency for infecting retinal or other ocular cell types, a property called tropism (Colella & Auricchio 2010). Mutant capsids sometimes allow better transduction efficiencies or different tropisms over the natural variants. Unfortunately, the tropism of a serotype is species dependent. Therefore, all serotypes must be tested in primates in vivo and, ideally, ex vivo in human retinal explants. However, we do not know for certain if any of these tests reliably predict good in vivo expression in humans: First, it is not clear if nonhuman primates have the same capsid-cell-type interactions as humans do; and second, due to dilution and other factors, ex vivo tests do not necessarily reflect the in vivo situation. Recent studies have found that some capsid residue (tyrosine) mutations increase gene expression by decreasing the intracellular elimination of AAV particles. These capsid modifications can be fashioned regardless of the capsid type used and could improve gene expression. The slow onset (~3 weeks) of transgene expression is considered to be a limitation of AAVs, but it could be circumvented by using self-complementary

AAVs (scAAVs), which ensure more rapid transgene expression (<1 week). It is not clear if, in the long term, scAAVs express more, and a limitation of scAAVs is that they can package only half the amount (2.4 kb) of DNA compared with normal AAVs (4.8 kb). The site of injection is an important determinant of the number of AAV particles that reach a given cell type. For retinal delivery, AAVs can be injected into the subretinal or the intravitreal space. When a normal AAV is injected into a primate eye intravitreally, the cells in the fovea are densely labeled but cells outside the fovea are sparsely transduced (Ivanova et al. 2010, Yin et al. 2011; D. Dalkara & J. Flannery, personal communication). This difference of infectivity in the fovea versus periphery is likely due to the very thick inner limiting membrane that forms a barrier between the intravitreal space and the retina. Subretinal injection leads to high-density labeling around the injection site. New AAV variants, with modified capsids, may be able to infect densely from the vitreal side outside of the fovea, but this has not yet been demonstrated in primates. Intravitreal injections have one more limitation: The intravitreal space is large and not well defined because the vitreous body that occupies this space is heterogeneous and changes in consistency with the subject's age, being gelatinous in young people and liquid in older people. The vitreous body can be removed, leaving an aqueous space several milliliters in size that could be filled either with a very large amount of virus or with a smaller amount, leading to significant AAV dilution. An advantage of intravitreal injection is that it is simple, and no detachment of the neural retina from the pigment epithelium is created. Subretinal injection is more difficult, leading to a temporary separation of the pigment epithelium from the photoreceptors; however, AAV concentrations can be tightly controlled. In gene-replacement therapy, where the photoreceptor outer segments are intact and need to interact with the pigment epithelium for outer segment renewal, subretinal injections must be done with care. We argue, however, that when the rods and cones do not respond to light, and therefore their interaction with the pigment epithelium may not be as important, subretinal injections are the preferred method. A limitation of AAVs is the restricted length of DNA (a total of 4.8 kb) that can be packaged into these vectors. Possible solutions to this limitation, as yet still theoretical, would be either to generate dual AAV vectors, each carrying one half of the transgene, which then reassembles in vivo or the packaging of oversized genomes (Colella & Auricchio 2012).

Restrictive Factors

The specificity of AAVs can be controlled by cell-type-specific promoters (Busskamp et al. 2012). The notion "specific promoter" must be treated with caution. First, gene expression driven by a specific promoter is likely detectable in nontargeted cell types. The ratio of expression between target and off-target cell types and the threshold for the biological effect are the relevant quantities to describe specificity. Second, whether a promoter drives specific expression in a cell type depends on the method of expression. The same promoter could behave differently when used in a transgenic animal, when electroporated, or when expressed from a virus. This variance is because the specific promoter is simply a DNA sequence that binds a combination of transcription factors, some of which are cell-type specific. However, when surrounded by different sequences in a viral or transgenic context, the binding can be modified. Using different expression methods, the copy number of the promoter-gene construct also varies significantly. Specificity and expression also depend on the species. Although finding promoters based on the gene-expression patterns of different retinal cell types (Siegert et al. 2012) and based on sequence conservation in mammals is a rational starting point, the process to select the stretch of DNA that fits the AAV and confers specificity is still trial and error. A key point here is that, ideally, the promoter should be tested in combination with all the other elements of the targeting vector in the final formulation, both in vivo in nonhuman primates and ex vivo in postmortem human retinas. Specificity could also be achieved by sensing the cell's gene-expression state combinatorially using cell-type classifiers built by synthetic biology approaches (Benenson 2012).

Control of Mean and Variance of Gene Expression

The number of AAV particles present in the different cells varies, even within cells of the same type. To our knowledge, none of the present preclinical or clinical gene-therapy trials controls for the variance of gene expression due to copy number variation. In some instances, the precise level of expression is not as important; the replacement of defective enzymes could be one such case. However, when precise stoichiometry is needed, regulation of the mean and variance of gene expression could be critical. The mean expression can be tuned using promoters of different strengths as well as using different combinations of permissive factors, whereas the control of cell-to-cell variation ("noise") in gene expression may require insights from synthetic biology (Benenson 2012).

FORMS OF GENE THERAPY

Gene therapy for hereditary eye diseases may take one of two different forms: One depends on the mutated gene or even on the mutation itself; the other is independent of the mutated gene and depends only on the structural and functional state of the retinal circuit. The first form is called gene supplementation or gene replacement. Although this approach is the most logical and straightforward to correct a genetic disease of the eye, one challenge in gene-replacement strategies is the large number of affected genes. Hundreds of disease-causing gene mutations and dozens of mutated genes have been reported. A second challenge is the limited packaging capabilities of gene-delivery vectors suitable for use in humans. Only a few disease-associated genes fit into these vectors. A third challenge is that the cell type that expresses the mutated gene must be alive at the time of therapy. A fourth challenge concerns dominant mutations in which the mutated allele is toxic to the cell. The eye diseases most suited to gene supplementation are recessive hereditary diseases caused by mutations in small genes in long-surviving retinal cell types such as the retinal pigment epithelium.

A number of mutation-independent approaches either attempt to slow down retinal degeneration (neuroprotection) or do not interfere with the intrinsic progression of retinal degeneration, but attempt to restore photosensitivity by creating new photosensors and coupling them to the remaining retinal circuitry. For this latter strategy, researchers have considered three approaches: Electronic implants (Humayun et al. 2012, Zrenner et al. 2011) gather light using a technology similar to that used by video cameras and communicate pixel intensity to retinal cells via injected currents; stem cell approaches (Ong & da Cruz 2012, Singh & MacLaren 2011, Tibbetts et al. 2012) attempt to derive new photoreceptor cells, which must then be integrated into existing retinal circuits; and finally, optogenetic approaches (Busskamp et al. 2012, Busskamp & Roska 2011) can be used to target light sensors genetically to strategically important retinal cells. Either nonphotoreceptor cells are turned into photosensors (Bi et al. 2006, Lagali et al. 2008), or the photosensitivity of native photoreceptors compromised by disease is restored (Busskamp et al. 2010). The key point in optogenetic therapy is that retinal cells are already connected to other retinal circuit elements, and therefore a critical technological problem, namely how to connect photosensors to existing retinal circuits in a biologically relevant way, has already been solved. Here we discuss neuroprotective and optogenetic therapies. Stem cell (Ong & da Cruz 2012, Singh & MacLaren 2011, Tibbetts et al. 2012) and electronic implant approaches (Weiland et al. 2005, Zrenner et al. 2011) have been reviewed elsewhere.

Which kind of gene therapy can be performed depends on the stage of retinal degeneration. **Figure 2** correlates the stage of degeneration with the possible forms of gene

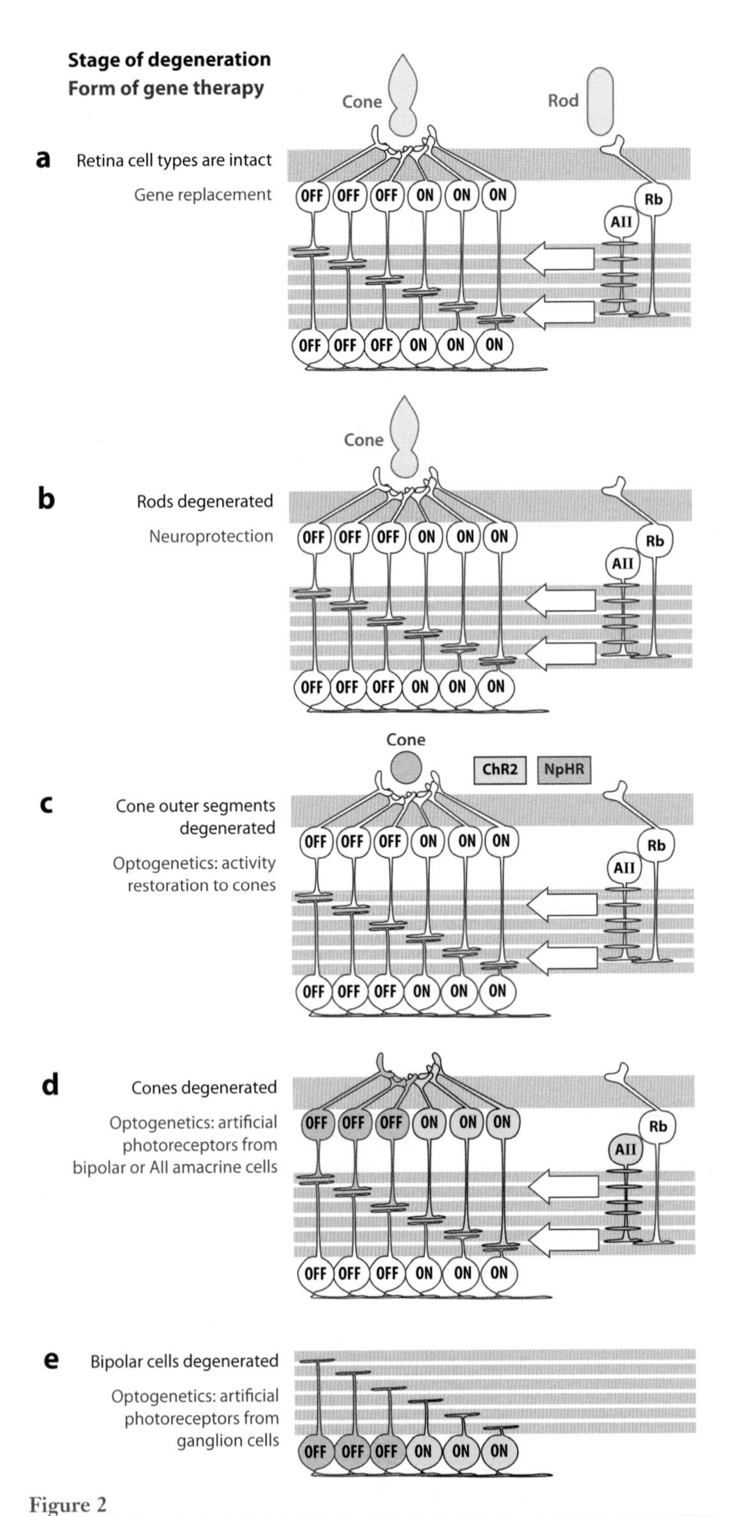

Figure 2

Stages of retinal degeneration and the forms of gene therapy that can be used in that stage.

therapy. Note that to what extent a stage at which no cones are left or only ganglion cells remain does exist in patients is not clear (**Figure 2**).

GENE-REPLACEMENT THERAPIES

Hereditary eye diseases can affect only the eye (nonsyndromic) or can affect other organs in addition to the eye (syndromic).

Nonsyndromic Retinitis Pigmentosa

Retinitis pigmentosa (RP) refers to a diverse group of progressive, hereditary diseases that often lead to incurable blindness and affect two million people worldwide. RP commonly starts with night blindness in young adults, reflecting early degeneration of the highly sensitive rod photoreceptors. This symptom is followed by a delay with a progressive decline in daylight central vision, owing to loss of function of the less-sensitive cone photoreceptors. Mutations in more than 44 genes have been demonstrated in different forms of RP, but in ~50% of cases the mutation has not yet been identified. Most known RP genes are expressed in rods or retinal pigment epithelium, and in such cases, degeneration of cones is thought to be a secondary consequence of the death of the rods.

In nonsyndromic, recessive RP, retinal pigment epithelium and photoreceptor cells are the targets of a large proportion of gene-replacement studies (Ali et al. 2000, Bennett et al. 1996, Takahashi et al. 1999, Vollrath et al. 2001). Recessive RP is particularly suitable for gene-replacement therapy if the gene of interest fits the targeting vector and if the cells involved, usually rods, are alive at the time of therapy.

The mutational heterogeneity (see **https://sph.uth.tmc.edu/retnet/**), together with the molecular gain-of-function leading to degeneration, represents a significant obstacle to the development of gene therapies for dominantly inherited RP. Several new studies try to address simultaneous silencing of the effect of dominant mutations and the substitution for normal function (Baulcombe 2002, Chadderton et al. 2009, Farrar et al. 2011, Mao et al. 2012, Tam et al. 2008).

Syndromic Retinitis Pigmentosa

The most frequent forms of syndromic RP are Usher syndrome and Bardet-Biedl syndrome. There is, as yet, no effective cure or prevention of the vision loss associated with these diseases, but recent studies have provided potential strategies for therapy of the retinal phenotype.

Bardet-Biedl syndrome affects the function of nonmotile cilia and leads to obesity, RP, the formation of extra digits, hypogonadism, and kidney problems (Forsythe & Beales 2012). The first successful gene-based treatment of Bardet-Biedl syndrome–associated retinal degeneration was demonstrated in a *Bbs4*-null mouse model, where an AAV-based vector expressing BBS4 prevented photoreceptor death, improved retinal function, and visually evoked behavior responses, providing hope for a promising future therapy in humans (Simons et al. 2011).

Usher syndrome is manifested in deafness and blindness. On the basis of symptom onset and severity, the disease is categorized into three groups: Usher I, II, and III. In Usher I syndrome, one of the major hurdles in the design, development, and validation of gene therapy is the discovery that those mutations that cause retinal degeneration in humans cause little or no retinal degeneration in mice. Sahly et al. (2012) recently reported clues to understanding this discrepancy and the function of Usher proteins in the primate retina. Despite the mild retinal phenotype in mice, Hashimoto et al. (2007) demonstrated the retinal expression of an Usher I gene, *MYO7A*, and the correction of some cellular defects. Owing to the large size of *MYO7A* (~7 kb), the researchers used LV for gene delivery. Researchers in this field still intensely debate the effectiveness of LV variants for transducing adult photoreceptors in vivo. Usher II mouse models better resemble the human phenotype, and a recent study using an AAV-based delivery system has demonstrated successful and specific expression of one of the Usher II genes, *DFNB31*, in both rods and cones (Zou et al. 2011).

Other Hereditary Retinal Diseases Amenable to Gene-Replacement Therapy

Although gene-therapy prospects for retinal degeneration have been most intensively studied in LCA and RP, studies have also demonstrated successful gene therapy for other degenerative conditions of the retina.

Achromatopsia is a genetic disorder of cones caused by mutations in a number of cone-expressed genes: It leads to the loss of cone function, but the cells often remain alive for a long time. In the Gnat2$^{-/-}$ mouse model of achromatopsia, the AAV-mediated *GNAT2* gene therapy targeted the cones and rescued the cone-mediated electroretinogram (ERG) and visual acuity (Alexander et al. 2007). In the Cngb3$^{-/-}$ mouse model, the AAV-delivered gene driven by a cone arrestin promoter showed a similar effect (Carvalho et al. 2011). Researchers have found that the robustness and stability of the observed treatment effect were independent of mutation but dependent on both promoter type and age. Komaromy et al. (2010) achieved a stable therapeutic effect (for at least 33 months) in younger animals. Because mutations in the *CNGB3* gene are the most prevalent cause of achromatopsia, accounting for more than 50% of all known cases of this disease, this study provides proof-of-concept for a potential therapy to treat the biggest subset of patients.

Stargardt disease, which affects cones and rods, is most often manifested in early-onset macular degeneration. In Abca4$^{-/-}$ mice, a model of the most common human recessive Stargardt disease, intraocular administration of an AAV encoding *ABCA4* resulted in protein localization to rod outer segments and stable morphological and functional improvement of the phenotype in one study (Allocca et al. 2008). The large size of the *ABCA4* gene (~7 kb) led other groups to use LVs. In the same mouse model, LVs ensured high transduction efficiency in rods and cones and significant reduction of lipofuscin pigment A2E accumulation, suggesting that LV gene therapy is potentially an efficient tool to treat ABCA4-associated diseases (Kong et al. 2008).

X-linked juvenile retinoschisis (XLRS) is another hereditary disease that leads to macular degeneration, although in some cases the peripheral retina is also affected. The gene that is responsible for most cases of XLRS is *RS1*, which codes for the protein retinoschisin. The lack of retinoschisin causes small tears between the layers of the retina. In the $Rs1h^{-/-}$ mouse model of XLRS, AAV-mediated intravitreal delivery of the normal *RS1* gene reduced the structural and functional loss of the retina when evaluated at 14 months of age; substantial (but variable) long-term rescue ERG amplitudes and waveforms were also reported (Kjellstrom et al. 2007, Min et al. 2005, Park et al. 2009).

Leber's hereditary optic neuropathy (LHON) is a mitochondrial disorder affecting ganglion cells that results in severe and usually irreversible visual loss in one eye. With some delay the fellow eye frequently suffers a similar loss of function. Unlike most retinal degenerations, which result in slow, progressive loss of vision over many years, LHON progresses quickly, although the timing of vision loss in the first eye is not predictable. Because vision loss in the second eye is highly likely a few months after the vision loss in the first eye, genetic correction of the mitochondrial defect for the second eye during this time window is logical. Investigators have recently undertaken important steps toward gene therapy for LHON. In a rat model, *ND1* expression restored vision, and signs of recovery appeared as early as 1–2 weeks after AAV-mediated *ND1* delivery into the superior colliculus (Marella et al. 2010). It should be emphasized that few practical methods for delivering genes to the mitochondria are currently available. To address this, Manfredi et al. (2002) developed an approach termed allotropic expression. Rescue of optic neuropathy (Ellouze et al. 2008) in an induced rat model of LHON suggested that allotropic *ND4* gene therapy could be effective in LHON patients with an *ND4* mutation. The authors optimized the allotropic expression for the mitochondrial genes *ATP6*, *ND1*, and *ND4* and obtained a complete and long-lasting rescue of mitochondrial dysfunction in human fibroblasts in which these genes were mutated. The same group has recently reported a substantial and long-lasting protection of retinal ganglion cell and optic nerve integrity after intravitreal administration of an AAV vector containing the full-length open reading frame and the 3′ untranslated region of the *AIF1* (apoptosis-inducing factor 1) gene in a spontaneous model of optic atrophy, the Harlequin mouse (Bouaita et al. 2012).

NEUROPROTECTION

Gene-therapy strategies can be used not only to correct the gene defect, but also to delay the degeneration independent of the mutation. One of the most frequently studied neuroprotective methods in animal models is the local expression of neurotrophic proteins to promote the survival of photoreceptors and the retinal pigment epithelium. Neuroprotection offers the possibility to treat not only genetic diseases but also a range of conditions, including acquired ocular disorders. Among the most extensively studied neuroprotective factors are ciliary neurotrophic factor (CNTF), brain-derived neurotrophic factor (BDNF), basic fibroblast growth factor (FGF), and glial cell–derived neurotrophic factor (GDNF) (Barnstable & Tombran-Tink 2006).

An attractive candidate to prevent and treat retinal degeneration is the protein rod-derived cone viability factor (RdCVF), specifically expressed and secreted by photoreceptors (Léveillard et al. 2004). RdCVF is a product of the nucleoredoxin-like 1 (*NXNL1*) gene homologous to the family of thioredoxins known to possess strong antioxidative properties. This trophic factor has directly induced cone survival in animal models of recessive and dominant RP, and studies have documented functional rescue independent of the

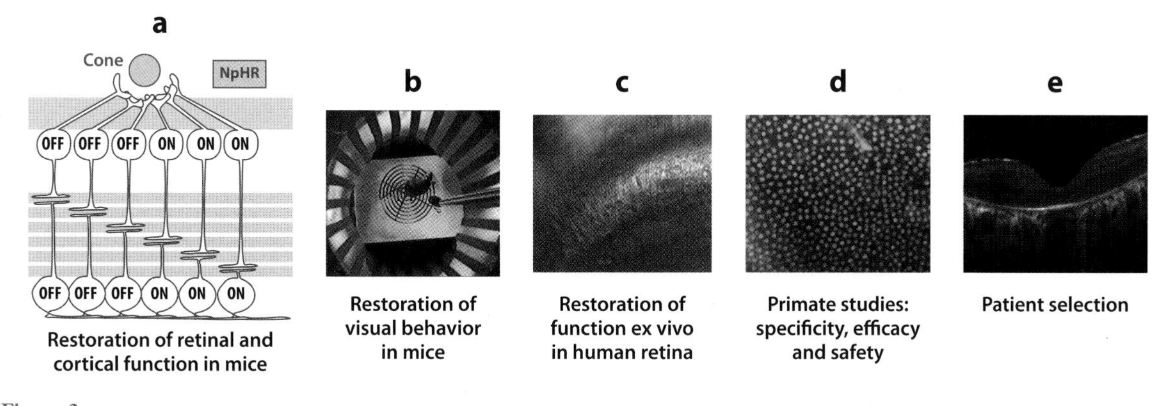

Figure 3

Translational route for one of the optogenetic therapies where cone activity is restored by the expression of halorhodopsin (NpHR).

mechanism and extent of rod degeneration. In treated animals, the functional effect observed appears to be related to the maintenance of cone outer segments (Yang et al. 2009). The gene encoding RdCVF also encodes for a second product with the characteristics of a thioredoxin-like enzyme, suggesting that RdCVF may play a physiological signaling role. It may be involved in the maintenance of photoreceptors by both autocrine and paracrine mechanisms, of importance during aging and exposure to oxidative stress. The two forms of RdCVF could be of particular interest as a therapeutic modality to prevent the secondary degeneration of cones in RP and, thereby, save vision (Fridlich et al. 2009, Léveillard et al. 2004, Léveillard & Sahel 2010).

Metabolic dysregulation and oxidative stress correlate with loss of visual activity, particularly in cones, and novel approaches are designed to modulate these pathways (Punzo et al. 2009, 2012). In addition, gene-based delivery can also be used to bring antiapoptotic, anti-inflammatory, or antiangiogenic molecules. Clinical trials are currently under way to investigate the effectiveness of the various neuroprotective therapies.

OPTOGENETIC THERAPY

Optogenetic therapy is relevant for treating forms of blindness in which some part of the retinal circuitry remains intact. Most diseases that fall into this category affect photoreceptors, rods, and cones. Initial attempts to restore vision using optogenetic methods have focused on RP, which is a collection of hereditary diseases that affect mostly rod-expressed genes and lead to rod degeneration. Later in the course of disease, due to a process that is not well understood, the cones also degenerate or lose their outer segments and become light insensitive, although, in most cases, they do not express the gene affected by the mutation. Optogenetic therapy is in the stage of preclinical studies. A translational route for optogenetic therapies is shown in **Figure 3**.

Optogenetic Tools

The two most well-known optogenetic tools are channelrhodopsin-2 (ChR2) (Boyden et al. 2005, Nagel et al. 2003), from the algae *Chlamydomona reinhardtii*, and halorhodopsin (NpHR) (Zhang et al. 2007), from the archaebacterium *Natronomas pharaonis*. These proteins are photosensitive and can be activated at specific light wavelengths. ChR2, a nonselective cation channel, provides neurons with excitatory currents and can therefore be used to activate cells (optogenetic activator), whereas NpHR, a chloride pump, generates inhibitory currents, which subsequently inactivate cells (optogenetic inhibitor). In recent years, a large number of mutant channelrhodopsins have been produced, with different functional properties, and new light-gated pumps have been isolated from different species (Chow et al. 2012; Kleinlogel et al. 2011a,b; Prigge et al. 2012; Tye & Deisseroth 2012). A key property of these light-gated proteins is that in most animals, including mammals, they do not require any externally supplied cofactors. Both optogenetic activators and inhibitors can be used to restore retinal photosensitivity because certain retinal cells, such as ON bipolar and ON ganglion cells, are naturally activated by light, whereas photoreceptors, OFF bipolar cells, and OFF ganglion cells are naturally inhibited by light (Masland 2001, Wässle 2004).

Apart from natural light-gated channels, pumps, and their mutants, synthetic light-gated actuators have also been described using azobenzene as the light sensor (Fehrentz et al. 2011, Kramer et al. 2009). One version of this approach uses a genetically expressed channel in combination with an organic component (Caporale et al. 2011). The second approach uses only an organic molecule to modulate intrinsic channels expressed in retinal neurons (Polosukhina et al. 2012). Finally, researchers have used intrinsic light-sensitive proteins, such as melanopsin (Lin et al. 2008). Melanopsin is expressed in intrinsically photosensitive retinal ganglion cells, which participate in non-image-forming visual functions such as the entrainment of the circadian rhythm and the pupillary reflex (Schmidt et al. 2011).

Functional State of the Retinal Circuit in Blinding Diseases

A large body of work has detailed the changes in the retinal architecture in the rd1 mouse model of RP (Jones et al. 2012, Marc et al. 2003). These changes are mostly localized to the rod circuitry, the cones, and, to a lesser degree, the cone circuitry. The degeneration is progressive: Despite complete blindness, the state of the retinal circuit is different in two-month-old and one-year-old mice. Work on a different mouse model, *rd10*, has revealed a different time course for the disease (Gargini et al. 2007, Phillips et al. 2010). From research on the anatomy of RP mouse models, we can draw at least two conclusions. First, the time course of degeneration can be different depending on the mutation. Second, the state of the retina can be different within a group of individuals affected by the same mutation depending on the time of intervention. Work on the anatomy of postmortem human retinas from RP patients (Milam et al. 1998) has revealed interesting details about the changes in human retinal architecture. Notably, in all foveal regions examined, at least one layer of cone cell bodies was alive, but these dormant cones lacked outer segments (Milam et al. 1998). The existence of dormant cones is not specific to the human fovea: Work by LaVail and his colleagues revealed surviving cones in retinal degenerations at stages when no visual function could be detected (LaVail 1981). Despite all this important RP research, the state of the retina of a particular patient at a given time cannot be predicted. In vivo imaging, including optical coherence tomography, autofluorescence, and adaptive optics, allows clinicians to assess each patient and, therefore, decide which optogenetic strategy is most appropriate.

Optogenetic Strategies

A basic question when choosing a strategy is whether to express the optogenetic sensor randomly, using broad-spectrum promoters (Caporale et al. 2011, Isago et al. 2012, Ivanova et al. 2010, Ivanova & Pan 2009, Lin et al. 2008, Nirenberg & Pandarinath 2012, Thyagarajan et al. 2010, Tomita et al. 2009, 2010), or to target certain cell types using cell-type-specific promoters (Busskamp et al. 2010, Doroudchi et al. 2011, Lagali et al. 2008). At least two factors affect this decision. The first is safety: After an injection into the eye, some viral particles will appear in the blood. Broad promoters may drive gene expression in many tissues, including reproductive organs. Special care is needed in preclinical tests to assess this possibility. Furthermore, if ganglion cells express the sensor, without a dendritic localization signal, the antigen will be present not only in the eye but also in the brain regions where the ganglion cells project (Bi et al. 2006). If there is an immune reaction, dealing with the eye alone is simpler than having to treat higher brain regions. Second, with respect to retinal function, broad promoters will make most cells that take up the virus light sensitive. This way, the function of any remaining retinal circuit is lost and the retinal output will simply signal changes in light intensity, just as do the intrinsically photosensitive ganglion cells but more quickly. Furthermore, unless the sensor is localized to the dendrites (Greenberg et al. 2011), retinal ganglion cell axons will be light sensitive; local light stimulation will therefore activate both local retinal cells and any retinal cells whose axons cross the stimulation point, which may result in the patient seeing randomly distributed dots instead of one localized spot. Finally, investigators have shown that broad promoters injected into primate eyes mostly drive expression in the fovea (Ivanova et al. 2010, Yin et al. 2011), where ganglion cells and bipolar cells are not organized in mosaics, which prevents the projected image from forming regular neural images. Despite these problems, some broad promoters are strong and often work across species; therefore, some light sensitivity will be conferred to patients. Because they are highly adaptive, the higher visual centers will likely use all possible information present in the concerted activity of ganglion cells to extract information about the visual world.

When the goal of therapy is to restore biologically relevant function, specificity at the periphery of visual circuits is more important than at other sensory circuits such as auditory circuits. This is because the physical arrangement of the front circuitry for audition is different from that for vision. In the auditory system, only the first synapse, between sensory and secondary cells, is positioned at the periphery. Therefore, only a limited amount of neuronal processing occurs in the inner ear. In the visual system, the second synapse is also embodied in the retina, and this second synapse is where complex neuronal computations, among amacrine, bipolar, and ganglion processes, are performed at parallel sites. The effect of nonspecific stimulation of the inner ear is similar to that of cell-type-specific cone or bipolar cell stimulation in the retina.

To restore biologically appropriate neuronal activity to the retina, using cell-type-targeted treatments, the key principles for designing the strategy are as follows: First, the closer to the photoreceptors we stimulate cells in the neuronal chain, the more natural the retinal processing will be; second, the optogenetic sensor-evoked activity should match the natural activity of the stimulated cell. ON cells are activated by light, whereas OFF cells are inhibited by light. However, the cell types that can be stimulated depend on which cells are alive in the patient.

At least four different strategies can be used, dictated by the functional organization of the mammalian retina. First, if cones are still alive but are nonfunctional, i.e., they lack outer segments, they can be targeted with optogenetic inhibitors (Busskamp et al. 2010). Second, ON (Lagali et al. 2008) and OFF bipolar cells can be targeted with activators and inhibitors, respectively. Third, AII amacrine cells, which could drive both the ON and OFF systems, can be targeted with activators. Fourth, ON and OFF ganglion cells can be targeted with activators and inhibitors, respectively.

Each strategy has its own advantages and disadvantages. Some of the limitations are "dynamic"; e.g., a lack of promoter or suitable serotype could be a problem today but be solved in the near future. Others are dictated by biology and therefore are unlikely to be resolved easily. One example is the special structure of the fovea. As described earlier, the fovea, which is the place for high-resolution color vision and is responsible for most of our visual perception, has a specialized structure in which the ganglion cells and the bipolar cells are not organized in special mosaics but are piled on top of each other in a ring around the fovea. As a consequence, if the ganglion or bipolar cells become light sensitive, any projected image will be severely distorted. The distortion itself could be corrected to some degree, but because the cells pile up, a single image pixel will illuminate many cells with unknown topology. The only cells that are organized in a mosaic are the cones or, more precisely, the cone outer segments. The cone cell bodies are also piled on top of each other but more regularly than bipolar and ganglion cells: If the outer segments of two cones are close, the cell bodies are also close. It is the axons of the cones that break the spatial continuity. From these arguments, one may conclude that, in the fovea, the only cell type that can be stimulated with an image is the cones; however, even for these cells, the light sensitivity of their axons may cause image irregularities. Furthermore, because cone cell bodies are also piled on top of each other, once the outer segments are lost, it is impossible to regain the same resolution as that in healthy humans. For optogenetically transduced cones with no outer segments, the physical resolution is dictated by the size of the cone cell bodies and is further diminished by the light-sensitive axons. However, the resolution probably will not decrease with the further loss of cone cell bodies until only one layer of cones is left; this situation was observed in all postmortem examinations of human RP retinas.

Goggles, Adaptation, and Sensitivity

Whereas rods and cones can adapt to intensity distributions across eight decades, optogenetic sensors respond in a narrow range of intensities across only two decades. One possibility to increase the dynamic range of optogenetically transformed cells is to express multiple sensors with different light sensitivities. A more practical solution is the use of an external device, embedded in goggles, which acquires images across a large range of intensities, ranging from dim indoor environments to bright outdoors, and projects an image with the intensity distribution of the light-sensitized retinal cell type. Such goggles are currently under development (Grossman et al. 2010). The need for goggles is independent of the sensitivity of the optogenetically transduced cell. Using the vectors and optogenetic sensors that have already been described, the sensitivity of transduced cells is low. However, recent developments in both sensors and vectors have resulted in new vectors, leading to significantly increased sensitivities (Kleinlogel et al. 2011a).

Sophisticated Control: Color and Fine Control of Cellular State

Optogenetic vision restoration could be qualitatively improved in at least two ways. First, none of the currently available approaches allows for color perception. Introducing sensors with different optimal wavelengths could enable at least rudimentary color discrimination. Second, the state of the optogenetically transduced cells is unknown; therefore, optogenetic stimulation with one polarity, activation or inhibition, may use only a fraction of the cells' dynamic range. As an example, a normal cone in the dark is depolarized and constantly releases glutamate. When the cone is illuminated with light, hyperpolarization occurs and glutamate release ceases. Cones can modulate the release of glutamate within a given voltage range and can therefore transmit information to the next cells, the bipolar cells. Imagine cones in a blind patient being transduced with NpHR and then stimulated with red light. As a response, the cone cells will hyperpolarize, but whether this process will modulate glutamate release will depend on the dark voltage of the diseased cones. In *rd1* mice, cones are depolarized enough to modulate glutamate release (Busskamp et al. 2010). The best way to drive cones through their full dynamic range would be to modulate the dark voltage of the cones, for example, by expressing a highly sensitive but slow-acting ChR2 together with NpHR. The dark voltage could be modulated using background blue light, and the light responses could be modulated with red light. This kind of push-pull technique would allow complete control of the cones, independent of their voltage state: This is important because the voltage state may change during the course of the disease.

Potential Complications: Immune Reaction, Eye Movement, Photophobia, and Interference with Remaining Central Vision

Restoring optogenetic vision presents at least four potential complications. First, an immune reaction against the sensor is a potential danger. In our preliminary preclinical work, we found no sign of inflammation or other forms of immune reaction. A number of people are naturally infected by AAVs and therefore have antibodies against particular serotypes. These serotypes could be neutralized during gene therapy; therefore, prescreening of patients may be necessary. Second, the presence of uncontrolled eye movement in some patients may interfere with image projection to the region surrounding the optogenetically transduced area. Third, some RP patients experience discomfort or pain in the eyes when exposed to higher light intensities, a symptom called photophobia. These last two difficulties can be controlled for by careful patient selection. A fourth possible complication is the pupillary reflexes mediated by intrinsically photosensitive ganglion cells. The peak sensitivity of the melanopsin-mediated response in these ganglion cells is ~490 nm (Berson et al. 2002); therefore, the more red shifted the sensor is, the less pupillary constriction it causes. A more conceptual difficulty is that many patients have some remaining central vision. The development of optogenetic sensors in the near-infrared region of the electromagnetic spectrum could enable the use of these remaining natural sensors together with injected optogenetic sensors.

ASSESSING ELIGIBILITY FOR GENE THERAPY

The advances in gene therapy in animal models require investigators to identify patients who would benefit from these treatments. Key steps for this process are human genetics and in vivo noninvasive imaging of the retina in patients.

Modern optical coherence tomography (OCT) devices can produce high-resolution images of the retina and optic nerve in a noninvasive manner. OCT is used to examine the retinal architecture and, in particular, the photoreceptor inner/outer segment border. OCT is particularly useful to correlate the retinal sensitivity and outer retina structure. With this technique, researchers can evaluate the preservation of the cone outer segments when the rods are already lost. For example, patients with existing outer segments would be eligible for treatment with neuroprotective agents such as RdCVF. Individuals with visual acuity below light perception and no visual field but with a preserved layer of cone bodies still shown on the OCT could be eligible for optogenetic functional restoration of cones (**Figure 3**). Autofluorescence imaging indicates the status of the interactions between photoreceptor and pigment epithelium cells.

The retinal imaging devices that are currently available do have limits, however. Irregular optical defects (ocular wave-front aberrations) may limit the resolution of these devices. The newly introduced adaptive optics-based laser-imaging technologies can correct the irregular optical defects and provide a retinal image with a lateral resolution of 2.5–3 μm, making it possible to visualize human cones noninvasively not only in the macula but all over the retina in a few minutes. Adaptive optics imaging to assess photoreceptor mosaic structure directly may have important implications for establishing functional correlates and studying gene-therapy outcomes.

In connection with such imaging studies, functional investigations should extend beyond testing visual acuity. Several protocols for visual field testing allow regional sensitivity thresholds to be determined. Microperimetry provides a direct mapping of function by projecting stimuli of variable size and luminance while observing the back of the eye, known as the fundus. Once an accurate map of retinal function has been obtained, together with high-resolution imaging, it will be easier to select patients for trials, and investigators can

Figure 4

Rehabilitation using the combination of real and virtual visual elements. To implement innovative adaptive rehabilitation strategies and to demonstrate the functional benefit of different visual restoration approaches relevant to real-life situations, new rehabilitation and testing platforms are being developed. The picture shows the Streetlab platform at the Vision Institute, Paris, which integrates real and virtual-reality visual elements to provide a place for rehabilitation as well as quantifiable tests for visual functions that are useful for performing tasks in everyday life.

identify the area of injection to optimize the ratio between the expected functional benefits and potential hazards resulting from subretinal injections.

Functional studies are also of paramount importance to establish the actual benefit of these novel therapies. Although visual acuity represents the gold standard for regulatory and funding bodies, no significant changes may be measured, although the visual field may stabilize or improve. In the LCA trials, even though visual acuity did not improve, testing for mobility in dark and even lighted environments indicated an obvious improvement (fewer bumps, shorter time to target). Because most quality-of-life questionnaires may not detect such changes, it is timely to develop standardized mobility and task-related tests that would provide both sensitivity and reliability (**Figure 4**).

REHABILITATION

Rehabilitation is an important component of vision restoration for patients who become blind both before and after the critical period of vision. Following rehabilitation in patients who became blind after the critical period, vision restoration using retinal prosthesis has led to some visual experience, including the localization of light sources and objects, and in some cases to the recognition of shapes and letters (Humayun et al. 2012, Zrenner et al. 2011). Even in early-onset blindness caused by LCA2, vision restoration using gene therapy and rehabilitation led to a measurable retinotopic map in the visual cortex (Ashtari et al. 2011, Sahel 2011). Finally, studies have demonstrated that the visual cortex can be activated in congenitally blind patients by stimulating other sensory pathways, showing the plasticity of the cortex (Pascual-Leone et al. 2005, Reich et al. 2012). Upon vision restoration to blind patients, innovative rehabilitation programs can recruit brain plasticity to teach the brain the novel “language” that the restored retina uses to communicate visual input (**Figure 4**).

CONCLUSION

Our capability to perform ocular gene therapy has increased substantially in the past decade owing to the enormous progress made in uncovering novel genetic causes and risks in blinding eye diseases, in developing and analyzing animal models, in developing in vivo imaging modalities in human patients, and in refining gene-delivery tools. Despite all this progress, many questions still remain unanswered: how to choose promoters and prepare vectors for clinical use; how to decide on the volume of injection and the location of intraocular vector administration; whether to use gene-replacement or rather mutation-independent gene therapy; how to choose and standardize patients for a given therapy; and how to evaluate visual function before and after gene therapy. To provide the most relevant therapy, we need further improvement in our understanding of genotype-phenotype correlations and in the diagnosis of the functional status of retinal cells in vivo in patients. The enthusiasm to provide therapy for such a major unmet medical need propels the field of ocular gene therapy forward to answer these questions. We believe that the recent major advances in gene therapy for the eye will pave the way for gene therapies in other parts of the brain.

DISCLOSURE STATEMENT

Both authors are among the founders of Gensight, Inc., a new company that specializes in innovative retinal gene therapy applications such as optogenetic vision restoration. José-Alain Sahel is also a consultant for Sanofi and a founder of Pixium Vision, a new company dedicated to the development of artificial retinas.

ACKNOWLEDGMENTS

We thank Katia Marazova, Dasha Nelidova, Deniz Dalkara, Serge Picaud, and Sara Oakeley for advice on and corrections to the manuscript.

LITERATURE CITED

Acland GM, Aguirre GD, Bennett J, Aleman TS, Cideciyan AV, et al. 2005. Long-term restoration of rod and cone vision by single dose rAAV-mediated gene transfer to the retina in a canine model of childhood blindness. *Mol. Ther.* 12:1072–82

Acland GM, Aguirre GD, Ray J, Zhang Q, Aleman TS, et al. 2001. Gene therapy restores vision in a canine model of childhood blindness. *Nat. Genet.* 28:92–95

Alexander JJ, Umino Y, Everhart D, Chang B, Min SH, et al. 2007. Restoration of cone vision in a mouse model of achromatopsia. *Nat. Med.* 13:685–87

Ali RR, Sarra GM, Stephens C, Alwis MD, Bainbridge JW, et al. 2000. Restoration of photoreceptor ultrastructure and function in retinal degeneration slow mice by gene therapy. *Nat. Genet.* 25:306–10

Allocca M, Doria M, Petrillo M, Colella P, Garcia-Hoyos M, et al. 2008. Serotype-dependent packaging of large genes in adeno-associated viral vectors results in effective gene delivery in mice. *J. Clin. Invest.* 118:1955–64

Ashtari M, Cyckowski LL, Monroe JF, Marshall KA, Chung DC, et al. 2011. The human visual cortex responds to gene therapy-mediated recovery of retinal function. *J. Clin. Invest.* 121:2160–68

Azeredo da Silveira R, Roska B. 2011. Cell types, circuits, computation. *Curr. Opin. Neurobiol.* 21:664–71

Bainbridge JW, Smith AJ, Barker SS, Robbie S, Henderson R, et al. 2008. Effect of gene therapy on visual function in Leber's congenital amaurosis. *N. Engl. J. Med.* 358:2231–39

Bainbridge JW, Tan MH, Ali RR. 2006. Gene therapy progress and prospects: the eye. *Gene Ther.* 13:1191–97

Barnstable CJ, Tombran-Tink J. 2006. Molecular mechanisms of neuroprotection in the eye. *Adv. Exp. Med. Biol.* 572:291–95

Bartel MA, Weinstein JR, Schaffer DV. 2012. Directed evolution of novel adeno-associated viruses for therapeutic gene delivery. *Gene Ther.* 19:694–700

Baulcombe D. 2002. RNA silencing. *Curr. Biol.* 12:R82–84

Benenson Y. 2012. Biomolecular computing systems: principles, progress and potential. *Nat. Rev. Genet.* 13:455–68

Bennett J, Ashtari M, Wellman J, Marshall KA, Cyckowski LL, et al. 2012. AAV2 gene therapy readministration in three adults with congenital blindness. *Sci. Transl. Med.* 4:120ra15

Bennett J, Tanabe T, Sun D, Zeng Y, Kjeldbye H, et al. 1996. Photoreceptor cell rescue in retinal degeneration (rd) mice by in vivo gene therapy. *Nat. Med.* 2:649–54

Berson DM, Dunn FA, Takao M. 2002. Phototransduction by retinal ganglion cells that set the circadian clock. *Science* 295:1070–73

Bhatia B, Jayaram H, Singhal S, Jones MF, Limb GA. 2011. Differences between the neurogenic and proliferative abilities of Müller glia with stem cell characteristics and the ciliary epithelium from the adult human eye. *Exp. Eye Res.* 93:852–61

Bi A, Cui J, Ma YP, Olshevskaya E, Pu M, et al. 2006. Ectopic expression of a microbial-type rhodopsin restores visual responses in mice with photoreceptor degeneration. *Neuron* 50:23–33

Bouaita A, Augustin S, Lechauve C, Cwerman-Thibault H, Bénit P, et al. 2012. Downregulation of apoptosis-inducing factor in Harlequin mice induces progressive and severe optic atrophy which is durably prevented by AAV2-AIF1 gene therapy. *Brain* 135:35–52

Boyden ES, Zhang F, Bamberg E, Nagel G, Deisseroth K. 2005. Millisecond-timescale, genetically targeted optical control of neural activity. *Nat. Neurosci.* 8:1263–68

Bringmann A, Wiedemann P. 2012. Müller glial cells in retinal disease. *Ophthalmologica* 227:1–19

Busskamp V, Duebel J, Balya D, Fradot M, Viney TJ, et al. 2010. Genetic reactivation of cone photoreceptors restores visual responses in retinitis pigmentosa. *Science* 329:413–17

Busskamp V, Picaud S, Sahel JA, Roska B. 2012. Optogenetic therapy for retinitis pigmentosa. *Gene Ther.* 19:169–75

Busskamp V, Roska B. 2011. Optogenetic approaches to restoring visual function in retinitis pigmentosa. *Curr. Opin. Neurobiol.* 21:942–46

Caporale N, Kolstad KD, Lee T, Tochitsky I, Dalkara D, et al. 2011. LiGluR restores visual responses in rodent models of inherited blindness. *Mol. Ther.* 19:1212–19

Carvalho LS, Xu J, Pearson RA, Smith AJ, Bainbridge JW, et al. 2011. Long-term and age-dependent restoration of visual function in a mouse model of CNGB3-associated achromatopsia following gene therapy. *Hum. Mol. Genet.* 20:3161–75

Caspi RR. 2010. A look at autoimmunity and inflammation in the eye. *J. Clin. Invest.* 120:3073–83

Chadderton N, Millington-Ward S, Palfi A, O'Reilly M, Tuohy G, et al. 2009. Improved retinal function in a mouse model of dominant retinitis pigmentosa following AAV-delivered gene therapy. *Mol. Ther.* 17:593–99

Chow BY, Han X, Boyden ES. 2012. Genetically encoded molecular tools for light-driven silencing of targeted neurons. *Prog. Brain Res.* 196:49–61

Cideciyan AV, Hauswirth WW, Aleman TS, Kaushal S, Schwartz SB, et al. 2009a. Human RPE65 gene therapy for Leber congenital amaurosis: persistence of early visual improvements and safety at 1 year. *Hum. Gene Ther.* 20:999–1004

Cideciyan AV, Hauswirth WW, Aleman TS, Kaushal S, Schwartz SB, et al. 2009b. Vision 1 year after gene therapy for Leber's congenital amaurosis. *N. Engl. J. Med.* 361:725–27

Colella P, Auricchio A. 2010. AAV-mediated gene supply for treatment of degenerative and neovascular retinal diseases. *Curr. Gene Ther.* 10:371–80

Colella P, Auricchio A. 2012. Gene therapy of inherited retinopathies: a long and successful road from viral vectors to patients. *Hum. Gene Ther.* 23:796–807

Dalkara D, Klimczak R, Visel M, Schaffer D, Flannery J. 2011. *Developing photoreceptor targeted AAV variant by directed evolution.* Presented at Am. Soc. Gene Cell Therapy's, Annu. Meet., 14th, Seattle, WA

den Hollander AI, Roepman R, Koenekoop RK, Cremers FP. 2008. Leber congenital amaurosis: genes, proteins and disease mechanisms. *Prog. Retin. Eye Res.* 27:391–419

Doroudchi MM, Greenberg KP, Liu J, Silka KA, Boyden ES, et al. 2011. Virally delivered channelrhodopsin-2 safely and effectively restores visual function in multiple mouse models of blindness. *Mol. Ther.* 19:1220–29

Ellouze S, Augustin S, Bouaita A, Bonnet C, Simonutti M, et al. 2008. Optimized allotopic expression of the human mitochondrial ND4 prevents blindness in a rat model of mitochondrial dysfunction. *Am. J. Hum. Genet.* 83:373–87

Falk MJ, Zhang Q, Nakamaru-Ogiso E, Kannabiran C, Fonseca-Kelly Z, et al. 2012. NMNAT1 mutations cause Leber congenital amaurosis. *Nat. Genet.* 44:1040–45

Farrar GJ, Millington-Ward S, Chadderton N, Humphries P, Kenna PF. 2011. Gene-based therapies for dominantly inherited retinopathies. *Gene Ther.* 19:137–44

Fehrentz T, Schönberger M, Trauner D. 2011. Optochemical genetics. *Angew. Chem. Int. Ed. Engl.* 50:12156–82

Fletcher EL, Jobling AI, Vessey KA, Luu C, Guymer RH, Baird PN. 2011. Animal models of retinal disease. *Prog. Mol. Biol. Transl. Sci.* 100:211–86

Forsythe E, Beales PL. 2012. Bardet-Biedl syndrome. *Eur. J. Hum. Genet.* 21:8–13

Fridlich R, Delalande F, Jaillard C, Lu J, Poidevin L, et al. 2009. The thioredoxin-like protein rod-derived cone viability factor (RdCVFL) interacts with TAU and inhibits its phosphorylation in the retina. *Mol. Cell Proteomics* 8:1206–18

Gargini C, Terzibasi E, Mazzoni F, Strettoi E. 2007. Retinal organization in the retinal degeneration 10 (rd10) mutant mouse: a morphological and ERG study. *J. Comp. Neurol.* 500:222–38

Gollisch T, Meister M. 2010. Eye smarter than scientists believed: neural computations in circuits of the retina. *Neuron* 65:150–64

Greenberg KP, Pham A, Werblin FS. 2011. Differential targeting of optical neuromodulators to ganglion cell soma and dendrites allows dynamic control of center-surround antagonism. *Neuron* 69:713–20

Grossman N, Poher V, Grubb MS, Kennedy GT, Nikolic K, et al. 2010. Multi-site optical excitation using ChR2 and micro-LED array. *J. Neural Eng.* 7:16004

Hashimoto T, Gibbs D, Lillo C, Azarian SM, Legacki E, et al. 2007. Lentiviral gene replacement therapy of retinas in a mouse model for Usher syndrome type 1B. *Gene Ther.* 14:584–94

Hauswirth WW, Aleman TS, Kaushal S, Cideciyan AV, Schwartz SB, et al. 2008. Treatment of Leber congenital amaurosis due to RPE65 mutations by ocular subretinal injection of adeno-associated virus gene vector: short-term results of a phase I trial. *Hum. Gene Ther.* 19:979–90

Hendrickson A. 1992. A morphological comparison of foveal development in man and monkey. *Eye* 6(Part 2):136–44

Humayun MS, Dorn JD, da Cruz L, Dagnelie G, Sahel JA, et al. 2012. Interim results from the international trial of Second Sight's visual prosthesis. *Ophthalmology* 119:779–88

Isago H, Sugano E, Wang Z, Murayama N, Koyanagi E, et al. 2012. Age-dependent differences in recovered visual responses in Royal College of Surgeons rats transduced with the channelrhodopsin-2 gene. *J. Mol. Neurosci.* 46:393–400

Ivanova E, Hwang G-S, Pan Z-H, Troilo D. 2010. Evaluation of AAV-mediated expression of Chop2-GFP in the marmoset retina. *Invest. Ophthalmol. Vis. Sci.* 51:5288–96

Ivanova E, Pan Z-H. 2009. Evaluation of the adeno-associated virus mediated long-term expression of channelrhodopsin-2 in the mouse retina. *Mol. Vis.* 15:1680–89

Jones BW, Kondo M, Terasaki H, Lin Y, McCall M, Marc RE. 2012. Retinal remodeling. *Jpn. J. Ophthalmol.* 56:289–306

Kjellstrom S, Bush RA, Zeng Y, Takada Y, Sieving PA. 2007. Retinoschisin gene therapy and natural history in the Rs1h-KO mouse: long-term rescue from retinal degeneration. *Invest. Ophthalmol. Vis. Sci.* 48:3837–45

Kleinlogel S, Feldbauer K, Dempski RE, Fotis H, Wood PG, et al. 2011a. Ultra light-sensitive and fast neuronal activation with the Ca2+-permeable channelrhodopsin CatCh. *Nat. Neurosci.* 14:513–18

Kleinlogel S, Terpitz U, Legrum B, Gökbuget D, Boyden ES, et al. 2011b. A gene-fusion strategy for stoichiometric and co-localized expression of light-gated membrane proteins. *Nat. Methods* 8:1083–88

Klimczak RR, Koerber JT, Dalkara D, Flannery JG, Schaffer DV. 2009. A novel adeno-associated viral variant for efficient and selective intravitreal transduction of rat Müller cells. *PLoS One* 4:e7467

Komaromy AM, Alexander JJ, Rowlan JS, Garcia MM, Chiodo VA, et al. 2010. Gene therapy rescues cone function in congenital achromatopsia. *Hum. Mol. Genet.* 19:2581–93

Kong J, Kim SR, Binley K, Pata I, Doi K, et al. 2008. Correction of the disease phenotype in the mouse model of Stargardt disease by lentiviral gene therapy. *Gene Ther.* 15:1311–20

Kramer RH, Fortin DL, Trauner D. 2009. New photochemical tools for controlling neuronal activity. *Curr. Opin. Neurobiol.* 19:544–52

Lagali PS, Balya D, Awatramani GB, Munch TA, Kim DS, et al. 2008. Light-activated channels targeted to ON bipolar cells restore visual function in retinal degeneration. *Nat. Neurosci.* 11:667–75

LaVail MM. 1981. Analysis of neurological mutants with inherited retinal degeneration. Friedenwald lecture. *Invest. Ophthalmol. Vis. Sci.* 21:638–57

Léveillard T, Mohand-Saïd S, Lorentz O, Hicks D, Fintz AC, et al. 2004. Identification and characterization of rod-derived cone viability factor. *Nat. Genet.* 36:755–59

Léveillard T, Sahel JA. 2010. Rod-derived cone viability factor for treating blinding diseases: from clinic to redox signaling. *Sci. Transl. Med.* 2:26ps16

Li S, Huang L. 2000. Nonviral gene therapy: promises and challenges. *Gene Ther.* 7:31–34

Lin B, Koizumi A, Tanaka N, Panda S, Masland RH. 2008. Restoration of visual function in retinal degeneration mice by ectopic expression of melanopsin. *Proc. Natl. Acad. Sci. USA* 105:16009–14

Maguire AM, Simonelli F, Pierce EA, Pugh EN Jr, Mingozzi F, et al. 2008. Safety and efficacy of gene transfer for Leber's congenital amaurosis. *N. Engl. J. Med.* 358:2240–48

Manfredi G, Fu J, Ojaimi J, Sadlock JE, Kwong JQ, et al. 2002. Rescue of a deficiency in ATP synthesis by transfer of MTATP6, a mitochondrial DNA-encoded gene, to the nucleus. *Nat. Genet.* 30:394–99

Mao H, Gorbatyuk MS, Rossmiller B, Hauswirth WW, Lewin AS. 2012. Long-term rescue of retinal structure and function by rhodopsin RNA replacement with a single adeno-associated viral vector in P23H RHO transgenic mice. *Hum. Gene Ther.* 23:356–66

Marc RE, Jones BW, Watt CB, Strettoi E. 2003. Neural remodeling in retinal degeneration. *Prog. Retin. Eye Res.* 22:607–55

Marella M, Seo BB, Thomas BB, Matsuno-Yagi A, Yagi T. 2010. Successful amelioration of mitochondrial optic neuropathy using the yeast *NDI1* gene in a rat animal model. *PLoS One* 5:e11472

Mariotti SP. 2012. *Global Data on Visual Impairments 2010*. Geneva: WHO. **http://www.who.int/blindness/GLOBALDATAFINALforweb.pdf**

Masland RH. 2001. The fundamental plan of the retina. *Nat. Neurosci.* 4:877–86

Milam AH, Li ZY, Fariss RN. 1998. Histopathology of the human retina in retinitis pigmentosa. *Prog. Retin. Eye Res.* 17:175–205

Min SH, Molday LL, Seeliger MW, Dinculescu A, Timmers AM, et al. 2005. Prolonged recovery of retinal structure/function after gene therapy in an Rs1h-deficient mouse model of x-linked juvenile retinoschisis. *Mol. Ther.* 12:644–51

Nagel G, Szellas T, Huhn W, Kateriya S, Adeishvili N, et al. 2003. Channelrhodopsin-2, a directly light-gated cation-selective membrane channel. *Proc. Natl. Acad. Sci. USA* 100:13940–45

Nirenberg S, Pandarinath C. 2012. Retinal prosthetic strategy with the capacity to restore normal vision. *Proc. Natl. Acad. Sci. USA* 109:15012–17

Ong JM, da Cruz L. 2012. A review and update on the current status of stem cell therapy and the retina. *Br. Med. Bull.* 102:133–46

Park TK, Wu Z, Kjellstrom S, Zeng Y, Bush RA, et al. 2009. Intravitreal delivery of AAV8 retinoschisin results in cell type-specific gene expression and retinal rescue in the Rs1-KO mouse. *Gene Ther.* 16:916–26

Pascual-Leone A, Amedi A, Fregni F, Merabet LB. 2005. The plastic human brain cortex. *Annu. Rev. Neurosci.* 28:377–401

Pawlyk BS, Smith AJ, Buch PK, Adamian M, Hong DH, et al. 2005. Gene replacement therapy rescues photoreceptor degeneration in a murine model of Leber congenital amaurosis lacking RPGRIP. *Invest. Ophthalmol. Vis. Sci.* 46:3039–45

Petrs-Silva H, Dinculescu A, Li Q, Deng WT, Pang JJ, et al. 2011. Novel properties of tyrosine-mutant AAV2 vectors in the mouse retina. *Mol. Ther.* 19:293–301

Petrs-Silva H, Dinculescu A, Li Q, Min S-H, Chiodo V, et al. 2009. High-efficiency transduction of the mouse retina by tyrosine-mutant AAV serotype vectors. *Mol. Ther.* 17:463–71

Phillips MJ, Otteson DC, Sherry DM. 2010. Progression of neuronal and synaptic remodeling in the rd10 mouse model of retinitis pigmentosa. *J. Comp. Neurol.* 518:2071–89

Polosukhina A, Litt J, Tochitsky I, Nemargut J, Sychev Y, et al. 2012. Photochemical restoration of visual responses in blind mice. *Neuron* 75:271–82

Prigge M, Schneider F, Tsunoda SP, Shilyansky C, Wietek J, et al. 2012. Color-tuned channelrhodopsins for multiwavelength optogenetics. *J. Biol. Chem.* 287:31804–12

Punzo C, Kornacker K, Cepko CL. 2009. Stimulation of the insulin/mTOR pathway delays cone death in a mouse model of retinitis pigmentosa. *Nat. Neurosci.* 12:44–52

Punzo C, Xiong W, Cepko CL. 2012. Loss of daylight vision in retinal degeneration: Are oxidative stress and metabolic dysregulation to blame? *J. Biol. Chem.* 287:1642–48

Reich L, Maidenbaum S, Amedi A. 2012. The brain as a flexible task machine: implications for visual rehabilitation using noninvasive versus invasive approaches. *Curr. Opin. Neurol.* 25:86–95

Sahel JA. 2011. Spotlight on childhood blindness. *J. Clin. Invest.* 121:2145–49

Sahly I, Dufour EA, Schietroma C, Michel V, Bahloul A, et al. 2012. Localization of Usher 1 proteins to the photoreceptor calyceal processes, which are absent from mice. *J. Cell Biol.* 199:381–99

Schmidt TM, Chen S-K, Hattar S. 2011. Intrinsically photosensitive retinal ganglion cells: many subtypes, diverse functions. *Trends Neurosci.* 34:572–80

Siegert S, Cabuy E, Scherf BG, Kohler H, Panda S, et al. 2012. Transcriptional code and disease map for adult retinal cell types. *Nat. Neurosci.* 15:487–95, S1–2

Simonelli F, Maguire AM, Testa F, Pierce EA, Mingozzi F, et al. 2010. Gene therapy for Leber's congenital amaurosis is safe and effective through 1.5 years after vector administration. *Mol. Ther.* 18:643–50

Simons DL, Boye SL, Hauswirth WW, Wu SM. 2011. Gene therapy prevents photoreceptor death and preserves retinal function in a Bardet-Biedl syndrome mouse model. *Proc. Natl. Acad. Sci. USA* 108:6276–81

Singh MS, MacLaren RE. 2011. Stem cells as a therapeutic tool for the blind: biology and future prospects. *Proc. Biol. Sci.* 278:3009–16

Streilein JW. 2003. Ocular immune privilege: The eye takes a dim but practical view of immunity and inflammation. *J. Leukoc. Biol.* 74:179–85

Streilein JW, Ksander BR, Taylor AW. 1997. Immune deviation in relation to ocular immune privilege. *J. Immunol.* 158:3557–60

Sun X, Pawlyk B, Xu X, Liu X, Bulgakov OV, et al. 2010. Gene therapy with a promoter targeting both rods and cones rescues retinal degeneration caused by AIPL1 mutations. *Gene Ther.* 17:117–31

Takahashi M, Miyoshi H, Verma IM, Gage FH. 1999. Rescue from photoreceptor degeneration in the rd mouse by human immunodeficiency virus vector-mediated gene transfer. *J. Virol.* 73:7812–16

Tam LC, Kiang AS, Kennan A, Kenna PF, Chadderton N, et al. 2008. Therapeutic benefit derived from RNAi-mediated ablation of IMPDH1 transcripts in a murine model of autosomal dominant retinitis pigmentosa (RP10). *Hum. Mol. Genet.* 17:2084–100

Thyagarajan S, van Wyk M, Lehmann K, Löwel S, Feng G, Wässle H. 2010. Visual function in mice with photoreceptor degeneration and transgenic expression of channelrhodopsin 2 in ganglion cells. *J. Neurosci.* 30:8745–58

Tibbetts MD, Samuel MA, Chang TS, Ho AC. 2012. Stem cell therapy for retinal disease. *Curr. Opin. Ophthalmol.* 23:226–34

Tomita H, Sugano E, Fukazawa Y, Isago H, Sugiyama Y, et al. 2009. Visual properties of transgenic rats harboring the channelrhodopsin-2 gene regulated by the thy-1.2 promoter. *PLoS One* 4:e7679

Tomita H, Sugano E, Isago H, Hiroi T, Wang Z, et al. 2010. Channelrhodopsin-2 gene transduced into retinal ganglion cells restores functional vision in genetically blind rats. *Exp. Eye Res.* 90:429–36

Tye KM, Deisseroth K. 2012. Optogenetic investigation of neural circuits underlying brain disease in animal models. *Nat. Rev. Neurosci.* 13:251–66

Vollrath D, Feng W, Duncan JL, Yasumura D, D'Cruz PM, et al. 2001. Correction of the retinal dystrophy phenotype of the RCS rat by viral gene transfer of *Mertk*. *Proc. Natl. Acad. Sci. USA* 98:12584–89

Wässle H. 2004. Parallel processing in the mammalian retina. *Nat. Rev. Neurosci.* 5:747–57

Weiland JD, Liu W, Humayun MS. 2005. Retinal prosthesis. *Annu. Rev. Biomed. Eng.* 7:361–401

Yang Y, Mohand-Said S, Danan A, Simonutti M, Fontaine V, et al. 2009. Functional cone rescue by RdCVF protein in a dominant model of retinitis pigmentosa. *Mol. Ther.* 17:787–95

Yin L, Greenberg K, Hunter JJ, Dalkara D, Kolstad KD, et al. 2011. Intravitreal injection of AAV2 transduces macaque inner retina. *Invest. Ophthalmol. Vis. Sci.* 52:2775–83

Zhang F, Wang LP, Brauner M, Liewald JF, Kay K, et al. 2007. Multimodal fast optical interrogation of neural circuitry. *Nature* 446:633–39

Zhong L, Li B, Mah CS, Govindasamy L, Agbandje-McKenna M, et al. 2008. Next generation of adeno-associated virus 2 vectors: point mutations in tyrosines lead to high-efficiency transduction at lower doses. *Proc. Natl. Acad. Sci. USA* 105:7827–32

Zou J, Luo L, Shen Z, Chiodo VA, Ambati BK, et al. 2011. Whirlin replacement restores the formation of the USH2 protein complex in whirlin knockout photoreceptors. *Invest. Ophthalmol. Vis. Sci.* 52:2343–51

Zrenner E, Bartz-Schmidt KU, Benav H, Besch D, Bruckmann A, et al. 2011. Subretinal electronic chips allow blind patients to read letters and combine them to words. *Proc. Biol. Sci.* 278:1489–97

Translating Birdsong: Songbirds as a Model for Basic and Applied Medical Research

Michael S. Brainard and Allison J. Doupe

Center for Integrative Neuroscience and Departments of Physiology and Psychiatry, University of California, San Francisco 94143-0444; email: msb@phy.ucsf.edu, ajd@phy.ucsf.edu

Annu. Rev. Neurosci. 2013. 36:489–517

The *Annual Review of Neuroscience* is online at neuro.annualreviews.org

This article's doi:
10.1146/annurev-neuro-060909-152826

Keywords

reinforcement learning, basal ganglia, speech, mirror neurons, hearing, neurogenesis

Abstract

Songbirds, long of interest to basic neuroscience, have great potential as a model system for translational neuroscience. Songbirds learn their complex vocal behavior in a manner that exemplifies general processes of perceptual and motor skill learning and, more specifically, resembles human speech learning. Song is subserved by circuitry that is specialized for vocal learning and production but that has strong similarities to mammalian brain pathways. The combination of highly quantifiable behavior and discrete neural substrates facilitates understanding links between brain and behavior, both in normal states and in disease. Here we highlight (*a*) behavioral and mechanistic parallels between birdsong and aspects of speech and social communication, including insights into mirror neurons, the function of auditory feedback, and genes underlying social communication disorders, and (*b*) contributions of songbirds to understanding cortical-basal ganglia circuit function and dysfunction, including the possibility of harnessing adult neurogenesis for brain repair.

Contents

INTRODUCTION

Humans are experts at vocal learning, an ability shared by few other vertebrates. No nonhuman primates or rodents have been shown to learn their vocal behavior, and there are only a few other known mammalian vocal learners—cetaceans, elephants, and some bats. Songbirds, therefore, represent an almost unique animal model: Song learning has remarkable parallels to human speech learning (**Figure 1**) (Brenowitz et al. 2010, Doupe & Kuhl 1999, Marler 1970, Mooney 2009) and provides an opportunity to investigate mechanistically both vocal learning and its disorders.

Because vocal learning involves both perceptual learning and motor skill learning, birdsong provides a model for investigating these general learning processes and their disturbances. Additionally, song learning may be shaped by social interactions and is limited to a "critical" period during which experience plays an especially crucial role in nervous system development. Such factors similarly shape and constrain many aspects of behavior in other vertebrates including humans, making song learning an attractive system for understanding these constraints and how they may be overcome to enable nervous system repair in disease and injury.

Birdsong is a complex motor sequence that is readily quantified, making it a sensitive assay for investigating the effects of behavioral and neural manipulations. Moreover, song depends on dedicated and accessible brain nuclei (**Figure 2**) (Nottebohm et al. 1976) that have strong homologies to mammalian circuits (see sidebar, Avian and Mammalian Brains). The specialization of this circuitry for a quantifiable, learned behavior facilitates investigation of links between brain and behavior, and it offers the opportunity to control experience and manipulate brain function in a manner impossible in humans. Thus, animal models such as songbirds can reveal basic mechanisms underlying normal and abnormal function. Accordingly, in this review, we take a "translational" stance and discuss ways in which birdsong can contribute to biomedical research and, ultimately, human medicine.

Here, we focus on contributions of songbird research to studies of (*a*) speech and social communication and (*b*) cortical-basal ganglia (CBG) circuitry and motor and reward learning. We refer the reader to recent reviews of additional areas of basic and translational relevance of birdsong, including general auditory processing (Knudsen & Gentner 2010, Theunissen & Shaevitz 2006, Woolley 2012), central and peripheral vocal motor production (Fee & Scharff 2010, Riede & Goller 2010, Schmidt et al. 2012), and the effects of steroid hormones on brain and behavior (Ball & Balthazart 2010, Pinaud & Tremere 2012, Remage-Healey et al. 2010).

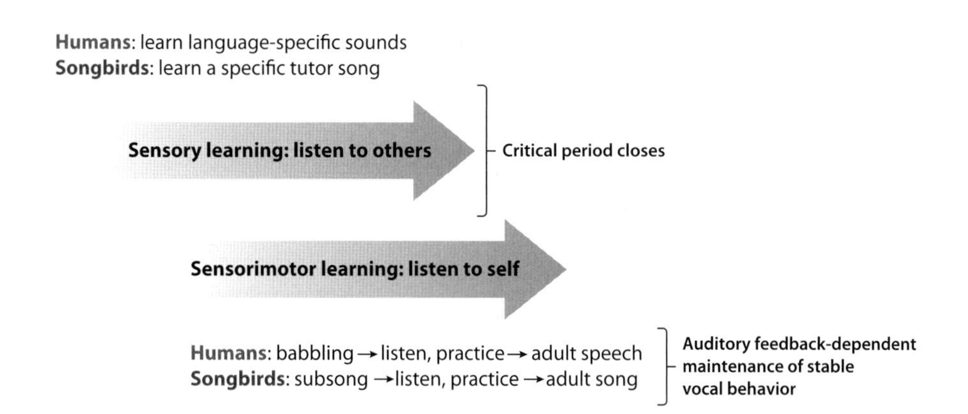

Figure 1

Similarities between human speech and birdsong learning. In both humans and birds, there is an early period of purely perceptual learning, when experiencing the sounds of others shapes the brain. This results in categorical perception of the sounds of one's own language or song, and in accents; such perception is hard to reverse after the critical period closes. Once youngsters begin to produce speech- or song-like sounds, sensorimotor learning has also begun, and the vocalizing animal must pay attention to its own sounds and gradually improve them to match the learned perceptual targets. This gradual process of vocal-motor shaping is dramatically disrupted by abnormal or absent auditory feedback. Following stabilization, adult language and song both remain acutely sensitive to disrupted auditory feedback, such as delayed auditory feedback, and slowly decay over time if auditory feedback is absent.

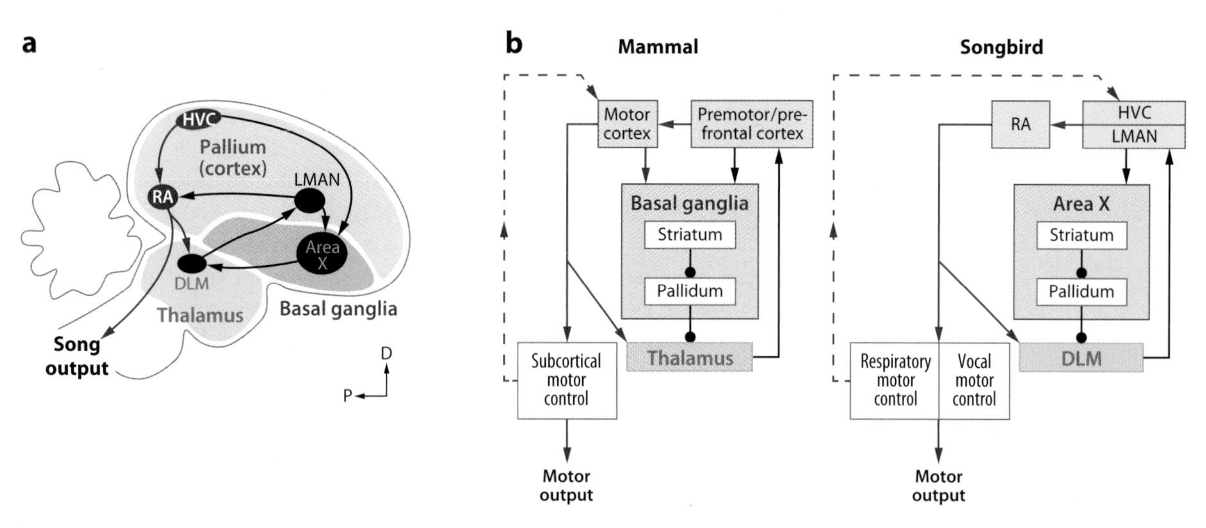

Figure 2

The song system. (*a*) Simplified diagram of the song system. The motor pathway (*red lines*) includes nuclei HVC and the robust nucleus of the arcopallium (RA) as well as downstream midbrain and brain stem centers (not shown) that control patterned respiration and the bird's syringeal vocal musculature. The anterior forebrain pathway (*black lines*) is a cortical-basal ganglia (CBG) loop that includes a basal ganglia homologue (Area X), a thalamic nucleus (DLM), and a frontal cortical nucleus (LMAN). Other abbreviations: P, posterior; D, dorsal. (*b*) Block diagram highlighting parallels between mammalian and birdsong circuitry. Also illustrated are the recurrent projections from the respiratory centers noted in panel *a* back to the HVC, which play an important role in song control, including bilateral coordination of the two hemispheres (Schmidt et al. 2012). Figure modified by S. Kojima from Fee & Goldberg (2011), Gale & Perkel (2010), and Schmidt et al. (2012).

AVIAN AND MAMMALIAN BRAINS

Many brain nuclei in avian telencephalon are derived from the pallial layer of embryos, which also gives rise to mammalian cortex. Basic connectivity and molecular markers specific to layer 4 input and layer 5 output cortical cells are also conserved across reptiles, birds, and mammals. Thus, many telencephalic nuclei in birds may correspond to different cortical layers (Dugas-Ford et al. 2012, Wang et al. 2010), despite differences in organization and laminar structure. As in mammals, the functionally "cortical" regions in birds lie above and adjacent to subpallial structures, including basal ganglia. Revised terminology, introduced in 2004, accurately reflects these homologies and eliminates the historical naming of most of the avian telencephalon as basal ganglia (Reiner et al. 2004). The comparative data all strengthen the parallels that can be drawn between avian and mammalian brains. For simplicity, in this article we mostly use the term cortical to refer to the pallial, cortical-equivalent structures that are part of the song circuit.

Background

Song, similar to speech, is learned in two, sometimes overlapping, phases (**Figure 1**). During a critical period of "sensory learning," birds hear and form a memory of a specific adult vocal model (the "tutor song") (for reviews, see Doupe & Kuhl 1999, Mooney 2009, Slater et al. 1988). Similarly, human infants progress from universal perception of speech sounds to perception that reflects exposure to a specific language. During subsequent "sensorimotor learning," birds gradually shape their initial rambling, immature "subsong" into mature song resembling that of the tutor, whereas humans turn their initial "babbling" into the mature adult sounds of the language surrounding them. Both birds and humans must be able to hear themselves to refine their vocalizations. In songbirds, the tutor no longer needs to be present during this phase, indicating that a memory of the adult model, often called the template, can guide learning. For song, as for speech, there must be circuitry for producing vocalizations, circuitry for hearing and learning sounds of self and others, and linkage between the two.

The motor control of both song and speech involves high-level structures that ultimately coordinate the patterned breathing and vocal muscle activity necessary for vocalization. In songbirds, these areas are collectively known as the song system and include a "motor pathway" as well as a CBG circuit for song, the anterior forebrain pathway (AFP) (**Figure 2**). The motor pathway includes the premotor cortical nucleus HVC (abbreviation used as proper name) and the robust nucleus of the arcopallium (RA), which is functionally equivalent to vocal motor cortex (we use the term "cortical" to refer to song-system nuclei that are equivalent to the mammalian cortex) (see sidebar, Avian and Mammalian Brains). HVC is involved in generating the timing and sequencing of song (Fee & Scharff 2010, Hahnloser et al. 2002, Long & Fee 2008, Yu & Margoliash 1996) (see sidebar, Motor Sequence Generation). The nucleus RA receives inputs from HVC and projects directly to respiratory centers and to brain stem motor neurons controlling the vocal organ; the respiratory centers send recurrent information back to HVC via the thalamus, reflecting the importance of bidirectional coordination between telencephalic and brain stem structures in vocal control (**Figure 2**) (Schmidt et al. 2012). HVC and RA are required for normal song production throughout life; complete bilateral lesions of either of these nuclei cause song disruptions or even muteness, akin to human aphasias (Nottebohm et al. 1976). The AFP indirectly connects HVC and RA (**Figure 2**) (Doupe et al. 2005, Fee & Scharff 2010, Gale & Perkel 2010, Bottjer & Johnson 1997), and like CBG circuitry in mammals, it functions importantly in motor learning (see Cortical Basal Ganglia Circuits and Reward and Motor Learning, section below).

The songbird Field L, which is analogous to the primary auditory cortex of mammals, projects to a complex set of higher auditory areas known broadly as NCM (caudo-medial nidopallium) and CM (caudal mesopallium), and these high-level areas are the likely sources of auditory inputs to the song system. Enhanced immediate early gene (IEG) expression and neurophysiological activity in response to songs in CM/NCM suggest that they, in addition to the song system, are sites of operations critical to vocal learning, with analogies to speech-related regions of the human superior temporal gyrus (Knudsen & Gentner 2010, Moorman et al. 2011, Woolley 2012).

SPEECH AND SOCIAL COMMUNICATION

Influence of Sensory Exposure on Vocal Learning and the Brain

Both humans and birds must hear adult vocal models to learn normal vocalizations. This requirement to imitate others results in different languages, dialects, and individual accents in humans, and in dialects within species of birds, as well as individual differences between birds tutored by different adults. The amount of early exposure to adult vocalizations contributes critically to vocal development. In humans, these data are largely correlative (Hurtado et al. 2008). In birds, controlled manipulations of experience demonstrate that insufficient exposure to vocal models can impair subsequent vocal learning (Catchpole & Slater 2008, Marler 1970). Interestingly, too much song model exposure can also impair learning (Tchernichovski et al. 1999), suggesting there may be an optimal balance between different kinds of experience (for example, between listening to others and listening to self). These effects of early experience can be remarkably rapid. In songbirds, after the first exposure to a tutor, vocal performance can improve in less than 1 day (Derégnaucourt et al. 2005, Shank & Margoliash 2009). This is analogous to the almost immediate appearance of new speech sounds in infants exposed to novel sounds (Kuhl & Meltzoff 1996).

The quality of acoustic features heard during this early period additionally shapes learning. For humans, exaggerated features adopted by adults in speaking to infants, such as slowing of speech or increased pitch (infant-directed speech), may preferentially shape perceptual learning for speech (Fernald 1985, Kuhl 2010). Similarly, for songbirds, some songs are more readily learned than others (Marler 1990), and key features within songs may enhance learning of associated sounds (Catchpole & Slater 2008, Soha & Marler 2000). Song learning also can be enhanced by factors such as hearing a greater variety of tutor songs (Tchernichovski et al. 1999) and impeded, for instance, by hearing tutor songs that are sung at too fast a tempo (Podos 1996). The controlled studies possible in songbirds enable investigation of such behavioral variables that may also be relevant to normal speech development and rehabilitation.

In songbirds, it is possible to investigate directly how and where sensory exposure influences the brain. For example, brain activity can

MOTOR SEQUENCE GENERATION

Because song is a complex yet often stereotyped sequence of motor gestures, it provides an attractive opportunity to understand the neural mechanisms of sequence learning and production. Studies of the motor pathway for song provide insights into motor sequence generation. HVC premotor neurons projecting to the robust nucleus of the arcopallium (RA) each fire sparsely at distinct times during short sequences of song (Amador et al. 2013, Hahnloser et al. 2002, Long & Fee 2008, Long et al. 2010). HVC neuron firing shapes the moment-by-moment pattern of activity of RA neurons (Dave & Margoliash 2000, Leonardo & Fee 2005, Sober et al. 2008), which in turn determine the spectrotemporal features of song via their projections to syringeal and respiratory premotor centers. Hence, HVC to RA synapses are one likely locus for plasticity underlying learned changes to syllable structure. The precise sequences of HVC firing extend across the duration of syllables or sub-syllabic features, and these sequences are likely linked together by recurrent feedback from brain stem vocal control regions (Amador et al. 2013, Long & Fee 2008, Schmidt et al. 2012, Wang et al. 2008). These findings were facilitated by specializations of the song system, but similar principles likely operate in mammalian systems. In particular, there are multiple examples of neurons, in mammalian prefrontal, parietal, and hippocampal cortices, each of which fire at discrete times during a sequence of movements but collectively tile the entire movement trajectory (e.g., Harvey et al. 2012).

be locally and reversibly disrupted specifically when young birds are hearing auditory models (and not when they are actively rehearsing). Such experiments have identified several areas required for learning of the tutor song. These include not only high-level auditory areas (London & Clayton 2008), but also the song sensorimotor nuclei HVC and LMAN (Basham et al. 1996, Roberts et al. 2012). Consistent with the importance of vocal premotor regions to sensory learning, exposure to the tutor song can drive rapid changes in these regions, including dramatic changes to patterns of neural activity and to the structure and motility of dendritic spines (Roberts et al. 2010, Shank & Margoliash 2009). These findings suggest that perceptual learning for vocal production both shapes neural selectivity in auditory areas and organizes sensorimotor structures involved in production. For song, and likely for speech, these data reinforce the notion of an intimate link between the mechanisms that contribute to sensory learning of vocal models and learning to produce those sounds.

Songbirds provide a particularly clear example of learning that is limited to a critical or sensitive period early in life, with special relevance to the human critical period for language acquisition (Kuhl 2010). Many species of birds must hear a tutor song while young and will not learn songs to which they are subsequently exposed (for reviews, see Doupe & Kuhl 1999, Mooney 2009). Around the time that the song sensitive period normally closes, changes also occur to many neuronal factors implicated in mammalian development and critical period regulation, including spine density, axonal arborization, NMDA receptor current decay, and perineuronal nets (e.g., Balmer et al. 2009, Heinrich et al. 2005, Livingston et al. 2000, Miller-Sims & Bottjer 2012, Roberts et al. 2010). Because the critical period for tutor song memorization can be altered or extended by manipulations of experience, it should be possible to identify neural mechanisms that are causally linked to its regulation.

Bidirectional Links Between Perception and Production

Although perceptual learning guides the development of vocal production, it can also constrain our ability to perceive communication sounds. For example, in humans, exposure to a specific language can result in categorical perception, in which acoustically distinct versions of a single phoneme become grouped so that the capacity to distinguish between them is reduced. Songbirds also develop categorical perception of song elements, for instance, for a particular range of syllable durations (**Figure 3*a*,*b***) (Nelson & Marler 1989, Prather et al. 2009). Such categorical perception is adaptive in creating an abstracted representation of sounds that enables recognition despite acoustic variation in production. However, in humans, such perceptual shaping also contributes to a difficulty in perceiving or producing non-native phonemes. Indeed, the difficulty in changing speech, both normally and in vocal pathologies, may stem primarily from the stability of perceptual targets, rather than an inflexibility of production mechanisms. Similarly, stability of perceptual targets contributes to the normal stability of adult song (Sober & Brainard 2009), suggesting that songbirds provide a useful model to investigate the neural basis of perceptual constraints on learning.

The tight, bidirectional link between perception and production is evident in the nervous system. Human imaging studies indicate that sensorimotor structures for speech, such as Broca's area, are active during both production and perception. Damage or disruption of these areas as well as of structures linked to comprehension, such as Wernicke's area, can cause both productive and receptive deficits (e.g., Hickok et al. 2011). Similarly, in adult birds, playback of the tutor song and the bird's own song (BOS) can activate both high-level auditory areas and neurons throughout the song system (Dave & Margoliash 2000; Moorman et al. 2011; Prather et al. 2008, 2010; Solis & Doupe 1999). Moreover, damage to sensorimotor song nuclei (such as HVC, LMAN, and Area X)

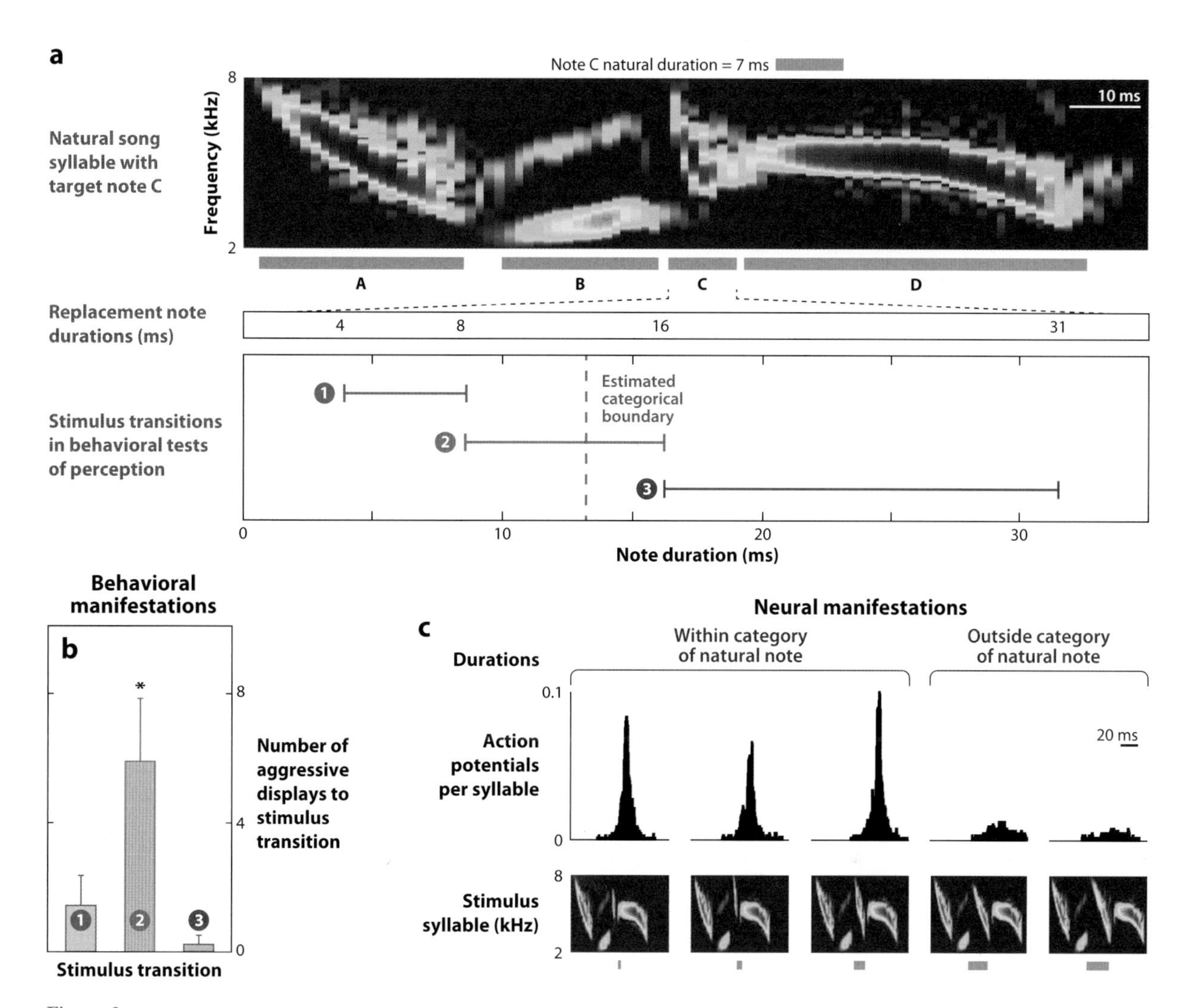

Figure 3

Behavioral and neural manifestations of categorical perception in songbirds. (*a*) Spectrogram of swamp sparrow song composed of syllables ABCD. Across a local New York dialect of swamp sparrow song, the naturally occurring durations of syllable C ranged up to ~13 ms, suggesting a potential correspondence with a categorical boundary for perception. To test this possibility, behavioral and neural sensitivities to changes in the duration of syllable C were tested for transitions in duration that either were within the hypothesized categories (*1*, *blue*; *3*, *red*) or spanned the boundary (*2*, *green*). (*b*) Birds were largely insensitive to transitions in durations that remained within the category (*1* and *3*) but responded strongly (with aggressive displays) to transitions that spanned the categorical boundary (*2*). (*c*) Neurons within HVC responded similarly to playback of different stimuli that had syllable durations within the naturally occurring range (within category), but they responded little to stimuli that had durations outside this range (outside category). Modified from Prather et al. (2009).

can cause difficulty in discriminating songs (Burt et al. 2000, Gentner et al. 2000, Scharff et al. 1998).

A particularly compelling example of the neural link between production and perception is provided by mirror neurons, which are found in human and nonhuman primate cortical motor areas, including speech areas. These neurons not only are active during production of movements but also exhibit similar firing patterns during observation or hearing of the same movements (Rizzolatti & Fabbri-Destro 2010).

This sensorimotor correspondence suggests that mirror neurons have been shaped both by the production of gestures and by the sensory stimuli that these same gestures produce. Such a correspondence may contribute to subjects' understanding of sensory inputs from others, based on their own production of gestures. Thus, mirroring may be important not only in learning, but also in subsequent mimicry and other social communication, as well as in disorders of the ability to understand and produce social gestures, such as autism (Rizzolatti & Fabbri-Destro 2010).

Songbirds provide striking examples of the kind of behavioral coordination and imitation that may require mirroring. Some species, such as lyrebirds and mockingbirds, can immediately produce imitations of newly heard sounds. Many species also engage in countersinging, in which birds sing songs in alternation, or duetting, in which two individuals cooperate to produce the alternating syllables of a single song (Catchpole & Slater 2008, Fortune et al. 2011). These behaviors indicate that songbirds can rapidly translate sensory inputs into the motor actions necessary to produce an imitation or the next gesture in a learned motor sequence.

At the neural level, the song system possesses song mirror neurons well suited to subserve learning, mimicry, and social coordination (Dave & Margoliash 2000, Fortune et al. 2011, Prather et al. 2008). In adult birds, these mirror neurons exhibit motor-related activity during song production and also fire more strongly in response to playback of the sound of the BOS than in response to other songs. The auditory activity of these BOS-selective neurons exhibits precise and complex feature selectivity for the spectral structure, relative ordering, and timing of song elements of the BOS (Dave & Margoliash 2000, Solis & Doupe 1999). For example, in swamp sparrows, HVC neurons fire most strongly for playback of song that matches the bird's behavioral boundaries for categorical perception of the same song (**Figure 3**) (Prather et al. 2009). Strikingly, in a duetting species, HVC neurons fire most when songs of both partners, in their normal alternation, are played back, and they fire much less when either song is played alone (Fortune et al. 2011). This suggests that neurons have encoded not only the BOS but also the duetting birds' combined, cooperative sensory output.

In birds, studying the development and learning-related properties of these mirror neurons is straightforward. Many aspects of BOS selectivity clearly emerge only during vocal production learning (Solis & Doupe 1999, Volman 1993). Damage to the vocal periphery in both adult and juvenile birds results in altered auditory tuning of BOS-selective neurons (Roy & Mooney 2007, Solis & Doupe 2000), and birds prevented from developing normal song by peripheral muting, but otherwise exposed to a normal acoustic environment, exhibit degraded song perception (Pytte & Suthers 1999). Intriguingly, songbird mirror neurons can encode more than the sound of the BOS. In both LMAN and HVC, some neurons also respond to the songs of tutors that have been heard and, thus, may be involved in encoding memory of the tutor song or in matching of the BOS to the tutor (Nick & Konishi 2005; Prather et al. 2010; Solis & Doupe 1999, 2000). The presence of such precise sensorimotor tuning in song-selective neurons provides an opportunity to investigate the mechanistic basis of interactions between perception and production as well as what happens when this interaction is disrupted.

Behavioral and Brain Mechanisms of Auditory Feedback

Speech and song are both motor skills that rely crucially on auditory feedback. Congenitally deaf individuals have extreme difficulty developing normal patterns of articulation, and hearing loss even in adulthood can lead to gradual deterioration of speech, suggesting that auditory feedback continues to play an important role in calibrating vocal output. Moreover, perturbation of feedback, by altering loudness, timing, or pitch, can drive both online disruptions and compensatory adaptations of speech (for a review, see Houde &

Nagarajan 2011). Correspondingly, deficits in central feedback processing mechanisms may contribute to a variety of speech abnormalities, including deficits in speech acquisition and in control of phonation and sequencing of speech sounds, as in stuttering.

Auditory feedback in birds is equally important for learning, controlling, and maintaining production. Juvenile birds that are deafened after exposure to a tutor song fail to develop normal songs, and adult birds deafened or subjected to disruption of auditory feedback exhibit gradual song deterioration (Konishi 1965, Leonardo & Konishi 1999, Nordeen & Nordeen 1992). As in humans, delayed auditory feedback results in slowing and mis-sequencing of syllables, with the most disruptive delay corresponding approximately to the duration of a single syllable (Cynx & von Rad 2001, Sakata & Brainard 2006). Furthermore, birds make gradual compensatory adjustments to the pitch at which they produce song to correct perceived errors in production (Sober & Brainard 2009).

Despite the importance of auditory feedback, we have relatively little understanding of the neural processes whereby information derived from auditory feedback is used to evaluate and adjust speech or song. Studies in humans suggest there are specialized substrates within vocal control and associated auditory regions that are specifically gated or modulated by the act of vocalizing. For example, responses in the human superior temporal gyrus are decreased when speakers hear themselves during speaking versus when the same speech sounds are played back during quiescence (Ford & Mathalon 2012, Hickok et al. 2011, Price 2012). This could represent damping of the responsiveness of auditory neurons by motor preparatory or "efference copy" signals, in advance of the predictable loud stimulus associated with vocalizing, a phenomenon widely observed across species (Ford & Mathalon 2012). However, in humans, alteration of the sound of the speaker's voice can increase neural activity during speaking to the level seen in response to playback (Price 2012). This suggests that efferent activity may specifically attenuate the expected sensory consequences of speech motor commands and thus enhance sensitivity to deviations from expectation.

Several related phenomena are present in birds. First, in some species, the responses to playback of the BOS within premotor nuclei are strongest in sleeping or anesthetized birds and attenuated or absent in awake birds (Cardin & Schmidt 2004, Castelino & Schmidt 2010, Coleman et al. 2007). Such context-dependent modulation is one indication that auditory feedback from vocalizations is processed differently from other sounds and is linked to mechanisms that control production. Second, transient perturbation of feedback during singing results in changes to the timing and sequencing of syllables at latencies of only a few tens of milliseconds. This indicates that sensory inputs must have rapid "online" access to song premotor circuitry (Sakata & Brainard 2006). One locus where such neural responses to feedback perturbation have been observed is the premotor nucleus HVC of the Bengalese finch (Sakata & Brainard 2008; but also see Kozhevnikov & Fee 2007, Leonardo 2004, Prather et al. 2008), supporting the possibility that auditory signals within premotor circuitry may normally contribute to online control of song production and to song learning. Finally, in areas conventionally thought of as primarily auditory regions of the brain, some neurons exhibit strong responses only when feedback is perturbed (Keller & Hahnloser 2009). Such responses, like those in HVC and in human auditory cortex, could indicate that the actual sensory input differs from the expected consequences of vocalizing and would provide ideal signals for correcting vocal performance.

In songbirds, it is possible to test experimentally both how the brain encodes sensory feedback and how it uses feedback to shape vocal production. For example, the state-dependent neural mechanisms that engage or disengage auditory responsiveness in song control regions include neuromodulatory tone. Direct manipulation of neuromodulators can enhance or eliminate auditory responses in central vocal control structures (Cardin & Schmidt 2004; Shea

& Margoliash 2010). Moreover, at a behavioral level, the presence of a female bird renders male song less sensitive to perturbations of auditory feedback (Sakata & Brainard 2009), perhaps also reflecting shifts in attentional and motivational systems arising from altered neuromodulatory tone.

For humans, some nonspeech disorders also involve abnormal processing of auditory signals both of others and of self. For instance, schizophrenic patients may have impaired efference copy mechanisms, evident as less damping of self-generated feedback. Similarly, these patients may misattribute delayed auditory feedback of their own voice to others (Ford & Mathalon 2012, Frith et al. 2000). These abnormalities are hypothesized to contribute to the generation of symptoms such as hallucinations and delusions. Thus, a mechanistic understanding of efference copy function during vocal behavior, and of how neuromodulatory systems act to alter neural and behavioral sensitivity to sensory feedback, may provide insights into self-monitoring systems relevant not only to speech but also more generally to neuropsychiatric disorders.

Importance of Social Factors in Vocal Learning and Brain Organization

Both speech and birdsong are highly social behaviors, and social interactions can strongly influence multiple phases of vocal learning. For instance, social factors can determine what models juveniles choose to learn. For human infants, live exposure to foreign-language phonemes preserves children's discriminatory ability for those phonemes, whereas exposure via video or audiotape does not (Kuhl 2010). Similarly, for highly social birds, like the zebra finch, song is better learned from live models than from passive playback of acoustic models (Catchpole & Slater 2008). However, if a young bird must press a key to elicit song playback interactively, the bird can learn well (Adret 1993, Tchernichovski et al. 2001). Moreover, birds will learn from interacting tutors beyond the age at which taped playback of song becomes ineffective, and they will learn normally nonpreferred (heterospecific) song if the heterospecific tutor is an interactive tutor (Catchpole & Slater 2008).

Social cues are also important in vocal motor learning. Juvenile white-crowned sparrows singing a variety of almost-mature songs stabilize and retain the variants that elicit countersinging from other males (Nelson & Marler 1994), and young male cowbirds preferentially retain song variants that trigger visual courtship responses from females (West & King 1988). Social factors in songbirds likely act in a variety of ways, including via changes in neuromodulators and steroid hormone levels (Riters 2011, Sasaki et al. 2006, White & Livingston 1999). They can also exert influences regardless of acoustic experience. For example, starlings prevented from forming social bonds, even if they can hear normal songs, develop highly abnormal auditory cortices (Cousillas et al. 2008).

Mechanistic and neurophysiological studies in songbirds have identified many brain areas affected by social signals. The familiarity and social relevance of songs can affect the degree of expression of immediate early genes (IEGs) in a number of high-level auditory areas (Knudsen & Gentner 2010, Moorman et al. 2011, Woolley & Doupe 2008), and in some cases, this induction can be prevented by blockade of norepinephrine receptors (Velho et al. 2012). The CBG circuit for song (the AFP; see **Figure 2**) is also strongly affected by social cues. IEG induction, which is high in Area X and LMAN when birds sing alone, is dramatically decreased when birds sing to a female (Jarvis et al. 1998), and this is accompanied by marked changes in neuronal firing properties (Hessler & Doupe 1999a,b; Kao et al. 2008). Midbrain neuromodulatory areas that project strongly to the AFP, such as the noradrenergic locus coeruleus and the dopaminergic ventral tegmental area (VTA), likely mediate these social influences. Lesions of the locus coeruleus eliminate the social modulation of IEG induction in Area X (Castelino & Schmidt 2010). VTA neurons, which project strongly to the AFP, increase their firing in the presence of a female, and the amount of this increase correlates with the amount of female-directed singing (Hara et al. 2007, Huang & Hessler 2008, Riters 2011, Yanagihara & Hessler 2006). Intriguingly, VTA neurons in young birds also show enhanced IEG induction in response to tutor song versus that of unfamiliar males, which may reflect the social salience of this signal or a possible function of the VTA in song memorization or evaluation during learning (Nordeen et al. 2009).

Interactions between social and vocal behavior are central to human disorders, such as autism spectrum disorders and frontotemporal dementias, in which disruptions of social and linguistic capacities frequently co-occur. Songbirds provide an opportunity to directly manipulate the relevant behavioral and neural variables to understand their role in normal learning. Studying songbirds, researchers can also explore the nature of social and vocal deficits that arise when experience or activation of neural circuitry is abnormal. Moreover, many bird species have complex social structures and strong social bonds, which are subject to modification and disruption by experience (Elie et al. 2011). This makes songbirds potentially informative not just for social-vocal interaction, but also for general brain mechanisms of social interaction, attention, and bonding.

Genetics of Speech and Social Communication

Numerous genes that are implicated in human diseases are enriched in song system nuclei and exhibit differential expression both at specific points in development and during behaviors such as listening to song or singing (Fee & Scharff 2010, Wada et al. 2006, White 2010). Accordingly, such genes may play conserved roles in circuitry that underlies complex behaviors such as vocal learning, and songbirds may also provide insights into the functions of these genes both normally and in disease.

The gene *FoxP2* provides a compelling example of parallels between songbirds and humans. *FoxP2* codes for a transcription factor that, when mutant in humans, leads to abnormalities in basal ganglia, cerebellum, and cortex, including Broca's area, and resultant language impairment with disordered sequencing of orofacial movements. In songbirds, as in humans, *FoxP2* is expressed strongly in the striatum (Haesler et al. 2004, Teramitsu et al. 2004). In addition, it is dynamically regulated during development and by adult behavioral state (Fee & Scharff 2010, White 2010). Knockdown of *FoxP2* in song striatum impairs song learning, is accompanied by decreased spine density of striatal spiny neurons, and results in a highly variable song reminiscent of the abnormal speech of humans with mutant *FoxP2* (Haesler et al. 2007, Schulz et al. 2010). Furthermore, study of the strong expression in several song nuclei of *FoxP1*, which dimerizes with *FoxP2*, led to the correct prediction that *FoxP1* would also play a role in human speech (Teramitsu et al. 2004).

Many additional genes relevant to speech or other human disorders are differentially expressed in the song system. For example, one of *FoxP2*'s interaction targets is *contactin-associated protein2* (*CNTNAP2*), a gene linked to autism and language impairment. In rodents, *CNTNAP2* expression is broad and relatively uniform, but in both human fetuses and songbirds, its expression is differentially enhanced in CBG circuits (Panaitof et al. 2010), suggesting a specialized function in vocal learners. In addition, an analysis of genes based on patterns of coexpression during singing identified ~2,000 genes in Area X that are regulated by singing, including known regulators of synaptic plasticity as well as novel genes (Hilliard et al. 2012). The investigation of these genes will be further aided by the recent sequencing of the zebra finch genome (Warren et al. 2010) and by the generation of songbird brain gene expression libraries (**http://www.zebrafinchatlas.org**). Thus, songbird researchers can contribute to human research both by identifying novel genes of translational relevance and by enabling disruption of such genes via transgenesis or viral transfection (Agate et al. 2009, Haesler et al. 2007) in a system with

HOMOLOGY OF AVIAN AND MAMMALIAN CORTICAL-BASAL GANGLIA CIRCUITRY

The anterior forebrain pathway (AFP) is an avian cortical-basal ganglia (CBG) circuit, with the special feature that most neurons within the AFP are specifically engaged during song production. There is an additional medial CBG loop with song-related activity, as well as shell structures around each of the AFP nuclei, which have been implicated in song production or learning (Jarvis et al. 1998, Bottjer & Altenau 2009), but the AFP has been most extensively investigated. It sits within more generalist CBG circuitry that likely subserves other nonsong aspects of motor control and learning (Feenders et al. 2008).

Area X is particularly well established as a homologue of mammalian striatum and pallidum, although there are some important differences (for more complete references and review of parallels, see Fee & Scharff 2010, Gale & Perkel 2010). Area X receives strong midbrain dopamine projections and contains the four major cell classes found in mammalian striatum, including spiny neurons; these classes resemble their mammalian homologues both neurochemically and neurophysiologically. The pallidal targets of Area X spiny neurons are intermingled with striatal neurons and include two classes of high-firing-rate neurons akin to the internal and external pallidal segments of mammals. Pallido-recipient thalamic (DLM) neurons in the AFP also exhibit properties typical of mammalian thalamus, including inputs from motor cortical areas (RA) and rebound burst firing from hyperpolarized potentials (Goldberg & Fee 2012, Person and Perkel 2005). Hence, the AFP preserves core features that enable investigation of broad issues relevant to CBG contributions to motor control and learning.

sensitive behavioral and neural assays of vocal and social communication.

CORTICAL-BASAL GANGLIA CIRCUITS AND REWARD AND MOTOR LEARNING

CBG circuits are critical for motor and cognitive processes, including control of movements as well as reward- and reinforcement-based learning. Accordingly, damage to CBG circuits can contribute to dysfunctions ranging from movement disorders to neuropsychiatric diseases. Yet, a detailed understanding of how these circuits function normally and are disrupted in disease is lacking. This may, in part, be attributable to the very importance of CBG circuits to multiple functions and the correspondingly complex and heterogeneous properties of neurons and subcircuits within the basal ganglia.

The AFP (**Figure 2**) is an avian CBG circuit that preserves core features of mammalian CBG circuits but is simplified and dedicated to song control and learning (see sidebar, Homology of Avian and Mammalian Cortical-Basal Ganglia Circuitry). Cortico-striatal neurons in HVC project to Area X (**Figure 2**), which has both striatal and pallidal components, and Area X pallidal neurons send inhibitory projections to the thalamic nucleus DLM. These thalamic neurons project onward to the frontal cortical nucleus LMAN, thus completing a cortico-basal ganglia-thalamo-cortical loop. LMAN projects both to the song motor cortical nucleus RA and back to Area X. One simplified but advantageous feature of this architecture is that the output of the AFP flows primarily through LMAN en route to song premotor nucleus RA. Hence, the output of the song-related basal ganglia circuitry can be monitored and manipulated at a single bottleneck for behaviorally relevant signals.

The Role of Cortical-Basal Ganglia Circuitry in Motor and Reinforcement Learning

The specific contributions of basal ganglia circuits to learning are particularly evident in songbirds. In juvenile zebra finches, lesions of the cortical outflow nucleus of this circuit (LMAN) cause an abrupt reduction in song variability and a failure of song to progress toward a good match with the tutor song (Bottjer et al. 1984, Scharff & Nottebohm 1991). Lesions of the striato-pallidal Area X also disrupt song learning but leave song structure highly variable (Scharff & Nottebohm 1991, Sohrabji et al. 1990). In adult finches, lesions of LMAN or Area X have little gross effect on song structure. However, such lesions subtly reduce rendition-by-rendition variation in song and prevent a variety of forms of adult song plasticity (Brainard & Doupe 2000, Kao et al. 2005, Thompson et al. 2007, Thompson & Johnson 2006, Williams & Mehta 1999). Similarly, for humans and mammalian models, disruptions of CBG circuits can cause more conspicuous deficits during learning than during execution of well-learned skills (Graybiel 2008).

In motor skill learning, which includes both song and speech, initially variable actions are refined over an extended period of rehearsal, during which sensory feedback is used to gradually improve performance relative to a goal. Although multiple processes may contribute to such learning, these likely include reinforcement learning using trial and error (Sutton & Barto 1998). Such reinforcement learning involves (*a*) the generation of behavioral variation, sometimes referred to as motor exploration, (*b*) the evaluation of which behavioral variants result in better versus worse outcomes, and (*c*) modifications to produce preferentially those variants that result in better outcomes. Studies of songbirds are helping to reveal the neural mechanisms that underlie these different processes.

Contributions of basal ganglia circuitry to the generation and regulation of behavioral variation. Motor exploration is a crucial substrate for learning, yet how such variability is generated has received relatively little attention from a biological perspective. Studies of songbirds indicate that such variability is actively generated by CBG circuitry. The observation that LMAN lesions cause an abrupt reduction in song variability prompted speculation that one function of basal ganglia circuitry may be to introduce variability or "motor exploration" into motor output for trial-and-error learning (Bottjer et al. 1984, Doya & Sejnowski 2000, Fiete et al. 2007, Kao & Brainard 2006, Kao et al. 2005, Ölveczky et al. 2005, Scharff & Nottebohm 1991). For example, reversible inactivation of LMAN in singing birds, or interfering with LMAN synaptic inputs to RA, decreases variability of both RA activity and song structure within minutes (Ölveczky et al. 2005, 2011; Stepanek & Doupe 2010, Warren et al. 2011). This suggests that LMAN introduces variability via synaptic influences on RA target neurons (Mooney 1992, Stark & Perkel 1999).

The ability to modulate behavioral variability is critical for trial-and-error learning, as too little variation can limit learning, and excess variability can interfere with performance. In songbirds, both behavioral and neural variability can be rapidly modulated by the social context in which a bird produces song. When a male sings courtship song (directed song) to a female, variability in syllable sequencing and in syllable structure is reduced relative to when males sing undirected song alone (Kao & Brainard 2006, Kao et al. 2005, Sakata et al. 2008, Sossinka & Bohner 1980). This is especially striking in juveniles, for which an immature and variable undirected song can be immediately transformed into nearly perfect adult song by the presence of a female (Kojima & Doupe 2011). These context-dependent changes in song variability are accompanied by dramatic changes in singing-related neural activity within LMAN of adult birds. During undirected song, neural activity is greater, more variable, and more likely to occur in bursts than it is during directed song (**Figure 4**) (Hessler & Doupe 1999a,b; Kao et al. 2008). In young birds singing alone, LMAN activity is also dominated by bursts, and these are correlated with sound production (Aronov et al. 2008, Ölveczky et al. 2005). Interestingly, IEG expression within the AFP is also greater following periods of undirected song than after periods of directed song (Jarvis et al. 1998). The elevated activity and variable bursting during undirected song are consistent with the idea that LMAN actively introduces variability into the motor pathway. Indeed, lesions or inactivation of LMAN reduce the variability normally present in undirected song to the level present during directed song (Hampton et al. 2009, Kao & Brainard 2006, Kao et al. 2005, Stepanek & Doupe 2010), suggesting that the default state of the adult motor pathway (in the absence of extrinsic

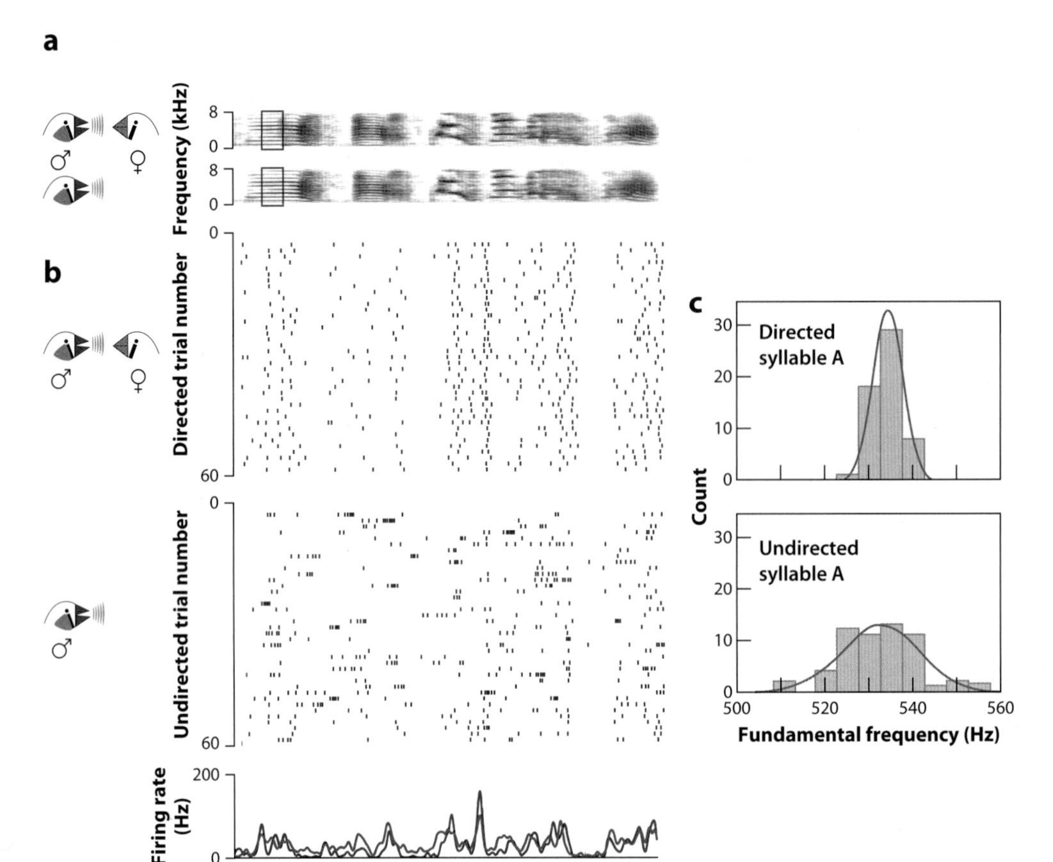

Figure 4

Rapid modulation of behavioral and neural variability by social context. (*a*) Spectrograms of song motifs produced by a male zebra finch singing directed song to a female (*top*) or undirected song alone (*bottom*). At a gross level, the acoustic structure of song is unaltered between these social contexts. (*b*) Neural activity recorded from a single neuron in LMAN during repeated renditions of the motif shown in panel *a*. When the male sings directed song (*top*), the pattern of neural activity is more stereotyped and less likely to occur in bursts than when the male sings undirected song (*middle*). Recordings are from interleaved renditions of directed versus undirected song, indicating that a single neuron can rapidly switch its firing pattern depending on social context. Despite the greater trial-by-trial variability during undirected song, the average pattern of adult LMAN neuron activity across many trials (*bottom*) is similar during directed (*red*) and undirected (*blue*) singing. (*c*) Distributions of fundamental frequency for typical renditions of a harmonic syllable (e.g., *red boxes* in panel *a*) during directed (*top*) or undirected (*bottom*) song. Although changes in social context do not grossly alter song structure, they consistently reduce the variability of fundamental frequency across renditions of directed song. Modified from Kao et al. (2005, 2008).

perturbations from LMAN) is to produce the low-variability songs characteristic of directed singing.

The active generation of behavioral variation during undirected song could reflect purposeful motor exploration. In principle, such motor exploration may enable trial-and-error refinement of vocalizations to match an acoustic target. Although the importance of such variability to juvenile song learning remains to be tested, this variability can subserve trial-and-error learning in adult birds. In particular, whereas the acoustic structure of "crystallized" adult finch song normally remains very

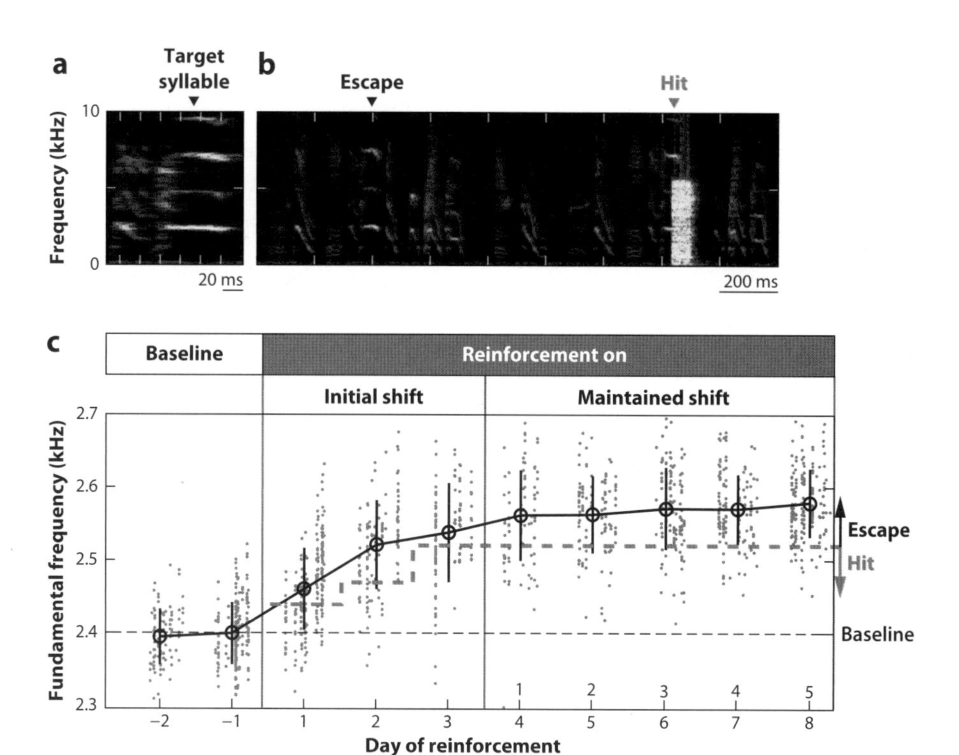

Figure 5

Trial-and-error learning in adult birdsong. Spectrogram of (*a*) a target syllable and (*b*) a song from an experiment in which white noise was delivered to targeted syllable renditions with low fundamental frequency (FF) ("hit") but not high FF ("escape"). (*c*) Delivering white noise to syllables with low FF (*below dashed green line*) elicited increases in FF. Each small point corresponds to one syllable rendition; black symbols indicate the mean and standard deviation of FF on each day of the experiment. Significant adaptive changes to FF could be driven within hours and were maintained as long as differential reinforcement with white noise continued. Figure modified from Warren et al. (2012).

stereotyped, differential reinforcement of subtle variations in adult song can drive rapid changes in song (Andalman & Fee 2009, Charlesworth et al. 2011, Tumer & Brainard 2007). For example, if an aversive white noise burst is played each time a bird sings a particular syllable at a lower pitch (fundamental frequency) but not when the syllable is produced at a higher pitch, the bird gradually adjusts the pitch of the targeted syllable upward so as to "escape" the white noise (**Figure 5**). These data reveal that even the very subtle rendition-to-rendition variations that are present in adult song can be utilized to enable adaptive changes to song structure, and they support the idea that undirected song in adult birds may reflect a state of motor "practice" in which motor exploration enables continued optimization of song. In the case of explicit reinforcement with white noise, this optimization entails shifting song to avoid a disruptive stimulus. Under normal circumstances, this optimization may entail the adjustment of premotor commands to maintain a close match to the tutor song, despite central and peripheral changes to components of the song system that occur during aging or injury.

A corollary of the idea that undirected song reflects a practice state, during which AFP-dependent motor exploration enables song fine-tuning, is the possibility that directed song reflects a performance state in which variability is reduced to produce the current "best version" of the song (in the context of courtship). Consistent with this possibility, males are less sensitive to altered auditory feedback when singing directed song than when singing alone (Sakata & Brainard 2009). Moreover, female birds tested with playback of recorded songs, without any additional cues, prefer directed songs over undirected songs (Woolley & Doupe 2008).

In mammalian systems, dopamine (DA) may play a role in the stability versus flexibility of behaviors. In songbirds, DA is one likely modulator of song variability via its actions on the AFP. Dopaminergic neurons send a strong projection to Area X (Gale & Perkel 2010, Person et al. 2008), and DA release in the AFP is enhanced by the presence of a female, as in rodent striatum (Sasaki et al. 2006). Moreover, both IEG induction and neural activity in the VTA are altered during directed singing, and VTA neurons show physiological changes following directed singing that are consistent with an elevated release of DA within the VTA (Hara et al. 2007, Huang & Hessler 2008, Riters 2011, Yanagihara & Hessler 2006). These data indicate that directed singing, which reduces song variability, engages neuromodulatory circuitry that encodes salience and reward. Finally, blockade of D1 receptors in the AFP decreases the excitability of Area X neurons and the strength of the synaptic inputs onto them (Ding & Perkel 2002, Ding et al. 2003). Correspondingly, such blockade interferes with modulation of variability by social context, pointing to a causal role of DA in modulating song variability (Leblois et al. 2010).

Together, these data suggest that the AFP is a useful model both to understand how CBG circuits contribute to behavioral variation important for learning as well as how such variability is dynamically regulated, for example, between practice and performance states. More broadly, the context-dependent changes that are readily apparent in songbirds may provide insights into human behavioral disruptions such as tremor, stuttering and other speech disorders, "choking" during performance, and difficulties with movement initiation, each of which can exhibit strong context-dependence in the severity of their manifestations.

Contributions of basal ganglia circuits to implementing adaptive changes in behavior. In mammalian systems, CBG circuits are broadly linked to reward-based action selection and learning (Graybiel 2008). In songbirds, studies indicate a central role of the AFP not only in generating variability, but also in the subsequent selection and implementation of more successful behavioral variants. These studies generally support an actor-critic model of basal ganglia function (Sutton & Barto 1998) (schematized in **Figure 6** in the context of reinforcement-driven changes to fundamental frequency). According to this model, at baseline (prior to learning), varying patterns of activity from LMAN "act" to drive rendition-to-rendition variation in RA activity and song structure (Andalman & Fee 2009; Doya & Sejnowski 2000; Kao et al. 2005; Ölveczky et al. 2005, 2011; Sober et al. 2008; Warren et al. 2011). During initial stages of learning, some patterns of LMAN activity cause syllable renditions that elicit more favorable feedback (i.e., escape from white noise bursts). These adaptive patterns of LMAN activity are then reinforced by favorable feedback from a "critic" so that over time they become more prevalent. Hence, LMAN activity is altered over the course of learning to bias vocal output in an adaptive direction.

Multiple lines of evidence support the actor-critic model. First, lesion experiments indicate that signals from LMAN are necessary for learning and can influence song structure (Bottjer et al. 1984, Brainard & Doupe 2000, Kao & Brainard 2006, Thompson et al. 2007, Thompson & Johnson 2006, Williams & Mehta 1999). Second, recording experiments indicate that LMAN activity during singing is temporally patterned, consistent with the possibility that LMAN provides input to RA that implements changes to song structure at specific times in song (Hessler & Doupe 1999a,b; Kao et al. 2008; Leonardo 2004). Third, lesions of adult Area X strip LMAN of its patterned

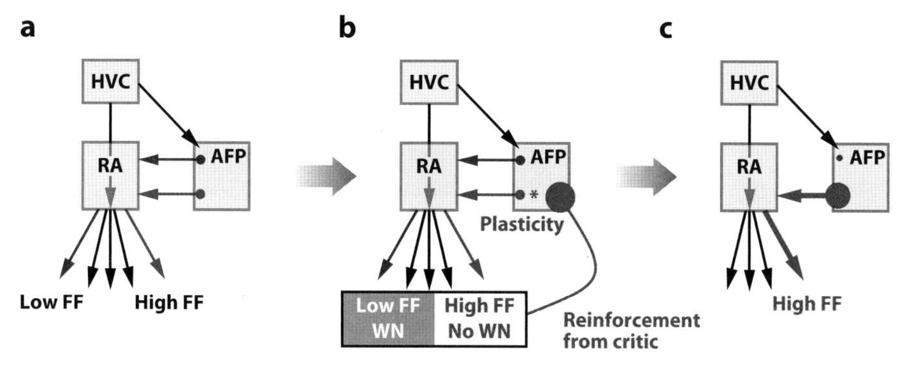

Figure 6

Actor-critic model of AFP contributions to reinforcement learning. (*a*) The AFP generates variation in performance (motor exploration); red and light blue indicate distinct activity patterns in the AFP that lead to distinct FF values on different renditions of the same syllable. (*b*) The AFP receives feedback about the behavioral variants that it generates, and this feedback strengthens patterns of AFP activity yielding better outcomes (*light blue*) (feedback and plasticity shown as *blue circle* and *asterisk*) and weakens patterns of AFP activity yielding worse outcomes (*red*). (*c*) This changes the output of the AFP so that it selectively implements more successful behaviors. Figure modified from Charlesworth et al. (2012). Abbreviations: AFP, anterior forebrain pathway; FF, fundamental frequency; RA, robust nucleus of the arcopallium; WN, white noise.

bursting during singing without decreasing its overall firing rate, simultaneously eliminating song plasticity in response to altered auditory feedback (Kojima et al. 2013). Finally, alterations in LMAN activity elicited by local microstimulation can acutely alter features of song, including the fundamental frequency and amplitude of individual syllables, indicating that signals from LMAN can acutely bias vocal output in a directed fashion (Kao et al. 2005).

Experiments in adult birds in which explicit reinforcement is used to direct changes to fundamental frequency provide further evidence that such biasing contributes to the initial expression of learning (Andalman & Fee 2009, Warren et al. 2011). During baseline singing (prior to learning), inactivation of LMAN causes a reduction in rendition-to-rendition variability of fundamental frequency for individual syllables, with little shift in the mean fundamental frequency. However, after the fundamental frequency of a targeted syllable is driven upward by differential reinforcement (**Figure 7**), inactivation of LMAN causes not only a reduction in variability, but also a reversion of the mean fundamental frequency toward the original baseline (Andalman & Fee 2009, Warren et al. 2011). Infusions of NMDA receptor antagonists into RA, which preferentially block LMAN synapses onto RA neurons, cause a quantitatively matched reversion of fundamental frequency toward baseline (Warren et al. 2011). Hence, a parsimonious account of these findings is that, during initial stages of learning, specific patterns of LMAN activity direct adaptive changes to song structure via their action on targets in RA, and that reversion following LMAN inactivation reflects the removal of this biasing signal.

Although important aspects of this model remain to be tested, including whether and how LMAN activity changes during adult reinforcement learning, these findings indicate crucial contributions of LMAN to the initial expression of such learning. It remains unclear whether similar contributions obtain for juvenile sensorimotor learning, in which auditory feedback guides developing vocalizations toward an acoustic target. However, in adults, vocal error correction in the absence of external instruction similarly depends on contributions of LMAN to the initial expression of learning

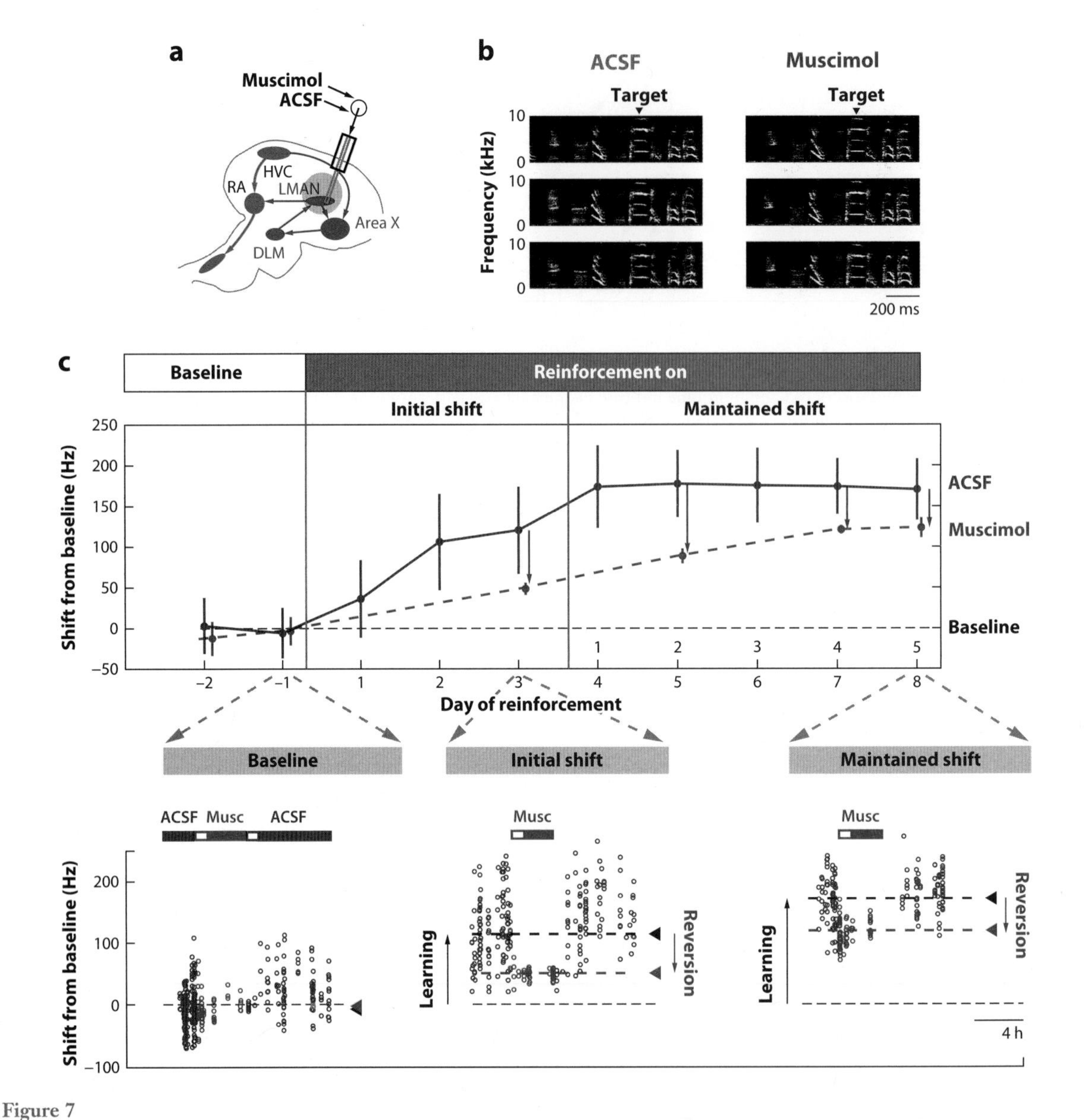

Figure 7

Contributions of LMAN to the expression of learning in adult song. (*a*) Dialysis probes were used to reversibly inactivate LMAN over the course of learning by periodically infusing the GABA agonist muscimol. (*b*) Spectrograms show little gross difference in the structure of songs produced under control conditions (ACSF) versus during LMAN inactivation (muscimol). (*c*) Effects of LMAN inactivation during different stages of learning in which differential reinforcement was used to drive an upward shift in fundamental frequency (FF) of the target syllable shown in panel *b*. Gray points indicate the mean and standard deviation of FF across renditions of the target syllable on each day of the experiment under conditions of saline infusion; red points indicate the mean FF during several hour periods when LMAN was inactivated. At baseline, prior to learning, LMAN inactivation reduced variability without altering mean FF. During initial learning, LMAN inactivation not only reduced variability, but also caused a large reversion of FF toward the baseline value. During a period of maintained learning, the effects of LMAN inactivation on the expression of learning progressively declined. Modified from Warren et al. (2011).

(Warren et al. 2011). Accordingly, the AFP may play a general role in directing adaptive changes to behavior during initial stages of learning.

This model of AFP contributions to the initial expression of learning implies that there is adaptive plasticity within the AFP or its upstream inputs. To achieve such adaptive plasticity, the nervous system must integrate information about neural activity patterns during a particular syllable rendition with feedback about resultant consequences (i.e., the presence or absence of reinforcement). In particular, because learned changes to song structure are precisely localized to reinforced features of song, there must be a correspondingly detailed record of the timing and structure of associated neural activity (Charlesworth et al. 2011, Tumer & Brainard 2007). The AFP is well positioned to receive information both about motor actions (an efference copy) from HVC and RA (Goldberg & Fee 2012, Kozhevnikov & Fee 2007, Prather et al. 2008, Vates et al. 1997) and about feedback regarding the outcomes resulting from those actions, either as auditory inputs via HVC or from neuromodulatory centers, such as the VTA, that may provide signals reflecting an evaluation of outcomes relative to expectation. Indeed, synaptic mechanisms of plasticity in Area X are modulated by dopaminergic tone, rendering Area X one likely locus for plasticity underlying initial learning (Ding & Perkel 2004, Fee & Goldberg 2011).

Basal ganglia–dependent learning without basal ganglia–dependent variation. The actor-critic model outlined above posits that basal ganglia circuitry monitors the consequences of behavioral variations that it generates and then implements the variations that result in better outcomes. However, experiments in adult songbirds indicate that basal ganglia circuitry can contribute to learning even when it is prevented from generating behavioral variation (Charlesworth et al. 2012). The AFP can be prevented from contributing to song variation by infusing NMDA receptor blockers into RA, which interrupts LMAN to RA synapses (Mooney 1992, Ölveczky et al. 2005, Stark & Perkel 1999). During this interruption, variation in the pitch of individual syllables is reduced, but some residual variation continues to arise from elsewhere, most likely from the motor pathway. If this residual variation is differentially reinforced, by playing white noise in response to lower versus higher pitched syllables, there is no expression of learning as long as the input from LMAN to RA remains interrupted. Surprisingly, however, as soon as input from LMAN to RA is restored, there is an immediate appearance of appropriate learned changes to song—an upward shift in pitch that is as large as would have occurred if AFP input to RA had remained intact. In contrast, if activity within the AFP is disrupted during differential reinforcement, no learning develops. These data indicate that the AFP is required for the acquisition and expression of learning, but that this learning does not require the AFP to act as the source of motor exploration for learning. Under normal conditions, the AFP may generate variation in song, but it can also integrate information about variation arising from elsewhere.

The immediate appearance of learning that is specific to reinforced features of song when the output of the AFP is unblocked indicates the importance of precise and bidirectional coordination between the AFP and the motor pathway in learning (Charlesworth et al. 2012). Furthermore, the specificity of this learning implies that the AFP receives detailed information about the song features produced by the motor pathway during reinforcement, likely as a copy of premotor commands in HVC and RA (Fee & Goldberg 2011, Goldberg & Fee 2012, Troyer & Doupe 2000a). These findings support the suggestion that efference copy signals from the motor cortex to basal ganglia circuitry play a fundamental role in mammalian skill learning (Crapse & Sommer 2008). These results also imply that the AFP can change its output to implement specifically the song features that were reinforced. Anatomically defined topographic loops that could mediate interactions between cortex and basal ganglia are well described in both mammals and songbirds (Johnson et al. 1995, Luo et al. 2001, Vates et al. 1997). However, the capacity of the AFP to direct precisely the activity within the motor pathway indicates surprisingly fine-scale functional coordination in the projections both from the motor pathway to the AFP and from the AFP back to the motor pathway. This coordination may be mediated by segregated functional loops between the AFP and the motor pathway, each encoding a particular feature of song, such as high fundamental frequency in a particular syllable. Under normal conditions with AFP output intact, such functional loops could enable the AFP to amplify and bias specific behavioral features, functions that have been attributed to mammalian basal ganglia circuits (Turner & Desmurget 2010). More generally, these results suggest that CBG circuits such as the AFP can act as a specialized hub that can both regulate and monitor variability in other brain regions and then direct those regions to produce patterns of variation that result in better outcomes.

Consolidation of motor skill learning. Numerous forms of learning, including motor skill learning, are stabilized or "consolidated" over time, often involving a shift in behavioral control from one region to another (Censor et al. 2012). Adult reinforcement learning in the songbird provides a clear example of such consolidation. Although the cortical outflow of the AFP (i.e., LMAN) must be active for the initial expression of adult learning, the degree to which learned changes revert toward baseline during LMAN inactivation gradually decreases over time (Andalman & Fee 2009, Warren et al. 2011). Thus, learning that initially requires the AFP is transferred to downstream structures in the motor pathway. One likely locus for this more slowly developing component of learning is at HVC-RA synapses within the motor pathway. These synapses influence the activity of RA neurons (Hahnloser et al. 2002, Long & Fee 2008) (see sidebar, Motor Sequence Generation), thereby shaping the spectral structure of ongoing song (Leonardo & Fee 2005, Sober et al. 2008). They are also a site where LMAN inputs to RA are well poised to drive synaptic plasticity (Mooney 1992, Sizemore & Perkel 2011). LMAN could instruct adaptive changes in motor circuitry by persistently imposing patterns of RA activity that eventually become engrained by spike-timing-dependent plasticity mechanisms acting on HVC to RA synapses (Troyer & Doupe 2000b, Swinehart & Abbott, 2005). In this case, LMAN would be a source of signals that drive the formation of longer-lasting habits but need not be the repository of those habits. Such song consolidation is consistent with other cases in which mammalian striatal activity initially implements adaptive changes in behavior but ultimately drives downstream consolidation of those changes (e.g., Censor et al. 2012, Pasupathy & Miller 2005).

DISORDERS OF BASAL GANGLIA CIRCUITRY

Given the importance of CBG circuits in motor and reinforcement learning, these pathways are also a locus of many neurological and psychiatric diseases. These include movement disorders, such as Parkinson's and Huntington's diseases, as well as psychiatric disorders, such as obsessive-compulsive disorder, psychoses, and addictions. As outlined above, the simplified CBG circuit for song has facilitated identification of ways in which such circuits contribute to normal learning. By extension, the avian CBG circuit may also be useful to investigate disorders of such circuits.

In Parkinson's disease and other human diseases of basal ganglia, disruptions of movement control (and perhaps also of cognitive function) reflect not only the absence of adaptive contributions of CBG activity to behavior, but also the addition of abnormal signals that actively corrupt behavior. In Parkinson's disease, neural activity in CBG circuits is excessively synchronized and is dominated by highly abnormal burst-firing (e.g., Hammond et al. 2007). Correspondingly, surgical elimination or alteration (via electrical stimulation) of components of human CBG circuitry can dramatically improve symptoms. Despite the clinical importance of these observations, how abnormal patterns of activity within CBG circuits interfere with normal function and how interventions such as deep-brain stimulation exert their beneficial effects (and deleterious side effects) are difficult to investigate in the human brain.

In songbirds, the consequences of abnormal activity within CBG circuits may similarly be much more disruptive than the complete absence of activity. Lesions of LMAN or Area X have relatively modest effects on the structure of adult song (though acute effects can be more disruptive in juvenile birds or in cases where song structure is more variable) (Aronov et al. 2008, Bottjer et al. 1984, Kobayashi et al. 2001, Nordeen & Nordeen 1992, Scharff & Nottebohm 1991, Sohrabji et al. 1990). However, if abnormal patterns of activity are artificially introduced into LMAN by electrical or pharmacological perturbations, song is more dramatically disrupted (Hamaguchi & Mooney 2012, Kao et al. 2005). Most strikingly, abnormalities of adult song that develop in response to a variety of insults to the nervous system (including peripheral damage to hearing or production mechanisms as well as central damage to HVC) can be prevented or even reversed by lesion or inactivation of LMAN (Brainard & Doupe 2000, Nordeen & Nordeen 2010, Thompson et al. 2007, Williams & Mehta 1999).

The importance of CBG circuitry in songbirds for regulating behavioral variability supports a view that aberrant regulation of variability may be an important component of many CBG diseases. For example, Huntington's disease is characterized by excessive and uncontrolled movement variability. Similarly, Parkinson's disease is characterized by decreased movement amplitude and speed. Abnormalities of speech in Parkinson's disease can include not only low volume and reduced amplitude of articulation, but also increased variability of articulation, which can contribute significantly to loss of intelligibility (Skodda 2011). Similarly, freezing of gait in Parkinson's disease may follow steps that have increased variability in their size and timing (e.g., Hausdorff et al. 2003). At the neuropsychiatric level, cognitive disturbances that may reflect abnormal basal ganglia and frontal function can be construed as expressing too little variability (as in obsessive-compulsive disorder and depression) or too much variability (as in psychosis or attention-deficit disorders).

Parkinson's disease and other basal ganglia disorders are also characterized by difficulty in initiating movements, which can depend strongly on the context of movement generation. Parkinsonian patients who are impaired in spontaneous movement generation are often much better at movements elicited by a sensory cue or an arousing stimulus (Berardelli et al. 2001, Carlsen et al. 2012). Parkinson's disease reflects, in part, a loss of dopaminergic inputs to striatum. In this regard, a bird's normal switch in performance between social contexts resembles a patients' switch between spontaneously generated and cued movements. Although birds can sing in isolation (uncued), presentation of a female often elicits immediate initiation of song (cued), consistent with the idea that neuromodulatory changes, such as elevation of DA release in Area X, may act to promote movement initiation.

Such similarities to findings in human diseases suggest that songbird studies can help reveal how pathological activity arising from CBG circuits contributes to movement and neuropsychiatric symptoms and can provide insights into mechanisms whereby these signals can be controlled or corrected. Moreover, in terms of speech and communication, songbirds can provide a sensitive assay for the cellular, circuit, and behavioral effects of the expression of genes identified in human neurodegenerative and neuropsychiatric diseases.

Adult Neurogenesis

One impediment to brain repair in the case of disease or injury, both in CBG circuits and more broadly, is our inability to replace neurons that are lost or damaged. Correspondingly, increasingly promising approaches to enabling adult brain repair include leveraging and extending the capacity for adult neurogenesis and the functional incorporation of new neurons into the adult brain. In songbirds, the phenomenon of adult neurogenesis is especially robust, with large-scale neurogenesis in HVC as well as in high-level auditory structures, striatum, and hippocampus (Nottebohm 2004). Such findings played a key role in overturning the dogma that the adult nervous system does not generate or incorporate new neurons.

There are numerous similarities between avian and mammalian adult neurogenesis, including lifespan and fate restrictions, progenitor cell types, and sensitivity to hormones and other experiential factors. However, songbird neurogenesis also exhibits several differences that may be usefully harnessed for human health (Gould 2007, Nottebohm 2004). These include much greater levels of neurogenesis, distributed over more brain regions, than are observed in mammals. Comparative molecular and genetic studies in mammals and birds could reveal how to upregulate cell division, migration, and functional incorporation rates in mammals in key additional areas. For instance, striatal spiny neurons that are the targets of neurodegenerative diseases such as Huntington's disease are specifically replaced in adult songbirds, raising the possibility that they could be renewed therapeutically (Rochefort et al. 2007). In addition, HVC neurons replaced in adult songbirds are premotor neurons that grow long projections into RA, through a heavily myelinated tract, and insert without disrupting already-learned song (Kirn et al. 1991). Understanding the molecular and cellular processes that enable this incorporation in adult songbirds has clear relevance to brain and spinal cord repair in humans.

Finally, songbirds are well suited to test how new neurons link to learning and behavior. The amount of neurogenesis in songbirds is correlated with periods of learning and repair (Nottebohm 2004), but there is also marked turnover and insertion of new neurons in birds that do not change their song (Pytte et al. 2012, Tramontin & Brenowitz 1999). The manipulation of neurogenesis in songbirds could provide sensitive behavioral and neural tests of the role of new neurons. An understanding of how new neurons contribute to sensory and sensorimotor learning may be particularly clear in birds and will be relevant to further understanding and therapeutic manipulation of neurogenesis in mammals.

CONCLUSIONS

No single animal model is likely to recapitulate all the features of complex neurological and psychiatric diseases, but we argue here that songbirds, with their easily quantifiable learned behavior and dedicated circuit, are a uniquely useful addition to more traditional model systems. Songbirds show remarkable parallels to humans in their links between sensory processing and motor output, their use and regulation of auditory feedback, their neural sensitivity to social signals, and the genes and circuits underlying vocal and motor skill learning. Songbird research has provided new insights into CBG circuitry, a major locus of human neural disease, including into the function of such circuitry in reinforcement learning and the social modulation of behavior. Finally, such research provides striking examples of time-limited neural plasticity and of functional insertion of newborn adult neurons into brain circuits, with great potential for enabling mechanistic understanding of these phenomena. It is our hope that enhanced dialogue between songbird and human researchers, with attention to both the similarities and differences between systems, can provide new therapeutic insights.

DISCLOSURE STATEMENT

The authors are not aware of any affiliations, memberships, funding, or financial holdings that might be perceived as affecting the objectivity of this review.

ACKNOWLEDGMENTS

The writing of this review and the cited work from the authors' labs were supported by NIH grant DC006636 and NSF grant IOS-0951348 (M.S.B.), and NIH grants MH055987 and DC011356, and a NARSAD Distinguished Investigator Award (A.J.D). We are grateful for thoughtful editing by Mimi Kao, Allison Hall, and Thomas Jessell, and to Satoshi Kojima for preparation of the illustration in **Figure 2**.

LITERATURE CITED

Adret P. 1993. Operant conditioning, song learning and imprinting to taped song in the zebra finch. *Anim. Behav.* 46:149–59

Agate RJ, Scott BB, Haripal B, Lois C, Nottebohm F. 2009. Transgenic songbirds offer an opportunity to develop a genetic model for vocal learning. *Proc. Natl. Acad. Sci. USA* 106:17963–67

Amador A, Perl YS, Mindlin GB, Margoliash D. 2013. Elemental gesture dynamics are encoded by song premotor cortical neurons. *Nature* 495:59–64

Andalman AS, Fee MS. 2009. A basal ganglia–forebrain circuit in the songbird biases motor output to avoid vocal errors. *Proc. Natl. Acad. Sci. USA* 106:12518–23

Aronov D, Andalman AS, Fee MS. 2008. A specialized forebrain circuit for vocal babbling in the juvenile songbird. *Science* 320:630–34

Ball GF, Balthazart J. 2010. Seasonal and hormonal modulation of neurotransmitter systems in the song control circuit. *J. Chem. Neuroanat.* 39:82–95

Balmer TS, Carels VM, Frisch JL, Nick TA. 2009. Modulation of perineuronal nets and parvalbumin with developmental song learning. *J. Neurosci.* 14(29):12878–85

Basham M, Nordeen E, Nordeen K. 1996. Blockade of NMDA receptors in the anterior forebrain impairs sensory acquisition in the zebra finch (*Poephila guttata*). *Neurobiol. Learn. Mem.* 66:295–304

Berardelli A, Rothwell JC, Thompson PD, Hallett M. 2001. Pathophysiology of bradykinesia in Parkinson's disease. *Brain* 124:2131–46

Bottjer SW, Altenau B. 2009. Parallel pathways for vocal learning in basal ganglia of songbirds. *Nat. Neurosci.* 13:153–55

Bottjer SW, Johnson F. 1997. Circuits, hormones, and learning: vocal behavior in songbirds. *J. Neurobiol.* 33:602–18

Bottjer SW, Miesner EA, Arnold AP. 1984. Forebrain lesions disrupt development but not maintenance of song in passerine birds. *Science* 224:901–3

Brainard MS, Doupe AJ. 2000. Interruption of a basal ganglia–forebrain circuit prevents plasticity of learned vocalizations. *Nature* 404:762–66

Brenowitz EA, Perkel DJ, Osterhout L. 2010. Language and birdsong: introduction to the special issue. *Brain Lang.* 115:1–2

Burt J, Lent K, Beecher M, Brenowitz E. 2000. Lesions of the anterior forebrain song control pathway in female canaries affect song perception in an operant task. *J. Neurobiol.* 42:487

Cardin JA, Schmidt MF. 2004. Noradrenergic inputs mediate state dependence of auditory responses in the avian song system. *J. Neurosci.* 24:7745–53

Carlsen AN, Almeida QJ, Franks IM. 2012. Using a startling acoustic stimulus to investigate underlying mechanisms of bradykinesia in Parkinson's disease. *Neuropsychologia* 51:392–99

Castelino CB, Schmidt MF. 2010. What birdsong can teach us about the central noradrenergic system. *J. Chem. Neuroanat.* 39:96–111

Catchpole CK, Slater PJB. 2008. *Bird Song: Biological Themes and Variations*. Cambridge, UK: Cambridge Univ. Press

Censor N, Sagi D, Cohen LG. 2012. Common mechanisms of human perceptual and motor learning. *Nat. Rev. Neurosci.* 13:658–64

Charlesworth JD, Tumer EC, Warren TL, Brainard MS. 2011. Learning the microstructure of successful behavior. *Nat. Neurosci.* 14:373–80

Charlesworth JD, Warren TL, Brainard MS. 2012. Covert skill learning in a cortical-basal ganglia circuit. *Nature* 486:251–55

Coleman MJ, Roy A, Wild JM, Mooney R. 2007. Thalamic gating of auditory responses in telencephalic song control nuclei. *J. Neurosci.* 27:10024–36

Cousillas H, George I, Henry L, Richard JP, Hausberger M. 2008. Linking social and vocal brains: Could social segregation prevent a proper development of a central auditory area in a female songbird? *PLoS One* 3:e2194

Crapse TB, Sommer MA. 2008. Corollary discharge across the animal kingdom. *Nat. Rev. Neurosci.* 9:587–600

Cynx J, von Rad U. 2001. Immediate and transitory effects of delayed auditory feedback on bird song production. *Anim. Behav.* 62:305–12

Dave AS, Margoliash D. 2000. Song replay during sleep and computational rules for sensorimotor vocal learning. *Science* 290:812–16

Derégnaucourt S, Mitra PP, Fehér O, Pytte C, Tchernichovski O. 2005. How sleep affects the developmental learning of bird song. *Nature* 433:710–16

Ding L, Perkel DJ, Farries MA. 2003. Presynaptic depression of glutamatergic synaptic transmission by D1-like dopamine receptor activation in the avian basal ganglia. *J. Neurosci.* 23:6086–95

Ding L, Perkel DJ. 2002. Dopamine modulates excitability of spiny neurons in the avian basal ganglia. *J. Neurosci.* 22:5210–18

Ding L, Perkel DJ. 2004. Long-term potentiation in an avian basal ganglia nucleus essential for vocal learning. *J. Neurosci.* 24:488–94

Doupe AJ, Kuhl PK. 1999. Birdsong and human speech: common themes and mechanisms. *Annu. Rev. Neurosci.* 22:567–631

Doupe AJ, Perkel DJ, Reiner A, Stern EA. 2005. Birdbrains could teach basal ganglia research a new song. *Trends Neurosci.* 28:353–63

Doya K, Sejnowski T. 2000. A computational model of avian song learning. In *The New Cognitive Neurosciences*, ed. M Gazzaniga, pp. 469–82. Cambridge, MA: MIT Press

Dugas-Ford J, Rowell JJ, Ragsdale CW. 2012. Cell-type homologies and the origins of the neocortex. *Proc. Natl. Acad. Sci. USA* 109:16974–79

Elie JE, Soula HA, Mathevon N, Vignal C. 2011. Same-sex pair-bonds are equivalent to male-female bonds in a life-long socially monogamous songbird. *Behav. Ecol. Sociobiol.* 65:2197–220

Fee MS, Goldberg JH. 2011. A hypothesis for basal ganglia–dependent reinforcement learning in the songbird. *Neuroscience* 198:152–70

Fee MS, Scharff C. 2010. The songbird as a model for the generation and learning of complex sequential behaviors. *ILAR J.* 51:362–77

Feenders G, Liedvogel M, Rivas M, Zapka M, Horita H, et al. 2008. Molecular mapping of movement-associated areas in the avian brain: a motor theory for vocal learning origin. *PLoS One* 3:e1768

Fernald A. 1985. Four-month-old infants prefer to listen to motherese. *Infant Behav. Dev.* 8:181–95

Fiete IR, Fee MS, Seung HS. 2007. Model of birdsong learning based on gradient estimation by dynamic perturbation of neural conductances. *J. Neurophysiol.* 98:2038–57

Ford JM, Mathalon DH. 2012. Anticipating the future: automatic prediction failures in schizophrenia. *Int. J. Psychophysiol.* 83:232–39

Fortune ES, Rodríguez C, Li D, Ball GF, Coleman MJ. 2011. Neural mechanisms for the coordination of duet singing in wrens. *Science* 334:666–70

Frith CD, Blakemore S, Wolpert DM. 2000. Explaining the symptoms of schizophrenia: abnormalities in the awareness of action. *Brain Res. Rev.* 31:357–63

Gale SD, Perkel DJ. 2010. Anatomy of a songbird basal ganglia circuit essential for vocal learning and plasticity. *J. Chem. Neuroanat.* 39:124–31

Gentner TQ, Hulse SH, Bentley GE, Ball GF. 2000. Individual vocal recognition and the effect of partial lesions to HVC on discrimination, learning, and categorization of conspecific song in adult songbirds. *J. Neurobiol.* 42:117–33

Goldberg JH, Fee MS. 2012. A cortical motor nucleus drives the basal ganglia–recipient thalamus in singing birds. *Nat. Neurosci.* 15:620–27

Gould E. 2007. How widespread is adult neurogenesis in mammals? *Nat. Rev. Neurosci.* 8:481–88
Graybiel AM. 2008. Habits, rituals, and the evaluative brain. *Annu. Rev. Neurosci.* 31:359–87
Haesler S, Rochefort C, Georgi B, Licznerski P, Osten P, Scharff C. 2007. Incomplete and inaccurate vocal imitation after knockdown of *FoxP2* in songbird basal ganglia nucleus Area X. *PLoS Biol.* 5:e321
Haesler S, Wada K, Nshdejan A, Morrisey EE, Lints T, et al. 2004. *FoxP2* expression in avian vocal learners and non-learners. *J. Neurosci.* 24:3164–75
Hahnloser RH, Kozhevnikov AA, Fee MS. 2002. An ultra-sparse code underlies the generation of neural sequences in a songbird. *Nature* 419:65–70
Hamaguchi K, Mooney R. 2012. Recurrent interactions between the input and output of a songbird cortico-basal ganglia pathway are implicated in vocal sequence variability. *J. Neurosci.* 32:11671–78
Hammond C, Bergman H, Brown P. 2007. Pathological synchronization in Parkinson's disease: networks, models and treatments. *Trends Neurosci.* 30:357–64
Hampton CM, Sakata JT, Brainard MS. 2009. An avian basal ganglia-forebrain circuit contributes differentially to syllable versus sequence variability of adult Bengalese finch song. *J. Neurophysiol.* 101:3235–45
Hara E, Kubikova L, Hessler NA, Jarvis ED. 2007. Role of the midbrain dopaminergic system in modulation of vocal brain activation by social context. *Eur. J. Neurosci.* 25:3406–16
Harvey CD, Coen P, Tank DW. 2012. Choice-specific sequences in parietal cortex during a virtual-navigation decision task. *Nature* 484:62–68
Hausdorff JM, Schaafsma JD, Balash Y, Bartels AL, Gurevich T, Giladi N. 2003. Impaired regulation of stride variability in Parkinson's disease subjects with freezing of gait. *Exp. Brain Res.* 2:187–94
Heinrich JE, Nordeen KW, Nordeen EJ. 2005. Dissociation between extension of the sensitive period for avian vocal learning and dendritic spine loss in the song nucleus lMAN. *Neurobiol. Learn. Mem.* 83:143–50
Hessler NA, Doupe AJ. 1999a. Singing-related neural activity in a dorsal forebrain–basal ganglia circuit of adult zebra finches. *J. Neurosci.* 19:10461–81
Hessler NA, Doupe AJ. 1999b. Social context modulates singing-related neural activity in the songbird forebrain. *Nat. Neurosci.* 2:209–11
Hickok G, Houde J, Rong F. 2011. Sensorimotor integration in speech processing: computational basis and neural organization. *Neuron* 69:407–22
Hilliard AT, Miller JE, Fraley ER, Horvath S, White SA. 2012. Molecular microcircuitry underlies functional specification in a basal ganglia circuit dedicated to vocal learning. *Neuron* 73:537–52
Houde JF, Nagarajan SS. 2011. Speech production as state feedback control. *Front. Hum. Neurosci.* 5:82
Huang YC, Hessler NA. 2008. Social modulation during songbird courtship potentiates midbrain dopaminergic neurons. *PLoS ONE* 3:e3281
Hurtado N, Marchman VA, Fernald A. 2008. Does input influence uptake? Links between maternal talk, processing speed and vocabulary size in Spanish-learning children. *Dev. Sci.* 11:F31–39
Jarvis ED, Scharff C, Grossman MR, Ramos JA, Nottebohm F. 1998. For whom the bird sings: context-dependent gene expression. *Neuron* 21:775–88
Johnson F, Sablan M, Bottjer SW. 1995. Topographic organization of a pathway involved with vocal learning in zebra finches. *J. Comp. Neurol.* 358:268–75
Kao MH, Brainard MS. 2006. Lesions of an avian basal ganglia circuit prevent context-dependent changes to song variability. *J. Neurophysiol.* 96:1441–55
Kao MH, Doupe AJ, Brainard MS. 2005. Contributions of an avian basal ganglia–forebrain circuit to real-time modulation of song. *Nature* 433:638–43
Kao MH, Wright BD, Doupe AJ. 2008. Neurons in a forebrain nucleus required for vocal plasticity rapidly switch between precise firing and variable bursting depending on social context. *J. Neurosci.* 28:13232–47
Keller GB, Hahnloser RH. 2009. Neural processing of auditory feedback during vocal practice in a songbird. *Nature* 457:187–90
Kirn JR, Alvarez-Buylla A, Nottebohm F. 1991. Production and survival of projection neurons in a forebrain vocal center of adult male canaries. *J. Neurosci.* 11:1756–62
Knudsen DP, Gentner TQ. 2010. Mechanisms of song perception in oscine birds. *Brain Lang.* 115:59–68
Kobayashi K, Uno H, Okanoya K. 2001. Partial lesions in the anterior forebrain pathway affect song production in adult Bengalese finches. *Neuroreport* 12:353–58

Kojima S, Doupe AJ. 2011. Social performance reveals unexpected vocal competency in young songbirds. *Proc. Natl. Acad. Sci. USA* 108:1687–92

Kojima S, Kao MH, Doupe AJ. 2013. Task-related "cortical" bursting depends critically on basal ganglia input and is linked to vocal plasticity. *Proc. Natl. Acad. Sci. USA* 110:4756–61

Konishi M. 1965. The role of auditory feedback in the control of vocalization in the white-crowned sparrow. *Z. Tierpsychol.* 22:770–83

Kozhevnikov AA, Fee MS. 2007. Singing-related activity of identified HVC neurons in the zebra finch. *J. Neurophysiol.* 97:4271–83

Kuhl PK. 2010. Brain mechanisms in early language acquisition. *Neuron* 67:713–27

Kuhl PK, Meltzoff AN 1996. Infant vocalizations in response to speech: vocal imitation and development change. *J. Acoust. Soc. Am.* 100:2425–38

Leblois A, Wendel BJ, Perkel DJ. 2010. Striatal dopamine modulates basal ganglia output and regulates social context–dependent behavioral variability through D1 receptors. *J. Neurosci.* 30:5730–43

Leonardo A. 2004. Experimental test of the birdsong error-correction model. *Proc. Natl. Acad. Sci. USA* 101:16935–40

Leonardo A, Fee MS. 2005. Ensemble coding of vocal control in birdsong. *J. Neurosci.* 25:652–61

Leonardo A, Konishi M. 1999. Decrystallization of adult birdsong by perturbation of auditory feedback. *Nature* 399:466–70

Livingston FS, White SA, Mooney R. 2000. Slow NMDA-EPSCs at synapses critical for song development are not required for song learning in zebra finches. *Nat. Neurosci.* 3:482–88

London SE, Clayton DF. 2008. Functional identification of sensory mechanisms required for developmental song learning. *Nat. Neurosci.* 11:579–86

Long MA, Fee MS. 2008. Using temperature to analyze temporal dynamics in the songbird motor pathway. *Nature* 456:189–94

Long MA, Jin DZ, Fee MS. 2010. Support for a synaptic chain model of neuronal sequence generation. *Nature* 468:394–99

Luo M, Ding L, Perkel DJ. 2001. An avian basal ganglia pathway essential for vocal learning forms a closed topographic loop. *J. Neurosci.* 21:6836–45

Marler P. 1970. Birdsong and speech development: Could there be parallels? *Am. Sci.* 58:669–73

Marler P. 1990. Innate learning preferences: signals for communication. *Dev. Psychobiol.* 23:557–68

Miller-Sims VC, Bottjer SW. 2012. Auditory experience refines cortico-basal ganglia inputs to motor cortex via remapping of single axons during vocal learning in zebra finches. *J. Neurophysiol.* 107:1142–56

Mooney R. 1992. Synaptic basis for developmental plasticity in a birdsong nucleus. *J. Neurosci.* 12:2464–77

Mooney R. 2009. Neural mechanisms for learned birdsong. *Learn. Mem.* 16:655–69

Moorman S, Mello CV, Bolhuis JJ. 2011. From songs to synapses: molecular mechanisms of birdsong memory. Molecular mechanisms of auditory learning in songbirds involve immediate early genes, including *zenk* and *arc*, the ERK/MAPK pathway and synapsins. *BioEssays* 33:377–85

Nelson DA, Marler P. 1989. Categorical perception of a natural stimulus continuum: birdsong. *Science* 244:976–78

Nelson DA, Marler P. 1994. Selection-based learning in bird song development. *Proc. Natl. Acad. Sci. USA* 91:10498–501

Nick TA, Konishi M. 2005. Neural song preference during vocal learning in the zebra finch depends on age and state. *J. Neurobiol.* 62:231–42

Nordeen EJ, Holtzman DA, Nordeen KW. 2009. Increased Fos expression among midbrain dopaminergic cell groups during birdsong tutoring. *Eur. J. Neurosci.* 30:662–70

Nordeen KW, Nordeen EJ. 1992. Auditory feedback is necessary for the maintenance of stereotyped song in adult zebra finches. *Behav. Neural Biol.* 57:58–66

Nordeen KW, Nordeen EJ. 2010. Deafening-induced vocal deterioration in adult songbirds is reversed by disrupting a basal ganglia–forebrain circuit. *J. Neurosci.* 30:7392–400

Nottebohm F. 2004. The road we travelled: discovery, choreography, and significance of brain replaceable neurons. *Ann. NY Acad. Sci.* 1016:628–58

Nottebohm F, Stokes TM, Leonard CM. 1976. Central control of song in the canary, *Serinus canarius*. *J. Comp. Neurol.* 165:457–86

Ölveczky BP, Andalman AS, Fee MS. 2005. Vocal experimentation in the juvenile songbird requires a basal ganglia circuit. *PLoS Biol.* 3:e153

Ölveczky BP, Otchy TM, Goldberg JH, Aronov D, Fee MS. 2011. Changes in the neural control of a complex motor sequence during learning. *J. Neurophysiol.* 106:386–97

Panaitof SC, Abrahams BS, Dong H, Geschwind DH, White SA. 2010. Language-related *Cntnap2* gene is differentially expressed in sexually dimorphic song nuclei essential for vocal learning in songbirds. *J. Comp. Neurol.* 18:1995–2018

Pasupathy A, Miller EK. 2005. Different time courses of learning-related activity in the prefrontal cortex and striatum. *Nature* 433:873–76

Person AL, Gale SD, Farries MA, Perkel DJ. 2008. Organization of the songbird basal ganglia, including area X. *J. Comp. Neurol.* 508:840–66

Person AL, Perkel DJ. 2005. Unitary IPSPs drive precise thalamic spiking in a circuit required for learning. *Neuron* 46:129–40

Pinaud R, Tremere LA 2012. Control of central auditory processing by a brain-generated oestrogen. *Nat. Rev. Neurosci.* 13:521–27

Podos J. 1996. Motor constraints on vocal development in a songbird. *Anim. Behav.* 51:1061–70

Prather JF, Nowicki S, Anderson RC, Peters S, Mooney R. 2009. Neural correlates of categorical perception in learned vocal communication. *Nat. Neurosci.* 12:221–28

Prather JF, Peters S, Nowicki S, Mooney R. 2008. Precise auditory-vocal mirroring in neurons for learned vocal communication. *Nature* 451:305–10

Prather JF, Peters S, Nowicki S, Mooney R. 2010. Persistent representation of juvenile experience in the adult songbird brain. *J. Neurosci.* 30:10586–98

Price CJ. 2012. A review and synthesis of the first 20 years of PET and fMRI studies of heard speech, spoken language and reading. *Neuroimage* 62:816–47

Pytte CL, George S, Korman S, David E, Bogdan D, Kirn JR. 2012. Adult neurogenesis is associated with the maintenance of a stereotyped, learned motor behavior. *J. Neurosci.* 32:7052–57

Pytte CL, Suthers RA. 1999. A bird's own song contributes to conspecific song perception. *Neuroreport* 10:1773–78

Reiner A, Perkel DJ, Mello CV, Jarvis ED. 2004. Songbirds and the revised avian brain nomenclature. *Ann. NY Acad. Sci.* 1016:77–108

Remage-Healey L, London SE, Schlinger BA. 2010. Birdsong and the neural production of steroids. *J. Chem. Neuroanat.* 39:72–81

Riede T, Goller F. 2010. Peripheral mechanisms for vocal production in birds: differences and similarities to human speech and singing. *Brain Lang.* 115:69–80

Riters LV. 2011. Pleasure seeking and birdsong. *Neurosci. Biobehav. Rev.* 35:1837–45

Rizzolatti G, Fabbri-Destro M. 2010. Mirror neurons: from discovery to autism. *Exp. Brain Res.* 200:223–37

Roberts TF, Gobes SMH, Murugan M, Ölveczky BP, Mooney R. 2012. Motor circuits are required to encode a sensory model for imitative learning. *Nat. Neurosci.* 15:1454–59

Roberts TF, Tschida KA, Klein ME, Mooney R. 2010. Rapid spine stabilization and synaptic enhancement at the onset of behavioural learning. *Nature* 463:948–52

Rochefort C, He X, Scotto-Lomassese S, Scharff C. 2007. Recruitment of *FoxP2*-expressing neurons to Area X varies during song development. *Dev. Neurobiol.* 67:809–17

Roy A, Mooney R. 2007. Auditory plasticity in a basal ganglia–forebrain pathway during decrystallization of adult birdsong. *J. Neurosci.* 27:6374–87

Sakata JT, Brainard MS. 2006. Real-time contributions of auditory feedback to avian vocal motor control. *J. Neurosci.* 26:9619–28

Sakata JT, Brainard MS. 2008. Online contributions of auditory feedback to neural activity in avian song control circuitry. *J. Neurosci.* 28:11378–90

Sakata JT, Brainard MS. 2009. Social context rapidly modulates the influence of auditory feedback on avian vocal motor control. *J. Neurophysiol.* 102:2485–97

Sakata JT, Hampton CM, Brainard MS. 2008. Social modulation of sequence and syllable variability in adult birdsong. *J. Neurophysiol.* 99:1700–11

Sasaki A, Sotnikova TD, Gainetdinov RR, Jarvis ED. 2006. Social context–dependent singing-regulated dopamine. *J. Neurosci.* 26:9010–14

Scharff C, Nottebohm F. 1991. A comparative study of the behavioral deficits following lesions of various parts of the zebra finch song system: implications for vocal learning. *J. Neurosci.* 11:2896–913

Scharff C, Nottebohm F, Cynx J. 1998. Conspecific and heterospecific song discrimination in male zebra finches with lesions in the anterior forebrain pathway. *J. Neurobiol.* 36:81–90

Schmidt MF, McLean J, Goller F. 2012. Breathing and vocal control: the respiratory system as both a driver and a target of telencephalic vocal motor circuits in songbirds. *Exp. Physiol.* 97:455–61

Schulz SB, Haesler S, Scharff C, Rochefort C. 2010. Knockdown of *FoxP2* alters spine density in Area X of the zebra finch. *Genes Brain Behav.* 9:732–40

Shank SS, Margoliash D. 2009. Sleep and sensorimotor integration during early vocal learning in a songbird. *Nature* 458:73–77

Shea SD, Margoliash D. 2010. Behavioral state–dependent reconfiguration of song-related network activity and cholinergic systems. *J. Chem. Neuroanat.* 39:132–40

Sizemore M, Perkel DJ. 2011. Premotor synaptic plasticity limited to the critical period for song learning. *Proc. Natl. Acad. Sci. USA* 108:17492–97

Skodda S. 2011. Aspects of speech rate and regularity in Parkinson's disease. *J. Neurol. Sci.* 310:231–36

Slater PJB, Eales LA, Clayton NS. 1988. Song learning in zebra finches (*Taeniopygia guttata*): progress and prospects. In *Advances in the Study of Behavior*, Vol. 18, pp. 1–34. New York: Academic

Sober SJ, Brainard MS. 2009. Adult birdsong is actively maintained by error correction. *Nat. Neurosci.* 12:927–31

Sober SJ, Wohlgemuth MJ, Brainard MS. 2008. Central contributions to acoustic variation in birdsong. *J. Neurosci.* 28:10370–79

Soha JA, Marler P. 2000. A species-specific acoustic cue for selective song learning in the white-crowned sparrow. *Anim. Behav.* 60:297–306

Sohrabji F, Nordeen EJ, Nordeen KW. 1990. Selective impairment of song learning following lesions of a forebrain nucleus in the juvenile zebra finch. *Behav. Neural Biol.* 53:51–63

Solis MM, Doupe AJ. 1999. Contributions of tutor and bird's own song experience to neural selectivity in the songbird anterior forebrain. *J. Neurosci.* 19:4559–84

Solis MM, Doupe AJ. 2000. Compromised neural selectivity for song in birds with impaired sensorimotor learning. *Neuron* 25:109–21

Sossinka R, Bohner J. 1980. Song types in the zebra finch *Poephila guttata castanotis*. *Z. Tierpsychol.* 53:123–32

Stark LL, Perkel DJ. 1999. Two-stage, input-specific synaptic maturation in a nucleus essential for vocal production in the zebra finch. *J. Neurosci.* 19:9107–16

Stepanek L, Doupe AJ. 2010. Activity in a cortical-basal ganglia circuit for song is required for social context–dependent vocal variability. *J. Neurophysiol.* 104:2474–86

Sutton RS, Barto AG. 1998. *Reinforcement Learning: An Introduction.* Cambridge, MA: MIT Press

Swinehart CD, Abbott LF. 2005. Supervised learning through neuronal response modulation. *Neural Comput.* 17:609–31

Tchernichovski O, Lints T, Mitra PP, Nottebohm F. 1999. Vocal imitation in zebra finches is inversely related to model abundance. *Proc. Natl. Acad. Sci. USA* 96:12901–4

Tchernichovski O, Mitra PP, Lints T, Nottebohm F. 2001. Dynamics of the vocal imitation process: how a zebra finch learns its song. *Science* 291:2564–69

Teramitsu I, Kudo LC, London SE, Geschwind DH, White SA. 2004. Parallel *FoxP1* and *FoxP2* expression in songbird and human brain predicts functional interaction. *J. Neurosci.* 24:3152–63

Theunissen FE, Shaevitz SS. 2006. Auditory processing of vocal sounds in birds. *Curr. Opin. Neurobiol.* 16:400–7

Thompson JA, Johnson F. 2006. HVC microlesions do not destabilize the vocal patterns of adult male zebra finches with prior ablation of LMAN. *J. Neurobiol.* 67:205–18

Thompson JA, Wu W, Bertram R, Johnson F. 2007. Auditory-dependent vocal recovery in adult male zebra finches is facilitated by lesion of a forebrain pathway that includes the basal ganglia. *J. Neurosci.* 27:12308–20

Tramontin AD, Brenowitz EA. 1999. A field study of seasonal neuronal incorporation into the song control system of a songbird that lacks adult song learning. *J. Neurobiol.* 40:316–26

Troyer TW, Doupe AJ. 2000a. An associational model of birdsong sensorimotor learning I. Efference copy and the learning of song syllables. *J. Neurophysiol.* 84:1204–23

Troyer TW, Doupe AJ. 2000b. An associational model of birdsong sensorimotor learning II. Temporal hierarchies and the learning of song sequence. *J. Neurophysiol.* 84:1224–39

Tumer EC, Brainard MS. 2007. Performance variability enables adaptive plasticity of 'crystallized' adult birdsong. *Nature* 450:1240–44

Turner RS, Desmurget M. 2010. Basal ganglia contributions to motor control: a vigorous tutor. *Curr. Opin. Neurobiol.* 20:704–16

Vates GE, Vicario DS, Nottebohm F. 1997. Reafferent thalamo-"cortical" loops in the song system of oscine songbirds. *J. Comp. Neurol.* 380:275–90

Velho TA, Lu K, Ribeiro S, Pinaud R, Vicario D, Mello CV. 2012. Noradrenergic control of gene expression and long-term neuronal adaptation evoked by learned vocalizations in songbirds. *PLoS One* 7:e36276

Volman SF. 1993. Development of neural selectivity for birdsong during vocal learning. *J. Neurosci.* 13:4737–47

Wada K, Howard JT, McConnell P, Whitney O, Lints T, et al. 2006. A molecular neuroethological approach for identifying and characterizing a cascade of behaviorally regulated genes. *Proc. Natl. Acad. Sci. USA* 103:15212–17

Wang CZ, Herbst JA, Keller GB, Hahnloser RH. 2008. Rapid interhemispheric switching during vocal production in a songbird. *PLoS Biol.* 6:e250

Wang Y, Brzozowska-Prechtl A, Karten HJ. 2010. Laminar and columnar auditory cortex in avian brain. *Proc. Natl. Acad. Sci. USA* 107:12676–81

Warren TL, Tumer EC, Charlesworth JD, Brainard MS. 2011. Mechanisms and time course of vocal learning and consolidation in the adult songbird. *J. Neurophysiol.* 106:1806–21

Warren WC, Clayton DF, Ellegren H, Arnold AP, Hillier LW, et al. 2010. The genome of a songbird. *Nature* 464:757–62

West MJ, King AP. 1988. Female visual displays affect the development of male song in the cowbird. *Nature* 334:244–46

White SA. 2010. Genes and vocal learning. *Brain Lang.* 115:21–28

White SA, Livingston F, Mooney R. 1999. Androgens modulate NMDA receptor–mediated EPSCs in the zebra finch song system. *J. Neurophysiol.* 82:2221–34

Williams H, Mehta N. 1999. Changes in adult zebra finch song require a forebrain nucleus that is not necessary for song production. *J. Neurobiol.* 39:14–28

Woolley SC, Doupe AJ. 2008. Social context–induced song variation affects female behavior and gene expression. *PLoS Biol.* 6:e62

Woolley SM 2012. Early experience shapes vocal neural coding and perception in songbirds. *Dev. Psychobiol.* 54:612–31

Yanagihara S, Hessler NA. 2006. Modulation of singing-related activity in the songbird ventral tegmental area by social context. *Eur. J. Neurosci.* 24:3619–27

Yu AC, Margoliash D. 1996. Temporal hierarchical control of singing in birds. *Science* 273:1871–75

Transient Receptor Potential Channels and Mechanosensation

Niels Eijkelkamp,[1] Kathryn Quick,[2] and John N. Wood[2]

[1]Laboratory of Neuroimmunology and Developmental Origins of Disease, University Medical Center Utrecht, 3584 EA Utrecht, The Netherlands; email: N.Eijkelkamp@umcutrecht.nl

[2]Wolfson Institute for Biomedical Research, University College London, London WC1 6BT, United Kingdom; email: J.Wood@ucl.ac.uk, K.quick@ucl.ac.uk

Annu. Rev. Neurosci. 2013. 36:519–46

The *Annual Review of Neuroscience* is online at neuro.annualreviews.org

This article's doi:
10.1146/annurev-neuro-062012-170412

Keywords

mechanotransduction, hearing, touch, pain

Abstract

Transient receptor potential (TRP) channels act as sensors for a range of stimuli as diverse as light, sound, touch, pheromones, and tissue damage. Their role in mechanosensation in the animal kingdom, identified by gene ablation studies, has raised questions about whether they are directly mechanically gated, whether they act alone or in concert with other channels to transduce mechanical stimuli, and their relative importance in various functions and disease states in humans. The ability of these channels to form heteromultimers and interact with other ion channels underlies a range of cell-specific functions in different cell types. Here we overview recent advances in this rapidly expanding field, focusing on somatosensation, hearing, the cardiovascular system, and interactions between TRP channels and other proteins involved in mechanoelectrical signaling.

Contents

INTRODUCTION

Transient receptor potential (TRP) channels contribute to intracellular signaling in many physiological systems and have been a focus of particular interest in studies of sensory transduction. The first clue that such channels existed came from an electrophysiological analysis of *Drosophila* by Cosens & Manning (1969), who found that a mutant fly showed impaired vision upon continuous illumination. It took 20 years before the mutant gene was cloned (Montell & Rubin 1989, Wong et al. 1989), and only in 1991 were TRP proteins defined as ion channels (Hardie 1991, 2011). The first mammalian TRP homologues were cloned by Montell (Wes et al. 1995) and by Birnbaumer (Zhu et al. 1995) in 1995, and a role for this class of ion channels in all types of sensory transduction has continued to accumulate (Zhu 2011). In this review we focus on TRP channels and mechanosensation.

TRP channels form multimeric cation-selective complexes (usually tetramers) and play an important role in calcium signaling. Some TRP channels are also important for magnesium homeostasis, for example, TRPM6 (Xie et al. 2011) and other magnesium permeant channels (TRPM2, TRPV1/2). TRP channels generally have six transmembrane domains with intracellular N and C termini, with a contribution by the S5–S6 transmembrane segments to the cation-selective pore. These features combined with voltage-dependent activity in some channels suggest an evolutionary link between TRP channels and the voltage-activated sodium and calcium channels that are central to the functioning of animal nervous systems. Thus, TRP channels may be not only the "doors of perception" through which external signals are detected at cellular and organismal levels but also the original building blocks of the electrical signaling system in animal nervous systems. The presence or absence of TRP channels is characteristic of animals or plants, respectively (Ramsey et al. 2006).

The subtypes of TRP channel are shown in **Figure 1**, where structural features of the channels are graphically represented. In mammals, there are broadly expressed structurally distinct subsets of channels, termed TRPC (canonical channels; seven members), TRPV (vanilloid; six members), TRPA (ankyrin; one member), TRPM (melastatin; seven members), TRPML (mucolipin; three members), and TRPP (polycystin; three members) (Wu et al. 2010). Channels both within and across structural subsets may be able to heteromultimerize to provide an extraordinary diversity of channels (**Figure 2**). For example, evidence indicates the existence of TRPC heteromultimers as well as TRPC1/TRPV4, TRPC1/TRPP2, and TRPV1/TRPA1 heteromultimers (Kobori et al. 2009, Ma et al. 2011a, Staruschenko et al. 2010). The genetic diversity of the channels is further enhanced by the existence of splice variants that could theoretically produce increasingly complex repertoires of different channel subtypes (**Figure 2**). Splice variants with distinct functions have been identified in TRPC3 (Kim et al. 2012). A practical consequence of the broad expression of a range of different TRP channels is that the characterization of the properties of a particular channel type expressed in a cell line may be affected by the presence of endogenous TRPs. Permissive factors that regulate TRP channel activity have also been found. For example, in pancreatic acinar cells, Stim1 translocates TRPC1 into the cell membrane to activate calcium influx (Yuan et al. 2007). The various discrepancies in the literature on the topic of TRP channel activation may reflect the complexities of TRP interactions in controlling activity.

TRP channels are found throughout the animal kingdom, in yeast, fungi, worms, flies, mice, and humans. In yeast, hyperosmotic stimuli cause calcium movement from the vacuole to the cytoplasm through the TRPY1 channel. Other fungi such as *Candida albicans* express related TRPY proteins that can functionally substitute for TRPY1 as mechanosensors in yeast (Zhou et al. 2005). Worms and flies also express a range of TRP channels. For example, *Drosophila* have the original TRP channel and related TRP-like protein (TRPL) as well as 11 other TRP channels involved in a variety of sensations, some of which involve interactions with rhodopsin outside the context of light sensing (Fowler & Montell 2013, Shen et al. 2011). Similarly, representatives of all types of known TRP channels (17 in total) are found in *Caenorhabditis elegans*, for which roles in mechanosensation have been described (Xiao & Xu 2011).

Genetic evidence of a contribution to mechanosensation of a variety of TRP channels is strong, as loss-of-function mutations in organisms ranging from invertebrates to zebrafish, to mice and humans have been recorded (Zhu 2011). However, formal proof of a role for these (or, indeed, any) channels is lacking, as mechanically gated ion channel activity in lipid bilayers of purified TRP proteins has not been observed. It is quite likely that such proof will never be obtained because a requirement for the integrity of the cytoskeleton has been observed with many mechanosensitive ion channel systems. Sensory neuron mechanically gated ion channels depend on the actin cytoskeleton

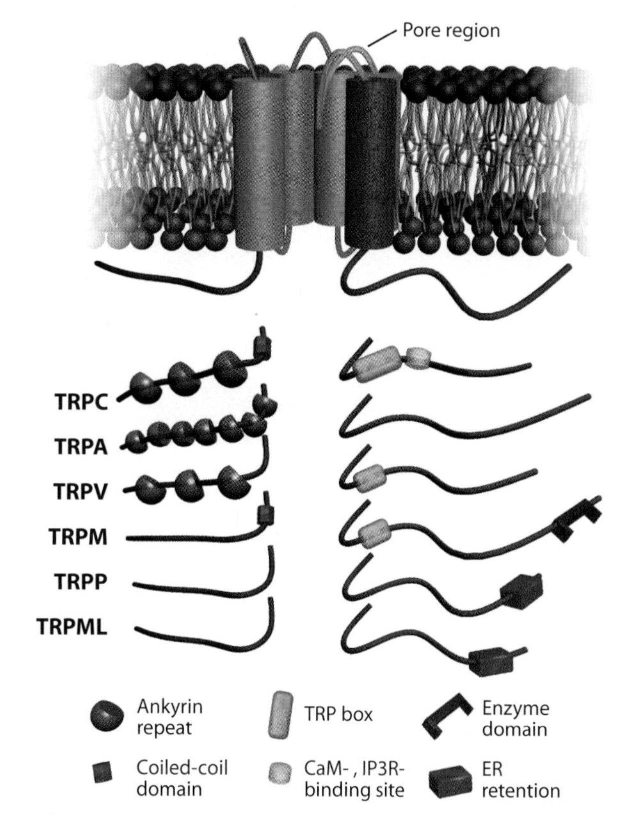

Figure 1

Structure of transient receptor potential (TRP) channel families. The six TRP families differ in their amino and carboxyl termini. The TRPC (TRP canonical channel), TRPA (TRP ankyrin), and TRPV (TRP vanilloid) subfamilies contain ankyrin repeat motifs present in tandem copies that contain elastic properties, which may be relevant to mechanosensation. The TRP box is thought to be involved in gating and is present in TRPV, TRPM (TRP melastatin), and TRPC subfamilies. TRPP (TRP polycystin) and TRPML (TRP mucolipin) specifically contain endoplasmic reticulum (ER) retention domains that may be important for their functional localization in intracellular organelles. Members of the TRPC subfamily also contain a calmodulin inositol-1,4,5-trisphosphate receptor binding domain at the carboxyl terminus that may be involved in channel gating.

(Drew et al. 2002), as does the non-TRP mechanotransducer Piezo2 (Coste et al. 2012, Eijkelkamp et al. 2013). Cold and ligand-gated TRP channel activity for TRPM8 in lipid bilayers conclusively demonstrates a role for this channel in cold transduction (Zakharian et al. 2010). By contrast, even though Piezo proteins form ion channels in isolation in lipid bilayers, they may also require the actin cytoskeleton to become mechanosensitive (Dubin et al. 2012).

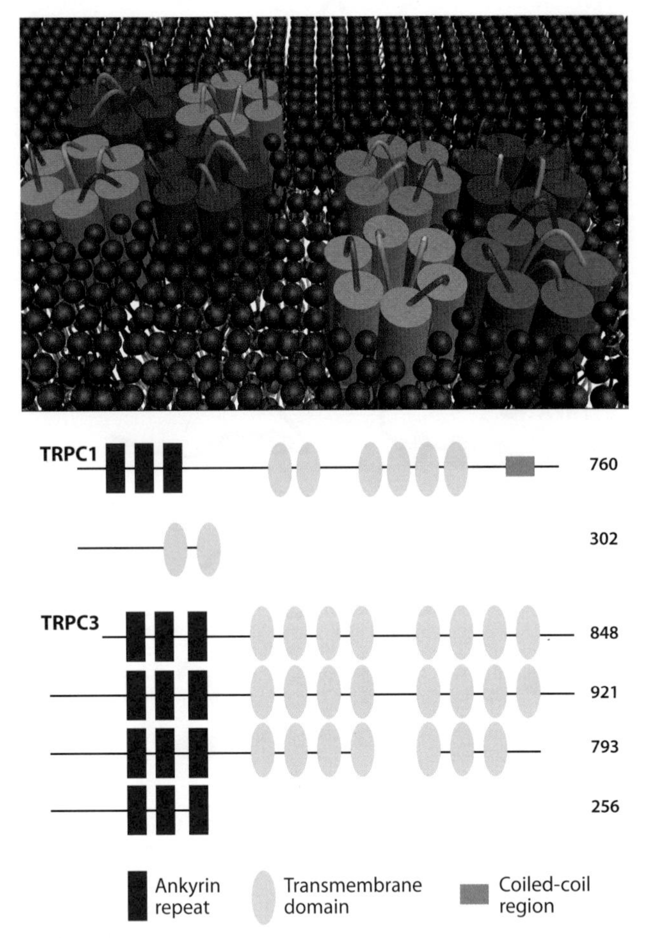

Figure 2

Multimerization and splice variants increase TRP diversity. Many closely related members of the TRP superfamily of proteins form tetramers. TRP channels form homomeric or heteromeric complexes (*upper panel*) depending on subunit types. Multimerization modulates the function, subcellular localization, and biophysical properties of the interacting channels. The C-terminal region has been proposed to contribute to subunit assembly. The different members of the TRP family can form many different combinations, which increases the functional diversity among TRP channels. Channel activity has indicated that TRPC channels function as heteromultimers, and we have recently shown that coexpression of TRPC6 with TRPC3 enhances mechanosensitivity. (*Lower panel*) A schematic representation of predicted splice variants of two members of the TRPC family, TRPC1 and TRPC3, to exemplify TRP diversity. The numbers indicate the total number of amino acids of each individual splice variant. This figure highlights that the existence of multiple splice variants of each TRP family member even further increases the possible number of TRP heteromeric complexes, which thus likely increases the functional diversity of TRPs.

To make a compelling case for a mechanotransducing role, a number of criteria need to be fulfilled:

1. Loss of expression of the channel should result in a loss of mechanosensitivity.
2. Heterologous expression of the candidate channel should result in the acquisition of mechanosensitivity in a number of different cellular environments.
3. The pharmacological and biophysical properties of the candidate channel should align with those of the endogenous mechanotransducing current studied.
4. Modulators of mechanosensation at the behavioral or electrophysiological levels should show parallel effects on channel activity.
5. The channel should, ideally, form a mechanotransducing complex in a lipid bilayer.

A critical role in mechanosensation does not necessarily imply a role as a direct mechanotransducer. Recent studies in *C. elegans* have shown an intimate relationship between the ENaC channel DEG-1 and the TRPV channel OSM9, where ENaC channels are mechanotransducers. By contrast, TRPV channels such as OSM9 play an important role in downstream signaling events that result in behavior effects on mechanosensation (Geffeney et al. 2011, Liedtke 2007). In contrast to the situation in *C. elegans*, there is compelling evidence against a direct role for the ENaC set of acid-sensing ion channels in mammalian somatosensation (Kang et al. 2012, Raouf et al. 2012). The evidence for ENaC channels as mechanotransducers in shear-stress sensing is indirect, and heterologous expression of mammalian ENaC channels as mechanotransducing channels has not been achieved. However, the evidence for ENaC family members as mechanotransducers in *C. elegans* is compelling (O'Hagan et al. 2005). Other *C. elegans* TRP family members such as TRP-4 also seem likely to be bona fide mechanotransducers, as indicated by mutagenesis studies of the putative pore region of this channel (Kang et al. 2010). Similarly, TRPP polycystins that are involved in kidney mechanosensation interact closely with K2P channels and respond to mechanical pressure to enhance opening of these channels and protect against apoptosis (Peyronnet et al. 2012).

TOUCH AND MECHANICAL PAIN

The ability to detect mechanical stimuli involves TRP channels in many cell types. In flies, most attention has focused on hearing rather than touch sensation, although *Painless*, a TRP gene first identified in the Benzer lab, does have a role in noxious mechanosensation in larvae and pressure sensing in the adult heart (Sénatore et al. 2010, Tracey et al. 2003). In *C. elegans*, a role for TRPVs in touch sensation has been established, and TRPA1 loss is associated with mechanosensory deficits. In contrast to human TRPA1, worm TRPA1 forms a mechanosensitive channel when expressed heterologously in eukaryotic cell lines (Kindt et al. 2007, Rugiero et al. 2010).

On the basis of an in-depth mutagenesis study, TRP-4 found in ciliated sensory neurons also seems very likely to be a bona fide mechanotransducer. TRP-4 (or TRP-N) channel loss-of-function mutants do not move, consistent with a mechanosensory role (Kang et al. 2010). Other channels, for example, *Drosophila* PKD2 involved in mating behavior and expressed in mechanosensitive sensory cells, are also candidate mechanosensors. Additionally, in zebrafish, inactivation of the unusual integral kinase TRP channel TRPM7 leads to a loss of escape responses upon mechanical stimulation in the mutant *touchdown*. However, sensory neuron integrity and function are retained in these animals, suggesting a downstream role for this particular channel (Low et al. 2011).

Somatosensation in mammals involves specialized sensory neurons that express different types of mechanosensitive channels associated with light touch or noxious mechanosensation (Delmas et al. 2011, Nilius & Honore 2012, Wood & Eijkelkamp 2012). Rapidly adapting channels seem to be a hallmark of light-touch transducing neurons, whereas noxious mechanosensation has been linked to slower classes of inactivating mechanosensitive currents (Delmas et al. 2011, Drew et al. 2007). Recently, two classes of mechanotransducing candidates have also been linked to touch. The novel mechanotransducer Piezo2 is required for expression of rapidly adapting channels in approximately 66% of the neurons that express such channels (Coste et al. 2012). Antisense knockdown of Piezo2 results in touch and allodynia deficits (Eijkelkamp et al. 2013).

Interestingly, a role for TRPC channels in light touch has also been described. When TRPC3 is deleted, mechanosensory behavior of knockout mice is normal, but there is a change in kinetics of the mechanosensitive channels exhibited by sensory neurons (Quick et al. 2012). However, deletion of TRPC3 together with TRPC6 results in a loss of sensitivity of 50% of the rapidly adapting mechanosensitive neurons and a deficit in light touch (**Figure 3**). Storch et al. (2012a) showed that TRPC1 does not express as a functional channel alone, although trafficking of TRPC1 by STIM1 results in a functional channel (Yuan et al. 2007). TRPC1 can also interact with TRPC3, -4, -5, -6, or -7, resulting in channels with lowered calcium permeability (Storch et al. 2012a). When TRPC1 is deleted in mice, light-touch behavioral responses are partially lost (Garrison et al. 2012). In addition, recordings of skin-nerve preparation showed that mechanical stimuli evoked 50% fewer action potentials in slowly adapting Aβ fibers and rapidly adapting Aδ fibers associated with touch-sensitive Merkel cells and down hairs, respectively. Thus, TRPC1 may play a role, probably as a complex with other channels in light touch. However, mechanotransduction deficits in isolated neurons have not yet been identified.

Noxious mechanosensation is associated with a slowly adapting class of mechanosensing channels in sensory neurons (Drew et al. 2007). Some evidence suggests that TRPA1, a known mechanosensor in *C. elegans* (Kindt et al. 2007), plays a related role in mice. There is a deficit in Randal-Sellito noxious mechanosensation in the TRPA1 null mutant (Andersson et al. 2009), and slowly adapting currents

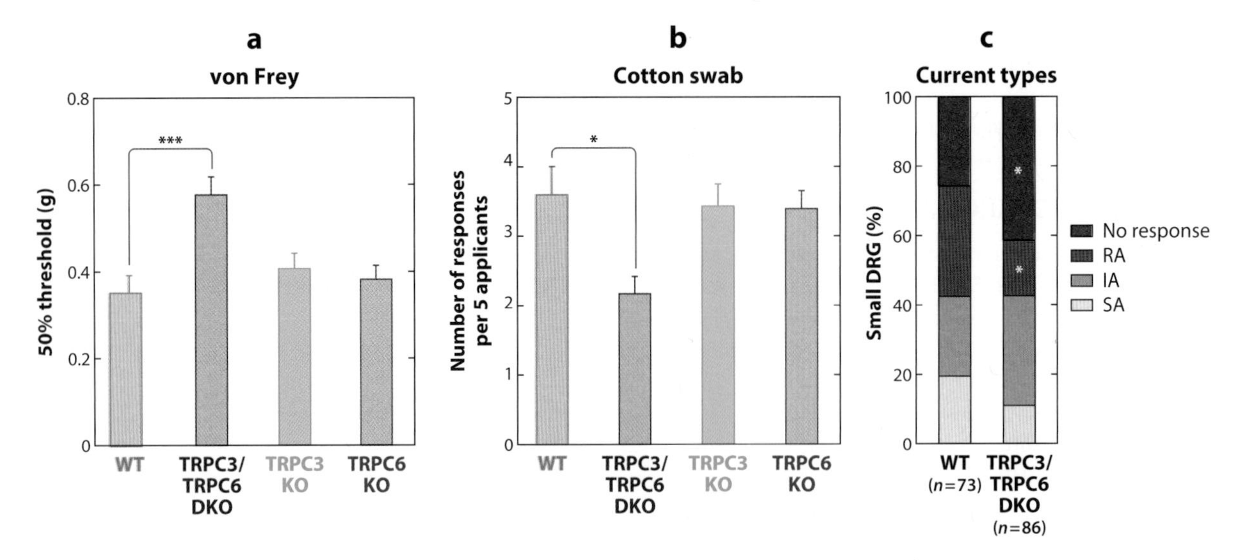

Figure 3

Innocuous touch sensation is attenuated in TRPC3/TRPC6 double knockout mice shown by (*a*) von Frey hairs and (*b*) response to a cotton bud. (*a*) 50% threshold to mechanical stimulation using von Frey hairs in wild-type (WT), TRPC3/TRPC6 double knockout (DKO), TRPC3 knockout (KO), and TRPC6 KO mice. (*b*) Responses of WT, TRPC3/TRPC6 DKO, TRPC3 KO, and TRPC6 KO mice to a cotton bud applied to the plantar surface of the hind paw. (*c*) Small-diameter dorsal root ganglion (DRG) neurons had mechanically activated currents, which could be classified on the basis of their adaptation kinetics to a static mechanical stimulus as rapidly adapting (RA), intermediately adapting (IA), and slowly adapting (SA). A significant reduction in the number of neurons displaying RA currents and an increase in the number of nonresponsive neurons were observed in TRPC3/TRPC6 DKO mice (χ-square test, $p < 0.05$). See Quick et al. 2012. *, $p < 0.05$; ***, $p < 0.01$.

in a subset of nociceptive neurons are lost (Vilceanu & Stucky 2010). Blockers of TRPA1 also inhibit mechanical firing (Kerstein et al. 2009), and TRPA1 contributes to intermediately adapting mechanosensory currents and mechanical hyperalgesia (Brierley et al. 2011). Further support for a role for TRPA1 in pain-related mechanosensation comes from Wei et al. (2012), who studied postoperative mechanical hyperalgesia in the rat, and Okubo et al. (2012), who established a role for TRPA1 in mouse mechanosensitization by hydrogen sulfide (Okubo et al. 2012). Interestingly, a heritable pain disorder has been mapped to a point mutation in TRPA1 (Kremeyer et al. 2010), and the overriding sensation associated with the gain-of-function channel is that of unbearable mechanical pain. The same channel shows enhanced activation by cold (one of the initiating factors of pain attacks).

Behavioral analysis of TRP mutants in mechanosensation has also identified roles for TRP channels in central information processing. Thus, spinal cord TRPV1 activation (Kanai et al. 2005) leads to allodynia in some neuropathic pain models, and specific antagonists can reverse this form of mechanical sensitization. Some research also claims that TRPC3 is centrally involved in synaptic transmission (Hartmann et al. 2008), which could have effects on some behavioral assays of mechanosensation.

HEARING

Hearing and balance are dependent on hair cells within the inner ear. The mechanically sensitive hair cell stereociliary bundle comprises 50–100 stereocilia. Hair cells are arranged along the length of the cochlea and vary in their frequency sensitivity; cells responding to the highest frequencies are found at the base of the cochlea (**Figure 4**). The kinetics of the mechanoelectrical transduction (MET) channels of cochlear outer hair cells also vary with an increase in single channel conductance and channel number from low to high frequencies (Beurg et al. 2006). Similar to TRP channels, mechanically activated channels are calcium permeant and can be blocked by Gd^{3+} and ruthenium red (Farris et al. 2004, Kroese et al. 1989, Marcotti et al. 1999, Rusch et al. 1994). Despite heroic efforts, the nature of the hair cell MET channels has still not been established.

INVERTEBRATES

In *Drosophila*, the genes *Nanchung* and *inactive* are related to worm ocr and osm-9 proteins, respectively. Both *Nanchung* and *inactive* are found in the mechanosensory cilia of the chordotonal organ (Kim et al. 2003), and *Nanchung* null *Drosophila* and *inactive* null *Drosophila* mutants have ablated sound-evoked potentials (Kim 2007, Kim et al. 2003). *Nanchung* and *inactive* are localized to the cilia only when both are present, indicating interactions between the two are required for functionality, even though each alone forms mechanosensitive channels when expressed in culture.

NOMPC or TRPN1 is found in the cilia of chordotonal neurons in *Drosophila* (Liang et al. 2011). Walker et al. (2000) reported that NOMPC mutants displayed nearly absent sound-evoked potentials but that a small nonadapting mechanosensory current remained, thus casting doubt over its role as the mechanotransducer. However, the remaining nerve potentials have recently been attributed to gravity receptor cells, which supports the possibility that NOMPC is a *Drosophila* auditory transducer (Effertz et al. 2011).

VERTEBRATES

Vertebrate hair cell transduction in the cochlea relies on tip links between stereocilia (Schwander et al. 2009) and mechanotransduction channels within these stereocilia that may be gated by a purely hypothetical elastic element termed "the gating spring" (Corey & Hudspeth 1983). The amplitude of the mechanically gated current and speed of adaptation increase from the apex to the basal region of the cochlea (Ricci et al. 2003). The tonotopic gradients in the cochlea could be explained by a channel consisting of multiple subunits that vary in composition along the cochlear duct. The nonselective permeability, large conductance, pharmacology, and ability to heteromultimerize suggest TRP channels as potential candidates for the MET channel; thus, a number of TRP channels have been investigated.

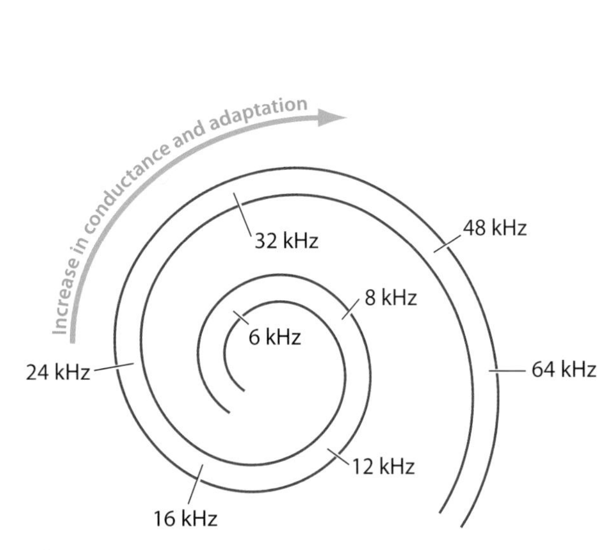

Figure 4

Frequency-place map for the mouse cochlea. Hair cells along the length of the basilar membrane respond to differing sound frequencies creating a tonotopic gradient. Approximate locations for the highest response to certain frequencies are indicated. The conductance and kinetics of the mechanotransducer also create a gradient from the apex to the basal end of the cochlea (figure adapted from Viberg 2004, original data from Muller 2004).

Of the mammalian TRPs, TRPA1 has the most ankyrin repeats, a property shared with TRPN channels and that could be the gating spring proposed to be involved in mechanotransduction in the hair cells. TRPA1 is expressed in mammalian hair cells and is blocked by the same series of blockers (amilorides, lanthanides, and ruthidium red) as found in the mechanotransduction channel in hair cells (Nagata et al. 2005). Although TRPA1 knockout mice do not show a behavioral auditory or vestibular deficit, auditory brain stem responses (ABRs) show that the hearing of $TRPA1^{-/-}$ mice is more sensitive than that of wild-type mice (Kwan et al. 2006). However, in *Drosophila*, TRPA1 mutants with a loss of mechanosensation normal mechanofunction was recovered with the introduction of a TRPA1 mutant that has all the ankyrin repeats removed (Hwang et al. 2012).

The vertebrate TRPV4 channel is expressed in hair cells of rodents (Liedtke et al. 2000, Shen et al. 2006) and zebrafish (Amato et al. 2012). TRPV4 homomers are mechanosensitive when expressed in cultured cells and respond to hypo-osmotic stimuli, although this activation is slower than would be expected of a directly gated mechanosensor. TRPV4 knockout mice have normal hearing at 8 weeks of age, as shown by ABR and distortion product otoacoustic emission, suggesting that TRPV4 is not an essential component of the mechanotransduction channel. However, at 24 weeks of age, these mice exhibit increased ABR thresholds across all frequencies (Tabuchi et al. 2005). Comparable age-related hearing loss in humans has been mapped to the chromosomal region containing the human *TRPV4* gene (Greene et al. 2001). No cell death or disruption of hair cells was found in *TRPV4* knockout mice at either 8 or 24 weeks, and the mechanism of this delayed-onset hearing loss is not known. TRPV4 mutations in humans cause an enormous variety of disorders, including many developmental disorders as well as a neuropathy that can also be associated with hearing loss. Conflicting studies report the mutant TRPV4 to be either gain-of- (Deng et al. 2010) or loss-of-function (Auer-Grumbach et al. 2010). However, as the mutations are in the ankyrin repeat, domain disruption of protein-protein interactions could be important (Zimon et al. 2010).

Gain-of-function mutations in TRPML3 found in the varitint-waddler mouse cause deafness and vestibular dysfunction (van Aken et al. 2008). However, TRPML3 knockout mice have normal Preyer reflexes (ear flattening in response to loud noise) and ABR thresholds (Jors et al. 2010). The hearing deficit is caused by stereocilia defects that arise because the mutant TRPML3 is constitutively active (Grimm et al. 2007). Similarly, TRPP1 is required for stereocilia structural integrity but is not required for mechanotransduction in inner hair cells of mice (Steigelman et al. 2011).

So far, TRPC3 and TRPC6 are the only TRP channels linked to a deficit in the MET currents in hair cells (Tadros et al. 2010, Quick et al. 2012). TRPC3 and TRPC6 are both expressed in the inner ear and localize to inner and outer hair cells as well as spiral ganglion cells. Behavioral and ABR threshold testing show that deletion of *TRPC3* or *TRPC6* genes alone has no effect on hearing. However, when both TRPC3 and TRPC6 are knocked out, high-frequency hearing is lost, as shown by ABR analysis and the fact that MET currents recorded from the basal, but not the apical, cochlear hair cells are lost in vitro (Quick et al. 2012). TRPC3/TRPC6 double-knockout mice have hearing deficits at frequencies over 32 kHz, which corresponds with the ablation of MET currents in the basal, but not the apical, region of the cochlea from knockout animals (**Figure 5**).

Like TRP channels, TMC proteins are thought to have six transmembrane domains and a pore loop. Functioning in a redundant manner, TMC1 and TMC2 loss have both been linked to deafness and hearing loss in humans as well as mice (Kawashima et al. 2011). The similarities in structure and expression patterns (high levels in the cochlea and sensory neurons) as well as the deafness associated with loss of TMC-1 all indicate that an interaction between TRP and TMC channels could contribute to mechanically sensitive channels in vivo.

CARDIOVASCULAR SYSTEM

Within the cardiovascular system, various mechanisms operate to sense changes in metabolic demand and hemodynamics that are required to generate adequate responses to maintain homeostasis. Transduction of mechanical forces such as shear stress, stretch, and (osmotic) pressure is key for the regulation of vascular tone and local blood flow. Ca^{2+} signaling is a central event in relaying these signals in the cardiovascular system. Endothelium and vascular smooth muscle cells are involved in the

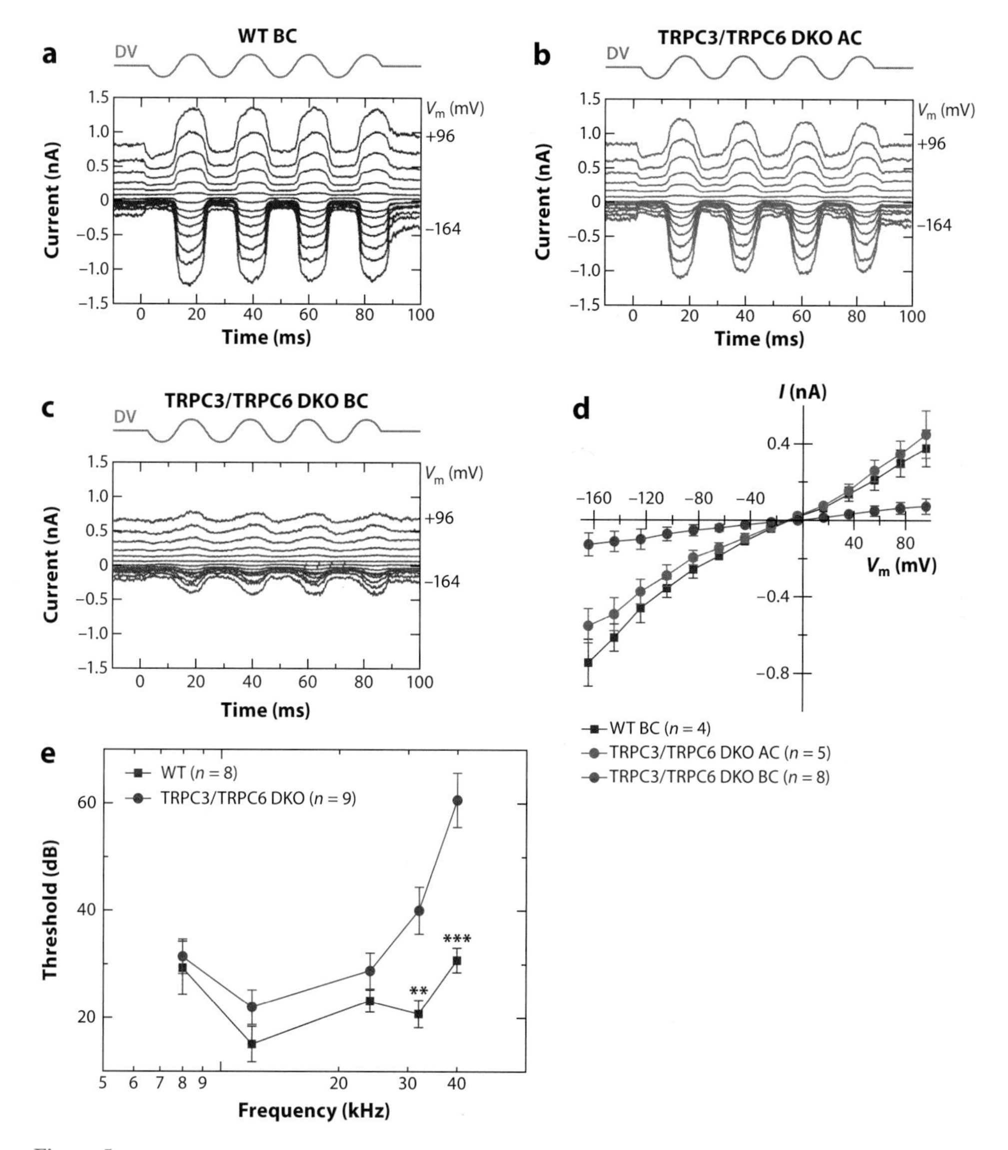

Figure 5

Hearing loss and deficits in mechanotransduction in hair cells from TRPC3/C6 knockout mice. (*a–c*) MET currents in response to 45 Hz sinusoidal force stimuli from a fluid jet in WT and TRPC3/TRPC6 outer hair cells (OHCs). Holding potential was −84 mV, and the membrane potential was stepped in 20 mV increments from −164 mV to +96 mV. Driver voltage (DV, amplitude 40 V) waveform to the fluid jet is shown above the current traces. Positive DV moves the hair bundles in the excitatory direction toward the kinocilium. Recordings in a–c are averages from two stimulus presentations each. (*a*) WT OHC, mid-basal coil P2+1, C_m 5.6 pF; R_s 0.80 MΩ. (*b*) TRPC3/TRPC6 DKO OHC, basal end of apical coil P2+2, C_m 7.4 pF; R_s 0.63 MΩ. (*c*) TRPC3/TRPC6 DKO OHC, mid-basal coil P2+2, C_m 6.9 pF; R_s 0.74 MΩ. (*d*) Current-voltage curves averaged from 4 WT OHCs from the basal coil, 5 apical-coil TRPC3/TRPC6 DKO OHCs and 8 basal-coil TRPC3/TRPC6 DKO OHCs. (*e*) Auditory brain stem recording (ABR) response thresholds to tone bursts of wild-type (WT) ($n = 8$) and TRPC3/TRPC6 double knockout (DKO) mice ($n = 9$). Data are expressed as mean ± SEM. *, $p < 0.05$; **, $p < 0.01$; ***, $p < 0.001$. AC, apical cochlea; BC, basal cochlea; dB, decibel; nA, nanoAmps. See Quick et al. (2012).

process of vascular mechanotransduction. Substantial evidence exists that pressure or stretch induces membrane depolarization in blood vessels and the heart (Kuipers et al. 2012, Patel et al. 2010). Mechanically activated cation channels triggered by stretch or osmotic swelling have been described in endothelial cells, vascular smooth muscle cells, and podocytes, which are epithelial cells essential to plasma ultrafiltration (Inoue et al. 2009, Patel et al. 2010). The properties of these mechanically activated channels resemble those of TRP channels; they are sensitive to blockage by Gd^{3+} and have similar ion selectivity and conductance. Thus, researchers have indicated that mechanosensitive TRP channels may play a role in cardiovascular function. Several members of the TRP family, including TRPV, TRPC, TRPP, and TRPM, have been implicated in sensing (osmo)mechanical stress (Jin et al. 2011).

Multiple TRP members, including TRPC, TRPA, TRPM, TRPV, and TRPP, have been identified in the heart and vasculature system (Inoue et al. 2009). Disturbed TRP channel functioning has been associated with several cardiovascular pathologies (Earley & Brayden 2010). TRP channels are also involved in regulating the cellular function of endothelial cells. Shear stress evokes an increase in intracellular calcium in endothelial cells that provokes the release of vasodilating agents such as nitric oxide that act on vascular smooth muscles. Although TRP channels in endothelial cells have been implicated in the detection of mechanical stimuli such as shear stress, they are also activated by vasoactive molecules (Kuipers et al. 2012). However, a unitary inward current was identified by Lansman et al. (1987) that was associated with an increased opening of a Ca^{2+}-permeable ion channel after the cell membrane patches of aortic endothelial cells had been stretched.

Endothelial TRPV4 is required for shear-stress-induced vasodilation. Hypotonic cell-swelling-induced currents are also lost in *TRPV4*$^{-/-}$ endothelial cells, and *TRPV4*$^{-/-}$ mice show complete absence of shear-stress-induced vasodilation, although acetylcholine-induced vasodilation is normal (Hartmannsgruber et al. 2007). More recently, Matthews et al. (2010) showed that TRPV4 can be activated by mechanical force via the relay of mechanical stretch through β1-integrin. A similar interplay between TRPV4 and α2β1-integrin has been found in mechanotransduction in the somatosensory system (Alessandri-Haber et al. 2008). Polycystins (TRPP1, TRPP2) have a key role in renal and vascular mechanosensory transduction (Patel & Honore 2010). In the primary cilium of renal, nodal, and endothelial cells, polycystins may act as flow sensors (Patel & Honore 2010). Moreover, in arterial myocytes, the ratio of TRPP1 and TRPP2 regulates pressure sensing (Sharif-Naeini et al. 2009). Polycystins also condition the stretch sensitivity of stretch-activated K2P channels in renal epithelial cells (Peyronnet et al. 2012). Although some research indicates that TRPC1 is intrinsically mechanosensitive (Maroto et al. 2005), this finding has been disputed as CHO cells overexpressing TRPC1 do not show enhanced mechanosensitive currents (Gottlieb et al. 2008).

Increasing evidence indicates that different TRP channels interact and cooperate to form mechanosensitive complexes in the cardiovascular system. Heteromeric TRPV4-TRPC1 channels display distinct electrophysiological properties that differ from homomeric TRPV4 channels (Ma et al. 2011b). TRPC1 interacts with TRPV4 to mediate shear-stress-induced calcium influx in endothelial cells (Ma et al. 2010). TRPV4 can also interact with other members of the TRP family (see **Table 1**). A common subunit arrangement may be present among heteromeric tetramers of TRP channels, as TRPP2 and TRPV4 assemble identically to TRPP2 and TRPC1 (Stewart et al. 2010). In blood-brain endothelial cells, calcium influx induced by mechanical stretch injury is probably mediated by TRPC1 and TRPP2 channels (Berrout et al. 2012b). Other proteins are also likely involved in the modulation of TRP channel complexes in endothelial function. For example, the protein Orai1, which spans four transmembrane domains, interacts

Table 1 TRP channels and mechanosensation

Species	Subfamily and channel		Mechanosensitivity	Mechanosensitive complexes	Cell type	Reference(s)
Worms	TRPV	**OSM-9**	Touch, osmolarity, auditory	Deg1	Neurons	Liedtke (2007), Geffeney et al. (2011)
		OCR-2	Touch, osmolarity		Neurons	Liedtke et al. (2000)
	TRPN	**TRP-4**	Stretch		Ciliated neurons	Kang et al. (2010)
Flies	TRPV	**NAN**	Auditory	IAV	Chordotonal organs	Kim (2007)
		IAV	Auditory	NAN	Chordotonal organs	Kim et al. (2003)
	TRPN	**NOMPC**	Auditory		Chordotonal neurons	Effertz et al. (2011)
Yeast		**TRPY1**	Osmolarity, stretch		Vacuolar membrane	Zhou et al. (2005), Denis & Cyert (2002)
Zebrafish	TRPM	**TRPM7**	Touch		Sensory neurons	Low et al. (2011)
Mammals	TRPV	**TRPV1**	Stretch, osmolarity?		Urothelial cells, colon afferent fibers	Ciura & Bourque (2006), Sharif et al. (2006)
		TRPV2	Stretch, osmolarity?		Epithelial cells, sensory neurons, smooth muscle cells	Muraki et al. (2003)
		TRPV4	Noxious pressure, shear stress, osmolarity, stretch	TRPP2, TRPC1, β1-integrin	Sensory neurons, endothelial cells	Wu et al. (2007)
	TRPC	**TRPC1**	Stretch, osmolarity, shear stress, touch	TRPP2, TRPV4, TRPC4	Endothelial cells, sensory neurons	Chen & Barritt (2003), Maroto et al. (2005)
		TRPC3	Touch, auditory	TRPC6	Sensory neurons	Quick et al. (2012)
		TRPC5	Pressure, osmolarity		Sensory neurons?	Gomis et al. (2008)
		TRCP6	Pressure, osmolarity, touch, auditory	TRPC3, Gq receptors (AT1R)	Sensory neurons, cardiac myocytes, smooth muscle cells	Quick et al. (2012), Inoue et al. (2009), Spassova et al. (2006)
	TRPP	**TRPP1**	Pressure, shear stress	TRPP2	Epithelial cells	Nauli et al. (2003)
		TRPP2	Pressure, shear stress	TRPC1, TRPP1, TRPV4	Epithelial and endothelial cells	Nauli et al. (2003)
		TRPP3	Osmolarity			Shimizu et al. (2009)
	TRPA	**TRPA1**	Pressure, stretch, osmolarity	TRPV1	Sensory neurons	Zhang et al. (2008)
	TRPM	**TRPM3**	Osmolarity			Grimm et al. (2003)
		TRPM4	Shear stress, stretch		Smooth muscle cells	Morita et al. (2007)
		TRPM7	Stretch, osmolarity, shear stress		Smooth muscle cells	Numata et al. (2007a)

with the TRPC4/TRPC1 channel complex, where it controls activation and channel-permeation characteristics (Cioffi et al. 2012). The calcium sensor STIM1 also interacts with TRPC channels and may be involved in the gating of these channels (Yuan et al. 2009). Finally, the interaction with cytoskeletal proteins or extracellular matrix proteins through molecules such as integrins may be essential to tune the mechanosensitivity of TRPs.

Smooth muscle TRP channels are implicated in the transduction of changes in pressure to changes in the vascular smooth muscle contractility of the cytoskeleton (Di & Malik 2010, Earley & Brayden 2010). Smooth muscle cells respond to changes in intraluminar pressure (termed the myogenic response). They are also affected by dilation in response to reduced intraluminar pressure and constriction in response to increases in intraluminar pressure. Vascular smooth muscle cells express multiple subunits of TRPs (Guibert et al. 2011), among which TRPC6, TRPV2, and TRPM4 exhibit some mechanosensitivity and, thus, could participate in pressure-induced depolarization (Kuipers et al. 2012). TRPV2 is a component of an osmo(mechanical)-sensitive channel. TRPV2 expression in CHO cells induces swelling-activated currents with similar characteristics to the endogenous channel in aortic smooth muscle cells that is activated by osmotic changes (Muraki et al. 2003). TRPC6 antisense oligonucleotides reduce the expression of TRPC6 in smooth muscle cells and attenuate hypo-osmotic-induced cation currents (Welsh et al. 2002). Overexpression of TRPC6 alone does not induce mechanosensitivity in different cell lines (Quick et al. 2012, Yin & Kuebler 2010). However, TRPC6 mechanosensitivity has been linked to $G_{q/11}$-coupled receptors (Inoue et al. 2009, Schnitzler et al. 2008, Sharif-Naeini et al. 2010). In cardiac myocytes, mechanical stress can activate TRPC channels indirectly through the angiotensin II type 1 receptor (AT1R) (Dietrich et al. 2005, Schnitzler et al. 2008). Whereas coexpression of AT1R with TRPC6 in HEK293 cells does not confer mechanosensitivity, coexpression of TRPC3 and TRPC6 in a sensory neuron line confers low-threshold mechanosensitivity (Quick et al. 2012). However, another study found no evidence that $G_{q/11}$-coupled receptor activation enhances the mechanosensitivity of TRPC-like currents or the myogenic responsiveness of anterior cerebral arteries (Anfinogenova et al. 2011). These differences may be inherently due to the fact that activity of the TRP-like channels depends on cell-dependent machinery. Indeed, TRP channel behavior is expression system dependent (Lev et al. 2012). TRPM7 may be a stretch- and swelling-activated channel. TRPM7 translocates to the membrane after shear stress that is associated with an increase in native TRPM7 current amplitude in smooth muscle cells (Oancea et al. 2006). In excised patches, TRPM7 displays single channel activity that is induced after membrane stretch (Numata et al. 2007a).

KIDNEY MECHANOTRANSDUCTION

All subfamilies of the TRP channels are expressed in the kidney, and several play a role in multiple kidney disorders (Hsu et al. 2007). Although TRP channels regulate many different functions in the kidney, sensing mechanical cues is an important function for normal renal function. Podocytes, the epithelial cells in the kidney, play an important role in ultrafiltration of plasma. They are exposed and sense different forms of mechanical stimuli, including mechanical stress due to glomerular capillary pressure, shear stress, and changes in osmolarity. Podocytes react and adapt to these mechanical cues via cytoskeletal rearrangements to maintain the glomerular barrier (Greka & Mundel 2012). TRPV4 mediates hypotonic cell-swelling-induced calcium influx (reviewed in Pedersen et al. 2011). Some reports showed that osmosensitivity of TRPV4 is mediated through archidonic acid metabolite 5′, 6′-epoxyeicosatrienoic acid (Nilius et al. 2004). However, in yeast that do not synthesize the metabolite, TRPV4 still responds to changes in osmolarity (Loukin et al. 2009). The precise mechanisms that underpin osmosensitivity of TRPV4 remain unclear, although TRPV4 function depends on cytoskeletal elements and can also interact with aquaporin-5 (Liu et al. 2006). Mutations in the *PDK1* and *PDK2* genes encoding TRPP1 and TRPP2 are the major cause of autosomal-dominant polycystic kidney disease. As mentioned above, TRPP1 and TRPP2 from the polycystin family could function as mechanosensors in renal epithelial cells (Patel & Honore 2010). Recently, Wang et al. (2012) showed that the actin cross-linking protein filamin regulates TRPP2 function. Podocyte injury is characterized by deregulated calcium homeostasis (Hunt et al. 2005), and TRPC6 has been linked to human genetic kidney disease (Reiser et al. 2005, Winn et al. 2005). To regulate AT1R-mediated cytoskeletal remodeling, TRPC5 and TRPC6 function in a protein complex with Rac and RhoA (Rho family of small guanosine triphosphatases), respectively (Tian et al. 2010). Finally, TRPM7 produces osmotic swelling and pressure-induced currents (Numata et al. 2007a).

VISCERAL MECHANOSENSATION

Mechanosensation in the bladder or intestinal tract can occur directly through activation of afferent nerve fibers or indirectly through mediators released by non-neuronal cells that act through afferent fibers. Some evidence suggests that TRP channels also play a role in mechanosensation in viscera. TRPM8 is detected in a small portion of primary sensory neurons that innervate the bladder (Hayashi et al. 2009) and colon (Harrington et al. 2011). Icilin, a TRPM8 superagonist, desensitizes colon afferent to mechanical stimulation and may couple to TRPV1 and TRPA1 to inhibit their downstream chemosensory and mechanosensory actions (Harrington et al. 2011). The behavioral responses to noxious colonic distension are reduced in *TRPA1*$^{-/-}$ mice, suggesting that TRPA1 is required for normal visceral mechanosensation (Brierley et al. 2009). Although TRPV1 is not considered to be mechanically gated, TRPV1-deficient mice display reduced mechanosensitivity of the bladder and urothelial cells display diminished hypoosmolarity-evoked ATP and NO release (Birder et al. 2002, Daly et al. 2007).

TRPV4 is highly expressed in bladder epithelium and in sensory neurons innervating the colon (Araki 2011, Brierley et al. 2008). Stretch-evoked Ca^{2+} influx is decreased in urothelial cells from TRPV4-deficient mice (Mochizuki et al. 2009), and a TRPV4 agonist induces bladder contraction and overreactivity (Thorneloe et al. 2008). In the colon, TRPV4 contributes to mechanically evoked visceral pain (Brierley et al. 2008). TRPV2 is expressed in nerve fibers within the muscularis and myenteric plexus, myenteric cell bodies, and epithelial cells in the stomach and intestine (Kashiba et al. 2004, Kowase et al. 2002, Zhang et al. 2004). In urothelial cells, TRPV2 is expressed and activation of TRPV2 induces a Ca^{2+} flux (Everaerts et al. 2010). Although TRPV2 may mediate stretch-induced calcium influx (Muraki et al. 2003), a role for TRPV2 in visceral organ mechanosensation has not been explored.

TRP INTERACTIONS WITH OTHER PROTEINS AND MECHANOTRANSDUCTION

The bewildering variety of TRP channel protein interactions has been comprehensively cataloged by the TRIP (Mammalian Transient Receptor Potential Channel-Interacting Protein) Database (Shin et al. 2011). As an example, we have selected the topic of interactions between TRPC3 and proteins involved in mechanotransduction because of the potential role of TRPC3 in somatosensation and hearing. Peer-reviewed papers from 1997 to 2012 show 58 interactions have been validated in vivo and/or in vitro studies. Interactions were determined via coimmunoprecipitation, patch clamp electrophysiology, fusion protein pull-down assay, calcium measurement, coimmunofluorescence staining, fluorescence probe labeling, FRET, yeast two-hybrid screens, in

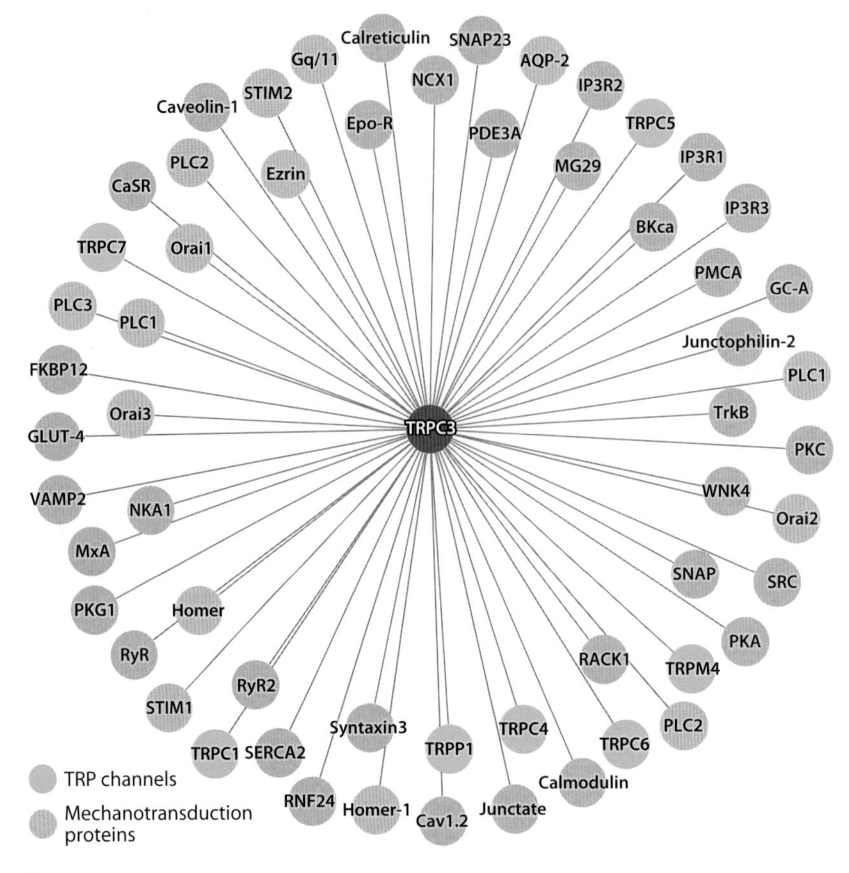

Figure 6

Protein-protein interactions of TRPC3 with cellular proteins taken from the TRIP (Mammalian Transient Receptor Potential Channel-Interacting Protein) Database (**http://www.tripdatabase.com**). Green circles indicate TRP channels; yellow circles indicate proteins implicated in mechanotransduction.

vitro post-translational modification assay, or cell surface biotinylation assays using human, mouse, rat, or bovine TRPC3 constructs (see **Figure 6**).

Aquaporins (AQP) have been implicated in hearing (Li & Verkman 2001) and are found in specialized mechanosensitive cells in human skin (Nandasena et al. 2007). AQP2 coimmunoprecipitates with TRPC3 in rat kidney cells (Goel et al. 2007) and is required for calcium influx in response to hypotonic shock (Galizia et al. 2008). Homer-1 coimmunoprecipitates with TRPC3 in HEK cells and regulates plasma membrane translocation of channels (Kim et al. 2006). In Homer-1 knockout mice, an increased cation influx can be blocked by GsMTx4, a known stretch-activated channel blocker (Stiber et al. 2008).

Phospholipases coimmunoprecipitated with TRPC3 in epithelial cells (Bandyopadhyay et al. 2005) and HEK cells (Tong et al. 2008). Phospholipases are activated by stretch in *Xenopus* oocytes (Montag et al. 2004), cardiomyocytes (Ruwhof et al. 2001), and aortic tissue (Matsumoto et al. 1997). Additionally, when phospholipase C (PLC) β2 is knocked down in cultured osteoblasts, responses to fluid shear stress are diminished (Hoberg et al. 2005). These enzymes may produce lipid activators of mechanosensitive TRPs.

STIM and Orai proteins are also important in regulating calcium entry. They interact with a number of TRPs including TRPC3, as shown by coimmunoprecipitation in HEK cells and platelets (Berna-Erro et al. 2012, Liao et al. 2007). Orai1 plays a role in stretch sensing in cardiomyocytes: siRNA knockdown for Orai1 impairs stretch-mediated responses (Volkers et al. 2012). In various systems, cytoskeletal interactions have also been implicated in mechanotransduction. Ezrin is a protein that serves as an intermediate between the plasma membrane and the cytoskeleton and interacts with TRPC3 (Lockwich et al. 2001). Ezrin also interacts with the mechanosensitive channel TREK-1 (Lauritzen et al. 2005).

These examples underline the potential diversity of TRP-dependent interactions that may play a role in mechanotransduction. Physical interactions may also define the cellular environment in which TRP channels can function in mechanotransducing complexes. Although neither TRPC3 nor TRPC6 expressed in HEK293 or CHO cells responds to mechanical pressure, rapidly adapting mechanosensitive currents are expressed in the neuronal cell line ND-C, expressing TRPC3 channels, which is potentiated by coexpression of TRPC6 (Quick et al. 2012) (**Figure 7**). Additional proteins in the ND-C cell line that traffic or contribute to mechanosensitive channels may explain the cell-type-specific expression of

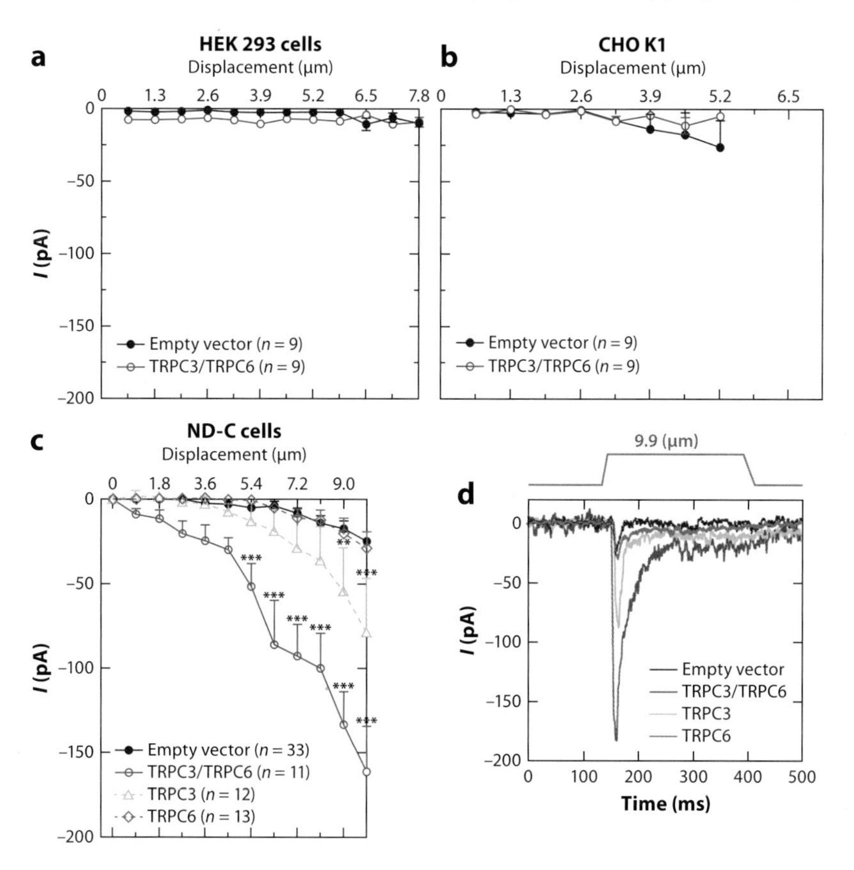

Figure 7

Heterologous expression of TRPC3 with or without TRPC6 confers mechanosensitivity on some cell lines. TRPC3 and TRPC6 alone or together do not confer mechanosensitivity on HEK 293 or CHO cells. Mechanical distension of the neuronal ND-C cell line expressing TRPC3 or TRPC3 and TRPC6 produces large rapidly adapting currents shown in panel *d* from Quick et al. 2012.

mechanosensitivity of TRPC3. The lipid environment in which the channels are expressed may also play an important role in channel gating mechanisms (Carrillo et al. 2012).

TRP CHANNEL GATING MECHANISMS

Activation of mechanotransduction channels by membrane perturbation phospholipids or lipid second messengers released by phospholipases have been investigated as potential mechanisms and are summarized below (**Figure 8**).

Membrane Perturbation and Potential Direct Activation

The sensitivity of TRP channels in response to membrane tension has been investigated in heterologous expression systems. TRPV2 expressed in CHO cells grown on an elastic membrane can be activated by membrane stretch and by pressure applied with a patch pipette (Muraki et al. 2003). Mammalian TRPA1 does not form a mechanically activated channel when transfected into mammalian cells (Rugiero & Wood 2009), but *C. elegans* TRPA1 expressed in CHO cells can be activated by 50 mm Hg suction in patch-clamp experiments (Kindt et al. 2007). TRPV4 can also be activated by suction when expressed in *Xenopus* oocytes (Loukin et al. 2009). TRPM4 and TRPM7 in HEK cells can be activated by negative pressure of approximately 20 mm Hg (Morita et al. 2007, Numata et al. 2007a), and TRPC5 in HEK2983 cells can be activated by positive pressure of approximately 21.3 mm Hg (Gomis et al. 2008). Although disputed in different cell lines, TRPC6-expressing HEK cells can also be activated by negative pressure (Gottlieb et al. 2008, Quick et al. 2012, Spassova et al. 2006).

TRPC5, TRPM3, and TRPV4 can be activated by osmotic pressure when transfected into HEK cells (Fan et al. 2009; Grimm et al. 2003, 2008). TRPV2 expressed in CHO cells as well as endogenously expressed TRPV2 in myocytes can also be activated by hypo-osmotic solutions (Muraki et al. 2003). Cardiomyocytes exposed to hypotonic solutions induce cationic currents that can be blocked by PLC inhibition or AT1R blockade, and they are not present in *TRPC1*$^{-/-}$ mice (Seth et al. 2009). Hypo-osmotic activation of TRPC5 appears to depend on phosphatidylinositol 4,5-bisphosphate (PIP_2) levels. PIP_2-depleted cells show low levels of activation that can be rescued by the addition of exogenous PIP_2. Temperature also plays a part in the sensation of hypertonic stimuli: TRPV1-transfected HEK293 cells show minimal response to hypertonic solutions at 24°C but a robust increase of Ca^{2+} at 36°C (Nishihara et al. 2011). TRPV4 can be activated by shear stress generated by fluid flow when transfected into HEK cells (Mendoza et al. 2010, Wu et al. 2007). TRPM7 expressed in HEK cells or found endogenously in HeLa cells can also be activated by fluid flow (Numata et al. 2007a,b).

Endogenously expressed TRP channels also show activation by membrane tension. In mouse kidney tissue, shear-stress flow induces a sustained rise in Ca^{2+} that is abolished in *TRPV4*$^{-/-}$ animals (Berrout et al. 2012a). TRPP1 is also implicated in fluid flow sensing, as mutations in TRPP1 result in abolition of fluid flow–induced Ca^{2+} rises in cultured kidney epithelial cells (Nauli et al. 2003). Brain endothelial cells grown on an elastic membrane exhibit an influx of Ca^{2+} when a 50-ms stretching pulse is applied. This current is decreased when TRPP2 and TRPC1 are knocked down by siRNA (Berrout et al. 2012b). Similarly, stretch-activated Ca^{2+} increases in myoblasts show a 90% reduction after knockdown of TRPC1 by siRNA (Formigli et al. 2009). Mechanical stretch also activates endogenous TRPC6 in cardiomyocytes, and expression of dominant-negative TRPC6 strongly reduces the Ca^{2+} increase in response to 20% transient stretch (Nishida et al. 2010). Nevertheless, all these examples of TRP activation could depend, not on membrane deformation, but on the actions of second messengers released from distinct mechanosensors (see **Figure 8**).

Indirect Activation

Trafficking, second messengers, and posttranslational modification are potential mechanisms of TRP channel regulation (**Figure 9**). Calcium homeostasis involving store-activated channel activity as well as STIM and Orai may contribute to TRP channel trafficking. They also enhance calcium fluxes, and mechanical currents in some cells seem to depend on this process (e.g., Volkers et al. 2012).

Second-messenger activation of TRP channels does not involve cyclic nucleotides, but a variety of lipid second messengers are effective agonists, raising the possibility that TRP channels are not directly gated by mechanical force in vivo but instead are activated downstream of uncharacterized mechanosensors (Makino et al. 2006). All seven families of the TRP channels contain members that can be activated by PLC-coupled GPCR receptors and downstream signaling cascades. The possibility of mechanosensation being downstream of a GPCR has been extensively explored with the AT1R implicated in the myogenic response (Storch et al. 2012b). Evidence that agonist activation of the AT1R is mimicked by agonist-independent mechanical stress on the membrane has been presented. However, coexpression of AT1R with TRPC6 does not confer mechanical sensitivity in a number of cell lines (e.g., Quick et al. 2012). Thus, this mechanism may not represent a general model for channel activation.

If GPCRs and lipids are required for mechanosensation by TRP channels, which second messengers are involved? Lipid regulation of TRP channel activity has been described in many systems (e.g., Kahn-Kirby et al. 2004). Studies of the lipid activation of TRP channels have focused on inositol phospholipid derivatives and PLC because of their important roles in regulating intracellular calcium levels. PIP_2 has been proposed to block TRPV1, with PLC-mediated hydrolysis releasing the inhibition. However, PIP_2 activates TRPV1 in isolated patches. The mechanisms involved in these two apparently contradictory events have been overviewed (Rohacs et al. 2008). The role of other phospholipids has received little attention. There is evidence for PLD activation of TRPC3 downstream of metabotropic glutamate receptor activation and translocation of the enzyme to the membrane. PLD releases phosphatidic acid and choline and may subsequently produce lysophosphatidic acid and diacylglycerol (DAG) (Glitsch 2010). Intriguingly, TRPC3-mediated calcium fluxes evoked by DAG in T cells depend on the fatty acid at the 2′ position of DAG. Thus, DAG-containing oleic acid is an effective agonist, whereas arachidonic acid at the 2′ position is much less effective (Carrillo et al. 2012). Sphingosine-1-phosphate is a product of ceramidase that acts through the GPCR SP1R1 and has been implicated in the gating of TRPC1 channels that are repeatedly associated with mechanosensation (Formigli et al. 2009).

Most studies of lipid activation of TRP channels have focused on inositol phospholipid derivatives because the function of these second messengers in intracellular calcium signaling is well known. PLC activation leads to the hydrolysis of PIP_2-releasing DAG, inositol trisphosphate (IP_3), which releases both calcium and protons from intracellular stores, as emphasized in recent studies of *Drosophila* TRP channels. For example, Huang et al. (2010) found that PIP_2 hydrolysis and intracellular acidification activated TRPL channels. Early work showed that polyunsaturated fatty acids (PUFAs) were also channel activators (Chyb et al. 1999). However, Lev et al. (2012) observed no activation of TRPL channels with PIP_2 hydrolysis and intracellular acidification in transfected HEK cells (Lev et al. 2012). Yet, PUFAs do activate the channel, and inhibitors of DAG lipase could block this effect—another example of different observations in distinct cellular contexts. PUFAs that activate TRP channels are likely to be produced by DAG lipases downstream of PLC or by PLA2 activation. In *C. elegans*, a subset of PUFAs with omega-3 and omega-6 acyl groups acts as endogenous modulators of TRPV signal transduction (Kahn-Kirby et al. 2004). Motter & Ahern (2012) also found potent agonist effects of PUFAs on mammalian TRPA1.

A recent study potentially reconciles the membrane deformation/second messenger debate on TRP mechanotransduction by providing compelling evidence that light sensing in *Drosophila* is directly mediated by membrane tension–mediated mechanical activation of TRP channels (Hardie & Franze 2012). The depletion of IP_3 catalyzed by PLC leaving DAG in the membrane can cause photoreceptor contraction that may be the result of altered membrane tension that is visible at the light-microscope level. Hypo-osmotic tension causing membrane deformation also results in increased TRP channel activity. Gramicidin, a cation selective mechanosensitive channel, shows enhanced channel activity when light

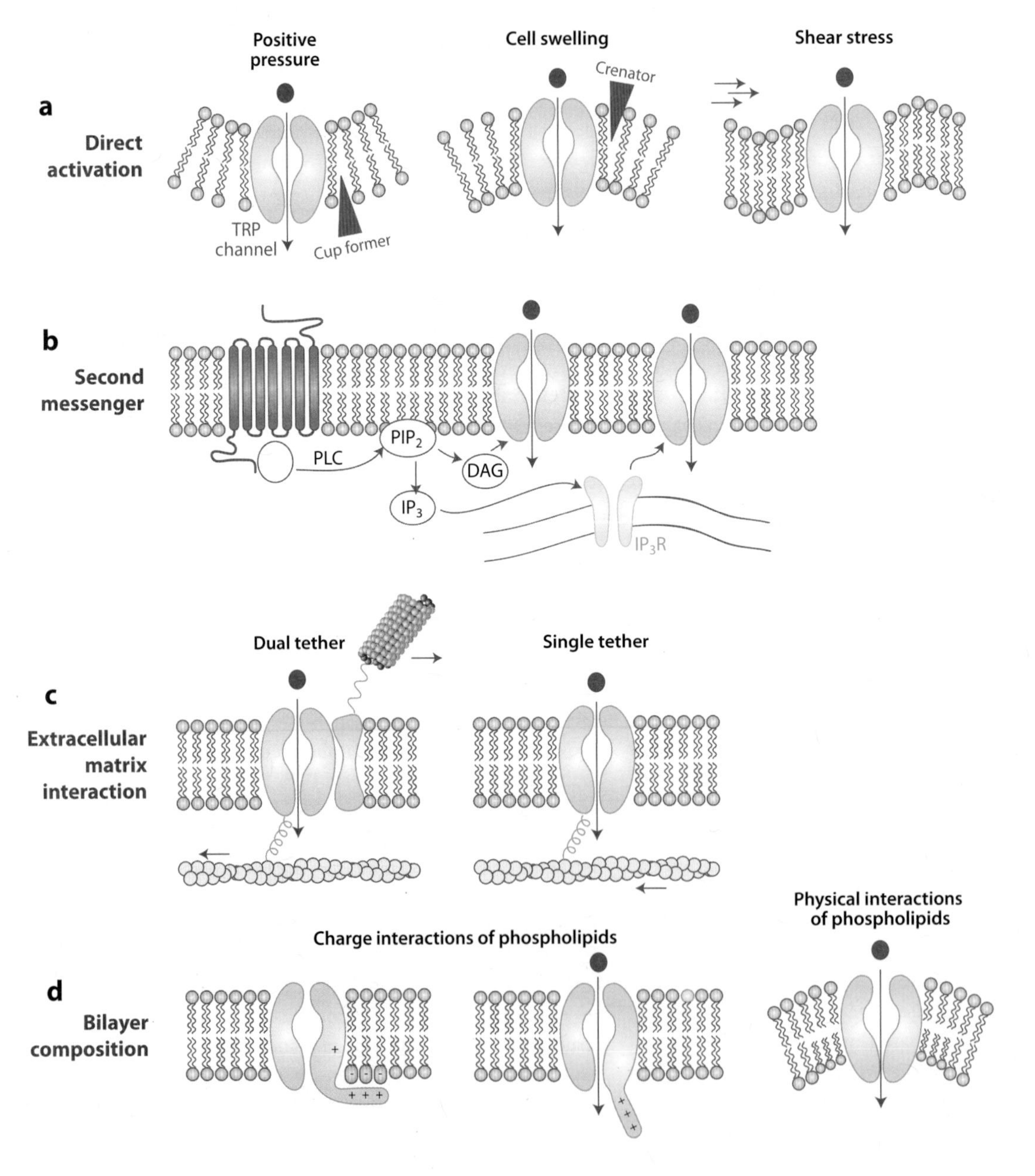

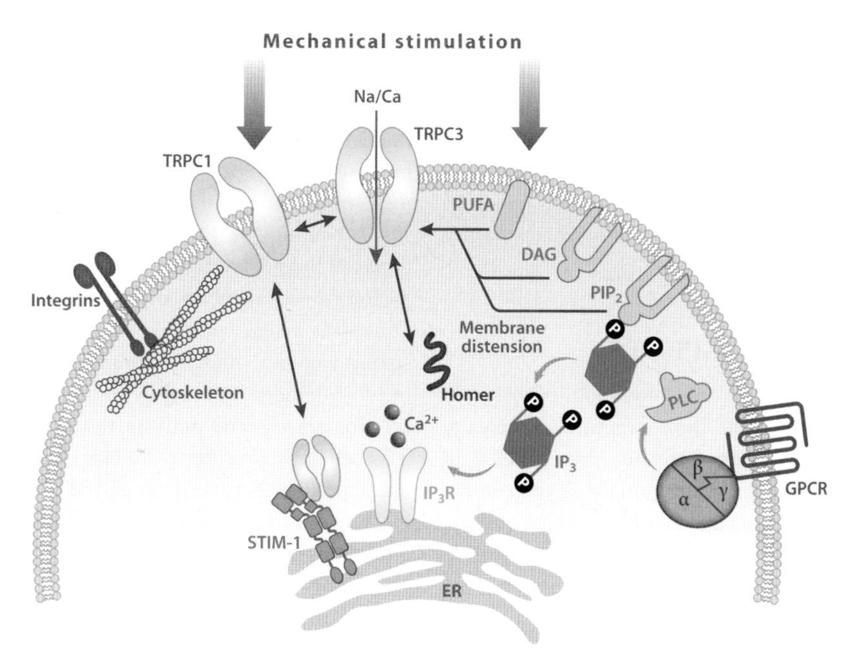

Figure 9

Mechanisms of mechanosensitive TRP activation. Intracellular proteins STIM and Homer have been linked to TRPC1 trafficking and activation. TRPC1 is known to require other TRP channels to function at the membrane. Membrane distension may itself activate the channels, whereas various lipids, particularly PUFAs, and the release of IP_3 from PIP_2 may alter membrane tension and activate TRPs in a way that is similar to the *Drosophila* visual system. Abbreviations: DAG, diacylglycerol; ER, endoplasmic reticulum; GPCR, G protein–coupled receptor; IP_3, inositol trisphosphate; PIP_2, phosphatidylinositol 4,5-bisphosphate; PLC, phospholipase C; PUFA, polyunsaturated fatty acids.

stimulates photoreceptors, clearly demonstrating that light acting via rhodopsin causes membrane deformation. This has interesting implications for the role of rhodopsins in heat sensing in *Drosophila* (Fowler & Montell 2013). In the Hardie scheme, PLC acts to alter membrane structure through the release of IP_3. DAG-derived negatively charged

Figure 8

Mechanisms for activating TRP channels. (*a*) Direct activation. Positive pressure or hypo-osmolarity cause the membrane to curve inward; this action can be replicated by cup formers, which insert into the inner leaflet of the bilayer, and also by cone-shaped lipids or lipids with large head groups. The opposite membrane curvature can be achieved by hyperosmotic cell swelling or application of crenators, which insert into the outer leaflet of the membrane. Shear stress occurs in tissues such as blood vessels and bone from fluid flow or friction; the membrane is put under stress by the movement and changes shape. (*b*) Second messengers. The ion channel may be activated indirectly by second-messenger signaling cascades from closely associated G protein–coupled receptors (GPCRs). GPCRs activate phospholipase C (PLC) leading to hydrolysis of PIP_2 to form diacylglycerol (DAG), which could activate the channel directly, and inositol trisphosphate (IP_3), which could activate the channel via IP_3R. (*c*) Extracellular matrix interaction. The dual tether model is proposed for vertebrate hair cell transduction. Intra- and extracellular tethers transfer the mechanical force to the channel to enable channel opening. The single tether model employs the same principle of mechanical forces being transferred to the channel but with just intra- or extracellular tethers. (*d*) Bilayer composition. The composition of the bilayer has a huge effect on the shape and structure of the membrane and its interactions with membrane proteins. Hypothetical models include those that utilize charge interactions between phospholipids and ion channels, mechanical force causing phosphorylation or hydrolysis of charged phospholipids; the modified lipid loses the charges, severing interactions and enabling the channel to change conformation and open. This principle could also work in reverse, with charged interactions favoring the open conformation of the channel. Alternatively, the modification of the lipids could simply exert membrane curvature by a difference in head group shape and the membrane curvature directly opening the channel, as in panel *a*.

PUFAs acting through the outer leaflet of the membrane could also enhance membrane tension. The use of gramicidin as a probe combined with studies of TRP channels to examine the effects of lipid mediators on membrane-mediated channel openings should be applicable to other mechanosensory systems.

FUTURE PROSPECTS

The diversity of the physiological roles of TRP channels and their structural complexity help explain some of the discrepancies in experimental observations made in vitro on the mechanisms associated with the gating of these channels. The multiple roles of TRP channels also suggests that exploiting them as drug targets will be a daunting task. For example, mammalian TRPA1 has been implicated not only in sensing environmental hazards, cold, and mechanical pressure, but also in regulating inhibitory synapse activity through the effects of astrocyte intracellular calcium (Shigetomi et al. 2012). The vast range of human developmental disorders linked to different mutations in TRPV4 also suggests that many functions of TRP channels remain unidentified (Cho et al. 2012). TRPV4 and TRPML3 knockouts are relatively normal, whereas gain-of-function mutants have dramatic phenotypes, suggesting that redundancy in TRP function masks some functions of TRP channels that are not detected in knockout mice.

Recent advances in the understanding of *Drosophila* visual transduction have provided strong evidence that membrane perturbation in rhabdomere microvilli caused by PLC activity can directly activate the prototypical TRP channel (Hardie & Franze 2012). This dramatic observation, if translated into mammalian systems, could explain the multiple effects of many lipid second messengers on mechanotransduction in terms of their ability to perturb membrane structure and indirectly gate TRP channels. Thus, the case for some TRP channels being direct mechanotransducers is becoming more convincing; unraveling the mechanisms through which lipids regulate TRP channel activity is likely to give new insights into mechanotransduction as well as other forms of sensation.

DISCLOSURE STATEMENT

The authors are not aware of any affiliations, memberships, funding, or financial holdings that might be perceived as affecting the objectivity of this review.

ACKNOWLEDGMENTS

We thank the BBSRC, the MRC, and the Wellcome Trust for generous support. J.N.W. was also supported by WCU grant R31-2008-000-10103-0 at SNU. We apologize for omissions caused by space constraints or oversights and thank many colleagues for helpful comments. N.E. was supported by a Rubicon fellowship of the Netherlands Organisation for Scientific Research.

LITERATURE CITED

Alessandri-Haber N, Dina OA, Joseph EK, Reichling DB, Levine JD. 2008. Interaction of transient receptor potential vanilloid 4, integrin, and SRC tyrosine kinase in mechanical hyperalgesia. *J. Neurosci.* 28:1046–57

Amato V, Vina E, Calavia MG, Guerrera MC, Laura R, et al. 2012. TRPV4 in the sensory organs of adult zebrafish. *Microsc. Res. Tech.* 75:89–96

Andersson DA, Gentry C, Moss S, Bevan S. 2009. Clioquinol and pyrithione activate TRPA1 by increasing intracellular Zn^{2+}. *Proc. Natl. Acad. Sci. USA* 106:8374–79

Anfinogenova Y, Brett SE, Walsh MP, Harraz OF, Welsh DG. 2011. Do TRPC-like currents and G protein-coupled receptors interact to facilitate myogenic tone development? *Am. J. Physiol. Heart. Circ. Physiol.* 301:H1378–88

Araki I. 2011. TRP channels in urinary bladder mechanosensation. *Adv. Exp. Med. Biol.* 704:861–79

Auer-Grumbach M, Olschewski A, Papic L, Kremer H, McEntagart ME, et al. 2010. Alterations in the ankyrin domain of TRPV4 cause congenital distal SMA, scapuloperoneal SMA and HMSN2C. *Nat. Genet.* 42:160–64

Bandyopadhyay BC, Swaim WD, Liu X, Redman RS, Patterson RL, Ambudkar IS. 2005. Apical localization of a functional TRPC3/TRPC6-Ca^{2+}-signaling complex in polarized epithelial cells. Role in apical Ca^{2+} influx. *J. Biol. Chem.* 280:12908–16

Berna-Erro A, Galan C, Dionisio N, Gomez LJ, Salido GM, Rosado JA. 2012. Capacitative and non-capacitative signaling complexes in human platelets. *Biochim. Biophys. Acta* 1823:1242–51

Berrout J, Jin M, Mamenko M, Zaika O, Pochynyuk O, O'Neil RG. 2012a. Function of transient receptor potential cation channel subfamily V member 4 (TRPV4) as a mechanical transducer in flow-sensitive segments of renal collecting duct system. *J. Biol. Chem.* 287:8782–91

Berrout J, Jin M, O'Neil RG. 2012b. Critical role of TRPP2 and TRPC1 channels in stretch-induced injury of blood-brain barrier endothelial cells. *Brain Res.* 1436:1–12

Beurg M, Evans MG, Hackney CM, Fettiplace R. 2006. A large-conductance calcium-selective mechanotransducer channel in mammalian cochlear hair cells. *J. Neurosci.* 26:10992–1000

Birder LA, Nakamura Y, Kiss S, Nealen ML, Barrick S, et al. 2002. Altered urinary bladder function in mice lacking the vanilloid receptor TRPV1. *Nat. Neurosci.* 5:856–60

Brierley SM, Castro J, Harrington AM, Hughes PA, Page AJ, et al. 2011. TRPA1 contributes to specific mechanically activated currents and sensory neuron mechanical hypersensitivity. *J. Physiol.* 589:3575–93

Brierley SM, Hughes PA, Page AJ, Kwan KY, Martin CM, et al. 2009. The ion channel TRPA1 is required for normal mechanosensation and is modulated by algesic stimuli. *Gastroenterology* 137:2084–95

Brierley SM, Page AJ, Hughes PA, Adam B, Liebregts T, et al. 2008. Selective role for TRPV4 ion channels in visceral sensory pathways. *Gastroenterology* 134:2059–69

Carrillo C, Hichami A, Andreoletti P, Cherkaoui-Malki M, Del Mar CM, et al. 2012. Diacylglycerol-containing oleic acid induces increases in $[Ca^{2+}]_i$ via TRPC3/6 channels in human T-cells. *Biochim. Biophys. Acta* 1821:618–26

Chen J, Barritt GJ. 2003. Evidence that TRPC1 (transient receptor potential canonical 1) forms a Ca^{2+}-permeable channel linked to the regulation of cell volume in liver cells obtained using small interfering RNA targeted against TRPC1. *Biochem. J.* 373:327–36

Cho TJ, Matsumoto K, Fano V, Dai J, Kim OH, et al. 2012. TRPV4-pathy manifesting both skeletal dysplasia and peripheral neuropathy: a report of three patients. *Am. J. Med. Genet. A* 158A:795–802

Chyb S, Raghu P, Hardie RC. 1999. Polyunsaturated fatty acids activate the *Drosophila* light-sensitive channels TRP and TRPL. *Nature* 397:255–59

Cioffi DL, Wu S, Chen H, Alexeyev M, St Croix CM, et al. 2012. Orai1 determines calcium selectivity of an endogenous TRPC heterotetramer channel. *Circ. Res.* 110:1435–44

Ciura S, Bourque CW. 2006. Transient receptor potential vanilloid 1 is required for intrinsic osmoreception in organum vasculosum lamina terminalis neurons and for normal thirst responses to systemic hyperosmolality. *J. Neurosci.* 26:9069–75

Corey DP, Hudspeth AJ. 1983. Kinetics of the receptor current in bullfrog saccular hair cells. *J. Neurosci.* 3:962–76

Cosens DJ, Manning A. 1969. Abnormal electroretinogram from a *Drosophila* mutant. *Nature* 224:285–87

Coste B, Xiao B, Santos JS, Syeda R, Grandl J, et al. 2012. Piezo proteins are pore-forming subunits of mechanically activated channels. *Nature* 483:176–81

Daly D, Rong W, Chess-Williams R, Chapple C, Grundy D. 2007. Bladder afferent sensitivity in wild-type and TRPV1 knockout mice. *J. Physiol.* 583:663–74

Delmas P, Hao J, Rodat-Despoix L. 2011. Molecular mechanisms of mechanotransduction in mammalian sensory neurons. *Nat. Rev. Neurosci.* 12:139–53

Deng HX, Klein CJ, Yan J, Shi Y, Wu Y, et al. 2010. Scapuloperoneal spinal muscular atrophy and CMT2C are allelic disorders caused by alterations in TRPV4. *Nat. Genet.* 42:165–69

Denis V, Cyert MS. 2002. Internal Ca^{2+} release in yeast is triggered by hypertonic shock and mediated by a TRP channel homologue. *J. Cell Biol.* 156:29–34

Di A, Malik AB. 2010. TRP channels and the control of vascular function. *Curr. Opin. Pharmacol.* 10:127–32

Dietrich A, Kalwa H, Rost BR, Gudermann T. 2005. The diacylgylcerol-sensitive TRPC3/6/7 subfamily of cation channels: functional characterization and physiological relevance. *Pflug. Arch.* 451:72–80

Drew LJ, Rugiero F, Cesare P, Gale JE, Abrahamsen B, et al. 2007. High-threshold mechanosensitive ion channels blocked by a novel conopeptide mediate pressure-evoked pain. *PLoS ONE* 2:e515

Drew LJ, Wood JN, Cesare P. 2002. Distinct mechanosensitive properties of capsaicin-sensitive and -insensitive sensory neurons. *J. Neurosci.* 22:RC228

Dubin AE, Schmidt M, Mathur J, Petrus MJ, Xiao B, et al. 2012. Inflammatory signals enhance piezo2-mediated mechanosensitive currents. *Cell Rep.* 2:511–17

Earley S, Brayden JE. 2010. Transient receptor potential channels and vascular function. *Clin. Sci.* 119:19–36

Effertz T, Wiek R, Gopfert MC. 2011. NompC TRP channel is essential for *Drosophila* sound receptor function. *Curr. Biol.* 21:592–97

Eijkelkamp N, Linley JE, Torres JM, Bee L, Dickenson AH et al. 2013. A role for Piezo2 in EPAC1-dependent mechanical allodynia. *Nat. Commun.* 4:1682

Everaerts W, Vriens J, Owsianik G, Appendino G, Voets T, et al. 2010. Functional characterization of transient receptor potential channels in mouse urothelial cells. *Am. J. Physiol. Renal. Physiol.* 298:F692–701

Fan HC, Zhang X, McNaughton PA. 2009. Activation of the TRPV4 ion channel is enhanced by phosphorylation. *J. Biol. Chem.* 284:27884–91

Farris HE, LeBlanc CL, Goswami J, Ricci AJ. 2004. Probing the pore of the auditory hair cell mechanotransducer channel in turtle. *J. Physiol.* 558:769–92

Formigli L, Sassoli C, Squecco R, Bini F, Martinesi M, et al. 2009. Regulation of transient receptor potential canonical channel 1 (TRPC1) by sphingosine 1-phosphate in C2C12 myoblasts and its relevance for a role of mechanotransduction in skeletal muscle differentiation. *J. Cell Sci.* 122:1322–33

Fowler MA, Montell C. 2013. *Drosophila* TRP channels and animal behavior. *Life Sci.* 92(8–9):394–403

Galizia L, Flamenco MP, Rivarola V, Capurro C, Ford P. 2008. Role of AQP2 in activation of calcium entry by hypotonicity: implications in cell volume regulation. *Am. J. Physiol. Renal. Physiol.* 294:F582–90

Garrison SR, Dietrich A, Stucky CL. 2012. TRPC1 contributes to light-touch sensation and mechanical responses in low-threshold cutaneous sensory neurons. *J. Neurophysiol.* 107:913–22

Geffeney SL, Cueva JG, Glauser DA, Doll JC, Lee TH, et al. 2011. DEG/ENaC but not TRP channels are the major mechanoelectrical transduction channels in a *C. elegans* nociceptor. *Neuron* 71:845–57

Glitsch MD. 2010. Activation of native TRPC3 cation channels by phospholipase D. *FASEB J.* 24:318–25

Goel M, Sinkins WG, Zuo CD, Hopfer U, Schilling WP. 2007. Vasopressin-induced membrane trafficking of TRPC3 and AQP2 channels in cells of the rat renal collecting duct. *Am. J. Physiol. Renal. Physiol.* 293:F1476–88

Gomis A, Soriano S, Belmonte C, Viana F. 2008. Hypoosmotic- and pressure-induced membrane stretch activate TRPC5 channels. *J. Physiol.* 586:5633–49

Gottlieb P, Folgering J, Maroto R, Raso A, Wood TG, et al. 2008. Revisiting TRPC1 and TRPC6 mechanosensitivity. *Pflug. Arch.* 455:1097–103

Greene CC, McMillan PM, Barker SE, Kurnool P, Lomax MI, et al. 2001. DFNA25, a novel locus for dominant nonsyndromic hereditary hearing impairment, maps to 12q21-24. *Am. J. Hum. Genet.* 68:254–60

Greka A, Mundel P. 2012. Cell biology and pathology of podocytes. *Annu. Rev. Physiol.* 74:299–323

Grimm C, Cuajungco MP, van Aken AF, Schnee M, Jors S, et al. 2007. A helix-breaking mutation in TRPML3 leads to constitutive activity underlying deafness in the varitint-waddler mouse. *Proc. Natl. Acad. Sci. USA* 104:19583–88

Grimm C, Kraft R, Sauerbruch S, Schultz G, Harteneck C. 2003. Molecular and functional characterization of the melastatin-related cation channel TRPM3. *J. Biol. Chem.* 278:21493–501

Guibert C, Ducret T, Savineau JP. 2011. Expression and physiological roles of TRP channels in smooth muscle cells. *Adv. Exp. Med. Biol.* 704:687–706

Hardie RC. 1991. Whole-cell recordings of the light induced current in dissociated *Drosophila* photoreceptors: evidence for feedback by calcium permeating the light-sensitive channels. *Proc. Biol. Sci.* 245:203–10

Hardie RC. 2011. A brief history of TRP: commentary and personal perspective. *Pflug. Arch.* 461:493–98

Hardie RC, Franze K. 2012. Photomechanical responses in *Drosophila* photoreceptors. *Science* 338:260–63

Harrington AM, Hughes PA, Martin CM, Yang J, Castro J, et al. 2011. A novel role for TRPM8 in visceral afferent function. *Pain* 152:1459–68

Hartmann J, Dragicevic E, Adelsberger H, Henning HA, Sumser M, et al. 2008. TRPC3 channels are required for synaptic transmission and motor coordination. *Neuron* 59:392–98

Hartmannsgruber V, Heyken WT, Kacik M, Kaistha A, Grgic I, et al. 2007. Arterial response to shear stress critically depends on endothelial TRPV4 expression. *PLoS ONE* 2:e827

Hayashi T, Kondo T, Ishimatsu M, Yamada S, Nakamura K, et al. 2009. Expression of the TRPM8-immunoreactivity in dorsal root ganglion neurons innervating the rat urinary bladder. *Neurosci. Res.* 65:245–51

Hoberg M, Gratz HH, Noll M, Jones DB. 2005. Mechanosensitivity of human osteosarcoma cells and phospholipase C β2 expression. *Biochem. Biophys. Res. Commun.* 333:142–49

Hsu YJ, Hoenderop JG, Bindels RJ. 2007. TRP channels in kidney disease. *Biochim. Biophys. Acta* 1772:928–36

Huang J, Liu CH, Hughes SA, Postma M, Schwiening CJ, Hardie RC. 2010. Activation of TRP channels by protons and phosphoinositide depletion in *Drosophila* photoreceptors. *Curr. Biol.* 20(3):189–97

Hunt JL, Pollak MR, Denker BM. 2005. Cultured podocytes establish a size-selective barrier regulated by specific signaling pathways and demonstrate synchronized barrier assembly in a calcium switch model of junction formation. *J. Am. Soc. Nephrol.* 16:1593–602

Hwang RY, Stearns NA, Tracey WD. 2012. The ankyrin repeat domain of the TRPA protein painless is important for thermal nociception but not mechanical nociception. *PLoS ONE* 7:e30090

Inoue R, Jensen LJ, Jian Z, Shi J, Hai L, et al. 2009. Synergistic activation of vascular TRPC6 channel by receptor and mechanical stimulation via phospholipase C/diacylglycerol and phospholipase A2/ω-hydroxylase/20-HETE pathways. *Circ. Res.* 104:1399–1409

Inoue R, Jian Z, Kawarabayashi Y. 2009. Mechanosensitive TRP channels in cardiovascular pathophysiology. *Pharmacol. Ther.* 123:371–85

Jin M, Berrout J, O'Neil RG. 2011. Regulation of TRP channels by osmomechanical stress. In *TRP Channels*, ed. MX Zhu, pp. 189–207. Boca Raton, FL: CRC Press

Jors S, Grimm C, Becker L, Heller S. 2010. Genetic inactivation of *Trpml3* does not lead to hearing and vestibular impairment in mice. *PLoS ONE* 5:e14317

Kahn-Kirby AH, Dantzker JL, Apicella AJ, Schafer WR, Browse J, et al. 2004. Specific polyunsaturated fatty acids drive TRPV-dependent sensory signaling in vivo. *Cell* 119:889–900

Kanai Y, Nakazato E, Fujiuchi A, Hara T, Imai A. 2005. Involvement of an increased spinal TRPV1 sensitization through its up-regulation in mechanical allodynia of CCI rats. *Neuropharmacology* 49:977–84

Kang L, Gao J, Schafer WR, Xie Z, Xu XZ. 2010. *C. elegans* TRP family protein TRP-4 is a pore-forming subunit of a native mechanotransduction channel. *Neuron* 67:381–91

Kang S, Jang JH, Price MP, Gautam M, Benson CJ, et al. 2012. Simultaneous disruption of mouse *ASIC1a*, *ASIC2* and *ASIC3* genes enhances cutaneous mechanosensitivity. *PLoS ONE* 7:e35225

Kashiba H, Uchida Y, Takeda D, Nishigori A, Ueda Y, et al. 2004. TRPV2-immunoreactive intrinsic neurons in the rat intestine. *Neurosci. Lett.* 366:193–96

Kawashima Y, Geleoc GS, Kurima K, Labay V, Lelli A, et al. 2011. Mechanotransduction in mouse inner ear hair cells requires transmembrane channel-like genes. *J. Clin. Investig.* 121:4796–809

Kerstein PC, del Camino D, Moran MM, Stucky CL. 2009. Pharmacological blockade of TRPA1 inhibits mechanical firing in nociceptors. *Mol. Pain* 5:19

Kim C. 2007. TRPV family ion channels and other molecular components required for hearing and proprioception in *Drosophila*. See Liedtke & Heller 2007, pp. 313–27

Kim J, Chung YD, Park DY, Choi S, Shin DW, et al. 2003. A TRPV family ion channel required for hearing in *Drosophila*. *Nature* 424:81–84

Kim JY, Zeng W, Kiselyov K, Yuan JP, Dehoff MH, et al. 2006. Homer 1 mediates store- and inositol 1,4,5-trisphosphate receptor-dependent translocation and retrieval of TRPC3 to the plasma membrane. *J. Biol. Chem.* 281:32540–49

Kim Y, Wong AC, Power JM, Tadros SF, Klugmann M, et al. 2012. Alternative splicing of the TRPC3 ion channel calmodulin/IP_3 receptor-binding domain in the hindbrain enhances cation flux. *J. Neurosci.* 32:11414–23

Kindt KS, Viswanath V, Macpherson L, Quast K, Hu H, et al. 2007. *Caenorhabditis elegans* TRPA-1 functions in mechanosensation. *Nat. Neurosci.* 10:568–77

Kobori T, Smith GD, Sandford R, Edwardson JM. 2009. The transient receptor potential channels TRPP2 and TRPC1 form a heterotetramer with a 2:2 stoichiometry and an alternating subunit arrangement. *J. Biol. Chem.* 284:35507–13

Kowase T, Nakazato Y, Yoko O, Morikawa A, Kojima I. 2002. Immunohistochemical localization of growth factor-regulated channel (GRC) in human tissues. *Endocr. J.* 49:349–55

Kremeyer B, Lopera F, Cox JJ, Momin A, Rugiero F, et al. 2010. A gain-of-function mutation in TRPA1 causes familial episodic pain syndrome. *Neuron* 66:671–80

Kroese AB, Das A, Hudspeth AJ. 1989. Blockage of the transduction channels of hair cells in the bullfrog's sacculus by aminoglycoside antibiotics. *Hear. Res.* 37:203–17

Kuipers AJ, Middelbeek J, van Leeuwen FN. 2012. Mechanoregulation of cytoskeletal dynamics by TRP channels. *Eur. J. Cell Biol.* 91:834–46

Kwan KY, Allchorne AJ, Vollrath MA, Christensen AP, Zhang DS, et al. 2006. TRPA1 contributes to cold, mechanical, and chemical nociception but is not essential for hair-cell transduction. *Neuron* 50:277–89

Lansman JB, Hallam TJ, Rink TJ. 1987. Single stretch-activated ion channels in vascular endothelial cells as mechanotransducers? *Nature* 325:811–13

Lauritzen I, Chemin J, Honore E, Jodar M, Guy N, et al. 2005. Cross-talk between the mechano-gated K_{2P} channel TREK-1 and the actin cytoskeleton. *EMBO Rep.* 6:642–48

Lev S, Katz B, Minke B. 2012. The activity of the TRP-like channel depends on its expression system. *Channels* 6:86–93

Li J, Verkman AS. 2001. Impaired hearing in mice lacking aquaporin-4 water channels. *J. Biol. Chem.* 276:31233–37

Liang X, Madrid J, Saleh HS, Howard J. 2011. NOMPC, a member of the TRP channel family, localizes to the tubular body and distal cilium of *Drosophila campaniform* and chordotonal receptor cells. *Cytoskeleton* 68:1–7

Liao Y, Erxleben C, Yildirim E, Abramowitz J, Armstrong DL, Birnbaumer L. 2007. Orai proteins interact with TRPC channels and confer responsiveness to store depletion. *Proc. Natl. Acad. Sci. USA* 104:4682–87

Liedtke W. 2007. TRPV channels' role in osmotransduction and mechanotransduction. *Handb. Exp. Pharmacol.* 179:473–87

Liedtke W, Choe Y, Marti-Renom MA, Bell AM, Denis CS, et al. 2000. Vanilloid receptor-related osmotically activated channel (VR-OAC), a candidate vertebrate osmoreceptor. *Cell* 103:525–35

Liedtke WB, Heller S, eds. 2007. *TRP Ion Channel Function in Sensory Transduction and Cellular Signaling Cascades*. Boca Raton, FL: CRC Press

Liu X, Bandyopadhyay BC, Nakamoto T, Singh B, Liedtke W, et al. 2006. A role for AQP5 in activation of TRPV4 by hypotonicity: concerted involvement of AQP5 and TRPV4 in regulation of cell volume recovery. *J. Biol. Chem.* 281:15485–95

Lockwich T, Singh BB, Liu X, Ambudkar IS. 2001. Stabilization of cortical actin induces internalization of transient receptor potential 3 (Trp3)-associated caveolar Ca^{2+} signaling complex and loss of Ca^{2+} influx without disruption of Trp3-inositol trisphosphate receptor association. *J. Biol. Chem.* 276(45):42401–8

Loukin SH, Su Z, Kung C. 2009. Hypotonic shocks activate rat TRPV4 in yeast in the absence of polyunsaturated fatty acids. *FEBS Lett.* 583:754–58

Low SE, Amburgey K, Horstick E, Linsley J, Sprague SM, et al. 2011. TRPM7 is required within zebrafish sensory neurons for the activation of touch-evoked escape behaviors. *J. Neurosci.* 31:11633–44

Ma X, Cheng KT, Wong CO, O'Neil RG, Birnbaumer L, et al. 2011a. Heteromeric TRPV4-C1 channels contribute to store-operated Ca^{2+} entry in vascular endothelial cells. *Cell Calcium* 50:502–9

Ma X, Nilius B, Wong JW, Huang Y, Yao X. 2011b. Electrophysiological properties of heteromeric TRPV4-C1 channels. *Biochim. Biophys. Acta* 1808:2789–97

Ma X, Qiu S, Luo J, Ma Y, Ngai CY, et al. 2010. Functional role of vanilloid transient receptor potential 4-canonical transient receptor potential 1 complex in flow-induced Ca^{2+} influx. *Arterioscler. Thromb. Vasc. Biol.* 30:851–58

Makino A, Prossnitz ER, Bunemann M, Wang JM, Yao W, Schmid-Schonbein GW. 2006. G protein-coupled receptors serve as mechanosensors for fluid shear stress in neutrophils. *Am. J. Physiol. Cell Physiol.* 290:C1633–39

Marcotti W, Geleoc GS, Lennan GW, Kros CJ. 1999. Transient expression of an inwardly rectifying potassium conductance in developing inner and outer hair cells along the mouse cochlea. *Pflug. Arch.* 439:113–22

Maroto R, Raso A, Wood TG, Kurosky A, Martinac B, Hamill OP. 2005. TRPC1 forms the stretch-activated cation channel in vertebrate cells. *Nat. Cell Biol.* 7:179–85

Matsumoto S, Takeda M, Saiki C, Takahashi T, Ojima K. 1997. Effects of tachykinins on rapidly adapting pulmonary stretch receptors and total lung resistance in anesthetized, artificially ventilated rabbits. *J. Pharmacol. Exp. Ther.* 283:1026–31

Matthews BD, Thodeti CK, Tytell JD, Mammoto A, Overby DR, Ingber DE. 2010. Ultra-rapid activation of TRPV4 ion channels by mechanical forces applied to cell surface beta1 integrins. *Integr. Biol.* 2:435–42

Mendoza SA, Fang J, Gutterman DD, Wilcox DA, Bubolz AH, et al. 2010. TRPV4-mediated endothelial Ca^{2+} influx and vasodilation in response to shear stress. *Am. J. Physiol. Heart. Circ. Physiol.* 298:H466–76

Mochizuki T, Sokabe T, Araki I, Fujishita K, Shibasaki K, et al. 2009. The TRPV4 cation channel mediates stretch-evoked Ca^{2+} influx and ATP release in primary urothelial cell cultures. *J. Biol. Chem.* 284:21257–64

Montag S, Kruger K, Madeja M, Speckmann EJ, Musshoff U. 2004. Contribution of the cytoskeleton and the phospholipase C signaling pathway to fluid stream-induced membrane currents. *Cell Calcium* 35:333–43

Montell C, Rubin GM. 1989. Molecular characterization of the *Drosophila trp* locus: a putative integral membrane protein required for phototransduction. *Neuron* 2:1313–23

Morita H, Honda A, Inoue R, Ito Y, Abe K, et al. 2007. Membrane stretch-induced activation of a TRPM4-like nonselective cation channel in cerebral artery myocytes. *J. Pharmacol. Sci.* 103:417–26

Motter AL, Ahern GP. 2012. TRPA1 is a polyunsaturated fatty acid sensor in mammals. *PLoS ONE* 7:e38439

Muraki K, Iwata Y, Katanosaka Y, Ito T, Ohya S, et al. 2003. TRPV2 is a component of osmotically sensitive cation channels in murine aortic myocytes. *Circ. Res.* 93:829–38

Nagata K, Duggan A, Kumar G, Garcia-Anoveros J. 2005. Nociceptor and hair cell transducer properties of TRPA1, a channel for pain and hearing. *J. Neurosci.* 25:4052–61

Nandasena BG, Suzuki A, Aita M, Kawano Y, Nozawa-Inoue K, Maeda T. 2007. Immunolocalization of aquaporin-1 in the mechanoreceptive Ruffini endings in the periodontal ligament. *Brain Res.* 1157:32–40

Nauli SM, Alenghat FJ, Luo Y, Williams E, Vassilev P, et al. 2003. Polycystins 1 and 2 mediate mechanosensation in the primary cilium of kidney cells. *Nat. Genet.* 33:129–37

Nilius B, Honore E. 2012. Sensing pressure with ion channels. *Trends Neurosci.* 35:477–86

Nilius B, Vriens J, Prenen J, Droogmans G, Voets T. 2004. TRPV4 calcium entry channel: a paradigm for gating diversity. *Am. J. Physiol. Cell Physiol.* 286:C195–205

Nishida M, Watanabe K, Sato Y, Nakaya M, Kitajima N, et al. 2010. Phosphorylation of TRPC6 channels at Thr69 is required for anti-hypertrophic effects of phosphodiesterase 5 inhibition. *J. Biol. Chem.* 285:13244–53

Nishihara E, Hiyama TY, Noda M. 2011. Osmosensitivity of transient receptor potential vanilloid 1 is synergistically enhanced by distinct activating stimuli such as temperature and protons. *PLoS ONE* 6:e22246

Numata T, Shimizu T, Okada Y. 2007a. Direct mechano-stress sensitivity of TRPM7 channel. *Cell Physiol. Biochem.* 19:1–8

Numata T, Shimizu T, Okada Y. 2007b. TRPM7 is a stretch- and swelling-activated cation channel involved in volume regulation in human epithelial cells. *Am. J. Physiol. Cell Physiol.* 292:C460–67

Oancea E, Wolfe JT, Clapham DE. 2006. Functional TRPM7 channels accumulate at the plasma membrane in response to fluid flow. *Circ. Res.* 98:245–53

O'Hagan R, Chalfie M, Goodman MB. 2005. The MEC-4 DEG/ENaC channel of *Caenorhabditis elegans* touch receptor neurons transduces mechanical signals. *Nat. Neurosci.* 8:43–50

Okubo K, Matsumura M, Kawaishi Y, Aoki Y, Matsunami M, et al. 2012. Hydrogen sulfide-induced mechanical hyperalgesia and allodynia require activation of both $Ca_V3.2$ and TRPA1 channels in mice. *Br. J. Pharmacol.* 166:1738–43

Patel A, Honoré E. 2010. Polycystins and renovascular mechanosensory transduction. *Nat. Rev. Nephrol.* 6:530–38

Patel A, Sharif-Naeini R, Folgering JR, Bichet D, Duprat F, Honoré E. 2010. Canonical TRP channels and mechanotransduction: from physiology to disease states. *Pflug. Arch.* 460:571–81

Pedersen SF, Kapus A, Hoffmann EK. 2011. Osmosensory mechanisms in cellular and systemic volume regulation. *J. Am. Soc. Nephrol.* 22:1587–97

Peyronnet R, Sharif-Naeini R, Folgering JH, Arhatte M, Jodar M, et al. 2012. Mechanoprotection by polycystins against apoptosis is mediated through the opening of stretch-activated K_{2P} channels. *Cell Rep.* 1:241–50

Quick K, Zhao J, Eijkelkamp N, Linley JE, Rugiero F, et al. 2012. TRPC3 and TRPC6 are essential for normal mechanotransduction in subsets of sensory neurons and cochlear hair cells. *Open Biol.* 2:120068

Ramsey IS, Delling M, Clapham DE. 2006. An introduction to TRP channels. *Annu. Rev. Physiol.* 68:619–47

Raouf R, Rugiero F, Kiesewetter H, Hatch R, Hummler E, et al. 2012. Sodium channels and mammalian sensory mechanotransduction. *Mol. Pain* 8:21

Reiser J, Polu KR, Möller CC, Kenlan P, Altintas MM, et al. 2005. TRPC6 is a glomerular slit diaphragm-associated channel required for normal renal function. *Nat. Genet.* 37:739–44

Ricci AJ, Crawford AC, Fettiplace R. 2003. Tonotopic variation in the conductance of the hair cell mechanotransducer channel. *Neuron* 40:983–90

Rohacs T, Thyagarajan B, Lukacs V. 2008. Phospholipase C mediated modulation of TRPV1 channels. *Mol. Neurobiol.* 37:153–63

Rugiero F, Drew LJ, Wood JN. 2010. Kinetic properties of mechanically activated currents in spinal sensory neurons. *J. Physiol.* 588:301–14

Rugiero F, Wood JN. 2009. The mechanosensitive cell line ND-C does not express functional thermoTRP channels. *Neuropharmacology* 56:1138–46

Rusch A, Kros CJ, Richardson GP. 1994. Block by amiloride and its derivatives of mechano-electrical transduction in outer hair cells of mouse cochlear cultures. *J. Physiol.* 474:75–86

Ruwhof C, van Wamel JT, Noordzij LA, Aydin S, Harper JC, van der Laarse A. 2001. Mechanical stress stimulates phospholipase C activity and intracellular calcium ion levels in neonatal rat cardiomyocytes. *Cell Calcium* 29:73–83

Schnitzler M, Storch U, Meibers S, Nurwakagari P, Breit A, et al. 2008. Gq-coupled receptors as mechanosensors mediating myogenic vasoconstriction. *EMBO J.* 27:3092–103

Schwander M, Xiong W, Tokita J, Lelli A, Elledge HM, et al. 2009. A mouse model for nonsyndromic deafness (DFNB12) links hearing loss to defects in tip links of mechanosensory hair cells. *Proc. Natl. Acad. Sci. USA* 106:5252–57

Sénatore S, Rami Reddy V, Sémériva M, Perrin L, Lalevée N. 2010. Response to mechanical stress is mediated by the TRPA channel painless in the *Drosophila* heart. *PLoS Genet.* 6:e1001088

Seth M, Zhang ZS, Mao L, Graham V, Burch J, et al. 2009. TRPC1 channels are critical for hypertrophic signaling in the heart. *Circ. Res.* 105:1023–30

Sharif-Naeini R, Folgering JH, Bichet D, Duprat F, Delmas P, et al. 2010. Sensing pressure in the cardiovascular system: Gq-coupled mechanoreceptors and TRP channels. *J. Mol. Cell Cardiol.* 48:83–89

Sharif-Naeini R, Folgering JH, Bichet D, Duprat F, Lauritzen I, et al. 2009. Polycystin-1 and -2 dosage regulates pressure sensing. *Cell* 139(3):587–96

Sharif-Naeini R, Witty MF, Séguéla P, Bourque CW. 2006. An N-terminal variant of Trpv1 channel is required for osmosensory transduction. *Nat. Neurosci.* 9:93–98

Shen J, Harada N, Kubo N, Liu B, Mizuno A, et al. 2006. Functional expression of transient receptor potential vanilloid 4 in the mouse cochlea. *NeuroReport* 17:135–39

Shen WL, Kwon Y, Adegbola AA, Luo J, Chess A, Montell C. 2011. Function of rhodopsin in temperature discrimination in *Drosophila*. *Science* 331:1333–36

Shigetomi E, Tong X, Kwan KY, Corey DP, Khakh BS. 2012. TRPA1 channels regulate astrocyte resting calcium and inhibitory synapse efficacy through GAT-3. *Nat. Neurosci.* 15:70–80

Shimizu T, Janssens A, Voets T, Nilius B. 2009. Regulation of the murine TRPP3 channel by voltage, pH, and changes in cell volume. *Pflug. Arch.* 457:795–807

Shin YC, Shin SY, So I, Kwon D, Jeon JH. 2011. TRIP Database: a manually curated database of protein-protein interactions for mammalian TRP channels. *Nucleic Acids Res.* 39:D356–61

Spassova MA, Hewavitharana T, Xu W, Soboloff J, Gill DL. 2006. A common mechanism underlies stretch activation and receptor activation of TRPC6 channels. *Proc. Natl. Acad. Sci. USA* 103:16586–91

Staruschenko A, Jeske NA, Akopian AN. 2010. Contribution of TRPV1-TRPA1 interaction to the single channel properties of the TRPA1 channel. *J. Biol. Chem.* 285:15167–77

Steigelman KA, Lelli A, Wu X, Gao J, Lin S, et al. 2011. Polycystin-1 is required for stereocilia structure but not for mechanotransduction in inner ear hair cells. *J. Neurosci.* 31:12241–50

Stewart AP, Smith GD, Sandford RN, Edwardson JM. 2010. Atomic force microscopy reveals the alternating subunit arrangement of the TRPP2-TRPV4 heterotetramer. *Biophys. J.* 99:790–97

Stiber JA, Zhang ZS, Burch J, Eu JP, Zhang S, et al. 2008. Mice lacking Homer 1 exhibit a skeletal myopathy characterized by abnormal transient receptor potential channel activity. *Mol. Cell. Biol.* 28:2637–47

Storch U, Forst AL, Philipp M, Gudermann T, Schnitzler M. 2012a. Transient receptor potential channel 1 (TRPC1) reduces calcium permeability in heteromeric channel complexes. *J. Biol. Chem.* 287:3530–40

Storch U, Schnitzler M, Gudermann T. 2012b. G protein-mediated stretch reception. *Am. J. Physiol. Heart. Circ. Physiol.* 302:H1241–49

Tabuchi K, Suzuki M, Mizuno A, Hara A. 2005. Hearing impairment in TRPV4 knockout mice. *Neurosci. Lett.* 382:304–8

Tadros SF, Kim Y, Phan PA, Birnbaumer L, Housley GD. 2010. TRPC3 ion channel subunit immunolocalization in the cochlea. *Histochem. Cell Biol.* 133:137–47

Thorneloe KS, Sulpizio AC, Lin Z, Figueroa DJ, Clouse AK, et al. 2008. *N*-((1*S*)-1-{[4-((2*S*)-2-{[(2,4-dichlorophenyl)sulfonyl]amino}-3-hydroxypropanoyl)-1-piperazinyl]carbonyl}-3-methylbutyl)-1-benzothiophene-2-carboxamide (GSK1016790A), a novel and potent transient receptor potential vanilloid 4 channel agonist induces urinary bladder contraction and hyperactivity: part I. *J. Pharmacol. Exp. Ther.* 326:432–42

Tian D, Jacobo SM, Billing D, Rozkalne A, Gage SD, et al. 2010. Antagonistic regulation of actin dynamics and cell motility by TRPC5 and TRPC6 channels. *Sci. Signal.* 3:ra77

Tong Q, Hirschler-Laszkiewicz I, Zhang W, Conrad K, Neagley DW, et al. 2008. TRPC3 is the erythropoietin-regulated calcium channel in human erythroid cells. *J. Biol. Chem.* 283:10385–95

Tracey WD Jr, Wilson RI, Laurent G, Benzer S. 2003. *painless*, a *Drosophila* gene essential for nociception. *Cell* 113:261–73

van Aken AF, Atiba-Davies M, Marcotti W, Goodyear RJ, Bryant JE, et al. 2008. TRPML3 mutations cause impaired mechano-electrical transduction and depolarization by an inward-rectifier cation current in auditory hair cells of varitint-waddler mice. *J. Physiol.* 586:5403–18

Vilceanu D, Stucky CL. 2010. TRPA1 mediates mechanical currents in the plasma membrane of mouse sensory neurons. *PLoS ONE* 5:e12177

Volkers M, Dolatabadi N, Gude N, Most P, Sussman MA, Hassel D. 2012. Orai1 deficiency leads to heart failure and skeletal myopathy in zebrafish. *J. Cell Sci.* 125:287–94

Walker RG, Willingham AT, Zuker CS. 2000. A *Drosophila* mechanosensory transduction channel. *Science* 287:2229–34

Wang Q, Dai XQ, Li Q, Wang Z, Cantero MR, et al. 2012. Structural interaction and functional regulation of polycystin-2 by filamin. *PLoS ONE* 7:e40448

Wei H, Karimaa M, Korjamo T, Koivisto A, Pertovaara A. 2012. Transient receptor potential ankyrin 1 ion channel contributes to guarding pain and mechanical hypersensitivity in a rat model of postoperative pain. *Anesthesiology* 117(1):137–48

Welsh DG, Morielli AD, Nelson MT, Brayden JE. 2002. Transient receptor potential channels regulate myogenic tone of resistance arteries. *Circ. Res.* 90:248–50

Wes PD, Chevesich J, Jeromin A, Rosenberg C, Stetten G, Montell C. 1995. TRPC1, a human homolog of a *Drosophila* store-operated channel. *Proc. Natl. Acad. Sci. USA* 92(21):9652–56

Winn MP, Conlon PJ, Lynn KL, Farrington MK, Creazzo T, et al. 2005. A mutation in the TRPC6 cation channel causes familial focal segmental glomerulosclerosis. *Science* 308:1801–4

Wong F, Schaefer EL, Roop BC, LaMendola JN, Johnson-Seaton D, Shao D. 1989. Proper function of the *Drosophila trp* gene product during pupal development is important for normal visual transduction in the adult. *Neuron* 3:81–94

Wood JN, Eijkelkamp N. 2012. Noxious mechanosensation: molecules and circuits. *Curr. Opin. Pharmacol.* 12:4–8

Wu L, Gao X, Brown RC, Heller S, O'Neil RG. 2007. Dual role of the TRPV4 channel as a sensor of flow and osmolality in renal epithelial cells. *Am. J. Physiol. Renal. Physiol.* 293:F1699–713

Wu LJ, Sweet TB, Clapham DE. 2010. International Union of Basic and Clinical Pharmacology. LXXVI. Current progress in the mammalian TRP ion channel family. *Pharmacol. Rev.* 62:381–404

Xiao R, Xu XZ. 2011. *C. elegans* TRP channels. *Adv. Exp. Med. Biol.* 704:323–39

Xie J, Sun B, Du J, Yang W, Chen HC, et al. 2011. Phosphatidylinositol 4,5-bisphosphate (PIP_2) controls magnesium gatekeeper TRPM6 activity. *Sci. Rep.* 1:146

Yin J, Kuebler WM. 2010. Mechanotransduction by TRP channels: general concepts and specific role in the vasculature. *Cell Biochem. Biophys.* 56:1–18

Yuan JP, Kim MS, Zeng W, Shin DM, Huang G, et al. 2009. TRPC channels as STIM1-regulated SOCs. *Channels* 3:221–25

Yuan JP, Zeng W, Huang GN, Worley PF, Muallem S. 2007. STIM1 heteromultimerizes TRPC channels to determine their function as store-operated channels. *Nat. Cell Biol.* 9:636–45

Zakharian E, Cao C, Rohacs T. 2010. Gating of transient receptor potential melastatin 8 (TRPM8) channels activated by cold and chemical agonists in planar lipid bilayers. *J. Neurosci.* 30:12526–34

Zhang L, Jones S, Brody K, Costa M, Brookes SJ. 2004. Thermosensitive transient receptor potential channels in vagal afferent neurons of the mouse. *Am. J. Physiol. Gastrointest. Liver Physiol.* 286:G983–91

Zhang XF, Chen J, Faltynek CR, Moreland RB, Neelands TR. 2008. Transient receptor potential A1 mediates an osmotically activated ion channel. *Eur. J. Neurosci.* 27:605–11

Zhou XL, Loukin SH, Coria R, Kung C, Saimi Y. 2005. Heterologously expressed fungal transient receptor potential channels retain mechanosensitivity in vitro and osmotic response in vivo. *Eur. Biophys. J.* 34:413–22

Zhu MX. 2011. *TRP Channels*. Boca Raton, FL: CRC Press

Zhu X, Chu PB, Peyton M, Birnbaumer L. 1995. Molecular cloning of a widely expressed human homologue for the *Drosophila* trp gene. 1995. *FEBS Lett.* 373(3):193–98

Zimon M, Baets J, Auer-Grumbach M, Berciano J, Garcia A, et al. 2010. Dominant mutations in the cation channel gene *transient receptor potential vanilloid 4* cause an unusual spectrum of neuropathies. *Brain* 133:1798–809

The Molecular Basis of Self-Avoidance

S. Lawrence Zipursky[1] and Wesley B. Grueber[2]

[1]Department of Biological Chemistry, Howard Hughes Medical Institute, David Geffen School of Medicine, University of California, Los Angeles, California 90095-1662; email: lzipursky@mednet.ucla.edu

[2]Department of Physiology and Cellular Biophysics, Department of Neuroscience, College of Physicians and Surgeons, Columbia University Medical Center, New York, NY 10032; email: wg2135@columbia.edu

Annu. Rev. Neurosci. 2013. 36:547–68

The *Annual Review of Neuroscience* is online at neuro.annualreviews.org

This article's doi:
10.1146/annurev-neuro-062111-150414

Keywords

Dscam, protocadherins, homophilic, synapses, dendrites

Abstract

Self-avoidance, the tendency of neurites of the same cell to selectively avoid each other, is a property of both vertebrate and invertebrate neurons. In *Drosophila*, self-avoidance is mediated by a large family of cell recognition molecules of the immunoglobulin superfamily encoded, via alternative splicing, by the Dscam1 locus. Dscam1 promotes self-avoidance in dendrites, axons, and prospective postsynaptic elements. Expression analysis suggests that each neuron expresses a unique combination of isoforms. Identical isoforms on sister neurites exhibit isoform-specific homophilic recognition and elicit repulsion between processes, thereby promoting self-avoidance. Although any isoform can promote self-avoidance, thousands are necessary to ensure that neurites readily discriminate between self and nonself. Recent studies indicate that a large family of cadherins in the mouse, i.e., the clustered protocadherins, functions in an analogous fashion to promote self-avoidance. These studies argue for the evolution of a common molecular strategy for self-avoidance.

Contents

INTRODUCTION

Important discoveries over the past two decades have transformed the study of neural circuit assembly from the descriptive to the molecular. In the late 1980s and early 1990s, researchers identified key axon guidance molecules, including netrins (Hedgecock et al. 1990, Ishii et al. 1992, Kennedy et al. 1994, Serafini et al. 1994), slits (Brose et al. 1999, Kidd et al. 1999, Li et al. 1999), and semaphorins (Kolodkin et al. 1993, Luo et al. 1993) and their receptors. These act in various combinations to control the directed motility of axons and dendrites in many different developmental contexts in both vertebrates and invertebrates. In the 1990s, investigators also determined that topographic maps, a prominent feature of the vertebrate visual system, form via gradients of Ephrins and Ephs on growth cones and their targets (Cheng et al. 1995, Drescher et al. 1995). In addition, other receptors such as cell adhesion molecules of the cadherin and immunoglobulin (Ig) superfamilies mediating interactions between opposing cell surfaces and integrins acting as receptors for extracellular matrix components are widely expressed on developing neurites (Neugebauer et al. 1988, Tomaselli et al. 1988). Together, these studies led to the view by the late 1990s that neural circuit assembly emerged as a result of a relatively small number of different signals and their receptors, some acting in a graded fashion and in different combinations (Tessier-Lavigne & Goodman 1996). Studies over the past decade have underscored the key role of this core set of intercellular signaling molecules in regulating circuit assembly in a wide variety of vertebrate and invertebrate systems.

The discovery some 10–15 years ago of two large families of cell surface proteins encoded by the *Drosophila Down syndrome cell adhesion1* (*Dscam1*) locus and the clustered protocadherin (Pcdh) loci in mammals raised the possibility that these proteins, with diverse extracellular domains and shared cytoplasmic presumptive intracellular signaling domains, could provide diverse recognition specificities to a vast array of different neurites and convert these to a common output (Kohmura et al. 1998, Wu & Maniatis 1999, Schmucker et al. 2000). One obvious and intriguing possibility was that these molecules provided recognition between pre- and postsynaptic neurites. Although a role in synaptic matching remains possible, a series of studies on Dscam1 in flies and recent studies on clustered Pcdhs in the mouse have uncovered a shared molecular strategy by which these families of cell recognition molecules endow neurons with a unique cell surface identity that, in turn, allows neurons to distinguish between self and nonself. This discrimination lies at the heart of a phenomenon called self-avoidance, a process proposed many years ago as crucial for patterning neural circuits but that has received little attention until recently.

The concept of self-avoidance arose roughly 40 years ago from studies of highly organized receptive fields of the medicinal leech, *Hirudo medicinalis*. Mapping of touch receptor axonal receptive fields revealed distinct types of boundaries created between "self" and "nonself" axons. At the outer margins of receptive fields neighboring cells showed overlapping innervation. By contrast, boundaries between stereotyped branches of the same cell were sharp and well-defined, indicating little or no overlap between them (Nicholls & Baylor 1968). Nicholls & Baylor (1968) suggested a mechanism for the spatial arrangement of axons in which "a fiber might repel other branches more strongly if they arise from the same cell than if they come from a homologue, and not at all if they come from a cell with a different modality," although other mechanisms involving interactions with skin or underlying tissue were not ruled out. Yau (1976) confirmed their findings and proposed that the branches of a cell recognized each other to avoid growing into the same territory.

Kramer & Stent (1985) provided the first experimental support for self-recognition and repulsion between branches of the same cell. They described a large mechanoreceptor neuron that extends three branches, each into one of three contiguous body segments where each then elaborates a complex branching pattern providing uniform coverage of the segments. The branches within each segment did not overlap with one another, nor did they overlap with the processes from the same neuron in the adjacent segment. In animals in which one axon and its terminal arborization were eliminated, neighboring branches from the same neuron grew into the territory that was vacated by the manipulation. This phenomenon was coined "self-avoidance" to indicate the involvement of selective recognition and avoidance between sister branches. Researchers envisioned that self-avoidance was a universal patterning mechanism for axons and dendrites, in vertebrates and invertebrates, and in both the central and peripheral nervous systems (Kramer & Kuwada 1983, Kramer & Stent 1985). Self-avoidance would ensure even spreading of arbors across their territory, and the ability to discriminate between self and nonself would permit coexistence of arbors, ensuring parallel streams of information processing for various types of neurons sharing the same receptive field.

Here we review the role of fly *Dscam1* in regulating self-avoidance. We also discuss recent studies demonstrating that clustered Pcdhs play a similar role in the mouse. These studies suggest that vertebrate and fly neurons have solved the self-recognition problem in fundamentally similar ways via a common molecular principle.

DROSOPHILA Dscam1 PROTEINS REGULATE SELF-AVOIDANCE

A molecular basis for neurite self-avoidance was first uncovered through biochemical and genetic studies of *Drosophila Dscam1*. As much of this work has been reviewed previously (Zipursky et al. 2006, Millard & Zipursky 2008, Hattori et al. 2009, Schmucker & Chen 2009, Shi & Lee 2012), here we review the key findings in an abbreviated form and focus on more recent studies.

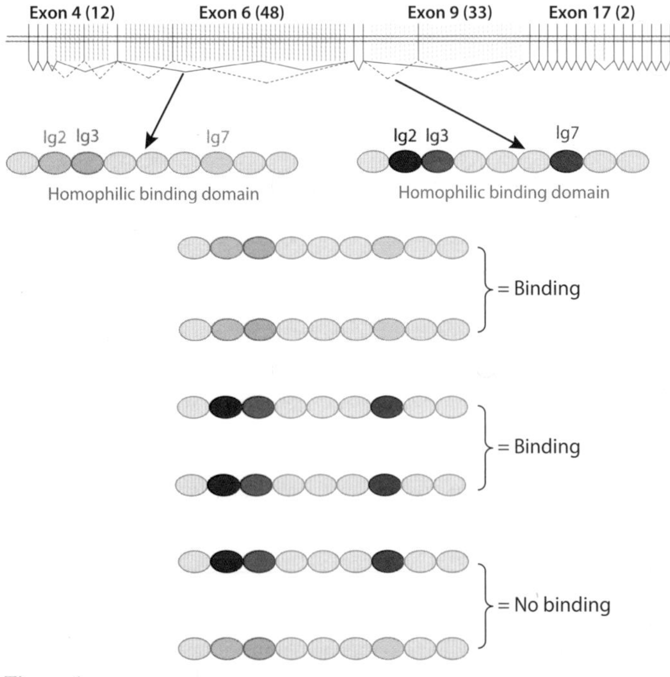

Figure 1

Alternative splicing of the *Drosophila Dscam1* gene generates multiple isoforms through alternative splicing. The *Dscam1* gene comprises four blocks of alternatively utilized exons. Each transcript contains a single exon from each block. Exons 4, 6, and 9 encode alternative versions of immunoglobulin (Ig) domain 2 (Ig2), Ig3, and Ig7, respectively, thus potentially generating 19,008 ectodomains. Two alternative versions of exon 17 encode different transmembrane domains. Alternative splicing gives rise to four different C-terminal tails (not shown). Biochemical studies support the view that 18,024 ectodomains exhibit isoform-specific homophilic binding. The binding properties of two isoforms are summarized. All three variable domains must match for binding to occur (see **Figure 2**).

Dscam1 Isoforms Exhibit Isoform-Specific Homophilic Recognition

Dscam1 encodes a family of cell surface receptors, each of which comprises an ectodomain with 10 Ig domains, 6 fibronectin type III repeats, a transmembrane domain, and an intracellular C-terminal tail. Alternative splicing of three large blocks of alternative exons arranged in a cassette-like fashion encodes variable Ig domains giving rise to as many as 19,008 isoforms that differ in one or more of three Ig domains, Ig2, Ig3, and Ig7 (**Figure 1**) (Schmucker et al. 2000). These diverse ectodomains lie at the center of the mechanism by which neurites of different cells discriminate between one another. In addition to these exons, a pair of alternatively spliced exons encode different transmembrane domains and some adjacent sequences; these differentially control localization of Dscam1 proteins to axons and dendrites (Wang et al. 2004, Shi et al. 2007, Yang et al. 2008). Finally, alternative splicing also generates different variants of the cytoplasmic domain, although how these variants contribute to Dscam1 function is not known (Yu et al. 2009).

The key to Dscam1 function lies in its isoform-specific homophilic binding, as has been demonstrated in vitro via studies ranging from traditional cell aggregation studies to a variety of biochemical assays (Wojtowicz et al. 2004, 2007; Matthews et al. 2007). Binding between isoforms relies on matching of the three variable Ig domains (i.e., Ig2 to Ig2, Ig3 to Ig3, and Ig7 to Ig7) (**Figure 2*a***). Typically, isoforms must match at all three variable domains for binding to occur, although some binding is seen between isoforms in which two domains match and the third is highly related. Isoform-specific binding has been shown for more than 100 isoforms. However, using analytical ultracentrifugation, investigators have assessed the binding affinities for only a few isoforms (Wu et al. 2012; G. Alsen & L. Shapiro, personal communication). The K_d values fall between 1 and 14 μM, a range similar to that observed for other cell adhesion molecules such as classical cadherins. Importantly, dimers of isoforms differing in only a single Ig2 domain were not detected (i.e., >500 μM), underscoring the importance of matching at all three Ig domains for binding to occur (Wu et al. 2012) (**Figure 2*a***).

Biochemical (Wojtowicz et al. 2004, 2007; Wu et al. 2012) and structural (Meijers et al. 2007, Sawaya et al. 2008) studies demonstrated that homophilic binding involves pairing of each variable domain with the same variable domain in its binding partner. The structure of a large fragment comprising all three variable domains (Ig1–8), which is sufficient to promote homophilic binding equivalent to the entire ectodomain, revealed that Dscam1 adopts a twofold symmetric S shape (**Figure 2*a***). Two tight turns, one between Ig2 and Ig3 and the other between Ig5 and Ig6, facilitate alignment of each variable domain in an antiparallel fashion; recognition between each cognate variable domain pair relies on a matching via charge and shape complementarity achieved through this antiparallel alignment of the two domains (for the Ig2 interface, see **Figure 2*b***). Extensive binding studies in which the binding specificity of each variant of a variable domain was tested for binding to all other variants of the same domain argued that 18,048 Dscam1 isoforms show isoform-specific homophilic recognition (Wojtowicz et al. 2007).

Dscam1 Isoform Expression in Individual Neurons

The expression of Dscam1 isoforms has been examined in only a few cell types, including mushroom body neurons located in the developing central brain, postmitotic neurons from the eye-antennal imaginal disc, and photoreceptors (R3/R4 and R7 subtypes). Thus, isoform expression analysis represents a major limitation in our current understanding of Dscam1 function. Neves et al. (2004) developed a custom microarray for probing all 93 alternative Dscam1 exons comprising the extracellular domain and examined expression of splice variants in populations of cells isolated by fluorescence activated cell sorting. These studies assessed the distribution of variable domains, but the distribution of different combinations of variable exons giving rise to distinct isoforms could not be determined. Although there was little difference in the distribution of exon 4 and exon 6 variants, some preferential utilization of exon 9 was observed. For instance, using clustering analysis, the researchers determined that the two populations of photoreceptors expressed significantly different exon 9 variants from one another and also from imaginal disc tissue. Different variants of exon 9 were also expressed in single neurons of the same cell type. These studies, together with those of Zhan et al. (2004) in mushroom body neurons, support the view that each neuron expresses between 10 and 50 different isoforms. Chess and colleagues characterized this pattern of Dscam1 as biased stochastic expression (Neves et al. 2004). These studies have led to the view that each neuron expresses a set of Dscam1 isoforms largely different from their neighbors. Additional studies are necessary to assess critically the expression of Dscam1 isoforms, particularly in dendritic arborization (da) neurons (see below), in which our understanding of how Dscam1 regulates self-avoidance is most advanced.

Sorting Out Dscam1 Function In Vivo

Dscam1 function has been studied using knock-out, knock-in, and isoform overexpression approaches in several different cell types in the central and peripheral nervous systems of *Drosophila* (Schmucker et al. 2000; Wang et al. 2002a; Wojtowicz et al. 2004; Zhan et al. 2004, Zhu et al. 2006; Chen et al. 2006). Much of this work has been reviewed extensively, so we summarize these findings only briefly, and focus primarily on the role of *Dscam1* in self-avoidance in da neurons where the most compelling case for its role in self-avoidance can be made. We then consider more recent studies in da neurons, addressing the requirement for homophilic recognition in vivo, the number of isoforms required for neurites to discriminate between self and nonself, and the interplay between self-avoidance and guidance factors. In the final section, we consider the role of Dscam proteins in regulating synaptic specificity in the visual system.

Da neurons are a group of sensory neurons in the embryo, larva, and adult with highly branched and extensively overlapping dendritic arbors. These neurons detect diverse stimuli and transmit this information from the body wall to the central nervous system. There are four classes of da neurons (classes I–IV) that differ markedly in dendritic complexity and axonal projection pattern (Grueber et al. 2002, 2007). The organization of da neuron dendrites

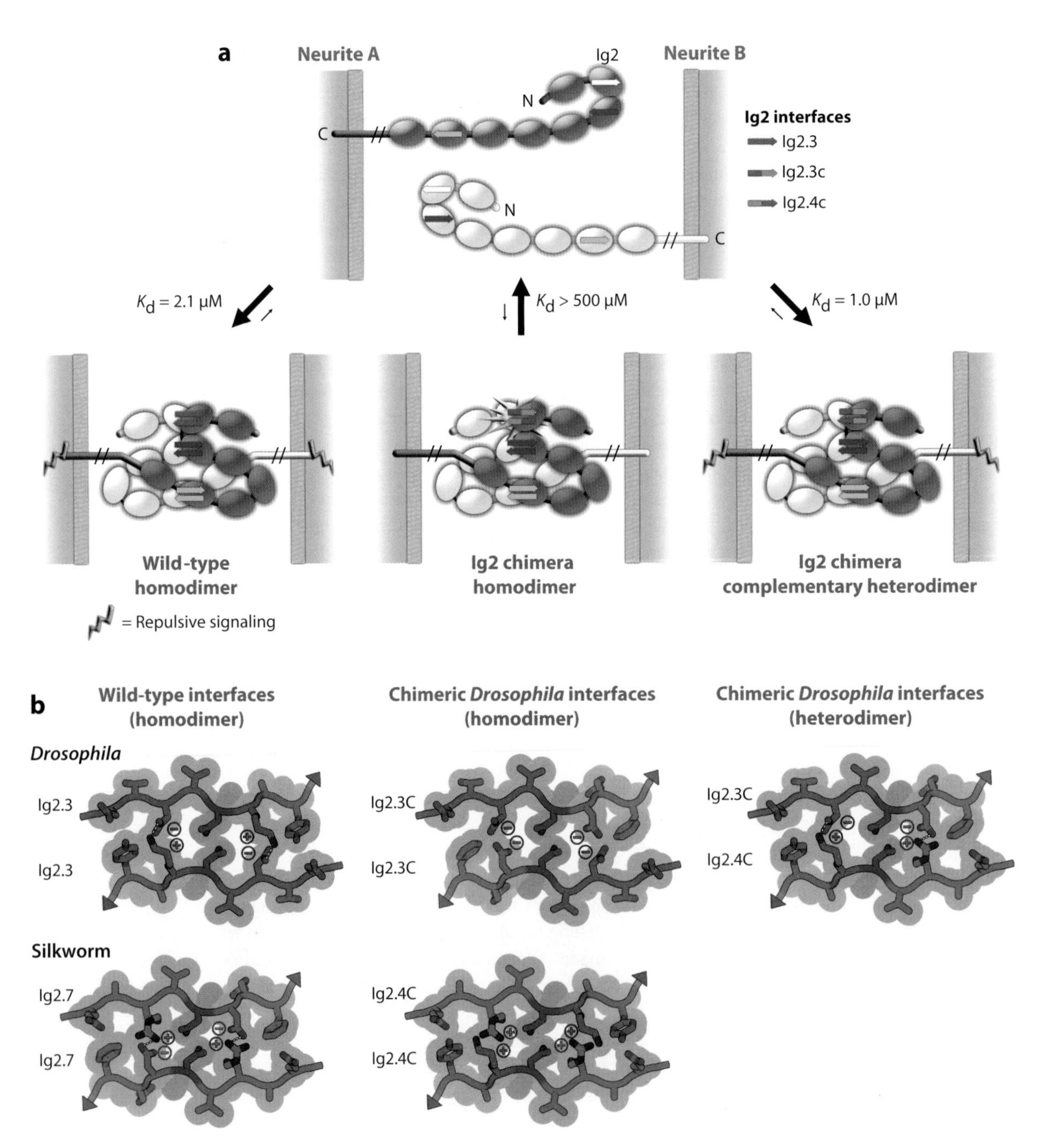

is defined by three basic features: (*a*) Dendrites of different classes of da neurons that share the same receptive field overlap (**Figure 3*c***). (*b*) Dendrites of the same class of cells typically do not overlap. For instance, the dendrites of class IV da neurons completely cover the body wall but do not overlap with each other (**Figure 4*a***). This phenomenon is typically referred to as tiling. (*c*) Dendrites from the same cell, or sister dendrites, avoid crossing each other and thus uniformly cover the receptive field (Grueber et al. 2002, 2003; Sweeney et al. 2002; Sugimura et al. 2003) (**Figures 3*c*** and **4*a***). All four classes of da neurons exhibit this self-avoidance property. Tiling and self-avoidance ensure that the body wall is covered completely and nonredundantly by dendrites of each cell class. Coexistence of dendrites of different cell classes ensures that redundant streams of sensory information are received and transmitted from each point on the body wall. Thus, self- versus nonself-discrimination is central to the patterning of da neuron dendritic fields.

Dscam1 Controls Self-Avoidance of Dendritic Arborization Neuron Dendrites

At the single cell level, da neuron dendrites are almost planar in organization, which makes the nearly perfect nonoverlapping pattern of self-dendrites very apparent (Grueber et al. 2002, Sweeney et al. 2002) and thus favorable for sorting out the molecular control of self-avoidance. The relative simplicity and planar arrangement of da neurons facilitate quantification of self-avoidance by measuring the incidence of self-dendrite crossing or fasciculation. Furthermore, researchers can quantify discrimination between self and nonself in different mutant backgrounds by assessing crossover between the processes of different da neurons (see below) (**Figure 3*a–d***). Loss of *Dscam1* in single da neurons (that are surrounded by normal neurons in genetically mosaic animals) causes severe cell-autonomous defects in self-avoidance (Hughes et al. 2007, Matthews et al. 2007, Soba et al. 2007). This phenotype was seen in all da sensory neurons examined, indicating a shared requirement for Dscam1 in self-recognition and repulsion. As a consequence of the self-avoidance phenotypes, coverage of territories was incomplete with unusually large gaps between branches (**Figure 3*b***).

The analysis of Dscam1's role in self-avoidance in da neurons has led to four key conclusions. First, no particular Dscam1 isoform is required for self-avoidance, as self-avoidance phenotypes were not observed in a series of alleles analyzed in which, in aggregate,

Figure 2

Binding between all three variable domains is required for Dscam1 function in vivo. (*a*) Schematic representation of interactions between isoforms with wild-type or chimeric ectodomains. Both biochemical and functional studies in vivo are summarized. Electron microscopic studies suggest that monomeric Dscam1 proteins exhibit a tight terminal horseshoe arrangement of the four N-terminal immunoglobulin (Ig) domains tethered to an otherwise highly flexible C terminus linked to the membrane (see Meijers et al. 2007). Matching at all three variable Ig domains promotes dimer formation through matching of each variable domain with its partner, leading to a marked conformational change to form the twofold symmetric S-shape structure. This, in turn, activates signaling to promote repulsion. Isoforms containing a chimeric Ig2 domain cannot pair and, hence, cannot form a twofold symmetric double S and are not active in self-avoidance. By contrast, if two opposed neurites display isoforms that can pair heterophilically through Ig2, then binding restores self-avoidance. K_d values were determined by analytical ultracentrifugation (see Wu et al. (2012)). (*b*) Chimeric Dscam1 proteins with altered binding specificity were generated via mutations at the Ig2 interface. (*Left column*) Interfaces from a wild-type *Drosophila* and Silkworm Ig2 domain are shown. These share the same symmetry center (*shaded*) but differ at other positions at the interface. (*Middle column*) The fly and silkworm interfaces were used as templates to generate chimeric interfaces of *Drosophila* Ig2.3 and Ig2.4 to generate Ig2.3C and Ig2.4C, respectively. Owing to charge incompatibility, a stable homodimeric interface does not form between monomers of each chimera. (*Right column*) On the basis of structural criteria, these chimeric interface segments were predicted to bind to each other (heterodimer). Biochemical and genetic studies indicate that the chimeric interface restores binding and biological activity.

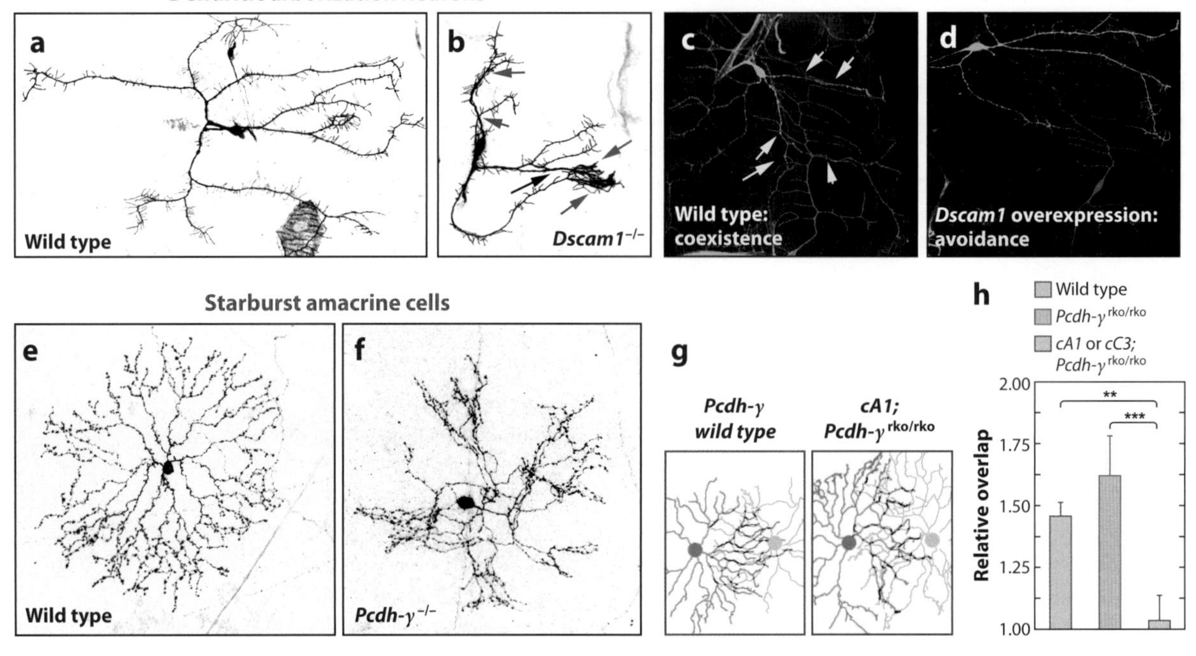

Figure 3

Dscam1 and *Pcdh-γ* genes regulate self-avoidance. (*a*, *b*) *Dscam1* is required for self-avoidance (reproduced with permission from Matthews et al. 2007). The branches of a single wild-type type III neuron in the body wall of a *Drosophila* larva seldom cross one another. By contrast, processes of *Dscam1* mutant cells cross and often fasciculate (*arrows*) with each other. (*c*, *d*) Ectopic expression prevents coexistence. (*c*) In wild type, cells of different classes of da neurons frequently share the same receptive field. Thus, dendrites of different cells cross one another: (*magenta*) type I dendritic arborization (da) neuron, (*green*) type III neuron. (*d*) Ectopic expression of the same isoform of Dscam1 in type I and type III neurons promotes repulsion between their dendrites. As a consequence, these two cells do not cover the same receptive field. (*e*, *f*) *Pcdh-γ* is required for self-avoidance. (*e*) Wild-type starburst amacrine cell (SAC) processes show little overlap, particularly in the regions proximal to the cell body. (*f*) SACs lacking all 22 Pcdh-γ isoforms are highly fasciculated. These images were kindly provided by J. Lefebvre and J.R. Sanes. (*g*, *h*) The relative overlap between the processes of SACs in cells ectopically expressing the same isoform is reduced compared with wild type (reproduced with permission from Lefebvre et al. 2012). (*g*) The relative overlap is shown between real images and between images in which one of the two cells is flipped (*left-hand image*; *blue bar in panel h*). The overlap is greater in the real image in the wild type than in the rotated images, arguing for a positive interaction between processes of the different SACs. A reduction in overlap is indicated between isoforms if both cells ectopically express the same Pcdh-γ isoform, in this case cA1 (*right-hand image*; *red bar in panel h*). An image between mutant cells lacking Pcdh-γ (*Pcdhg*$^{rko/rko}$) is not shown.

all 12 versions of exon 4, and thus all possible isoforms, were removed. Second, Dscam1 diversity is not required in individual neurons for self-avoidance; single arbitrarily chosen isoforms rescue the Dscam1 null self-avoidance phenotype. Third, Dscam1 diversity is essential for discriminating between self- and nonself-dendrites, because overexpression of single isoforms in neurons whose dendrites overlap in wild-type animals led to their segregation into nonoverlapping territories (Hughes et al. 2007, Matthews et al. 2007, Soba et al. 2007) (**Figure 3*d***). Fourth, expression of single Dscam1 molecules lacking most of their cytoplasmic tail prevented ectopic branch segregation and instead led to apparently stable adhesion between dendrites. How the cytoplasmic domain promotes repulsive interactions remains unknown. In sum, isoform identity between branches of the same neuron leads to recognition via the extracellular region and repulsion mediated by the intracellular tail.

As the Dscam1 isoforms expressed in different da neurons are likely to be different, dendrites of different da neurons do not inappropriately recognize nonself as self. Thus, Dscam1 proteins are required for self-avoidance and provide the molecular code by which neurites discriminate between self-dendrites and those of neighboring cells.

Homophilic Recognition Provides the Molecular Basis for Self-Avoidance

To test whether homophilic binding of Dscam1 isoforms is required for self-avoidance, Wu and coworkers generated homophilic binding defective mutants (Wu et al. 2012). In this study, pairs of mutant isoforms were engineered to lose homophilic binding and to simultaneously acquire heterophilic binding specificity to each other (**Figure 2**). Efforts to alter specificity focused on modifying alternative versions of exon 4, which encode variable Ig2 domains. By comparing 89 Ig2 interface segments of Dscam and Dscam1 proteins from 39 species, researchers generated pairs of chimeric Ig2 domains with desired specificity in vitro and introduced them into flies to test for their ability to support self-avoidance (Wu et al. 2012) (**Figure 2**).

The ability of single chimeric isoforms ($Dscam1^{single\ chimera}$) knocked into the endogenous Dscam1 locus to support self-avoidance was compared to single isoforms that retained homophilic binding properties ($Dscam1^{single}$). $Dscam1^{single}$ isoforms shared the same Ig3 and Ig7 domains with mutants, but they encoded wild-type Ig2 domains. Whereas $Dscam1^{single}$ isoforms were sufficient for self-avoidance in different classes of da neurons, $Dscam1^{single\ chimera}$ isoforms showed reduced ability to support self-avoidance. $Dscam1^{single\ chimera}$ isoforms were also compromised in their ability to induce ectopic repulsion between different da neuron dendrites when overexpressed. Nevertheless, nonmatching isoforms still retained some activity both when expressed from the endogenous locus and in overexpression experiments, arguing that weak binding between isoforms, matching at only two variable domains, is sufficient to induce low levels of repulsion (Wu et al. 2012). Weak binding may arise from avidity between oligomerized proteins on the cell surface (J.J. Flanagan & S.L. Zipursky, unpublished results), as these proteins showed no detectable binding affinity using analytical ultracentrifugation (i.e., a measure of *trans*-dimerization). In contrast to the weak ability of $Dscam1^{single\ chimera}$ isoforms to promote repulsion, coexpression of two complementary chimeric isoforms in da neurons with overlapping dendrites in wild type promoted ectopic repulsion between them, resulting in the establishment of nonoverlapping dendritic fields (Wu et al. 2012). Homophilic binding is also important for axonal branching in mushroom body neurons. Indeed, coexpression of complementary chimeras in single *Dscam1* null mutant mushroom body neurons rescued self-avoidance defects. Together, these results provide strong evidence that binding between Dscam1 isoforms on opposing surfaces of neurites of the same cell is required for self-avoidance (Wu et al. 2012).

Diversity at the Dscam1 Locus Is Essential for Discrimination Between Self- and Nonself-Neurites

Although analysis of mutants in which the ectodomain diversity was reduced to a single isoform through a knock-in replacement strategy clearly showed that Dscam1 diversity was required for discrimination between self and non-self (Hattori et al. 2007), researchers did not know how many isoforms were necessary to ensure that neurites do not inappropriately recognize and avoid nonself-neurites. To address this question, Hattori et al. (2009) took a genomic replacement strategy in which alternative exons were replaced with cDNAs encoding exons 7–11 (i.e., to fix exon 9 and limit the number of potential isoforms to 576) or exons 5–11 (i.e., to fix exons 6 and 9 and limit the number of potential isoforms to 12). By combining two different $Dscam1^{576\text{-}isoform}$ alleles, the number of isoforms was fixed at 1152; likewise, use of two different Dscam1$^{12\text{-isoform}}$ alleles fixed maximum diversity at 24 isoforms. In addition, two deletion alleles that removed blocks of 3 and 9 alternative exon 4s were used to limit the number of isoforms to 14,256 and 4,752, respectively. Thus, these alleles together allowed testing discrimination between self versus nonself in animals carrying no isoforms (null) and those carrying 12, 24, 576, 1,152, 4,752, and 14, 256 isoforms.

Self-recognition was normal with all knock-ins (i.e., from 12 or more isoforms). By contrast, inappropriate recognition of nonself as self was seen in knock-ins up to 1,152 isoforms; da neurons discriminated between self and nonself with 4,752 isoforms. Thus, thousands of Dscam1 isoforms are required for discrimination between self- and nonself-dendrites. Similar requirements for thousands of isoforms were observed in other neurons. For example, separation of mushroom body lobes, taken as a measure of self-avoidance between sister axon branches segregating to form the two lobes, depends on *Dscam1* and is defective when the number of potential isoforms falls to 1,152 (Hattori et al. 2009). Mathematical modeling based on the number of isoforms expressed in neurons and an estimate of the number of shared isoforms allowed (i.e., that do not activate inappropriate repulsion) supports

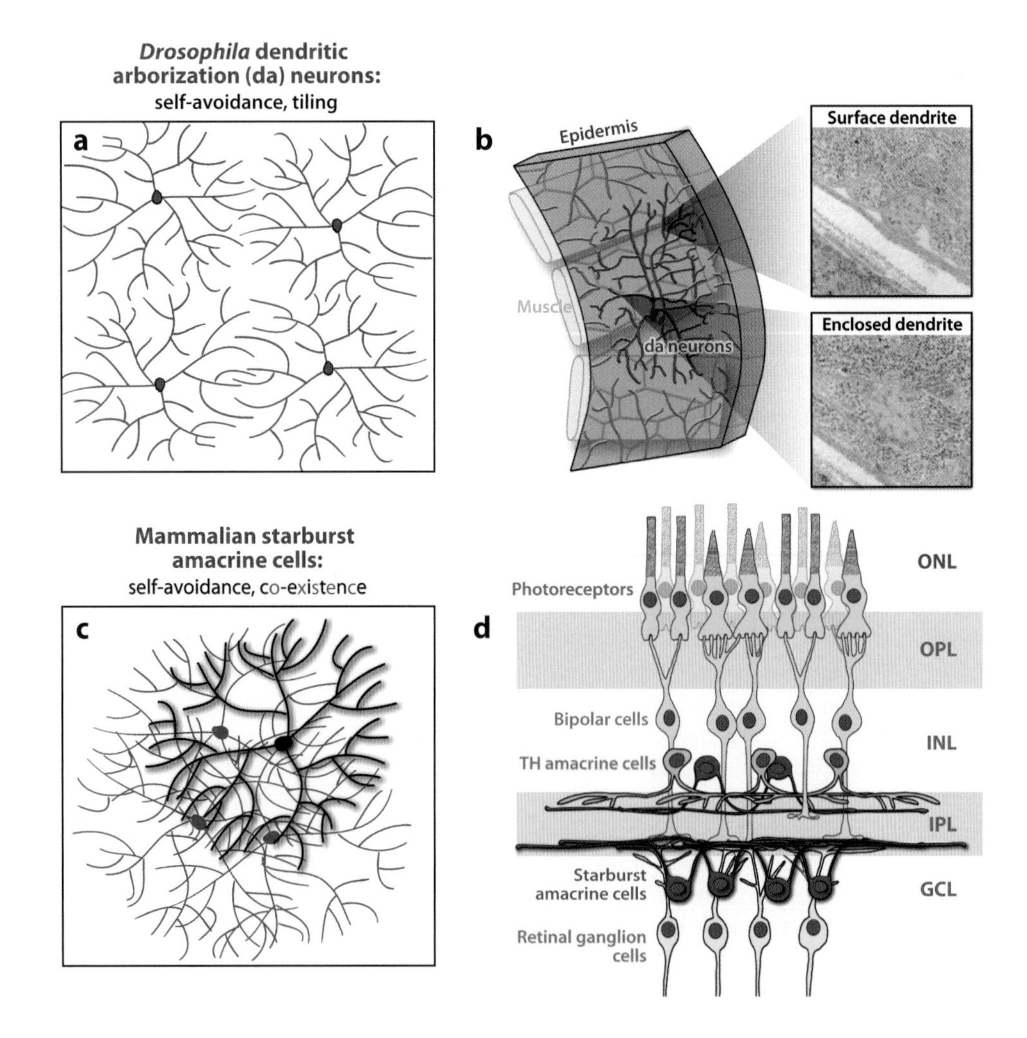

the conclusion that thousands of isoforms at a minimum are required to produce on the order of tens of distinct neuronal identities via a stochastic mechanism (Hattori et al. 2009, Forbes et al. 2011).

INTEGRATING SELF-AVOIDANCE WITH OTHER DENDRITIC PATTERNING PROCESSES

Self-Avoidance and Netrins

Self-recognition and repulsion occur as dendrites are responding to other extracellular cues that determine their targeting and position; thus these various signals must be integrated by developing dendritic branches. Indeed, studies on da neurons have identified interplay between Dscam1-mediated repulsion and other guidance signals. In *Dscam1* null mutant clones, not only did dendritic processes form clumps, but clumping also often occurred at stereotyped positions (Matthews et al. 2007). These clumps corresponded in position to Netrin-B expressing cells that promote specific dendritic targeting of da neurons (Matthews & Grueber 2011). In the absence of either Netrin-B or the netrin receptor Frazzled, dendrites fail to target their normal territory. By contrast, in the absence of Dscam1, dendrites show an opposite "hypertargeting" to the source of Netrin-B. Removal of both Frazzled and Dscam1 signaling causes strong reductions in dendritic fields, and sister dendrites become highly tangled. Thus, integration of Netrin/Frazzled and Dscam1 systems promotes spreading of dendrites from the cell body to more distant targets and prevents sister dendrites from targeting in unison to the source of guidance cues. In this manner, dendrite self-avoidance and guidance cues act together to promote territory coverage. However, the mechanism by which Dscam1-dependent repulsion is integrated with

Figure 4

Organization of dendritic arborization (da) neurons and starburst amacrine cells (SACs), contrasting self-avoidance, tiling, and coexistence. (*a*, *c*) Both da neurons and SACs elaborate highly branched dendrites that self-avoid within restricted layers. For da neurons, mechanisms that constrain dendrites to a common surface facilitate self-avoidance. To our knowledge, the detailed surface interactions between developing processes of SACs within a layer have not been described. (*a*) *Drosophila* da neurons show both self-avoidance and tiling. The schematic shows a field of class IV da neurons. The processes of these cells do not overlap with processes of the same cell (self-avoidance) or with process of da neurons of the same class (tiling). They do overlap with the processes of different classes of da neurons (not shown). (*b*) Da neurons arborize dendrites in a narrow plane sandwiched between muscle (*green*) and epidermis (*gray*). This planar organization promotes self-avoidance. Complex class IV arbors (*red*) and a single overlapping class I arbor (*blue*) are represented. Electron microscopic analysis shows that dendrites in cross section (*shaded red*) either reside on the basal surface of the epidermis (*shaded yellowish-brown in the two photo details*) in contact with the extracellular matrix (ECM) (*shaded blue in the two photo details*) or are enclosed within invaginations of the epidermal membrane. In wild type, tight association of dendrites with the ECM facilitates contact between neurites, thereby promoting repulsion. Envelopment within the epidermal membrane would physically isolate processes and thus antagonize direct contact between them, leading to crossing in a different three-dimensional plane (see text). (*c*) Mammalian SACs elaborate dendrites that show self-avoidance (*visible in black neuron*) but that coexist, or overlap, with many neighboring SACs (*red*). (*d*) Organization of mammalian retina. Rod and cone photoreceptors reside in the outer nuclear layer (ONL) and extend into the outer plexiform layer (OPL) where they synapse with rod and cone bipolar cells. Cell bodies of bipolar cells, horizontal cells (not shown), tyrosine-hydroxylase (TH)-immunoreactive cells (*orange*), and SACs (*red*) reside in the inner nuclear layer (INL). Bipolar and amacrine cells project to different layers of the inner plexiform layer (IPL). SACs reside both in the INL and in the ganglion cell layer (GCL), and they send processes to distinct narrow layers within the IPL, where they overlap extensively with neighboring SAC dendrites. Mammalian DSCAM and DSCAML1 regulate interactions between dendrites of many of these cell types. *Pcdh-*γ proteins regulate interactions between self-processes of SACs. Drawing in panel *d* modified with permission from Wassle (2004).

other guidance cues that give dendrites and axons their proper shape and trajectory remains unknown. Counterbalancing roles for Dscam1 could be particularly important in complex neuropils where dendrites respond to a multitude of extracellular cues during targeting and could conceivably help to explain clumping and fasciculation phenotypes seen in several different contexts in *Dscam1* mutant brains.

Self-Avoidance and Integrins

An integrin-dependent pathway that constrains da dendrites to a common surface plays a key role in self-avoidance (Han et al. 2012, Kim et al. 2012). Without such a constraint, for instance, in a three-dimensional field, arbors might simply inch past one another. Integrin receptors, which mediate interactions between cell surfaces and the extracellular matrix, normally restrict da dendrites to a single plane (Han et al. 2012, Kim et al. 2012). In the absence of neuronal integrins, some arbors become enclosed within segments of the overlying epidermis and dendrites show excessive overlap (see **Figure 4*b***). By contrast, in response to co-overexpression of α- and β-integrin subunits, more dendrites attach to the extracellular matrix, thus decreasing enclosure. Other components of this two-dimensional restriction pathway include the Tricornered kinase and Furry (Emoto et al. 2004) and components of the TOR complex 2 (Koike-Kumagai et al. 2009, Han et al. 2012). The enclosure of dendrites prevents Dscam1-dependent repulsion, because arbors do not come into contact and, as a consequence, are unable to self-repel. These mechanisms may also function within the neuropil. Here, processes may be constrained to grow along the surface of other cells, such as glia, thereby enforcing interaction between neurites of the same cell.

Self-Avoidance and Synapses

Using electron micrographic studies of developing synapses in the fly visual system, investigators determined that repulsive interactions between prospective postsynaptic elements of the same cell regulate the correct assembly of synapses between photoreceptor neurons and their postsynaptic targets. Photoreceptor cells in the fly retina form multiple contact synapses, or tetrads, with four distinct elements, lamina interneurons L1 and L2, and two other elements (either from a glial cell, a lamina L3, or a lamina amacrine cell) (Meinertzhagen & O'Neil 1991) (**Figure 5*a*,*b***). Each lamina microcircuit contains only one L1 and one L2 cell. Each cell forms many small dendritic processes, which contribute postsynaptic elements to multiple contact synapses on both the same and six different photoreceptor cell axons (**Figure 5*b***). Pairing ensures that L1 and L2 receive matched inputs from photoreceptor neurons. Such pairing appears to arise from synaptic exclusion and refinement of intermingled pairs of L1/L1 and L2/L2 seen during synapse assembly (Frohlich & Meinertzhagen 1983) (**Figure 5*b*,*c***), raising the possibility of a self-avoidance mechanism at work to ensure proper tetrad pairing.

Tetrad composition is normal in *Dscam1* mutants, and *Dscam2* mutant retinas exhibit a small fraction of tetrads with two L2 elements (Millard et al. 2010). In the absence of both Dscam1 and Dscam2, however, tetrad composition is randomized, indicating that Dscam1 and Dscam2 act redundantly to exclude inappropriate synaptic pairing at tetrads (Millard et al. 2010). This phenotype may arise through a defect in self-avoidance, such that during normal development L1-L1 and L2-L2 pairs are eliminated by Dscam1/Dscam2-dependent repulsion leaving L1-L2 pairs at mature tetrad synapses (**Figure 5*d***). As multiple contact synapses are seen throughout the fly visual system, and indeed elsewhere within the fly central nervous system, Dscam-mediated self-avoidance likely contributes in this way to the precise assembly of synaptic connectivity.

Summary

On the basis of classical work largely conducted in invertebrates, researchers proposed that self-avoidance plays a widespread role in patterning neural circuits. Nevertheless, the phenomenon received little attention, perhaps in large part owing to the difficulty of studying the development of neuronal arbors, particularly in the central nervous system, at the level of single cells. Studies in *Drosophila* over the past decade have not only underscored the importance of self-avoidance in the developing nervous system but also uncovered an underlying molecular strategy regulating this process. In this system, neurons acquire a unique cell surface identity through the expression of different combinations of a vast array of homophilic cell recognition molecules. This enables each neuron to distinguish its own neurites from the neurites of other neurons that it encounters. Binding between identical isoforms signals self and promotes repulsion of processes away from one another.

Importantly, Dscam1 endows each neuron, whether of the same or a different cell type, with a unique identity and thus facilitates neuronal individualization. One consequence of the vast number of different identities is coexistence, either among dendrites of different cell types that must cover overlapping receptive territories or in contexts in which neurites of a single cell must distinguish between sister neurites and neurites of other cells of the same cell type. This latter scenario arises in mushroom body axons in the fly central brain (Wang et al. 2002a, Zhan et al. 2004, Hattori et al. 2009) and, as we discuss below, during dendritic patterning of starburst amacrine cells (SACs) in the vertebrate retina.

SELF-AVOIDANCE BUT NOT SELF/NONSELF-DISCRIMINATION IS MEDIATED BY VERTEBRATE DSCAMS

Self-avoidance has only recently been explored in vertebrate brain development and largely in the context of patterning neurites in the inner plexiform layers (IPLs) of the vertebrate retina. Studies from Burgess and colleagues have demonstrated that mouse DSCAM and DSCAM-like-1 (DSCAMLl) mediate self-avoidance (Fuerst et al. 2008, 2009). By contrast to *Drosophila*, however, mouse DSCAMs are typical cell surface molecules, lacking the massive alternative splicing their fly cousin Dscam1 undergoes. So although DSCAMs may retain a conserved function in vertebrates, it is unclear, in the absence of molecular diversity, how neurites of different cells of the same type would discriminate between self and nonself. More recent studies demonstrated that mice use a different family of cell recognition molecules, clustered Pcdhs, in a fly Dscam1-like strategy to regulate self-avoidance. In the following sections, we first discuss the role of mammalian DSCAM proteins in regulating the interaction between neurites and then describe the role of Pcdhs in regulating self-avoidance.

DSCAM Regulates Interactions Between Dendrites of the Same Cell Classes in the Mammalian Retina

The first clue that DSCAM proteins regulate interactions between processes of the same cell in mammals came from studies on retinal amacrine cells. These cells arborize within a layered neuropil called the IPL where they make synapses with both bipolar neuron axons and the dendrites of retinal ganglion cells (RGCs) (**Figure 4*d***). DSCAM is selectively expressed in two of approximately two dozen classes of amacrine cells whose cell bodies form evenly spaced mosaics across the retina (Fuerst et al. 2008): The tyrosine-hydroxylase-positive and bNOS-expressing amacrine cells arborize in the S1 and S3 layers of the IPL, respectively. Loss of DSCAM leads to fasciculation of tyrosine-hydroxylase and bNOS amacrine cell dendrites and abnormal cell body spacing. In contrast to the selective expression of DSCAM in segregated amacrine cell populations, DSCAM is expressed in the dendrites of virtually all RGCs that also extend into the IPL (Fuerst et al. 2009). Removal of DSCAM results in the fasciculation of dendrites with sister dendrites as well as selective fasciculation with the dendrites of cells of the same class of RGCs. These

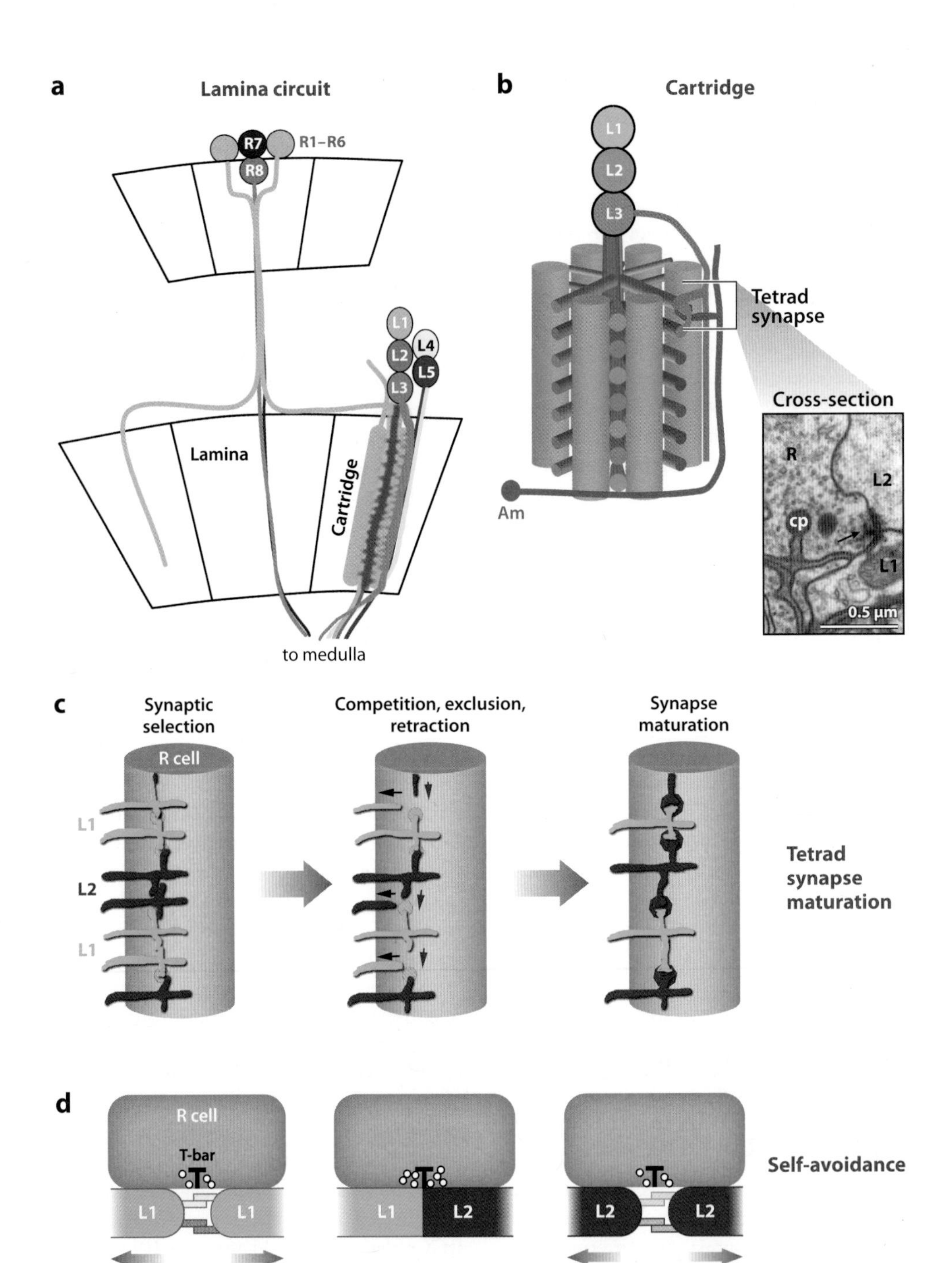
a
Lamina circuit
R7
R1–R6
R8
L1
L2
L4
L5
L3
Lamina
Cartridge
to medulla
b
Cartridge
L1
L2
L3
Tetrad synapse
Cross-section
R
L2
cp
L1
0.5 µm
Am
c
Synaptic selection
Competition, exclusion, retraction
Synapse maturation
R cell
L1
L2
L1
Tetrad synapse maturation
d
R cell
T-bar
L1
L1
L1
L2
L2
L2
Self-avoidance
Matching Dscams (1 and 2)
Repulsion
Nonmatching Dscams
Synapse formation
Matching Dscams (1 and 2)
Repulsion

findings led to the conclusion that DSCAM may function primarily to mask cell-type-specific adhesive interactions between dendrites of the same cell class, rather than to actively promote repulsion between them. In this way, DSCAM acts to negate cell-type-specific interactions that are promoted by other, as-yet unidentified, recognition molecules. Such a mechanism would not, however, allow dendrites of one cell to distinguish between self-dendrites and dendrites belonging to other cells of the same cell or different cell types. Thus, although DSCAM mediates self-avoidance for the amacrine cell subclasses expressing it and all RGC dendrites, it does not allow processes of one cell to discriminate between self-processes and the processes of other cells.

The expression of DSCAML1, versus DSCAM, is far more restricted within the mouse retina. It is expressed in cells in the rod circuit, including rod, rod bipolar cells (RBCs), and AII amacrine cells, but it does not appear to be expressed in RGCs (Fuerst et al. 2009). Like DSCAM, loss of DSCAML1 results in cell clustering and neurite fasciculation, a function consistent with self-avoidance. For instance, the dendrites of RBCs appear to be highly fasciculated, and the AII amacrine cells form clumps (Fuerst et al. 2009). Though largely normal synaptic structures were observed between cells in the rod circuit that lacked DSCAMLl, some abnormalities were also observed. For instance, there was a marked increase in synaptic vesicles in RBC/AII/A17 synapses suggestive of a defect in exocytosis. In addition, there was a nearly fourfold decrease in synapses between rods and RBCs, which may be a consequence of the ~40% decrease in average length of RBC dendrites, their fasciculation, or both. These defects in RBC dendrite structure and organization may prevent normal extension of RBC dendrites to their rod presynaptic targets. Thus, DSCAML1 also appears to regulate self-avoidance. As with DSCAM, DSCAML1 defects are also associated with an increase in cell number. Although experiments on DSCAM support the view that the defects in the organization of neurites are independent

Figure 5

Dscam1 and Dscam2 play redundant roles in regulating the composition of postsynaptic elements at multiple contact synapses. (*a*) The R1–R6 photoreceptor neurons in the *Drosophila* retina form synapses with neurons in the lamina within microcircuits called cartridges. The dendrites of target neurons, as is typical of arthropods, extend from the proximal region of the axon. There is only a single cell of each L1–L5 neuron in a cartridge. (*b*) R1–R6 axons form tetrad synapses. Each R cell axon (*gray*) forms ~40 presynaptic release sites adjacent to four postsynaptic elements, including an invariant pair from L1 and L2 and a variable pair, in the example shown here an L3 neuron and an amacrine cell (amacrine cells form synapses in multiple cartridges) (not shown). Note each L1 and L2 elaborates many dendrites, yet there are no L1 or L2 pairs. The invariant pairing of L1 and L2 is thought to be important for motion detection. (*Inset*) A transmission electron microscopy section through a tetrad synapse shows the T-bar (*arrow*), thus indicating the presynaptic release site or active zone, and the invariant L1 and L2 postsynaptic elements. (*c*) Model for the formation of invariant pairs of L1 and L2 postsynaptic elements based on studies by Meinertzhagen et al. (2000). The multiple L1 (*green*) and L2 (*blue*) dendrites are each from the same L1 and L2, respectively. Pink circle indicates the prospective presynaptic sites. At some sites, two L1 or two L2 processes pair transiently. These are resolved through interactions between processes leading to the establishment of mature synapses (*red dot* indicates mature presynaptic site) resulting in the invariant pairing of L1 and L2. Black and red arrows indicate retraction and extension, respectively, of postsynaptic elements to establish L1/L2 pairing. The other postsynaptic elements in the tetrad are not shown. (*d*) Model of Dscam1 and Dscam2 function in regulating tetrad assembly. Dscam1 and Dscam2 are required in a redundant fashion to prevent inappropriate pairing through self-avoidance. Orange and yellow bars represent different Dscam1 isoforms expressed on the surface of L1 and L2. Light blue and purple bars indicate different Dscam2 isoforms (either the A or B isoform) proposed to be expressed on L1 and L2. Because Dscam1 and Dscam2 homophilic binding activates repulsion, only the L1/L2 pairing is permitted. Drawings and inset image reproduced with permission from Millard et al. (2010).

of the cell-death phenotypes (Keeley et al. 2012), whether the morphological phenotypes observed in DSCAML1 retinas are an indirect consequence of this increase has not been addressed.

In summary, Burgess and coworkers have proposed that the mammalian DSCAMs promote an anti-adhesive function, preventing the adhesion between processes of cells of the same type rather than actively promoting repulsion between them (Fuerst & Burgess 2009, Garrett et al. 2012b). Although it remains unclear what a general antiadhesive function might look like at a mechanistic level, in the absence of diversity, mammalian DSCAMs would not provide cells with the ability to distinguish between their own processes and the processes of all other cells, including processes from cells of the same type. As we describe in the next section, clustered Pcdhs provide SACs with precisely this function, thereby utilizing a strategy remarkably similar to *Drosophila* Dscam1, albeit in the guise of a different protein superfamily.

PCDH-γS REGULATE SELF-AVOIDANCE

Diversity of Pcdh-γs Provides a Basis for the Identity of Vertebrate Neurons

Just prior to the discovery of the diversity in the *Drosophila Dscam1* locus, two studies in vertebrates led to the finding of a large cluster of closely related putative cell recognition molecules of the cadherin superfamily. In 1998, Yagi and colleagues reported the discovery of eight structurally related transmembrane proteins of the cadherin superfamily termed cadherin-related neuronal receptors (Kohmura et al. 1998). These proteins are expressed broadly in the developing nervous system and are localized to synapses. Wu & Maniatis (1999) identified the genomic loci encoding these proteins and demonstrated that these were tightly linked and organized within three distinct clusters. These are now referred to as the clustered Pcdhs with each cluster referred to as Pcdh-α (including the cadherin-related neuronal receptors identified by Yagi and colleagues) and the related Pcdh-β and Pcdh-γ proteins (**Figure 6**). The number of Pcdh isoforms varies between different vertebrate species, but in aggregate, there are typically on the order of 50 isoforms. In the mouse, for instance, there are 14 Pcdh-α, 22 Pcdh-β, and 22 Pcdh-γ isoforms. Studies of clustered Pcdhs indicate diverse roles in the nervous system, including promotion of neuronal survival (Wang et al. 2002c, Lefebvre et al. 2008, Prasad et al. 2008), synaptic development (Weiner et al. 2005), axon target recognition (Prasad & Weiner 2012), and dendritic arborization (Garrett et al. 2012a).

Clustered Pcdhs have been found only in vertebrates; conversely, vertebrate DSCAMs lack the diversity of their arthropod counterparts. Such findings raised the intriguing possibility that these protein families could play analogous roles in the developing nervous system. Although both clustered Pcdhs and *Dscam1* genes generate families of proteins with diverse ectodomains joined to a common cytoplasmic domain, the organization and mode of generating clustered Pcdh and fly Dscam1 diversity are markedly different. Pcdh diversity is largely generated by alternative promoter choice, as opposed to alternative splicing (Tasic et al. 2002, Wang et al. 2002b). Alternative ectodomains are encoded by single large exons arranged in a head to tail fashion (see **Figure 6**), and each exon is preceded by a transcription start site. For the Pcdh-α and Pcdh-γ clusters, each ectodomain-encoding exon is spliced to a common cytoplasmic tail found within the 3′ end of the cluster. By contrast, each Pcdh-β is encoded in its entirety by a single exon. In addition, the Pcdh-α and Pcdh-γ clusters contain five exons that are highly related to one another; these are referred to as C1 and C2 within the α cluster and as C3–C5 within the γ cluster. Molecular studies argue that variable Pcdh-α and Pcdh-γ are expressed in a largely stochastic fashion (Kaneko et al. 2006). Single-cell polymerase chain reaction analyses were

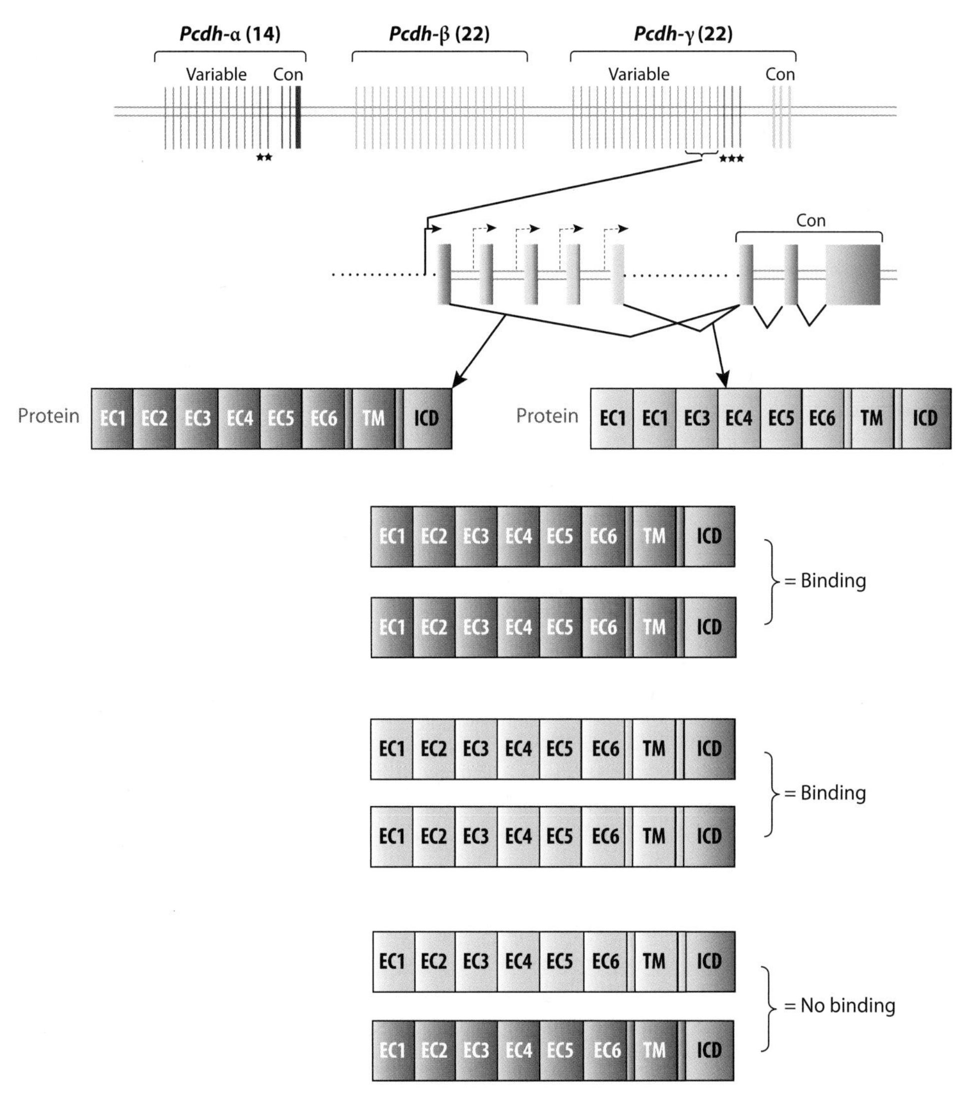

Figure 6

Alternative transcription initiation generates multiple isoforms of *Pcdh-γ*. Three clusters of protocadherin (Pcdh) genes are tightly linked to one another. Fourteen *Pcdh-α* and 22 *Pcdh-γ* alternative exons encode different ectodomains. These each comprise six extracellular cadherin (EC1–EC6) repeats and a single transmembrane (TM) domain. These variable ectodomain-encoding exons are spliced to a common C-terminal tail [intracellular domain (ICD)]. Each Pcdh-β exon encodes an entire protein isoform tethered to the membrane. Each exon is preceded by a transcription initiation site. Like alternative isoforms of Dscam1, different Pcdhγ ectodomains exhibit isoform-specific homophilic binding. Pcdhγ proteins form tetramers (not shown) that exhibit preferential binding for tetramers of the same composition. Abbreviation: Con, conserved.

undertaken to study Pcdh-α and Pcdh-γ expression in Purkinje cells [Pcdh-β expression was also examined, which was noted as unpublished results discussed in Yagi (2012)]. Each Purkinje cell expresses all C1–C5 isoforms and different combinations of Pcdh-αs (~2 of 12), Pcdh-βs (~4 of 22) (Yagi 2012), and Pcdh-γs (~4 of 19). Thus, a vast number of different combinations of Pcdhs is possible, and as with Dscam1, this would have the potential to endow neurons with a unique cell surface identity.

Pcdh-γ Isoforms Exhibit Isoform-Specific Homophilic Recognition

Although the clustered Pcdhs were discovered in the late 1990s, compelling evidence for discrete binding specificities of different isoforms was not uncovered until 2010. By devising an elegant and quantitative assay and a suitable cell line for protein expression, Schreiner & Weiner (2010) convincingly demonstrated that Pcdhs promote isoform-specific homophilic recognition. In their assay, the activity of seven Pcdh-γs was assessed. Of critical importance, of the many cell lines tested in various Pcdh-γ transfection experiments, only the human leukemia cell line K562 supported Pcdh-γ-dependent cell aggregation. To compare homophilic and heterophilic binding, the authors further assessed interactions between two cell populations, one expressing an isoform tagged with a hemagglutinin (HA) epitope and the other expressing either the same or a different isoform along with the enzyme β-galactosidase. These cells were mixed and adsorbed to an anti-HA antibody; bound β-galactosidase enzymatic activity was then measured.

The studies of Schreiner & Weiner (2010) led to several important findings. First, each of the seven Pcdh-γ isoforms tested exhibited isoform-specific homophilic recognition. Second, the Pcdh-γ isoforms oligomerize and appear to form tetramers preferentially. Third, Pcdh-γs form hetero-oligomers in cells cotransfected with different isoforms, and cells expressing the same combination of isoforms preferentially bind to themselves rather than to cells expressing different combinations (even if some isoforms are shared between the cells). Studies also suggest that Pcdh-γ isoforms may associate with Pcdh-α and Pcdh-β isoforms. Should these Pcdhs share similar biochemical properties, the recognition specificities would be further diversified.

Thus, while the number of Pcdh isoforms pales in comparison to the number of Dscam1 isoforms, hetero-oligomerization of Pcdhs markedly increases the number of discrete binding specificities encoded by the locus. With the shared properties of clustered Pcdhs and Dscam1—isoform diversity, homophilic binding specificity, and stochastic combinatorial expression—Pcdhs have emerged as attractive candidates for regulating self-avoidance in vertebrates.

Pcdh-γs Regulate Self-Avoidance and Self/Nonself-Discrimination in the Mouse Brain

Lefebvre et al. (2012) set out to assess whether *Pcdh-γ* genes regulate self-avoidance in the mouse retina. They focused on the stereotyped relationship between dendrites of the same SACs and between dendrites of different SACs. SACs arborize extensively within specific laminae (designated S2 and S4) in the IPL of the mouse retina. SAC dendrites make contacts with dendrites of other SACs and form transient contacts with sister dendrites at early postnatal stages. These self-contacts are promptly eliminated to generate a radial arbor with minimal overlap among self-dendrites and no autapses. Thus, by these criteria, SACs distinguish between self and nonself.

By using the Cre-Lox system, Lefebvre and colleagues selectively deleted all 22 variable domains of the Pcdh-γ locus in the developing retina and demonstrated that SAC processes from the same cell crossed each other extensively, effects that looked remarkably similar to the removal of Dscam1 from da neurons (see **Figure 3*a,b,e,f***) (Lefebvre et al. 2012). They also demonstrated that single arbitrarily chosen isoforms rescued the mutant phenotype and that expression of the same isoform in neighboring SACs reduces the overlap between them (**Figure 3*g,h***). Furthermore, although single isolated wild-type SACs exhibited self-avoidance in vitro, dendrites from single mutant SACs formed clumps. These data indicate that Pcdh-γs act in a cell-autonomous fashion and that they mediate interactions between Pcdh-γ isoforms on dendrites of the same cell. These data further argue that, as with Dscam1, self-avoidance in SACs does not rely on a specific isoform, but rather requires that isoform use differs among neighboring cells.

Genetic analyses have also revealed that subsets of the Pcdh-γs, the A/B Pcdh-γs, and C Pcdh-γs have diverse functions. Deletion of all 22 Pcdh-γs led to massive cell death. Self-avoidance is not an indirect effect of this cell death because cell death, but not the self-avoidance phenotypes, was suppressed in $Bax^{-/-}$ mice (Lefebvre et al. 2012). Furthermore, this cell death was seen in animals lacking C3–C5 but not in animals with deletions of A1–A3 (Chen et al. 2012). Thus, these data argue that variable Pcdh-γs are essential for self-avoidance in SACs, whereas the constant domains are required for cell survival. Finally, Lefebvre et al. (2012) also demonstrated that Pcdh-γs promote self-avoidance between dendrites of cerebellar Purkinje cells. Future experiments will be important to determine whether the Pcdhα and Pcdhβ proteins also contribute to self-avoidance in these or other neurons. In summary, clustered Pcdh-γs allow neurites of different cells to discriminate between self and nonself, thereby promoting self-avoidance.

CONCLUDING REMARKS

As proposed approximately 30 years ago, studies over the past 5–10 years have demonstrated that self-avoidance plays a widespread role in patterning axons and dendrites in both vertebrates and invertebrates. A common molecular strategy underlies this process, albeit through the utilization of different large families of cell recognition molecules. In each system, neurons express multiple isoforms of isoform-specific homophilic cell recognition molecules. Although the mechanisms by which specific isoforms are selected for expression within each cell remain unclear, each cell appears to express a unique population of isoforms endowing it with a cell surface identity significantly different from its neighboring neurons. This difference and the isoform-specific binding properties of different isoforms endow each neurite with the ability to discriminate between a sister neurite and the neurite of other cells, even cells of the same type. Through activation of a common signaling domain, via homophilic binding, sister neurites are repelled from one another. Given the complexity of both the insect and mammalian nervous systems, the diverse and highly specific binding properties of different isoforms may also contribute to other aspects of neuronal patterning in different developmental contexts.

DISCLOSURE STATEMENT

The authors are not aware of any affiliations, memberships, funding, or financial holdings that might be perceived as affecting the objectivity of this review.

LITERATURE CITED

Brose K, Bland KS, Wang KH, Arnott D, Henzel W, et al. 1999. Slit proteins bind Robo receptors and have an evolutionarily conserved role in repulsive axon guidance. *Cell* 96:795–806

Chen BE, Kondo M, Garnier A, Watson FL, Puettmann-Holgado R, et al. 2006. The molecular diversity of Dscam is functionally required for neuronal wiring specificity in *Drosophila*. *Cell* 125:607–20

Chen WV, Alvarez FJ, Lefebvre JL, Friedman B, Nwakeze C, et al. 2012. Functional significance of isoform diversification in the protocadherin gamma gene cluster. *Neuron* 75:402–9

Cheng HJ, Nakamoto M, Bergemann AD, Flanagan JG. 1995. Complementary gradients in expression and binding of ELF-1 and Mek4 in development of the topographic retinotectal projection map. *Cell* 82:371–81

Drescher U, Kremoser C, Handwerker C, Loschinger J, Noda M, Bonhoeffer F. 1995. In vitro guidance of retinal ganglion cell axons by RAGS, a 25 kDa tectal protein related to ligands for Eph receptor tyrosine kinases. *Cell* 82:359–70

Emoto K, He Y, Ye B, Grueber WB, Adler PN, et al. 2004. Control of dendritic branching and tiling by the Tricornered-kinase/Furry signaling pathway in *Drosophila* sensory neurons. *Cell* 119:245–56

Forbes EM, Hunt JJ, Goodhill GJ. 2011. The combinatorics of neurite self-avoidance. *Neural Comput.* 23:2746–69

Frohlich A, Meinertzhagen IA. 1983. Quantitative features of synapse formation in the fly's visual system. I. The presynaptic photoreceptor terminal. *J. Neurosci.* 3:2336–49

Fuerst PG, Bruce F, Tian M, Wei W, Elstrott J, et al. 2009. DSCAM and DSCAML1 function in self-avoidance in multiple cell types in the developing mouse retina. *Neuron* 64:484–97

Fuerst PG, Burgess RW. 2009. Adhesion molecules in establishing retinal circuitry. *Curr. Opin. Neurobiol.* 19:389–94

Fuerst PG, Koizumi A, Masland RH, Burgess RW. 2008. Neurite arborization and mosaic spacing in the mouse retina require DSCAM. *Nature* 451:470–74

Garrett AM, Schreiner D, Lobas MA, Weiner JA. 2012a. γ-Protocadherins control cortical dendrite arborization by regulating the activity of a FAK/PKC/MARCKS signaling pathway. *Neuron* 74:269–76

Garrett AM, Tadenev AL, Burgess RW. 2012b. DSCAMs: restoring balance to developmental forces. *Front. Mol. Neurosci.* 5:86

Grueber WB, Jan LY, Jan YN. 2002. Tiling of the *Drosophila* epidermis by multidendritic sensory neurons. *Development* 129:2867–78

Grueber WB, Ye B, Moore AW, Jan LY, Jan YN. 2003. Dendrites of distinct classes of *Drosophila* sensory neurons show different capacities for homotypic repulsion. *Curr. Biol.* 13:618–26

Grueber WB, Ye B, Yang CH, Younger S, Borden K, et al. 2007. Projections of *Drosophila* multidendritic neurons in the central nervous system: links with peripheral dendrite morphology. *Development* 134:55–64

Han C, Wang D, Soba P, Zhu S, Lin X, et al. 2012. Integrins regulate repulsion-mediated dendritic patterning of *Drosophila* sensory neurons by restricting dendrites in a 2D space. *Neuron* 73:64–78

Hattori D, Chen Y, Matthews BJ, Salwinski L, Sabatti C, et al. 2009. Robust discrimination between self and non-self neurites requires thousands of Dscam1 isoforms. *Nature* 461:644–48

Hattori D, Demir E, Kim HW, Viragh E, Zipursky SL, Dickson BJ. 2007. Dscam diversity is essential for neuronal wiring and self-recognition. *Nature* 449:223–27

Hedgecock EM, Culotti JG, Hall DH. 1990. The *unc-5*, *unc-6*, and *unc-40* genes guide circumferential migrations of pioneer axons and mesodermal cells on the epidermis in *C. elegans*. *Neuron* 4:61–85

Hughes ME, Bortnick R, Tsubouchi A, Baumer P, Kondo M, et al. 2007. Homophilic Dscam interactions control complex dendrite morphogenesis. *Neuron* 54:417–27

Ishii N, Wadsworth WG, Stern BD, Culotti JG, Hedgecock EM. 1992. UNC-6, a laminin-related protein, guides cell and pioneer axon migrations in *C. elegans*. *Neuron* 9:873–81

Kaneko R, Kato H, Kawamura Y, Esumi S, Hirayama T, et al. 2006. Allelic gene regulation of Pcdh-α and Pcdh-γ clusters involving both monoallelic and biallelic expression in single Purkinje cells. *J. Biol. Chem.* 281:30551–60

Keeley PW, Sliff B, Lee SC, Fuerst PG, Burgess RW, et al. 2012. Neuronal clustering and fasciculation phenotype in Dscam- and Bax-deficient mouse retinas. *J. Comp. Neurol.* 520:1349–64

Kennedy TE, Serafini T, de la Torre JR, Tessier-Lavigne M. 1994. Netrins are diffusible chemotropic factors for commissural axons in the embryonic spinal cord. *Cell* 78:425–35

Kidd T, Bland KS, Goodman CS. 1999. Slit is the midline repellent for the Robo receptor in *Drosophila*. *Cell* 96:785–94

Kim ME, Shrestha BR, Blazeski R, Mason CA, Grueber WB. 2012. Integrins establish dendrite-substrate relationships that promote dendritic self-avoidance and patterning in *Drosophila* sensory neurons. *Neuron* 73:79–91

Kohmura N, Senzaki K, Hamada S, Kai N, Yasuda R, et al. 1998. Diversity revealed by a novel family of cadherins expressed in neurons at a synaptic complex. *Neuron* 20:1137–51

Koike-Kumagai M, Yasunaga KI, Morikawa R, Kanamori T, Emoto K. 2009. The target of rapamycin complex 2 controls dendritic tiling of *Drosophila* sensory neurons through the Tricornered kinase signalling pathway. *EMBO J.* 28:3879–92

Kolodkin AL, Matthes DJ, Goodman CS. 1993. The *semaphorin* genes encode a family of transmembrane and secreted growth cone guidance molecules. *Cell* 75:1389–99

Kramer AP, Kuwada JY. 1983. Formation of the receptive fields of leech mechanosensory neurons during embryonic development. *J. Neurosci.* 3:2474–86

Kramer AP, Stent GS. 1985. Developmental arborization of sensory neurons in the leech *Haementeria ghilianii*. II. Experimentally induced variations in the branching pattern. *J. Neurosci.* 5:768–75

Lefebvre JL, Kostadinov D, Chen WV, Maniatis T, Sanes JR. 2012. Protocadherins mediate dendritic self-avoidance in the mammalian nervous system. *Nature*. doi:10.1038/nature11305

Lefebvre JL, Zhang Y, Meister M, Wang X, Sanes JR. 2008. γ-Protocadherins regulate neuronal survival but are dispensable for circuit formation in retina. *Development* 135:4141–51

Li HS, Chen JH, Wu W, Fagaly T, Zhou L, et al. 1999. Vertebrate Slit, a secreted ligand for the transmembrane protein Roundabout, is a repellent for olfactory bulb axons. *Cell* 96:807–18

Luo Y, Raible D, Raper JA. 1993. Collapsin: a protein in brain that induces the collapse and paralysis of neuronal growth cones. *Cell* 75:217–27

Matthews BJ, Grueber WB. 2011. Dscam1-mediated self-avoidance counters netrin-dependent targeting of dendrites in *Drosophila*. *Curr. Biol.* 21:1480–87

Matthews BJ, Kim ME, Flanagan JJ, Hattori D, Clemens JC, et al. 2007. Dendrite self-avoidance is controlled by Dscam. *Cell* 129:593–604

Meijers R, Puettmann-Holgado R, Skiniotis G, Liu JH, Walz T, et al. 2007. Structural basis of Dscam isoform specificity. *Nature* 449:487–91

Meinertzhagen IA, O'Neil SD. 1991. Synaptic organization of columnar elements in the lamina of the wild type in *Drosophila melanogaster*. *J. Comp. Neurol.* 305:232–63

Meinertzhagen IA, Piper ST, Sun XJ, Frohlich A. 2000. Neurite morphogenesis of identified visual interneurons and its relationship to photoreceptor synaptogenesis in the flies, *Musca domestica* and *Drosophila melanogaster*. *Eur. J. Neurosci.* 12:1342–56

Millard SS, Lu Z, Zipursky SL, Meinertzhagen IA. 2010. *Drosophila* Dscam proteins regulate postsynaptic specificity at multiple-contact synapses. *Neuron* 67:761–68

Millard SS, Zipursky SL. 2008. Dscam-mediated repulsion controls tiling and self-avoidance. *Curr. Opin. Neurobiol.* 18:84–89

Neugebauer KM, Tomaselli KJ, Lilien J, Reichardt LF. 1988. N-cadherin, NCAM, and integrins promote retinal neurite outgrowth on astrocytes in vitro. *J. Cell Biol.* 107:1177–87

Neves G, Zucker J, Daly M, Chess A. 2004. Stochastic yet biased expression of multiple Dscam splice variants by individual cells. *Nat. Genet.* 36:240–46

Nicholls JG, Baylor DA. 1968. Specific modalities and receptive fields of sensory neurons in CNS of the leech. *J. Neurophysiol.* 31:740–56

Prasad T, Wang X, Gray PA, Weiner JA. 2008. A differential developmental pattern of spinal interneuron apoptosis during synaptogenesis: insights from genetic analyses of the protocadherin-γ gene cluster. *Development* 135:4153–64

Prasad T, Weiner JA. 2012. Direct and indirect regulation of spinal cord Ia afferent terminal formation by the γ-Protocadherins. *Front. Mol. Neurosci.* 4:54

Sawaya MR, Wojtowicz WM, Andre I, Qian B, Wu W, et al. 2008. A double S shape provides the structural basis for the extraordinary binding specificity of Dscam isoforms. *Cell* 134:1007–18

Schmucker D, Chen B. 2009. Dscam and DSCAM: complex genes in simple animals, complex animals yet simple genes. *Genes Dev.* 23:147–56

Schmucker D, Clemens JC, Shu H, Worby CA, Xiao J, et al. 2000. *Drosophila* Dscam is an axon guidance receptor exhibiting extraordinary molecular diversity. *Cell* 101:671–84

Schreiner D, Weiner JA. 2010. Combinatorial homophilic interaction between γ-protocadherin multimers greatly expands the molecular diversity of cell adhesion. *Proc. Natl. Acad. Sci. USA* 107:14893–98

Serafini T, Kennedy TE, Galko MJ, Mirzayan C, Jessell TM, Tessier-Lavigne M. 1994. The netrins define a family of axon outgrowth-promoting proteins homologous to *C. elegans* UNC-6. *Cell* 78:409–24
Shi L, Lee T. 2012. Molecular diversity of Dscam and self-recognition. *Adv. Exp. Med. Biol.* 739:262–75
Shi L, Yu HH, Yang JS, Lee T. 2007. Specific *Drosophila* Dscam juxtamembrane variants control dendritic elaboration and axonal arborization. *J. Neurosci.* 27:6723–28
Soba P, Zhu S, Emoto K, Younger S, Yang SJ, et al. 2007. *Drosophila* sensory neurons require Dscam for dendritic self-avoidance and proper dendritic field organization. *Neuron* 54:403–16
Sugimura K, Yamamoto M, Niwa R, Satoh D, Goto S, et al. 2003. Distinct developmental modes and lesion-induced reactions of dendrites of two classes of *Drosophila* sensory neurons. *J. Neurosci.* 23:3752–60
Sweeney NT, Li W, Gao FB. 2002. Genetic manipulation of single neurons *in vivo* reveals specific roles of Flamingo in neuronal morphogenesis. *Dev. Biol.* 247:76–88
Tasic B, Nabholz CE, Baldwin KK, Kim Y, Rueckert EH, et al. 2002. Promoter choice determines splice site selection in protocadherin α and γ pre-mRNA splicing. *Mol. Cell* 10:21–33
Tessier-Lavigne M, Goodman CS. 1996. The molecular biology of axon guidance. *Science* 274:1123–33
Tomaselli KJ, Neugebauer KM, Bixby JL, Lilien J, Reichardt LF. 1988. N-cadherin and integrins: two receptor systems that mediate neuronal process outgrowth on astrocyte surfaces. *Neuron* 1:33–43
Wang J, Ma X, Yang JS, Zheng X, Zugates CT, et al. 2004. Transmembrane/juxtamembrane domain-dependent Dscam distribution and function during mushroom body neuronal morphogenesis. *Neuron* 43:663–72
Wang J, Zugates CT, Liang IH, Lee CH, Lee T. 2002a. *Drosophila* Dscam is required for divergent segregation of sister branches and suppresses ectopic bifurcation of axons. *Neuron* 33:559–71
Wang X, Su H, Bradley A. 2002b. Molecular mechanisms governing *Pcdh-γ* gene expression: evidence for a multiple promoter and *cis*-alternative splicing model. *Genes Dev.* 16:1890–905
Wang X, Weiner JA, Levi S, Craig AM, Bradley A, Sanes JR. 2002c. Gamma protocadherins are required for survival of spinal interneurons. *Neuron* 36:843–54
Wassle H. 2004. Parallel processing in the mammalian retina. *Nat. Rev. Neurosci.* 5:747–57
Weiner JA, Wang X, Tapia JC, Sanes JR. 2005. Gamma protocadherins are required for synaptic development in the spinal cord. *Proc. Natl. Acad. Sci. USA* 102:8–14
Wojtowicz WM, Flanagan JJ, Millard SS, Zipursky SL, Clemens JC. 2004. Alternative splicing of *Drosophila* Dscam generates axon guidance receptors that exhibit isoform-specific homophilic binding. *Cell* 118:619–33
Wojtowicz WM, Wu W, Andre I, Qian B, Baker D, Zipursky SL. 2007. A vast repertoire of Dscam binding specificities arises from modular interactions of variable Ig domains. *Cell* 130:1134–45
Wu Q, Maniatis T. 1999. A striking organization of a large family of human neural cadherin-like cell adhesion genes. *Cell* 97:779–90
Wu W, Ahlsen G, Baker D, Shapiro L, Zipursky SL. 2012. Complementary chimeric isoforms reveal Dscam1 binding specificity in vivo. *Neuron* 74:261–68
Yagi T. 2012. Molecular codes for neuronal individuality and cell assembly in the brain. *Front. Mol. Neurosci.* 5:45
Yang JS, Bai JM, Lee T. 2008. Dynein-dynactin complex is essential for dendritic restriction of TM1-containing *Drosophila* Dscam. *PLoS One* 3:e3504
Yau KW. 1976. Receptive fields, geometry and conduction block of sensory neurones in the central nervous system of the leech. *J. Physiol.* 263:513–38
Yu HH, Yang JS, Wang J, Huang Y, Lee T. 2009. Endodomain diversity in the *Drosophila* Dscam and its roles in neuronal morphogenesis. *J. Neurosci.* 29:1904–14
Zhan XL, Clemens JC, Neves G, Hattori D, Flanagan JJ, et al. 2004. Analysis of Dscam diversity in regulating axon guidance in *Drosophila* mushroom bodies. *Neuron* 43:673–86
Zhu H, Hummel T, Clemens JC, Berdnik D, Zipursky SL, Luo L. 2006. Dendritic patterning by Dscam and synaptic partner matching in the *Drosophila* antennal lobe. *Nat. Neurosci.* 9:349–55
Zipursky SL, Wojtowicz WM, Hattori D. 2006. Got diversity? Wiring the fly brain with Dscam. *Trends Biochem. Sci.* 31:581–88

Cumulative Indexes

Contributing Authors, Volumes 27–36

D

E

F

G

H

I

J

S

T

V

W

Y

Z

Article Titles, Volumes 27–36

Cognitive Neuroscience

Comparative Neuroscience

Computational Approaches to Neuroscience

Degeneration and Repair

Developmental Neurobiology

Neurogenetics

Neuronal Membranes

Neuroscience Techniques

Vestibular System

Vision and Hearing

Visual System